Lecture Notes in Computer Science 6715

Javier Lopez Gene Tsudik (Eds.)

Applied Cryptography and Network Security

9th International Conference, ACNS 2011
Nerja, Spain, June 7-10, 2011
Proceedings

Springer

Volume Editors

Javier Lopez
University of Malaga
Computer Science Department
29071 Malaga, Spain
E-mail: jlm@lcc.uma.es

Gene Tsudik
University of California
Computer Science Department
Irvine, CA 92697, USA
E-mail: gts@ics.uci.edu

ISSN 0302-9743
ISSN 1611-3349
ISBN 978-3-642-21553-7
ISBN 978-3-642-21554-4 (eBook)
DOI 10.1007/978-3-642-21554-4
Springer Heidelberg Dordrecht London New York

Library of Congress Control Number: 2011928562

CR Subject Classification (1998): E.3, E.4, K.6.5, D.4.6, C.2, K.4.4

LNCS Sublibrary: SL 4 – Security and Cryptology

Typesetting: Camera-ready by author, data conversion by Scientific Publishing Services, Chennai, India

Printed on acid-free paper

Springer is part of Springer Science+Business Media (www.springer.com)

Javier Lopez   Gene Tsudik (Eds.)

Applied Cryptography and Network Security

9th International Conference, ACNS 2011
Nerja, Spain, June 7-10, 2011
Proceedings

Springer

Volume Editors

Javier Lopez
University of Malaga
Computer Science Department
29071 Malaga, Spain
E-mail: jlm@lcc.uma.es

Gene Tsudik
University of California
Computer Science Department
Irvine, CA 92697, USA
E-mail: gts@ics.uci.edu

ISSN 0302-9743
ISSN 1611-3349
ISBN 978-3-642-21553-7
ISBN 978-3-642-21554-4 (eBook)
DOI 10.1007/978-3-642-21554-4
Springer Heidelberg Dordrecht London New York

Library of Congress Control Number: 2011928562

CR Subject Classification (1998): E.3, E.4, K.6.5, D.4.6, C.2, K.4.4

LNCS Sublibrary: SL 4 – Security and Cryptology

Typesetting: Camera-ready by author, data conversion by Scientific Publishing Services, Chennai, India

Printed on acid-free paper

Springer is part of Springer Science+Business Media (www.springer.com)

Preface

These proceedings contain 31 papers selected for presentation at the 9th International Conference on Applied Cryptography and Network Security (ACNS 2011) held June 7-10, 2011 in Nerja (Malaga), Spain, and hosted by the Computer Science Department of the University of Malaga.

Since 2003, ACNS is an annual conference that focuses on cutting-edge advances and results in applied cryptography and systems/network security. ACNS is a forum for research of academic as well as industrial/technical nature.

This year, a total of 172 papers were submitted. They were evaluated on the basis of research significance, novelty, and technical quality. Each submission was reviewed by at least three members of the Program Committee (PC). The PC meeting was held electronically and involved intensive discussions. In the end, 31 papers were selected for presentation at the conference, corresponding to an 18% acceptance rate. A further nine papers (not included in these proceedings) were selected for the industrial track of the conference.

Many people deserve acknowledgment for having volunteered their time and energy to make ACNS 2011 a resounding success. Many thanks are due to General Co-chairs, Roberto di Pietro and Rodrigo Roman, for their valuable help with the conference organization. We are also very grateful to Cristina Alcaraz and Claudio Soriente (Publicity Co-chairs), Ersin Uzun and Pablo Najera (Web Support) and Noelia Campos (Local Organization). Clearly, we are greatly indebted to all members of the PC and external reviewers for their selfless dedication and hard work during the review and selection process. We would also like to express our appreciation to the invited/keynote speakers: Refik Molva and Ed Dawson. Last, but certainly not least, our sincere gratitude goes to all submission authors as well as to all conference attendees.

We hope that you will find the program stimulating and that it will serve as a source of inspiration for future research.

June 2011

Javier Lopez
Gene Tsudik

ACNS 2011
9th International Conference on Applied Cryptography and Network Security

Nerja (Malaga), Spain
June 7–10, 2011

Organized by
Computer Science Department
University of Malaga
Spain

Program Co-chairs

Javier Lopez	University of Malaga, Spain
Gene Tsudik	University of California, Irvine, USA

General Co-chairs

Roberto di Pietro	University of Roma Tre, Italy
Rodrigo Roman	University of Malaga, Spain

Publicity Co-chairs

Cristina Alcaraz	University of Malaga, Spain
Claudio Soriente	Madrid Polytechnic University, Spain

Web Chair

Ersin Uzun	PARC, USA

Program Committee

Michel Abdalla	ENS, France
Elli Androulaki	Columbia University, USA
N. Asokan	Nokia Research, Finland
Giuseppe Ateniese	Johns Hopkins University, USA

Ahmad-Reza Sadeghi	Ruhr-Universität Bochum, Germany
Nitesh Saxena	Polytechnic Institute of New York University, USA
Abdullatif Shikfa	Alcatel-Lucent Bell Labs, France
Jessica Staddon	Google, USA
Angelos Stavrou	George Mason University, USA
Carmela Troncoso	KU Leuven, Belgium
Serge Vaudenay	EPFL, Switzerland
Ivan Visconti	University of Salerno, Italy - UCLA, USA
Shouhuai Xu	University of Texas at San Antonio, USA
Jianying Zhou	I2R, Singapore

External Reviewers

Seung Geol Choi
Ang Cui
Christophe De Canniere
Sharath Hiremagalore
Ivica Nikolic
Rahul Murmuria
Vasilis Pappas
Bart Preneel
Christian Rechberger
Jorge Villar

Table of Contents

Session 1: Malware and Intrusion Detection

Session 2: Attacks I

Session 3: Applied Crypto I

Session 4: Signatures and Friends

Session 5: Eclectic Assortment

Session 6: Theory

Session 7: Encryption

Session 8: Broadcast Encryption

Session 9: Security Services

Session 10: Attacks II

Session 11: Applied Crypto II

Inferring Protocol State Machine from Network Traces: A Probabilistic Approach*

Yipeng Wang[1,3], Zhibin Zhang[1], Danfeng (Daphne) Yao[2], Buyun Qu[1,3], and Li Guo[1]

[1] Institute of Computing Technology, Chinese Academy of Sciences, Beijing, China
[2] Department of Computer Science, Virginia Tech, Blacksburg, VA, USA
[3] Graduate University, Chinese Academy of Sciences, Beijing, China
yipeng.wang1@gmail.com, {zhangzhibin,guoli}@ict.ac.cn, danfeng@cs.vt.edu, qubuyun@software.ict.ac.cn

Abstract. Application-level protocol specifications (i.e., how a protocol should behave) are helpful for network security management, including intrusion detection and intrusion prevention. The knowledge of protocol specifications is also an effective way of detecting malicious code. However, current methods for obtaining unknown protocol specifications highly rely on manual operations, such as reverse engineering which is a major instrument for extracting application-level specifications but is time-consuming and laborious. Several works have focus their attentions on extracting protocol messages from real-world trace automatically, and leave protocol state machine unsolved.

In this paper, we propose *Veritas*, a system that can automatically infer protocol state machine from real-world network traces. The main feature of *Veritas* is that it has no prior knowledge of protocol specifications, and our technique is based on the statistical analysis on the protocol formats. We also formally define a new model – probabilistic protocol state machine (P-PSM), which is a probabilistic generalization of protocol state machine. In our experiments, we evaluate a text-based protocol and two binary-based protocols to test the performance of *Veritas*. Our results show that the protocol state machines that *Veritas* infers can accurately represent 92% of the protocol flows on average. Our system is general and suitable for both text-based and binary-based protocols. *Veritas* can also be employed as an auxiliary tool for analyzing unknown behaviors in real-world applications.

Keywords: Protocol Model Inference and Analysis; Probabilistic Protocol State Machine; Network Security.

1 Introduction

Detailed knowledge of protocol specifications is helpful in many network security applications, such as intrusion detection systems [16], vulnerability discovery [14], and protocol analyzers for Internet traffic monitoring [17]. Furthermore,

* This work is supported by the National Basic Research Program "973" of China (Grant No. 2007CB311100).

J. Lopez and G. Tsudik (Eds.): ACNS 2011, LNCS 6715, pp. 1–18, 2011.

given a protocol specification, it is important for application fingerprinting [15] and mapping traffic to applications [7]. However, many network protocols, such as private and non-standard protocols, have no public protocol specifications available. Therefore, it is a crucial network security problem for Internet Service Providers (ISP) to find out these unknown protocol specifications. In the context of obtaining protocol specifications, inferring protocol state machines plays a more important role in practice. Generally, the target of protocol specification discovery concerns not only protocol message formats (i.e., the packet encapsulation mechanism), but also the protocol state machine. The protocol state machine is a finite state automaton illustrating the states in the protocol and their transitions (i.e., the state transition manner). Discovering message format is useful in identifying protocols in monitored network traffic and building intrusion detection systems; and discovering protocol state machine can depict the behavior of an application. Much previous work [5,6,7,8,9] was focused on extracting the protocol format information, without resolving any protocol state machine. However, there are a few exceptions [3,4]. For example, Prospex [3] is an elegant solution for both protocol format and state machine inferencing, which is useful for malware analysis. Prospex's analysis is based on observing the dynamic execution of the program and thus requires the binary code.

Our paper provides a novel technique for inferring protocol state machine *solely* based on the real-world network trace of an application. There are several advantages associated with our approach. First, analyzing network traffic can be easily automated and requires less manual effort. The analysis does not require the distinction between client and server applications. Second, up to 40% of Internet traffic belongs to unknown applications [19], many of them ran by botnets. The binary code of these applications may not be available for reverse engineering. Inferring the state machine of unknown protocol from its real world trace can help ISPs have a better understanding to the behaviors of traffic passing through their networks.

We propose *Veritas*, a system that automatically extracts protocol state machine for stateful network protocols from Internet traffic. The input to *Veritas* is the network trace of a specific application. Our output is a *probabilistic* description of the protocol state machine. This probabilistic protocol state machine (P-PSM) is a new and powerful model that we define for capturing and representing any protocols with incomplete knowledge. In order to test and verify *Veritas*, we apply our system to several real-world applications, including a client-server protocol SMTP, two peer-to-peer protocols PPLIVE [21] and XUNLEI [20]. The experiment results show that our system is capable of correctly recognize and classify 86% SMTP flows, 100% PPLIVE and 90% XUNLEI flows. Our tool has the following features: (a) requiring no knowledge of protocol specifications, (b) based on the statistics of protocol formats, and (c) effective for both text and binary protocols. Our contributions are summarized as follows.

- We introduce and formalize a new model – probabilistic protocol state machine (P-PSM) – for describing the protocol state machine in a probabilistic fashion when there is incomplete knowledge about the protocol. P-PSM

model is general and can be used for representing any stateful protocol with uncertain information.
- We design a system called *Veritas*, which can automatically infer the protocol state machine of a specific protocol from its real-world trace with no prior knowledge about the protocol specification. We propose a new technique to extract protocol messages formats that is independent of the type of the target protocol.
- We apply our system to verify real world applications. The applications contain both text-based and binary-based protocols, which are quite complex. Our results demonstrate that *Veritas* is capable of inferring protocol state machine of good quality in practice.

The rest of the paper is organized as follows. Section 2 is dedicated to the related work. In Section 3, we introduce the architecture of *Veritas* and present each portion of the system. In Section 4, we make use of *Veritas* for protocol inference and evaluate the whole system with different protocols. Finally, we conclude our work with future research directions in Section 5.

2 Related Work

We divide our discussion of related work into three areas, namely automatic protocol reverse engineering, protocol message format extraction, and inferring protocol state machine.

Automatic protocol reverse engineering. Accurately reversing protocols typically involves manual efforts, such as in the cases of Gaim [23] and [22]. There are several proposals on automating this process. Lim *et al.* [1] proposed a method, which automatically extracted the format from files and application data output. Their works depend on some parameters, such as the output functions, which may not be available. Polyglot [5] proposed a new approach for reverse engineering a protocol by using dynamic analysis of program binaries. In our work, we assume that the program binary is not available; thus our work is orthogonal to the above.

Protocol message format extraction. Much work in the current literature is focused on protocol message format extraction. Kannan *et al.* [8] presented algorithms on extracting the structure of application sessions embedded in the connections. Haffner [7] automated the construction of protocol signatures on traffic that contains the known instance of each protocol. Ma [9] proposed an unexpected means of protocol inference without relying on the port numbers. His method classify the network data belonging to the same protocol automatically. Cui *et al.* [6] introduced a tool, which is for automatically reverse engineering the protocol message format from its network trace. His method divided protocol formats into different tokens by some experiential delimiters. In those studies, inferencing protocol state machine was not investigated.

Inferring protocol state machine. Inferring protocol state machine plays an important role in protocol specifications. The works that are closest to ours include ScriptGen and Prospex. ScriptGen [4] aims to infer protocol state machine

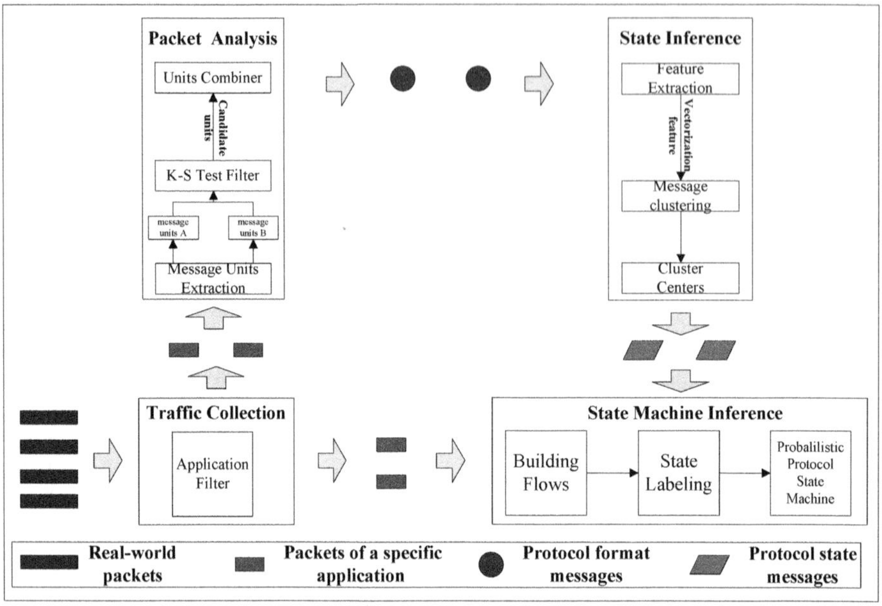

Fig. 1. The Architecture of *Veritas*

from network traffic. However, ScriptGen has to rebuilds TCP flows based on several assumptions, and it can not handle each TCP session precisely. So those limitations prevent it from emulating all possible protocols. Prospex [3] infers protocol state machine by means of analyzing the execution trace of a program on a stand-alone host. In comparison, our inference is based on observed network traffic that can be performed by ISPs.

3 Architecture of *Veritas*

The objective of our system is to infer the specifications of a protocol that is used for communication between different hosts. More specifically, given the packet sequences of flows of a specific application, we investigate how protocol state changes from one state to another in the flow. Our approach is to perform machine learning and probabilistic/statistical analysis on the syntax of observed network traces. In this section, we give the definitions used in *Veritas* and an overview of *Veritas* architecture.

We define a protocol as a Markov chain model on sessions of length at most n, which has a discrete state space. The Markov property states that the conditional probability distribution of a system at the next time period depends *only* on the current state of the system, i.e., not depending on the state of the system at previous time periods.

Definition 1. *The message that identifies the state of a protocol is referred to by us as a* ***protocol state message****.*

Protocol state messages are important for understanding the behaviors of the protocol. However, one may not always be able to obtain the state message directly from network traces. Our approach is to infer or estimate protocol state messages by observing and analyzing messages that frequently occur next to them.

Definition 2. *The* ***protocol format message*** *refers to the most frequent string (i.e., the keyword) in the protocol format.*

Protocol format messages are useful for both protocol format inference and obtaining the protocol state message. Protocol format messages include protocol state messages, which we will give more description in the following subsection.

In comparison, research on protocol format extraction typically regards a protocol as a distribution on sessions of length at most n [9], which is static. In other words, the existing solutions work only for extracting the format of a protocol, and cannot be used to describe the protocol states and their transitions.

The input to our system is the network trace of an application. In application-layer packet headers, there may be some protocol state messages. Each of these messages has a message *type*, which indicates the protocol state of the packet. The sequence of packets (belonging to the same flow) is determined by the protocol state machine of a specific application. Meanwhile, the protocol state machine describes how packets of different message types are transmitted in the traces.

Our assumptions In our work, we assume that the network trace is not encrypted. In addition, we assume that the network trace is only composed of flows from the application to be investigated. That is, there is no mixed traffic of multiple protocols.

Veritas has several components as shown in Figure 1, including network data collection, packet analysis, state message inference, and state machine inference, which we describe in more details next.

Network data collection. In this phase, network traffic of a specific application (such as SMTP, DNS, BitTorrent *etc.*) is collected carefully. There are several ways to get packets of a specific protocol, that is the ground truth. For example, the GT [2] method, capturing packets on a specific transport layer port, by means of reverse engineering and so on are all widely used. In this paper, the method of collecting packets on a specific transport layer port is adopted.

Packet analysis. During the phase of packet analysis, we first look for high frequency message units from off-line application-layer packet headers. Then, we employ Kolmogorov - Smirnov (K-S) test [11] to determine the optimal number of message units. Finally, we replay each application-layer packet header and construct protocol message formats with candidate message units.

State message inference. In this phase, we extract the feature from each protocol format message. The feature is used to measure the similarity between format messages. Then, the partitioning around medoids (PAM) clustering algorithm [12] is applied to group similar messages into a cluster. Finally, the medoid message of a cluster will be a protocol state message.

State machine inference. In order to infer protocol state machine, we should be aware of the protocol state sequence of flows. In order to label protocol state, firstly our system builds flows for a specific protocol. Then, each packet under analysis (if it has a state) will be assigned with a state. Afterwards, by constructing the relationship between different states, a protocol state machine is constructed. Moreover, in each flow the transitions probabilities of diverse states are counted. Finally, together with the protocol state messages, the probabilistic state machine is constituted.

3.1 Packet Analysis

The first stage of *Veritas* is to acquire the formats of protocol messages. In *Veritas*, we extract message formats by applying statistical learning methods on protocol packets. Protocol format messages can be extracted from application-layer packet headers by searching for frequently occurring strings (i.e., keywords). These keywords are typically encapsulated at the beginning of application-layer packets. Taking SMTP (Simple Mail Transfer Protocol) for example, both of the strings "MAIL FROM:" and "RCPT TO:" are its format messages, which usually reside in the SMTP protocol application headers. We assume that the protocol specifications is not available to us. Next, we describe in details how we analyze collected packets in order to infer protocol format messages.

Message Units Extraction. Protocol format messages are defined by us as the most frequently occurred strings in the traces of a protocol. From a statistical perspective, if each protocol format can be partitioned into a set of all possible subsequences with fixed lengths, the frequency of these subsequences can be counted precisely.

However, there are two practical problems. The first issue is how to choose the length l of these subsequences, which is critical to the performance of *Veritas*. The second issue is that given a packet how to determine the number n of bytes that are protocol related (i.e., not payload). The latter problem arises, as it is unnecessary to the payload of a packet. Thus, the problem message units extraction turns to determining proper values for l and n.

Definition 3. *The l-byte subsequence originated from the first few bytes of each packet header in network traces is referred to by us as a* ***message unit****.*

We investigate several common application-layer protocols, and find that the minimum field length of those protocols, both text and binary, is at least three bytes. In addition, it is easy to see that the subsequences with length three will be more differentiable than those with length two. The sequence set with three bytes is larger than that with two characters, so high frequency of three-byte sequences is more prominent in special subsequence seeking. On the other hand, the subsequences with length four or more will weaken its occurrence frequency. Therefore, in *Veritas* we set $l = 3$. Furthermore, if the packet length s is smaller than three bytes, we regard $l = s$. For the other parameter n, we just give a tentative value 12.

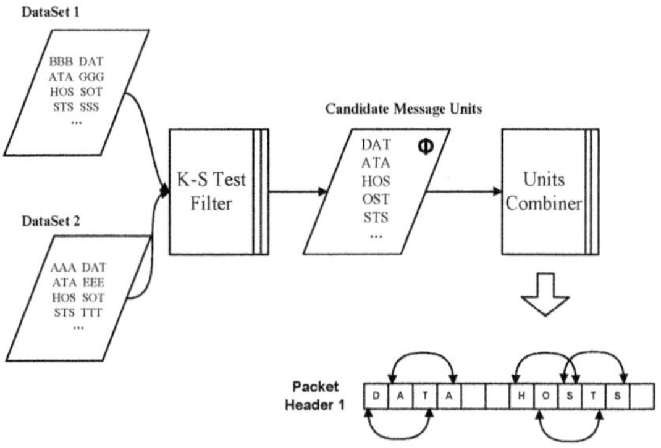

Fig. 2. Packet Analysis of Veritas

It should be noticed that not all of the message units obtained from the method aforementioned are protocol relative. In order to get the high frequency units helpful in characterizing application layer protocol, *Veritas* introduces the two-sample Kolmogorov-Smirnov statistical testing method (abbr. K-S test) [11] to tackle the resulting message units set.

K-S Test Filter. In this part, we employ a K-S test filter to obtain message units, which are associated with protocol formats. A concrete example is illustrated in Figure 2. The input to K-S test filter is two groups of message units which is obtained by packet analysis. Before conducting K-S test, *Veritas* will turn message units into numeric values. The output to K-S test filter is candidate units which are used for constructing protocol format messages.

The essence of K-S test is to estimate the similarity of two samples according to their empirical probability distributions in a nonparametric way. Given two samples, say S_n and $S_{n'}$, where the subscripts represent the sample sizes, their empirical distribution functions (denote F_n and $F_{n'}$ respectively) can be calculated as $F(X) = \frac{1}{n}\sum_{i=1}^{n} I_{X_i \leq x}$, where $I_{X_i \leq x}$ is the indicator function, equal to 1 if $X_i \leq x$ and equal to 0 otherwise. Then K-S test conducts the similarity measurement by quantifying the distance between F_n and $F_{n'}$ as a statistic $D_{n,n'} = \sup_x |F_n(x) - F_{n'}(x)|$. The null hypothesis is that the samples are drawn from the same distribution, without specifying what that common distribution is. The null hypothesis is rejected (at level α) if $\sqrt{nn'/(n+n')}D_{n,n'} > K_\alpha$, where K_α is the critical value which can be found from $Pr(K \leq K_\alpha) = 1 - \alpha$ under the Kolmogorov distribution.

So, in order to apply K-S test, the packet collection of a specific protocol should be randomly partitioned into two disjoint groups $\mathbb{A}$ and $\mathbb{B}$ with approximately the same size by utilizing the units extraction strategy described in Section 3.1, two groups will yield two message-unit sets, A and B respectively. Then after turning message units into numeric values, the frequency f_x of element x in each set can be counted easily. Then for set A, we partition those elements with frequency higher than or equal to λ into a subset A_λ. Here, λ

is a frequency threshold. Doing the same thing on B generates B_λ. Now the application of K-S test on A_λ and B_λ is just to choose a suitable value λ under which the null hypothesis is acceptable. The rejection of K-S test on A_λ and B_λ means the threshold λ is not high enough to cut off useless units.

In the circumstance of *Veritas*, the aim of K-S test is to filter out message units that is not relevant to protocol formats. That is, it requires the result sets A_λ and B_λ responsible for the reflection of protocol formats. Put it in the way of statistical testing, the K-S test on A_λ and B_λ should be accepted at a extremely low level (i.e., $1-\alpha$ should be small enough). In *Veritas*, $1-\alpha$ is valued less than 10^{-8}. Then, for the purpose of accepting the K-S test under the chosen reject level α, *Veritas* manipulates λ in a progressive way: it is initialized as 10^{-5} and gradually increases by 10^{-5} till K-S test accepts.

Once the K-S test finished (i.e., been accepted), the elements in $A_\lambda \bigcap B_\lambda$ will be called **candidate message units**. Then *Veritas* attempts to recover the protocol format messages from these candidate message units.

Protocol Format Message Inference. Obtaining protocol format messages is important, as these messages are used for inferring protocol state messages, which is described in Section 3.2. We design a units combiner, which is employed to recover protocol format messages from candidate message units obtained. Here we give a concrete example to explain the process of protocol format messages reconstruction. As shown in Figure 2, the candidate units set ϕ is comprised of five message units (DAT, ATA, HOS, OST and STS). The possible protocol format message can be checked out as follows,

1. Randomly selecting a group of packets from the traffic collection, say Packet Header 1 in Figure 2.
2. With candidate message units, the units combiner tries to rebuild all sequences as long as possible (maybe more than one) for each packet header. So all of these sequences only contain possible three-byte subsequences which are lying in ϕ.
3. All of these obtained sequences are regarded as protocol format messages, such as '$DATA\quad HOSTS$' in Figure 2.

Furthermore, not all packets contain protocol format message, since some packets only transmit data. Next, we describe our machine-learning methods for inferring protocol state messages.

3.2 Inferencing Protocol State Messages

As defined earlier, protocol state messages are important and can be used to represent states of a protocol. From our packet analysis, our system obtains a set of protocol format messages from application-layer packet headers, based on which we derive protocol state messages. This derivation is based on a statistical approach, namely using a clustering algorithm.

We need to assign a *type* to each protocol format message, which can be accomplished in two steps. First, we define the features of a protocol format message,

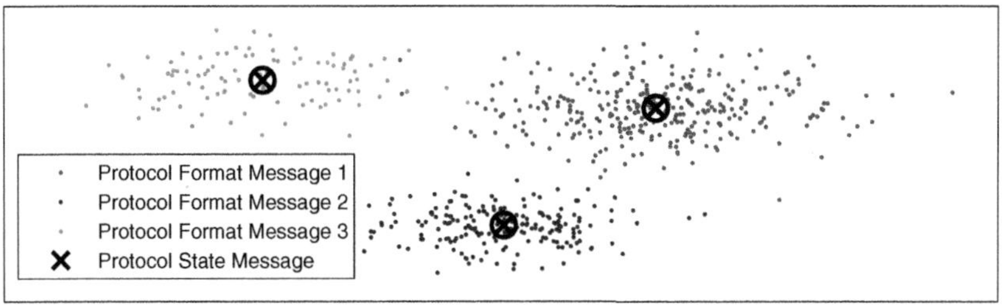

Fig. 3. The Relationship between Protocol Format Messages and Protocol State Messages

as well as a similarity metric between messages. We assume that messages of the same type share a similar message format. Second, our system will group similar format message into a cluster using machine learning methods. Similar messages are likely to be placed in the same cluster, which is labeled by us with the proper type. We define the center of each cluster as a *protocol state message*, which can be used to represent other messages in the cluster. The relationship between protocol state messages and protocol format messages is illustrated in Figure 3. In Figure 3, each dot represents one protocol format message. Furthermore, as shown in Figure 3, protocol state messages are part of protocol format messages, and a protocol state messages is the center message of a cluster.

However, there are two technical challenges in message clustering. First, we have no knowledge of the similarities between messages or their types. Second, we get no prior knowledge about how many state messages are in a certain protocol under analysis. Next, we describe the details of our similarity computation, and how we realize clustering and address the challenges.

Feature Extraction and Similarity Calculation. In order to group similar format messages, we need to extract the feature from each format message. In *Veritas*, the feature of a protocol format message is expressed by a vector in $\in \mathbb{R}^{256}$, with the i-th (starting from 0) component counting the number of one-byte character (values i) in that message. Meanwhile, we regard that two format messages of the same type should have a similar character composition. Afterwards, our system carries out similarity calculation between different format messages.

For the purpose of comparing the similarity between two format messages, we make use of the Jaccard index [13], which is defined as follows:

$$J(a, b) = \frac{|a \cap b|}{|a \cup b|}, \tag{1}$$

where, a is the set of elements associated with the feature of the first message, while b is the set that stands for the same feature of the second message. $J(a, b)$ gains its maximum value 1 when all the items in the given set are the same and it will achieve its minimum value 0 when all the items in the given set are distinct.

Message Clustering. Based on the feature and the similarity function introduced in the previous subsection, we define the distance between two protocol format messages, which is crucial in clustering. The distance of two protocol format messages a and b is defined as $d(a,b) = 1 - J(a,b)$, where J is the Jaccard index in this paper.

In order to group protocol format messages, we make use of the Partitioning Around Medoids (PAM) algorithm [12]. In contrast to the k-means algorithm, the partitioning around medoids algorithm is more robust to noise and outliers. Therefore, PAM algorithm are suitable for protocol state messages inferring. Just like most other clustering algorithms, the partitioning around medoids algorithm needs an integer value k (the number of clusters) as the input. In order to find out a proper k value, we use a generalization of the Dunn index [18] as a measurement. The Dunn index is a standard intrinsic measurement of clustering quality, defined as follows.

$$D(k) = \frac{\min_{1 \leq i \leq k}\{\min_{1 \leq j \leq k}\{\delta(C_i, C_j)\}\}}{\max_{1 \leq i \leq k}\{\Delta(C_i)\}}, \tag{2}$$

where $C_i, .., C_k$ are the clusters, $\Delta(C_i)$ is the diameter of cluster C_i, and $\delta(C_i, C_j)$ is the distance between two clusters. According to Equation 2, we may see in a clear way that the numerator of Equation 2 is a measure of cluster separation and denominator is a measure of cluster compactness. In our experiment, the k, which maximizes the Dunn index, would be chosen. Finally, the format message of each cluster center is regarded as a protocol state message, and the *type* of the protocol state message is represented by π.

3.3 Probabilistic Protocol State Machine

Because our analysis is based on statistical methods, *Veritas* is able to represent protocol state relations probabilistically. In this section, we introduce a novel expression of protocol state machine – probabilistic protocol state machine (P-PSM). P-PSM can be used to describe both protocol state transitions and their probabilities. Moreover, the probabilistic protocol state machine is helpful for identifying critical paths of a protocol.

Notation. Let Σ be the set of characters (256 possibilities) and Σ^* be the set of protocol state messages that can be built from Σ. In Σ, symbols can be denoted as (`\00`, `\01`, `\02`, ... , `\ff`) and protocol state messages in Σ^* will be represented by alphabet letters (`a`, `b`, ... , `z`). Therefore, a protocol state transition $\mathcal{T}_{ij}$ can be denoted by (σ_i,σ_2,... ,σ_j) from a starting state i to an accepting state j, where $\forall \sigma \in \Sigma^*$. $\Pr(\mathcal{T}_{ij})$ is a probability $\prod_{k=i,\sigma_k \in \Sigma^*}^{j} \sigma_k$. Moreover, the distribution must satisfy the equation $\sum_{ij \in \Sigma^*} \Pr(\mathcal{T}_{ij}) = 1$. The distribution can be modeled by a probabilistic protocol state machine $\mathcal{A}$ (defined next). The protocol under analysis will be described by $\mathcal{A}$ in a probabilistic manner.

Formal Definition of P-PSM. We give the formal definition for probabilistic protocol state machine (P-PSM). P-PSM is a specialization of the general probabilistic finite-state machine [10] in the (network) protocol context.

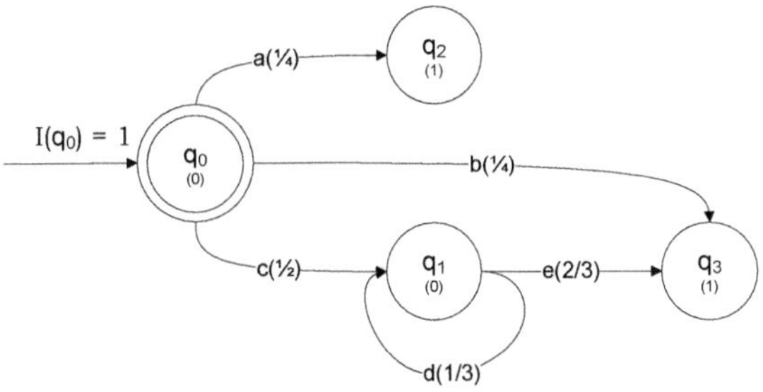

Fig. 4. An Example of P-PSM

Definition 4. *A P-PSM is a tuple $\mathcal{A}$.*

$\mathcal{A} = \langle Q_{\mathcal{A}}, \Sigma^*, \delta_{\mathcal{A}}, I_{\mathcal{A}}, F_{\mathcal{A}}, P_{\mathcal{A}} \rangle$, where:

- $Q_{\mathcal{A}}$ is a finite set of states;
- Σ^* is the set of protocol state messages;
- $\delta_{\mathcal{A}} \subseteq Q_{\mathcal{A}} \times \Sigma^* \times Q_{\mathcal{A}}$ is a set of transitions;
- $I_{\mathcal{A}} : Q_{\mathcal{A}} \longrightarrow \mathbb{R}^+$ (initial-state probabilities);
- $F_{\mathcal{A}} : Q_{\mathcal{A}} \longrightarrow \mathbb{R}^+$ (final-state probabilities);
- $P_{\mathcal{A}} : \delta_{\mathcal{A}} \longrightarrow \mathbb{R}^+$ (transition probabilities).

$I_{\mathcal{A}}, F_{\mathcal{A}}, P_{\mathcal{A}}$ are function such that:

$$\sum_{q \in Q_{\mathcal{A}}} I_{\mathcal{A}} = 1, \tag{3}$$

and

$$\forall q \in Q_{\mathcal{A}}, F_{\mathcal{A}}(q) + \sum_{x \in \Sigma^*, q' \in Q_{\mathcal{A}}} P_{\mathcal{A}}(q, x, q') = 1. \tag{4}$$

By convention, P-PSMs are illustrated by directed labeled graphs. In Figure 4, we give a concrete example of P-PSM. In what follows, the subscript $_{\mathcal{A}}$ will be dropped when there is no ambiguity. Typically, a protocol description by means of P-PSM begins with starting states (q_0 in Figure 4) and finishes with accepting states (q_2, q_3 in Figure 4). In Figure 4, there are four states $Q = \{q_0, q_1, q_2, q_3\}$, only one initial-state ($I(q_0) = 1$) and the real numbers in the states are the final-states probabilities. In addition, there are five protocol state messages, $\Sigma^* = \{a, b, c, d, e\}$, and real numbers in the arrows are transition probabilities.

3.4 State Machine Inference

Veritas constructs the protocol state machine based on protocol state messages, which are obtained from message clustering. Since each stateful packet has its own *type*, a TCP or UDP flow f_i, can be represented as a sequence $F_i = (\pi_1, .., \pi_h)$, where $\pi_1, .., \pi_h \in M$ and M is the set of protocol state messages. Next, we give details on how to associate a network packet with a state label and to construct the protocol state machine in a probabilistic fashion.

State Labeling. We describe how to label each network packet with a state *type*, which is assigned at message clustering. Since our state labeling method is entirely based on a flow model, a 5-tuple of a flow, (source address, destination address, source port, destination port, timeout), is needed as a distinction of different flows. In a 5-tuple, the timeout value indicates the duration of a flow. In our work, several timeout values (16s, 32s, 64s) have been examined in our experiments. From our experiment results, we find that the timeout value is not sensitive in our system, and different timeout values will yield the same experiment results. As a result, in the following experiments, the timeout value will be set to 64s.

As it is defined in previous section, $\pi_i, .., \pi_k$ are cluster center messages (protocol state messages). In this phase, after aligning the two messages to be compared, we denote the feature of the packet header under analysis with ρ, and the feature of the cluster center message π_i with θ_i. For each packet, our system calculates the distance between ρ and θ_i, and labels the packet header ρ with the type of π_i, which satisfies that $\arg\min d(\rho, \theta_i)$, where $i \in [1, k]$.

However, not each packet header have a state *type*. For example, some data transmission packet do not contain any protocol format message, so it will be not marked with any state *type*. Assuming that $\Delta(C_i)$ is the diameter of cluster C_i, $d_{\max}$ can be defined as follows, $d_{\max} = \max_{1 \le i \le k}\{2\Delta(C_i)\}$. C_h is the cluster that is nearest to the packet header ρ' under analysis. If $d(\theta_h, \rho') > d_{\max}$, the packet header ρ' will be assigned with an unknown state *type*.

After labeling all packets of a specific protocol, *Veritas* constructs a probabilistic protocol state machine, as explained next.

Obtaining Probabilistic Protocol State Machine. After the phase of state labeling, we are aware of the state *type* π of the packet in each flow. And then in each TCP or UDP flow F_i, $F_i = (\pi_1, .., \pi_h)$, our system calculate the frequency of each state type pair, such as (π_i, π_{i+1}). Therefore, *Veritas* will obtain both the order of different state types and the transition probability from state *type* π_i to π_{i+1}. For the reason that network packets may be out of order in real-world transmission environment, we employ a threshold value as a filter, which can get rid of state *type* pairs that is out of order. The system only keeps the state *type* pairs with a frequency above 0.005.

According to the set of state *type* pairs, our system is able to depict the linkage of each state *type* pair with a directed labeled graph. And all linkages of state *type* pairs are employed to construct a deterministic finite automaton (DFA) of the protocol under analysis, T. Afterwards, we find the minimal DFA that is consistent with T. In the end, probabilistic protocol state machine (P-PSM) will be the combination of minimal DFA and the set of state transition probabilities.

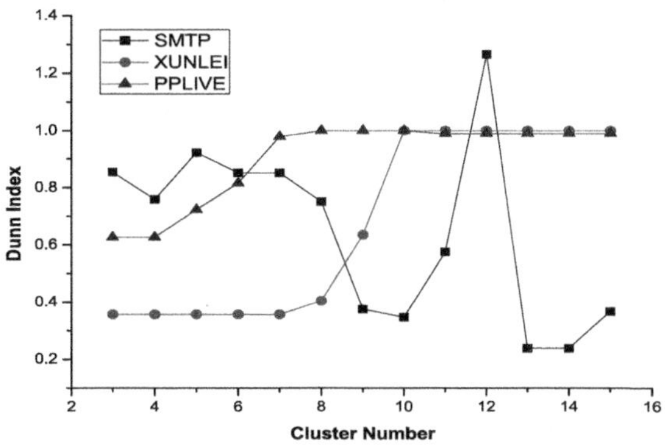

Fig. 5. The changes of Dunn indices with the number of clusters for the three protocols

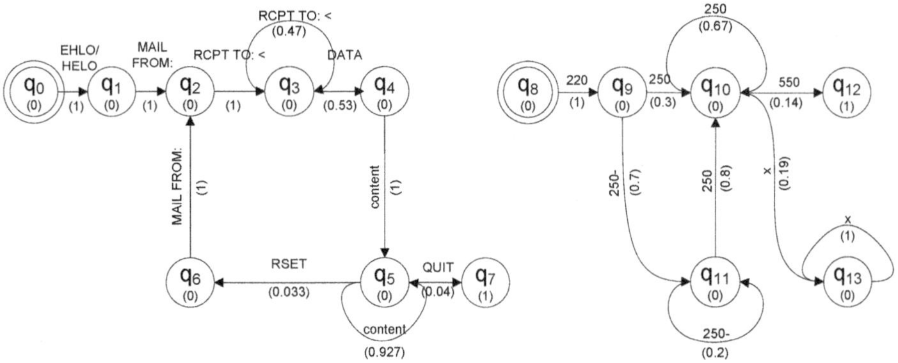

Fig. 6. Probabilistic protocol state machine of SMTP

4 Experimental Evaluation

In evaluation section, in order to verify the effectiveness of *Veritas*, we use two kinds of protocols, text and binary. For each protocol under analysis, the input to our system is real-world trace of the protocol, and the output to the system is the protocol state machine described in a probabilistic mode.

4.1 Text Protocol

In this paper, we choose SMTP (Simple Mail Transfer Protocol), which is a stateful and text-based protocol, as a verification of text protocol for our system. In order to infer the P-PSM of SMTP, we capture real-world packets of SMTP protocol. In this paper, the data source of SMTP is real-world trace, which is obtained from a backbone router on TCP port 25.

In message clustering phase, as it is shown in Figure 5, the optimal cluster number k for SMTP is 12. Moreover, after several iterative clustering experiments, we find that EHLO and HELO messages are grouped into a cluster. And the probability of EHLO and HELO being the medoid of the cluster is equal. This is due to the fact that mail clients may send an EHLO or a HELO command to initialize a connection.

After state machine minimization, the final P-PSM of SMTP protocol we inferred is shown in Figure 6. As it is shown in Figure 6, the P-PSM of SMTP protocol contains two parts, one may be the state transition of client to server, and the other is the state transition of server to client. In addition, from State q_4 to State q_5, the state machines only carry on SMTP data transmission, which does not contain any state information. Furthermore, from State q_{10} to State q_{13}, unknown protocol state message is represented by x currently.

4.2 Binary Protocols

To test the validation of our system to the binary protocol, in this part we choose PPLIVE and XUNLEI, which are peer-to-peer and binary-based protocols.

Analysis on a P2P Streaming Video Application. PPLIVE is a famous peer-to-peer streaming video application in China. The data source of PPLIVE protocol is obtained from our server which only runs an entertainment channel of PPLIVE on UDP port 3987. After state message inference phase, as it is shown in Figure 5, the optimal cluster number k for PPLIVE protocol is 8.

After state machine inference, the ultimate P-PSM of PPLIVE protocol is shown in Figure 7. Moreover, the set of protocol state messages are illustrated in Table 1.

Table 1. PPLIVE Protocol State Messages

Sign	Protocol State Message
a	0xe9 0x03 0x62 0x01 0x98 0xab 0x01 0x02 0x01
b	0xe9 0x03 0x61 0x01 0x98 0xab 0x01 0x02 0x01
c	0xe9 0x03 0x63 0x01 0x98 0xab 0x01 0x02 0x01
d	0xe9 0x03 0x53 0x00 0x98 0xab 0x01 0x02 0x5b
e	0xe9 0x03 0x49 0x01 0x98 0xab 0x01 0x02 0x98
f	0xe9 0x03 0x51 0x01 0x98 0xab
g	0xe9 0x03 0x50 0x00 0x98 0xab 0x01 0x02 0x9b
h	0xe9 0x03 0x4a 0x01 0x98 0xab 0x01 0x02 0x01

Analysis on a P2P File-Sharing Application. XUNLEI is a popular P2P application in China, and it holds a significant UDP Internet traffic. The data source of XUNLEI protocol is obtained from backbone routers on UDP port 15000. In message clustering phase, as it is shown in Figure 5, the optimal cluster number k for XUNLEI protocol is 10. However, 10 is not the final number of protocol state messages. In the next step, the system will construct DFA and find the minimal DFA that is consistent with it.

After state machine minimization, the ultimate P-PSM of XUNLEI protocol is shown in Figure 8. Moreover, the set of protocol state messages is illustrated in Table 2. Furthermore, sign f is not depicted in Figure 8 for the reason that

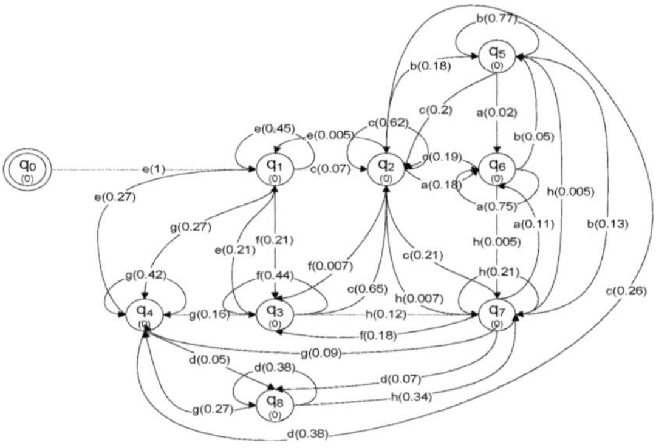

Fig. 7. Probabilistic protocol state machine of PPLIVE protocol

Table 2. XUNLEI Protocol State Messages

Sign	Protocol State Message
a	0x32 0x00 0x00 0x00 0x06 0x00 0x00
b	0x32 0x00 0x00 0x00 0x07
c	0x32 0x00 0x00 0x00 0x08
d	0x32 0x00 0x00 0x00 0x11
e	0x32 0x00 0x00 0x00 0x12
f	0x32 0x00 0x00 0x00 0x09

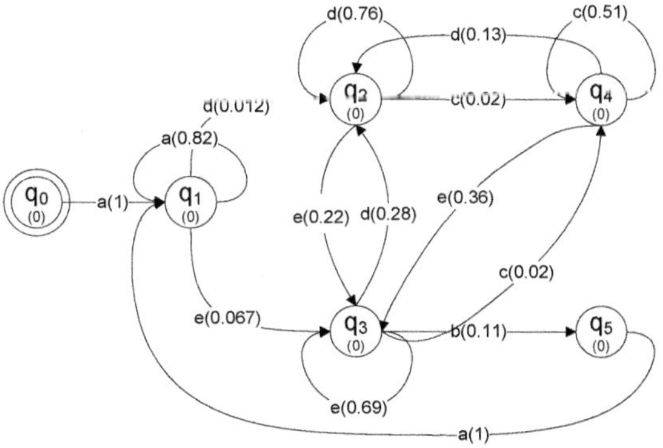

Fig. 8. Probabilistic protocol state machine of XUNLEI protocol

state *type* pairs correspond with f are of very small probability. As far as we know, f is an old version of XUNLEI protocol state message. If we analyze flows correlated with f respectively, we will get a more comprehensive experiment result, which we do not show here.

4.3 Quality of Protocol Specification

In order to evaluate the quality of protocol specification inferred by *Veritas*, we make use of real-world network traces to test P-PSMs we inferred. In the following experiments, we will demonstrate that the P-PSMs we inferred are complete. Completeness is a measurement of protocol specifications accepting valid protocol sessions.

For SMTP, there are about 100,000 SMTP flows captured from the backbone router. Out of those flows, the SMTP protocol state machine are able to parse the state transitions of about 86% flows successfully. The remaining SMTP flows may use an encryption transmission, which we can not handle properly as one of the limitations of our system is its incompetence to deal with encrypted traffic.

PPLIVE peers always employ UDP packets to communicate and transmit data with each other. For the purpose of testing the quality of the PPLIVE specification, about 200,000 UDP flows of PPLIVE are captured from a server which runs a news channel of PPLIVE on September 9th, 2009. For PPLIVE flows, we are able to parse the state transitions of all flows successfully.

In order to test and verify XUNLEI protocol specification, there are about 500,000 UDP flows of XUNLEI obtained from a backbone router. For XUNLEI flows, we are able to parse the state transitions of about 90% flows successfully. The flows we parsed take up more than 99% XUNLEI protocol packets under analysis. Since our method is based on high probability sets, it will not be sensitive to the event of small probability.

From the above experiment results, we can find that the probabilistic protocol state machines we inferred are of good quality. The whole system can be employed as an auxiliary tool for analyzing unknown behaviors in real-world applications.

4.4 Summary

Our technique for inferring protocol state machine is based on a statistical model, and it is sensitive to states which are statistically significant. Therefore, maybe *Veritas* cannot cover all the paths of a protocol state machine. However, our method is suitable for analyzing critical paths in a protocol, which is very important in intrusion detection. Moreover, our experiment results show that the our inference method has a high degree of accuracy in practice.

5 Conclusions and Future Work

Inferring protocol state machine from Internet traffic is a fundamental network security problem, solutions to which have many practical applications. In this paper, we presented a new solution to this problem. We proposed *Veritas*, a system that can automatically extract protocol state machine for stateful network protocols solely from Internet traffic. The input to *Veritas* is the real-world trace of a specific application, and the output is the protocol state machine of that application with a probabilistic description. Our technique proceeds mainly in the following steps. First, the real-world trace of a specific application is extracted from Internet traffic. Then, by analyzing each packet header, we capture the protocol message format from packet headers. Afterwards, by means of clustering algorithms, protocol state messages will be obtained. Based on the clusters, we assign a type to each packet of flows. Finally, we obtain the probabilistic protocol state machine. Our verification experiments show that *Veritas* is general and suitable for both text and binary protocols. The P-PSM inferred by our system reflects the actual applications with high degrees of accuracy.

For future work, we plan to work on semantic inference with *Veritas* for better understanding of protocol specifications. Moreover, *Veritas* can only deal with real-world network trace of a single application. In the future, we would like to make it fit for the multi-protocol environment.

References

1. Lim, J., Reps, T., Liblit, B.: Extracting Output Formats from Executables. In: WCRE 2006: Proceedings of the 13th Working Conference on Reverse Engineering (2006)
2. Gringoli, F., Salgarelli, L., Dusi, M., Cascarano, N., Risso, F., Claffy, K.C.: GT: picking up the truth from the ground for internet traffic. In: SIGCOMM Comput. Commun. Rev. (2009)
3. Comparetti, P.M., Wondracek, G., Kruegel, C., Kirda, E.: Prospex: Protocol Specification Extraction. In: SP 2009: Proceedings of the 2009 30th IEEE Symposium on Security and Privacy (2009)
4. Leita, C., Mermoud, K., Dacier, M.: Scriptgen: an automated script generation tool for honeyd. In: Annual Computer Security Applications Conference (2005)
5. Caballero, J., Yin, H., Liang, Z., Song, D.: Polyglot: automatic extraction of protocol message format using dynamic binary analysis. In: CCS 2007: Proceedings of the 14th ACM conference on Computer and Communications Security (2007)
6. Cui, W., Kannan, J., Wang, H.J.: Discoverer: automatic protocol reverse engineering from network traces. In: SS 2007: Proceedings of 16th USENIX Security Symposium on USENIX Security Symposium (2007)
7. Haffner, P., Sen, S., Spatscheck, O., Wang, D.: ACAS: automated construction of application signatures. In: MineNet 2005: Proceedings of the 2005 ACM SIGCOMM workshop on Mining network data (2005)
8. Kannan, J., Jung, J., Paxson, V., Koksal, C.E.: Semi-automated discovery of application session structure. In: IMC 2006: Proceedings of the 6th ACM SIGCOMM conference on Internet measurement (2006)
9. Ma, J., Levchenko, K., Kreibich, C., Savage, S., Voelker, G.M.: Unexpected means of protocol inference. In: IMC 2006: Proceedings of the 6th ACM SIGCOMM conference on Internet measurement (2006)
10. Vidal, E., Thollard, F., de la Higuera, C., Casacuberta, F., Carrasco, R.C.: Probabilistic Finite-State Machines-Part I. IEEE Trans. Pattern Anal. Mach. Intell. (2005)
11. Kendall, M.G., Stuart, A., Ord, J.K.: Kendall's advanced theory of statistics. Oxford University Press, Inc., Oxford (1987)
12. Kaufman, L., Rousseeuw, P.: Finding Groups in Data: An Introduction to Cluster Analysis. Wiley, Chichester (1990)

13. Jaccard, P.: The distribution of the flora in the alpine zone. The New Phytologist (1912)
14. Brumley, D., Caballero, J., Liang, Z., Newsome, J., Song, D.: Towards automatic discovery of deviations in binary implementations with applications to error detection and fingerprint generation. In: SS 2007: Proceedings of 16th USENIX Security Symposium on USENIX Security Symposium (2007)
15. Caballero, J., Venkataraman, S., Poosankam, P., Kang, M.G., Song, D., Blum, A.: FiG: Automatic Fingerprint Generation. In: Annual Network and Distributed System Security Symposium (2007)
16. Dreger, H., Feldmann, A., Mai, M., Paxson, V., Sommer, R.: Dynamic application-layer protocol analysis for network intrusion detection. In: USENIX-SS 2006: Proceedings of the 15th conference on USENIX Security Symposium (2006)
17. Borisov, N., Brumley, D.J., Wang, H.J.: A Generic Application-Level Protocol Analyzer and its Language. In: Network and Distributed System Security Symposium (2007)
18. Dunn, J.C.: Well separated clusters and optimal fuzzy-partitions. Journal of Cybernetics (1974)
19. Internet2 netflow statistics, `http://netflow.internet2.edu`
20. XUNLEI, `http://www.xunlei.com/`
21. PPLIVE, `http://www.pptv.com/`
22. How Samba Was Written, `http://samba.org/ftp/tridge/misc/french_cafe.txt`
23. Gaim Instant Messaging Client, `http://gaim.sourceforge.net/`

A Specification Based Intrusion Detection Framework for Mobile Phones

Ashwin Chaugule, Zhi Xu, and Sencun Zhu

Department of Computer Science and Engineering,
The Pennsylvania State University, University Park, PA 16802
{avc114,zux103,szhu}@cse.psu.edu

Abstract. With the fast growth of mobile market, we are now seeing more and more malware on mobile phones. One common pattern of many commonly found malware on mobile phones is that: the malware always attempts to access sensitive system services on the mobile phone in an unobtrusive and stealthy fashion. For example, the malware may send messages automatically or stealthily interface with the audio peripherals on the device without the user's awareness and authorization. To detect the unauthorized malicious behavior, we present *SBIDF*, a *Specification Based Intrusion Detection Framework*, which utilizes the keypad or touchscreen interrupts to differentiate between malware and human activity. Specifically, in the proposed framework, we use an application independent specification, written in *Temporal Logic of Causal Knowledge (TLCK)*, to describe the normal behavior pattern, and enforce this specification to all third party applications on the mobile phone during runtime by monitoring the inter-component communication pattern among critical components. Our evaluation of simulated behavior of real world malware shows that we are able to detect all forms of malware that attempts to access sensitive services without possessing user's permission. Furthermore, the SBIDF incurs a negligible overhead (20 μ secs) which makes it very feasible for real world deployment.

Keywords: Mobile Phone, Intrusion Detection, Messaging Attack, Audio Attack.

1 Introduction

With the fast growth of mobile market, surveys and research show that there is an increasing number of mobile phone malware. There are over 400 mobile phone viruses detected so far [12]. Over 17% of manufacturers reported more than 1 million attacks on mobile phones in 2008 [2]. Infected phones are even capable of bringing down the GSM infrastructure of a whole city by exploiting the SMS/MMS messaging protocols [7]. With the advent of newer more capable mobile phone platforms, this security risk will only increase.

Most commonly found malware [11,23,25,29] on mobile phones share a common pattern, in which the malware always attempts to access sensitive services stealthily without authorization from the mobile phone user. For example,

J. Lopez and G. Tsudik (Eds.): ACNS 2011, LNCS 6715, pp. 19–37, 2011.

- In the *Messaging Attack*, such as Cabir [8], Commwarrior [9] and Viver [10], the attacker interfaces directly with the exported serial port of the GSM engine and implement his own messaging framework, thus bypassing all the phone stack components. This can allow him to send messages and interfere with GSM calls. The attacker's intention may be to deplete the battery charge on the phone, spread the malware to other devices, or affect the monthly bill of the user[23,13,25].
- In the *Audio/Video Attack*, such as the research presented by Xu et al. [29], the attacker may access the audio/video peripherals and use them to covertly record an ongoing conversation, interfere in a conversation by playing an audio file or even record an ambient conversation.

To detect those stealthy unauthorized service accesses, the main challenge is *how to differentiate between a purely software generated action and a user initiated action.* Software generated actions without user awareness or interactions will have a high probability of being malicious. Another challenge is *how to detect the unauthorized access efficiently.* Due to the limited battery and computing resources, security solutions for desktops, such as the machine learning based misuse detection approach in [24] and mandatory access control based solutions [20], are not suitable for mobile phones.

In this paper, we propose *Specification Based Intrusion Detection Framework (SBIDF)*, a framework designed specifically for detecting the unauthorized access to sensitive services on mobile phones, specifically for SMS service and audio service. In the proposed framework, we solve the differentiation challenge by observing hardware interrupts. Mobile phones come with touchscreens and/or keypads which generate hardware interrupts for each key press event. This asynchronous notification mechanism is the key to differentiate between a user generated activity and a purely software generated one, since the latter cannot explicitly generate a hardware interrupt. To solve the efficiency challenge, the proposed SBIDF only monitors the Inter-Process Communication (IPC) for the critical components of the userspace stack, which include a *finite* and *small* set of applications *always* involved for performing any function of the device.

Briefly, SBIDF consists of two phrases: training phrase and enforcing phrase. In the training phrase, the vendor defines specifications, written in *Temporal Logic of Causal Knowledge(TLCK)* language, before shipping to the user. These specifications are independent to applications, thus no change on specifications is needed when new applications are installed on mobile phone. In the enforcing phrase, these specifications are then converted into a precise sequence of inter-component communication events in the system. During the usage of mobile phone, the SBIDF enforces the predefined specifications at runtime. Whenever a deviation from specifications is detected, the SBIDF will alert user immediately. The main contributions of this paper are summarized as follows.

- We define concise and precise specifications for legitimate behavior in the system for the events of messaging and calling. The specifications describe system activities related to calling and messaging originating from a hardware interrupt of a corresponding key press event.

- Our framework, SBIDF, resides in the kernel and enforces the behavior of system according to the specification and detects any malicious activity that tries to subvert SBIDF.
- We simulate behavior of real world malware and successfully prevent its malicious attempts to send multiple SMS's and control the audio hardware which interferes with an on going call or stealthily records ambient conversations.
- SBIDF is able to detect all malware with negligible runtime overhead on the system (20μ secs).

Here is the roadmap of this paper. Section 2 describes related research work, Section 3 describes a typical phone stack, Section 4 outlines the security analysis of design, Section 5 describes a design overview, Section 6 describes the framework in detail and Section 7 shows the promising feasibility of our work. We end with laying out future work in Section 8 and conclusions in Section 9.

2 Related Work

Most of the previous research work focused on optimizing desktop solutions for embedded devices. Very few of these propositions started the design from the ground up with an exclusive focus on mobile devices.

Enck et al. [22] and Ontang et al. [21] look at individual application requirements on the Android operating system and define mechanisms to enforce policies defined by the application provider. This approach is very application centric and thus implies numerous policies as number of applications increases. On the other hand, our approach is application agnostic and enforces policies defined only for critical components that provide services to other applications. Thus the number of applications doesn't affect the complexity of the framework.

Xie et al. [15] propose a behavior-based malware detection system named pBMDS, which uses a probabilistic approach to detect anomalous activities in phones via monitoring system calls. In this paper, our detection relies on the inter-components communications within the phone, including IPC events, system calls, and hardware interrupts. Moreover, the specification in SBIDF is predefined thus prevent potential false positive detections caused by false learning.

Bose et al. [3] propose a solution to logically order the events caused by applications on the device. They use machine learning theory to detect the pattern of these events and compare them to a whitelist of behavior signatures. However, their scheme requires a vulnerable complex framework in userspace to detect and monitor these learning patterns and they depend on remote analysis of behavior to reduce the overhead of computation on the device. Their framework is prone to mimicry attacks, so they can have false negatives with their approach.

Cheng et al. [4] proposes a collaborative detection and alert system where neighboring phones collect and analyze system data for intrusions. These designs may also need a proxy server in case collaborative analyses is not possible. They require additional user space components that constantly run in the background collecting this data, which are an expensive overhead for these devices.

Vigna et al. [18] show a way of labeling processes and data to prevent cross service attacks. By tagging resources used by processes upon network activity they monitor the flow of data in between processes. Their scheme is effective in achieving their goal, but not without overheads. Labeling resources requires several rules, transition of labeling incurs monitoring overheads and false positives can easily occur when legitimate processes initiate network activity.

Vogel et al. [28] ported SELinux on the mobile phone to prevent the SMS/MMS related attacks. On the same lines Divya et al. [19] use a stripped down SELinux policy infrastructure based on the PRIMA model to define policies for applications running on the mobile phone to prove to remote verifiers that the system is safe to run their third party software. Desmet et al. [6] describe a way to securely run third party applications on mobile phones without the conventional sandboxing techniques. Their design uses secure execution techniques like run-time monitoring, static verification and proof carrying code. The run time monitors insert hooks into the applications and enforce correctness according to the policies and rely on defining complex policies manually, where improper settings could easily compromise the system. Also, addition of SELinux on mobile phones shows significant overheads in the kernel as shown by Nakamura [20].

Venugopal [26] came up with a faster way to lookup signatures using hashes. He focused on the overhead of detecting which signature matches the current system behavior. However, these anomaly detection techniques still have the downside of false positives and cannot detect zero day attacks. Venugopal et al. [27] describe a virus detection system for the Symbian platform which monitors the DLL functions used by applications. Using Bayesian decision theory and past virus samples, they observe malicious activity. Although they claim a 0% false negative rate, they are only able to detect 95% of the viruses.

3 Background

3.1 Cellphone Platform

For this research we used the Qtopia userspace stack (Qt-Extended-4.4.3) on the Openmoko Neo1973 handset which contains an ARMv7 based CPU (Samsungs S3c2410). This stack is widely used in many of Nokia phones which use the Linux kernel. This stack's design represents many other stack implementations very well. Although there will be differences, we believe our framework can be extended to other variants easily. We describe the implementation of SBIDF on Android platform in the Discussion Section.

Figure 1 shows the components of the Qtopia phone stack and the key interactions. Qtopia contains a critical component called QPE. This is the main server that interfaces with the operating system through device nodes and sockets for IPC communication. QPE is the first application to start when the stack executes. It opens all the necessary sockets and device nodes and then initiates other critical components like Message Server and Media Server. The Message Server controls the messaging and emailing functions. The Media Server controls the voice and audio related functions. However, QPE makes the final system calls

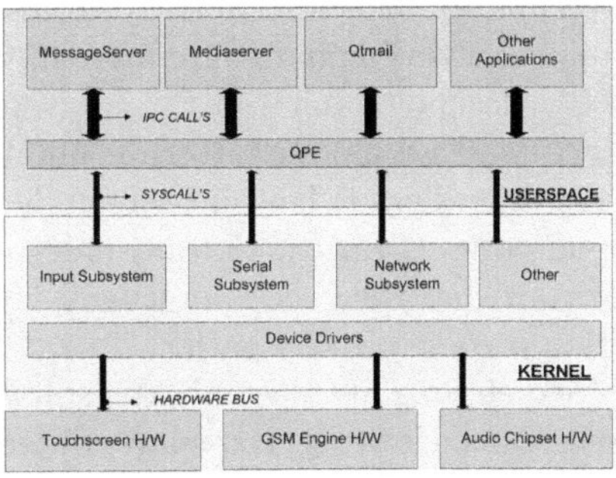

Fig. 1. Qtopia Stack Design

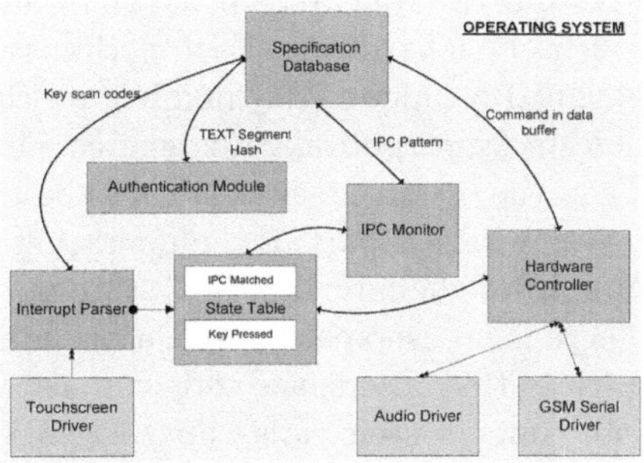

Fig. 2. Specification Based Intrusion Detection Framework (SBIDF)

to the kernel for submitting the SMS/MMS packet and the calls to alter the microphone state. QPE, Message Server and Media Server are started up when the device boots and remain resident till reboot or shutdown. There are several other applications like Qtmail, Games, Browser etc. that are invoked as plugins. These plugins are separate binaries that link with the QPE server at runtime and are executed only when needed. They terminate after their function is over or when the user closes them explicitly. The plugins communicate with each other via QPE by using IPC mechanisms.

In the case of Qtopia, the communication channels are implemented as Sockets and Pipes. These are created by QPE during startup and then the connection at the other end is completed by the other components when they are executed. Different components communicating over the same named socket can still be identified uniquely in the kernel. *The key concept in SBIDF is that we only need to monitor activity between critical components and as such define specifications only for their interactions.* So, the number of specifications does not increase as the number of applications downloaded to the phone increases.

3.2 Design Motivation

Our framework is based on a key observation which is: a user generated activity usually includes a hardware interrupt generated via touchscreens and/or keypads; however, a purely software generated activity cannot explicitly generate a hardware interrupt. This is true because the interrupt generated by the touchscreen hardware is received directly by the CPU. The CPU then responds by calling the interrupt handler of the touchscreen device driver. As the interrupt handler is part of operating system, userspace code cannot directly call it unless the operating system has been tampered with. e.g. via malicious system calls. In our trust model, we assume that the operating system is within the TCB. Therefore, the hardware interrupt handler can not be called directly by userspace code.

In mobile devices, the touchscreen interrupt out line is connected to the CPU interrupt [17] in line via an interrupt controller (which is part of the CPU).

Pressing the touchscreen, triggers an interrupt to the CPU directly from the touchscreen hardware. Each peripheral capable of raising hardware interrupts is assigned a unique IRQ number by the CPU designers. The operating system, detects which peripheral triggered the interrupt via a unique interrupt number. Then the operating system calls the interrupt handler of the touchscreen device driver which is part of the operating system code. Clearly, the legitimate path of flow for the interrupt is all through hardware and then the operating system. There is no userspace application interaction involved in the path.

Note that, userspace code can also raise exceptions like data aborts, software interrupts (system calls), floating point exceptions. However, these have different handlers in the operating system because they are not "hardware interrupts". As explained in the ARM Architecture Reference Manual [17], the CPU jumps to different addresses for these type of exceptions and thus calls different handlers. So unless the operating system is tampered with, there is no direct way in which the touchscreen interrupt handler will be called from userspace code.

4 Security Model

4.1 Threat Model

We assume that attackers use malware to exploit vulnerable components in userspace. Malware is considered as third party applications. Attackers may even compromise the integrity of existing components in the platform, such as the Message Server and Qtmail. For example, in the messaging attacks, we assume that the attacker can interface directly with the exported serial port of the GSM engine and implement his own messaging framework, thus bypassing all the phone stack components. In audio attacks, peripherals are exported through device nodes and can be configured through system calls (IOCTL's etc). The telephony stack on the device registers with these nodes and provides the respective service to other applications. But an attacker can easily open them by himself and use them to alter the audio configurations.

There are other threats where the attacker can interface directly to the network interfaces like BT or WiFi or even hijack the browser or mailing applications to spread malware. If the mobile phone's operating system is not protected (e.g., operating system memory maps are exposed to userspace), then he may even install rootkits that maliciously alter the kernel control flow. This later category is only commonly found on desktops where the user typically has to install device drivers and as such is not a major threat to mobile devices.

In this research, we address the messaging and audio attacks due to their significant implications and widespread occurrence. SMS is an ubiquitous and reliable method to communication among mobile phones. It exists on almost all mobile phones with more than 2.4 billion active users. Also, the mature telecommunication infrastructure makes the message delivery very reliable even when the receiver is offline at the moment of sending. Thus, it is one of favorite choices for spreading malware or building command-and-control channels. Audio attacks are chosen because of its serious consequences. Leaking the audio information directly invades the user's privacy. Video attacks can be detected in a similar way as that of audio attack. We discuss other types in future work section.

4.2 Trust Model

In our work, the operating system running on the device mainly forms the Trusted Computing Base (TCB). The bootloader of phone should correctly load the kernel which contains the SBIDF implementation. Kernel rootkits which may compromise the kernel are not the focus of this research.

We assume that the kernel memory interfaces are not exported to userspace. This requirement is a necessary to prevent userspace applications from writing into kernel memory and altering kernel control flow[14]. For example, in Linux these are exported as /dev/mem and /dev/kmem. These interfaces can be easily disabled during kernel configuration time.

In the userspace, the integrity of critical userspace components should be guaranteed, such as *Message Server* and *Qtmail.* This requirement is needed to prevent a malicious application from bypassing the monitoring of SBIDF. To provide such guarantee, we authenticate the critical components by setting up a TEXT segment hash of the critical userspace component. We only need to trust the critical components during the training phase. After the phone is deployed, the attacker could tamper with them, but as we explain later, our SBIDF can easily prevent any damage.

Note that, those critical components are bundled with the kernel and shipped by the phone vendor. The user and third party applications lack permissions to modify those components. Therefore, their authentication only needs to be done once when the phone is shipped. If there is a need of update , the vendor will redo the authentication during the update.

5 Design Overview

SBIDF utilizes the keypad or touchscreen interrupts to differentiate between malware and human activity. Here we assume that specifications defining correct pattern have been defined and stored in SBIDF. We will explain how to create these specifications in details in Section 6.

Figure 2 shows the overview of our framework. Whenever these critical components start up for execution, the *Authentication Module* computes an md5 hash over the TEXT segment of the component. The TEXT segment is the only read-only segment of a process and any modifications to it during runtime will be detected by the operating system. However, the attacker may replace or infect the binary file of the critical component when it is stored on the flash device. In order to detect this, the *Authentication Module* compares the hash with a pre-computed copy of the hash of that component. This pre-computed copy is present in the *Specification Database.* It could be calculated for the first time by the phone stack provider during a training phase and then statically stored in the *Specification Database* for future use. When the *Authentication Module* finds a match of the hash with the pre-computed copy, it sets an authentication bit in that component's Process Control Block (PCB), which is an in-kernel representation of the userspace application, to signify the successful authentication. This avoids recalculation of the hash for later stages.

In order to detect the messaging or calling activity in userspace from within the kernel, the *IPC Monitor* observes all read and write IPC calls and it checks the *State Table* for the *Key Pressed* flag. If the flag is unset, then it simply returns, since the current IPC transaction was not triggered by a hardware interrupt. If it is set, the *IPC Monitor* checks the PCB of the process that has made this IPC call. If it finds the authentication bit to be set, then it knows that the *Authentication Module* has authenticated this process previously. Similarly, *IPC Monitor* also checks if the communicating peer involved in the current IPC has been authenticated. Following this, it checks if the processes are communicating over a specific named socket. This socket information is present in the specifications in the *Specification Database*. Now the *IPC Monitor* knows which authenticated critical components are communicating over a specific socket and that this activity was triggered by a hardware interrupt. For each type of the attacks (i.e., messaging or audio attacks) we aim to prevent, the *Specification Database* contains unique specifications which define the expected IPC pattern between the critical components. Amongst all the critical components, there is usually a single component that finally sends out an SMS/MMS message from userspace to the kernel. If all the IPC transactions occur as specified, then the *IPC Monitor* sets a permission bit of that critical component in its PCB, granting it the permission to submit the SMS/MMS, and it also sets the flag *IPC Matched* in the *State Table*.

The *Hardware Controller* mediates all the read and write calls that occur on the GSM serial port and audio interface. When the *Hardware Controller* detects that there is a GSM command (e.g., *Submit SMS*) in the write call, it checks the *IPC Matched* flag in the *State Table*. If this flag is set, then the *Hardware Controller* checks if a permission bit is set for the critical component invoking the write. If both these conditions are true, it allows the message to go through the GSM hardware. If either of them is false, the message is denied.

SBIDF detects audio attacks in a similar way. The *Interrupt Parser* checks if the ACCEPT CALL or END CALL keys are pressed. The *IPC Monitor* checks if the expected IPC pattern occurs. The *Hardware Controller* checks if an expected authenticated critical component as defined by the specification has been granted permission by the *IPC Monitor* in the write call to the audio driver to switch the microphone ON . If the *IPC Matched* flag is set, and the critical component has been granted the permission, then the MIC is turned ON. The *Hardware Controller* parses the data buffer of the write call to the GSM serial driver so as to determine when to switch the MIC OFF. If it finds a *CALL HANGUP* GSM command, then it knows the call is over and then switches the MIC OFF.

Compare the above scenarios with what the malware would do. There will be no hardware interrupt generated. So, if the malware directly tries to access the hardware, the *Hardware Controller* will deny access, because it will not see the permission bit set by the *IPC Monitor*. If the malware tries to masquerade as one of the critical components, it will be detected by the *Authentication Module*, which mediates all application's startups. If the malware tries to mimic the IPC pattern, the *IPC Monitor* will detect this, because it will not get the interrupt information from the *Interrupt Parser*, or it will not get the authentication information from the *Authentication Module* and therefore will not set the permission bit. Therefore, the *Hardware Controller* will always deny any malicious access.

6 Specification Design and Enforcement

In SBIDF, the specifications define the precise expected behavior of the system for specific events like messaging and calling. Each specification contains the $TEXT$ segment hash of the critical components involved in the activity, the key scan codes that trigger the IPC activity, the expected IPC pattern and the action for the *Hardware Controller.* The *Specification Database* is available to all the above components of the *SBIDF.* Each specification is third party application independent.

6.1 Specification Formalization

We use the TLCK (Temporal Logic of Causal Knowledge) described by Bose et al. [3] to describe the sequence of events in the system and the interactions in the SBIDF's state machine. Temporal events in the system are described by following notations.
$\odot_t$ is an event true at time t.
$\triangle_t$ is an event true before time t.
$\square_t^{t-k}$ is an event true in the interval $[t-k, t]$.
The other operators such as $\wedge$ and $\vee$ etc. carry their usual meanings. Next we define some propositional variables.

- **KeyPressed(S):** Where, $S = \{s_1, s_2, s_3, .., s_n\}$ which is a set of *'n'* scan code interrupts which we need to monitor. e.g.*SEND, RECORD, STOP* etc. $KeyPressed(S)$returns $TRUE$ if any of the monitored keys is pressed.
- **AuthApp(T):** Where, $T = \{t_1, t_2, ..., t_m\}$ which is a set of *'m'* applications that we need to authenticate. e.g. *QPE, Qtmail, Mediaserver, Messageserver* etc. *AuthApp*(T) returns $TRUE$ if the hash of the running application matches with a pre stored digest of that application.
- **IPC(T, Sockname):** IPC encapsulates both IPC read and write functions over the AF_UNIX socket. The socket name is defined by *Sockname*. The IPC variable returns $TRUE$ iff all the IPC transactions occur over the specified socket in the specified sequence and within a timeframe. The pattern for the expected IPC is defined by a truth table as shown in the next section.
- **ParseATCMD(C):** Where, $C = \{AT + CMGS, AT + CHLD = 1\}$ is the set of AT commands to be monitored. $ParseATCMD(C)$ returns $TRUE$ when the *Hardware Controller* finds any of these commands in the *data* buffer that is passed to the GSM engine over the serial port.

- **Perm(X):** $X = \{QPE\}$. $Perm(X)$ returns $TRUE$ if QPE has its permission bit set as explained previously. For other stacks, X will contain the critical components which make the final system calls to access the hardware.

Now, we define specifications using the TLCK and the aforementioned notations for messaging attack and audio attack.

For Messaging Attack. Let
$A \xrightarrow{r} B$ denote an IPC Read of application A from application B.
$A \xrightarrow{w} B$ denote an IPC Write of A to B.
As a first example let us consider the specification for submitting an SMS.
The set $S = \{SEND\}$
The set $T = \{Qpe, Qtmail, Messageserver\}$
The set $C = \{AT + CMGS\}$
Sockname = *"/tmp/qtembedded-0/QtEmbedded-0"*

The truth table for $IPC(T, Sockname)$ is shown in Table 1. The IPC variable is $TRUE$ iff all the other variables are $TRUE$.

Table 1. Messaging Truth Table

Variable	Value
$Qpe \xrightarrow{r} Qtmail$	T
$Qtmail \xrightarrow{r} Qpe$	T
$Qpe \xrightarrow{w} Qtmail$	T
$Qtmail \xrightarrow{w} Qpe$	T
$Qpe \xrightarrow{r} Msgserver$	T
$MsgServer \xrightarrow{r} Qpe$	T
$Qpe \xrightarrow{w} MsgServer$	T
$Msgserver \xrightarrow{w} Qpe$	T
IPC(T, Sockname)	**T**

Table 2. Audio Call Truth Table

Variable	Value
$Qpe \xrightarrow{w} Mediaserver$	T
$Mediaserver \xrightarrow{r} Qpe$	T
$Mediaserver \xrightarrow{w} Qpe$	T
$Qpe \xrightarrow{r} Mediaserver$	T
IPC(T, Sockname)	**T**

The specification is defined as follows:
$\odot_t(SubmitSMS) = \Delta_t(KeyPressed(S) \wedge AuthApp(T)) \wedge (\Box_t^{t-k}(IPC(T, Sockname) \wedge Perm(X) \wedge ParseATCMD(C)))$

Here, $SubmitSMS$ is true only when a real user pressed the $SEND$ key on the touchscreen/keypad, the applications in set T were authenticated by the *Authentication Module* and there was an expected IPC transaction over the socket defined by *Sockname* between these authenticated components within a time frame, the *IPC Monitor* set the permission bit for QPE and the *Hardware Controller* received an $AT + CMGS$ command to submit the SMS from QPE in the data buffer of the GSM serial port. The time frame for IPC can be customized depending upon the overheads of the IPC calls. When $SubmitSMS$ evaluates to $TRUE$, then the outgoing SMS is allowed, else denied.

For Audio Attack. In a similar way, the audio attack specification can be constructed as follows.

The set $S = \{CALL, ENDCALL\}$
The set $T = \{Qpe, Mediaserver\}$
The set $C = \{AT + CHLD = 1\}$
Sockname="/tmp/qt-embedded/valuespace_applayer"

The truth table for IPC(T, Sockname) is shown in Table 2. The IPC variable is $TRUE$ iff all the other variables are $TRUE$. The specification is defined as :

$\odot_t(AllowAudio) = \Delta_t(KeyPressed(S) \wedge AuthApp(T)) \wedge (\Box_t^{t-k}(IPC(T, Sockname) \wedge Perm(X)))$

Here, *AllowAudio* is true when a real user presses the CALL key, the *IPC Monitor* confirms the IPC pattern amongst authenticated processes on the specified socket within the specified time frame and the authenticated QPE is given the permission to toggle the microphone, then the *Hardware Controller* turns the microphone ON. At all other times, any command to alter the microphone state is denied. Since the *Hardware Controller* controls the microphone, it needs to know when to turn it OFF again. For this we have another specification.

$\odot_t(DenyAudio) = \Delta_t(KeyPressed(S) \wedge AuthApp(T)) \wedge (\Box_t^{t-k}(IPC(T, Sockname) \wedge ParseATCMD(C) \wedge Perm(X)))$

Here the *Interrupt Parser* looks for the $ENDCALL$ scan code. The truth table for this IPC operation in case of the Qtopia stack is the same as the case for $CALL$. In order to ensure that a call is being dropped or ended, the *Hardware Controller* parses the AT commands passed to the GSM engine. So, when it detects the command $AT + CHLD = 1$ after the IPC operations, and QPE has the permission, it switches the microphone OFF.

6.2 Specification Enforcement

The description of specification enforcement in the *SBIDF* is described with a State Machine. In Figure 3, we show the state machine for specifications we introduced in previous subsection.

State INT: This is the Start state of the machine, where the *Interrupt Parser* is just parsing the Input key events. The set of key scan codes to be monitored is dependent upon the specification. When either of these key presses are detected, the *SBIDF* transitions to the *IPC State.*

State IPC: This is where the *IPC Monitor* begins to monitor the ensuing IPC communication between processes after a specific key press event. In this state, it also checks if the communication is being performed over a specified named socket as defined by the specification and the communicating peer processes are authenticated. If authentic, it proceeds to either the $T1$ or $T2$ *States.*

State T1: Here, the *IPC Monitor* detects there was an IPC Send operation from Qpe to Qtmail ($Qpe \xrightarrow{w} Qtmail$). This is an indication that there could be a event to form an outgoing SMS. Then the SMS timer is started. After this timer expires, the *IPC Monitor* checks how many expected IPC operations were performed. Here, the timer is set to expire in 1 second. The timer expiry value is decided by observing the overheads involved in a normal IPC call. Upon timer

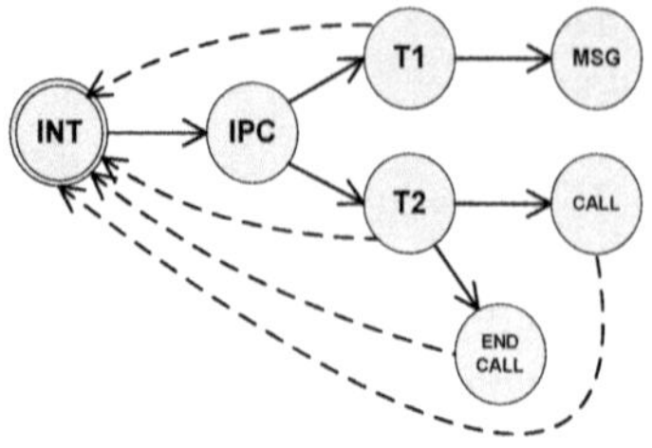

Fig. 3. State Machine

expiry, if all the expected IPC occurred and the *Hardware Controller* gets an $AT + CMGS$ command in the *data* buffer of the GSM serial driver from an authenticated QPE with the permission granted, then the GSM transmit path is unlocked and the SMS is let though the GSM engine in $STATE\ MSG$.

State T2: Here the *IPC Monitor* detects if there was an IPC Send operation from QPE to Mediaserver ($Qpe \xrightarrow{w} Mediaserver$). This signifies that an outgoing GSM call could be in progress. The MIC timer is triggered to expire in 1 second. After this timer expires, the *IPC Monitor* checks if the expected IPC pattern occurs, then the *Hardware Controller* checks if the $AT + CHLD = 1$ command is found in the *data* buffer of the GSM serial driver written by an authenticated QPE with the permission granted. If this command is found, then it signifies an end of call, so the *Hardware Controller* switches the microphone OFF, else it is turned ON. This way, the microphone is kept *ON* only during the duration of a call and when the call is in progress, any modification to reroute the audio is denied by the *Hardware Controller.*

Reverse Edges: The dashed back arrows in the State Machine in Figure 3 are explained here. These edges characterize the false conditions of the propositional variables described previously.

INT $\rightarrow$ INT: This means the *Interrupt Parser* has not encountered any key press events for the ones it is monitoring.

IPC $\rightarrow$ INT: This could happen when

- The *Interrupt Parser* falsely recognized a key press event but the *IPC Monitor* did not find a matching IPC pattern.
- Unauthenticated entities tried to communicate over the specified socket. This could be when malware is masquerading as legitimate applications, or when malware is trying to mimic the IPC pattern.
- Authenticated entities sent data over some other socket than the one specified. This could happen when the ongoing pattern represents other system activity that is legitimate but has no relation to the attack prevention.
- When the *IPC Monitor* detected that the ensuing IPC activity was not triggered by an expected hardware interrupt.

T1 → INT: The *IPC Monitor* started the SMS timer, but the ensuing IPC transactions didn't match with the specification or didn't complete in time due to system noise.

T2 → INT: The *IPC Monitor* started the Call timer, but the ensuing IPC transactions didn't match with specification or didn't complete due to system noise.

(MSG/CALL/ENDCALL) → INT: In any of these cases, the *Hardware Controller* did not get the commands in the data buffer from an authenticated QPE component. Or, the *IPC Matched* flag was not set, which implies QPE did not get permission. The latter case would arise when some userspace application tried to directly interface with hardware. This could also happen when the system noise mimicked the events in the specification but it was a false alarm.

Since we are only concerned with the interactions amongst the select few critical components in the phone stack, the complexity of the state machine does not increase with the number of applications running in the system. The number of nodes and state transitions will only increase if we aim to prevent more attacks than the ones we have described.

7 Evaluation

All instrumentation was performed completely in the kernel of the Openmoko device. The hardware of the device consists of a Samsung S3c2410 ARMv7 based CPU running at 266MHz, with 128MB SDRAM, 64MB NAND Flash. The Linux kernel version used was v2.6.24 with modifications to add the SBIDF code. Only 784 lines of kernel code were added and no userspace code was changed.

7.1 Security Evaluation

For the simulation of malware, we coded two representative applications to demonstrate the Messaging and Audio attacks that replicate the behavior of commonly found malware. The SMS attack program implemented its own message server and interfaced directly with the GSM serial port. Without the SBIDF, this program automatically sends SMSes. The SBIDF, denied all SMSes from this program, since the events in the system did not match with the specification therefore the *Hardware Controller* denied access to the hardware.

Note that although the attack we implemented is a specific one, to our knowledge all the existing messaging attacks are similar. Indeed, our framework is able to prevent any variant of the Messaging attacks since all these attacks finally try to use the GSM hardware either directly or indirectly. For example, such malware may exploit another application to interface with the GSM hardware on its behalf. Since the SBIDF uses precise specifications that define the behavior amongst critical applications to send SMSes, any malware that deviates from this behavior will be detected. Also, even if the malware attempts to mimic the specification behavior, the *Authentication Module* detects malware that masquerades as legitimate applications, the *IPC Monitor* detects any malicious activity since it checks the authenticity of the communicating peers, the *Hardware Controller* denies any access to hardware if the *IPC Matched* flag is not set and finally since there will be no hardware interrupt to trigger the IPC, the *Interrupt Parser* will not set the *Key Pressed* flag.

The audio attack, firstly implemented by us, tries to interface with the audio subsystem node to configure the audio chipset. This malware tries two attacks on the audio interface. The first one tries to play a file during a call, with an intention to interfere with the on-going conversation by making the other peer hear the sound from the file. The second one tries to record an ongoing conversation with and without a call in progress. In the second case, the intention of the malware is to covertly record the conversations during a call and to record ambient conversations when the user is not on a call. The recorded audio is routed to a file, which can then be transferred to the attacker via an SMS or any other network interface. This part of transferring the recorded audio was not implemented. However, with the SBIDF, any attempt to alter the microphone configuration was denied by the SBIDF. Using the similar principles behind preventing the SMS/MMS attack, our framework is able to prevent any variation of the audio malware. Moreover, the video attack as described in [29] can also be easily prevented by the SBIDF, because the audio attack program closely matches with their video capture malware.

7.2 Overhead Evaluation

Application Text Segment Sizes. The *Authentication Module* only calculates the hashes of the $TEXT$ segment of the critical applications. Their sizes as stored in the respective PCBs are as follows: QPE(176 KB), MediaServer(192KB), MessageServer(596KB), Qtmail(28KB).

Application Load Time. The SBIDF affects the load time of only the critical components. All the other applications being loaded in the system are not considered by the SBIDF. Since *do_execv* is the system call to load the application into main memory for the Linux kernel, the table shown below shows the overheads for this call with and without the SBIDF.

The time taken for these applications to load through the *Authentication Module* interface may seem very high. But this is because it scans through the whole $TEXT$ segment of the application, calculates an md5 hash and then

Table 3. Application Load Overheads

Application	Time	
	W/ SBIDF(ms)	**W/O SBIDF**(ms)
QPE	374	2
MediaServer	210	78
MessageServer	561	66
Qtmail	40	10

Table 4. Training Run Overheads

Application	**Time** (ms)
QPE	375
MediaServer	185
MessageServer	816
Qtmail	43

Table 5. IPC Overheads

With SBIDF(μs)		
	Send	**Receive**
No Key Press	1.0	1.0
Not Authenticated	1.0	1.0
Authenticated	20	17

compares this hash with a pre-stored digest that was computed in the training run. This scan may include a page table walk to fetch the respective pages into memory. However, this is only a one-time overhead, since the *Authentication Module* sets a bit in that applications PCB after comparing the hashes to signify whether the application has been authenticated. This bit is then checked during the IPC transactions thus avoiding re-calcuation of the hashes.

Training Run Time. For the training run, the SBIDF uses the same logic to calculate the hash of the running applications, but stores the hashes into a file, which may then be statically added to the *Specification Database* for run-time usages. Table 4 includes the implicit file I/O overheads.

Input Event Overheads. The *Interrupt Parser* parses the scan code of specific keys such as the *SEND, CALL, ENDCALL keys.* For the OpenMoko device, the touchscreen hardware produces screen co-ordinates which then map to scan codes. Each button on screen is represented by a range of $[x, y]$ co-ordinates. Hence the *Interrupt Parser* decodes the keys according to the range of co-ordinates for each button. It takes only $1.1\mu sec$ to check if an expected key is pressed. Thus, the SBIDF can easily detect when to proceed in the state machine even with random key presses.

IPC Overheads. The *IPC Monitor* monitors each IPC Send and Receive operation over the AF_UNIX sockets. However, it only mediates IPC operations over specific sockets as defined by the specification. Also, the time taken by the IPC operation depends upon the data being transferred between the two processes. This varies for every run. Hence the results listed in Table 5 show the average time over 5 runs of sending an SMS and making a GSM call. For each IPC operation, first the *IPC Monitor* checks to see if any of the monitored keys was pressed. If not it simply returns, because that IPC operation was not triggered as a result of a hardware interrupt. Similarly, it then checks if the application that initiated the IPC operation was authenticated previously. If not, then it returns. This prevents malware from sending an SMS or modifying the microphone state. If the applications are authentic and some monitored key was pressed, then it checks which applications sent or received data and on which socket. Accordingly it decides which timer to trigger as per the specification. The cases shown here imply that the SBIDF takes on average just $20\mu s$ more while trying to prevent malware activity and negligible overhead for all other cases. This shows how the SBIDF can easily account for false alarms due to random system noise.

8 Discussions

8.1 Scalability

Specification Database. We define specifications for legitimate sequence of events in the system for the attacks we aim to prevent. For each type of attack, the proposed framework defines a specification written in TLCK and enforces the specification with state machines. The number of specifications only increase if the number of attacks to be prevented increases. The length of the specification depends on the number of critical components in the system, which by design in most phone stacks is a small number. Unlike the context-related policies [5], specifications in SBIDF focus on low-level system events.

Application to other platforms. The SBIDF is a framework that uses observation of inter-component communications to detect abnormal behaviors of third party applications. Our preliminary experiments show that this framework can be implemented for the Android platform on the Android ADP1 developer phone. Android uses the Binder framework [1] for its IPC mechanism. A trace of system events in the kernel shows that the binder kernel driver is able to detect both peers involved in the IPC transaction. The critical components in this case are *rild, com.android.phone, system_server, media_server*. Since the binder framework does not use named sockets, we found that there were node identifiers for each object involved in the transaction. These objects are represented as nodes of a red-black tree per process/thread. The nodes in the tree act as senders or receivers of data. The *logcat* service on the *radio* log shows the *AT* commands exchanged with the GSM core. These commands could be parsed in the kernel driver that maps the memory for the shared memory bridge between the application processor and GSM core (smd_qmi.c). SBIDF will need to include additional hooks in the sliding keypad interface driver to mediate inputs other than the touchscreen.

8.2 Limitations and Future Work

Given the great variety of mobile phone platforms and the sophistication of attacks, we do not think any single or a few defense techniques will be sufficient for all cases. SBIDF focuses on detecting unauthorized access to sensitive services. Below we discuss some limitations as well as some ideas for future research.

Message Integrity. At this stage, we do not consider the integrity of the outgoing messages. So, if there is a vulnerability in the data segment of the messaging application, once exploited, a malware could piggy-back malicious data along with an outgoing message. Here, the SBIDF will not be able to detect the alteration, but still let it pass since the message was initiated through the *SEND* key interrupt followed by the specified IPC pattern. We think this is a challenging research problem using our framework. We might need to always trust the message server to protect the integrity of the message.

Reducing False Positives. We notice that a few normal applications which automatically send SMS messages (e.g., with location data) in the background have emerged. By our design principle our system would not be able to differentiate these automated applications from malware which have similar behavior. Their difference is on the intention, not on the techniques. So we need to attest the intention of such automated messages. For an automated message detected by our system, we are considering to validate the application originating it through a graphic Turing test [16]. If the user confirms it is from an authorized application, we may white-list the application (assuming the application is not compromised yet at its first-time use). Such white-listed applications can be authenticated using our *Authentication Module*. If the application is authenticated and authorized, the message claimed to be from it will be allowed and otherwise denied.

Network Interfaces. Parsing data communication via BT and WiFi to detect malicious activity without incurring high overheads and false positives is challenging. However, we think we can use SBIDF to mediate requests to power ON and OFF these interfaces, or alter their configuration at runtime. This is possible since, a real user will have to press a sequence of keys on screen. This is typically a system configuration screen with one button for BT/WiFi power ON/OFF. But we will not be able to parse events after these interfaces are switched ON, using current techniques. We plan to leverage the work of Bose et al. [3] to detect malware that tries to send sensitive files via these interfaces.

Component Authentication. To protect the integrity of critical components, we use the *Authentication Module* to compute an md5 hash over the TEXT segment of the component. This authentication approach works well to protect the static part of the component. However, a component may also contain dynamic part which changes during run time. Attacker may take advantage of this dynamic part and bypass the authentication check. Verifying the dynamic part is still a hard problem requiring more effort and investigations.

9 Conclusions

The framework described in this research shows a specification based intrusion prevention approach to detect unauthorized access to sensitive services, such as SMS, audio, and video services. We believe our framework is the first of its kind to address mobile phone malware using hardware interrupts. For each type of attack, the proposed framework defines a specification written in TLCK and enforces the specification with state machines. The number of specifications only increase if the number of attacks to be prevented increases. Through evaluations, we show how the framework is capable of detecting unauthorized access to sensitive services at runtime with neglectable overhead.

Acknowledgment

We thank the reviewers for the valuable comments.This work was supported in part by NSF CAREER 0643906 and ARL CTA 2010-2015. The views and conclusions contained in this document are those of the author(s) and should not be interpreted as representing the official policies, either expressed or implied, of the U.S. Army Research Laboratory or the U.S. Government.

References

1. Android developer phone, `http://developer.android.com/`
2. F-secure report, `http://www.vnunet.com/vnunet/news/2230481/f-secure-launches-mobile`
3. Bose, A., Hu, X., Shin, K.G., Park, T.: Behavioral detection of malware on mobile handsets. In: MobiSys 2008 (2008)
4. Cheng, J., Wong, S.H.Y., Yang, H., Lu, S.: Smartsiren: virus detection and alert for smartphones. In: MobiSys 2007 (2007)
5. Conti, M., Nguyen, V.T.N., Crispo, B.: CRePE: Context-related policy enforcement for android. In: Burmester, M., Tsudik, G., Magliveras, S., Ilić, I. (eds.) ISC 2010. LNCS, vol. 6531, pp. 331–345. Springer, Heidelberg (2011)
6. Lieven, D., Wouter, J., Fabio, M., Katsiaryna, N., Pieter, P., Frank, P., Dries, V.: A flexible security architecture to support third-party applications on mobile devices. In: CSAW 2007 (2007)
7. Enck, W., Traynor, P., McDaniel, P., La Porta, T.: Exploiting open functionality in sms-capable cellular networks. In: CCS 2005 (2005)
8. F-Secure. Cabir, `http://www.f-secure.com/v-descs/cabir.shtml`
9. F-Secure. Commwarrior, `http://www.f-secure.com/v-descs/commwarrior.shtml`
10. F-Secure. Viver, `http://www.f-secure.com/v-descs/trojan_symbos_viver_a.shtml`
11. Gostev, A.: Mobile malware evolution: An overview, `http://www.viruslist.com/en/analysis?pubid=200119916`
12. Kidman, A.: Mobile viruses set to explode mobile viruses set to explode, `http://www.zdnet.com.au/news/security/soa/mobile-viruses-set-to-explode/`
13. Kim, H., Smith, J., Shin, K.G.: Detecting energy-greedy anomalies and mobile malware variants. In: MobiSys 2008 (2008)
14. Kruegel, C., Robertson, W., Vigna, G.: Detecting kernel-level rootkits through binary analysis. In: ACSAC 2004 (2004)
15. Seifert, J., Xie, L., Zhang, X., Zhu, S.: pbmds: A behavior-based malware detection system for cellphone devices. In: WiSec 2010 (2010)
16. Liang, X., Xinwen, Z., Sencun, Z., Trent, J., Chaugule, A.: Designing system-level defenses against cellphone malware. In: SRDS (2009)
17. ARM Limited, Arm architecture reference manual
18. Mulliner, C., Vigna, G., Dagon, D., Lee, W.: Using labeling to prevent cross-service attacks against smart phones (2006)
19. Muthukumaran, D., Sawani, A., Schiffman, J., Jung, B.M., Jaeger, T.: Measuring integrity on mobile phone systems. In: SACMAT 2008 (2008)
20. Nakamura, Y.: Selinux for consumer electric devices. In: Linux Symposium
21. Ongtang, M., McLaughlin, S., Enck, W., McDaniel, P.: Semantically rich application-centric security in android. In: ACSAC 2009 (2009)
22. Ongtang, W.E.M., McDaniel, P.: On lightweight mobile phone app certification. In: CCS 2009 (2009)
23. Chen Racic, M.: Exploiting mms vulnerabilities to stealthily exhaust mobile phone's battery. In: SecureComm 2006 (2006)

24. Sekar, R., Gupta, A., Frullo, J., Shanbhag, T., Tiwari, A., Yang, H., Zhou, S.: Specification-based anomaly detection: a new approach for detecting network intrusions. In: CCS 2002 (2002)
25. Traynor, P., Enck, W., McDaniel, P., La Porta, T.: Mitigating attacks on open functionality in sms-capable cellular networks. In: MobiCom 2006, pp. 182–193. ACM, New York (2006)
26. Venugopal, D.: An efficient signature representation and matching method for mobile devices. In: WICON 2006 (2006)
27. Venugopal, D., Hu, G., Roman, N.: Intelligent virus detection on mobile devices. In: PST 2006 (2006)
28. Vogel, B., Steinke, B.: Using selinux security enforcement in linux-based embedded devices. In: MOBILWARE 2008 (2007)
29. Xu, N., Zhang, F., Luo, Y., Jia, W., Teng, J.: Stealthy video capturer: A new video-based spyware in 3g smartphones. In: WiSec 2009 (2009)

Misuse Detection in Consent-Based Networks

Mansoor Alicherry[1] and Angelos D. Keromytis[2]

[1] Bell Labs, Alcatel-Lucent, Murray Hill, NJ, USA
[2] Department of Computer Science, Columbia University, New York, USA

Abstract. Consent-based networking, which requires senders to have permission to send traffic, can protect against multiple attacks on the network. Highly dynamic networks like Mobile Ad-hoc Networks (MANETs) require destination-based consent networking, where consent needs to be given to send to a destination in any path. These networks are susceptible to multipath misuses by misbehaving nodes.

In this paper, we identify the misuses in destination-based consent networking, and provide solution for detecting and recovering from the misuses. Our solution is based on our previously introduced DIPLOMA architecture. DIPLOMA is a *deny-by-default* distributed policy enforcement architecture that can protect the end-host services and network bandwidth. DIPLOMA uses capabilities to provide consent for sending traffic. In this paper, we identify how senders and receivers can misuse capabilities by using them in multiple paths, and provide distributed solutions for detecting those misuses. To that end, we modify the capabilities to aid in misuse detection and provide protocols for exchanging information for distributed detection. We also provide efficient algorithms for misuse detection, and protocols for providing proof of misuse. Our solutions can handle privacy issues associated with the exchange of information for misuse detection. We have implemented the misuse detection and recovery in DIPLOMA systems running on Linux operating systems, and conducted extensive experimental evaluation of the system in Orbit MANET testbed. The results show our system is effective in detecting and containing multipath misuses.

Keywords: network capability, MANET security, misuse detection.

1 Introduction

Consent-based networking is emerging as a "clean-slate" design for providing security against multiple attacks in the Internet [12,5,17,4]. In consent-based networking, a sender needs to have permission to send traffic to a destination. Consent-based architectures may support permission to send to a destination on a particular path (*path-based*) or on any path (*destination-based*). In path-based consent architectures, every node (or realm) in the path from a source to a destination need to give consent to send traffic. This gives the nodes control over the traffic passing through them, making it suitable for networks like the Internet, where there are multiple providers (or administrative domains) and the

J. Lopez and G. Tsudik (Eds.): ACNS 2011, LNCS 6715, pp. 38–56, 2011.

paths are mostly static. In destination-based consent architectures, permission is given to send traffic to a destination on any of the available paths; intermediate nodes honor those permissions and forward the traffic. This architecture is useful for networks where the paths are dynamic, as in mobile ad-hoc networks.

In a consent-based system, senders are given the permission to send traffic in the form of verifiable proofs of consent (capabilities). The nodes perform bandwidth enforcement by rate controlling the bandwidth used for the flows that are part of the capability. In destination-based consent architectures, it is possible to use the capability to reach a destination on multiple paths. In those cases, not all traffic corresponding to a capability may go through a node. Hence, any single node may not be able to enforce the bandwidth constraints of the capability. Furthermore, a node that has authority over multiple destination nodes may assign permission to reach those destinations in a single capability. Hence, the same capability may be used for multiple unrelated flows. Even if all the traffic passes through a node, the node may be unable to enforce the bandwidth constraints across unrelated flows due to high processing required to account for traffic across the flows. It is also possible for certain nodes to collude with senders allowing for larger bandwidth than the one allocated in the capability. When misuse prevention is not feasible, we need a detection mechanism. Once misuse is detected, the capability may be revoked or temporarily not honored, or the node misusing the capability may be isolated.

Recently proposed DIPLOMA [4,3] is a destination-based consent architecture for MANETs based on the concept of network capabilities [5]. A capability is a cryptographically sealed token of authority that has associated rights. DIPLOMA capabilities propagate both access control rules and traffic-shaping parameters that should govern a node's traffic. All the nodes in the path from a source to a destination can verify and enforce the capability. The architecture is based on deny-by-default paradigm, where nodes can only access the services and hosts they are authorized for by the capabilities given to them. Thus, DIPLOMA provides two main features: access control of the end-host services, and protection of network bandwidth.

In this paper, we identify the sources of misuse in DIPLOMA and provide solutions for detecting those misuses. A misuse may constitute either the use of a capability in multiple paths to a destination, or the use of the same capability to multiple destinations. The detection of misuse may be done based on the information locally available to the node (local detection), or based on the information exchanged among the nodes (distributed detection).

To provide solutions for detecting misuses, we enhance the capability establishment protocol to enable nodes to detect the misuses. We also describe the protocols for communicating the information about the flows going through the nodes to enable distributed detection. We also provide efficient algorithms for detecting misuses.

The node detecting a misuse should be able to provide the proof of the same, so that other nodes can take action based on the misuse. Our solution can provide the proof of the misuse, so that rogue nodes cannot exploit the misuse detection algorithms itself. Our solution also handles privacy issues associated with the exchange of information about the flows.

We implemented our algorithms in the Orbit lab testbed [1]. We show that the algorithms require minimum processing and memory. We also show that the amount of information exchanged for the misuse detection algorithm is minimal. We also conduct extensive experiments on capability misuses, and show that our system effectively detects and contains these misuses.

2 Misuses in Consent-Based Architectures

The misuses in consent-based architectures depend on the type of the architecture and the resources it is trying to protect. It depends whether the consent is for a particular path or on any path to the destination. The misuse also depends on whether the consent has any bandwidth constraint or it is just an access constraint (i.e. unlimited bandwidth).

In architectures like network capabilities [5,16,17] and visas [7], consent is given to access the receiver in any path. In network capabilities, all the nodes from a source to destination participate in the protocol. In visas, on other hand, only the source and destination networks are involved in the protocol. In ICING architecture [12], consent is given for a particular path. ICING also requires that all the intermediate nodes from the source to destination give explicit consent for the packet to go through. In DIPLOMA, consent is given for any path, but enforcement is done at all the intermediate nodes from source to destination.

Another factor that influences the misuse is whether the consent based architecture depends on the trusted nature of the routers or intermediate nodes and security of the communication medium. If the protocol assumes trusted intermediate nodes, it may be possible to overcome the protection by compromising the routers. In general, protocols designed for wired networks assume trusted routers. DIPLOMA is designed for wireless networks where the routers may not be trusted and the communication medium is broadcast in nature.

A consent-based architecture may provide only access control or may additionally provide bandwidth limitations. It is easier to enforce bandwidth constraints on the path-based architectures. In destination based architecture like DIPLOMA, which also provides bandwidth constraints, it is a challenge to enforce the bandwidth constraints due to use of multiple paths to a destination; the constraints has to be enforced across all the paths. The focus of this paper is to detect and recover from misuses involving multiple paths.

3 DIPLOMA Overview

We assume wireless ad-hoc network setting where the nodes have limited mobility. In DIPLOMA architecture, the resources needed to access a service are allocated by the *group controller(s)* (GCs) of the MANET. Group controllers are nodes responsible for maintaining the group membership for a set of MANET nodes, and *a priori* authorize communications within the group. This means

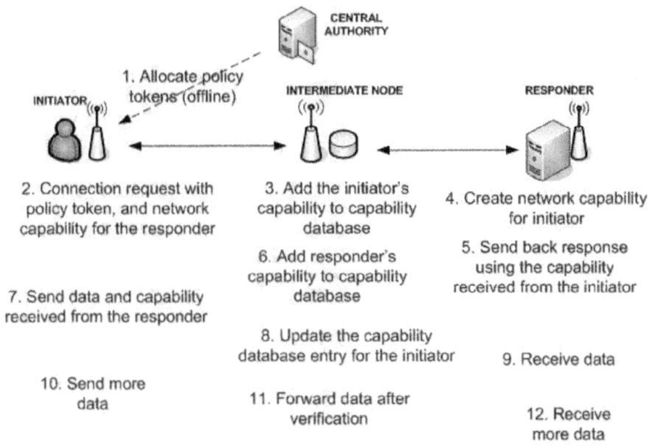

Fig. 1. System overview

that GCs do not participate in the actual communications, nor do they need to be consulted by nodes in real time; in fact, if they distribute the appropriate policies ahead of time, they need not even be members of the MANET. In some cases, the GC may be reachable through a high-energy-consumption, high-latency, low-bandwidth long-range link (*e.g.*, a satellite connection); interactions in such an environment should be kept to a minimum, and only for exceptional circumstances (*e.g.*, for revoking access for compromised nodes). The resource allocation by GC to a node is represented as a credential (capability) called *policy token*, and it can be used to express the services and the bandwidth a node is allowed to access. They are cryptographically signed by the GC, which can be verified any node in the MANET.

When a node (initiator) requests a service from another MANET node (responder) using the policy token assigned to the initiator, the responder can provide a capability back to the initiator. This is called a *network capability*, and it is generated based on the resource policy assigned to the responder and its dynamic conditions (*e.g.*, level of utilization).

Figure 1 gives a brief overview of DIPLOMA. All nodes in the path between an initiator to a responder (*i.e.*, nodes relaying the packets) enforce and abide by the resource allocation encoded by the GC in the policy token and the responder in the network capability. The enforcement involves both access control and bandwidth allocation. A responder accepts packets (except for the first) from an initiator only if the initiator is authorized to send, in the form of a valid network capability. It accepts the first packet only if the initiator's policy token is included. An intermediate node will forward the packets from a node only if they have an associated policy token or network capability, and if they do not violate the conditions contained therein. Possession of a capability does not imply resource reservation; they are the maximum limits a node can use. Available resources are allocated by the intermediate nodes in a fair manner, in proportion to the allocations defined in the capability.

The capability need not be contained in all packets. The first packet carries the capability, along with a transaction identifier (TXI) and a public key. Subsequent packets contain only the TXI and a packet signature based on that public key. Intermediate nodes cache policy tokens and network capabilities in a *capability database*, treating them as soft state. A capability database entry contains the source and the destination addresses, TXI, the capability, public key for the packet signature and packet statistics. Capability retransmissions update the soft state of intermediate nodes when the route changes due to node mobility. The soft state after a route change is also updated using an on-demand query for the capability database entry from the upstream nodes.

3.1 Misuses in DIPLOMA

In this section, we identify ways of misusing capabilities in destination-based consent systems like DIPLOMA. These includes simultaneous use of a capability on multiple paths to get more than allocated bandwidth, or misusing a policy to create network capabilities more than the policy is entitled to.

Misuse of policy tokens: Policy tokens are capabilities allocated by the group controllers to the nodes to access the services running on other nodes in MANET. The node for which the policy token is allocated is called *owner* of that policy token. A policy token contains the owner, the destination node, the type of service, the allocated bandwidth, and the signature of the group controller. The destination field of a policy token may correspond to a specific host or a group of hosts. A sender (*i.e.,* the owner) can send traffic to multiple receivers simultaneously using a policy token that has authorization to access those receivers. While accessing multiple receivers, the sender should not exceed the total bandwidth allocated to that policy token. A misbehaving sender may try to exceed this allocation by deliberately communicating with multiple receivers without satisfying the overall bandwidth constraints of the capability. We call this misuse as *concurrent-destination misuse.*

Another way to misuse the capabilities is to use multiple paths to the receiver. The sender may use the same capability on multiple paths, and may claim the bandwidth allocation of the capability in each of the paths. This way the sender can bypass the bandwidth enforcement that is performed by the intermediate nodes. Though the receiver can easily detect this kind misuse, it might be collaborating with the sender to receive a larger bandwidth. We call this misuse as *multi-path misuse.*

Misuses of network capabilities: Network capabilities are the capabilities issued by the receiver nodes to the senders that authorize sending traffic to those receivers. They are similar to policy tokens, except that the destination field cannot be arbitrary; it has to be the receiver that issued the capability. The capabilities also need to contain a signed policy issued by the group controller authorizing the receiver to issue such a capability.

Nodes can misuse a network capability in two ways: either by a receiver issuing more than it is entitled to, or by a sender sending more than the network capability. A sender could misuse the network capability by sending the capability over multiple paths. This is same as the multi-path misuse. Note that concurrent destination misuse is not possible with network capabilities, since those capabilities have fixed destination.

A receiver creating network capabilities may perform another form of misuse. The receiver needs to conform to the policy while creating network capabilities. A policy puts an upper bound on the amount of bandwidth a receiver may allocate to capabilities simultaneously. A receiver might not abide by this policy, and might allocate more network capabilities than it is entitled to. We call this misuse as *policy misuse.*

4 System Architecture

Figure 2 shows the architecture of the misuse detection system in DIPLOMA. The DIPLOMA engine, which is responsible for packet processing and capability enforcement, collects the information about the capabilities going through the node and provides them to the misuse detection engine. This information is stored in the local records table. The detection engine may also receive the information about the communication flows and the associated capabilities from other nodes, which are stored in the external records table. The detection engine periodically runs the misuse detection algorithm described in Section 5 on these records. Whenever the algorithm detects a misuse, it informs the local DIPLOMA engine as well as the misuse detection engines of the other nodes. The DIPLOMA engine makes use of this information while accepting the capabilities for connection establishment and packet forwarding.

Based on where the flow information is obtained, the misuse detection algorithm is classified as local or distributed.

Local detection: In many cases, we can detect capability misuse using the local information a node has, received either through the packets passing through that node, or by listening to the channel and snooping on the packets in its neighborhood. For example, a receiver can detect any misuse by a sender directed towards it. Nodes in the sender's neighborhood may be able to hear all the packets by listening to the channel. In those cases, a neighboring node will be able to detect any misuse by a sender, and provide a proof of the misuse.

Distributed detection: When it is not possible to detect the misuse based on local information, we resort to distributed detection. For example, if the sender is using a directional antenna, then its neighboring nodes may not be able to hear all the packets it has sent. A misuse may be targeted towards multiple receivers; hence, a single receiver cannot detect it. Even if a misuse involves a single receiver, the receiver may be colluding with the sender and may not report the misuse. In distributed detection, the nodes periodically exchange information about the flows passing through them. The misuse detection algorithm is run using the combination of the information a node collected locally and what it received from other nodes. In distributed detection, one or more nodes in the MANET are designated as *verifier nodes.* All the other nodes (called *collector nodes*) send information about the flows going through them to one or more

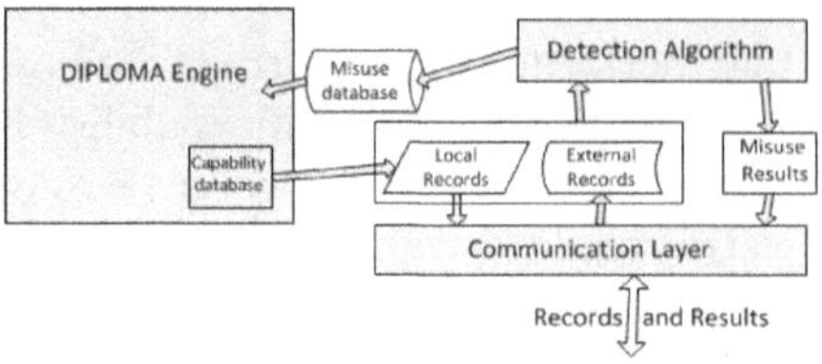

Fig. 2. Misuse detection architecture

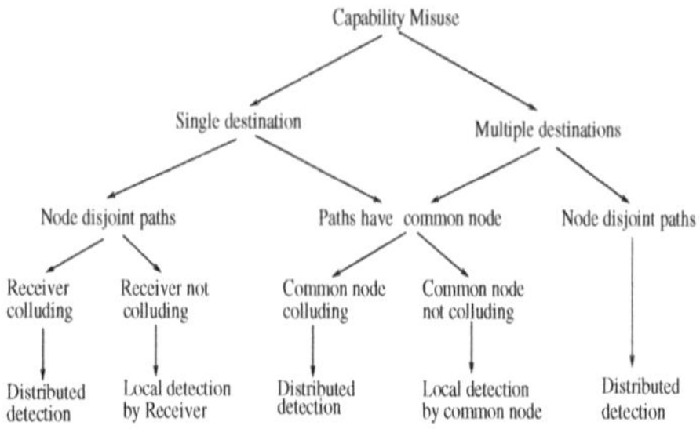

Fig. 3. Various types of capability reuse and detection algorithm

of these verifier nodes. The verifier nodes run the misuse detection algorithm, and inform the collector nodes about any misbehaving nodes and the associated capabilities, along with proof of misuse.

Figure 3 illustrates the types of the detection methods that are useful in various misuse scenarios. Whether distributed detection is required or not is dictated by whether there exists any common node in the misuse paths, and whether the common node is willing to co-operate. A common node can run the detection algorithm based on the local information alone to identify, and report the misuse. For the misuse involving multiple paths to a single destination, the receiver is always a common node. If all the common nodes are colluding with the sender, then a distributed detection is required.

Our solution consists of the following: a capability-encoding scheme that aids detection, protocols for exchanging information for distributed detection, and detection algorithms.

4.1 Capability Encoding

In the DIPLOMA architecture, it is permissible to use a capability for multiple communication sessions concurrently. For example, a node possessing a policy token to communicate with a group of destinations may be simultaneously communicating with multiple nodes in that destination group. Similarly, it is possible to use a policy authorizing the issue of the network capability to create multiple capabilities simultaneously. For example, a node may be receiving packets from multiple source nodes, and may want to allocate network capabilities to those senders based on a single policy authorizing the allocations. Both of these concurrent uses of policies are valid as long as the nodes do not use (or allocate) more bandwidth than allowed by the policies. When the nodes split the bandwidth of a policy into multiple capabilities, we need protocols that enable other nodes to check if these capabilities are within the limit.

While sending the policy token or while creating receiver capabilities, the owner has to decide on how to split the available bandwidth. The protocol allows dividing the available bandwidth into 32 or 64 equal sized slots, which are represented using a bitmask of 32 or 64 bits. We call this bitmask as the *allocation vector*. A bit in the allocation vector is set, if the corresponding bandwidth slot is used. The allocation vector is included in the capability request packets, as well as on the network capabilities created by the receivers. For a policy token that a sender is using to communicate with multiple destinations, the allocation vector on a capability request indicates the portion of the available bandwidth allocated to that communication session. When a sender uses multiple paths to reach a destination, the allocation vectors on the capability requests on each of the paths indicate the portion of the available bandwidth from the capability allocated to that path. If a receiver node creates multiple network capabilities, based on a policy, the allocation vector field in the capability indicates the portion of the available bandwidth allocated to that capability. Note that there could be multiple bits set in the allocation vector indicating bandwidth allocation proportional to number of bits set in the vector.

It is permissible to allocate the same bandwidth slot to different capabilities, derived from the same policy, at different times. Every capability request and network capabilities contain a start time stamp and a validity duration, which indicates the time until they are valid. To extend its validity, the owner needs to create a new request. Hence, a misuse constitute the existence of two capability requests for the same capability, or two network capabilities for the same policy that has a common bit set in their allocation vector at the same time.

Our misuse detection algorithms do not depend on the data structure used for dividing the allocated bandwidth. Allocation vectors have easy representation, and allow for easy unions and intersection operations using bitwise operators. If finer granularity is needed in dividing the bandwidth, one could use other representations like slab allocation.

4.2 Communication Protocol

When a sender node wants to communicate with a receiver node, it uses the *capability request* packet to inform the intermediate nodes about the capability that will be used for the communication [4]. This request is signed by the sender's private key. To avoid any non-repudiation of the capability request by the sender, the nodes participating in the misuse detection are required to store the capability request and the signature. The capability request have many information that are not necessary for the misuse detection. Storing this information puts undue burden on the misuse detection nodes. One way to solve this problem is to sign the information that are essential for misuse detection (called *misuse detection block*), separately from that which are not useful for misuse detection (called *capability establishment block*). Unfortunately, this requires senders to perform two expensive signature operations. As a compromise between the amount of information stored by nodes and the processing needed at the sender, we sign the request in two steps. First, the sender computes the hash of the capability establishment block. Then, it signs the combination of this hash and the misuse detection block. To prove capability misuse by a sender, the nodes only need to keep track of the hash, the misuse detection block, and the signature.

A capability request packet contains the capability establishment block and misuse detection block. A node can verify the capability request by first computing the hash of the capability establishment block, and verifying the signature for the combination of hash and the misuse detection block. A valid signature indicates that the packet was not tampered with. We also use a similar scheme for signing the networks capabilities created by the receiver nodes.

Information exchange: For the distributed detection, the detector nodes send information about the communication sessions and the capabilities passing through them to the verifier nodes. The detector nodes use the underlying MANET unicast or multicast routing protocol to reach the verifier nodes. The information required to detect misuse consists of the identity of the sender node, the transaction identifier, serial number and the issuer of the capability, the allocation vector, the time stamps, and the previous and next nodes in the path. This information about the communication session is called a *record*. A node can send multiple records in a packet. The nodes sign the packet using their private keys. A similar record is also send for the network capabilities for detecting misuses in them. The algorithms used for detection of the misuse by the senders, and the receivers are similar. Hence, we will deal only with the sender misuse in rest of the paper.

5 Detection Algorithms

In this Section, we describe the DIPLOMA misuse detection algorithms that are used for both local and distributed detection. Then we describe how a verifier node can provide a proof of misuse. Finally, we provide solution for handling the privacy issues in our misuse detection architecture.

Recall that there is a misuse if there are two communication sessions that use the same capability and have a common bit set in the allocation vectors with overlapping validity periods. Hence, the goal of the algorithm is to find such communication sessions. To that end, the algorithm first groups the records corresponding to a communication session. This is because there could be multiple records for a communication, received from different collector nodes. Once the records of communication sessions are grouped together, the algorithm look at records across the communication sessions to identify misuse.

In DIPLOMA, a communication session can be uniquely identified by the (transaction identifier, sender identity) pair. If the sender uses multiple paths to a destination, the sender is required to use different transaction identifiers for each path.

The misuse detection algorithm has two phases. In the first phase, it removes the duplicate records for each communication sessions from the collection of records it gathered locally and from other nodes. It also detects if a sender uses the same transaction identifier on multiple paths. The output of the first phase is a set of records, consisting of at most one record for a transaction identifier per sender. This phase is not required if all the records are obtained from the capability database of the local DIPLOMA engine, as the engine already prevents duplicates. In the second phase, the algorithm detects if there is any misuse on the filtered records output by the first phase.

5.1 Phase 1 – Duplicate Removal and Multipath Detection

The removal of duplicate records for a communication session is performed by sorting the records based on (sender, transaction id) pair and keeping only one record per pair. However, this step will not detect use of the same transaction identifier by a sender on multiple paths. To detect the multiple path misuse, we use the following property of the paths.

If the same transaction id is used in two different paths to a destination, then there will be two records for it that has a common previous node or a common next node. This is because since the source and the destination nodes are common in both paths, the paths need to bifurcate at some node and join at another node. If the paths are bifurcating at any node other than the sender, or joins anywhere other than the receiver, then the DIPLOMA engine at the common nodes in the path can detect the misuse during the connection establishment stage. If the paths are node disjoint and the receiver is not colluding with the misbehaving sender, then the receiver can detect misuse. If all the common nodes are colluding with the sender, then the phase 1 algorithm can detect the misuse looking for the common nodes. This is depicted in Figure 4.

Algorithm 1 describes the phase 1 algorithm. It goes through the records corresponding to the same (sender, transaction id) pair and verifies that all the records use the same capability, allocation vector and time stamps. It also stores the previous nodes and the next nodes of each record in temporary arrays. The presence of duplicates in these arrays indicates a misuse.

Analysis: The algorithm will fail to detect a multipath misuse if certain nodes collude with the sender and do not provide the relevant records to the verifier. If the common nodes, including the receiver, collude with the sender, then local detection of the multipath misuse will fail. If at least one of the nodes in the path next to the common node colludes, where the forking of the paths has occurred, then the algorithm will fail to detect that common node. Similarly if one of the nodes before the common node at which the joining of the paths take place, then also the algorithm will fail to detect the common next node. The algorithm will fail to detect a multipath to a destination, when it cannot detect both the common previous and next nodes. This is depicted in Figure 4. It is still possible

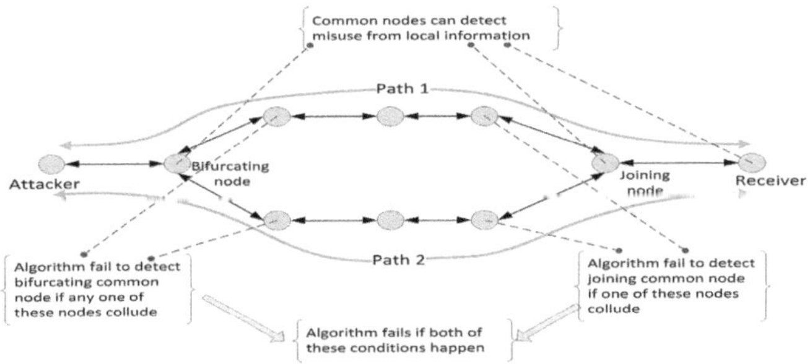

Fig. 4. Properties of multiple paths in aiding misuse detection

for a verifier to detect that it has not received records from some of the nodes (which may be colluding with the sender), because of the existence of two path fragments (as opposed to one path) for the transaction identifier. However, we cannot use this against the sender, because of the possibility of packet losses, or the possibility of a node deliberately not sending the records to the verifier.

5.2 Phase 2 – Reuse of the Capability Detection

The second phase, the algorithm detects misuse of capabilities across the communication sessions. Recall that a misuse is identified by a common bit set in the allocation vectors at overlapping validity periods. The input to phase 2 is the records output by phase 1. Hence, there is only one record per communication session. The algorithm goes through all the records corresponding to each of the capabilities and detects misuse.

Algorithm 1. Duplicate-record removal & multi-path detection

$L_i \leftarrow$ List of all (source node, transaction id) pair
$L_u \leftarrow$ NULL {Output list of unique records}
for all $id \in L_i$ **do**
    $L_r \leftarrow$ List of records for id
    $H_{prev}, H_{next} \leftarrow$ NULL
    Add first record of L_r to L_u
    for all $rec \in L_r$ **do**
        if attributes of rec different from $head(L_r)$ **then**
            print Misuse. Different attributes for same transaction
        else if $prevhop(rec) \in H_{prev}$ **or** $nexthop(rec) \in H_{next}$ **then**
            print Misuse. Same transaction in different paths
        end if
        Add $prevhop(rec)$ to H_{prev} and $nexthop(rec)$ to H_{next}
    end for
end for
return L_u

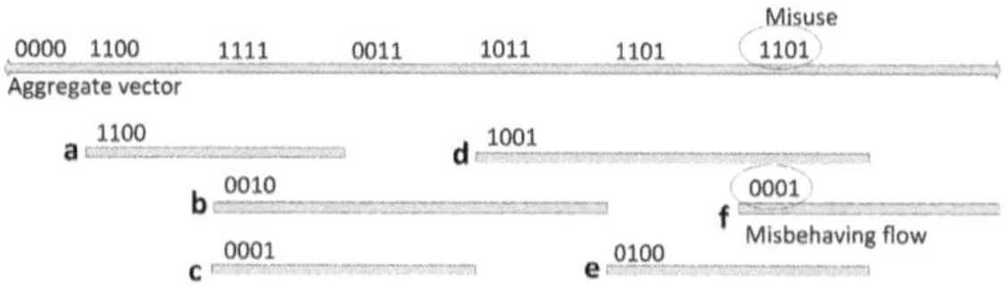

Fig. 5. Computation of aggregate allocation vector and misuse detection using interval graphs

The algorithm treats the records as an interval graph, where each record corresponds to an interval for which they are active. There is an allocation vector associated with each interval, which is the allocation vector of the corresponding record. We define the *aggregate allocation vector* at any point of time as the union of the allocation vectors of the intervals passing through it. There is misuse at any point in time if the intersection of any two intervals passing through it is not empty. This is depicted in Figure 5.

Once the interval graph is formed, we can detect any misuse in linear time in the number of intervals. The algorithm goes through the end points of the intervals in increasing time and updates the aggregate allocation vector. At the beginning of the algorithm, this vector is set to NULL. Whenever the algorithm considers the beginning of an interval, it checks if the intersection of the aggregate allocation vector and the allocation vector of that interval is non-empty. If it is not empty, then there is misuse. Otherwise, the allocation vector of that interval is added to the aggregate vector. Similarly, when the algorithm considers the end of an interval, its allocation vector is subtracted from the aggregate allocation vector. If there are both entering and leaving intervals at any point, then the leaving operation is considered before the entering operation. This is because the new entering interval could use slots from the leaving interval, without causing misuse.

Figure 5 illustrates the computation of aggregate allocation vector and the misuse detection. There are six flows labeled as $a, b, \ldots, f$. They are represented as the intervals in which they are active. Their corresponding allocation vectors are also shown. For simplicity of illustration, we use the allocation vector of 4 bits. The vector on the top line shows the aggregate allocation vector when the flows enter or leave the system. The aggregate vector at any point is the union of the allocation vectors of the interval going through that point. There is no misuse for flows $a, b, \ldots, e$. The flow f uses one of the slots of flow d, hence there is misuse. The aggregate allocation vector before f entered the system was 1101. The flow f uses one bandwidth slot. The sender assigned it the slot 0001, which is a reuse of the existing slot. If the sender had assigned it the slot 0010, then there would not be any misuse. Hence, it is important that senders allocate the right bandwidth slot for flows.

Creating an interval graph from the records is performed by sorting the end-points of the interval. In fact, our algorithm maintains two sorted lists: one for the starting points of the intervals and the second for the ending points of the intervals. The algorithm is given in Algorithm 2.

5.3 Privacy Issues

In the protocol presented so far, the detector nodes send the information about all the flows going through them to the verifier. Even though the records contain only the sender node identity and does not have the receiver identities of the flow, it is still possible to deduce the receiver identity by following the path using the previous and next hop information. Hence, the verifier can know about the source and destination of all the flows. Another privacy concern is the knowledge about the number of flows a sender is sending, even if the verifier is not interested in knowing the receivers.

Algorithm 2. Phase 2 - checking for reuse of bandwidth slots

```
L_i ← List of all (capability id, issuer) pair
for all id ∈ L_i do
    L_r ← List of records for id
    L_s ← Records in L_r sorted on start time
    L_e ← Records in L_r sorted on end time
    aggregate ← φ
    while L_s not empty do
        time_s ← starttime(head(L_s))
        time_e ← endtime(head(L_e))
        if time_e ≤ time_s then
            rec ← head(L_e)
            L_e ← L_e − rec
            aggregate ← aggregate − allocvector(rec)
        else
            rec ← head(L_s)
            L_s ← L_s − rec
            if aggregate ∩ allocvector(rec) ≠ NULL then
                print Misuse. Reuse of bandwidth slots.
            end if
            aggregate ← aggregate ∪ allocvector(rec)
        end if
    end while
end for
```

We can modify the protocol to honor privacy, and still detect the misuse as follows. The detection algorithm continues to function even if all the fields in the records, except the time stamps and the allocation vector, were encrypted with a key that is common across all the flows corresponding to a capability. The detector nodes can create such a key by taking a known function (*e.g.*, hash) of the capability. Since the flows are going through the detector nodes, they know about the complete capability associated with the flow. However, verifier knows only about their serial numbers, and cannot recreate the keys. Hence, verifier cannot decrypt the records for the flows not going through it. In this scheme, the verifier can still get information about all the flows that use any of the capabilities passing through it.

6 Experimental Evaluation

In this section, we study the effectiveness of the misuse detection algorithms. The experiments were conducted in the Orbit Lab Testbed [1]. The algorithms were implemented on DIPLOMA systems running on Debian Linux with kernel 2.6.30. We analyzed memory, bandwidth and processing overheads, and found them to be minimal; due to lack of space we omit the results.

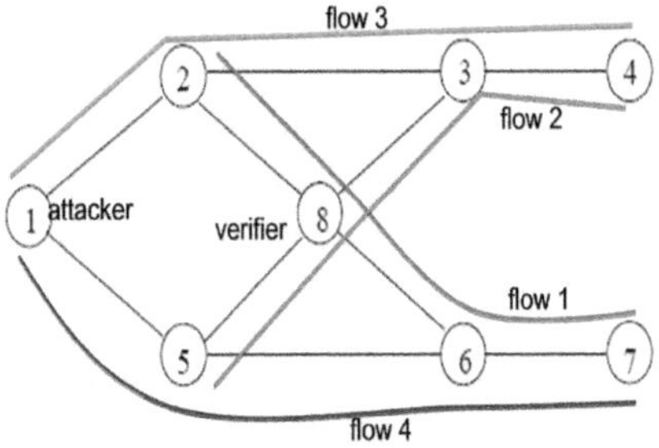

Fig. 6. Topology to study the performance of detection algorithm

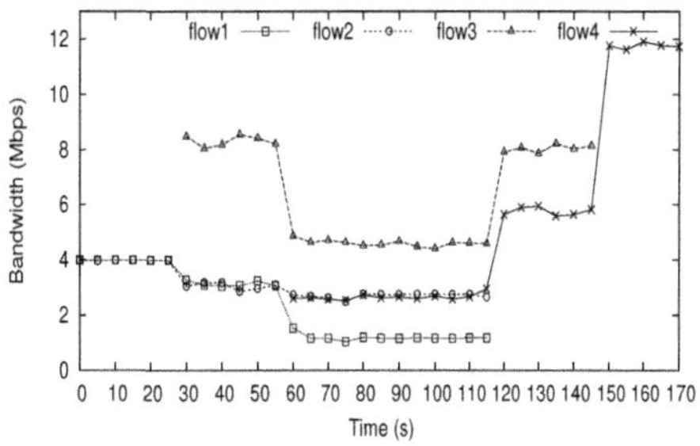

Fig. 7. Bandwidth of flows in a system that does not require consent to send

6.1 Effectiveness in Containing Attacks

Now we study the effectiveness of misuse detection algorithm in detecting and containing the attacks. We used the topology given in Figure 6, created by assigning non-overlapping channels of 802.11b and 802.11a to the links [3].

In this set of experiments, there are four flows: two by good nodes and two by the attacker. Each of the senders is allocated a capability of 4 Mbps. All the flows were created using UDP *iperf*. The good nodes, 2 and 5, send traffic at 4 Mbps each to the destinations nodes 7 and node 4 respectively. We denote those flows as flow 1 and flow 2 respectively and call them good flows. Flow 1 takes the path $2-8-6-7$ and flow 2 takes the path $5-8-3-4$. These flows are started at time 0 seconds and last for 120 seconds. The attacker, which is the node 1, sends flows to nodes 4 and 7, which we denote by flows 3 and 4. We call these attack flows. Each of the attack flows are 12 Mbps, even though the attacker has one capability with the allocated bandwidth of 4 Mbps, which can be used for either destination. Flow 3 is started at time 30 seconds and takes the path $1-2-3-4$. Flow 4 is started at time 60 seconds and takes the path $1-5-6-7$. Both the flows last for 120 seconds, and use the same capability with all the bits in the allocation vector set. Hence, the attacker launches two types of attacks. First, it is sending higher bandwidth than that is allocated in the capability, which starts at time 30 seconds. Secondly, it uses the same capability to talk to multiple destinations simultaneously using the same bandwidth slot. This attack starts at time 60 seconds.

We conduct three sets of experiments. The first is called the *original*, and does not require any consent for sending the traffic. This scheme cannot protect against both the attacks. Then we use the consent-based scheme, where DIPLOMA requires capabilities for sending traffic. This DIPLOMA without misuse detection, can handle the first bandwidth hogging attack but cannot prevent reuse of the capability across the flows. Finally, we use DIPLOMA with misuse detection to handle both types of attacks. For each of the experiments, we report the bandwidth of each of the flow over a period of time. All experiments were run 6 times, and we show the average bandwidth. The iperf servers (receivers) measured the bandwidths at 5 second intervals.

Figure 7 shows the results for the original system. In this system, until 30 seconds, where there are only two flows from the good nodes, both the flows get their requested 4 Mbps bandwidth. At 30 seconds, when the first attacker flow (flow 3) arrives, the bandwidth of the good flows drop as the attacker flow takes up most of the bandwidth. The attacker receives a bandwidth between 8 and 8.5 Mbps, whereas the bandwidth of good flows drop to 3 Mbps. At 60 seconds, when the second attack flow arrives (flow 4), the bandwidth of the good flows further drops along with the bandwidth of the first attacker flow. The bandwidth of the flow 1 and flow 2 drops to 1.1 Mbps and 2.7 Mbps respectively. The bandwidth of first attacker flow drops to 4.6 Mbps. The new attacker flow receives 2.6 Mbps. This trend continues until 120 seconds, when the good flows end. At that point, the first attacker flow bandwidth recovers to its previous levels (8 Mbps) and the second attacker flow receives a higher bandwidth of 5.8 Mbps. At 150 seconds, when the first attacker flow ends, the second attacker flow receives 11.8 Mbps, which is close to the requested bandwidth of that flow.

Figure 8 shows the results for the DIPLOMA scheme when misuse detection is not in effect. Until 30 Seconds, where the attacker had not started sending any traffic, the good flows 1 and 2 get the bandwidth that are allocated in their capability. The bandwidth reported by the flows is 3.74 Mbps, which is slightly less than the allocated 4 Mbps due to the additional headers present in DIPLOMA packets. At 30 seconds, when the attacker starts sending the first attack flow (flow 3) at the rate of 12 Mbps, the bandwidth of the good flows drops only slightly to 3.71 Mbps. The attacker gets a bandwidth of 3.5 Mbps, which is closer to its allocated bandwidth. Hence, the consent-based DIPLOMA scheme is able to protect the good flows and contain the attacker to its allocated bandwidth. At 60 seconds, when the second attack flow (flow 4) starts, the bandwidth of the good flows and the attacker drops. This drop for the good flow is not as drastic as the original scheme. Here the bandwidth of the flow 1 drops to 2.3 Mbps and that of flow 2 drops to 3 Mbps. The existing attacker flow drops to 2.9 Mbps, and the new attacker flow receives 1.9 Mbps bandwidth. This drop in bandwidth is due to limited available bandwidth on the network. The attacker is reusing the capability at this point, and DIPLOMA ends up honoring the same capability in two node disjoint paths. At 120 s, the genuine flow ends. At that point, the first attack flow bandwidth moves back to its original level of 3.5 Mbps. The second attacker flow bandwidth ends up at 3.3 Mbps. This increase is due to the freed up capacity from the good flows. At 150 seconds, the first attack flow ends and the second attack bandwidth increases slightly to 3.6 Mbps. Even then, the bandwidth of the individual attack flows does not go above the allocated bandwidth of 4 Mbps, because DIPLOMA enforces the bandwidth. Therefore, in DIPLOMA without the misuse detection, an attacker cannot go above the allocated bandwidth in a single path. However, it can bypass that check by sending traffic to multiple destinations in disjoint paths, if the capability permits it. When this happens, genuine traffic is affected due to capacity sharing.

Figure 9 shows the results for the DIPLOMA system with misuse detection. The behavior of the system is same as DIPLOMA without misuse detection, until

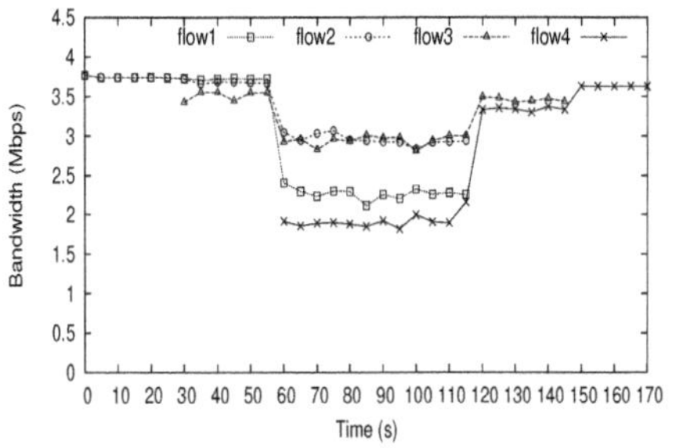

Fig. 8. Bandwidth of the flows in DIPLOMA without misuse detection

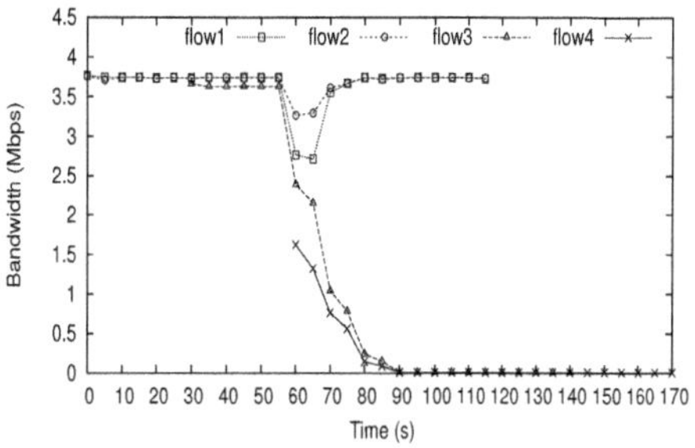

Fig. 9. Bandwidth of the flows in DIPLOMA with misuse detection

the misuse happens at 60 seconds. At 60 seconds, when the attacker sends the second flow (flow 4), which constitutes a capability reuse, the behavior changes from DIPLOMA without the misuse detection. In this case, the nodes send the record to the verifier (node 8), which detects the misuse. In this experiment, the period at which nodes send the record is 10 seconds. Hence, the verifier detects the misuse before 70 seconds, and informs the forwarding nodes. The forwarding nodes starts to block the attack flows. The drop of the attack bandwidth is gradual due the nature of our implementation [3]. The bandwidth of the good flows drops to 2.8 Mbps and 3.3 Mbps for a short duration (10 seconds) at 60 seconds while the misuse detection and recovery takes place. At 85 seconds, the recovery is complete and the bandwidth of the good flows moves back to the levels before the reuse attack. Even after the genuine flow ends at 120 s, the attacker flows continue to be blocked due to their misuse action. This continues even after the attacker stops the misuse at 150 seconds, when the first attacker flow ends. Hence, DIPLOMA can effectively contain capability misuse.

6.2 Speed of Detecting Misuse

Next, we study how fast our scheme can detect the misuses and communicate with the affected nodes. The time to detect the misuse will depend on the frequency at which records are sent to the verifiers. Hence, we study the misuse detection speed as a function of that period, using the same set of flows as in the previous experiment.

Figure 10 plots the time required at the detector nodes to get the misuse notification after the misuse happened for various record reporting periods. For the periods up to 30 seconds, the time to detect is the same as the period. Hence, the misuse is detected as soon as the record is received to the verifier. For the periods 40 and 50 seconds the time to detect misuse was less than the period, and for 60 seconds the detection time was the same as the period. This is because for 40 and 50 seconds experiments, the start of the attack and the start of the period may not have been synchronized. Hence, it is possible for the detectors to send the record to the verifier in time less than the period after the attack has happened. The verifier detects the misuse upon receiving the records.

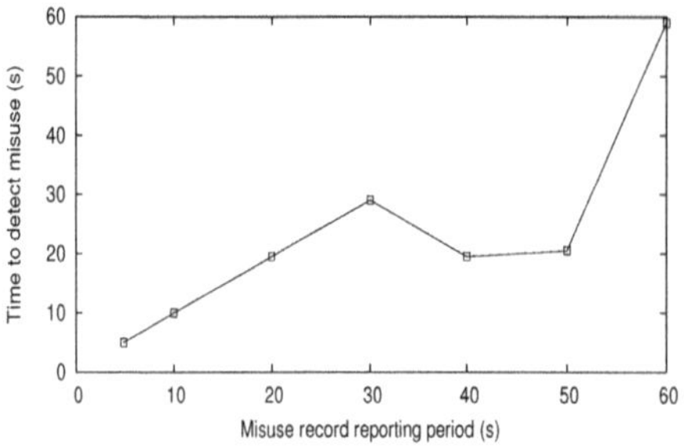

Fig. 10. Time to detect the misuse for different record reporting periods

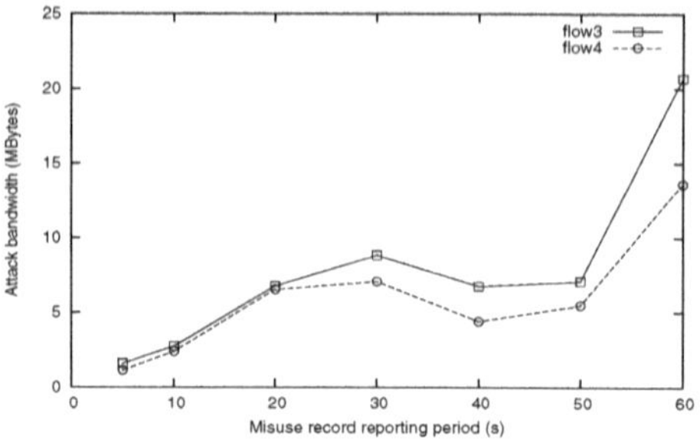

Fig. 11. Additional attack traffic after misuse for different record reporting periods

6.3 Attack Bandwidth after Misuse

Now we study the amount of additional traffic the attacker is able to send after the misuse. This quantity depends on how fast the system is able to detect the misuse and take action against the attacker. Hence, we study the additional traffic as a function of record reporting period. We use the same flows and attack scenario as the previous experiment and measure the data received at the receiver for the attack flows (flows 3 and 4), after the misuse has started.

Figure 11 plots the amount of traffic received at the receivers for the attack flows after the misuse, for different record reporting periods. Up to 30 seconds, the attack traffic increases as the record reporting time increases. This is because the misuse traffic will be treated as the legitimate traffic and allowed to pass through until the misuse is detected; and the misuse detection time is proportional to the record reporting period. Note that even if the attacker is trying to send the traffic at 12 Mbps, the throughput the flows receive is less than 4 Mbps due to bandwidth enforcement by the intermediate nodes. In this set of experiments, the flow 3 had slightly higher bandwidth than flow 4, similar to the experiment in Figure 9. For 40 and 50 seconds, the attack traffic drops is less than that of 30 seconds, as the misuse is detected before the complete period as explained in the previous experiment.

7 Related Work

The concept of capabilities was used in operating system for securing resources [15]. Follow-on work investigated the controlled exposure of resources at the network layer using the concept of "visas" for packets [7], which is similar to network capabilities. More recently, network capabilities were proposed to prevent DDoS attacks [5]. We extend the concept to MANETs and use it for both access control and traffic shaping [4,3]. All these works represent destination-based consent architectures. Path-based consent architectures in the context of Internet have also been proposed [12].

Intrusion detection systems (IDS) for MANETs is an active area of research [10]. In the Local Intrusion Detection Architecture [2] communities of nodes are formed, which exchange various security data and intrusion alerts. The nodes can also place mobile agents in the other nodes, to do a specific mission in an autonomous and asynchronous manner. Distributed IDS architecture [18] is another proposed IDS architecture for MANETs. It uses a local detection engine, with the input from the local data collection, and a co-operative detection engine, with the input from the neighboring nodes to detect intrusions.

Solutions are also proposed to specifically detect attacks on routing protocols based on the protocol specification. [8] detects violations in the protocol specification based on an extended finite state automation (EFSA) of AODV. Distributed Evidence-Driven Message Exchange Intrusion Detection Model (DEMEM) [13] detects inconsistency among the routing messages in OLSR. [6] models the attacks on AODV protocol using attack tree and identify the damages. Proposals are also made when to isolate a misbehaving node based on the criticality of that node in maintaining the connectivity [14]. There is also a rich literature on detecting DoS and node replication attacks in sensor networks [9,11].

8 Conclusions and Future Work

We identified sources of misuse in destination-based consent architectures and provided a distributed solution for detecting misuses. Our solution is demonstrated for DIPLOMA, a consent-based architecture for MANETs. DIPLOMA is a *deny-by-default* distributed policy enforcement architecture based on capabilities. We provided capability encodings and protocols for exchanging information for distributed misuse detection, and presented efficient algorithms for misuse detection. We showed through experimental evaluations that our algorithms are effective in identifying and containing the misuse. In the future, we plan to measure the energy consumed for detecting misuses and the energy savings due to preventing attacks, and to look at misuse in path-based consent architectures.

Acknowledgements. This work was supported in part by the National Science Foundation through Grant CNS-07-14277. Any opinions, findings, conclusions, and recommendations expressed in this paper are those of the authors and do not necessarily reflect the views of the NSF.

References

1. Orbit Lab Test bed, `http://www.orbit-lab.org`
2. Albers, P., Camp, O., Jouga, B., Me, L., Puttini, R.: Security in ad hoc networks: A general id architecture enhancing trust based approaches. In: IEEE Network, WIS (2002)
3. Alicherry, M., Keromytis, A.D.: DIPLOMA: Distributed Policy Enforcement Architecture for MANETs. In: International Conference on Network and System Security (September 2010)

4. Alicherry, M., Keromytis, A.D., Stavrou, A.: Deny-by-default distributed security policy enforcement in mobile ad hoc networks. In: Chen, Y., Dimitriou, T.D., Zhou, J. (eds.) SecureComm 2009. Lecture Notes of the Institute for Computer Sciences, Social Informatics and Telecommunications Engineering, vol. 19, pp. 41–50. Springer, Heidelberg (2009)
5. Anderson, T., Roscoe, T., Wetherall, D.: Preventing Internet Denial-of-Service with Capabilities. In: Proc. of Hotnets-II (2003)
6. Ebinger, P., Bucher, T.: Modelling and analysis of attacks on the manet routing in aodv. LNCS. Springer, Heidelberg (2006)
7. Estrin, D., Mogul, J.C., Tsudik, G.: Visa protocols for controlling interorganizational datagram flow. IEEE J. on Selected Areas in Comm. (May 1989)
8. Huang, Y., Lee, W.: Attack analysis and detection for ad hoc routing protocols. In: Jonsson, E., Valdes, A., Almgren, M. (eds.) RAID 2004. LNCS, vol. 3224, pp. 125–145. Springer, Heidelberg (2004)
9. McCune, J., Shi, E., Perrig, A., Reiter, M.: Detection of denial-of-message attacks on sensor network broadcasts. IEEE Security and Privacy (2005)
10. Mishra, A., Nadkarni, K., Pactha, A.: Intrusion detection on wireless ad hoc networks. IEEE Wireless Communications (2004)
11. Parno, B., Perrig, A., Gligor, V.: Distributed detection of node replication attacks in sensor networks. IEEE Security and Privacy (2005)
12. Seehra, A., Naous, J., Walfish, M., Mazires, D., Nicolosi, A., Shenker, S.: A policy framework for the future internet. In: ACM Workshop on Hot Topics in Networks (HotNets) (October 2009)
13. Tseng, C., Wang, S., Ko, C., Levitt, K.: Demem: Distributed evidence-driven message exchange intrusion detection model for manet. In: Zamboni, D., Krügel, C. (eds.) RAID 2006. LNCS, vol. 4219, pp. 249–271. Springer, Heidelberg (2006)
14. Wang, S., Tseng, C.H., Levitt, K., Bishop, M.: Cost-sensitive intrusion responses for mobile ad hoc networks. In: Kruegel, C., Lippmann, R., Clark, A. (eds.) RAID 2007. LNCS, vol. 4637, pp. 127–145. Springer, Heidelberg (2007)
15. Wobber, E., Abadi, M., Burrows, M., Lampson, B.: Authentication in the Taos Operating System. ACM Trans. on Computer Systems 12 (February 1994)
16. Yaar, A., Perrig, A., Song, D.: SIFF: A Stateless Internet Flow Filter to Mitigate DDoS Flooding Attacks. In: IEEE Symposium on Security and Privacy (2004)
17. Yang, X., Wetherall, D., Anderson, T.: A DoS-limiting network architecture. In: In Proceedings of ACM SIGCOMM, pp. 241–252 (2005)
18. Zhang, Y., Lee, W.: Intrusion detection for wireless ad-hoc networks. In: MOBICOM (2000)

Cold Boot Key Recovery by Solving Polynomial Systems with Noise

Martin Albrecht* and Carlos Cid

[1] INRIA, Paris-Rocquencourt Center, SALSA Project
UPMC Univ Paris 06, UMR 7606, LIP6, F-75005, Paris, France
CNRS, UMR 7606, LIP6, F-75005, Paris, France
malb@lip6.fr
[2] Information Security Group,
Royal Holloway, University of London
Egham, Surrey TW20 0EX, United Kingdom
carlos.cid@rhul.ac.uk

Abstract. A method for extracting cryptographic key material from DRAM used in modern computers has been recently proposed in [9]; the technique was called *Cold Boot attacks*. When considering block ciphers, such as the AES and DES, simple algorithms were also proposed in [9] to recover the cryptographic key from the observed set of round subkeys in memory (computed via the cipher's key schedule operation), which were however subject to errors due to memory bits decay. In this work we extend this analysis to consider key recovery for other ciphers used in Full Disk Encryption (FDE) products. Our algorithms are also based on closest code word decoding methods, however apply a novel method for solving a set of non-linear algebraic equations with noise based on Integer Programming. This method should have further applications in cryptology, and is likely to be of independent interest. We demonstrate the viability of the Integer Programming method by applying it against the Serpent block cipher, which has a much more complex key schedule than AES. Furthermore, we also consider the Twofish key schedule, to which we apply a dedicated method of recovery.

1 Introduction

The structure of block cipher key schedules has received much renewed attention, since the recent publication of high-profile attacks against the AES [4] and Kasumi [3] in the related-key model. While the practicality of such attacks is subject of debate, they clearly highlight the relevance of the (often ignored) key schedule operation from a cryptanalysis perspective. An unrelated technique, called *Cold Boot* attacks, was proposed in [9] and also provided an insight into the strength of a particular key schedule against some forms of practical attacks. The method is based on the fact that DRAM may retain large part of its content

* This author was supported by the Royal Holloway Valerie Myerscough Scholarship.

J. Lopez and G. Tsudik (Eds.): ACNS 2011, LNCS 6715, pp. 57–72, 2011.

for several seconds after removing its power, with gradual loss over that period. Furthermore, the time of retention can be potentially increased by reducing the temperature of memory. Thus contrary to common belief, data may persist in memory for several minutes after removal of power, subject to slow decay. As a result, data in DRAM can be used to recover potentially sensitive information, such as cryptographic keys, passwords, etc. A typical application is to defeat the security provided by disk encryption programs, such as Truecrypt [16]. In this case, cryptographic key material is maintained in memory, for transparent encryption and decryption of data. One could apply the method from [9] to obtain the computer's memory image, potentially extract the encryption key and then recover encrypted information.

The *Cold Boot* attack has thus three stages: (a) the attacker physically removes the computer's memory, potentially applying cooling techniques to reduce the memory bits decay, to obtain the memory image; (b) locate the cryptographic key material and other sensitive information in the memory image (likely to be subject to errors due to memory bits decay); and (c) recover the original cryptographic key from this information in memory. While all stages present the attacker with several challenges, from the perspective of the cryptologist the one that poses the most interesting problems is the latter stage; we therefore concentrate in this work on stage (c). We refer the reader to [9,10] for discussion on stages (a) and (b).

A few algorithms were proposed in [9] to tackle stage (c), which requires one to recover the original key based on the observed key material, probably subject to errors (the extent of which will depend on the properties of the memory, lapsed time from removal of power, and temperature of memory). In the case of block ciphers, the key material extracted from memory is very likely to be a set of round subkeys, which are the result of the cipher's key schedule operation. Thus the key schedule can be seen as an *error-correcting code*, and the problem of recovering the original key can be essentially described as a decoding problem.

The paper [9] contains methods for the AES and DES block ciphers (besides discussion for the RSA cryptosystem, which we do not consider in this work). For DES, recovering the original 56-bit key is equivalent to decoding a repetition code. Textbook methods are used in [9] to recover the encryption key from the *closest code word* (i.e. valid key schedule). The AES key schedule is not as simple as DES, but still contains a large amount of linearity (which has also been exploited in recent related-key attacks, e.g. [4]). Another feature is that the original encryption key is used as the initial whitening subkey, and thus should be present in the key schedule. The authors of [9] model the memory decay as a binary asymmetric channel, and recover an AES key up to error rates of $\delta_0 = 0.30, \delta_1 = 0.001$ (see notation in Section 2 below). The results against the AES – under the same model – were further improved in [17,11].

Contribution of this paper. We note that other block ciphers were not considered in [9]. For instance, the popular FDE product Truecrypt [16] provides the user with a choice of three block ciphers: SERPENT [2], Twofish [14] (both formerly AES candidates) and AES. The former two ciphers present much more complex key schedule operations than DES and AES. Another feature is that the original encryption key does not *explicitly* appear in the expanded key schedule material (but rather has its bits non-linearly combined to derive the several round subkeys). These two facts led to the belief that these ciphers were not susceptible to the attacks in [9], and could perhaps provide an *inherently* more secure alternative to AES when protecting against *Cold Boot* attacks[1].

In this work, we extend the analysis from [9] and demonstrate that one can also recover the encryption key for the SERPENT and Twofish ciphers up to some reasonable amount of error. We propose generic algorithms which apply a novel method for solving a set of non-linear algebraic equations with noise based on Integer Programming[2]. Our methods also allow us to consider different noise models and are not limited to the binary asymmetric channel setting usually considered in the *Cold Boot* scenario; in particular we improve the results against the AES from [9] and extend the range of scenarios in which the attack can be applied when compared to [9,17,11]. Finally, we note that our methods can in principle be applied to any cipher and thus provide a *generic* (but possibly impractical for some ciphers) solution to the *Cold Boot* problem.

2 The *Cold Boot* Problem

Cold Boot attacks were proposed and discussed in detail in the seminal work [9]. The authors of [9] noticed that bit decay in DRAM is usually asymmetric: bit flips $0 \to 1$ and $1 \to 0$ occur with different probabilities, depending on the "ground state". To motivate our work, we model more formally the cold boot problem for block ciphers below.

We define the *Cold Boot* problem (for block cipher) as follows. Consider an efficiently computable vectorial Boolean function $\mathcal{KS} : \mathbb{F}_2^n \to \mathbb{F}_2^N$ where $N > n$, and two real numbers $0 \leq \delta_0, \delta_1 \leq 1$. Let $K = \mathcal{KS}(k)$ be the image for some $k \in \mathbb{F}_2^n$, and K_i be the i-th bit of K. Now given K, compute $K' = (K'_0, K'_1, \ldots, K'_{N-1}) \in \mathbb{F}_2^N$ according to the following probability distribution:

$$\begin{array}{ll} Pr[K'_i = 0 \mid K_i = 0] = 1 - \delta_1 \ , & Pr[K'_i = 1 \mid K_i = 0] = \delta_1, \\ Pr[K'_i = 1 \mid K_i = 1] = 1 - \delta_0 \ , & Pr[K'_i = 0 \mid K_i = 1] = \delta_0. \end{array}$$

Thus we can consider such a K' as the output of $\mathcal{KS}$ for some $k \in \mathbb{F}_2^n$ except that K' is noisy, with the probability of a bit 1 in K flipping to 0 is δ_0 and the

[1] In fact, a message in one of the most popular mailing lists discussing cryptography, commenting at the time about the contributions of [9]: "While they did have some success with recovering an entire AES key schedule uncorrupted, it seems important to note that the simplistic nature of the AES and DES key schedules allowed them to recover the entire original key even after the state had been somewhat degraded with only moderate amounts of work. A cipher with a better key schedule (Blowfish or Serpent, for instance) would seem to offer some defense here"[12], was one of the initial motivations for our work in this problem.

[2] We note that a similar method for key recovery on a different model of leakage, namely side-channel analysis, was independently proposed in [13].

probability of a bit 0 in K flipping to 1 is δ_1. It follows that a bit $K'_i = 0$ of K' is correct with probability

$$Pr[K_i = 0 \mid K'_i = 0] = \frac{Pr[K'_i = 0|K_i = 0]Pr[K_i = 0]}{Pr[K'_i = 0]} = \frac{(1-\delta_1)}{(1-\delta_1+\delta_0)}.$$

Likewise, a bit $K'_i = 1$ of K' is correct with probability $\frac{(1-\delta_0)}{(1-\delta_0+\delta_1)}$. We denote these values by Δ_0 and Δ_1 respectively.

Now assume we are given a description of the function $\mathcal{KS}$ and a vector $K' \in \mathbb{F}_2^N$ obtained by the process described above. Furthermore, we are also given a control function $\mathcal{E} : \mathbb{F}_2^n \rightarrow \{True, False\}$ which returns *True* or *False* for a candidate k. The task is to recover k such that $\mathcal{E}(k)$ returns $True$. For example, $\mathcal{E}$ could use the encryption of some known data to check whether k is the original key.

In the context of this work, we can consider the function $\mathcal{KS}$ as the key schedule operation of a block cipher with n-bit keys. The vector K is the result of the key schedule expansion for a key k, and the noisy vector K' is obtained from K due to the process of memory bit decay. We note that in this case, another goal of the adversary could be recovering K rather than k (that is, the expanded key rather than the original encryption key), since with the round subkeys one could implement the encryption/decryption algorithm. In most cases, one should be able to efficiently recover the encryption key k from the expanded key K. However it could be conceivable that for a particular cipher with a highly non-linear key schedule, the problems are not equivalent.

Finally, we note that the *Cold Boot* problem is equivalent to decoding (potentially non-linear) binary codes with biased noise.

3 Block Cipher Key Expansion

In this section we briefly describe some of the relevant features of the key schedule operation of the target ciphers.

3.1 AES

For details of the key schedule of the AES block cipher we refer the reader to [7]. In this work, we are interested in its description as a system of polynomial equations over $\mathbb{F}_2$, see [6]. We note that the non-linearity of the key schedule is provided by four S-box operations in the computation of each round subkey. The explicit degree of the S-box Boolean functions is 7, while it is well known that the key schedule can be described as a system of quadratic equations.

3.2 Serpent

SERPENT, designed by Anderson et al. [2], was one of the five AES finalists. The cipher key schedule operation produces 132 32-bit words of key material as follows. First, the user-supplied key k is padded to 256 bits using known constants, and written as eight 32-bit words $w_{-8}, \ldots, w_{-1}$. This new string is then expanded into the prekey words $w_0, \ldots, w_{131}$ by the following affine recurrence:

$$w_i = (w_{i-8} \oplus w_{i-5} \oplus w_{i-3} \oplus w_{i-1} \oplus \psi \oplus i) \lll 11,$$

where ψ is some known constant. Finally the round keys are calculated from the prekeys w_i using the S-boxes S_i in bitslice mode in the following way:

$$\begin{aligned}
\{k_0, k_1, k_2, k_3\} &= S_3(w_0, w_1, w_2, w_3) \\
\{k_4, k_5, k_6, k_7\} &= S_2(w_4, w_5, w_6, w_7) \\
\{k_8, k_9, k_{10}, k_{11}\} &= S_1(w_8, w_9, w_{10}, w_{11}) \\
\{k_{12}, k_{13}, k_{14}, k_{15}\} &= S_0(w_{12}, w_{13}, w_{14}, w_{15}) \\
\{k_{16}, k_{17}, k_{18}, k_{19}\} &= S_7(w_{16}, w_{17}, w_{18}, w_{19}) \\
&\ldots \\
\{k_{124}, k_{125}, k_{126}, k_{127}\} &= S_4(w_{124}, w_{125}, w_{126}, w_{127}) \\
\{k_{128}, k_{129}, k_{130}, k_{131}\} &= S_3(w_{128}, w_{129}, w_{130}, w_{131}).
\end{aligned}$$

The rounds subkeys are then $K_i = \{k_{4i}, k_{4i+1}, k_{4i+2}, k_{4i+3}\}$.

We note the following features of the cipher key schedule which are of relevance to *Cold Boot* key recovery: the user-supplied key does not appear in the output of the SERPENT key schedule operation, the explicit degree of the S-box Boolean functions is three, and every output bit of the key schedule depends non-linearly on the user-supplied key.

3.3 Twofish

Twofish, designed by Schneier et al. [14], was also one of the five AES finalists. The cipher is widely deployed, e.g. it is part of the cryptographic framework in the Linux kernel and is also available in Full Disk Encryption products. Twofish has a rather complicated key schedule, which makes it a challenging target for *Cold Boot* key recovery attacks. We note that while Twofish is defined for all key sizes up to 256 bits, we will focus here on the 128-bit version. We also follow the notation from [14].

The Twofish key schedule operation generates 40 32-bit words of expanded key $K_0, \ldots, K_{39}$, as well as four key-dependent S-boxes from the user-provided key M. Let $k = 128/64 = 2$, then the key M consists of $8k = 16$ bytes $m_0, \ldots, m_{8k-1}$. The cipher key schedule operates as follows. Each four consecutive bytes are converted into 32-bit words in little endian byte ordering. That is, the leftmost byte is considered as the least significant byte of the 32-bit word. This gives rise to four words M_i. Two key vectors M_e and M_o are defined as $M_e = (M_0, M_2)$ and $M_o = (M_1, M_3)$. The subkey words K_{2i} and K_{2i+1} for $0 \leq i < 20$ are then computed from M_e and M_o by the routine `gen_subkeys` given in Algorithm 2 in the Appendix.

Algorithm 1 (also in the Appendix) defines the function h used in the key schedule. There, we have that q_0 and q_1 are applications of two 8-bit S-boxes defined in [14] and $MDS(Z)$ is a multiplication of Z interpreted as a 4 element vector over the field $\mathbb{F}_{2^8} \cong \mathbb{F}_2[x]/\langle x^8 + x^6 + x^5 + x^3 + 1\rangle$ by a 4×4 MDS matrix. The explicit degree of the S-boxes' Boolean functions is also seven.

Finally, a third vector S is also derived from the key. This is done by combining the key bytes into groups of eight (e.g. $m_0, \ldots, m_7$), interpreting them as a vector over the field $\mathbb{F}_{2^8} \cong \mathbb{F}_2[x]/\langle x^8 + x^6 + x^3 + x^2 + 1\rangle$, which is multiplied by a 4×8 matrix RS. Each resulting four bytes are then interpreted as a 32-bit word S_i. These words make up the third vector $S = (S_1, S_0)$. The key-dependent S-Box g maps 32 bits to 32 bits and is defined as $g(X) = h(X, S)$.

Full disk encryption products use infrequent re-keying, and to provide efficient and transparent access to encrypted data, applications will in practice precompute the key schedule and store the expanded key in memory. For the Twofish block cipher, this means that the subkey words K_0, ..., K_{39} as well as the key dependent S-boxes are typically precomputed.

Storing 40 words $K_0, \ldots, K_{39}$ in memory is obviously straightforward (we note however that this set of words does not contain a copy of the user-supplied key). To store the key dependent S-box, the authors of [14] state: "Using 4 Kb of table space, each S-box is expanded to a 8-by-32-bit table that combines both the S-box lookup and the multiply by the column of the MDS matrix. Using this option, a computation of g consists of four table lookups, and three XORS. Encryption and decryption speeds are constant regardless of key size." We understand that most software implementations choose this strategy to represent the S-box (for instance, the Linux kernel chooses this approach, and by default Truecrypt also implements this technique, which can however be disabled with the C macro `TC_MINIMIZE_CODE_SIZE`); we assume this is the case in our analysis.

4 Solving Systems of Algebraic Equations with Noise

In this section we model a new family of problems – solving systems of multivariate algebraic equations with noise – and propose a first method for solving problems from this family. We use the method to implement a *Cold Boot* attack against ciphers with key schedule with a higher degree of non-linearity, such as SERPENT.

Polynomial system solving (*PoSSo*) is the problem of finding a solution to a system of polynomial equations over some field $\mathbb{F}$. We consider the set $F = \{f_0, \ldots, f_{m-1}\} \subset \mathbb{F}[x_0, \ldots, x_{n-1}]$. A solution to F is any point $x \in \mathbb{F}^n$ such that $\forall f \in F$, we have $f(x) = 0$. Note that we restrict ourselves to solutions in the base field in the context of this work.

Moreover, denote by *Max-PoSSo* the problem of finding any $x \in \mathbb{F}^n$ that satisfies the maximum number of polynomials in F. Likewise, by *Partial Max-PoSSo* we denote the problem of finding a point $x \in \mathbb{F}^n$ such that for two sets of polynomials $\mathcal{H}, \mathcal{S} \subset \mathbb{F}[x_0, \ldots, x_{n-1}]$, we have $f(x) = 0$ for all $f \in \mathcal{H}$, and the number of polynomials $f \in \mathcal{S}$ with $f(x) = 0$ is maximised. Max-PoSSo is Partial Max-PoSSo with $\mathcal{H} = \varnothing$.

Finally, by *Partial Weighted Max-PoSSo* we denote the problem of finding a point $x \in \mathbb{F}^n$ such that $\forall f \in \mathcal{H} : f(x) = 0$ and $\sum_{f \in \mathcal{S}} \mathcal{C}(f, x)$ is minimised where $\mathcal{C} : f \in \mathcal{S}, x \in \mathbb{F}^n \rightarrow \mathbb{R}_{\geq 0}$ is a cost function that returns 0 if $f(x) = 0$ and some value $v > 0$ if $f(x) \neq 0$. Partial Max-PoSSo is Partial Weighted Max-PoSSo where $\mathcal{C}(f, x)$ returns 1 if $f(x) \neq 0$ for all f.

The family of "Max-PoSSo" problems defined above is analogous to the well-known Max-SAT family of problems. In fact, these problems could well be reduced to their SAT equivalents. However, the modelling as polynomial systems seems more natural in this context since more algebraic structure can be preserved.

4.1 Cold Boot as Partial Weighted Max-PoSSo

We can consider the *Cold Boot* Problem as a Partial Weighted Max-PoSSo problem over $\mathbb{F}_2$. Let $F_{\mathcal{K}}$ be an equation system corresponding to $\mathcal{KS}$ such that the only pairs (k, K) that satisfy $F_{\mathcal{K}}$ are any $k \in \mathbb{F}_2^n$ and $K = \mathcal{K}\mathcal{f}(k)$. In our task however, we need to consider $F_{\mathcal{K}}$ with k and K'. Assume that for each noisy output bit K'_i there is some $f_i \in F_{\mathcal{K}}$ of the form $g_i + K'_i$ where g_i is some polynomial. Furthermore assume that these are the only polynomials involving the output bits ($F_{\mathcal{K}}$ can always be brought into this form) and denote the set of these polynomials by $\mathcal{S}$. Denote the set of all remaining polynomials in $F_{\mathcal{K}}$ as $\mathcal{H}$, and define the cost function $\mathcal{C}$ as a function which returns

$$\begin{array}{cr} \frac{1}{1-\Delta_0} & \text{for } K'_i = 0, f(x) \neq 0, \\ \frac{1}{1-\Delta_1} & \text{for } K'_i = 1, f(x) \neq 0, \\ 0 & \text{otherwise.} \end{array}$$

This cost function returns a cost proportional to the inverse of the probability that a given value is correct. Finally, let $F_{\mathcal{E}}$ be an equation system that is only satisfiable for $k \in \mathbb{F}_2^n$ for which $\mathcal{E}$ returns *True*. This will usually be an equation system for one or more encryptions. Add the polynomials in $F_{\mathcal{E}}$ to $\mathcal{H}$. Then $\mathcal{H}, \mathcal{S}, \mathcal{C}$ define a Partial Weighted Max-PoSSo problem. Any optimal solution x to this problem is a candidate solution for the *Cold Boot* problem.

In order to solve Max-PoSSo problems, we propose below an approach which appears to better capture the algebraic structure of the underlying problems (compared to SAT-solvers), and should thus have further applications.

4.2 Mixed Integer Programming

Integer optimisation deals with the problem of minimising (or maximising) a function in several variables subject to linear equality and inequality constraints, and integrality restrictions on some or all the variables. A linear mixed integer programming problem (MIP) is defined as a problem of the form

$$\min_x \{c^T x | Ax \leq b, x \in \mathbb{Z}^k \times \mathbb{R}^l\},$$

where c is an n-vector ($n = k + l$), b is an m-vector and A is an $m \times n$-matrix. This means that we minimise the linear function $c^T x$ (the inner product of c and x) subject to linear equality and inequality constraints given by A and b. Additionally $k \geq 0$ variables are restricted to integer values while $l \geq 0$ variables are real-valued. The set S of all $x \in \mathbb{Z}^k \times \mathbb{R}^l$ that satisfy the linear constraints $Ax \leq b$, that is

$$S = \{x \in \mathbb{Z}^k \times \mathbb{R}^l \mid Ax \leq b\},$$

is called the feasible set. If $S = \varnothing$ the problem is infeasible. Any $x \in S$ that minimises $c^T x$ is an optimal solution.

Efficient MIP solvers use a branch-and-cut algorithm as one of their core components. The main advantage of MIP solvers compared to other branch-and-cut solvers (e.g. SAT-solvers) is that they can relax the problem to a floating point linear programming problem in order to obtain lower and upper bounds. These bounds can then be used to cut search branches. This relaxation also allows one to prove optimality of a solution without exhaustively searching for all possible solutions.

Moreover we can convert the PoSSo problem over $\mathbb{F}_2$ to a mixed integer programming problem using the Integer Adapted Standard Conversion [5] as follows. Consider the square-free polynomial $f \in \mathbb{F}_2[x_0, \dots, x_{n-1}]$. We interpret the Boolean equation $f = 0$ as an equation over the integers by replacing XOR by addition and AND by multiplication. All solutions of f over $\mathbb{F}_2$ will correspond to multiples of 2 when considered over the integers. Let ℓ be the mininum and u the maximum value of these multiples of two. We introduce an integer variable m and restrict it between $\frac{\ell}{2}$ and $\frac{u}{2}$ (inclusive). Finally we linearise $f - 2m$ and add equations relating the new linear variables to the original monomials. More details of this modelling[3] can be found in [5]. It then follows that solving the resulting MIP problem for any objective function will recover a value x that also solves the PoSSo problem.

Moreover we can convert a Partial Weighted Max-PoSSo problem into a Mixed Integer Programming problem as follows. Convert each $f \in \mathcal{H}$ to linear constraints as before. For each $f_i \in \mathcal{S}$ add some new binary slack variable e_i to f_i and convert the polynomial $f_i + e_i$ as before. The objective function we minimise is $\sum c_i e_i$, where c_i is the value of $\mathcal{C}(f, x)$ for some x such that $f(x) \neq 0$. This way, the penalty for having $e_i = 1$ is proportional to the cost implied by the cost function $\mathcal{C}$. Any optimal solution $x \in S$ will be an optimal solution to the Partial Weighted Max-PoSSo problem.

We note that in our modelling of Cold Boot key recovering as a Mixed Integer Programming problem, we are using a linear objective function, which we expect to be a first order approximation of the true noise model. Our results will however demonstrate that this approximation is sufficient. Finally, we note that the approach discussed above is essentially the non-linear generalisation of decoding random linear codes with linear programming [8].

5 Cold Boot Key Recovery against Block Ciphers

The original approach proposed in [9] is to model the memory decay as a binary asymmetric channel (with error probabilities δ_0, δ_1), and recover the encryption key from the *closest code word* (i.e. valid key schedule) based on commonly used decoding techniques. The model of attack used in [9] often assumes $\delta_1 \approx 0$ (that

[3] We note that the MIP solver SCIP [1] used in this work generates linear constraints for (among others) AND clauses automatically and thus we do not need to model these explicitly in our experiments.

is, the probability of a 0 bit flipping to 1 is negligible), which appears to be a reasonable assumption to model memory decay in practice. We will sometimes do the same in our discussions below. However, all experimental data in this work was generated with $\delta_1 > 0$, even where our algorithms assume $\delta_1 = 0$, in order to estimate the success rate in practice more precisely. Thus, contrary to prior work, when we construct experimental data we do not consider the asymmetry to be perfect.

Under the adopted model, recovering the original 56-bit key for DES is equivalent to decoding a repetition code, as discussed in [9]. In this section we will discuss potential methods for recovering the user-supplied key for the key schedules of Twofish, Serpent and the AES, under the *Cold Boot* attack scenario. We note that the attack's main parameters (the error probabilities δ_0, δ_1) obviously affect the effectiveness (and the viability) of the methods discussed below; in particular, while some methods may have a superior performance for a certain range of δ_0, δ_1, they may however not be viable outside this particular range. For instance, the technique presented in [11] relies on $\delta_1 = 0$.

5.1 Prior Work on AES

The AES key schedule is not as simple as the one from DES, but still contains a large amount of linearity. Furthermore, the original encryption key is used as the initial whitening subkey, and thus should be present in the key schedule output. The method proposed in [9] for recovering the key for the AES-128 divides this initial subkey into four subsets of 32 bits, and uses 24 bits of the second subkey as redundancy. These small sets are then decoded in order of likelihood, combined and the resulting candidate keys are checked against the full schedule. The idea can be easily extended to the AES with 192- and 256-bit keys. The authors of [9] recover up to 50% of keys for error rates of $\delta_0 = 0.30, \delta_1 = 0$ within 20 minutes. In [17] an improved algorithm making better use of the AES key schedule structure was proposed which allows one to recover the vast majority of keys within 20 minutes for $\delta_0 = 0.70, \delta_1 = 0$. It is noted in [17] that the algorithm can be adapted for the case $\delta_1 > 0$.

In [11] an alternative algorithm is proposed which models the *Cold Boot* problem as a SAT problem by ignoring all output bits equal to zero. In the considered model this implies that the remaining output bits are correct (since $\delta_1 = 0$ implies $\Delta_1 = 1$). Thus a set of correct SAT clauses can be constructed, which – due to the amount of redundant information available in the AES key schedule – still allows one to recover the encryption key. The authors of [11] report recovery of encryption keys for $\delta_0 = 0.80, \delta_1 = 0$ with an average running time of about 30 minutes.

Below we discuss the different methods we have considered for cold boot key recovery, and our results when applied to the AES, Serpent and Twofish. We first discuss a naïve decoding technique in order to have a base line to compare our algebraic technique to. Our algebraic technique is then discussed in Section 5.4.

5.2 Generic Combinatorial Approach

Assuming that the cipher key schedule operation is invertible, we can still consider a somewhat naïve combinatorial approach, even when the user-supplied key does not explicitly appear in the expanded key schedule material. In order to recover the full n-bit key, we will consider at least n bits of key schedule output. We assume that bit-flips are biased towards zero with overwhelming probability (i.e., we assume the asymmetry is perfect) and assume the distribution of bits arising in the original key schedule material is uniform. Then for an appropriate n-bit segment K in the noisy key schedule, we can expect approximately $\frac{n}{2} + r$ zeros, where $r = \lceil \frac{n}{2}\delta_0 \rceil$. We have thus to check

$$\sum_{i=0}^{r} \binom{n/2 + r}{i}$$

candidates for the segment K. Each check entails to correct the selected bits, invert the key schedule and verify the candidate for k using for example $\mathcal{E}$. For $n = 128$ and $\delta_0 = 0.15$ we would need to check approximately 2^{40} candidates; for $\delta_0 = 0.30$ we would have to consider approximately 2^{64} candidates; for $\delta_0 = 0.50$ we would have to consider approximately 2^{85} candidates. By focusing the search around the expected error rate, we may be able to improve these times. This approach is applicable to both SERPENT and the AES. However we need to adapt it slightly for Twofish.

5.3 Adapted Combinatorial Approach for Twofish

We recall that for the Twofish key schedule, we assume that the key dependent S-boxes are stored as a lookup table in memory. In fact, each S-box is expanded to a 8-by-32-bit table holding 32-bit values combining both the S-Box lookup and the multiplication by the column of the MDS matrix (see Section 3.3) Thus, we will have in memory a 2-dimensional 32-bit word array $s[4][256]$, where $s[i][j]$ holds the result of the substitution for the input value j for byte position i. The output of the complete S-Box for the word $X = (X_0, X_1, X_2, X_3)$ is $s[0][X_0] \oplus s[1][X_1] \oplus s[2][X_2] \oplus s[3][X_3]$.

Each array $s[i]$ holds the output of an MDS matrix multiplication by the vector X, with three zero entries and all possible values $0 \leq X_i < 256$, with each value occurring only once. Thus, we have exactly 256 possible values for the 32-bit words in $s[i]$ and we can simply adjust each disturbed word to its closest possible word. Furthermore, we do not need to consider all values, we can simply use those values with low Hamming distance to their nearest candidate word but a large Hamming distance to their second best candidate. We can thus implement a simple decoding algorithm to eventually recover an explicit expression for each of the four key-dependent S-boxes.

Using this method, we can recover all bytes of S_0 and S_1. More specifically, if we assume that $\delta_0 = \delta_1$, we can recover the correct S_0, S_1 with overwhelming probability if 30% of the bits have been flipped. If we assume an asymmetric channel ($\delta_1 \approx 0$) then we can recover the correct values for S_0, S_1 with overwhelming probability if 60% of the bits have been flipped. This gives us 64-bit of *information* about the key.

In order to recover the full 128-bit key, we can adapt the combinatorial approach discussed above. In the noise-free case, we can invert the final modular addition and the MDS matrix multiplication. Since these are the only steps in the key schedule where diffusion between S-box rows is performed, we should get eight 8-bit equation systems of the form $C_1 = Q_0(C_0 \oplus M_0) \oplus M_1$, where Q_0 is some S-box application and C_0 and C_1 are known constants. Each such equation restricts the number of possible candidates for M_0, M_1 from 2^{16} to 2^8. Using more than one pair C_0, C_1 for each user-supplied key byte pair M_0, M_1 allows us to recover the unique key. Thus, although the Twofish key schedule is not as easily reversed as the SERPENT or AES key schedule, the final solving step is still very simple. Thus, the estimates given for the combinatorial approach ($\delta_0 = 0.15 \rightarrow 2^{36}$ candidates and $\delta_0 = 0.30 \rightarrow 2^{62}$ candidates) also apply to Twofish.

Alternatively, we may consider one tuple of C_0, C_1 only and add the linear equations for S. This would provide enough information to recover a unique solution; however S does mix bytes from M_0 across S-box rows, which makes the solving step more difficult.

5.4 Algebraic Approach Using Max-PoSSo

If the algebraic structure of the key schedule permits, we can model the *Cold Boot* key recovery problem as a Partial (Weighted) Max-PoSSo problem, and use the methods discussed earlier to attempt to recover the user-supplied key or a noise-free version of the key schedule. We applied those methods to implement a *Cold Boot* attack against the AES and SERPENT. We focused on the 128-bit versions of the two ciphers.

For each instance of the problem we performed 100 experiments with randomly generated keys. In the experiments we usually did not consider the full key schedule but rather a reduced number of rounds of the *key schedule* in order to improve the running time of our algorithms. We note however that this does not mean we are attacking a reduced-round version of the algorithm: considering a reduced amount of data (less redundancy) in the key schedule still allows us to recover the *entire* encryption key of the full-round version of the block cipher. The running time is increased when more data is considered due to the increase of terms in the objective function. Furthermore, we did not include equations for $\mathcal{E}$ explicitly. This is again to reduce the amount of data the solver has to consider. Finally, we also considered at times an "aggressive" modelling, where our algorithm assumes $\delta_1 = 0$ instead of $\delta_1 = 0.001$. In this case all values $K_i' = 1$ are considered correct by the algorithm (since $\Delta_1 = 1$), and as a result all corresponding equations are promoted to the set $\mathcal{H}$. We stress, however, that the input data in our experiments was always generated with $\delta_1 > 0$. Thus, increasing the amount of data considered also increases the chances of including an equation for $K_i' = 1$ which is not correct. We note that in the "aggressive" modelling our problem reduces to

Table 1. AES considering N rounds of key schedule output

N	δ_0	aggr	limit t	r	min t	avg. t	max t
2	0.05	–	3600.00	59%	50.80 s	2124.90 s	3600.00 s
3	0.15	+	60.0s	63%	1.38 s	8.84 s	41.66 s
4	0.15	+	60.0s	70%	1.78 s	11.77 s	59.16 s
4	0.30	+	600.0s	66%	4.81 s	116.07 s	600.00 s
4	0.30	+	3600.0s	69%	4.86 s	117.68 s	719.99 s
4	0.35	+	600.0s	65%	4.66 s	185.14 s	600.00 s
4	0.35	+	3600.0s	68%	4.45 s	207.07 s	1639.55 s
4	0.40	+	600.0s	47%	4.95 s	284.99 s	600.00 s
4	0.40	+	3600.0s	61%	4.97 s	481.99 s	3600.00 s
5	0.40	+	3600.0s	62%	7.72 s	704.33 s	3600.00 s
4	0.50	+	3600.0s	8%	6.57 s	3074.36 s	3600.00 s
4	0.50	+	7200.0s	13%	6.10 s	5882.66 s	7200.00 s

Partial Max-PoSSo and that the specific weights assigned in the cost function are irrelevant, since all weights are identical.

Running times for the AES and Serpent using the MIP solver SCIP [1] are given in Tables 1 and 2 respectively. For each cipher dedicated tuning parameters were used and we also made use of advanced features in SCIP such as the support for AND constraints which are not available in other MIP solvers. The column "a" denotes whether we chose the aggressive ("+") or normal ("–") modelling. The column "cutoff t" denotes the time we maximally allowed the solver to run until we interrupted it. The column r gives the success rate, i.e. the percentage of instances we recovered the correct key for.

For the Serpent key schedule we consider decays up to $\delta_0 = 0.50, \delta_1 = 0.001$. We also give running times and success rates for the AES up to $\delta_0 = 0.50, \delta_1 = 0.001$ in order to compare our approach with previous work. We note that a success rate lower than 100% may still allow a successful key recovery since the algorithm can be run using other data from the key schedule if it fails for the first few rounds. Considering later rounds of the key schedule has no performance penalty for the AES, but does decrease the performance for Serpent as indicated in the row $16 \ll 8$ which considers 16 words of key schedule output starting from the 8-th word. Our attacks were implemented using the Sage mathematics software [15].

We note from the results in Table 1 that the Max-PoSSo based method proposed in this work compares favourably to the results in [9]. However, it offers poorer results when compared to the ones in [17,11]. While there is no prior work to compare Table 2 with, we note that our technique compares favourably to the generic combinatorial approach discussed earlier. Furthermore, our method has some attractive features and flexibility which allow its application to more extended scenarios and in principle to any block cipher[4].

[4] We note however that this does not imply the technique is *practical* for all block ciphers as demonstrated by the lack of success against Twofish.

Table 2. SERPENT considering $32 \cdot N$ bits of key schedule output

N	δ_0	aggr	limit t	r	min t	avg. t	max t
12	0.05	–	600.0s	37%	8.22 s	457.57 s	600.00 s
12	0.15	+	60.0s	84%	0.67 s	11.25 s	60.00 s
16	0.15	+	60.0s	79%	0.88 s	13.49 s	60.00 s
$16 \ll 8$	0.15	+	1800.0s	64%	95.52 s	1089.80 s	1800.00 s
16	0.30	+	600.0s	74%	1.13 s	57.05 s	425.48 s
16	0.50	+	1800.0s	21%	135.41 s	1644.53 s	1800.00 s
16	0.50	+	3600.0s	38%	136.54 s	2763.68 s	3600.00 s

Table 3. SERPENT considering $32 \cdot N$ bits of key schedule output (symmetric noise)

N	$\delta_0 = \delta_1$	limit t	r	min t	avg. t	max t
12	0.01	3600.0	96%	4.60 s	256.46	3600.0 s
12	0.02	3600.0	79%	8.20 s	1139.72	3600.0 s
8	0.03	3600.0	41%	3.81 s	372.85 s	3600.0 s
12	0.03	7200.0	53%	24.57 s	4205.34 s	7200.0 s
12	0.05	3600.0	18%	5.84 s	1921.89 s	3600.0 s

Finally, we note that our technique does not rely on $\delta_1 = 0$ and can thus be applied if perfect asymmetry cannot be assumed. To demonstrate this feature, we give results against SERPENT for our technique when considering symmetric noise (i.e., $\delta_0 = \delta_1$) in Table 3. For comparison, for $\delta_0 = \delta_1 = 0.05$ a combinatorial approach similar to Section 5.2 would have to check rougly $\binom{128}{\lceil 0.05 \cdot 128 \rceil} \approx 2^{36.4}$ candidates. In order to make the comparison fair, we aim for a success rate of $\approx 20\%$ and thus only have to consider roughly 1/5 of those candidates. If each of those checks costs at least $15 \cdot 10^{-8}$ seconds – which ammounts to ≈ 390 CPU cycles on a 2.6 Ghz CPU – then the overall running time would be greater than the average running time reported in Table 3.

6 Conclusion and Discussions

In this paper we followed up from the original work on *Cold Boot* key recovery attacks in [9,11,17], and extended the analysis to consider other block ciphers such as Twofish and SERPENT. Our algorithms apply a novel method for solving a set of non-linear algebraic equations with noise based on Integer Programming. Besides improving some existing results and extending the range of scenarios in which the attacks can be applied, this paper also brings into attention two topics which in our opinion should be of enough interest for future research in cryptology:

Block Cipher Key Schedule: the structure of the key schedule of block ciphers has recently started attracting much attention from the cryptologic research community. Traditionally, the key schedule operation has perhaps received much less consideration from designers, and other than for efficiency and protection against some known attacks (e.g. slide attacks), the key schedule was often designed in a somewhat ad-hoc way (in contrast to the usually well-justified and motivated cipher round structure). However the recent attacks against the AES and Kasumi have brought this particular operation to the forefront of block cipher cryptanalysis (and as a result, design). While one can argue that some of the models of attack used in the recent related-key attacks may be far too generous to be of practical relevance, it is clear that resistance of ciphers against these attacks will from now on be used as another form of measure of security of block ciphers.

In this spirit, we propose in this paper a further measure of security for key schedule operations, based on the *Cold Boot* attack scenario. These attacks are arguably more practical than some of the other attacks targeting the key schedule operation. More importantly, we believe the model can be used to further evaluate the strength of the key schedule operation of block ciphers. Our results show however that it is not trivial to provide high security against *Cold Boot* attacks. In fact, by proposing generic algorithms for solving the *Cold Boot* problem, we showed that, contrary to general belief, several popular block ciphers are also susceptible to attack under this model. How to come up with design criteria for a secure key schedule under this model (while preserving other attractive features such as efficiency) remains a topic for further research.

Polynomial System Solving with Noise: another contribution of this paper, which is very likely to be of independent interest, is the treatment of the problem of solving non-linear multivariate equations with noise. In fact, several interesting problems in cryptography such as algebraic attacks, side-channel attacks and the cryptanalysis of LPN/LWE-based schemes can be naturally modeled as Max-PoSSo problems. However, so far this problem was not considered in its general form. This paper presents a formalisation of this problem and a novel method, based on Integer Programming, which proved to be a powerful technique in some situations. We expect that this will bring MIP solvers further to the attention of the cryptography research community and consider studying and improving (MIP-based) Max-PoSSo methods an interesting area for future research.

Acknowledgements

We would like to thank Stefan Heinz, Timo Berthold and Ambros M. Gleixner for helpful discussions on mixed integer programming and for providing tuning parameters for SCIP suitable for our problems. We would also like to thank Kenny Paterson for helpful comments on this work.

References

1. Achterberg, T.: Constraint Integer Programming. PhD thesis, TU Berlin (2007), `http://scip.zib.de`
2. Biham, E., Anderson, R.J., Knudsen, L.R.: SERPENT: A New Block Cipher Proposal. In: Vaudenay, S. (ed.) FSE 1998. LNCS, vol. 1372, pp. 222–238. Springer, Heidelberg (1998)

3. Biham, E., Dunkelman, O., Keller, N.: A Related-Key Rectangle Attack on the Full KASUMI. In: Roy, B. (ed.) ASIACRYPT 2005. LNCS, vol. 3788, pp. 443–461. Springer, Heidelberg (2005)
4. Biryukov, A., Khovratovich, D., Nikolić, I.: Distinguisher and related-key attack on the full AES-256. In: Halevi, S. (ed.) CRYPTO 2009. LNCS, vol. 5677, pp. 231–249. Springer, Heidelberg (2009)
5. Borghoff, J., Knudsen, L.R., Stolpe, M.: Bivium as a Mixed-Integer Linear Programming Problem. In: Parker, M.G. (ed.) Cryptography and Coding 2009. LNCS, vol. 5921, pp. 133–152. Springer, Heidelberg (2009)
6. Cid, C., Murphy, S., Robshaw, M.J.B.: Algebraic Aspects of the Advanced Encryption Standard. Springer, Heidelberg (2007)
7. Daemen, J., Rijmen, V.: The Design of Rijndael. Springer, Heidelberg (2002)
8. Feldman, J.: Decoding Error-Correcting Codes via Linear Programming. PhD thesis, Massachusetts Institute of Technology (2003)
9. Alex Halderman, J., Schoen, S.D., Heninger, N., Clarkson, W., Paul, W., Calandrino, J.A., Feldman, A.J., Appelbaum, J., Felten, E.W.: Lest We Remember: Cold Boot Attacks on Encryption Keys. In: USENIX Security Symposium, USENIX Association, pp. 45–60 (2009)
10. Heninger, N., Shacham, H.: Reconstructing RSA Private Keys from Random Key Bits. Cryptology ePrint Archive, Report 2008/510 (2008)
11. Kamal, A.A., Youssef, A.M.: Applications of SAT Solvers to AES key Recovery from Decayed Key Schedule Images. In: Proceedings of The Fourth International Conference on Emerging Security Information, Systems and Technologies – SECURWARE 2010, Venice/Mestre, Italy, July 18-25 (2010)
12. Lloyd, J.: Re: cold boot attacks on disk encryption. Message posted to The Cryptography Mailing List on February 21 (2008), archived at `http://www.mail-archive.com/cryptography/metzdowd.com/msg08876.html`
13. Oren, Y., Kirschbaum, M., Popp, T., Wool, A.: Algebraic Side-Channel Analysis in the Presence of Errors. In: Mangard, S., Standaert, F.-X. (eds.) CHES 2010. LNCS, vol. 6225, pp. 428–442. Springer, Heidelberg (2010)
14. Schneier, B., Kelsey, J., Whiting, D., Wagner, D., Hall, C., Ferguson, N.: Twofish: A 128-Bit Block Cipher (1998), `http://www.schneier.com/paper-twofish-paper.pdf`
15. Stein, W., et al.: Sage Mathematics Software (Version 4.4.1). The Sage Development Team (2010) `http://www.sagemath.org`
16. TrueCrypt Project, `http://www.truecrypt.org/`.
17. Tsow, A.: An Improved Recovery Algorithm for Decayed AES Key Schedule Images. In: Jacobson Jr., M.J., Rijmen, V., Safavi-Naini, R. (eds.) SAC 2009. LNCS, vol. 5867, pp. 215–230. Springer, Heidelberg (2009)

Appendix

Input: Z – a 32-bit word
Input: L – a list of two 32-bit words
Result: a 32-bit word
begin
$L_0, L_1 \longleftarrow L[0], L[1]$;
$z_0, z_1, z_2, z_3 \longleftarrow$ split Z into four bytes;
$z_0, z_1, z_2, z_3 \longleftarrow q_0[z_0], q_1[z_1], q_0[z_2], q1[z_3]$;
$z_0, z_1, z_2, z_3 \longleftarrow z_0 \oplus L_1[0], z_1 \oplus L_1[1], z_2 \oplus L_1[2], z_3 \oplus L_1[3]$;
$z_0, z_1, z_2, z_3 \longleftarrow q_0[z_0], q_0[z_1], q_1[z_2], q_1[z_3]$;
$z_0, z_1, z_2, z_3 \longleftarrow z_0 \oplus L_0[0], z_1 \oplus L_0[1], z_2 \oplus L_0[2], z_3 \oplus L_0[3]$;
$z_0, z_1, z_2, z_3 \longleftarrow q_1[z_0], q_0[z_1], q_1[z_2], q_0[z_3]$;
$z_0, z_1, z_2, z_3 \longleftarrow MDS(z_0, z_1, z_2, z_3)$;
return the 32-bit word consisting of the four bytes z_0, z_1, z_2, z_3;
end

Algorithm 1. h

Input: i – an integer
Input: M_e – a list of 32-bit words
Input: M_o – a list of 32-bit words
Result: two 32-bit words
begin
$\rho \longleftarrow 2^{24} + 2^{16} + 2^8 + 2^0$;
$A_i \longleftarrow h(2i\rho, M_e)$;
$B_i \longleftarrow h((2i+1)\rho, M_o) \lll 8$;
$K_{2i} \longleftarrow A_i + B_i \bmod 2^{32}$;
$K_{2i+1} \longleftarrow (A_i + 2B_i \bmod 2^{32}) \lll 9$;
return K_{2i}, K_{2i+1};
end

Algorithm 2. gen_subkeys

Exponent Blinding Does Not Always Lift (Partial) Spa Resistance to Higher-Level Security

Werner Schindler[1] and Kouichi Itoh[2]

[1] Bundesamt für Sicherheit in der Informationstechnik (BSI)
Godesberger Allee 185–189, 53175 Bonn, Germany
Werner.Schindler@bsi.bund.de
[2] Fujitsu Laboratories Ltd., 211–8588, 4-1-1, KamiKodanaka, Nakahara-ku, Kawasaki, Japan
kito@labs.fujitsu.com

Abstract. Exponent blinding is known as a secure countermeasure against side-channel attacks. If single power traces reveal some exponent bits, an attack by Fouque et al. applies that recovers the exponent. However, this attack becomes infeasible if some of the guessed bits are incorrect. Thus, the attack was not assumed to be a realistic threat. In this paper we present two variants of a novel generic attack, which works for considerable error rates at each bit position, disproving the hypothesis that mere exponent blinding is always sufficient. We confirmed experimentally that our attack permits up to 28% (RSA case) or 23% (ECC case) error bits.

1 Introduction

Blinding mechanisms such as base and exponent blinding [8] have been effective algorithmic countermeasures against side-channel attacks. Both seem to prevent pure global timing attacks [9], since no pure timing attack is presently known defeating either base or exponent blinding.

However, base blinding does not protect against local timing attacks [11,1], which exploit timing information on the particular elementary operations: modular multiplications and squarings in the RSA case, point doubling and point addition (or arithmetic sub-operations thereof) in the ECC case. Attacks on table-based RSA implementations [11,1] exploit the pre-computation phase of modular exponentiation and the fact that the ith elementary operation in the exponentiation phase is always of the same type (either squaring or a multiplic ation by a particular table value).

A cryptographic device must be also secure against power analysis [9], which in particular comprises Simple Power Analysis (SPA) and Differential Power Analysis (DPA). However, exponent blinding prevents the alignment of power traces corresponding to elementary operations of the same type. This shall prevent an attacker from combining information from several RSA exponentiations/ECC

J. Lopez and G. Tsudik (Eds.): ACNS 2011, LNCS 6715, pp. 73–90, 2011.

scalar point multiplications. It might thus be assumed that exponent blinding protects against power attacks if the device is resistant against SPA attack, or more generally, against attacks on single exponentiations/scalar point multiplications. The assumption that exponent blinding automatically lifts SPA or partial SPA resistance (i.e. the device is secure enough to protect some private key bits against SPA) to higher-level security assertions (e.g., DPA resistance) should be valid in many cases of practical relevance. However, we present a new generic attack that shows that this conclusion is not true in general. We assume that the targeted device applies the square and always multiply (S&aM) algorithm [4], resp. a double and always add (D&aA) algorithm, as an SPA resistance algorithm. At first we sketch related previous results.

Applying exponent blinding is a powerful solution even when the device is partially SPA-resistant. In theory, a device with partial SPA resistance that only applies exponent blinding is not secure against side-channel attacks [2]. However this attack assumes a strong assumption: *all observed bits* of the randomized exponent must be perfectly revealed from a single power trace *with no error bits*, which is not easy to achieve in real environment because noise is included. There are some known attacks on the S&aM algorithm. First is the address-bit power analysis technique [6] because the address differs between the real and dummy operations. However, large hardware multipliers cause strong noise, which should hide the power consumption of the memory addresses. Second attack is a chosen message technique using $c = -1 \pmod N$ proposed by Yen et al [12]. Itoh et al.'s result [7] shows clear waveforms using auto-correlation techniques. However, filtering the messages, checking whether $c = -1 \pmod N$, prevents this attack. If $c = -1 \pmod N$ simply the lowest bit of the exponent is used to calculate an output. Third is an attack proposed by Courrège et al. [5] that utilizes the power consumption of partial multiplications on word length. It is effective if messages can be chosen, which minimize partial hamming weights. However, S&aM reduces its effectivity because the attacker can use only $c = -1 \pmod N$), which is easily prevented by the filtering. That is, we do not know concrete attacks, which fulfil the above assumption in real-world environments when a simple filtering technique is used.

Can the attacker reveal the private key even when the above assumption, consolidated knowledge on some key bits, is not satisfied? The scenario is the following: the attacker guesses the blinded exponent bits on basis of the corresponding waveforms. Some of the guessed bits may be wrong but the attacker does not know their positions. This scenario is realistic if the waveforms of real and dummy operations are similar, with noisy power traces. If there are 5% unnoticed error bits a brute-force attack costs $\approx \sum_{j=0}^{51} \binom{1024}{j} = 2^{288}$ steps for 1024-bit RSA, or $\approx \sum_{j=0}^{13} \binom{256}{j} = 2^{71}$ steps for 256-bit ECC.

This motivates the following idea: if the operations cannot perfectly be distinguished, the device should be secure against side-channel attack by asserting exponent blinding and input filtering. However, our attack shows this idea does not suffice in general. We show that the attacker is able to identify exponents with identical exponent blinding factors (basic attack) or sums of exponents with identical sums of exponent blinding factors (enhanced attack), respectively, undermining the declared goal of exponent blinding. The basic attack searches for t exponents with identical exponent blinding factors, which allows to simply apply the 'best of t' rule (majority decision). The enhanced attack applies more sophisticated techniques. Both variants tolerate if all exponent bit positions allow only uncertain guesses. In our analysis we assume identical error rate ϵ_b for all bit positions and traces. Interestingly, even additionally applied base blinding might not prevent our attack. Fortunately, generic countermeasures exist.

We mention that reference [3] also considers the recovery of a secret RSA key from defective information. It is a theoretical attack that tolerates certain errors rates in the binary representations of the particular components of (p, q, d, d_p, d_q), (p, q, d), or (p, q), respectively. Results and techniques from [3] cannot be transferred to our situation where an attacker does not get any direct information on the primes p and q. Our attack works against RSA (with and without CRT) and ECC without any restrictions on the private key nor on the public key.

In Section 2 and Section 3 we describe the basic version and an enhanced version of our new attack and present experimental results. Unlike the basic attack, the enhanced attack needs only a small number of power traces to find the private key and cannot effectively be prevented by limiting the number of operations with the targeted key. Instead, large blinding factors are necessary.

2 Basic Attack

The setting of the attack is as follows. A target device (typically a smart card) executes RSA or ECC operations. In both the basic and enhanced attacks we assume that S & aM (RSA)/ D& aA (ECC) are used within exponent/scalar blinding. The attacker performs a power attack on the target device. From single power traces he derives information on the corresponding exponentiations or scalar point multiplications with the secret key.

2.1 Notation

Our attack is generic and applies identically to RSA with CRT, RSA without CRT, and ECC. To avoid clumsy formulations we will simply speak of 'multiplication', 'square' and 'exponent' in the following, which correspond to 'addition', 'doubling' and 'scalar' in the ECC context. The blinded exponents are

$$v_j := d + r_j y \qquad \text{for } j = 1, 2, \ldots \tag{1}$$

where d denotes a long term key. For RSA d is the secret exponent, resp. the secret exponent $(\bmod\, p)$ (usually denoted by d_p in this context) if the CRT is used. For ECC d equals the secret scalar. Further, $y = \phi(n)$, $y = \phi(p)$, or y equals the order of an elliptic curve, respectively. The integer r_j is the exponent blinding factor used for exponentiation j.

Remark 1. We point out that for RSA with CRT it suffices to recover d_p if at least one (plaintext / ciphertext) pair (x, y) is known, e.g. a digital signature. Then $\gcd(y - x^{d_p} (\bmod\, n), n) = p$ factors the modulus n.

We assume $d, y < 2^k$ where the integers d and y (resp. only d for ECC applications) are selected randomly. Hence we may assume $\log_2(d), \log_2(y) \approx k$. The blinding factor r_j assumes values in $\{0, 1, \ldots, 2^R - 1\}$. The binary representation of v_j is $(v_{j;k+R-1}, \ldots, v_{j;0})_2$ where leading zero digits are allowed.

On basis of the SPA information provided by power trace j the attacker guesses the randomized exponent with SPA to obtain $\widetilde{v}_j = (\widetilde{v}_{j;n+R-1}, \ldots, \widetilde{v}_{j;0})_2$, which will usually differ from the correct binary representation $(v_{j;k+R-1}, \ldots, v_{j;0})_2$ of v_j. (Otherwise, the jth power trace alone would suffice for a successful SPA.)

The attacker may commit two types of guessing errors: Although $v_{j;i} = 0$ he might guess $\widetilde{v}_{j;i} = 1$ ('10-error'), or despite of $v_{j;i} = 1$ he might guess $\widetilde{v}_{j;i} = 0$ ('01-error'). We denote the integer that corresponds to $(\widetilde{v}_{j;n+R-1}, \ldots, \widetilde{v}_{j;0})_2$ by $\widetilde{v}_j$. Using this notation the attacker guesses

$$\widetilde{v}_j := (d + r_j y) \oplus e_j = v_j \oplus e_j \qquad \text{for } j = 1, 2, \ldots \tag{2}$$

where '$\oplus$' denotes the bitwise XOR operation while the integer e_j expresses the guessing error for exponent v_j. For $j \neq m$ we consider the hamming weight of

$$\widetilde{v}_j \oplus \widetilde{v}_m = (v_j \oplus v_m) \oplus (e_j \oplus e_m). \tag{3}$$

We use the relationship

$$\text{ham}(\widetilde{v}_j \oplus \widetilde{v}_m) = \begin{cases} \text{ham}(e_j \oplus e_m) & (\text{if } r_j = r_m) \\ \text{ham}(v_j \oplus v_m \oplus e_j \oplus e_m) & (\text{if } r_j \neq r_m) \end{cases} \tag{4}$$

to distinguish the cases $r_j = r_m$ and $r_j \neq r_m$.

2.2 Entire Process of the Basic Attack

The attack consists of two phases. In Phase 1 the attacker guesses $\widetilde{v}_1, \widetilde{v}_2, \ldots$ on basis of the corresponding power traces (independent SPA attacks). The goal of Phase 1 is to group the guessed exponents to classes with identical (though unknown) blinding factors ('blinding classes'). Phase 2 considers only the class $\mathcal{C}_w$, which at first contains t elements with (presumably) identical blinding factors('winning class'). The generic approach clearly is to apply the 'majority decision' to each elementary operation guess, based on the guessed binary representations $(\widetilde{v}_{j;n+R-1}, \ldots, \widetilde{v}_{j;0})_2$ for all $\widetilde{v}_j \in \mathcal{C}_w$. We point out that the attacker does not know d, y, r (RSA case), resp. d, r (ECC case) for any blinding class. Pseudoalgorithm 1 describes our attack in more detail. The integer t should be selected with regard to the error probabilities (cf. below). Typically, t is an odd number ($t = 3, 5, \ldots$), cf. Table 2 and Table 3. The attacker decides that $\widetilde{v}_j$ belongs to class $\mathcal{C}_m$ iff $\text{ham}(\widetilde{v}_j \oplus \widetilde{v}) < \gamma$ (a suitably selected threshold) for all $\widetilde{v} \in \mathcal{C}_m$. The algorithm stops when the some class $\mathcal{C}_w$ contains t elements (a so-called 't-birthday').

Pseudoalgorithm 1. (The basic attack)

```
   Select an integer t ≥ 3 and a suitable threshold γ > 0
1) Phase 1: Find a class C_m with t elements ( = 'winning class')
   j:=1; s:=0; stop:=0;
   while (stop = 0) {
     attacker estimates ṽ_j;
     if (j = 1) then { s := 1; C_1 := {ṽ_1}; }
     else { m:=1; found:=0;
       while ( (m ≤ s) && (found = 0) ) {
         if ( ham((ṽ_j ⊕ ṽ)) < γ ) for all ṽ ∈ C_m then {
           C_m := C_m ∪ {ṽ_j} ; found:=1;
           if ( |C_m| = t ) then { w:=m; stop:=1; }
            /* t-birthday, winning class found */  } }
       if (found = 0) then { s++; C_s := {ṽ_j}; }
   } }
2) Phase 2: Apply the majority decision to the elements
            of the winning class C_w
```

Theoretical Background of Phase 1. We show how to distinguish $r_j = r_m$ or $r_j \neq r_m$. This is based on Eqn. (4) which reads

$$\mathrm{ham}(\widetilde{v}_j \oplus \widetilde{v}_m) = \sum_{i=0,1,\ldots,n+R-1} (v_{j;i} \oplus v_{m;i} \oplus e_{j;i} \oplus e_{m;i}) \ , \tag{5}$$

$$\text{which simplifies to} \quad \sum_{i=0,1,\ldots,n+R-1} (e_{j;i} \oplus e_{m;i}) \qquad \text{if } r_j = r_m \tag{6}$$

where $e_{j;i}, e_{m;i}$ are the ith elements of error vectors $e_j = (e_{j;n+R-1}, \ldots, e_{j;0})_2$ and $e_m = (e_{m;n+R-1}, \ldots, e_{m;0})_2$. Let ϵ_b be the error rate in Phase 1, i.e. $e_{j;i}$ or $e_{m;i}$ is '1' with probability ϵ_b, and is '0' with probability $1 - \epsilon_b$. Hence $e_{j;i} \oplus e_{m;i} = 1$ is satisfied with probability $2\epsilon_b(1 - \epsilon_b)$. If $r_i = r_m$ we assume that $\widetilde{v}_j \oplus \widetilde{v}_m$ and $\mathrm{ham}(\widetilde{v}_j \oplus \widetilde{v}_m)$ are values that are taken on by random variables X' and $Y' := \mathrm{ham}(X')$, where the components of X' are independent and binomially $B(2\epsilon_b(1 - \epsilon_b), 1)$-distributed. Hence Y' is binomially $B(2\epsilon_b(1 - \epsilon_b), n + R)$-distributed and thus has mean $\mu_{X'} = 2\epsilon_b(1 - \epsilon_b)(n + R)$ and standard deviation $\sigma_{Y'} = \sqrt{Var(\mathrm{ham}(X'))} = \sqrt{2\epsilon_b(1 - \epsilon_b)(\epsilon_b^2 + (1 - \epsilon_b)^2)(n + R)}$.

In case $r_j \neq r_m$, we can assume that $v_{j;i} \oplus v_{m;i} = 1$ holds with probability $1/2$. Since $(v_{j;i}, v_{m;i})$ and $(e_{j;i}, e_{m;i})$ are independent $v_{j;i} \oplus v_{m;i} \oplus e_{j;i} \oplus e_{m;i} = 0$ is satisfied with probability $1/2$. Therefore, $\widetilde{v}_j \oplus \widetilde{v}_m$ and $\mathrm{ham}(\widetilde{v}_j \oplus \widetilde{v}_m)$ are assumed to be values that are taken on by random variables X and $Y := \mathrm{ham}(X)$, where the components of X are independent and binomially $B(1/2, 1)$-distributed. Hence Y is binomially $B(1/2, n + R)$-distributed and thus has mean $\mu_X = 0.5(n + R)$ and standard deviation $\sigma_Y = \sqrt{Var(\mathrm{ham}(X))} = \sqrt{(n + R)/4}$. Table 2 illuminates the decision step ($R = 16$, $\gamma = \mu_Y - 4\sigma_Y$, sample size $= 10000$ for each parameter set). The column 'success rate' provides the empirical probability for the decision $r_j = r_m$ when $r_j = r_m$ is indeed true whereas 'miss ratio' shows the empirical probability for $r_j = r_m$ when $r_j \neq r_m$ is true. An appropriate choice of γ, which should consider R is important for the success of the attack. Lowering

Table 1. ham(·) distributions for different sets of parameters

			$\text{ham}(\widetilde{v}_j \oplus \widetilde{v}_m)$							$\text{ham}(\widetilde{v}_j \oplus \widetilde{v}_m)$			
			$r_j = r_m$		$r_j \neq r_m$					$r_j = r_m$		$r_j \neq r_m$	
k	R	ϵ_b	$\mu_{Y'}$	$\sigma_{Y'}$	μ_Y	σ_Y	k	R	ϵ_b	$\mu_{Y'}$	$\sigma_{Y'}$	μ_Y	σ_Y
256	16	0.10	48.96	6.34	136	8.25	1024	16	0.15	265.2	14.06	520	16.12
256	16	0.15	69.36	7.19	136	8.25	1024	16	0.20	332.80	15.04	520	16.12
256	16	0.20	87.04	7.69	136	8.25	1024	16	0.25	390.0	15.61	520	16.12
256	16	0.25	102.00	7.98	136	8.25	1024	16	0.30	436.80	15.92	520	16.12

Table 2. Experimental results for $\gamma = \mu_Y - 4\sigma_Y$

			$r_j = r_m$					$r_j = r_m$	
k	R	ϵ_b	Success ratio	Miss ratio	k	R	ϵ_b	Success ratio	Miss ratio
256	16	0.10	10000/10000	0/10000	1024	16	0.15	10000/10000	1/10000
256	16	0.15	10000/10000	0/10000	1024	16	0.20	10000/10000	0/10000
256	16	0.20	9762/10000	1/10000	1024	16	0.25	10000/10000	1/10000
256	16	0.22	8793/10000	2/10000	1024	16	0.30	8576/10000	0/10000

γ reduces the probability that $\widetilde{v}_j$ is erroneously assumed to belong to blinding class $\mathcal{C}_m$ while the probability for a false rejection $\widetilde{v}_j \notin \mathcal{C}_m$ (and opening a new blinding class) increases. Increasing γ clearly has opposite effects.

Discussion and Experimental Result of Phase 2. Let r_w^* denote the (unknown) common blinding factor of all elements in the winning blinding class $\mathcal{C}_w$. In Phase 2 the majority decision rule yields a false guess for the ith elementary operation of the blinded exponent $v_w^* := d + r_w^* y$ if the majority of the respective error bits $e_{\cdot,i}$ in $\mathcal{C}_w$ is 1. This occurs with probability

$$q_{t,\epsilon_b} := \sum_{s=u+1}^{2u+1} \binom{t}{s} \epsilon_b^s (1-\epsilon_b)^{t-s} \quad \text{for } t = 2u+1\,,$$

and we expect $e_{t,\epsilon_b} := (k+R) q_{t,\epsilon_b}$ false elementary operation estimates in average. (7)

If there are $\leq \lceil e_{t,\epsilon_b} \rceil$ false operation guesses a brute force attack costs at most

$$w_{k+R,t,\epsilon_b} := \sum_{i=0}^{\lceil e_{t,\epsilon_b} \rceil} \binom{k+R}{i} \quad \text{trials (worst case estimate)} \tag{8}$$

to find and correct them. Table 3 supports the choice of an appropriate parameter t if the majority decision rule shall be applied in Phase 2. For $(k+R = 272, \epsilon_b = 0.05)$, for instance, 3-birthdays are sufficient while for $(k+R = 1040, \epsilon_b = 0.20)$ the parameter $t = 17$ seems appropriate. (Of course, smaller t is possible at cost of higher correction workload $w_{k+R,t,p}$; note further that $> \lceil e_{t,p} \rceil$ errors may

Table 3. Majority decision in Phase 2: Expected number of false operation estimates and guessing workload due to (8)

$k+R$	ϵ_b	t	q_{t,ϵ_b}	e_{t,ϵ_b}	$\log_2(w_{k+R,t,\epsilon_b})$	$k+R$	ϵ_b	t	q_{t,ϵ_b}	e_{t,ϵ_b}	$\log_2(w_{k+R,t,\epsilon_b})$
272	0.05	3	0.007	2.0	15.2	1040	0.05	5	0.001	1.2	19.1
272	0.10	5	0.008	2.3	21.7	1040	0.10	7	0.003	2.8	27.5
272	0.15	7	0.012	3.3	27.8	1040	0.15	11	0.003	2.8	27.5
272	0.20	11	0.012	3.2	27.8	1040	0.20	17	0.003	2.7	27.5
272	0.22	13	0.012	3.3	27.8	1040	0.25	27	0.002	2.5	27.5
272	0.24	15	0.013	3.7	27.8	1040	0.30	45	0.002	2.5	27.5

Table 4. Empirical results (The last column refers to the correct blinding classes)

					average number of			
							correct	different
t	k	R	ϵ_b	success rate	exponentiations	blinding classes	blinding classes	r-values
3	256	10	0.05	10/10	150	138	138	138
5	256	10	0.10	10/10	672	483	475	475
7	256	10	0.15	10/10	1435	758	748	748
13	256	10	0.22	10/10	5253	1419	1246	1004
15	256	10	0.23	9/10	7425	1919	1228	989
15	256	10	0.24	1/10	6154	2017	1064	905
17	1024	10	0.20	10/10	6890	1037	1022	1022
27	1024	10	0.25	10/10	13833	1044	1026	1024
37	1024	10	0.28	6/10	23682	2004	1971	1024
45	1024	10	0.30	0/10	10742	2047	1807	1024

occur.) Table 4 collects experimental results. In order to reduce the workload we used the small blinding parameter $R = 10$. An attack was counted successful iff the (first) winning class was correct. Table 4 shows that for $R = 10$ errors rates up to 23% (256-bit ECC) and 28% (1024-bit RSA) are tolerable. Depending on ϵ_b we used decision boundaries $\gamma \in [\mu_Y - 4\sigma_Y, \mu_Y - 3\sigma_Y]$.

Remark 2. Possible improvements
(i) Resuming Phase 1 if the winning class $\mathcal{C}_w$ is incorrect might increase the success rate and thus the tolerable error rate ϵ_b.
(iii) Phase 1: For large t it might be an option to apply the decision strategy "if $(\text{ham}(\widetilde{v}_j \oplus \widetilde{v}) < \gamma)$" to, let's say, at most 3 elements of $\mathcal{C}_m$ to reduce the probability of an erroneous decision $\widetilde{v}_j \notin \mathcal{C}_m$.
(iii) Phase 2: Depending on the target implementation for large t more efficient strategies than the majority decision rule might exist (e.g. DPA techniques), reducing t, thereby increasing the success rate and the maximal tolerable ϵ_b.

2.3 Efficiency, Scalability, and Limits

The term $N := 2^{\alpha R}$ with $\alpha := 1 + (\log(t!) + 1 - R\log(2))/(Rt\log(2) - 1)$ provides a rough estimate for the average number of traces needed until the first t-birthday occurs if all decisions are correct. (This formula is a result of a fruitful discussion with E. Schulte-Geers.) For moderate R the value α ranges from 0.70 to 0.75 for $t = 3$, while $\alpha \approx 1$ and $\alpha > 1$ for medium-sized, resp. for large t, the exact value depending on (t, R). If wrong decisions occur, either spoiling existing blinding classes with wrong elements or opening new blinding classes unnecessarily (which might occur for large ϵ_b, cf. Table 4), more power traces are needed. For large t after $\approx 0.29 \cdot 2^R$ traces $0.25 \cdot 2^R$ blinding classes exist so that even in absence of wrong decisions the overall number of decisions on whether an exponent belongs to a particular blinding class is roughly in the order of 2^{2R}. Thus the number of traces and the number of decisions are limiting factors for $R \geq 32$ with large t. Since more correct decisions are necessary until $\widetilde{v}_j$ has been assigned to the correct blinding class, then γ has to be lowered to guarantee comparable overall success rate. Coarse heuristic considerations suggest that for $R = 16$, compared to $R = 10$, the tolerable error rate should drop down by roughly $\approx 3\%$ for ECC and $\approx 2\%$ for RSA. In fact, the setup $(\epsilon_b, k, R) = (20\%, 256, 16)$ showed 100% empirical success rate, matching with this heuristics. For $R = 32$ clearly further percentage points are lost.

3 Enhanced Attack Variant

The basic attack aims at t-birthdays of exponent blinding factors $r_1, r_2, \ldots$ where the choice of t depends on ϵ_b. This costs clearly more than $\sqrt{2^R}$ exponentiations, for large t even more than 2^R. A designer thus might feel secure just by selecting some value R for which the maximum number of exponentiations within the life cycle of the device (e.g., RSA signatures on a smart card) is clearly below $\sqrt{2^R}$. In contrast to the basic attack the enhanced variant does not aim at collisions of blinding factors but on collisions of sums of blinding factors, which can be obtained from substantially fewer exponentiations than $\sqrt{2^R}$.

3.1 Enhanced Attack: Overview

We begin with a definition. As in Section 2 the letter N denotes the sample size, i.e. the number of observed exponentiations, and $r_1, \ldots, r_N \in \{0, 1, \ldots, 2^R - 1\}$ are the blinding factors.

Definition 1. *For $u > 1$ let $M_u := \{(j_1, \ldots, j_u) \mid 1 \leq j_1 \leq \cdots \leq j_u \leq N\}$. For each u-tuple $(j_1, \ldots, j_u) \in M_u$ we define the 'u-sum' $S_u(j_1, \ldots, j_u) := r_{j_1} + \cdots + r_{j_u}$. For $(j_1, \ldots, j_u), (i_1, \ldots, i_u) \in M_u$ we write $(j_1, \ldots, j_u) \sim (i_1, \ldots, i_u)$ iff $S_u(j_1, \ldots, j_u) = S_u(i_1, \ldots, i_u)$.*

Observation. $\sim$ defines an equivalence relation on M_u. For $0 \leq m \leq u(2^R - 1)$ the equivalence class E_m consists of all u-tuples $(j_1, \ldots, j_u) \in M_u$ with $S_u(j_1, \ldots, j_u) = m$. Next we summarize the particular steps of the enhanced attack variant. With regard to N and error rate ϵ_b one first selects the parameter u. The parameters $u = 2, 3, 4$ should be most relevant for applications.

- Step 1: Identify several pairs of u-tuples $(j_1, \ldots, j_u), (j'_1, \ldots, j'_u) \in M_u$ with $S_u(j_1, \ldots, j_u) = S_u(j'_1, \ldots, j'_u)$, i.e. with $(j_1, \ldots, j_u) \sim (j'_1, \ldots, j'_u)$. This yields a system of linear equations over Z in the blinding factors $r_1, \ldots, r_N$.

- Step 2: Solve the system of linear equations gained in Step 1. More precisely, find $(r_1^*, \ldots, r_N^*)$ such that $(r_1, \ldots, r_N) = (r_1^*, \ldots, r_N^*) + (c, \ldots, c)$ for some unknown integer c much smaller than 2^R.
- Step 3:
  - ECC case: Compute $\widetilde{v}_j - r_j^* y = d^* + e_j$ with $d^* = d + c$ for $j \leq N$. Task: 1.) Find d^*. 2.) Determine d by exhaustive search in c.
  - RSA case: Compute $\widetilde{v}_j - \widetilde{v}_i = (r_j^* - r_i^*)y + (e_j - e_i)$ for several pairs (j, i). Task: Find y $(=\phi(n)$, resp. $= \phi(p))$.

3.2 NAF Representations

This subsection collects relevant properties of NAF representations, which will be needed in Step 1 of the enhanced attack.

Definition 2. $\mathrm{Z_m} := \{0, 1, \ldots, \mathrm{m} - 1\}$, *and if* $z = \sum_{j=0}^{t} \beta_j 2^j$ *with* $\beta_j \in \{1, 0, -1\}$ *for all* j *then* $(\beta_t, \ldots, \beta_0)_{SD}$ *is called a* binary signed-digit representation *of* z. *A* signed *representation is said to be* non-adjacent *if consecutive non-zero coefficients are always separated by at least one zero. We briefly speak of* NAF representations *where* 'NAF' *abbreviates 'non-adjacent form'. The* Hamming weight ham$(\cdots)$ *of a signed-digit representation is the number of non-zero digits. The term* $\Phi(\cdot)$ *denotes the cumulative distribution function of the standard normal distribution.*

Every integer z has a unique non-adjacent representation, noted as NAF(z). In particular, NAF(z) has minimal hamming weight under all binary signed-digit representations of z. For $z < 0$ we set NAF$(z) = -$NAF$(|z|)$. Due to the lack of space we omit the proof of Lemma 1. Example 1 illustrates that the variance of ham(NAF(X)) is surprisingly small, which will be crucial for our attack.

Lemma 1. *For any integers* z_1 *and* z_2 *we have*

$$\mathrm{ham}(\mathrm{NAF}(z_1 + z_2)) \leq \mathrm{ham}(\mathrm{NAF}(z_1)) + \mathrm{ham}(\mathrm{NAF}(z_2)), \tag{9}$$

and the right-hand side is $\leq$ *the sum of the Hamming weights for any binary signed-digit representations* z_1 *and* z_2.
(ii) Let X *be a random variable uniformly distributed on* Z_{2^n} *with* $n \geq 2$. *Then*

$$\mathrm{Prob}\,(\mathrm{ham}(\mathrm{NAF}(X)) = k) =$$

$$2^{-n} \cdot \begin{cases} 1 & \text{for } k = 0 \\ n & \text{for } k = 1 \\ 2^{k-2}\left(\binom{n+1-k}{k} + \binom{n+2-k}{k}\right) & \text{for } k \in \{2, \ldots, \lfloor \frac{n+1}{2} \rfloor\} \\ 2^{\frac{n}{2}-1} & \text{for } k = \frac{n}{2} + 1, \text{ if } n \text{ is even} \end{cases} \tag{10}$$

(iii) Unless n *is too small, the random variable* $Y :=$ ham(NAF(X)) *is approximately normally distributed with expectation* E$(Y) \approx 0.333n$ *and variance* Var$(Y) \approx 0.075n$.

Example 1. $n = 1040$ (corresponds to 1024-bit RSA with $R = 16$), $Y :=$ ham(NAF(X)), E$(Y) \approx 0.334n = 347.4$ $\sigma_Y := \sqrt{\mathrm{Var}(Y)} \approx \sqrt{0.075n} = 8.8$.

3.3 Enhanced Attack: Step 1

In Step 1 we have to decide whether $(j_1,\ldots,j_u) \sim (i_1,\ldots,i_u)$, i.e. whether $S_u(j_1,\ldots,j_u) = S_u(i_1,\ldots,i_u)$. Therefore, we apply the decision rule

$$\begin{aligned} &\text{Decide for } (j_1,\ldots,j_u) \sim (i_1,\ldots,i_u) \text{ iff} \\ &\text{ham}(\text{NAF}(\widetilde{v}_{j_1} + \cdots + \widetilde{v}_{j_u} - (\widetilde{v}_{i_1} + \cdots + \widetilde{v}_{i_u}))) < b_0 \quad \text{for some suitable } b_0. \end{aligned} \tag{11}$$

Motivation for the decision strategy (11). Clearly, $(\widetilde{v}_{j_1} + \cdots + \widetilde{v}_{j_u}) - (\widetilde{v}_{i_1} + \cdots + \widetilde{v}_{i_u}) = ry + e_{j_1} + \cdots + e_{j_u} - e_{i_1} + \cdots - e_{i_u}$ with $r = 0$ iff $(j_1,\ldots,j_u) \sim (i_1,\ldots,i_u)$ and $r \in \mathrm{Z} \setminus \{0\}$ else. By Lemma 1(iii) for a uniformly distributed random variable X on $\mathrm{Z}_{2^{k+R}}$ the variance $\text{Var}(\text{ham}(\text{NAF}(X))) \approx 0.075(k+R)$ is very small, and thus $\text{E}(\text{ham}(\text{NAF}(X)))$ is a 'typical' value. Hence for randomly selected v_j and random e_j one may expect $\text{ham}(\text{NAF}(\widetilde{v}_{j_1} + \cdots + \widetilde{v}_{j_u} - (\widetilde{v}_{i_1} + \cdots + \widetilde{v}_{i_u}))) \approx \text{E}(\text{ham}(\text{NAF}(X))) \approx 0.333(k+R)$ if $(j_1,\ldots,j_u) \not\sim (i_1,\ldots,i_u)$ although the absolute differences of u-sums are not uniformly distributed on $Z_{2^{k+R}}$. Numerical experiments confirm the intuition (cf. Table 5). On the other hand, if $(j_1,\ldots,j_u) \sim (i_1,\ldots,i_u)$ by Lemma 1(i)

$$\begin{aligned} &\text{ham}(\text{NAF}(\widetilde{v}_{j_1} + \cdots + \widetilde{v}_{j_u} - \widetilde{v}_{i_1} - \cdots - \widetilde{v}_{i_u})) \leq \\ &\text{ham}(\text{NAF}(e_{j_1})) + \cdots + \text{ham}(\text{NAF}(e_{j_u})) - \cdots - \text{ham}(\text{NAF}(e_{i_u})) \leq \\ &\text{total number of guessing errors for } v_{j_1},\ldots,v_{j_u},v_{i_1},\ldots,v_{i_u}. \end{aligned} \tag{12}$$

If the error rate is sufficiently small for $(j_1,\ldots,j_u) \sim (i_1,\ldots,i_u)$ the term $\text{ham}(\text{NAF}(\widetilde{v}_{j_1} + \cdots + \widetilde{v}_{j_u} - \widetilde{v}_{i_1} - \cdots - \widetilde{v}_{i_u}))$ is significantly smaller than for $(j_1,\ldots,j_u) \not\sim (i_1,\ldots,i_u)$, justifying decision rule (11). Each decision for $(j_1,\ldots,j_u) \sim (i_1,\ldots,i_u)$ gives a linear equation

$$S_u(j_1,\ldots,j_u) - S_u(i_1,\ldots,i_u) = r_{j_1} + \cdots + r_{j_u} - r_{i_1} - \cdots - r_{i_u} = 0. \tag{13}$$

Number of linear equations in Step 1. We derive a coarse estimator for the average number of (not necessarily independent) linear equations that one gains from N exponentiations if all decisions are correct. Neglecting those pairs of u-tuples $\in M_u$, which have at least two identical components, we conclude

$$\begin{aligned} \text{E}(\#\text{linear equations}) &\approx \binom{\binom{N}{u}}{2} \text{Prob}(S_u(1,\ldots,u) = S_u(u+1,\ldots,2u)) \\ &\approx \frac{N^{2u}}{2u!u!} \sum_{t=0}^{(2^R-1)u} \text{Prob}(S_u(1,\ldots,u) = t)^2 \quad \text{for } u \ll N \end{aligned} \tag{14}$$

since for $u \ll N$ for the great majority of pairs $(j_1,\ldots,j_u),(i_1,\ldots,i_u)$ the components $j_1,\ldots,j_u,i_1,\ldots,i_u$ are mutually distinct. Approximately, the right-hand sum in (14) (without pre-factor $N^{2u}/(2u!u!)$) equals

$$\frac{c(u)}{2^R} \quad \text{with } c(2) \approx \frac{2}{3},\ c(3) \approx \frac{11}{20},\ \text{and } c(u) \approx \frac{\sqrt{3}}{\sqrt{\pi u}} \text{ for } u \geq 4. \tag{15}$$

For $R = 16$, for instance, $N = 116$ for $u = 2$, resp. $N = 28$ for $u = 3$, resp. $N = 16$ (coarse estimate since $4 \not\ll 16$) for $u = 4$, exponentiations provide $\approx 2N$ linear equations (equate (14) with $2N$). Numerical experiments agree with these estimates, and these $N's$ roughly give the necessary number of power traces needed for the overall attack (cf. Phase 2, Table 6). Similarly, for $R = 32$ we need $N = 4689$ ($N = 141$, $N = 79$) power traces for $u = 2$ ($u = 3$, $u = 4$) to obtain $\approx 2N$ linear equations. In particular, $2N$ equations cost roughly

$$\approx \frac{N^{2u}}{2u!u!} \approx \frac{\left(\frac{2^R \cdot 4u!u!}{c(u)}\right)^{\frac{2u}{2u-1}}}{2u!u!} \approx 10 \cdot (2^R)^{1+\frac{1}{2u-1}} \quad \text{comparisons for } u \leq 4. \qquad (16)$$

Discussion on the appropriate decision parameter. When applying decision rule (11) to u-tuples $(j_1, \ldots, j_u), (i_1, \ldots, i_u) \in M_u$ the attacker might commit two types of errors: False positives (decision for '$\sim$' although '$\not\sim$' is true) and false negatives (decision for '$\not\sim$' although '$\sim$' is true). A false negative just 'loses' one linear equation (13) whereas a false positive yields a wrong equation, making the solution of the system of linear equations (cf. Subsec. 3.4) meaningless.

Unlike in the basic version a false positive does not only slow down the attack (affecting a single blinding class) but the whole attack fails. False positives thus should definitely be prevented although the fact that $\sim$ defines an equivalence relation on M_u might allow to detect (and to cancel) some false positives Moreover, for the great majority of mutual comparisons '$\not\sim$' is the correct decision. More precisely, (14) to (16) says that only for the fraction $c(u)/2^R$ of all comparisons '$\sim$' is true. Assume that the random variables W and W' describe the distribution of $\mathrm{ham}(\mathrm{NAF}(\widetilde{v}_{j_1} + \cdots + \widetilde{v}_{j_u} - \widetilde{v}_{i_1} - \cdots - \widetilde{v}_{i_u}))$ under the condition $(j_1, \ldots, j_u) \not\sim (i_1, \ldots, i_u)$, resp. under the condition $(j_1, \ldots, j_u) \sim (i_1, \ldots, i_u)$. The empirical studies below consider $R = 16$ where we selected $b_0 := E(W) - 6.5\sigma_W$ in (11).

If we assume that W is at least approximately normally distributed the probability for a false positive in a single decision is $< 10^{-9}$. In Table 5 μ_W and $\mu_{W'}$ denote the mean values of W and W', while '$\sim$' and '$\not\sim$' briefly stand for $(j_1, \ldots, j_u) \sim (i_1, \ldots, i_u)$ and $(j_1, \ldots, j_u) \not\sim (i_1, \ldots, i_u)$, respectively. Each quintuple $(\mu'_W, \mu_W, \sigma'_W, \sigma_W, b_0)$ in Table 5 was figured on basis of 40000 pseudorandomly generated pairs $(\widetilde{v}_{j_1}, \ldots, \widetilde{v}_{j_u}), (\widetilde{v}_{i_1}, \ldots, \widetilde{v}_{i_u})$, fulfilling '$\sim$', resp. '$\not\sim$'. Table 5 verifies that the empirical values μ_W and σ_W are indeed very close to the expectation and standard deviation when X is uniformly distributed on $Z_{2^{k+R}}$ (cf. Lemma 1(iii) and Example 1), confirming our heuristics discussed at the beginning of this subsection. Table 5 shows for several pairs $(k+R, u)$ what error rates ϵ_b Step 1 of the enhanced attack can tolerate. As Phase 1 of the basic attack also Step 1 of the enhanced variant is more efficient for larger key sizes since the ratio $\sigma_W/(\mu_W - \mu_{W'})$ decreases as $(k+R)$ increases. We applied decision rule (11) in many simulation experiments. As predicted false positives did not occur for this choice of b_0. For $(k+R, u, \epsilon_b) = (1040, 3, 0.07)$ false negatives essentially do not occur whereas for $(k+R, u, \epsilon_b) = (272, 4, 0.04)$ about 29%

Table 5. ham(NAF(·))-u-sum-distributions for several sets of parameters

		$\mathrm{ham}(\mathrm{NAF}(\widetilde{v}_{j_1}+\cdots-\widetilde{v}_{i_u}))$							$\mathrm{ham}(\mathrm{NAF}(\widetilde{v}_{j_1}+\cdots-\widetilde{v}_{i_u}))$						
				$\sim$		$\not\sim$					$\sim$		$\not\sim$		
R	u	k	ϵ_b	$\mu_{W'}$	$\sigma_{W'}$	μ_W	σ_W	b_0	k	ϵ_b	$\mu_{W'}$	$\sigma_{W'}$	μ_W	σ_W	b_0
16	2	256	0.06	49.2	5.4	90.5	4.5	61.2	1024	0.11	272.7	10.4	347.0	8.8	289.8
16	2	256	0.07	54.7	5.4	90.1	4.5	61.0	1024	0.12	283.8	10.2	346.8	8.8	289.7
16	3	256	0.08	59.6	5.4	90.8	4.5	61.6	1024	0.13	293.4	10.1	346.8	8.8	289.9
16	3	256	0.04	48.7	5.4	90.5	4.5	61.2	1024	0.07	262.8	10.4	346.9	8.8	289.6
16	3	256	0.05	56.6	5.4	90.7	4.5	61.2	1024	0.08	279.8	10.2	346.9	8.8	290.1
16	3	256	0.06	63.0	5.4	90.8	4.5	61.7	1024	0.09	293.5	10.0	347.0	8.8	290.0
16	4	256	0.04	58.5	5.4	90.9	4.5	61.5	1024	0.06	277.7	10.3	347.0	8.8	290.0

($=100(1-\Phi((61.5-58.5)/5.4))\% = 100\cdot(\Phi(0.56))\%$) of the existing (correct) equations get lost due to false negatives, increasing the required number of power traces N somewhat (cf. (15)).

Remark 3. At cost of efficiency (more decisions and more power traces) Step 1 of the enhanced attack also applies to larger error rates than discussed in Table 5. From (15) one concludes that $N \approx 116\sqrt[3]{a}$ for $u=2$, $N \approx 28\sqrt[5]{a}$ for $u=3$, and $N \approx 16\sqrt[7]{a}$ for $u=4$ provide $\approx 2aN$ linear equations ($a>1$), thus tolerating a fraction of $(a-1)/a$ lost equations. Numerical Example: For $(k+R, u, \epsilon_b) = (1040, 2, 0.12)$ we expect $\approx 28\%$ false negatives but $N=129$ in place of $N=116$ should compensate this defect.

3.4 Enhanced Attack: Step 2

Basic idea. In Step 1 the attacker has gained a homogeneous system of q linear equations

$$Br' = \mathbf{0} \quad \text{with } B \in \mathrm{Mat}(q,N;\mathrm{Z}), \boldsymbol{r}' \in \mathrm{Z}^{\mathrm{N}}, \mathbf{0} = (0,\ldots,0)^{\mathrm{t}} \in \mathrm{Z}^{\mathrm{q}}. \qquad (17)$$

In the following we assume that all equations are correct. This is the case if all decisions for '$\sim$' were correct, or if all false positives could be eliminated due to consistency checks within the equivalence classes $\mathcal{E}_m$. Then $\boldsymbol{r} := (r_1,\ldots,r_N)^t$ clearly is a solution of (17). Moreover, $z(1,\ldots,1)^t \in \mathrm{Z}^{\mathrm{N}}$ is also a solution of (17) for all $z \in \mathrm{Z}$. As it is extremely unlikely that $\boldsymbol{r}$ is an integer multiple of $(1,\ldots,1)^t$ (which can easily be checked) the kernel $\ker B$, the solution of (17), is a proper superset of $\{z(1,\ldots,1)^t \mid z \in \mathrm{Z}\}$. Since $\dim(\ker(B)) \geq 2$ would be pointless to consider (17) as a system of linear equations over the real numbers.

Preliminary on Step 2

Definition 3. *Let $A \neq \{0\}$ be a commutative ring that operates on an abelian group $M \neq \{0\}$ via $(a,m) \mapsto am \in M$. Assume that besides $(ab)m = a(bm)$ and $1m = m$ also $a(m+n) = am + an$ and $(a+b)m = am + bm$ for all $a, b \in A$ and $m, n \in M$. Then M is called an A-module. The module M is called free if*

M admits an A-basis, i.e. if there exists a subset $B \subseteq M$ such that any $m \in M$ can be expressed in a unique way as an A-linear combination of elements of B. If M has a finite basis $\{m_1, \ldots, m_n\}$ then M has dimension n.

Note that if A is a field then M is a vector space. Lemma 2 illuminates the structure of $\ker B$ as a Z-module. We point out that for linear equations over $\mathbb{R}$ its assertion would be trivial while it is not obvious over Z.

Remark 4. (i) Example: Z^{n} becomes a Z-module via $y(z_1, \ldots, z_n) := (yz_1, \ldots, yz_n)$. (ii) Modules may not admit bases (e.g., the Z-module Z_{m}), and a set of independent elements may not be extendable to a basis: Consider, for example, the Z-module Z^2. The vector $(2, 0) \in \mathrm{Z}^2$ cannot be extended to a basis, and $(2, 0), (1, 0)$, for instance, only spans the sub-module $2\mathrm{Z} \times \mathrm{Z}$.
(iii) The definition of $\dim(M)$ tacitly uses the fact that all bases of M have the same cardinality since A is commutative (cf. [10], Chap. XIII).

Lemma 2. *$\ker B$ is a free sub-module of Z^{N}. In particular, $\ker B$ admits a module basis $\boldsymbol{s}_{(1)} := (1, \ldots, 1)^t, \boldsymbol{s}_{(2)}, \ldots, \boldsymbol{s}_{(\dim(\ker B))}$.*

Proof. Z^{N} is a free module over Z. As $\ker B$ is a sub-module of Z^{N} and since Z is a principal entire ring, $\ker B$ is also free and admits a basis with $\dim(\ker B)$ elements ([10], Chap. III, Thm. 7.1). It is $(1, \ldots, 1)^t, (0, 1, 0, \ldots, 0)^t, \ldots (0, 0, \ldots, 0, 1)^t$ is a basis of Z^{N}, and Z^{N} is a direct sum of the sub-modules T_1 and T_2 where T_1 is generated by $(1, \ldots, 1)^t$ and T_2 by basis vectors $(0, 1, 0, \ldots, 0)^t, \ldots (0, 0, \ldots, 0, 1)^t$. The mapping $\phi\colon T_1 \times T_2 \to \mathrm{Z}^{\mathrm{N}}, (\mathrm{t}_1, \mathrm{t}_2) \mapsto \mathrm{t}_1 + \mathrm{t}_2$ defines a module isomorphism. Then $\ker(B \circ \phi) = \{(t_1, t_2) \in T_1 \times T_2 \mid B(t_1 + t_2) = B(t_2) = 0\} = T_1 \times \{t_2 \in T_2 \mid B(t_2) = 0\}$. The second summand is a sub-module of $T_2 \cong \mathrm{Z}^{\mathrm{N}-1}$, is thus free and admits a basis $\boldsymbol{s}'_{(2)}, \ldots$ ([10], Chap. III, Thm. 7.1). Then $\boldsymbol{s}_{(1)} := \phi((1, \ldots, 1)^t, 0), \boldsymbol{s}_{(2)} := \phi(0, \boldsymbol{s}'_{(2)}), \ldots$ is a basis of $\ker B$, which proves the lemma.

If the number q of linear equations is sufficiently large, $\dim(\ker B) = 2$ may be expected. Experimental results confirm this conjecture (cf. Table 6). Larger parameter u allows smaller N but on the negative side the admissible error rate ϵ_b in Step 1 decreases. In the following we assume $\dim(\ker B)) = 2$. Then

$$\boldsymbol{r} = \mu_1 \boldsymbol{s}_{(1)} + \mu_2 \boldsymbol{s}_{(2)} \quad \text{with unknown } \mu_1, \mu_2 \in \mathrm{Z}. \tag{18}$$

Table 6. Enhanced attack: Experimental results for Step 2

	$u = 2$		$u = 3$		$u = 4$	
N	116	128	28	32	16	20
$\dim(\ker B) = 2$	43/50	49/50	49/50	50/50	12/50	50/50

Finding the basis. Finding μ_1 and μ_2 might seem to be a $O(2^{2R})$ problem, but this is not the case. Clearly, $|\mu_2| \cdot \max\{|\boldsymbol{s}_{(2)i} - \boldsymbol{s}_{(2)j}| : 1 \le i,j \le N\} = \max\{|r_i - r_j| : 1 \le i,j \le N\} < 2^R$. If the blinding factors $r_1, \ldots, r_N$ have been generated randomly it is very likely that the second term is $\approx 2^R$, and thus it is very unlikely that there is an integer μ_1 with $\gcd(r_1 - \mu_1, \ldots, r_N - \mu_1) > 1$. Thus almost certainly $\mu_2 \in \{-1, 0, 1\}$.

The case $\mu_2 = 0$ is very unlikely and easily checked. If $\mu_2 \ne 0$ order the indices $1, \ldots, N$ twice, first according to the size of the components $\boldsymbol{s}_{(2)1}, \ldots, \boldsymbol{s}_{(2)N}$ and secondly according to $\widetilde{v}_1, \ldots, \widetilde{v}_N$. If $\mu_2 = 1$ is correct both ordered sets should be similar while for $\mu_2 = -1$ both sets should become more similar if one order is reversed. Once μ_2 has been determined for μ_1 only

$$ca := 2^R - (\max\{|\boldsymbol{s}_{(2)i} - \boldsymbol{s}_{(2)j}| : 1 \le i,j \le N\} + 1) \text{ candidates remain} \quad (19)$$

to be checked. We assign to μ_1^* the minimal integer for which

$$\boldsymbol{r}^* = \mu_1^* \boldsymbol{s}_{(1)} + \mu_2 \boldsymbol{s}_{(2)} \quad \text{has only non-negative components.} \quad (20)$$

We point out that finding a Z-basis of $\ker B$ is more difficult than finding a basis of a vector space. In a first step we select some $\boldsymbol{v}_{(2)} \in \ker B$ that is no integer multiple of $\boldsymbol{s}_{(1)}$, where $\boldsymbol{v}_{(2)}$ is obtained by following steps.

1. Let $B' = \binom{B}{11\ldots1}$. Since B' is made by adding $\boldsymbol{s}_{(1)}$ to B, $\dim(\ker B') = 1$.
2. Use the Gaussian elimination to B', to obtain a reduced row echelon form

$$B'' = \begin{pmatrix} 1\,0 \cdots 0 & v'_{(2)1} \\ \vdots\ \vdots\ \ \vdots\ \vdots & \vdots \\ 0\,0 \cdots 1 & v'_{(2)N-1} \\ 0\,0 \cdots 0 & 0 \\ \vdots\ \vdots\ \ \vdots\ \vdots & \vdots \\ 0\,0 \cdots 0 & 0 \end{pmatrix}.$$

3. $\boldsymbol{v}_{(2)}$ is obtained by multiplying some integer to $(-v'_{(2)1}, \ldots, -v'_{(2)N-1}, 1)^t$.

If the greatest common divisor of its components exceeds 1, we divide $\boldsymbol{v}_{(2)}$ by this number. Lemma 3(i) provides a simple criterion to check whether $\{\boldsymbol{s}_{(1)}, \boldsymbol{v}_{(2)}\}$ is already a basis of $\ker B$. If not, Lemma 3(ii) explains how to get one, i.e. how to determine $\boldsymbol{s}_{(2)} = \boldsymbol{v}_{(3)}$. Due to lack of space we omit its proof.

Lemma 3. *Let* $\dim(\ker B) = 2$.*Assume further that that* $v_{(2)}$ *is no integer multiple of* $\boldsymbol{s}_{(1)}$, *and that the gcd over the components of* $\boldsymbol{v}_{(2)}$ *is* 1.
(i) Assertion (∗) and Assertion (∗∗) are equivalent:
(∗) $\{\boldsymbol{s}_{(1)}, \boldsymbol{v}_{(2)}\}$ *is basis of* $\ker B$.
(∗∗) There is no integer $1 < m \le 2\max\{|\boldsymbol{v}_{(2)1}|, \ldots, |\boldsymbol{v}_{(2)N}|\}$ *with*

$$\boldsymbol{v}_{(2)1} \equiv \cdots \equiv \boldsymbol{v}_{(2)N} (\bmod m) \quad (21)$$

(ii) Assume that $\{\boldsymbol{s}_{(1)}, \boldsymbol{v}_{(2)}\}$ *is not a basis of* $\ker B$ *and that* m^* *is the largest integer with property (21). Then* $\{\boldsymbol{s}_{(1)}, \boldsymbol{v}_{(3)}\} := ((m^* - \boldsymbol{v}_{(2)1})\boldsymbol{s}_{(1)} + \boldsymbol{v}_{(2)})/m^*\}$ *is basis of* $\ker B$.

3.5 Enhanced Attack: Step 3

The goal of Step 3 is to find the secret key d. Recall that r^* denotes solution (20). Step 3 differs between ECC and RSA.

The ECC Case. The attacker first computes

$$a_j := \widetilde{v}_j - r_j^* y = d + (\mu_1 - \mu_1^*)y + e_j = d^* + e_j \qquad \text{for } j = 1, \ldots, N. \quad (22)$$

He knows all a_j while d^* and $e_j = (e_{j,n+R-1}, \ldots, e_{j,0})_2$ are unknown.
Algorithm ECC (finds d^*):

1. Clearly, $e_{j;0} = 0$ (correct guess of elementary operation 0 for exponent v_j) iff $d^* \equiv a_j (\text{mod } 2)$. This leads to the following decision rule: (Majority Decision) For $i = 0, 1$ let $S_i := \{j \leq N \mid a_j \equiv i (\text{mod} 2)\}$. If $|S_0| \geq |S_1|$ the attacker assumes that $e_{j',0} = 0$ for $j' \in S_0$ and $e_{j',0} = 1$ for $j' \in S_1$. The case $|S_0| < |S_1|$ is treated accordingly.
   (Error Correction): $e_{j,0} = 1$ either means that the attacker has guessed $\widetilde{v}_{j,0} = 1$ although $v_{j,0} = 0$ is true (Case I) or that the attacker has guessed $\widetilde{v}_{j,0} = 0$ although $v_{j,0} = 1$ is true (Case II). The attacker clearly can distinguish between both cases since he knows his guess $\widetilde{v}_{j,0}$.
   For Case I he replaces $\widetilde{v}_j$ by $\widetilde{v}_j - 1$ and thus in (22) a_j changes to $a_j - 1$. (Note: Since we are not interested in the e_j's we simply keep the notation e_j.) Similarly, for Case II one gets the new equation (22) $d^* + e_j = a_j + 1$. (Note that after error correction $d^* \equiv a_j (\text{mod } 2)$ for all $j \leq N$.)
2. (Induction step): Assume that the guessing errors $e_{j,i}$ have already been corrected for $0 \leq i \leq k-1$. Thus $d^* \equiv a_j (\text{mod } 2^k)$ for $j \leq N$. Now we are going to correct the $e_{j,k}$.
   (Majority Decision) Since $e_j \equiv 0 (\text{mod } 2^k)$ equation $d^* + e_j \equiv a_j (\text{mod } 2^{k+1})$ implies $d_k^* + e_{j,k} \equiv a_{j,k} (\text{mod} 2)$. As in Step 1 the attacker applies the majority decision to identify those equations with a guessing error for elementary operation k.
   (Error Correction) as in Step 1, with $\pm 2^k$ in place of $\pm 1 = \pm 2^0$.

After this algorithm has terminated $d^* = a_1 = \cdots = a_N$ if all majority decisions had been correct. Since $d = d^* - (\mu_1 - \mu_1^*)y$ one known pair (P, dP) for some point P should suffice to identify d after at most ca trials (cf. (19)).

Remark 5. For error rate $\epsilon_b = \alpha$ one expects $e_{j,0} = 1$ in $\approx \alpha N$ equations but $e_{j,0} = 0$ in $\approx (1-\alpha)N$ equations. For $\alpha \leq 0.12$, for instance, $e_{1,i} + \cdots + e_{N,i} < N/2$ with overwhelming probability unless N is extremely small, and false majority decisions are very unlikely. For $N = 20$, $\alpha \leq 0.12$, the error probability is $\approx 10^{-5}$.

The RSA Case. At first the attacker re-labels $r_1^*, \ldots, r_N^*$ (and, in the same way, $\widetilde{v}_j, e_j$ etc.) such that $r_1^*, \ldots, r_{N'}^*$ have alternating parity (odd, even, odd, ... or even, odd, even,...) with maximal possible $N' \leq N$. The exponentiations $N'+1, \ldots, N$ are neglected in the following. Instead, we concentrate on a system of $N' - 1$ linear equations

$$b_j := \widetilde{v}_{j+1} - \widetilde{v}_j = \underbrace{(r^*_{j+1} - r^*_j)}_{=r_{j+1}-r_j} y + e_{j+1} - e_j := \Delta_j y + e_{j+1} - e_j \text{ for } 1 \le j < N'. \tag{23}$$

The integers b_j and Δ_j are known while y, e_j, e_{j+1} are unknown. By construction, $\Delta_j \equiv 1 (\bmod 2)$ for all $j \le N' - 1$. The algorithm below applies the same techniques as the ECC algorithm but it is more complicated in detail.
Algorithm RSA (finds y ($= \phi(n)$ or $= \phi(p)$)):

1. Analogously to the ECC case the equations (23) are considered (mod 2). This gives modular equations $b_j \equiv y + e_{j+1} - e_j (\bmod 2)$.
(Majority Decision) For $i = 0, 1$ let $S_i := \{j < N' \mid b_j \equiv i (\bmod 2)\}$. If $|S_0| \ge |S_1|$ the attacker assumes $b_{j'} \equiv y (\bmod 2)$ and thus $e_{j'+1,0} - e_{j',0} \equiv 0 (\bmod 2)$ for all $j' \in S_0$, and thus $e_{j'+1,0} + e_{j',0} \equiv 1 (\bmod 2)$ for $j' \in S_1$. In particular, he guesses $y_0 := b_j \pmod 2$. The case $|S_0| < |S_1|$ is treated accordingly.
   (Error Correction): Majority Decision gives a system of $N' - 1$ linear equations $e_{j'+1,0} + e_{j',0} \equiv c_{j'} (\bmod\ 2)$ where $c_{j'} = 0$ if j' belongs to the larger set S_i and $c_{j'} = 1$ else. Since the homogeneous system has 1-dimensional kernel $\{(0, \ldots, 0), (1, \ldots, 1)\}$ the non-homogeneous system has two solutions with mutually flipped components. The attacker decides for the solution with smaller Hamming weight. (If both Hamming weights are identical the solution is selected, which comes first in lexicographical order.)
   As for ECC $e_{j,0} = 1$ either means that the attacker has guessed $\widetilde{v}_{j,0} = 1$ although $v_{j,0} = 0$ is true (Case I) or that the attacker has guessed $\widetilde{v}_{j,0} = 0$ although $v_{j,0} = 1$ is true (Case II). The attacker replaces $\widetilde{v}_j$ by $\widetilde{v}_j - 1$ for CASE I, and by $\widetilde{v}_j + 1$ for Case II. Then he recalculates the left-hand sides of (23). (After error correction $y \equiv b_j (\bmod 2)$ and $e_{j,0} = 0$ for all $j \le N'$.)
2. (Induction step) Assume that we have already guessed $y_0, \ldots, y_{k-1}$, and that we have corrected the guessing errors for the k least significant elementary operations, i.e. $e_j \equiv 0 (\bmod 2^k)$ for all $j \le N'$. The next task is to guess y_k and to correct all guessing errors in $e_{1,k}, \ldots, e_{N',k}$. From (23) we conclude

$$b_j \equiv \widetilde{v}_{j+1} - \widetilde{v}_j = \Delta_j y + e_{j+1} - e_j \equiv$$
$$1 \cdot y_k 2^k + (\Delta_j - 1) \sum_{i=0}^{k-1} y_i 2^i + e_{j+1,k} 2^k - e_{j,k} 2^k (\bmod 2^{k+1}). \tag{24}$$

   Since Δ_j is known and $y_0, \ldots, y_{k-1}$ have already been guessed we obtain

$$b'_j \equiv 2^k (y_k + e_{j+1,k} - e_{j,k}) (\bmod 2^{k+1}) \tag{25}$$

   with $b'_j := b_j - (\Delta_j - 1) \sum_{i=0}^{k-1} y_i 2^i (\bmod 2^{k+1})$. The right-hand side implies $b'_j = b'_{j,k} 2^k$, which yields

$$y_k + e_{j+1,k} - e_{j,k} \equiv b'_{j,k} (\bmod 2) \tag{26}$$

   Analogously to Step 1 one estimates y_k (Majority Decision) and corrects the $e_{j,k}$ by modification of the $\widetilde{v}_j$ and recalculation of (23) (Error correction). The correction factor clearly is $\pm 2^k$ in place of ± 1.

Table 7. Success rates in Step 3 (computed from 300 trials each)

algo	$n+R$	ϵ_b	N	success rate	algo	$n+R$	ϵ_b	N	success rate
ECC	272	0.12	20	100%	RSA	1040	0.12	70	95%
ECC	272	0.12	16	96%	RSA	1040	0.12	60	85%
ECC	272	0.08	16	100%	RSA	1040	0.08	45	95%
ECC	272	0.08	10	91%	RSA	1040	0.08	35	74%

If all majority decisions have been correct the algorithm delivers y, which immediately gives the factorization of $n = pq$. We note that unlike (22) equations (23) are 'disturbed' by two error terms e_{j+1} and e_j, and some exponentiations are neglected. Consequently, for identical error rates the RSA algorithm requires larger sample size N than the ECC algorithm. Experiments (Table 7) verify that both error correction algorithms work. Note that the success rate depends only on (N, ϵ_b) but not on parameter u from Step 1. Table 7 shows that for both ECC and RSA for large error rate ϵ_b (e.g., $\epsilon_b = 0.12$, demanding $u = 2$) Step 1 and Step 2 determine the minimum number of power traces, which is needed for the overall attack. For RSA with small error rates ϵ_b (allowing $u > 2$) Step 3 defines the minimum number of power traces. Step 3 virtually does not depend on R, and thus for large R the minimum number of traces depends on Step 1 and Step 2.

3.6 Efficiency, Scalability, and Limits

The empirical results from Subsections 3.3 to 3.5 indicate that for $R = 16$ the enhanced attack tolerates error rates of 13%. Since the number of traces remains moderate even for large R the number of comparisons (16) is the limiting factor. However, the enhanced attack scales much better than the basic attack, and thus e.g. $R = 32$ is definitely feasible. For $R = 64$ even $u = 4$ requires roughly $\approx 2^{76}$ comparisons (NAF computations) in absence of false negatives, which might be viewed impractical for side-channel attacks, in particular since the attacker does not know in advance whether the error rate ϵ_b is indeed sufficiently small. Moreover, for increasing R the boundary b_0 has to be decreased (more comparisons per gained equation); replacing $6.5\sigma_W$, e.g. by $8.0\sigma_W$ (for $R = 32, u \geq 3$) or $8.5\sigma_W$ (for $R = 32, u = 2$), decreases the tolerable error rate by ≈ 1.5 percentage points. For $R = 48$ the reduction is ≈ 3 percentage points.

4 Conclusion

We introduced two variants of a new generic attack on exponent blinding (RSA without CRT, RSA with CRT, ECC). Experiments with simulated data confirmed the theoretical results. For small blinding factor $R = 10$ the basic attack was successful for error rates of $\epsilon_b \leq 0.23$ (256-bit ECC), resp. for $\epsilon_b \leq 0.28$ (1024-bit RSA). The enhanced attack works for $\epsilon_b \leq 0.13$ (for $R = 16$) but requires by orders of magnitudes less power traces and scales much better for large R. Moreover, the enhanced variant applies new techniques, which are interesting by themselves. Of course, for security implementations power traces should not provide any useable information at all. However, selecting $R = 64$, or maybe even better $R > 64$, should make both variants impractical anyway. Limiting the use of the key clearly below $2^{R/2}$ even prevents 2-birthdays (basic attack).

Acknowledgement. The authors would like to thank David Galindo for his valuable support and the anonymous referees for their comments, which helped to improve the editorial quality of the paper.

References

1. Acıiçmez, O., Schindler, W.: A Vulnerability in RSA Implementations due to Instruction Cache Analysis and Its Demonstration on OpenSSL. In: Malkin, T. (ed.) CT-RSA 2008. LNCS, vol. 4964, pp. 256–273. Springer, Heidelberg (2008)
2. Fouque, P., Kunz-Jacques, S., Martinet, G., Muller, F., Valette, F.: Power Attack on Small RSA Public Exponent. In: Goubin, L., Matsui, M. (eds.) CHES 2006. LNCS, vol. 4249, pp. 339–353. Springer, Heidelberg (2006)
3. Henecka, W., May, A., Meurer, A.: Correcting Errors in RSA Private Keys. In: Rabin, T. (ed.) CRYPTO 2010. LNCS, vol. 6223, pp. 351–369. Springer, Heidelberg (2010)
4. Coron, J.S.: Resistance against Differential Power Analysis for Elliptic Curve Cryptosystems. In: Koç, Ç.K., Paar, C. (eds.) CHES 1999. LNCS, vol. 1717, pp. 292–302. Springer, Heidelberg (1999)
5. Courrège, J.C., Feix, B., Roussellet, M.: Simple Power Analysis on Exponentiation Revisited. In: Gollmann, D., Lanet, J.-L., Iguchi-Cartigny, J. (eds.) CARDIS 2010. LNCS, vol. 6035, pp. 65–79. Springer, Heidelberg (2010)
6. Itoh, K., Izu, T., Takenaka, M.: Address-bit Differential Power Analysis of Cryptographic Schemes OK-ECDH and OK-ECDSA. In: Kaliski Jr., B.S., Koç, Ç.K., Paar, C. (eds.) CHES 2002. LNCS, vol. 2523, pp. 129–143. Springer, Heidelberg (2003)
7. Itoh, K., Yamamoto, D., Yajima, J., Ogata, W.: Collision-Based Power Attack for RSA with Small Public Exponent. IEICE Transactions on Information and Systems E92-D5, #5, 897–908 (2009)
8. Kocher, P.: Timing Attacks on Implementations of Diffie-Hellman, RSA, DSS, and Other Systems. In: Koblitz, N. (ed.) CRYPTO 1996. LNCS, vol. 1109, pp. 104–113. Springer, Heidelberg (1996)
9. Kocher, P., Jaffe, J., Jun, B.: Differential Power Analysis. In: Wiener, M. (ed.) CRYPTO 1999. LNCS, vol. 1666, pp. 388–397. Springer, Heidelberg (1999)
10. Lang, S.: Algebra, 3rd edn. Addison-Wesley, Reading (1993)
11. Schindler, W.: A Combined Timing and Power Attack. In: Naccache, D., Paillier, P. (eds.) PKC 2002. LNCS, vol. 2274, pp. 263–279. Springer, Heidelberg (2002)
12. Yen, S., Lien, W., Moon, S., Ha, J.: Power Analysis by Exploiting Chosen Message and Internal Collisions - Vulnerability of Checking Mechanism for RSA-Decryption. In: Dawson, E., Vaudenay, S. (eds.) Mycrypt 2005. LNCS, vol. 3715, pp. 183–195. Springer, Heidelberg (2005)

Cryptanalysis of the Atmel Cipher in SecureMemory, CryptoMemory and CryptoRF

Alex Biryukov, Ilya Kizhvatov, and Bin Zhang

University of Luxembourg
Faculty of Science, Technology and Communication
6, rue Richard Coudenhove-Kalergi, L-1359 Luxembourg
{alex.biryukov, ilya.kizhvatov, bin.zhang}@uni.lu

Abstract. SecureMemory (SM), CryptoMemory (CM) and CryptoRF (CR) are the Atmel chip families with wide applications in practice. They implement a proprietary stream cipher, which we call the Atmel cipher, to provide authenticity, confidentiality and integrity. At CCS'2010, it was shown that given 1 keystream frame, the secret key in SM protected by the simple version of the cipher can be recovered in $2^{39.4}$ cipher ticks and if 2640 keystream frames are available, the secret key in CM guarded by the more complex version of the cipher can be restored in 2^{58} cipher ticks. In this paper, we show much more efficient and practical attacks on both versions of the Atmel cipher. The idea is to dynamically reconstruct the internal state of the underlying register by exploiting the different diffusion speeds of the different cells. For SM, we can recover the secret key in $2^{29.8}$ cipher ticks given 1 keystream frame; for CM, we can recover the secret key in 2^{50} cipher ticks with around 24 frames. Practical implementation of the full attack confirms our results.

Keywords: Stream ciphers, RFID, Frame, SecureMemory, CryptoMemory.

1 Introduction

The Atmel Cipher. The Atmel chips AT88SC153 and AT88SC1608, called SecureMemory (SM) family, were introduced in 1999 [3]. The CryptoMemory (CM) family including the AT88SCxxxxC chips was introduced in early 2002 [2] with more advanced cryptographic features. These two families are ISO/IEC 7816 smart cards that communicate through a contact interface; in late 2003, the CryptoRF (CR) family (ISO/IEC 14443-B smart cards) [10], which is a variant of the CM family with the AT88SCxxxxCRF chips was introduced with a RF interface. These chips are widely used in practice all over the world, e.g. in smart cards (ID and access cards, health care cards, loyalty cards, Internet kiosks, energy meters and e-government) [4], in the widely sold NVIDIA graphics cards [12], the Microsoft's Zune Player [7] and SanDisk's Sansa Connect [13]. For more complete and detailed information, please see [4,8].

A proprietary stream cipher, which we call the Atmel cipher, lies at the core of the chips to provide authenticity, confidentiality and integrity. The SM family

J. Lopez and G. Tsudik (Eds.): ACNS 2011, LNCS 6715, pp. 91–109, 2011.

uses the simple version of the cipher, while in CM and CR, the more complex version of the cipher is adopted. The difference between the simple version and the complex one is that there is 1 byte feedback of the output into the other shift registers every cipher tick in the complex version. Besides, in CM, the initialization phase and the generation of the authenticators is much more complicated. There are more mixing rounds in CM before the output is produced. It is commonly believed that the complex version of the Atmel cipher provides much stronger security.

Previous work. At CCS'2010, the Atmel cipher is described and analyzed in the authentication mechanism [8]. In such a scenario, the tag and the reader exchange their 64-bit nonces and use the shared 64-bit key to generate some keystream nibbles as authenticators. The attacker is assumed to be able to capture some keystream frames produced by the same shared key, but with different nonces. Throughout this paper, we call the result of a single authentication session (with 128-bit keystream) a keystream frame, or briefly a frame. In [8], it was shown that there exists a key recovery attack on SM in $2^{39.4}$ cipher ticks with probability 0.57, given 1 frame and a key recovery attack on CM in 2^{58} cipher ticks[1], if 2640 frames are available. Hence, in theory, both versions of the cipher do not provide the full security with respect to their key length. However, in practice, the challenge is how to efficiently break the cipher with as few frames as possible? Are there any cryptanalytic techniques that could be used in such a *restricted* setting? Note that in our scenario, the attacker can only capture some *random known* frames with random nonces, he cannot choose the frames with the nonces satisfying some specific properties, e.g. some special differences. Thus, the techniques requiring chosen nonces, e.g. the differential-like chosen nonces attacks and the cube attacks [5,6] will not work in this realistic setting, neither will fast correlation attacks [11] which usually require large amounts of keystream.

Our contribution. In this paper, we present practical *random known* nonces attacks on both version of the Atmel cipher. In contrast to the attack in [8], which had to exhaustively search the left and right registers for each captured frame, our attack only makes an exhaustive search of the shortest right-most register and uses the optimal Viterbi-like decoding techniques [14] to recover the internal states of the other registers. We exploit the differences in diffusion speeds of the cells of the registers to restore the internal state efficiently. For SM, by starting from the most dense part of the known keystream segment of the left register, our technique can fill the gap of 2-step update for adjacent known keystream nibbles very well, resulting in a key recovery attack in $2^{29.8}$ cipher ticks with success probability 0.75, given 1 frame. This is about 1000 times faster than that in [8]. For CM, by a careful analysis of the state update function and the output function of the underlying register, we can partially determine chunks of the state with low complexity. The positions of the recovered chunks are chosen

[1] In [8], this complexity is claimed to be 2^{52} cipher ticks, however, the complexity of unrolling the cipher 64 steps is ignored [9].

Table 1. Key recovery attacks on SecureMemory

	data, frames	time	success probability	running time
attack of [8]	1	$2^{39.4}$	0.57	minutes
this paper	1	$2^{29.8}$	0.75	seconds

Table 2. Key recovery attacks on CryptoMemory (success probability 0.5)

	Theoretical			Practical	
	data, frames	time	memory	running time (200 CPU cores)	memory
attack of [8]	2640	2^{58}	$O(2^{32})$	several weeks (estimated)[9]	16 GB
this paper	30	2^{50}	$O(2^{24})$	several days	530 MB

in such a way that we can determine the maximum keystream information solely based on these states. By starting from the carefully chosen point in time, we mount an attack on CM in 2^{50} cipher ticks with around 24 frames, which is 2^8 times faster than that in [8] and uses much less frames. The extremely low data complexity of our attack makes it more threatening in practice, an attacker can easily get such a number of frames to mount the attack.

We have fully implemented our attack. It takes several minutes to find a good frame among the 30 given frames and recover the possible left-right state pairs subsequently. Then roughly $2-6$ days are needed to restore the full internal state of a good frame just after the initialization using the 200 CPU cores and another 2 hours on a single core are taken for the full key recovery. The short running time allowed us to run the full attack several times for different keys.

Tables 1 and 2 present a comparison of our new attacks and the attacks of [8][2] on SM and CM respectively. We note that due to the properties of our attack the use of bit-slicing techniques, which according to [9] were employed in [8], is not efficient in its implementation. Our original estimates were also very optimistic: the recovery of just the middle register took only 6 hours and complexity of the total attack seemed around $2^{45}-2^{47}$ cipher ticks, but full implementation has shown dependencies between all the phases of the attack, raising the total complexity to about $2^{49}-2^{50}$ cipher ticks.

Organization of the paper. We describe the two versions of the Atmel cipher in Section 2 together with the concrete authentication protocols. Our attack on SM is provided in Section 3 and the attack on CM is given in Section 4. We describe our practical implementation of the full key recovery for CM in Section 5 and give our conclusions in Section 6.

[2] The authors have corrected their original complexity of 2^{52} (for which the attack runs 2 days) to 2^{58} which probably means increase to several months. However, the authors of [9] used bit-slice implementation which offers some speedup. Our implementation currently is not bit-sliced.

2 Description of the Atmel Cipher and the Authentication Protocol

In this section, we present a description of the two versions of the Atmel cipher together with the concrete authentication protocols. Let us first specify the notations used hereafter.

- $\mathbb{F}_2^n = \{0, 1, \ldots, 2^n - 1\}$.
- $(x_0x_1 \ldots x_{n-1}) \in \mathbb{F}_2^n$ with x_0 being the most and x_{n-1} being the least significant bits.
- suc is the state transition function.
- $a \in \mathbb{F}_2^8$ is the input to the state.
- the cipher state s and the successor state $s' = suc^1(a, s) = suc(a, s)$.
- $suc^n(a, s) = suc^{n-1}(a, suc(a, s))$ for $n > 1$.
- the left register $l = (l_0, l_1, \ldots, l_6) \in (\mathbb{F}_2^5)^7$.
- the middle register $m = (m_0, m_1, \ldots, m_6) \in (\mathbb{F}_2^7)^7$.
- the right register $r = (r_0, r_1, \ldots, r_4) \in (\mathbb{F}_2^5)^5$.
- the feedback register $f = (f_0, f_1) \in (\mathbb{F}_2^4)^2$ for CM.
- $L : \mathbb{F}_2^n \to \mathbb{F}_2^n$ is the bitwise left rotation defined by $L(x_0x_1 \ldots x_{n-1}) = (x_1 \ldots x_{n-1}x_0)$.
- $+$ is the integer addition.
- the modified modular addition $\boxplus : \mathbb{F}_2^n \times \mathbb{F}_2^n \to \mathbb{F}_2^n$ is defined as:

$$x \boxplus y = \begin{cases} x + y \pmod{2^n - 1} & \text{if } x = y = 0 \text{ or } x + y \neq 0 \pmod{2^n - 1} \\ 2^n - 1 & \text{otherwise} \end{cases}$$

2.1 Specification of the Atmel Cipher

Both versions consist of 3 shift registers, i.e., the left register l, the middle register m and the right register r. The complex version in CM has an additional feedback register f to store the last 8 bits of output. The cipher structure is illustrated in Fig. 1. At each tick, a cipher state $s = (l, m, r, f) \in \mathbb{F}_2^{117}$ (for SM, ignore f and $s \in \mathbb{F}_2^{109}$) is converted into a successor state $s' = (l', m', r', f')$ as follows. First inject the input a into s at several cell positions, resulting in an intermediate state $\hat{s}$. For CM, let $b = a \oplus f_0 f_1$; while for SM, let $b = a$. Then, $\hat{l}_i = l_i$, $\hat{m}_j = m_j$ and $\hat{r}_k = r_k$ for $i \neq 2$, $j \neq 4$ and $k \neq 1$. For $i = 2$, $j = 4$ and $k = 1$,

$$\hat{l}_2 := l_2 \oplus b_3b_4b_5b_6b_7, \quad \hat{m}_4 := m_4 \oplus b_4b_5b_6b_7b_0b_1b_2, \quad \hat{r}_1 := r_1 \oplus b_0b_1b_2b_3b_4.$$

Second, shift the left, right and middle registers one cell to the right and compute the new 0th terms by the 1-bit left rotation L and the modified modular addition $\boxplus$.

$$\begin{aligned} &l'_{i+1} := \hat{l}_i, \ m'_{i+1} := \hat{m}_i, \ \text{for } i \in \{0, 1, \ldots, 5\}, \\ &r'_{i+1} := \hat{r}_i \ \text{for } i \in \{0, 1, \ldots, 3\}, \\ &l'_0 := \hat{l}_3 \boxplus L(\hat{l}_6), \ m'_0 := \hat{m}_5 \boxplus L(\hat{m}_6), \ r'_0 := \hat{r}_2 \boxplus \hat{r}_4. \end{aligned}$$

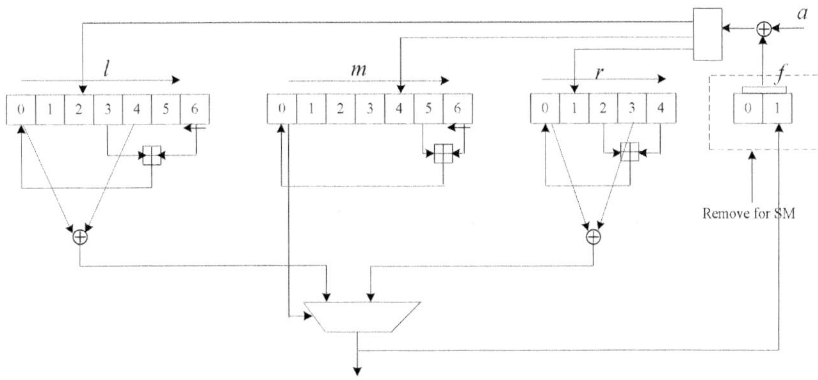

Fig. 1. The ciphers

Finally, generate the keystream and shift the feedback register f one cell to the left and set a new 1st entry as the output nibble for CM. Let $||$ be the concatenation operation, denote by $outputl(l') = l'_{0,1} \oplus l'_{4,1} \; || \; l'_{0,2} \oplus l'_{4,2} \; || \; l'_{0,3} \oplus l'_{4,3} \; || \; l'_{0,4} \oplus l'_{4,4}$ the rightmost 4 bits of $l'_0 \oplus l'_4$ and $outputr(r') = r'_{0,1} \oplus r'_{3,1} \; || \; r'_{0,2} \oplus r'_{3,2} \; || \; r'_{0,3} \oplus r'_{3,3} \; || \; r'_{0,4} \oplus r'_{3,4}$ the rightmost 4 bits of $r'_0 \oplus r'_3$. The output of s', denoted by $output(s')$, is given by

$$output(s')_i = \begin{cases} outputl(l')_i, & \text{if } m'_{0,i+3} = 0 \\ outputr(r')_i, & \text{if } m'_{0,i+3} = 1. \quad i \in \{0, \ldots, 3\}. \end{cases} \tag{1}$$

Note that the rightmost 4 bits of m'_0 selects either a bit from $outputl(l')$ or a bit from $outputr(r')$ to be output as the keystream bit. For CM, let $f'_0 = \hat{f}_1 = f_1$ and $f'_1 = output(s')$.

2.2 The Authentication Protocol

In the protocol, the tag and the reader exchange the nonces (depicted in Fig. 2) and use the cipher to generate keystream that will be used as authenticators for both sides.

Let $nt \in (\mathbb{F}_2^8)^8$ be a tag nonce, $nr \in (\mathbb{F}_2^8)^8$ a reader nonce and $k \in (\mathbb{F}_2^8)^8$ be the shared key between the tag and the reader. First initialize all the registers l, m, r and f (for SM, ignore f) to be zero, then the cipher is clocked as follows.

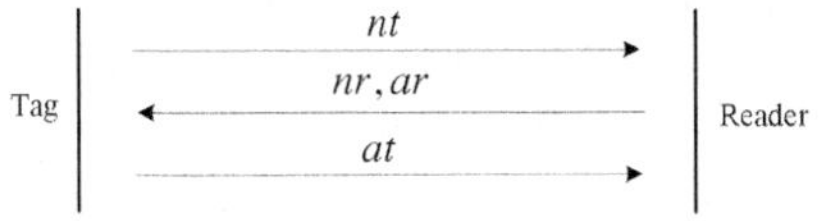

Fig. 2. The authentication protocol

Table 3. The injection procedures

	SM	CM
s_0	nt_0, nt_1, nr_0	$nt_0, nt_0, nt_0\ nt_1, nt_1, nt_1, nr_0$
s_1	nt_2, nt_3, nr_1	$nt_2, nt_2, nt_2\ nt_3, nt_3, nt_3, nr_1$
s_2	nt_4, nt_5, nr_2	$nt_4, nt_4, nt_4\ nt_5, nt_5, nt_5, nr_2$
s_3	nt_6, nt_7, nr_3	$nt_6, nt_6, nt_6\ nt_7, nt_7, nt_7, nr_3$
s_4	k_0, k_1, nr_4	$k_0, k_0, k_0\ k_1, k_1, k_1, nr_4$
s_5	k_2, k_3, nr_5	$k_2, k_2, k_2\ k_3, k_3, k_3, nr_5$
s_6	k_4, k_5, nr_6	$k_4, k_4, k_4\ k_5, k_5, k_5, nr_6$
s_7	k_6, k_7, nr_7	$k_6, k_6, k_6\ k_7, k_7, k_7, nr_7$
s_8		

$$s_0 := 0,$$
$$s_{i+1} := suc(nr_i, suc^v(nt_{2i+1}, suc^v(nt_{2i}, s_i))), \qquad i \in \{0, \ldots, 3\}$$

$$s_{i+5} := suc(nr_{i+4}, suc^v(k_{2i+1}, suc^v(k_{2i}, s_{i+4}))), \quad i \in \{0, \ldots, 3\}$$

where $v = 1$ for SM and $v = 3$ for CM. Table 3 shows the schematic view of the input in the initialization phase. Note that the states $s_0, \ldots, s_7, s_8$ are non-consecutive. There are 24 setup rounds for SM and 56 setup rounds for CM respectively. Let $at \in (\mathbb{F}_2^4)^{16}$ be the tag authenticators and $ar \in (\mathbb{F}_2^4)^{16}$ the reader authenticators. The precise definitions of the authentication process are given as follows.

SM Authentication. Define the following states and outputs:

$$s_i := suc^2(0, s_{i-1}), \quad i \in \{9, \ldots, 40\}.$$
$$at_i := output(s_{2i+9}), \quad at_{i+1} := output(s_{2i+10}), \quad i \in \{0, 2, \ldots, 14\},$$
$$ar_i := output(s_{2i+11}), \quad ar_{i+1} := output(s_{2i+12}), \quad i \in \{0, 2, \ldots, 14\}.$$

CM Authentication. Define the following states and outputs:

$$s_9 := suc^5(0, s_8), \quad s_{10} := suc(0, s_9), \quad s_i := suc^6(0, s_{i-1})\ i \in \{11, 13, \ldots, 23\};$$
$$s_i := suc(0, s_{i-1})\ i \in \{12, 14, \ldots, 24\}; \quad s_i := suc(0, s_{i-1})\ i \in \{25, 26, \ldots, 38\};$$
$$ar_i := output(s_{i+9})\ i \in \{0, 1, \ldots, 15\}; \quad at_0 := 0xf, \qquad at_1 := 0xf,$$
$$at_i := output(s_{i+23})\ i \in \{2, 3, \ldots, 15\}.$$

Note that there are 16 consecutive keystream nibbles in the frame, i.e., ar_{14}, ar_{15}, at_2, $at_3, \ldots, at_{15}$. Since CM and CR use the same version of the cipher, we will ignore this distinction hereafter. The attacker's aim is to restore the shared 64-bit key from a number of captured frames, for each of which only a short keystream segment is known.

3 Our Attack on SecureMemory

First note that from (1), for $i \in \{0, 1, 2, 3\}$ we have

$$output(s')_i = m'_{0,i+3} \cdot outputr(r')_i \oplus (1 \oplus m'_{0,i+3}) \cdot outputl(l')_i. \qquad (2)$$

This suggests that $P\big(output(s')_i = outputr(r')_i\big) = \frac{3}{4}$ and $P\big(output(s')_i = outputl(l')_i\big) = \frac{3}{4}$. Thus, we can make an exhaustive search of all the possible s_8 states of the right register r and use a classical correlation test to find the correct state. This is feasible because we can run r independently of l and m. Note that there are 128 bits of known keystream from 1 SM frame. The correct state of r could pass the test with probability $\sum_{i=T_r}^{128} \binom{128}{i} (\frac{3}{4})^i (\frac{1}{4})^{128-i}$ and a wrong guess would pass with probability $\sum_{i=T_r}^{128} \binom{128}{i} (\frac{1}{2})^{128}$, where T_r is the threshold value of the correlation test.

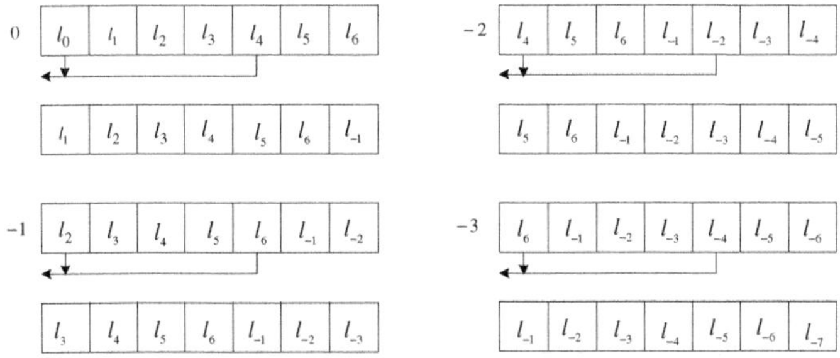

Fig. 3. The backward diffusion of l in SM

Then, unlike the attack in [8], we do not make a second exhaustive search of the left register l. To get the candidate states of l matching the keystream bits where the intermediate output $outputr(r')$ of r does not generate the correct bit (There are 32 such bits on average), we regard the known keystream bits generated by l as the observed events of the internal hidden states, as in the classical Viterbi decoding scenario [14]. However, we found that it is not easy to directly use the Viterbi decoding algorithm here. Instead, we exploit the different diffusion speeds of the different cells in l to enumerate the possible candidates dynamically. Fig. 3 and 4 show the diffusion process of the cells in l. Let l_i $(i \geq 0)$ be the 5-bit content of the corresponding cell, then for every second step i, we know some bits of the xor of the 0th cell and the 4th cell. In fact, the bits are distributed according to the xor between $outputr(r')$ and at and ar. At some instance, it may happen that there are no bits known; while at other step, it also may happen that we know the whole nibble.

Let $lk_i = outputl(l')$ at step i, though sometimes lk_i is unknown. Our observation is that some cells of the initial state affect the output more often than others. Hence, if we isolate the low-effect cells and first determine the cells that have the most extensive effect on the output, it is expected that in this way, we need not try all the possible states one-by-one. Besides, we can shift the starting point of our decoding algorithm. The chosen criterion is determined by the problem that which cells we choose to determine first. We have the following theorem (proved in the full version of the paper) on the latter problem. For any starting point, define the starting state to be $ls_0 = \{l_0, l_1, l_2, l_3, l_4, l_5, l_6\}$. The counting of the predecessor and successor states and the lk_is follow Fig. 3 and 4.

Theorem 1. *For any starting state ls_0, if we choose $A = \{l_0, l_1, l_3, l_4, l_6\}$ to determine first, then $\{lk_3, lk_5, lk_7, lk_9, lk_{11}\}$ depend on l_8 and A, $\{lk_{-3}, lk_0, lk_2, lk_4\}$ only depend on A, $\{lk_{-1}, lk_6\}$ depend on A and l_2, and $\{lk_1, lk_8, lk_{10}\}$ depend on A, l_2 and l_8.*

Theorem 1 shows that if we know A, then we can reduce the possible values of l_8 to a large extent, for there are 5 equations that could be used for check. After

Fig. 4. The forward diffusion of l in SM

determining l_8 and A, other cells of the state ls_0 could be restored easily, for we can just run l forward and backward from ls_0 to get the candidates and then clock back from ls_0 to restore the real initial state. This theorem also indicates how to choose the starting state. Let $N_H(lk_i)$ be number of known bits in lk_i for $0 \le i \le 31$, then we define the following function:

$$\Psi(i) = \begin{cases} \sum_{j\in\{0,2,3,4,5,7,9,11\}} N_H(lk_{i+j}) & \text{for } 0 \le i \le 2 \\ \sum_{j\in\{-3,0,2,3,4,5,7,9,11\}} N_H(lk_{i+j}) & \text{for } 3 \le i \le 20 \\ \sum_{j\in\{-3,0,2,3,4,5,7,9\}} N_H(lk_{i+j}) & \text{for } 21 \le i \le 24 \\ \sum_{j\in\{-3,0,2,3,4,5\}} N_H(lk_{i+j}) & \text{for } 25 \le i \le 26 \\ \sum_{j\in\{-3,0,2,3,4\}} N_H(lk_{i+j}) & \text{for } i = 27 \\ \sum_{j\in\{-3,0,2,3\}} N_H(lk_{i+j}) & \text{for } i = 28 \\ \sum_{j\in\{-3,0,2\}} N_H(lk_{i+j}) & \text{for } i = 29 \\ \sum_{j\in\{-3,0\}} N_H(lk_{i+j}) & \text{for } 30 \le i \le 31 \,. \end{cases}$$

Let $I = \max_{0\le i\le 31}\Psi(i)$ and $\Psi(J) = I$, then we can start from the state s_{8+J} in the real SM authentication, which will have a maximum reduction effect on the possible candidates. Table 4 shows the distributions of I and J obtained from experiments. From Table 4, we get $P(I \ge 10, J < 25) \approx 0.94$.

Table 4. The distributions of I and J in SM.

I	$I \ge 10$ 0.95	$I \ge 11$ 0.87	$I \ge 12$ 0.73	$I \ge 13$ 0.53	$I \ge 14$ 0.34
J	$J \ge 21$ (0.0297, 0.0344)	$J \ge 22$ (0.0205, 0.028)	$J \ge 23$ (0.003, 0.014)	$J \ge 24$ (0.003, 0.007)	$J \ge 25$ (0.0006, 0.0018)

Now we are ready to enumerate the possible states of A consistent with $\{lk_0, lk_2, lk_4\}$. We first guess l_0 in ls_0, there are 2^5 possibilities. From lk_0, we can get 2^3 candidates of l_4 on average for each l_0. Then from lk_2, we can get around 2 candidates of l_3 for each l_0. Knowing l_0 and l_3, we could derive l_{10}. From lk_4, we could get 2 candidates on average of l_7 for each pair of (l_0, l_3). For each l_7, we could derive one or two (due to the fact that $\boxplus$ is not injective) candidates of l_6. There are 2^5 candidates of l_1, so with a complexity of $2^5 \cdot 4 = 2^7$, we could get $2^5 \cdot 2^5 \cdot 2^3 \cdot 2 \cdot 2 = 2^{15}$ possible combinations of A. Then using the 5 check equations for l_8, we can determine several candidates of l_8 conditioned on A, sometimes even 1. For each possible combination of A and l_8, we can derive the corresponding l_2 from other lk_is. Given l_8 and l_2, we know l_5. Finally, we run l backwards and forwards from ls_0 to further reduce the possibilities. For each surviving candidate for ls_0, run l backwards to recover the real initial state s_8 in the authentication. This procedure has a complexity of $2^{15} \cdot 2^5 \cdot 2^3 \cdot \frac{64}{3} = 2^{27.4}$ cipher ticks. The number of candidates of s_8 of l is directly determined by the coincidence bits between the intermediate output, $outputr(r')$, of r and the keystream. More coincidence bits, more candidates of s_8. From experiments, we found that with probability around 0.92, there are less than or equal to 100 candidates of s_8 recovered. For each right-left candidate pair, we can run the meet-in-the-middle attack in Section 4.2 of [8] to recover the secret key with a complexity of around $2^{24.5}$ cipher ticks. So the total time complexity of our attack is $2^{27.4} + 2^{24.5} + 2^{25} \cdot \frac{64}{3} = 2^{29.8}$ cipher ticks. And the success probability of our attack is around $\sum_{i=T}^{128} \binom{128}{i} (\frac{3}{4})^i (\frac{1}{4})^{128-i} \cdot 0.92 \cdot 0.94$, which is around 0.75 if we set $T = 91$. Given 1 frame, our attack on SM is about 1000 times faster than that in [8] with a higher success rate. This attack is verified on a single CPU core in C. In experiments, it takes tens of seconds to restore the s_8 state of l and r in the real authentication.

4 Our Attack on CryptoMemory

The starting point of our attack on CM is the 16 consecutive keystream nibbles, i.e., we first use the correlation test to find some candidates of the right-most register r. Because of the existence of the feedback register f in CM, we cannot run l or r independently in general. But when CM generates the 16 consecutive nibbles, we can run either l or r independently, for in this case we know the feedback bytes, which is the same as the last 64-bit keystream in one frame.

Conditioned on r, we can derive 16-bit information of the intermediate output of l on average by selecting those keystream bits where the intermediate output

Fig. 5. The forward diffusion of l in CM

of r does not generate the correct keystream bit. Again, we regard the known intermediate output of l as the observed events of the corresponding internal hidden states. The good thing for us is that now we know the intermediate output of l for consecutive steps, instead of for every second step. Besides, we need to exploit the properties of the state update function and the output function of l to find the most likely internal states that generate the observed keystream bits. Since both the state update function and the output function of l only depend on very few variables, some of which are shared, we can partially determine chunks of the state with low complexity. Several overlapping partial states could be restored in this way. Then we take an intersection of the overlapping states and use this subset to further reduce the candidates of other parts. The positions of the recovered states are chosen in such a way that we can determine the maximum keystream information solely based on these states. The same techniques are also applied to the middle register m.

4.1 Recovering the Right and Left Registers r and l

Precisely, we first make an exhaustive search of all the possible s_{24} states of the right register r and use a correlation test to filter out the wrong guesses to some extent. The correct candidate could pass the test with probability $\sum_{i=T_r}^{64} \binom{64}{i} (\frac{3}{4})^i (\frac{1}{4})^{64-i}$ and a wrong guess would pass with probability $\sum_{i=T_r}^{64} \binom{64}{i}$ $\cdot(\frac{1}{2})^{64}$, where T_r is the threshold value in correlation test. Since in CM we can only use the 64 consecutive bits for correlation test, instead of 128 bits as in the SM case, we select T_r in such a way that we need not call the following parts of our attack for each captured frame. Our attack is continued only if there are some candidates of r pass the correlation test. If for one frame, there are no output from the correlation test, we discard the frame and try another one.

Now we look at the left register l. We want to restore the s_{24} state of l without exhaustively searching all the possibilities. Still, we did not find a way to use the Viterbi decoding algorithm directly. The following steps have some similarities with the Viterbi algorithm in the sense that we determine the most likely hidden state candidates up to a certain point depending only on the current observable events and the most likely state candidates at the last point. Fig. 5 depicts the forward diffusion process of l. In fact, we can shift the beginning point of our counting, i.e., we can make a chosen time point in the middle to be 0. The chosen criterion is determined by the diffusion properties of the different cells. Again, let lk_i be the intermediate output of l, though sometimes lk_i is unknown. For any starting point, define the starting state to be $ls_0 = \{l_0, l_1, l_2, l_3, l_4, l_5, l_6\}$. We have

$$lk_1 = (l_3 \boxplus L(l_6)) \oplus l_3, \quad lk_2 = (l_2 \boxplus L(l_5)) \oplus l_2, \tag{3}$$

$$lk_3 = (l_1 \boxplus L(l_4)) \oplus l_1, \quad lk_4 = (l_0 \boxplus L(l_3)) \oplus l_0, \tag{4}$$

$$lk_5 = (l_7 \boxplus L(l_2)) \oplus l_7, \quad lk_6 = (l_8 \boxplus L(l_1)) \oplus l_8. \tag{5}$$

The above equations indicate that the output function and the state update function of l depends on very few variables. For example, the 3rd cell of l is used in the current state update function and in the next step output function. Hence, with a complexity of 2^{10}, we can determine the solution set $H_0 = \{l_3, l_6\}$ to (3). Similarly, we derive $H_1 = \{l_2, l_5\}$, $H_2 = \{l_1, l_4\}$, $H_3 = \{l_0, l_3\}$, $H_4 = \{l_7, l_2\}$, $H_5 = \{l_8, l_1\}$ and $H_6 = \{l_9, l_0\}$ from (4) and (5). The cardinality of H_i depends on $N_H(lk_i)$. Note that $H_{0,3} = H_0 \cap H_3 = \{l_3\}$, $H_{1,4} = H_1 \cap H_4 = \{l_2\}$, $H_{2,5} = H_2 \cap H_5 = \{l_1\}$ and $H_{3,6} = H_3 \cap H_6 = \{l_0\}$. Hence, we can reduce the cardinalities of H_i and H_j by keeping only those solutions that have the value patterns existing in the corresponding intersection set $H_{i,j}$. Experiments show that the averaged value of $|H_i|$ is $2^{9.25}$. After reducing the cardinality by intersection, the cardinalities sometimes reduce to half or more, sometimes remain. Then, we combine H_0 with H_3 to get the possible values for $H_{0,3,6} = \{l_0, l_3, l_6\}$. The averaged number of solutions of $H_{0,3,6}$ is $2^{10.96}$ in 2^{20} times of experiments. Note that we have to xor the feedback byte b with the recovered l_0 in H_6 to get the original value of l_0 in ls_0. Similar operation has to be done for the recovered l_1 in H_5 too. To get a maximum reduction effect on the number of possible candidates, we define the following function:

$$\Psi(i) = \sum_{j \in \{1,3,4,8\}} N_H(lk_{i+j}) \quad \text{for } 1 \leq i \leq 7 .$$

This function considers the reduction effect on $A = \{l_0, l_1, l_3, l_4, l_6\}$ of the chosen starting state, as can be seen from the diffusion process in Fig. 5. Let $I = \max_{1 \leq i \leq 7} \Psi(i)$ and $\Psi(J) = I$, then we can start from the state s_{24+J} in the real CM authentication to have a maximum reduction effect on A. Since in CM, we know the 16 consecutive lk_is and $\Psi(i)$ is defined over $1 \leq i \leq 7$, we need not build a table similar to Table 4 to list the distributions of I and J, we just use I and J directly following their definitions. From Fig. 5, we have the following theorem, proved in the full version of the paper.

Theorem 2. *For any starting state ls_0, if we choose $A = \{l_0, l_1, l_3, l_4, l_6\}$ to determine first in CM, then $\{lk_0, lk_1, lk_3, lk_4, lk_7, lk_8, lk_{11}, lk_{15}\}$ depend only on A, $\{lk_{-1}\}$ depends on $\{l_5\}$ and A, and $\{lk_5, lk_{12}\}$ depend on A and l_2.*

Due to the fact that J is not necessarily equal to 1, some of the lk_i could not be used in practice. For example, if $J = 7$, we can only use the lk_is up to $i = 8$. Since we define $\Psi(i)$ only over $1 \leq i \leq 7$, we can use at least $\{lk_0, lk_1, lk_3, lk_4, lk_7, lk_8\}$ for the reduction check of A. Then we use lk_{-1} and lk_5 to reduce the possibilities of H_1. After that, we combine the remaining candidates of H_1 with those of A and run l backwards and forwards to cover the other lk_is. Finally, we clock l back from ls_0 to recover the original initial state, i.e., the state s_{24} in the real CM authentication. Note that the correct candidate of the s_{24} state of l will pass the above procedures with probability 1.

Let n_l be the number of candidates of the s_{24} state of l restored in the above way. Experiments show that n_l is determined by the threshold value T_r in the correlation test of the right-most register r. For example, if we set $T_r = 54$, the averaged value of n_l is approximately $2^{25.4}$, which is very close to $2^{35-64+T_r}$. To further reduce the possible candidates, we resort to the correlation test of l, but restrict ourselves only to the n_l candidates. This step has a complexity of $n^l \cdot \frac{16-2}{3}$, instead of $2^{35} \cdot \frac{16-2}{3} \approx 2^{37.2}$. Let T_l be the threshold value used in the correlation test of the n_l candidates, the correct candidate could pass this test with probability around $P_c = \sum_{i=T_l}^{64} \binom{64}{i}(\frac{3}{4})^i(\frac{1}{4})^{64-i}$, but the wrong candidates would pass with the probability much larger than $\sum_{i=T_l}^{64} \binom{64}{i}(\frac{1}{2})^{64}$. The reason is that the n_l candidates are more likely to be the correct candidate than the ones in the random case, which has 2^{35} possibilities. We use experiments to determine this probability. Let P_w be this false alarm probability, we found that if $T_r = 54$ and $T_l = 45$, then the averaged value of P_w is about 2^{-5}; if $T_r = 54$ and $T_l = 53$, then the averaged value of P_w is about 2^{-19}. We conjecture that in general, P_w is around 2^5 times larger than $\sum_{i=T_l}^{64} \binom{64}{i}(\frac{1}{2})^{64}$. After the correlation test of the n_l candidates, we get about $(n_l - 1) \cdot P_w + 1 \cdot P_c$ candidates for the left register l. Then we turn to the middle register m.

4.2 Recovering the Middle Register m

Now we know the candidate s_{24} (in the real authentication) states of l, r and the real feedback register f, our aim is to restore the candidate s_{24} states of m. By xoring the intermediate outputs $outputl(l')$ of l and those of r, we could know the exact values of the intermediate output of m from (1). Let mk_i denote such

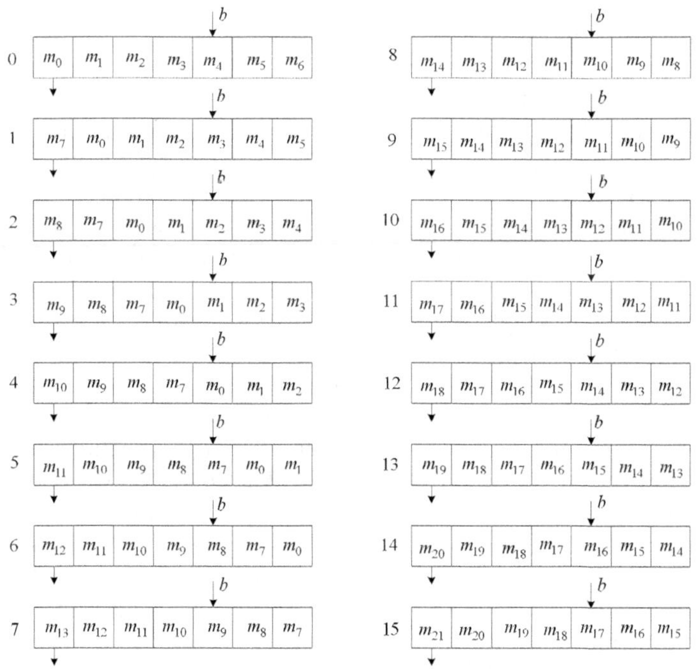

Fig. 6. The forward diffusion process of m in CM

value for the step i. Fig. 6 shows the forward diffusion process of m in CM. To make a full use of the known information, we start from the state $s_{24+6} = s_{30}$ in the real authentication. From Fig. 6, we have

$$mk_7 = m_7 \boxplus L(m_0), \tag{6}$$

$$mk_i = m_i \boxplus L(m_{i-1}). \ (8 \leq i \leq 15) \tag{7}$$

The above equations indicate that the output function and the state update function of the middle register m depend on even fewer variables than l. We can see the content of the feedback cell directly. Hence, with a complexity of 2^{14}, we can determine the candidates of $Q_0 = \{m_7, m_0\}$ satisfying (6). Similarly, we can determine $Q_1 = \{m_8, m_7\}$, $Q_2 = \{m_9, m_8\}$, $Q_3 = \{m_{10}, m_9\}$, $Q_4 = \{m_{11}, m_{10}\}$, $Q_5 = \{m_{12}, m_{11}\}$, $Q_6 = \{m_{13}, m_{12}\}$, $Q_7 = \{m_{14}, m_{13}\}$ and $Q_8 = \{m_{15}, m_{14}\}$ from (7). The cardinality of Q_i depends on $N_H(mk_i)$. Note that $Q_{0,1} = Q_0 \cap Q_1 = \{m_7\}$, $Q_{1,2} = Q_1 \cap Q_2 = \{m_8\}$, $Q_{2,3} = Q_2 \cap Q_3 = \{m_9\}$, $Q_{3,4} = Q_3 \cap Q_4 = \{m_{10}\}$, $Q_{4,5} = Q_4 \cap Q_5 = \{m_{11}\}$, $Q_{5,6} = Q_5 \cap Q_6 = \{m_{12}\}$, $Q_{6,7} = Q_6 \cap Q_7 = \{m_{13}\}$ and $Q_{7,8} = Q_7 \cap Q_8 = \{m_{14}\}$. Experiments show that the averaged value of $|Q_i|$ is $2^{9.1}$ in the random case.

We can reduce the cardinalities of Q_i and Q_{i+1} by keeping only those solutions that have the value patterns existing in the corresponding intersection set $Q_{i,i+1}$. To get a maximum reduction effect, let I $=\min_{0 \leq i \leq 8} |Q_i|$. Then we start the reduction process from $Q_I = \{m_j, m_k\}$ with $k = 0$ or $k = j - 1$. More precisely, if $I < 8$, we reduce $Q_{I+1} = \{m_{j+1}, m_j\}$ by keeping only those solutions that have the m_j value patterns existing in Q_I. If $I > 0$, we reduce $Q_{I-1} = \{m_k, m_{k-1}\}$ by keeping only those solutions that have the m_k value patterns existing in Q_I. This reduction process is continued to cover Q_0 and Q_8, i.e., we make the reduction step for each Q_i. After reducing each Q_i $(0 \leq i \leq 8)$, we combine Q_0, Q_1 and Q_2 together to get the possible values for the ith cells $(3 \leq i \leq 6)$ of state s_{24+6} in the real authentication, i.e., we first fill the 5th and the 6th cells of s_{30} from Q_0, then we fill the 4th cell from the 'm_8' item of Q_1 with the corresponding 'm_7' item equal to the 'm_7' item of Q_0 just filled in. The same procedure applies to the 3rd cell of s_{30}. This is something like a chain. Still, we have to xor the corresponding feedback byte with the recovered m_9 item of Q_2 to get the value of the 3rd cell of state s_{30}.

Then we can check the 3rd to 6th cells of s_{30} immediately by $mk_{20} = m_{14} \boxplus L(m_{13}) = (m_8 \boxplus L(m_7)) \boxplus L(m_7 \boxplus L(m_0))$ and $mk_{21} = m_{15} \boxplus L(m_{14}) = (m_9 \boxplus L(m_8)) \boxplus L(m_8 \boxplus L(m_7))$. Experiments show that with probability around 0.85, the number of the combined 3rd to 6th cells of state s_{30} is less than 2^{18} and in this case the averaged number is $2^{14.6}$, which is much less than 2^{28}. The reduction effect is obvious. Then, we continue the construction of the chain by filling the 0th to 2nd items of the state s_{30} from Q_3, Q_4 and Q_5 respectively. Here we can see the similarity of our method with the Viterbi decoding algorithm. After filling in, we run m from s_{30} backwards and forwards to check all the known mk_is for $0 \leq i \leq 15$. Finally, we clock back m from s_{30} to recover the s_{24} state in the real CM authentication. Let n_m be the number of candidates of the s_{24} state of m obtained in the above way, our experiments show that with probability around

0.72, n_m is less than 2^{21} and in this case, the averaged number is $2^{18.8}$, which is close to the theoretical value $2^{17} = 2^{49-32}$. Note that in [8], this number is at least $1.43 \cdot 10^9 = 2^{30.41}$. The gain of our method is obvious.

To further reduce the possible candidates, we now run the whole 117-bit state of the cipher backwards to cover the authenticators $ar_{13}, ar_{12}, ar_{11}, ar_{10}, ar_9, ar_8$. Here we have to clock the cipher back 21 steps, i.e., we check those candidates of the middle register that are with the common right-left pair, instead of checking the candidates of the right-left-middle triple. This batch treatment results in a good complexity. In fact, what we do is to regroup all the candidates of the right-left-middle triple of the s_{24} state in the real CM authentication into subgroups which has a common right-left pair and check these subgroups one-by-one. In most cases, we get 1 or 2 candidates of the middle register left corresponding to a right-left state pair, for the above reduction factor is 2^{-24}. If there are no surviving candidates of the middle register we conclude that the corresponding right-left pair is a wrong pair. After this step, there are around $n_m \cdot 2^{-24}$ candidates of m left corresponding to a common right-left state pair.

Next, for each survived candidate for m with the common right-left pair we run the cipher backwards to the s_8 state in the real CM authentication. In this process, we can use the authenticators ar_i for $0 \leq i \leq 7$ to further reduce the possible candidates of m and the right-left pairs. In general, only the correct right-left pair with the correct candidate for the middle register could pass this test, since the reduction factor is 2^{-32}.

4.3 Complexity Analysis

In summary, our attack works in three phases. First, we exhaustively search the shortest register r. In our attack, we set the threshold value T_r of the correlation test in such a way that for some frames, there are no output of the test. Thus, we need not run the following phases of our attack for each frame. Second, we use our method to get the candidates of l without trying all the possible values and for each candidate, we again call the correlation test with a threshold value T_l to further reduce the number of candidates. In this way, it is possible that the correct candidate of l is filtered out. We have to compensate this with more frames.

Definition 1. *We say a frame is a good frame, if both the correct candidate of l and that of r pass the correlation tests in our attack.*

Let $P_l = \sum_{i=T_l}^{T_l'} \binom{64}{i}(\frac{3}{4})^i(\frac{1}{4})^{64-i}$ and $P_r = \sum_{i=T_r}^{64} \binom{64}{i}(\frac{3}{4})^i(\frac{1}{4})^{64-i}$. In theory, with

$$F = \frac{1}{P_l \cdot P_r} = \frac{1}{\sum_{i=T_l}^{T_l'} \binom{64}{i}(\frac{3}{4})^i(\frac{1}{4})^{64-i} \cdot \sum_{i=T_r}^{64} \binom{64}{i}(\frac{3}{4})^i(\frac{1}{4})^{64-i}} \quad (8)$$

frames, we could encounter a good frame to mount our attack. In practice, we usually need more than F frames to get a good frame with probability around 0.5. This is mainly caused by the fact that the left and the right registers are not fully independent from each other. Third, we recover the middle register conditioned on the candidates of l and r. Since there is an unrolling check in this phase, we can determine either the correct right-left-middle state triple or there is no candidate survived, indicating that the frame is not good.

The time complexity of the first phase is $2^{25} \cdot \frac{16-2}{3} = 2^{27.8}$ cipher ticks and we have to repeat this step from each frame until we meet a good frame. Hence, in total, the time complexity in the first phase is $2^{27.8} \cdot F$ cipher ticks. For about $w = \frac{F}{2} \cdot (P_r + (2^{25} - 1) \cdot \sum_{i=T_r}^{64} \binom{64}{i}(\frac{1}{2})^{64})$ frames, we have about $n_r = 1 \cdot P_r + (2^{25} - 1) \cdot \sum_{i=T_r}^{64} \binom{64}{i}(\frac{1}{2})^{64}$ candidates passed from the first phase. In such cases, we invoke the second phase of our attack to find some candidates for the left register. The time complexity of a single run of the second phase of our attack is $C_l = n_l \cdot \frac{16-2}{3} + \frac{2^{10.96} \cdot 2^{10}}{2^{\Psi(J)}} \cdot \frac{2^{10}}{2^{N_H(lk_{-1}) + N_H(lk_5)}}$. We will get about $(n_l - 1) \cdot P_w + 1 \cdot P_c$ candidates after the second phase which is mainly determined by T_l. For each possible combination of the right-left state candidates pair, we check whether the underlying frame is a good one or not in the third phase. The time complexity of a single run of the third phase is at most $C_m = 2^{14} \cdot 9 + 2^{18} \cdot 4 \cdot \frac{16-2}{3} + n_m \cdot 21 + 2 \cdot (64 - 21)$ cipher ticks. Therefore, the time complexity of our attack so far is around

$$C_{total} = 2^{27.8} \cdot F + w \cdot (n_r \cdot C_l + n_r \cdot ((n_l - 1) \cdot P_w + 1 \cdot P_c) \cdot C_m) \qquad (9)$$

cipher ticks. If we set $T_r = 54$, $T_l = 45$ and $T_l' = 48$, we have $F = 24$ and $C_{total} \approx 2^{50}$ cipher ticks

The success probability of our attack depends on the number of captured frames. Table 5 shows the relation got from 10^6 experiments with randomly generated frames.

Table 5. The success probability of our attack if $T_r \geq 54$ and $T_l = 45$, $T_l' = 48$

F	24	30	40	45	60	90	120
P_{succ}	0.431	0.505	0.608	0.653	0.759	0.877	0.943

5 Practical Implementation

Here we describe the practical implementation of the full key recovery attack on the CryptoMemory. We have fully implemented our attack. The implementation consists of three stages:

1. finding a good frame and recovering the left-right pairs;
2. recovering the full internal state s_8;
3. recovering the full key from s_8.

The first stage is implemented on a single core (of an Intel Core 2 Duo 6600, 2.4 GHz). It takes about 10 minutes to find a possible good frame and recover the possible left-right state pairs subsequently. The second stage is the most time-consuming and is implemented on a computing cluster with 200 cores (of Intel Xeon L5640, 2.26 GHz). It takes roughly $2-6$ days to find the full internal state (this requires trying several possible good frames found in stage 1). The last stage is implemented on a single core. It takes on average 2 hours to recover the full secret key from s_8. Note that stage 2 can return several candidates for s_8; in this case we launch stage 3 in parallel on several cores. We will describe our low-memory (compared to that of [8]) key recovery algorithm in more details in the full version of the paper.

The low complexity of our attack allowed us to run it several times with different keys. The speed of our programm is about 2^7 clock cycles per inverse cipher tick. Here we present the flow of the attack for one of the runs. We obtained 30 authentication frames from the reference implementation of the Atmel cipher (which was verified against the hardware according to [8]) with the secret key `0xf7fb3e25ab1c74d8`. After this, we proceeded as if we had not known the key. We set $T_r = 54$, $T_l = 45$ and $T_l' = 48$. Then $P_r \approx 0.05$ and $P_l \approx 0.84 - 0.45 = 0.39$. Among the 30 frames we found one possible good frame

$$nr = \texttt{0xa8becfc790ce1272}, \quad nt = \texttt{0x8bd5987bdf33aec7},$$
$$ar = \texttt{0x2e0ba95f84eb0a50}, \quad at = \texttt{0xff3f26fab2fb809e},$$

for which there were around $2^{20.73}$ left-right state pairs. For each left-right state pair, $2^{27.2}$ inverse cipher ticks are done on average to reduce the number of possible candidates. We launched the second stage of our attack on 200 CPU cores. The attack succeeded during the 4th frame. Analysis of the 4th frame took about 20.4 hours to find 1 possible candidate state of s_8, while analysis of the 3 other frames took several days in total. For this s_8 state, we use our key-recovery technique to restore the key. The secret key `0xf7fb3e25ab1c74d8` was found for the state

$$s_8 = (\texttt{0x071d0308081a0e}, \texttt{0x1627033e566b74}, \texttt{0x1e1a100e1b}, \texttt{0x0109})$$

(each two hexadecimal digits in this notation represent a single register cell).

We note that due the properties of our attack (namely, frequent checking of the cipher output against the keystream while clocking back) its implementation cannot be significantly sped-up by employing a bit-sliced implementation of the cipher, as it was the case in [8] according to [9]. During the experiments, we found an inherent property of CM, i.e., the number of non-coincidence bits between the two intermediate outputs generated by one possible left-right state pair is a fixed constant, if the sum of the numbers of coincidence bits between each one of the intermediate output and the 64-bit keystream is a constant. It is checked 10^6 times, the experiments show it holds all the time. This property indicates that we cannot further reduce the time complexity of our attack by setting a larger T_l. Since in such cases, the entropy of the middle register also increases. This property also explains why we set $T_l = 45$ and $T_l' = 48$, for we have to discard the pairs resulting in high entropy middle register.

6 Conclusions

In this paper, we have shown practical key recovery attacks on both versions of the Atmel cipher. By using the optimal Viterbi-like decoding techniques to recover the internal states of the left and middle registers and exploiting the different diffusion speeds of the different cells of the underlying registers, our attacks significantly improved the best previously known results [8]. Our analysis shows that even the strongest version of the Atmel cipher succumbs to practical attacks using relatively few captured authentication frames. Our practical implementation recovers the full 64-bit secret key from 30 captured authentication frames in about 2 – 6 days using 200 CPU cores. Table 6 shows the comparison of the attack on the Atmel CryptoMemory cipher presented in this paper and that on another proprietary cipher KeeLoq in [1]. One can again conclude that such proprietary ciphers fail to provide enough security even from a practical point of view.

Table 6. Comparison of our attack on the Atmel cipher in CryptoMemory and the attack [1] on KeeLoq

	key length, bits	data complexity	time complexity
KeeLoq	64	2^{16} known plaintexts	$2^{44.5}$
CryptoMemory	64	30 known frames	2^{50}

Acknowledgements. We would like to thank Flavio D. Garcia of Radboud University Nijmegen, the Netherlands for useful discussions.

References

1. Aerts, W., Biham, E., de Moitie, D., de Mulder, E., Dunkelman, O., Indesteege, S., Keller, N., Preneel, B., Vandenbosch, G., Verbauwhede, I.: A practical attack on KeeLoq. Journal of Cryptology (to appear)
2. Atmel. CryptoMemory specification, 5211A-SMIC-04/07 (2007)
3. Benhammou, J.P., Colnot, V.C., Moore, D.J.: Secure memory device for smart cards, US Patent 7395435 B2 (July 2008)
4. Benhammou, J.P., Jarboe, M.: Security at an affordable price. Atmel Applications Journal 3, 29–30 (2004)
5. Dinur, I., Shamir, A.: Cube Attacks on Tweakable Black Box Polynomials. In: Joux, A. (ed.) EUROCRYPT 2009. LNCS, vol. 5479, pp. 278–299. Springer, Heidelberg (2009)
6. Dinur, I., Shamir, A.: Breaking Grain-128 with Dynamic Cube Attacks. In: Fast Software Encryption-FSE 2011. Springer, Heidelberg (2011) (to appear)
7. Dipert, B.: The Zune HD: more than an iPod touch wanna-be? EDN, p. 20 (October 2009)
8. Garcia, F.D., van Rossum, P., Verdult, R., Schreur, R.W.: Dismantling SecureMemory, CryptoMemory and CryptoRF. In: 17th ACM Conference on Computer and Communications Security-CCS 2010, pp. 250–259. ACM Press, New York (2010), `http://eprint.iacr.org/2010/169`
9. Garcia, F.D.: Private communication
10. Jarboe, M.: Introduction to CryptoMemory. Atmel Applications Journal 3, 28 (2004)

11. Meier, W., Staffelbach, O.: Fast correlation attacks on certain stream ciphers. Journal of Cryptology, 159–176 (1989)
12. http://download.nvidia.com/downloads/pvzone/Checklist_for_Building_a_HDPC.pdf
13. http://www.rockbox.org/wiki/SansaConnect
14. Viterbi, A.J.: Error bounds for convolutional codes and an asymptotically optimum decoding algorithm. IEEE Transactions on Information Theory 13(2), 260–269 (1967)
15. Amazon Elastic Compute Cloud (Amazon EC2), http://aws.amazon.com/ec2/#pricing (accessed January 22, 2010)

Cache Timing Analysis of RC4

Thomas Chardin[1], Pierre-Alain Fouque[2], and Delphine Leresteux[3]

[1] DGA Engineering and Integration, 7 rue des Mathurins, 92221 Bagneux Cedex
[2] Département d'informatique, École normale supérieure, 45 rue d'Ulm, F-75230 Paris Cedex 05
[3] DGA Information Superiority, BP7, 35998 Rennes Armées
thomas.chardin@dga.defense.gouv.fr,
pierre-alain.fouque@ens.fr,
delphine.leresteux@dga.defense.gouv.fr

Abstract. In this paper we present an attack that recovers the whole internal state of RC4 using a cache timing attack model first introduced in the cache timing attack of Osvik, Shamir and Tromer against some highly efficient AES implementations. In this model, the adversary can obtain some information related to the elements of a secret state used during the encryption process. Zenner formalized this model for LFSR-based stream ciphers.

In this theoretical model inspired from practical attacks, we propose a new state recovery analysis on RC4 using a belief propagation algorithm. The algorithm works well and its soundness is proved for known or unknown plaintext and only requires that the attacker queries the RC4 encryption process byte by byte for a practical attack. Depending on the processor, our simulations show that we need between 300 to 1,300 keystream bytes and a computation time of less than a minute.

Keywords: cryptanalysis, stream cipher, RC4, cache timing analysis.

1 Introduction

Some side channel attacks have been recently formalized in theoretic work by modelling powerful adversaries that can learn a bounded amount of arbitrary information on the internal state by Dziembowski and Pietrzak in [9]. Here we consider information coming from cache attacks which is of the same kind but more practical since they correspond to real attacks which have been experimented on AES implementation [18,7,4,22,3]. Concretely, when the cipher is looking for a value in a table, a whole line of cache is filled in, containing but not limited to the value looked for in the table. This mechanism allows to achieve better performance since in general when a program needs some data, it also requests the successive ones soon after. Osvik, Shamir and Tromer proposed in 2006 an attack on some AES implementations that use look-up tables to implement the S-box and showed that the adversary can learn the high order bits of the index looked for, but neither the whole index itself nor the corresponding

J. Lopez and G. Tsudik (Eds.): ACNS 2011, LNCS 6715, pp. 110–129, 2011.

value of the table. These attacks are rather practical since they have been implemented [18,7] on classical implementations used in the OpenSSL library. Others cache attacks target DSA [1] or ECDSA [8] operations in the OpenSSL library due to branch prediction on instructions.

To gain more information from cache monitoring, Osvik *et al.* propose to run a concurrent process at the same time as the encryption process. Attackers can evict data from the cache using the second process which begins by reading a large table to flush the cache. Then, the encryption process is run; the attacker finally tries to read again the elements of his table. If the element is in the cache, the access is fast (cache hit) and in the other case, the access is slow (cache miss) since the information has been evicted from the cache. Consequently, the adversary is not allowed to read the cache, but since the cache lines correspond to lines in the memory, if the adversary knows how the encryption process organizes the data in the memory (the address of the whole table for instance), the information of which cache line has been removed from the cache allows to recover the index (or a part of it) of the value looked for by the encryption process. Indeed, we do not recover the whole index since the cache is filled in line by line, so we know that the encryption process has read some element of the whole line but not exactly which element. Moreover, if the encryption process performs many table lookups, we do not have the order of the indexes since we perform timing on our own process which is run after the encryption process. These practical analyses allow us to consider such attacks on encryption schemes through a new security model. For example, Zenner *et al.* propose to study security of LFSR-based stream ciphers in [23,15].

RC4 is a stream cipher designed in 1987 by Ron Rivest and widely used in many standards such as in SSL, WEP, WPA TKIP, etc. The internal state of RC4 is composed of two indexes and of a permutation over $\mathbf{F}_{256}$. The initialization of the permutation table depends on the secret key (which size varies between 0 and 256 bits); the table is then updated during the generation of the keystream. Many attacks have been proposed on RC4 since its design was published in 1994 but none of them really breaks RC4. The bad initialization used in the WEP protocol and the key schedule algorithm of RC4 have been attacked by Fluhrer, Mantin and Shamir in [10]. Recent improvement has revealed new linear correlations in RC4 in order to mount key retrieve attacks on WEP and WPA [21]. Since then, from a cryptographic point of view, this scheme has not been broken despite many statistical properties. Finally, more powerful attacks have been taken into account, for instance fault attacks by Hoch and Shamir, Biham *et al.* in [12,6]. However, the number of faults is rather high, 2^{16} for the most efficient attack.

Previous Work. Our analysis is related to the one published in 1998 by Knudsen *et al.* in [14], which try to recover the internal state from the keystream. Once the internal state is recovered, it is possible to run the algorithm backward and efficient algorithms allow to recover the key [5]. Though an improvement was proposed in 2000 [11] and another in 2008 [16], such attacks remain impractical, having a time complexity of 2^{241} operations for the full RC4 version. The basic idea of the "deterministic" attacks (section 4 of [14] and [16]) is to guess some values of the table and then check if these guesses are valid with the output keystream. These algorithms perform a clever search by guessing bytes when they need them and then use a backtracking approach when a contradiction appears. However, a huge number of values have to be guessed so that the complexity is relatively high in the end. This is basically the algorithm of [14]. Maximov and Khovratovich in [16] improve this algorithm by looking at the equations of RC4:

$$\begin{aligned} i_t &= i_{t-1} + 1 \\ j_t &= j_{t-1} + S_{t-1}[i_t] \\ S_t[i_t] &= S_{t-1}[j_t], \quad S_t[j_t] = S_{t-1}[i_t] \\ Z_t &= S_t[k] \quad \text{where} \quad k = S_t[i_t] + S_t[j_t] \end{aligned}$$

In the algorithm of [14], the number of unknowns is 4 (j, $S[i]$, $S[j]$ and $k = S[i] + S[j]$) even if they are related). Maximov and Khovratovich solve these equations by noting that if j is known for different times t, then $S[i]$ also, and the number of unknowns is reduced to 2. Then, they show that it is possible to have the value of j for consecutive times t, and also to detect such patterns from the keystream. The attack begins by locating in the keystream a good pattern which gives information about the internal state and j, and then since the equations are simpler the complexity is lower. Solving such linear systems with *non-linear terms* has also been recently extended by Khovratovich *et al.* to more complex equations system in [13] in the context of differential trail for hash functions.

Finally, Knudsen *et al.* propose a "probabilistic" algorithm in section 5 of [14], which is different from the deterministic one since the idea is that the output keystream gives conditions on the internal secret state which leads to conditional probability distribution $\Pr(S[i] = v | Z_t = z)$. Now, the internal state is represented with a probabilistic distribution table: to each element $S[i]$ in the table is associated a probability distribution on the 256 possible values. At the beginning, for all, i and v, $\Pr(S[i] = v) = 1/256$. Then according to the output keystream byte Z_t, an a posteriori distribution is computed using Bayes rules and the previous values in the distribution table, and finally the algorithm accordingly updates the distribution table. This probabilistic algorithm does not work if no more information is used. Knudsen *et al.* *partially* fulfill the table at the beginning with correct values. Their experiments show that they need 170 values so that the algorithm converges. They use the same idea (used later by Maximov and Khovratovich): they fulfill the table such that consecutive values of j can be found which makes the equations easier.

The algorithm we propose here is different from the one described in [14]; however they have in common the manner of using the structure of PRGA, acronym of Pseudo Random Generation Algorithm, to propagate constraints on the values of elements of the secret state used by RC4.

Our Results. RC4 is a good candidate to study cache timing analysis since it uses a rather large table and indexes of the lookups give information about the table. In this paper, we present a probabilistic algorithm that recovers the current state of the permutation table. Contrary to previous state recovery attacks [14,16] whose complexity is about 2^{241} only based on cryptanalysis, our algorithm is more efficient in practice and its soundness has been proved due to information from cache timing analysis, although same cryptanalysis techniques are used. The experiments are derived using our algorithm and by simulating the information obtained in the model designed by Osvik *et al.* in [18]. The idea is the following: to generate a cipher byte, the generation algorithm uses three lookups in the permutation table and then updates the corresponding elements. Thus, the knowledge of the elements used to generate this byte allows the attacker to eliminate some candidate values for these elements, leading little by little to the recovery of the entire permutation.

The key recovery algorithm we developed is probabilistic as the one of Knudsen *et al.* (section 5 of [14]). However, we use a belief propagation algorithm to use the cache attack information. Belief propagation algorithms have been used in information theory and coding theory and are related to Bayesian networks and Hidden Markov Models, when a state is hidden and has to be recovered given some information. Recently, such algorithms have been successfully used to prove convergence and complexity results on the random assignment problem [20]. The algorithm propagates the partial information on the indexes by modifying the distribution table. According to the distribution table $\Pr(S[i] = v)$ and the partial indexes, the algorithm computes for all possible guesses, the values of the probability of such guesses. Since one guess is the correct one, we then normalize all these probabilities, and we update the distribution table according guesses modify or not the value of $\Pr(S[i] = v)$.

The algorithm gives good results but we improve the data complexity and the success probability by using time to time an assignment algorithm, such as the Hungarian algorithm, because we know that the table S is a permutation. Without cache analysis information, the algorithm cannot be effective and the complexity is too large as the probabilistic algorithm of Knudsen *et al.*. In this paper, we show that in practice, we can recover the secret permutation using only 300 bytes of the keystream in the best case. In the case of ciphertext only attack, our analysis works in practice only when the cache lines are half of the real size. Using partially known plaintext, we can recover the internal table using 3,000 bytes with probability 95%. This attack shows that RC4 must be used with some care when attackers can monitor the cache and query the cipher stream byte by byte.

Organization of the paper. The paper is organized as follows: after having presented the information expected to get from cache monitoring in section 2, we describe in section 3 a first algorithm, that only works on idealized cases. Then our main algorithm is detailed in section 4, which is able to take information from real cache monitoring to recover the permutation table through an example based on OpenSSL implementation. Finally we conclude and point out some practical thoughts on data collection in section 5.

2 The Context of the Attack

Our algorithm is designed to attack RC4 from a side-channel, the cache memory. We will briefly describe RC4 and the structure and the mechanism of cache memory. Then we explain how this cache memory can be considered as a side-channel and how to exploit the resulting data leakage.

2.1 Description of RC4

The design of RC4 has been kept secret until it was leaked anonymously in 1994 on the Cypherpunks mailing list [2]. It consists of two algorithms. The first one, named KSA (key-scheduling algorithm), is used to initialize the permutation table according to the secret key. The other PRGA is used to generate a keystream byte and update the permutation. While the details of KSA are not relevant to our subject, PRGA is described in algorithm 1.

Algorithm 1. Pseudo-random generation algorithm

$i \leftarrow i + 1$
$j \leftarrow j + S[i]$
swap($S[i]$, $S[j]$)
$k \leftarrow S[i] + S[j]$
return $S[k]$

2.2 Structure and Use of Cache Memory

To make up for the increasing gap between the latency of microprocessors and memories, some little but low-latency memory modules have been included in modern microprocessors. The aim of these cache memories is to preload frequently accessed data, reducing the latency of load/store instructions (an average code uses one such instruction out of three). There are often two caches, named L1 and L2. The L1 cache is closer to the CPU but smaller than the L2 cache. These memories are organized in Z sets. Each set contains W cache lines of B bytes; a memory block which address is a in memory is stored in cache at the cache set $a/B \bmod Z$, starting at the byte $a \bmod B$ of a random cache line. The main placement policy is to evict the most ancient stored data to store a new one. The values of these parameters for Pentium 4 processors can be found in table 1.

The use of cache is as follows:

- when the processor needs data from memory, it sends a request to L1 cache;
- If the L1 cache contains the required data (cache hit), they are sent to the CPU; otherwise (cache miss), the request is transmitted to the L2 cache;

Table 1. Cache parameters for the Pentium 4

	B	Z	W
L1	64	32	8
L2	64	4096	8

- If the L2 cache contains the data, they are transmitted to L1 cache (for further use) and CPU; if not, the request is transmitted to the main memory, and the data are copied in both caches.

This behavior allows us to make out four different time scales (the figures are given for a Pentium 4 CPU):

- the average execution time of an instruction: 1 CPU cycle (1 nanosecond on a 1 GHz CPU);
- the latency of the L1 cache: 3 cycles;
- the latency of the L2 cache (and time to write the data in the L1 cache): 18 cycles;
- the latency of the RAM: approximately 50 nanoseconds.

Because of this mechanism, the execution time of a process may vary significantly according to the relative number of cache hits and misses during its execution. Such variations can be used as a side-channel, either for covert communication or for cryptanalytic purposes.

2.3 Cache Timing Analysis

The first side-attack using cache memory was described by Page in 2002 [19]. It uses the fact that, in DES, if two rounds use the same element in some S-box, then the global encryption time will be reduced (the second lookup resulting in a cache hit). In 2006, Bonneau and Mironov [7] adapted this attack to AES, allowing to recover a secret key in an average of 10 minutes and needing approximately 2^{20} encryptions.

An even more powerful attack was published in 2006 by Osvik, Shamir and Tromer [18]. The use of a test table filling the cache memory allowed them to know which cache lines were used by the encryption process (causing the eviction from cache of the corresponding lines of the test table, hence a greater latency to access again to these elements). The knowledge of the position of the encryption tables in memory allowed them to know which elements of these tables were used, helping them to recover a secret key with less than 16,000 encryptions (depending on the processor on which the attack was implemented).

2.4 Prerequisites

The context of the analyses presents here is the same as the one of [18]. We assume that the attacker is able to monitor the cache memory and learns partial information on which element of the permutation table is used. Our algorithm is particularly fitted for "synchronous attacks", where the attacker can trigger encryption himself. Osvik *et al.* [18] first present the concept of synchronous cache attacks and Zenner *et al.* [23,15] establish a cache model based on this concept. They define an adversary having access to two oracles. The first oracle allows to obtain a list of output keystream bytes. At the same time, the second oracle delivers a truthful list of all cache requests realized by the first oracle. This cache attack in Zenner's model has several properties like noise-free and there is no order on the cache accesses. However, if the attacker can produce the same cipher operation with the same input and output many times, the noise and wrong cache accesses could be removed using many samples. Finally, one must note that monitoring the cache does not allow us to know exactly which element of the table is used. In fact, when a request causes a cache miss, the whole cache line corresponding to the requested data is loaded into the cache. As the elements of the RC4 permutation table are stored as 32-bit integers (in 32-bit CPUs), a cache line contains (in the case of a Pentium 4) 16 elements. Thus, using a test table to monitor the cache only allows us to know the index modulo 16 of the element used.

2.5 Conventions and Experimental Set-Up

We study the current version of RC4 with bytes even though some authors have tried to attack weaker versions with smaller permutation table. Consequently, unless explicitly stated, all additions and subtractions will be done modulo 256.

We denote by δ the number of integers per cache line, and S the permutation table used for encryption.

We assume that the attacker can trigger the generation of the stream cipher byte by byte; the data are collected as in [18].

We have simulated cache accesses on OpenSSL library and we have experimented several cache models, from idealized one to real one.

Finally, all the experimental results are presented as follows: "time" represents the time needed by the attack to succeed, "requests" the number of needed cipher byte requests, and "success" the number of times when the table given by the attack is equal to the permutation of RC4. All results are given on average, over 50 random secret keys.

3 A First Algorithm for Idealized Cases

In this section we assume for the sake of simplicity that monitoring the cache allows us to know exactly which element of the permutation has been used by the encryption process. It allows us to introduce our attack on a simple case which is presented in appendix A. We will see further how to adapt the attack to a more complex cache model.

Data Structures Used by the Algorithm. For each table element i we have to keep a trace of the remaining candidates for its value. To do so we use a 256x256-boolean table S_{values}, where $S_{values}[i][v] = 0$ if and only if we are sure that $S[i] \neq v$. Another 256-boolean table is used for j, with the same conventions. We do not need to keep an equivalent structure for i, as its value is known through the whole encryption process.

Exploitation of the Data Collected During the Generation of One Cipher Byte. The generation of one cipher byte gives us the following data:

- the first index i;
- the unknown internal index j;

- three indexes a, b and c (possibly equal) of elements used during the byte generation;
- the cipher byte, which we will call *out*.

We invite the reader to note that we do not know the order of use of the three elements of the table, except for the first (because we know i). We will suppose from now on that $a = i$. Two orders remain: (a, b, c) and (a, c, b). We then use the structure of PRGA, which imposes the following constraints in the case (a, b, c):

$$\begin{aligned} j + S[a] &= b \\ S[a] + S[b] &= c \\ S[c] &= out \end{aligned}$$

Algorithm 2. Algorithm for one step with order (a, b, c) when $a \neq b \neq c \neq a$

1. **for** $v \leftarrow 0$ **to** $2^n - 1$ **do**
2.   **for** $t \leftarrow 0$ **to** $2^n - 1$ **do**
3.   **if** $v \neq a$ **and** $v \neq b$ **and** $v \neq c$ **then**
4. $S_{values\_bis}[v][t] \leftarrow S_{values}[v][t]$
5.   **else**
6. $S_{values\_bis}[v][t] \leftarrow 0$
7.   **end if**
8.   **end for**
9. $j_{values\_bis}[v] \leftarrow 0$
10. **end for**
    {Exploit cache data, order (a, b, c)}
11. **for** $j \leftarrow 0$ **to** $2^n - 1$ **do**
12.   **if** $j_{values}[j] = 1$ **and** $S_{values}[a][b - j] = 1$ **and** $S_{values}[b][c - b + j] = 1$ **and** $S_{values}[c][out] = 1$ **then**
13. $S_{values\_bis}[a][c - b + j] \leftarrow 1$
14. $S_{values\_bis}[b][b - j] \leftarrow 1$
15. $S_{values\_bis}[c][out] \leftarrow 1$
16. $j_{values\_bis}[b] \leftarrow 1$
17.   **end if**
18. **end for**
    {Repeat steps 11. to 18. for order (a, c, b)}
    {Write new tables}
19. **for** $v \leftarrow 0$ **to** $2^n - 1$ **do**
20.   **for** $t \leftarrow 0$ **to** $2^n - 1$ **do**
21. $S_{values}[v][t] \leftarrow S_{values\_bis}[v][t]$
22.   **end for**
23. $j_{values}[v] \leftarrow j_{values\_bis}[v]$
24. **end for**

We deduce from these equations that the values of $S[a]$, $S[b]$ and $S[c]$ are completely determined given the order of the lookups and the values of j and *out*. We then do the following steps. For each order and each value v of j, we check if the corresponding values of $S[a]$, $S[b]$ and $S[c]$ remain possible. If not, we are sure that these order and values are not possible, which allows us to remove these candidates for $S[a]$, $S[b]$ and $S[c]$. We reduce the number of candidates. Finally, we update the S_{values} and j_{values} tables to take the swap of PRGA into account. Algorithm 2. gives the details of one step of the attack; a special care has to be taken when a, b or c are equal, because the swap operation is performed before reading the last element $S[c]$. A special care has to be taken when a, b or c are equal, because the swap operation is performed before reading the last element $S[c]$.

We continue to use such data until every table element has only one possible candidate value. Other candidates are eliminating due to contradiction. We have then found the current permutation table and value of j.

4 Adaptation to Real Caches

There exist a main difference between the idealized framework used above and the monitoring cache memory. Even if we did not know the order of use of the elements, we knew exactly which elements of the table were used. In particular, we were sure not to modify during an attack step the candidate values of an unused table element. When we monitor real cache memory, we only know some most significant bits of the index of used table elements. Therefore we cannot be sure that the candidate values, on which we are trying to study, really correspond to an effectively used table element.

To take this issue into account, we use the probabilistic frame presented in [14]. Instead of the boolean table S_{values}, we now use a 256x256-floating point number table S_{probas}. We now noted $\mathbf{P}(S[i] = v)$ where $S_{probas}[i][v]$ is the estimated probability $S[i] = v$. At the beginning of the attack, we suppose that every value is equiprobable for any element of S and for j so these two tables are filled with the value $1/256$.

4.1 The Known-Plaintext Attack

We assume that the attacker knows, for each step:

- the first index i;
- three indexes $a \wedge$ `mask`, $b \wedge$ `mask` and $c \wedge$ `mask`, where `mask` corresponds to the known bits and is computed from the value of δ: if $\delta = 16$, then `mask` $=$ `0xf0`;
- the cipher byte out.

A Probabilistic Version of the Algorithm. The belief propagation algorithm we design is exactly the same as the one for the idealized case, with the exception that we do not know b and c for sure. As in the previous section, we will assume that $a = i$ and that $a_\delta = a$ `&& mask`, $b_\delta = b$ `&& mask` and $c_\delta = c$ `&& mask` are different (the case with some equalities is treated in a similar way but taking care of the possible collisions in the permutation table).

We know that the indexes b and c used for the PRGA step are of the form $b = b_\delta + o_1$ and $c = c_\delta + o_2$ where $0 \leq o_1, o_2 < \delta$, so the equations giving $S[a]$,

$S[b]$ and $S[c]$ are the same as in the previous section, except that we have two more unknowns, o_1 and o_2.

Our algorithm is hence modified as following:

1. we make two guesses for the values of o_1 and o_2 as well as the guess for the value v of j and for the order of use of a, b and c;
2. we compute the corresponding values of $S[a]$, $S[b]$ and $S[c]$ similarly to what was done in the idealized case;
3. we evaluate the probability of this guess to be exact, assuming for the sake of simplicity that all probabilities are independent:

$$\begin{aligned}\mathbf{P}(guess) = \mathbf{P}(j = v) \cdot \mathbf{P}(S[a] = b_\delta + o_1 - v) \\ \cdot \mathbf{P}(S[b_\delta + o_1] = c_\delta + o_2 - b_\delta - o_1 + v) \\ \cdot \mathbf{P}(S[c_\delta + o_2] = out)\end{aligned}$$

4. having done these computations for all possible guesses, we normalize the corresponding probabilities (these guesses were the only ones having a chance to describe what really happened during the keystream byte generation);
5. finally each guess contributes to modify the tables S_{probas} and j_{probas}, since for each guess we know exactly the action of the PRGA step on the three elements looked up, the others remaining unmodified. Adding all these contributions we obtain the global transformation of the table S_{probas}:

$$\forall i, \forall v, \mathbf{P}(S[i] = v)_{\text{after attack step}} = \left(\sum_{\text{guesses imposing } S[i]=v} \mathbf{P}(\text{guess}) \right) \cdot 1$$

$$+ \left(1 - \sum_{\text{guesses imposing } S[i]} \mathbf{P}(\text{guess}) \right) \cdot \mathbf{P}(S[i] = v)_{\text{before attack step}}$$

(the first term of the sum gives the action of all transformations where the equation $S[i] = v$ is guaranteed by the action of the swap; the second step gives the action of all transformations where $S[i]$ is not affected, i.e. $S[i] \neq v$). The update of j_{probas} is much simpler since for each guess the new value of j is imposed by the collected data; the formula is similar to the one for S_{probas} without the second term on the right.

How to Know if the Attack Succeeds? Two parameters remain to be set: the time when we decide that the attack has succeeded (or failed), and the processing of the solution. We first use a simple criterion to stop the attack: we consider that it succeeds when for each table element a candidate value has a greater probability than 1/2. The solution is thus "built" by local optimization, retaining for each element the value with greatest probability. We consider that the attack failed if this event does not occur after a certain number of steps.

Once the solution is found, we must first reverse it back to the initial permutation, then test it. The reversion is easy, given that a step of PRGA is invertible if the values of i and *out* are known. To test it, several options can be chosen:

- a first and very simple test is to make sure that the solution given by local search on each element is a permutation. If this is the case, we have a good hint that this solution will be the good one: using the Stirling formula, we can evaluate the probability of a random application on $\mathbf{F}_{256}$ to be a permutation to approximately $2 \cdot 10^{-110}$;
- we can then verify that the solution is the good permutation trying to predict some following cipher bytes.

First Experimental Results. We use to implement our attack a simulation of the cache monitoring to collect the data necessary to the attack. The experimental results obtained are reproduced in Table 2 - the case where $\delta = 1$ being given only for comparison purposes, as the concerned results are similar to those obtained in idealized cases.

Table 2. Known-plaintext cache attack of RC4, first version

δ	1	2	4	8	16
Time	0.393 s	0.542 s	0.962 s	2.836 s	25.48 s
Requests (maximum)	417	498	498	594	898
Requests (average)	326	384	400	456	666
Requests (minimum)	268	310	344	387	556
Success	100%	78%	66%	44%	56%

We can bring out from these results that our intuition on the convergence of the probabilities is correct. However, the results for $\delta = 16$ make this attack nearly impractical for real CPUs.

4.2 An Improvement: Searching Permutations

The main issue concerning the previous test of success is that we never use the fact that what we search is not any application on $\mathbf{F}_{256}$ but a permutation. We take this crucial constraint into account reducing the search of a solution to a constraint-solving problem: once we have the densities of probability for each table element, searching the best permutation (which probability is given by the product of the probabilities of the values of all its elements) is equivalent to solve an assignment problem.

An assignment problem takes the following canonical formulation: given an integer n, a set $\mathcal{V}$ of size n (historically representing n workers), another set $\mathcal{D}$ of size n (representing n tasks) and an application $c : \mathcal{V} \times \mathcal{D} \to \mathbf{R}$ (representing the cost of assigning a worker to a task), find a bijection $f : \mathcal{V} \to \mathcal{D}$ that minimizes the global cost given by:

$$\sum_{v \in \mathcal{V}} c(a, f(a))$$

Our problem is easily reducible to an assignment problem, using for $0 \leq i, v < 256$ the cost function $c(i, j) = -\log(\mathbf{P}(S[i] = j))$.

Table 3. Known-plaintext cache attack of RC4 with global search of the solution

δ	1	2	4	8	16
Time	0.704 s	1.320 s	2.587 s	5.750 s	33.53 s
Requests (maximum)	300	400	400	500	700
Requests (average)	297	318	377	413	597
Requests (minimum)	268	300	300	368	533
Success	100%	98%	100%	100%	98%

The assignment problem can easily be written as a linear programming problem, but we prefer to solve it using the Hungarian algorithm, designed by Kuhn and Munkres in 1955–1957 [17] and which time complexity is cubic. We decide to stop the attack when the probability of the best found permutation is greater than 2^{-16}, value experimentally optimized according to a compromise between the economy of encryption requests and the improvement of the success rate. As the cost for the resolution of this problem is far greater than the cost of the previous local search, this criterion is used for one step amongst 100. Furthermore, it is not used at the beginning of the attack (for the 300 first steps). The previous criterion is used again in the latter cases.

The attack is carried out in the same conditions as before. The results are shown in Table 3.

As guessed, the global search allows to reduce significantly the number of needed encryption requests and greatly improves the success rate. We have no decisive argument so far to prove the termination or estimate the convergence of this algorithm, except for hints from information theory; however, its proofs of soundness can be found in appendix B.

4.3 The Unknown-Plaintext Attack

In this paragraph the considered hypothesis does not use the properties of probability distribution for each language characters and their probability to appear. We suppose that the attacker does not know the language or the byte code employed. The known-plaintext attack is easily adapted into an unknown plaintext one, simply by considering that all values are uniformly possible for the cipher byte (this is not exactly true, as most of the existing attacks against RC4 lie on the non-uniformity of the distribution of the first cipher bytes; however, this remains a good approximation). As a matter of consequence:

- the probability for each order, j value and offsets is computed as above, simply summing on all possible values for the output;

$$\begin{aligned}\mathbf{P}(guess) = \sum_{out=0}^{255} \mathbf{P}(j = v) \cdot \mathbf{P}(S[a] = b_\delta + o_1 - v) \\ \cdot \mathbf{P}(S[b_\delta + o_1] = c_\delta + o_2 - b_\delta - o_1 + v) \\ \cdot \mathbf{P}(S[c_\delta + o_2] = out)\end{aligned}$$

$$\sum_{out=0}^{255} \mathbf{P}(S[c_\delta + o_2] = out) = 1$$

$$\mathbf{P}(guess) = \mathbf{P}(j = v) \cdot \mathbf{P}(S[a] = b_\delta + o_1 - v) \\ \cdot \mathbf{P}(S[b_\delta + o_1] = c_\delta + o_2 - b_\delta - o_1 + v)$$

- the probabilities of the table element corresponding to the output (that is to say, the third one in the chosen order) are not updated anymore, since we know nothing on its value and the PRGA does not modify it.

Results. We test the resulting attack in the same conditions as before, except the number of steps without global search of the solution, which we adapt using the minimal number of needed requests; the results are shown in Table 4.

Table 4. Unknown plaintext cache attack of RC4 (with global search of the solution)

δ	1	2	4	8
Time	1.512 s	6.395 s	3.409 s	12.06 s
Requests (maximum)	350	600	900	1,287
Requests (average)	318	490	765	1,099
Requests (minimum)	300	429	700	950
Success	100%	96%	92%	96%

For $\delta = 16$, we do not succeed to make the probabilities converge.

4.4 A Partially-Known-Plaintext Attack

We finally develop another kind of the attack adapted to the main use of RC4: communications. In this case, the attacker partially knows the stream cipher, which corresponds for example to the case of TCP packets of which one attacker can guess some header bytes. If the output is known, we obtain the $\mathbf{P}(guess)$ of the known-plaintext attacks. On the contrary, if the output is unknown, the $\mathbf{P}(guess)$ formula of the unknown-plaintext attacks is used. The results are shown in Figure 1, for $\delta = 16$; the average time does not vary much, and the success rate is better than 47 over 50 trials.

5 Remarks on Collecting Data

In order to mount a practical attack on existing implementations from the above analysis, we make some comments on the data collection:

- in some implementations of RC4, as in OpenSSL, the internal state is stored as fields of the secret key; in particular, the two integers containing the value of i and j are stored just before the permutation table, which may have two effects:

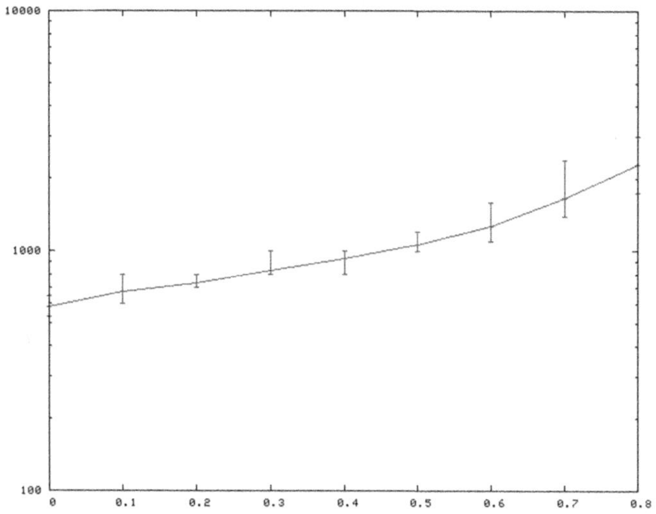

Fig. 1. Partially-known-plaintext cache attack on RC4: number of requests (logarithmic scale) as a function of the rate of unknown plaintext

  - first, there is high probability that the beginning of the permutation table does not correspond to the beginning of a cache line. On the one hand, the attacker is not supposed to know where the table begins, so he must use the attack algorithm for each possible offset between the first element of the table and the beginning of the nearest cache line. On the other hand, this offset causes the first and last cache lines covered by the table to contain less than δ elements, which gives the attacker a little more information,
  - besides, the PRGA reads the value of both index registers to generate each cipher byte, so the attacker cannot distinguish if the elements stored in the first covered cache line have been really used, which leads to a loss of information;
- last but not least, the attacker must be able to use the PRGA byte by byte; if he cannot, all the given information he will get will be a set of used cache line numbers through a large amount, say k, of cipher byte generation. He will then have to test all ordered sets of $3k$ cache line numbers using all the elements of the read set (but knowing with certainty the element out of three corresponding to the current value of i), which number is at least $(2k)!$, instead of k times the two orders used above.

6 Conclusion

We have detailed an efficient way of recovering the internal state of a RC4 process from cache monitoring. The presented algorithm is efficient in both known- and unknown-plaintext contexts. It allows to recover the internal permutation using the generation of on average about 550 keystream bytes and less than a minute of computation time in an OpenSSL implementation. The only prerequisite is the possibility for the attacker to run a process on the attacked machine and to trigger the keystream generation by itself.

To avoid cache attacks, many countermeasures have been proposed in [18]. The main thing consists of removing or masking data links to memory access. Cache could be replaced by registers or several copies of the lookup table could be used. These countermeasures cost more in time and resources required. Shuffling memory and adding random allow to avoid cache timing analysis and execution branch analysis too.

The recovering algorithm is based on a belief propagation mechanism to infer information on a hidden state which evolves in a deterministic manner and output some values of this state. Our algorithm is rather simply but very efficient, and its soundness has been proved. However, we are not able to prove its termination and complexity, which we leave as an open problem.

This article draws attention to RC4 implementation has to be carefully used to avoid cache attacks. Indeed, it would prevent from monitoring the cache and querying ciphertexts, from accessing cache from remote computer.

References

1. Aciiçmez, O., Brumley, B.B., Grabher, P.: New results on instruction cache attacks. In: Mangard, S., Standaert, F.-X. (eds.) CHES 2010. LNCS, vol. 6225, pp. 110–124. Springer, Heidelberg (2010)
2. Anonymous: RC4 source code. Cypherpunks mailing list (September 1994), `http://cypherpunks.venona.com/date/1994/09/msg00304.html`
3. Bernstein, D.J.: Cache-timing attacks on AES. Technical report (2005)
4. Bertoni, G., Zaccaria, V., Breveglieri, L., Monchiero, M., Palermo, G.: Aes power attack based on induced cache miss and countermeasure. In: ITCC, vol. (1), pp. 586–591. IEEE Computer Society, Los Alamitos (2005)
5. Biham, E., Carmeli, Y.: Efficient reconstruction of rc4 keys from internal states. In: Nyberg, K. (ed.) FSE 2008. LNCS, vol. 5086, pp. 270–288. Springer, Heidelberg (2008)
6. Biham, E., Granboulan, L., Nguyen, P.Q.: Impossible Fault Analysis of RC4 and Differential Fault Analysis of RC4. In: Gilbert, H., Handschuh, H. (eds.) FSE 2005. LNCS, vol. 3557, pp. 359–367. Springer, Heidelberg (2005)
7. Bonneau, J., Mironov, I.: Cache-Collision Timing Attacks against AES. In: Goubin, L., Matsui, M. (eds.) CHES 2006. LNCS, vol. 4249, pp. 201–215. Springer, Heidelberg (2006), `http://www.springerlink.com/content/v34t50772r87g851/fulltext.pdf`
8. Brumley, B.B., Hakala, R.M.: Cache-Timing Template Attacks. In: Matsui, M. (ed.) ASIACRYPT 2009. LNCS, vol. 5912, pp. 667–684. Springer, Heidelberg (2009)
9. Dziembowski, S., Pietrzak, K.: Leakage-resilient cryptography. In: FOCS, pp. 293–302. IEEE Computer Society, Los Alamitos (2008)
10. Fluhrer, S.R., Mantin, I., Shamir, A.: Weaknesses in the Key Scheduling Algorithm of RC4. In: Vaudenay, S., Youssef, A.M. (eds.) SAC 2001. LNCS, vol. 2259, pp. 1–24. Springer, Heidelberg (2001)

11. Golić, J.D.: Iterative Probabilistic Cryptanalysis of RC4 Keystream Generator. In: Clark, A., Boyd, C., Dawson, E.P. (eds.) ACISP 2000. LNCS, vol. 1841, pp. 220–233. Springer, Heidelberg (2000), http://www.springerlink.com/content/11510525523352p4/fulltext.pdf
12. Hoch, J.J., Shamir, A.: Fault Analysis of Stream Ciphers. In: Joye, M., Quisquater, J.-J. (eds.) CHES 2004. LNCS, vol. 3156, pp. 240–253. Springer, Heidelberg (2004)
13. Khovratovich, D., Biryukov, A., Nikolic, I.: Speeding up collision search for byte-oriented hash functions. In: Fischlin, M. (ed.) CT-RSA 2009. LNCS, vol. 5473, pp. 164–181. Springer, Heidelberg (2009)
14. Knudsen, L.R., Meier, W., Preneel, B., Rijmen, V., Verdoolaege, S.: Analysis Methods for (Alleged) RC4. In: Ohta, K., Pei, D. (eds.) ASIACRYPT 1998. LNCS, vol. 1514, pp. 327–341. Springer, Heidelberg (1998), http://www.springerlink.com/content/tyqqary0p5kfw7tp/fulltext.pdf
15. Leander, G., Zenner, E., Hawkes, P.: Cache Timing Analysis of LFSR-Based Stream Ciphers. In: Parker, M.G. (ed.) Cryptography and Coding 2009. LNCS, vol. 5921, pp. 433–445. Springer, Heidelberg (2009)
16. Maximov, A., Khovratovich, D.: New State Recovery Attack on RC4. In: Wagner, D. (ed.) CRYPTO 2008. LNCS, vol. 5157, pp. 297–316. Springer, Heidelberg (2008)
17. Munkres, J.: Algorithms for the Assignment and Transportation Problems. Journal of the Society for Industrial and Applied Mathematics 5, 32–38 (1957), http://www.jstor.org/stable/2098689
18. Osvik, D.A., Shamir, A., Tromer, E.: Cache Attacks and Countermeasures: The Case of AES. In: Pointcheval, D. (ed.) CT-RSA 2006. LNCS, vol. 3860, pp. 1–20. Springer, Heidelberg (2006), http://www.springerlink.com/content/f52x1h55g1632l17/fulltext.pdf
19. Page, D.: Theoretical Use of Cache Memory as a Cryptanalytic Side-Channel. Technical Report CSTR-02-003, Department of computer science, university of Bristol (2002), http://www.cs.bris.ac.uk/Publications/Papers/1000625.pdf
20. Salez, J., Shah, D.: Belief propagation: An asymptotically optimal algorithm for the random assignment problem. Math. Oper. Res. 34(2), 468–480 (2009)
21. Sepehrdad, P., Vaudenay, S., Vuagnoux, M.: Discovery and exploitation of new biases in rc4. In: Biryukov, A., Gong, G., Stinson, D.R. (eds.) SAC 2010. LNCS, vol. 6544, pp. 74–91. Springer, Heidelberg (2011)
22. Tromer, E., Osvik, D.A., Shamir, A.: Efficient cache attacks on aes, and countermeasures. J. Cryptology 23(1), 37–71 (2010)
23. Zenner, E.: A Cache Timing Analysis of HC-256. In: Avanzi, R.M., Keliher, L., Sica, F. (eds.) SAC 2008. LNCS, vol. 5381, pp. 199–213. Springer, Heidelberg (2009)

A A Numerical Example of the Algorithm With Idealized Cache

Let us imagine that the generation of a cipher byte gives the following information:

$$a = \texttt{0xf3}, \quad b = \texttt{0xd9}, \quad c = \texttt{0x43}, \quad out = \texttt{0x1c}$$

The candidates values for j, $S[a]$, $S[b]$ and $S[c]$, given by the preceding attack steps, are:

$$\begin{aligned} j &\in \{\texttt{0x2f}, \texttt{0xc6}\} \\ S[a] &\in \{\texttt{0x7d}, \texttt{0xaa}\} \\ S[b] &\in \{\texttt{0x14}, \texttt{0x1c}, \texttt{0x5d}, \texttt{0x99}\} \\ S[c] &\in \{\texttt{0x1c}, \texttt{0x5c}, \texttt{0xd4}\} \end{aligned}$$

Let us suppose first that the table elements are used in the order (a, b, c):

– if the value of j is `0x2f`, the constraints given above are:

$$\texttt{0x2f} + S[a] = \texttt{0xd9}, \quad S[a] + S[b] = \texttt{0x43}, \quad S[c] = \texttt{0x1c}$$

  The imposed values are therefore $S[a] = \texttt{0xaa}$, $S[b] = \texttt{0x99}$ and $S[c] = \texttt{0x1c}$. At this step of the attack, these three values are considered possible: this may correspond to what really happened when generating the keystream byte;

– using analogous computations for $j = \texttt{0xc6}$, we deduce that the value of $S[a]$ must be `0x13`, which is not possible.

We make similar computations for the order (a, c, b) and finally obtain two candidates for what happened during the step of PRGA:

– if the elements are used in the order (a, b, c), the corresponding values must have been $j = \texttt{0x2f}$, $S[a] = \texttt{0xaa}$, $S[b] = \texttt{0x99}$ and $S[c] = \texttt{0x1c}$;
– otherwise, they must be $j = \texttt{0xc6}$, $S[a] = \texttt{0x7d}$, $S[b] = \texttt{0x1c}$ and $S[c] = \texttt{0x5c}$.

We can now update the candidate values for j and the three table elements: after this pseudo-random generation step, the value of j can be `0xd9` or `0x43` (as we were not able to determine the order of use of the table elements with certainty); we also take the swap operation into account, which leads us to:

$$\begin{aligned} j &\in \{\texttt{0x43}, \texttt{0xd9}\} \\ S[a] &\in \{\texttt{0x5c}, \texttt{0x99}\} \\ S[b] &\in \{\texttt{0x1c}, \texttt{0xaa}\} \\ S[c] &\in \{\texttt{0x1c}, \texttt{0x7d}\} \end{aligned}$$

We proceed like above to determine which values are possible or not and so on.

B Proofs of Soundness

B.1 Algorithm for the Idealized Case

The proof of soundness of the algorithm for the idealized case is very straightforward:

- at the beginning of the attack, all the elements of the tables $S_{values}[i]$ have the value 1. In particular, for each i, the value v corresponding to the secret permutation table used by PRGA is associated to the value 1. Moreover, the value of j is well-known;
- we now suppose that before an attack step the tables $S_{values}[i]$ and j_{values} contain the value 1 for the value v corresponding to the corresponding value of the secret state of the encryption process. Consequently, when trying the good order for a, b and c and the good value for j, the imposed values $S[a]$, $S[b]$ and $S[c]$, which correspond to the values of the secret state, all will have the value 1 in S_{values}: this try will be considered as successful. Finally, the attack algorithm will update S_{values} and j_{values} to take this attack step into account. Since the guess corresponding to the real secret state has been considered as successful, the values for j, $S[a]$, $S[b]$ and $S[c]$ will be marked with the boolean 1, all other values of S_{values} remaining untouched as in PRGA: after the attack step, the tables S_{values} and j_{values} also associate the boolean 1 to the values of $S[i]$ and j corresponding to the secret state used by PRGA.

Using the axiom of induction, we obtain that during each step of the attack, the tables S_{values} and j_{values} associate the boolean 1 to the values corresponding to the secret state used by PRGA, so that the $S_{values}[i]$ and j_{values} always have at least one boolean with value 1, and that when all of these tables contain only one boolean with value 1, this boolean is the one corresponding to the value of the secret stage used by PRGA.

B.2 Probabilistic Algorithm for Realistic Caches

An Impractical Straightforward Algorithm. The proof of soundness of the probabilistic algorithm is a little more complicated. To do so, we will introduce the following algorithm:

- we consider the space of couples (S, j), where S is a permutation over $\mathbf{F}_{256}$ and j an element of $\mathbf{F}_{256}$;
- we use a `256.256!`-boolean table C_{values}, using for each couple (S, j) the same meaning as in the algorithm for the idealized case: if the value is 1, the couple is considered as probable as it explains the successive observed cipher bytes and cache lookups; if it is 0, the couple has failed to explain the observations and we know it to be impossible;
- during an attack step, we begin a new table C^*_{values}, filled with the value 0. For each couple (S, j) that has the value 1 in the table C_{values} and that explains the observed cipher byte and cache lookups, we compute the updated couple (S^*, j^*) with PRGA and give it the value 1 in C^*_{values};

– at the end of the attack step, we replace the table C_{values} by the table C^*_{values}.

As for the algorithm of the idealized case, the proof of soundness of this algorithm is very simple to do using induction, so we will not discuss it further. However, this algorithm is very inefficient, as the number of possible couples in the first steps if far too high to allow a computer to enumerate them.

The Probabilistic Algorithm, an Improvement of the Straightforward Algorithm. We introduce a 256x256-floating point table S_p and a 256-floating point table j_p, which values are computed as follows:

$$\forall i, \forall v, S_p[i][v] = \frac{\#\{(S,j)|C_{values}[(S,j)] = 1, S[i] = v\}}{\#\{(S,j)|C_{values}[(S,j)] = 1\}}$$

$$\forall v, j_p[v] = \frac{\#\{(S,j)|C_{values}[(S,j)] = 1, j = v\}}{\#\{(S,j)|C_{values}[(S,j)] = 1\}}$$

We will now prove that the evolution of S_p and j_p is the same as the one of S_{probas} and j_{probas}. First, it is clear that, at the beginning of the attack:

$$\forall i, \forall v, S_p[i][v] = \frac{256 \cdot 255!}{256 \cdot 256!} = 1/256$$

$$\forall v, j_p[v] = \frac{1 \cdot 256!}{256 \cdot 256!} = 1/256$$

Now, choosing an order (a, b, c), a value for j and for o_1, o_2 and being given the values of i, b_δ, c_δ and out, we can compute the number P_g of probable couples (S, j) concerned by this guess. We obtain, assuming that the different elements of the permutation are independent (the same assumption as what was done in the probabilistic algorithm) and using the constraints on $S[a]$, $S[b]$ and $S[c]$ given by the structure of PRGA, an equation analogous to the one giving $\mathbf{P}(\text{guess})$:

$$P_g = j_p[v] \cdot S_p[a][b - v] \cdot S_p[b][c - b + v] \cdot S_p[c][out] \cdot N$$

where N is the number of probable couples.

For given values of i and v, the couples (S', j') of C^*_{values} can have two origins:

– either probable couples where PRGA does not affect $S[i]$. Assuming again the independence of all concerned random variables, their number is:

$$\left(N - \sum_{\text{guesses imposing } S[i]} P_g \right) \cdot S_p[i][v]$$

– or couples where PRGA imposes $S[i] = v$. Their number is given by:

$$\sum_{\text{guesses imposing } S[i]=v} P_g$$

Adding these two contributions, we obtain that the evolution of S_p is the same as that of S_{probas}, except for a coefficient $\sum_{\text{any guesses}} P_g/N$, which is the proportion of couples (S, j) that are compatible with the observed cipher byte and cache lookups, and that corresponds to the renormalization of $\mathbf{P}(\text{guess})$. Using the axiom of induction, this achieves proving that the behavior of S_p is the same as the one of S_{probas} in the probabilistic algorithm.

For the behavior of j_p the proof is the same, as previously. In conclusion, the behavior of the tables S_{probas} and j_{probas} in the probabilistic algorithm is a consequence of the behavior of the table C_{values} in the algorithm introduced here. Furthermore, we know that if this algorithm only has one probable couple (S, j), then this couple is the right one (we also know that the right couple is always considered as probable). Consequently, we can deduce that if the tables $S_{probas}[i]$ and j_{probas} contain only one value 1.0 and the other elements have the value 0.0, then the value with probability 1 is the one of the secret state used by PRGA, which concludes this proof.

Further Thoughts on the Practical Results. First, it is important to stress that despite the previous proofs of soundness of the algorithms presented in this paper, it was neither possible to estimate the speed of convergence to the solution, nor to prove the termination of these algorithms. The only hints according to this problem are given considering arguments from information theory, since the entropy of the secret state is easy to evaluate to the first order, as is the entropy of every cipher byte and cache lookup. Furthermore, the number of cases when the probabilistic algorithm fails to give the right secret state do not go against the soundness of the algorithm, but rather show that the choice of stopping the attack before only one probable state remains is done with the risk of giving a wrong answer. Finally, as the number of possible couples is less than `256x256!`, the number of bits needed to store the floating-point values of the tables S_{probas} and j_{probas} is $\log_2(256 \cdot 256!) \approx 2056$. To improve the speed of the attack, we used basic floating-point numbers, which is also a probable cause for the failure of the attack in some cases.

Secure Efficient Multiparty Computing of Multivariate Polynomials and Applications

Dana Dachman-Soled[1], Tal Malkin[1], Mariana Raykova[1], and Moti Yung[2]

[1] Columbia University
[2] Google Inc. and Columbia University

Abstract. We present a robust secure methodology for computing functions that are represented as multivariate polynomials where parties hold different variables as private inputs. Our generic efficient protocols are fully black-box and employ threshold additive homomorphic encryption; they do not assume honest majority, yet are robust in detecting any misbehavior. We achieve solutions that take advantage of the algebraic structure of the polynomials, and are polynomial-time in all parameters (security parameter, polynomial size, polynomial degree, number of parties). We further exploit a "round table" communication paradigm to reduce the complexity in the number of parties.

A large collection of problems are naturally and efficiently represented as multivariate polynomials over a field or a ring: problems from linear algebra, statistics, logic, as well as operations on sets represented as polynomials. In particular, we present a new efficient solution to the multi-party set intersection problem, and a solution to a multi-party variant of the polynomial reconstruction problem.

Keywords: secure multiparty computation, multivariate polynomial evaluation, additive homomorphic encryption, threshold cryptosystems, secret sharing, multiparty set intersection.

1 Introduction

Secure multiparty computation (MPC) allows mutually distrustful parties to jointly compute a functionality while keeping their inputs private. Seminal feasibility results for the two-party and multi-party settings have been demonstrated in [34] [35] [19] [16] [1] [4] [31] [24]. These results show that any functionality can be securely computed in time polynomial in the size of its Boolean or arithmetic circuit representation.

While the works above yield strong feasibility results, these generic approaches typically lead to inefficient implementations since the circuit size representation of a functionality may be very large. Thus, an important open problem in MPC is designing highly efficient protocols for smaller, yet large enough to be interesting, sets of functionalities, taking advantage of the domain specific mathematical structure.

J. Lopez and G. Tsudik (Eds.): ACNS 2011, LNCS 6715, pp. 130–146, 2011.

Problem Statement. We consider the problem of secure multiparty computation functions that can be represented by *polynomial-size multivariate polynomials*. Each party's inputs correspond to some subset of the variables in the polynomial representation. There is a designated party receiving output that learns only the output of the polynomial evaluation while all other parties receive no output.[1] We assume a broadcast channel and that the private keys for the threshold encryption scheme are distributed in a preprocessing stage.

Our Results: The General Protocol. We present a protocol that allows multiple parties to compute the above functionalities, assuring security against a dishonest majority and robustness (detection misbehavior). Our protocol is fully black-box assuming any threshold additive homomorphic encryption with a natural property that we specify later, (instantiated by Paillier scheme, say). The protocol utilizes a "round table" structure where parties are nodes in a ring network (which means that frequently a party only communicates with its two neighboring parties around the table). This structure (employed already in past protocols) has two benefits: first, it allows each party to be offline for the majority of the execution of the protocol and to be involved only when it needs to contribute its inputs at its turn. Second, it allows a division of the communication complexity into two types: "round table" communication complexity including messages exchanged between two neighboring parties, and broadcast communication complexity including messages sent simultaneously to all parties. Our simulation-based proofs of security are given in the Ideal/Real (standard) Model as per definitions in [20].

To the best of our knowledge, the only paper that has considered secure computation of multivariate polynomials is [13]. This recent independent work has focused on multivariate polynomials of degree 3 but points out that the proposed protocols can be generalized to higher degree polynomials, however, with communication complexity that is no longer optimal, leaving as an open question improvements of this complexity. Their protocol is based on the compiler of [23], but with the difference being that the outer and the inner protocols are instantiated with efficient constructions tailored for multivariate polynomials. Their protocol's communication complexity is (sub)-exponential in the number of variables t: $O(poly(k)d^{\lfloor t/2 \rfloor})$ for polynomials of degree d and security parameter k. Our work, in turn, improves their communication complexity to be fully polynomial (i.e., polynomial in all parameters of the problem). Clearly, one can take a poly-size multivariate polynomial and translate it to a circuit with poly time secure computation soultion, but this will have a huge polynomial factor expansion and will lose the structure enabling the special-purpose speedups. We achieve "round-table" complexity $10kDn(m-1)$ and broadcast complexity $k(10D+1)(\sum_{j=1}^{m}\sum_{t=1}^{l_j}\log\alpha_{j,t}+1)$ for m parties where party i has l_i inputs of degrees $\alpha_{i,1},\ldots,\alpha_{i,l_i}$, D being the sum of the logarithms of the variable degrees for polynomials consisting of n monomials. Next, since every polynomial

[1] We note that our protocol can be generalized to allow any subset of the parties to receive output.

can be easily converted into an arithmetic circuit, our protocol is also a protocol for MPC of a subclass of all arithmetic circuits. From this point of view, the work of [24] addresses a comparable problem to ours (constructing a MPC protocol for all poly-size arithmetic circuits, using a black-box construction and assuming no honest majority). The work of [13] already improves in the worst case the complexity results of [24] (for proper set of multivariate polynomials), and as we noted above we bring additional improvement (intuitively, our amortized broadcast complexity is linear in the size of the representation of the largest term of the polynomial, and does not depend on the number of terms in the representation, which contributes to the size of the arithmetic circuit). Further, the protocol of [24] requires as many rounds (involving all the parties) as the depth of the circuit and communication complexity depending on the size of the circuit. In contrast, we achieve a number of rounds independent of the depth and the size of the arithmetic circuit of the polynomial (and our round-complexity is actually constant when either counting a round-table round as one round or when considering only a constant number of parties).

Our Results: Special Cases. The class of polynomial size multivariate polynomials contains a wide range of efficiently representable functionalities with special structure that enables further optimizations. Most of the commonly used statistics functions can either be represented as polynomials or approximated with polynomials using Taylor series approximation for trigonometric functions, logarithms, exponents, square, etc. Examples include average, standard deviation, variance, chi-square test, Pearson's correlation coefficients, and central moment of statistical distributions. Matrix operations (i.e., linear algebra) can also be translated to polynomial evaluations.

As a special case of the general protocol, we implement *secure multiparty set intersection* against a malicious adversary controlling a majority of the parties; we note that the set intersection question in the two party case has been addressed in many works [15] [22] [26] [25] [8] [7] while there are fewer works that have considered the multiparty version. Two works adress the issue in the computational protocol setting. First, Kissner et al. [26] present a semi-honest protocol and suggests using generic zero communication complexity $O(m^2d^2)$ for m parties with input sets of size d. The work of [32] improves this complexity by a factor of $O(m)$ for m party protocols, using more efficient ZK based on pairings. (In addition, relatively inefficient information theoretic solutions are presented in [29,30]).) Our protocol achieves communication complexity $O(md + 10d\log^2 d)$ improving the existing works. We achieve linear complexity in the number of parties m due to the round table communication paradigm, whereas even the recent work [5] is quadratic in the number of parties.

Finally, when polynomial's coefficients correspond to the input of the designated receiver, our method is turned into a *multi-party oblivious multivariate polynomial evaluation*, a generalization of the problem of oblivious polynomials evaluation [27] to inputs from multiple parties.

Techniques. Many of our techniques exploit the "nice structure" of multivariate polynomials as well as various interactions of this structure with other algebraic and cryptographic primitives. First, we crucially utilize the fact that multivariate polynomials are linear operators which are easy to combine with additive homomorphic encryption and polynomial secret sharing. We formalize this property by presenting a commutativity property between the evaluation of multivariate polynomials and reconstruction of Shamir's secret sharing [33]. Intuitively, this allows us to evaluate a given polynomial on multiple (modified) Shamir secret shares in parallel and obtain the final evaluatation of the polynomial by reconstructing the resulting secret shares. This technique allows us to apply (black box) "cut-and-choose" techniques to verify the correctness of the evaluation, without revealing information about the shared inputs or outputs. We note that analogous techniques were used in a different context by [6, 8].

A second property of multivariate polynomials is that they can be computed over additive homomorphic encryption non-interactively in a round-table type protocol where each participant incrementally contributes his inputs to the encryption of a monomial outputted by the previous participant (note that a participant's contribution to a given monomial amounts to multiplication of the encrypted monomial by a scalar).

We additionally use the polynomial structure of a variant of Shamir's threshold sharing in zero knowledge protocols proving that inputs were shared correctly and committed under homomorphic encryption. We utilize Lagrange interpolation combined with what we call vector homomorphic encryption (where the homomorphic properties hold for both the plaintexts and the encryption randomness; which is true for many of the known homomorphic encryption schemes [28, 12, 10, 21]). This is used to verify that inputs were shared and encrypted correctly, provided that the randomness for the encryptions was chosen in a specific way. This encrypted interpolation technique combined with the large minimum distance of Reed-Solomon codes allows us to guarantee the correctness of an entire computation on encrypted codewords based on the verification that a small random subset of shares were computed correctly. Finally, we use the linear operator properties of the sharing polynomials for share re-randomization under additive homomorphic encryption.

We note that when we instantiate our protocol with homomorphic encryption over a ring, we apply the technique of Feldman ([11]) also used, e.g., in Fouque et al. ([12]) for Paillier sharing that transforms interpolation over an RSA-composite ring to an interpolation over (an interval of) the integers (where computing inverses, i.e., division, is avoided and finding uninvertible elements is hard, assuming factoring is hard).

2 Protocol Overview

Semi-honest structure: As described above, multivariate polynomials can be computed over additive homomorphic encryption by a round-table protocol. This constitutes our underlying semi-honest evaluation protocol.

Robustness ideas: To achieve security against malicious adversaries, we employ the commutativity between evaluation of multivariate polynomials and Shamir's secret sharing reconstruction described above. Consider the following simplified example that illustrates our basic techniques: Let us have m parties that wish to evaluate the univariate polynomial $\mathbf{Q}(x) = x^5 + 10x^3 + 6x + 9$, at point x, where x is the committed input of Party 1. Note that allowing Party 1 to do the entire computation will not ensure that the outcome is consistent with the committed input. One possible solution is to require Party 1 to commit to its input x by encrypting x with a homomorphic encryption scheme, and have all parties compute the encrypted result using the homomorphic properties of the encryption, which is then decrypted. However, in order to compute all polynomial functions we will need a threshold doubly (or fully)-homomorphic encryption scheme. Although Gentry, [18], recently introduced the first (highly expensive) known doubly-homomorphic encryption scheme, a threshold analogue is not yet known.

Instead, we take the following approach: Party 1 computes a Shamir secret-sharing of its input x by choosing a polynomial P_x of degree k uniformly at random conditioned on $P_x(0) = x$. Now, instead of committing to the value x, Party 1 commits to, say, $20k$ input shares of $P_x : P_x(1), \ldots, P_x(20k)$. Next, Party 1 commits to $20k$ output shares of $\mathbf{Q} \circ P_x(i) : \mathbf{Q}(P_x(1)), \ldots, \mathbf{Q}(P_x(20k))$. Notice that $\mathbf{Q} \circ P_x(i)$ is a polynomial of degree $5k$ and that $\mathbf{Q} \circ P_x(0) = Q(P_x(0)) = Q(x)$. Thus, by reconstructing $\mathbf{Q} \circ P_x(0)$ we obtain the output value $\mathbf{Q}(x)$. After Party 1 sends the input and output commitments, the parties verify efficiently that the input and output shares indeed lie on a polynomial of degree k and $5k$ respectively using an interpolation algorithm we define below. Now, the parties run a cut-and-choose step where a set $I \subset [20k]$ of size k is chosen at random. For each index $i \in I$, Party 1 must open the commitments to reveal $P_x(i)$ and $\mathbf{Q} \circ P_x(i)$. The remaining parties now verify privately that $\mathbf{Q} \circ P_x(i)$ was computed correctly. Note that due to the secret-sharing properties of the commitment scheme, the cut-and-choose step reveals no information about $P_x(0) = x$ or $\mathbf{Q} \circ P_x(0) = \mathbf{Q}(x)$. Now, let us assume that Party 1 acted maliciously. Since the set I was chosen at random, and due to the large distance of Reed-Solomon codes, we show that if Party 1 is able to open all the shares corresponding to I correctly, then with very high probability Party 1 must have computed *all* of the output shares correctly. We note that the above description leaves out important **re-randomization techniques** (that are described in the full protocol) whose goal is to prevent parties from learning during the incremental evaluation and robustness checking.

Efficient Robustness: Although the technique described above is sufficient to ensure that the parties behave honestly, it induces a huge blow-up in the number of required shares. Indeed, in order to reconstruct the zero coefficient of a polynomial of degree deg, we must have at least $deg + 1$ secret shares. Thus, when evaluating a polynomial such as $\mathbf{Q} = x^{2^n}$, we would require an exponential number of shares. To prevent this blow-up, we employ an input preprocessing step (described in Section 3).

Secure output reconstruction: Finally, we use a threshold decryption algorithm to ensure that no subset of the parties can decrypt the intermediate messages exchanged. The threshold decryption is needed in the case where more than one party contributes its inputs to the polynomial (and is actually not necessary in our toy example above). Any additive homomorphic threshold encryption scheme (with one additional natural property, which we describe later) would suffice for the correctness of our protocol. Examples of such schemes are the El Gamal threshold encryption scheme [17] and the Paillier threshold encryption scheme [12]. Note that additive El Gamal does not allow efficient decryption over a large domain, but it sufficies for our Set Intersection applications. We use the Paillier threshold encryption scheme to instantiate our general polynomial evaluation protocols. To obtain the final output, the designated party reconstructs the encryption of the final output value using Lagrange interpolation over encrypted values and decrypts with the help of the other parties.

3 Definitions and Building Block Protocols

We use a standard simulation-based definition of security (e.f., [3]), and follow the definitions of zero knowledge proofs of knowledge and commitment schemes ([20]). We denote by Com_B a perfectly binding commitment scheme and by Com_H a perfectly hiding commitment scheme. Given $d+1$ evaluation points $(x_0, y_0), \dots (x_d, y_d)$ on a polynomial of degree d, we denote the interpolation value at the point x as $L_{x_0,\dots,x_d}(y_0,\dots,y_d,x)$.

3.1 Vector Homomorphic Encryption

We require threshold additive homomorphic encryption scheme with the following additional property, capturing the fact that the homomorphism applies also to the randomness.[2] This property is satisfied by most known homomorphic encryption schemes: Paillier [28] and threshold Paillier [12], ElGamal [10], and Goldwasser-Micali [21].

Property 1. Let $E = (\text{GEN}, \text{ENC}, \text{DEC})$ be an encryption scheme where the plaintexts come from a ring R_1 with operations $(+,\cdot)$, the randomness comes from a ring R_2 with operations $(\oplus,\odot)$, and the ciphertexts come from a ring R_3 with operations $(\otimes, \hat{\ })$. We say that E is *vector homomorphic* if the following holds: $\text{ENC}(m_1;r_1) \otimes \text{ENC}(m_2;r_2) = \text{ENC}(m_1+m_2; r_1 \oplus r_2)$ and $\text{ENC}(m;r)^c = \text{ENC}(c\cdot m; r\odot c)$.

[2] We actually only need a slightly weaker property, but to simplify the presentation we assume our encryption scheme possesses the stronger property defined here.

3.2 Polynomial Code Commutativity

Shamir secret sharing [33] or Reed-Solomon codes are commutative with respect to polynomial evaluations, which we formalize as follows:

Property 2 (Polynomial Code Commutativity). Let $\mathbf{Q}(x_1, \ldots, x_m)$ be a multivariate polynomial. Let $P_x(1), \ldots, P_x(t+1)$ be Shamir secret shares of a value x where P_x is a polynomial of degree t such $P_x(0) = x$. We can reconstruct x from its secret shares using Lagrange interpolation L. The evaluation of $\mathbf{Q}$ commutes with L in the sense that we can compute the value $\mathbf{Q}(x_1, \ldots, x_m)$ with either of the following algorithms:

$$\begin{aligned}&(\mathbf{Q} * L)(P_{x_1}(1), \ldots P_{x_1}(t+1), \ldots P_{x_m}(1), \ldots P_{x_m}(t+1), 0) = \\ &= \mathbf{Q}((L(P_{x_1}(1), \ldots P_{x_1}(t+1), 0), \ldots, L(P_{x_m}(1), \ldots P_{x_m}(t+1), 0))) = \\ &= \mathbf{Q}(x_1, \ldots x_m),\end{aligned}$$

where we first use L to retrieve the secrets and then evaluate $\mathbf{Q}$, or

$$\begin{aligned}&(L * \mathbf{Q})(P_{x_1}(1), \ldots P_{x_1}(t+1), \ldots P_{x_m}(1), \ldots P_{x_m}(t+1), 0) = \\ &= L(\mathbf{Q}(P_{x_1}(1), \ldots, P_{x_m}(1)), \ldots, \mathbf{Q}(P_{x_1}(t+1), \ldots, P_{x_m}(t+1)), 0) = \\ &= L(w_1, \ldots, w_{t+1}, 0) = \mathbf{Q}(x_1, \ldots x_m),\end{aligned}$$

where we evaluate $\mathbf{Q}$ on each set of shares of $x_1, \ldots, x_m$ to obtain shares of $\mathbf{Q}(x_1, \ldots, x_m)$ and then use L to reconstruct the final secret.

3.3 Incremental Encrypted Polynomial Evaluation

We will use homomorphic encryption to allow multiple parties to evaluate a multivariate polynomial depending on their inputs by incrementally contributing their inputs to partial encrypted evaluations of its monomials. This is facilitated by the following property:

Property 3 (Incremental Encrypted Polynomial Evaluation). Let m be the number of parties evaluating a multivariate polynomial $\mathbf{Q}$ defined by

$$\mathbf{Q}(x_{1,1}, \ldots, x_{1,l_1}, \ldots, x_{m,1}, \ldots, x_{m,l_m}) = \sum_{s=1}^{n} c_s(\prod_{j=1}^{m} h_{j,s}(x_{j,1}, \ldots, x_{j,l_j})),$$

where $h_{j,s}$ represents the inputs of party j to the s-th monomial of $\mathbf{Q}$. Let $E = (\mathrm{GEN}, \mathrm{ENC}, \mathrm{DEC})$ be an additive homomorphic encryption. We define the partial evaluations $b_{j,s}$ (including the contributions of parties 1,..., j) of the monomials $s, 1 \leq s \leq n$ of $\mathbf{Q}$ as follows:

$$b_{0,s} = \mathrm{ENC}(c_j) \text{ for } 1 \leq j \leq n, \text{ and } b_{j,s} = b_{j-1,s}^{h_{j,s}(x_{j,1},\ldots,x_{j,l_j})} \text{ for } 1 \leq j \leq m$$

3.4 Polynomial Interpolation over Encrypted Values

In this section we present a protocol that allows a verifier to verify (without help from the prover) that the prover's encrypted points lie on a polynomial of low degree, assuming the prover constructed the encryptions in a predetermined manner. Recall that Lagrange interpolation allows us, given $d+1$ points, to reconstruct the polynomial of degree d that interpolates the given points. In Figure 1, we use the fact that Lagrange interpolation can, in fact, be carried out over encrypted points when the known encryption used possesses the vector homomorphic Property 1. Since the encryption is over a ring we use Feldman's technique for shift interpolation by factorial [11].

Lagrange Interpolation Protocol Over Encrypted Values (LIPEV)

Input: $(1, \mathrm{ENC}_{pk}(y_1, r_1)), \ldots (A, \mathrm{ENC}_{pk}(y_A, r_A))$, d where $d+1 < A$,
Output: Verifier outputs Accept if there are polynomials $P_1 \in R_1[x], P_2 \in R_2[x]$ of degree at most d such that $y_j = P_1(j)$ for $1 \leq j \leq A$ and $r_j = P_2(j)$ (P_1 and P_2 are defined with respect to the operations in the respective rings) for $1 \leq j \leq A$.
Verification Protocol:

1. Let $\Delta = A!$.
2. Let $l_j(x) = \Delta \cdot \prod_{i=1, i \neq j}^{d+1} \frac{x-i}{j-i}$ for $1 \leq j \leq d+1$.
3. Verifier checks whether $\mathrm{ENC}_{pk}(y_i, r_i)^{\Delta} = \prod_{j=1}^{d+1} (\mathrm{ENC}_{pk}(y_j, r_j))^{l_j(i)}$, and rejects otherwise.

Fig. 1.

Using the LIPEV protocol, a prover can prove to a verifier that A encrypted points lie on one polynomial of degree d, provided that the randomness for the encryptions was chosen in a specific way; namely, the random values chosen must also lie on a polynomial of degree d. For completeness, we describe next how to compute the random values for the encryptions so that they lie on a polynomial $P_2 \in R_2[x]$ of degree d (see Figure 2). We note that even though the randomness for all A encrypted points are not chosen uniformly at random, semantic security is still preserved since the randomness for $d+1$ of the points is chosen uniformly at random and the remaining $A-d-1$ encryptions can be computed given only the first $d+1$ encryptions due to Property 1.

3.5 Input Sharing via Enhanced Shamir Scheme

One of the ideas that we employ in our main protocol is to share function evaluation by secret sharing the arguments of the function via polynomials of degree k and evaluating the function on the corresponding shares in order to obtain shares of the final value of the function. The above can be implemented straightforwardly using Shamir's secret sharing ([33]) and evaluating the polynomial on corresponding shares. The problem with this approach is that if the degree of the

Randomness Interpolation

Input: $(1, y_1), \ldots, (A, y_A), r_1, \ldots, r_{d+1}$ for $d+1 < A$
Output: $r^{\Delta}_{d+2}, \ldots, r^{\Delta}_A$ such that the Lagrange Interpolation Protocol Over Encrypted Values outputs accept on the input: $[i, \mathrm{ENC}_{pk}(y_i, r_i)^{\Delta}]_{1 \le i \le A}, d$.

Protocol:

1. Compute $l_j(x) = \Delta \prod_{i=0, i \neq j}^{d+1} \frac{x-i}{j-i}$ for $1 \le j \le d+1$.
2. Compute $r^{\Delta}_i = \prod_{j=0}^{d+1} (r_j)^{l_j(i)}$ for $d+2 \le i \le A$.

Fig. 2.

output function is Deg, then we require at least $k \cdot Deg$ shares to reconstruct the output value. We avoid this blow-up in number of required shares by applying the following transformation on the inputs.

We consider a multivariate polynomial $\mathbf{Q}(x_1, \ldots, x_\ell)$ over the input set $\overline{X} = \{x_1, \ldots, x_\ell\}$ which we would like to evaluate on shares of the input variables in order to obtain shares of the output value. Since the number of output shares that we will need to compute will depend on the degree of $\mathbf{Q}$ and the degrees of the sharing polynomials P_{x_i} that we use to share each of the input variable, we employ techniques that allow us to decrease the degree of the final polynomial [8]. The main idea is to introduce new variables that represent higher degrees of the current variables. For each variable x_i that has maximum degree d_i in $\mathbf{Q}$ we substitute each power $x_i^{2^j}$ with a new variable for $0 \le j \le \lfloor \log d_i \rfloor$. Note that if we view the original polynomial $\mathbf{Q}$ as a polynomial over these new variables, we have that each variable has degree at most one. Let M^t be the t-th monomial in $\mathbf{Q}$ and let $d_{M^t,i} \in M^t$ be the degrees of the variables in monomial M^t. Thus, the original degree of $\mathbf{Q}$ was $\max_t\{\sum_{i \in M^t} d_{M^t,i}\}$, whereas the degree of the transformed polynomial over the new variables is only $\max_t\{\sum_{i \in M^t} \log d_{M^t,i}\}$. In our protocol, each party will pre-process its inputs to compute its new input variables and their shares and will prove that the new shares are consistent with the initial inputs.

4 Multiparty Polynomial Evaluation

The multiparty polynomial evaluation has the following setup:

- Each party T_j has l_j inputs $\overline{X}_j = \{x_{j,1}, \ldots, x_{j,l_j}\}$ for $1 \le j \le m$.
- A designated output receiver T^* (one of the parties $T_1, \ldots, T_m$).
- A polynomial $\mathbf{Q}(x_{1,1}, \ldots, x_{1,l_1}, \ldots, x_{m,1}, \ldots, x_{m,l_m})$, which depends on the inputs of all parties.

We use the following representation of the polynomial $\mathbf{Q}$:

$$\mathbf{Q}(x_{1,1},\ldots,x_{1,l_1},\ldots,x_{m,1},\ldots,x_{m,l_m}) =$$
$$= \sum_{s=1}^{n} c_s x_{1,1}^{\alpha_{1,1,s}} \ldots x_{1,l_1}^{\alpha_{1,l_1,s}} \ldots x_{m,1}^{\alpha_{m,1,s}} \ldots x_{m,l_m}^{\alpha_{m,l_m,s}} = \sum_{s=1}^{n} c_s (\prod_{j=1}^{m} h_{j,s}).$$

where $h_{j,s}(x_{j,1},\ldots,x_{j,l_j}) = x_{j,1}^{\alpha_{j,1,s}} x_{j,2}^{\alpha_{j,2,s}} \ldots x_{j,l_j}^{\alpha_{j,l_j,s}}$ and c_s is a known coefficients for $1 \le j \le m$. If $x_{j,v}$ does not participate in the s-th monomial of $\mathbf{Q}$, then $\alpha_{j,v,s} = 0$. Alternatively, we view $h_{j,s}$ for $1 \le j \le m$, $1 \le s \le n$ in the following way:

$$h_{j,s}(x_{j,1}, x_{j,1}^2, \ldots, x_{j,1}^{2^{\lfloor \log \alpha_{j,1,s} \rfloor}}, \ldots, x_{j,l_j}, x_{j,l_j}^2, \ldots, x_{j,l_j}^{2^{\lfloor \log \alpha_{j,l_j,s} \rfloor}}) \quad (1)$$

in which each variable is of degree at most one.

Notation. In the protocol we will use the following variables: $D_{h,s} = \sum_{j=1}^{m} deg(h_{j,s})$ for $1 \le s \le n$; $D_{h,j,s} = k \sum_{v=1}^{j} deg(h_{v,s})$ for $1 \le j \le m$ where $h_{v,s}$ is defined as in Equation 1 (variables of degree at most 1); $D = \max_{s=1}^{n} D_{h,s}$. Also we let $\Delta = 10kD!$ be a public parameter, and $E = (\text{GEN}, \text{ENC}, \text{DEC})$ be a threshold encryption scheme that possesses Property 1 with public key pk and secret keys $sk_1, \ldots, sk_m$ for the m parties $T_1, \ldots, T_m$.

Protocol Intuition. The protocol consists of four phases: *Input Preprocessing, Round-Table Step, Re-randomization, Verification and Reconstruction.* During the *Input Preprocessing* phase, the parties use the technique from Section 3.5 to transform $\mathbf{Q}$ from polynomial over the variables $x_{1,1}, \ldots, x_{1,l_1}, \ldots, x_{m,1}, \ldots, x_{m,l_m}$ into a polynomial of lower degree over the variables $x_{j,1}, x_{j,1}^2, \ldots, x_{j,1}^{2^{\max_{s=1}^{n} \lfloor \log \alpha_{j,1,s} \rfloor}}, \ldots, x_{j,l_j}, x_{j,l_j}^2, \ldots, x_{j,l_j}^{2^{\max_{s=1}^{n} \lfloor \log \alpha_{j,l_j,s} \rfloor}}$ for $1 \le j \le m$. Each party T_i commits to shares of its new inputs via the Efficient Preprocessing protocol described in the full version of the paper [9]. In the *Round Table Step* the parties compute the encrypted evaluations of the monomials in $\mathbf{Q}$ in a round-table fashion. Next, in the *Re-Randomization* phase, each party helps to re-randomize the output shares. Honest behavior of the parties is checked during the *Verification* step via cut-and-choose and a *Preprocessing Verification* protocol for the committed inputs, which is described in the full version of the paper [9]. If the verification passes, the parties jointly decrypt the output shares and the output receiver reconstructs the final polynomial evaluation result in the *Reconstruction* phase. We now present the detailed protocol and state our main theorem[3].

[3] We note that for each intermediate monomial $h_{j,s}$ passed between the parties in the round-table step, each Party j needs to transmit only $D_{h,j,s} + 1$ shares to Party $j+1$ since the rest of the shares may be constructed by the receiving party via Lagrange interpolation over committed values. This may yield significant savings in the communication complexity, which we assumed in our discussion in the introduction.

Multiparty Polynomial Evaluation Protocol Π_{poly_eval}

Inputs: $T_1 : \overline{X}_1, sk_1; \ldots; T_m : \overline{X}_m, sk_m$
Outputs: $T^* : \mathbf{Q}(x_{1,1}, \ldots, x_{1,l_1}, \ldots, x_{m,1}, \ldots, x_{m,l_m}); \{T_1, \ldots, T_m\} \backslash T^* : \perp$

Input Preprocessing:

1. For $1 \leq j \leq m$, Party T_j converts each $h_{j,s}$ for $1 \leq s \leq n$ in the form of Equation 1 (each variable is of degree at most one).
2. For $1 \leq j \leq m$, Party T_j runs the Efficient Preprocessing protocol (see [9]) to generate $10kD$ shares for each of its new inputs $x_{j,1}, x_{j,1}^2, \ldots, x_{j,1}^{2^{\lfloor \log \alpha_{j,1} \rfloor}}, \ldots, x_{j,l_j}, x_{j,l_j}^2, \ldots, x_{j,l_j}^{2^{\lfloor \log \alpha_{j,l_j} \rfloor}}$ where $\alpha_{j,t} = \max_{i=1}^n \alpha_{j,t,i}$ for $1 \leq t \leq l_j$ and commits to the shares.

Round-Table Step:

3. Party T_1 computes encryptions of the polynomials coefficients $b_{0,i} = \text{ENC}_{pk}(c_s; 1)$ for $1 \leq s \leq n$.
4. For $1 \leq s \leq n$, T_1 chooses $D_{h,1,s} + 1$ random numbers $r_1^{1,s}, \ldots, r_{D_{h,1,s}+1}^{1,s}$. T_1 uses the Randomness Interpolation protocol to compute $(r_{D_{h,1,s}+2}^{1,s})^{\Delta}, \ldots, (r_{10kD}^{1,s})^{\Delta}$.
5. For $1 \leq i \leq 10kD$, $1 \leq s \leq n$ T_1 uses the values chosen above to compute

$$h_{1,s}(i) = h_{1,s}(P_{x_{1,1}}(i), \ldots, P_{x_{1,1}^{2^{\lfloor \log \alpha_{1,1,j} \rfloor}}}(i), \ldots, P_{x_{1,l_1}}(i), \ldots, P_{x_{1,l_1}^{2^{\lfloor \log \alpha_{1,l_1,s} \rfloor}}}(i))$$

and $b_{1,s,i} = b_{0,j}^{\Delta \cdot h_{1,s}(i)} \cdot \text{ENC}pk(0; r_i^1)^{\Delta}$, which he sends to party T_2.
6. For each $2 \leq j \leq m$:
   (a) Party T_j receives from party T_{j-1} coefficients $b_{j-1,1,i}, \ldots, b_{j-1,n,i}$ for $1 \leq i \leq 10kD$.
6. (b) For $1 \leq s \leq n$, T_j chooses $D_{h,j,s} + 1$ random numbers $r_1^{j,s}, \ldots, r_{D_{h,j,s}+1}^{j,s}$. P_j uses the Randomness Interpolation protocol to compute $(r_{D_{h,j,s}+2}^{j,s})^{\Delta}, \ldots, (r_{10kD}^{j,s})^{\Delta}$.
   (c) For $1 \leq s \leq n$ T_j uses the values $r_i^{j,s}$ chosen above to compute

$$h_{j,s}(i) = h_{j,s}(P_{x_{j,1}}(i), \ldots, P_{x_{j,1}^{2^{\lfloor \log \alpha_{j,1} \rfloor}}}(i), \ldots, P_{x_{j,l_j}}(i), \ldots, P_{x_{j,l_j}^{2^{\lfloor \log \alpha_{j,l_j} \rfloor}}}(i))$$

   and $b_{j,s,i} = b_{j-1,s,i}^{\Delta \cdot h_{j,s}(i)} \cdot \text{ENC}_{pk}(0; r_i^{j,s})^{\Delta}$.
   (d) If $j < m$ T_j sends all $b_{j,s,i}$ to T_{j+1}. If $j = m$ for each $1 \leq i \leq 10kD$ T_m computes $S_i' = \Pi_{s=1}^n b_{m,s,i}$ and sends them to all parties on the broadcast channel.

Re-Randomization Step:

7. For $1 \leq j \leq m$, Party T_j computes polynomial $P_{j,0}$ of degrees kD such that $P_{j,0}(0) = 0$.
8. For $1 \leq j \leq m$, Party T_j chooses $kD + 1$ random values $r_{j,1}, \ldots, r_{j,kD+1}$ and uses the Randomness Interpolation protocol to compute $r^{\Delta}_{j,kD+2}, \ldots, r^{\Delta}_{j,10kD}$. T_j commits to shares $Z_{j,0} = \text{ENC}_{pk}(P_{j,0}(i); r_{j,i})^{\Delta}$ for $1 \leq i \leq 10kD$
9. All parties run the LIPEV protocol and a zero knowledge proof protocol (HEPKPV, see [9]) to ensure that each $[Z_{j,i}]_{1 \leq i \leq 10kD}$ is an encryption of a polynomial with constant coefficient 0.
10. The final encryptions are: $S_i = S'_i \cdot \Pi^m_{j=1} Z_{j,0}$ for $1 \leq i \leq 10kD$.

Verification:

11. All parties verify using the Lagrange encrypted interpolation protocol that the values S_i lie on a polynomial of degree kD. Otherwise reject.
12. All parties run a multi-party coin-tossing protocol (see [9]) to choose a random subset I of size k from $[1, 10kD]$.
13. For each $i \in I$ parties $T_1, \ldots, T_m$ decommit their corresponding shares from the Efficient Input Preprocessing.
14. All parties run the Preprocessing Verification for their inputs (see [9]).
15. For each $i \in I$ each party T_j decommits the i-th shares of its inputs as well as the i-th share of the polynomials $P_{j,0}$. Additionally, each party T_j reveals the randomness $r^{j,s}_i$ for $1 \leq s \leq n$ and $r_{j,i}$ used for the corresponding shares. To verify, each party recomputes the entire share S^*_i, using the inputs and randomness revealed and checks that $S_i = S^*_i$. If any verification fails the protocol is aborted.

Reconstruction:

16. For each $1 \leq i \leq 10kD$ each party computes its partial decryption $s_{i,j}$ of S_i and sends it to the designated output receiver T^*.
17. Party T^* uses the partial decryptions $s_{i,j}$ for $1 \leq j \leq m$ to completely decrypt S_i. T^* reconstructs the value of $\mathbf{Q}(x_{1,1}, \ldots, x_{1,l_1}, \ldots, x_{m,1}, \ldots, x_{m,l_m})$ via interpolation and division by Δ^m.

Theorem 1. *If the Decisional Composite Residuosity problem is hard in $\mathbf{Z}^*_{n^2}$, where n is a product of two strong primes, and protocol Π_{poly_eval} is instantiated with the threshold Paillier encryption scheme TP^m_{enc} such that $E = TP^m_{enc}$, then Π_{poly_eval} securely computes the Polynomial Evaluation functionality in the presence of malicious adversaries.*

5 Communication and Computational Complexity

Our protocol computes the polynomial functionality in a constant number of rounds (counting round-table rounds as one, or otherwise when we have a constant number of players). The communication complexity of the protocol is divided into two types: broadcast messages and "round-table" (neighbors only) communication; we note that the "round-table" communication can be done off-line. The broadcast communication consists of the commitments of the inputs shares, the decommitments used in the final verification phase, the encrypted and decrypted output shares as well as the messages used in the coin tossing and HEPKPV protocols. These messages add up to $k(10D+1)(\sum_{j=1}^{m}\sum_{t=1}^{l_j}\log\alpha_{j,t}+1)$. km Note that the communication complexity may be much smaller than the size of the polynomial representation. For example, if party P_j with input $x_{j,1}$ must contribute $\alpha_{j,t}$ consecutive powers of x_i: $x_i^1,\dots,x_i^{\alpha_{j,t}}$ to $\alpha_{j,t}$ different terms, the broadcast communication complexity for this party will still only be $k(10D+1)\log\alpha_{j,t}+1$. round-table messages passed between consecutive parties include all intermediate messages in the computation that are sent by all the parties except the last one, which in total are $10kDn(m-1)$. The computational complexity (where we count number of exponentiations) for all m parties in total is $O(kDnm)$. Further, if we apply the share packing optimization from [14] over k executions of the protocol we can drop k factor for the new amortized complexities.

In summary, our protocol runs in constant number of "round table" rounds, in which every party is involved in order, while the protocol for secure computation of arithmetic circuits [24] requires as many rounds as the depth of the arithmetic circuit. It also requires fewer broadcasted messages compared to techniques proving the polynomial evaluation via zero knowledge proofs such as [2] since any ZK proof will have to be broadcasted. Additionally, a ZK protocol will require runs of a multiparty coin tossing protocol to generate randomness for each ZK proof.

6 Protocol Optimizations and Application to Multiparty Set Intersection

We apply several optimizations to the protocol given in Section 4 for polynomials with specific structures. First, if we have a monomial that is computed only from the inputs of a subset of the parties, then clearly, we can evaluate it in a round-table fashion that only includes parties in this subset and proceed to the Re-Randomization Step.

Additionally, in some cases, we can remove the requirement of a party to share all of its inputs. Recall that we require the input-sharing in order to enable the cut-and-choose verification of honest behavior of the parties In the case when an input is used only once in the polynomial, this type of proof may not be necessary. We can avoid sharing an input if it belongs to the first party in the round table computation of the corresponding monomial as long as we can verify that the encryption itself is valid with a ZKPOK and extract the encrypted value. We notice that the requirements imposed on the structure of the polynomial in order to be able to apply this optimization substantially limit the range of possible polynomials. However, in the next section we will see how the problem of multiparty set intersection can be reduced to the evaluation of exactly this type of polynomials.

Finally, we use the approach of multi-secret sharing from [14] that allows us to use the same polynomials to share the input values for multiple parallel executions of the protocol, which lowers the amortized communication complexity of our protocol. Intuitively, we choose a set of points on the sharing polynomials to represent the input values for each of the different executions of the protocol, say points 1 to k for each of k different executions. The shares that will be used in the computation will be those corresponding to points not in this set. As a result, the final output polynomial will evaluate to each of the different output values corresponding to each execution at the points 1 to k respectively.

In the setting of our protocol in Section 4 we assume that the multivariate polynomial is known to all parties. By removing this requirement and assuming that the polynomial coefficients are the inputs of one of the parties, we reduce the problem to oblivious multivariate polynomial evaluation (introduced by [27] in the single-variable case) for a small class of multivariate polynomials.

6.1 Multiparty Set Intersection

We apply the techniques introduced in Section 4 to the problem of multiparty set intersection. We give here a brief sketch. In the multiparty set intersection problem, there are m parties $T_1, \ldots, T_m$ who have input sets $X_1, \ldots, X_m$ and wish to jointly compute $X_1 \bigcap \ldots \bigcap X_m$.

Recall that a set $X = \{x_1, \ldots, x_d\}$ can be represented as a polynomial $P(x) = (x - x_1) \ldots (x - x_d)$. Now if we consider the polynomial $P'(x) = r \cdot P(x) + x$, where r is random, we have that if $x' \in X$ then $P'(x') = x'$ and if $x' \notin X$ then $P'(x')$ is uniformly distributed (see [15]). In the multiparty case we have m parties with input sets $X_1, \ldots, X_m$, represented by polynomials $P_{X_1}(x), \ldots, P_{X_m}(x)$. Thus the polynomial $\mathbf{R}(x) = \mathbf{r} \cdot \sum_{i=1}^{m-1} P_{X_i}(x) + x$, where $\mathbf{r} = r_1 + r_2 + \cdots + r_m$ and each r_i is a randomly chosen input contributed by Party i, will have the same property mentioned above: if $x' \in X_1 \bigcap \ldots \bigcap X_m$ then $\mathbf{R}(x') = x'$ and if $x' \notin X_1 \bigcap \ldots \bigcap X_m$ then $\mathbf{R}(x')$ is uniformly distributed. Now in the setting of polynomial evaluation we let a designated party P_m evaluates $\mathbf{R}(x)$ on its own inputs, and thus the output is exactly the intersection of all sets with some additional random values. This problem now reduces to the Multiparty Polynomial Evaluation problem.

Theorem 2. *If the Decisional Composite Residuosity problem is hard in $\mathbf{Z}_{n^2}^*$, where n is a product of two strong primes, protocol Π_{poly_eval} is instantiated with the threshold Paillier encryption scheme TP_{enc}^m such that $E = TP_{enc}^m$, and*

$\mathbf{Q} = \mathbf{R}$, then Π_{poly_eval} securely computes the Set Intersection functionality[4] *in the presence of malicious adversaries.*

Using the optimizations described in this section, we have that the broadcast communication complexity of the Set Intersection protocol is $O(md + 10d \log^2 d)$ (there is no round-table communication) and the computational complexity is $O(md^2)$, where $d >> k$ is the maximum input set size of each party.

References

1. Ben-Or, M., Goldwasser, S., Wigderson, A.: Completeness theorems for non-cryptographic fault-tolerant distributed computation. In: Proceedings of the Twentieth Annual ACM Symposium on Theory of Computing (1988)
2. Camenisch, J., Michels, M.: Proving in zero-knowledge that a number is the product of two safe primes. In: Stern, J. (ed.) EUROCRYPT 1999. LNCS, vol. 1592, pp. 107–122. Springer, Heidelberg (1999)
3. Canetti, R.: Security and composition of multiparty cryptographic protocols. Journal of Cryptology 13 (2000)
4. Chaum, D., Crépeau, C., Damgard, I.: Multiparty unconditionally secure protocols. In: STOC 1988: Proceedings of the Twentieth Annual ACM Symposium on Theory of Computing, pp. 11–19. ACM, New York (1988)
5. Cheon, J.H., Jarecki, S., Seo, J.H.: Multi-party privacy-preserving set intersection with quasi-linear complexity. Cryptology ePrint Archive, Report 2010/512 (2010), `http://eprint.iacr.org/`
6. Choi, S., Dachman-Soled, D., Malkin, T., Wee, H.: Black-box construction of a non-malleable encryption scheme from any semantically secure one. In: Canetti, R. (ed.) TCC 2008. LNCS, vol. 4948, pp. 427–444. Springer, Heidelberg (2008)
7. De Cristofaro, E., Kim, J., Tsudik, G.: Linear-complexity private set intersection protocols secure in malicious model. In: Abe, M. (ed.) ASIACRYPT 2010. LNCS, vol. 6477, pp. 213–231. Springer, Heidelberg (2010)
8. Dachman-Soled, D., Malkin, T., Raykova, M., Yung, M.: Efficient robust private set intersection. In: Abdalla, M., Pointcheval, D., Fouque, P.-A., Vergnaud, D. (eds.) ACNS 2009. LNCS, vol. 5536, pp. 125–142. Springer, Heidelberg (2009)
9. Dachman-Soled, D., Malkin, T., Raykova, M., Yung, M.: Multiparty secure computation over multivariate polynomials. Technical Report CUCS-024-10 (2010)
10. ElGamal, T.: A public key cryptosystem and a signature scheme based on discrete logarithms. In: CRYPTO 1984, pp. 10–18. Springer-Verlag New York, Inc., New York (1985)
11. Feldman, P.: A practical scheme for non-interactive verifiable secret sharing. In: FOCS, pp. 427–437. ACM, New York (1987)
12. Fouque, P.A., Poupard, G., Stern, J.: Sharing decryption in the context of voting or lotteries. In: Frankel, Y. (ed.) FC 2000. LNCS, vol. 1962, pp. 90–104. Springer, Heidelberg (2001)

[4] We consider here a slight variant of the Set Intersection functionality where Party i for $1 \leq i \leq m-1$ submits the polynomial P_{X_i} to the Trusted Party, Party m submits X_m and the Trusted Party returns the intersection of $X_1, \ldots, X_m$. In order to compute the standard Set Intersection functionality, we must use the threshold El Gamal encryption scheme.

13. Franklin, M., Mohassel, P.: Efficient and secure evaluation of multivariate polynomials and applications. In: Zhou, J., Yung, M. (eds.) ACNS 2010. LNCS, vol. 6123, pp. 236–254. Springer, Heidelberg (2010)
14. Franklin, M., Yung, M.: Communication complexity of secure computation (extended abstract). In: STOC 1992: Proceedings of the Twenty-fourth Annual ACM Symposium on Theory of Computing, pp. 699–710 (1992)
15. Freedman, M., Nissim, K., Pinkas, B.: Efficient private matching and set intersection. In: Cachin, C., Camenisch, J.L. (eds.) EUROCRYPT 2004. LNCS, vol. 3027, pp. 1–19. Springer, Heidelberg (2004)
16. Galil, Z., Haber, S., Yung, M.: Cryptographic computation: Secure fault tolerant protocols and the public-key model. In: Pomerance, C. (ed.) CRYPTO 1987. LNCS, vol. 293, pp. 135–155. Springer, Heidelberg (1988)
17. Gennaro, R., Jarecki, S., Krawczyk, H., Rabin, T.: Secure distributed key generation for discrete-log based cryptosystems. J. Cryptol. 20(1), 51–83 (2007)
18. Gentry, C.: Fully homomorphic encryption using ideal lattices. In: STOC 2009: Proceedings of the 41st Annual ACM Symposium on Theory of Computing, pp. 169–178. ACM, New York (2009)
19. Goldreich, O., Micali, S., Wigderson, A.: How to play any mental game. In: STOC 1987: Proceedings of the Nineteenth Annual ACM Symposium on Theory of Computing, pp. 218–229. ACM, New York (1987)
20. Goldreich, O.: Foundations of cryptography: a primer. Found. Trends Theor. Comput. Sci. 1(1), 1–116 (2005)
21. Goldwasser, S., Micali, S.: Probabilistic encryption and how to play mental poker keeping secret all partial information. In: STOC 1982: Proceedings of the Fourteenth Annual ACM Symposium on Theory of Computing, pp. 365–377. ACM, New York (1982)
22. Hazay, C., Lindell, Y.: Efficient protocols for set intersection and pattern matching with security against malicious and covert adversaries. In: Canetti, R. (ed.) TCC 2008. LNCS, vol. 4948, pp. 155–175. Springer, Heidelberg (2008)
23. Ishai, Y., Prabhakaran, M., Sahai, A.: Founding cryptography on oblivious transfer efficiently. In: Wagner, D. (ed.) CRYPTO 2008. LNCS, vol. 5157, pp. 572–591. Springer, Heidelberg (2008)
24. Ishai, Y., Prabhakaran, M., Sahai, A.: Secure arithmetic computation with no honest majority. In: Reingold, O. (ed.) TCC 2009. LNCS, vol. 5444, pp. 294–314. Springer, Heidelberg (2009)
25. Jarecki, S., Liu, X.: Efficient oblivious pseudorandom function with applications to adaptive OT and secure computation of set intersection. In: Reingold, O. (ed.) TCC 2009. LNCS, vol. 5444, pp. 577–594. Springer, Heidelberg (2009)
26. Kissner, L., Song, D.: Privacy-preserving set operations. In: Shoup, V. (ed.) CRYPTO 2005. LNCS, vol. 3621, pp. 241–257. Springer, Heidelberg (2005)
27. Naor, M., Pinkas, B.: Oblivious polynomial evaluation. SIAM J. Comput. 35(5), 1254–1281 (2006)
28. Paillier, P.: Public-key cryptosystems based on composite degree residuosity classes. In: Stern, J. (ed.) EUROCRYPT 1999. LNCS, vol. 1592, pp. 223–238. Springer, Heidelberg (1999)
29. Patra, A., Choudhary, A., Rangan, C.: Information theoretically secure multi party set intersection re-visited. In: Jacobson Jr., M.J., Rijmen, V., Safavi-Naini, R. (eds.) SAC 2009. LNCS, vol. 5867, pp. 71–91. Springer, Heidelberg (2009)

30. Patra, A., Choudhary, A., Rangan, C.P.: Round efficient unconditionally secure mpc and multiparty set intersection with optimal resilience. In: Roy, B., Sendrier, N. (eds.) INDOCRYPT 2009. LNCS, vol. 5922, pp. 398–417. Springer, Heidelberg (2009)
31. Rabin, T., Ben-Or, M.: Verifiable secret sharing and multiparty protocols with honest majority (extended abstract). In: STOC 1989: Proceedings of the Twenty-first Annual ACM Symposium on Theory of Computing, pp. 73–85 (1989)
32. Sang, Y., Shen, H.: Efficient and secure protocols for privacy-preserving set operations. ACM Trans. Inf. Syst. Secur. 13, 9:1–9:35 (2009)
33. Shamir, A.: How to share a secret. Commun. ACM 22(11), 612–613 (1979)
34. Yao, A.C.C.: Protocols for secure computations. In: FOCS, pp. 160–164 (1982)
35. Yao, A.C.C.: How to generate and exchange secrets (extended abstract). In: FOCS, pp. 162–167 (1986)

Private Discovery of Common Social Contacts

Emiliano De Cristofaro[1], Mark Manulis[2], and Bertram Poettering[2]

[1] Computer Science Department, University of California, Irvine
edecrist@uci.edu
[2] Cryptographic Protocols Group, TU Darmstadt & CASED, Germany
mark@manulis.eu, bertram.poettering@cased.de

Abstract. The increasing use of computing devices for social interactions propels the proliferation of online social applications, yet, it prompts a number of privacy concerns. One common problem occurs when two unfamiliar users, in the process of establishing social relationships, want to assess their social proximity by discovering mutual contacts. In this paper, we introduce *Private Contact Discovery*, a novel cryptographic primitive that lets two users, on input their respective contact lists, learn their common contacts (if any), and nothing else. We present an efficient and provably secure construction, that (i) prevents arbitrary list manipulation by means of contact certification, and (ii) guarantees user authentication and revocability. Following a rigorous cryptographic treatment of the problem, we define the privacy-protecting *contact-hiding* property and prove it for our solution, under the RSA assumption in the Random Oracle Model (ROM). We also show that other related cryptographic techniques, such as Private Set Intersection and Secret Handshakes, are unsuitable in this context. Experimental analysis attests to the practicality of our technique, which achieves computational and communication overhead (almost) linear in the number of contacts.

1 Introduction

The increasing volume of electronic social interactions motivates the need for efficient privacy-enhancing techniques. One interesting problem occurs when two unfamiliar users want to *privately* discover their common contacts: in doing so, they would like to reveal to each other only the contacts that they share.

We focus, for instance, on the discovery of mutual social-network *friends*. Online social networks provide friends with services to share interests, activities, or pictures, and help them build and reflect social relations. Popular social network websites, such as Facebook, Linkedin, or MySpace, involve millions of active users, who access their services *ubiquitously* — e.g., 250 of 500 million Facebook users access it from their mobile devices [39]. Other projects, such as Nokia's Awarenet [1], aim at letting users in physical proximity interact using their mobile phones, without relying on a central server or using an Internet connection.

One of the first steps toward establishing social-network relationships is to verify the existence of *common* friends. Consider the following settings: (1) a social network user wants to extend her network and is willing to establish friendships with other users with whom she has mutual friends; (2) a mobile phone

J. Lopez and G. Tsudik (Eds.): ACNS 2011, LNCS 6715, pp. 147–165, 2011.

user would like to interact with other users in physical proximity (e.g., in a bar or on the subway), given that they have some common friends on a given social network, e.g., Facebook. One crucial problem, in these scenarios, is discovering mutual friends in a *privacy-preserving* manner.

A naïve solution would require users to reveal their friends to each other. Clearly, this would not preserve users' privacy, since their complete lists would be exposed. Another trivial solution would employ and trust a central server to find and output the common friends. However, a central server is not necessarily trusted and not always available. Also, such a server would learn not only users' friends, but also which users become friends, how, when, and where. For instance, in scenario (2) above, mobile phone users might be willing to discover their mutual friends on a social network (e.g., Facebook) but may not be connected to the Internet or they may want to operate outside the social network website.

In order to protect privacy, we are faced with a couple of fundamental issues. First, we need to prevent a malicious user from manipulating her list of friends, e.g., by populating it with her best guesses of other user's list to maximize the amount of information learned, or by "impersonating" unwarranted friendships. Then, as friend relationships may vary over time, we need an efficient mechanism allowing to *revoke* friendships.

In this paper, we introduce the concept of **Private Contact Discovery**, a novel general construct geared to preserve user privacy, not only in social network interactions, but also in any other application that uses personal contact lists. We design a cryptographic primitive involving two users, e.g., Alice and Bob, on input their contact lists, that outputs only the list of mutual contacts (if any). The protocol prevents users from claiming unwarranted friendships by introducing *contact certification*. For instance, in order to include Carol in her contact list, Alice needs to obtain a certificate from Carol attesting this friendship. Then, when Alice interacts with Bob, not only the entries in her contact list are hidden from Bob, but also the possession of corresponding certificates with respect to non-common friends (and vice-versa). Note that, in our solution, these certificates are specific to individual users, i.e. malicious transfer of certificates, e.g. to enable others to claim unwarranted friendship, is impossible. Our protocol does not require any trusted server nor is bound to a specific network infrastructure, and can be used in both centralized and distributed environments.

1.1 Private Contact Discovery with Available Tools?

The problem of Private Contact Discovery bears some resemblance with several cryptographic constructs. We review them below and discuss why they are inappropriate for the problem of Private Contact Discovery.

Private Set Intersection (PSI). PSI techniques, e.g. [2, 13, 14, 15, 20, 23, 24, 29, 30], allow two parties to compute the intersection of their input sets, such that they learn nothing beyond the intersection (and the set sizes). However, PSI does not prevent parties from manipulating their inputs, thus, in the context of contact discovery, it would not prevent users from claiming unwarranted friendships.

Authorized PSI (APSI). APSI [14, 15] extends PSI by ensuring that inputs are authorized by an appropriate Certification Authority (CA). Thus, unless they hold authorizations on their inputs (typically, in the form of digital signatures), parties do not learn whether the corresponding input belongs to the set intersection. Similarly, Private Intersection of Certified Sets [10] allows a trusted CA to ensure that all protocol inputs in PSI are valid and bound to each protocol participant. Note, however, that these constructs involve one single CA, whereas, every user in the context of Contact Discovery would have to act as an independent "CA" for her contacts. Also, one may think that the social network provider could certify friendships, but such a solution would incur a fundamental problem. In fact, in APSI, authorizations are signatures: assuming that Alice and Bob are both friends with Carol, Alice would have a signature on a message in the form of "Carol→Alice", while Bob would have one on "Carol→Bob". Therefore, since (Authorized) Private Set Intersection techniques only output matching elements, we cannot use them for privately discovering common contacts (as messages representing common friendships would not match).

Secure Two-Party Computation. Generic secure two-party computation, e.g. [21, 32, 42], allows two parties with respective private inputs x and y to compute a functionality $f(x, y) = (f_1(x, y), f_2(x, y))$ such that: one party obtains $f_1(x, y)$ and the other receives $f_2(x, y)$, while neither learns more than its own input and the output. The functionality underlying Private Contact Discovery would require malicious model and could, possibly, be expressed and solved using general secure two-party computation techniques. However, the expected computational and communication overhead would be too large (possibly thousands of rounds of interaction) and thus not be applicable in practice.

Anonymous Credentials (AC). AC schemes, e.g. [7, 9], allow a provider to issue to a user an anonymous credential on various attributes. The user can then prove to a third party that she possesses valid credentials issued by that provider, yet without revealing further information about credentials and attributes. AC schemes do not seem to offer an immediate solution to Private Contact Discovery. Indeed, one could think that user's friends may act as providers issuing friendship credentials, however, AC proofs would disclose information about credential issuers. For the same reason, Credential-Authenticated Key Exchange [8] does not provide an immediate solution to the Private Contact Discovery problem.

Affiliation-Hiding Authentication (AHA). AHA protocols, also called *Secret Handshakes* (SH) [3, 11, 26, 28, 41], allow two parties with membership credentials issued by the same organization — called *Group Authority* (GA) — to *privately* authenticate each other. Specifically, one party can prove to the other that it has a valid credential, yet this proof hides the identity of the issuing organization, unless the other party also has a valid credential from the same organization. Some protocols [6, 27, 33] efficiently support *multiple* credentials, i.e, multiple Group Authorities, and are more closely related to the Contact Discovery problem. Specifically, Jarecki and Liu [27] introduced a multiple-credential Affiliation-Hiding Authentication scheme, with overhead (almost) linear in the number of credentials, secure under GapDH assumption. Recently, Manulis, Pinkas, and Poettering [33] proposed another efficient multiple-credential AHA scheme, secure under RSA assumption. One could think that Private Contact Discovery can be solved using efficient multiple-credential AHAs. For instance, every user could act as a GA and issue credentials as a contact certification: whenever two users want to discover whether or not they have common contacts, they execute AHA on input their credentials. However, this approach would incur several problems. In fact, multiple-credential AHA schemes, such as [27, 33], assume that GAs are unconditionally trusted and always follow protocol specification. While this assumption might be realistic in classic AHA scenarios (where GAs are courts or investigation agencies), it is not reasonable, in the context of Contact Discovery, to trust all users, e.g., of a social network. Consider the case of [27]: in the process of obtaining credentials from GAs, users need to surrender all their secret keys which are not dependent on the specific group but are valid for all of them. If Eve certifies Bob to be her friend, she would obtain Bob's secret keys, thus, she would be able to impersonate Bob and/or test Bob's friendship with other users. Although recent results in [35, 36] relax some of the trust assumptions on GAs in AHA protocols, it is not clear how to efficiently extend them to the multiple-credential setting.

Friend-of-Friend. Prior work has attempted to solve problems similar to the one considered in this paper. Von Arb et al. [40] present a mobile social networking platform which enables *Friend-of-Friend* (FoF) detection in physical proximity. Matching of friend lists is provided using PSI techniques [25, 29]. As discussed earlier, this approach fails to effectively guarantee privacy, as contact lists can be artificially expanded. Freedman and Nicolosi [19] propose two solutions for the FoF problem, in the context of trust establishment in email whitelisting. One solution is based on hash functions and symmetric encryptions, the other on bilinear maps. Both solutions leverage friendship attestation but do not support user revocation — a necessary requirement in our context. Also note that, as opposed to our protocol: (1) their solution based on symmetric encryption allows users to maliciously transfer attestations to other users, and (2) their technique using bilinear maps is inefficient, as it involves a quadratic number of bilinear map operations.

Non-Cryptographic Techniques. Besides the work focusing on protecting privacy by means of cryptographic techniques, some solutions targeting the discovery of common contacts have been proposed in different and broader contexts. Some techniques address the Friend-of-Friend problem with none or unclear privacy properties [12, 31]. Other solutions analyze, to a higher extent, social relationships, without focusing on privacy. For instance, [38] uses random walks to discover *communities* in large social-network graphs, [43, Chapter 12] formalizes the problem of dynamically identifying core communities (i.e., sets of entities with frequent and consistent interactions), [44] builds a prediction model to identify certain social structures, e.g., friendship ties and family circles, while [17] attempts at identifying communications that substantiate social relationship types.

Remark 1. Private Contact Discovery can be used as an important building block for privacy-preserving social interactions. Indeed, although prior work has focused on privacy concerns in this context, we highlight the need for a cryptographic treatment of them to obtain clear guarantees. This includes formal definition of privacy goals and design of provably secure and practical solutions. Also, Günther, Manulis, and Strufe [22] recently proposed a cryptographic model and solutions for Private User Profiles, another building block for privacy in social interactions.

1.2 Contribution and Organization

Our contributions are manifold: First, we define Private Contact Discovery, a novel cryptographic tool that allows two users to discover their common contacts, without leaking information on any other contacts, and without relying on any (trusted) third parties. Second, we provide rigorous privacy definitions and security model for this new notion. In particular, we define its main privacy goal called *Contact-Hiding.* Finally, we propose a very efficient solution, secure under the RSA assumption in ROM, which also supports efficient revocation. Performance analysis attests to the practicality of our protocol, which incurs almost linear computational and communication complexities in the number of alleged contacts. This efficiency stems from the use of the recent Index-Hiding Message Encoding (IHME) scheme [33, 34].

Paper Organization. After preliminaries in Section 2, we introduce Private Contact Discovery and present our solution, alongside its performance analysis, in Section 3. Next, in Section 4, we formalize the security model for Private Contact Discovery and state Contact-Hiding security of our scheme. Section 5 concludes the paper and provides an outlook into further research directions. In Appendix, we present the proof of Contact-Hiding security of our solution.

2 Preliminaries: Assumptions and Building Blocks

Definition 1 (RSA Assumption on Safe Moduli). *Let* $\mathsf{RSA\text{-}G}(\kappa')$ *be a probabilistic algorithm that outputs pairs* (N, e)*, where* $N = PQ$ *for random* κ'*-bit primes* $P \neq Q$ *such that* $P = 2P' + 1, Q = 2Q' + 1$ *for primes* P', Q'*, and* $e \in \mathbb{Z}_{\varphi(N)}$ *is coprime to* $\varphi(N)$*. The RSA-success probability of a PPT solver* $\mathcal{A}$ *is defined as*

$$\mathsf{Succ}^{\mathsf{rsa}}_{\mathcal{A}}(\kappa') = \Pr\big[(N, e) \leftarrow \mathsf{RSA\text{-}G}(\kappa');\ z \overset{\$}{\leftarrow} \mathbb{Z}_N;\ m \leftarrow \mathcal{A}(N, e, z);\ m^e = z \pmod N\big].$$

The RSA assumption on Safe Moduli *states that the maximum RSA-success probability* $\mathsf{Succ}^{\mathsf{rsa}}(\kappa')$ *(defined over all PPT solvers* $\mathcal{A}$*) is negligible in* κ'*.*

One important building block of our protocol is an Index-Hiding Message Encoding (IHME) scheme, recently introduced by Manulis, Pinkas, and Poettering [33], which we review below. Definition 2 recalls the underlying concept of Index-*Based*

Message Encoding (IBME), also introduced in [33]. It is an encoding technique that combines a set of *indexed* input messages, $m_1, \ldots, m_n \in \mathcal{M}$ (where $\mathcal{M}$ is a message space), into a single data structure $\mathcal{S}$. Any message can be individually recovered from $\mathcal{S}$ using its index, which is arbitrarily chosen from an index set $\mathcal{I}$, and specified at encoding-time.

Definition 2 (Index-Based Message Encoding). *An* index-based message encoding *scheme* (iEncode, iDecode) *over an index space $\mathcal{I}$ and a message space $\mathcal{M}$ consists of two efficient algorithms:*

iEncode($\mathcal{P}$): *On input a tuple of index/message pairs $\mathcal{P} = \{(i_1, m_1), \ldots, (i_n, m_n)\} \subseteq \mathcal{I} \times \mathcal{M}$, with distinct indices $i_1, \ldots, i_n$, this algorithm outputs an encoding $\mathcal{S}$.*

iDecode($\mathcal{S}, i$): *On input of an encoding $\mathcal{S}$ and an index $i \in \mathcal{I}$ this algorithm outputs a message $m \in \mathcal{M}$.*

An index-based message encoding scheme is correct *if* iDecode(iEncode($\mathcal{P}$), i_j) $= m_j$ *for all $j \in \{1, \ldots, n\}$ and all tuples $\mathcal{P} = \{(i_1, m_1), \ldots, (i_n, m_n)\} \subseteq \mathcal{I} \times \mathcal{M}$ with distinct indices i_j.*

Further, [33] defines IBME schemes that guarantee *index-hiding* security as *'Index-Hiding Message Encoding'* (IHME) schemes. Informally, an IHME scheme guarantees that no adversary, by inspecting an IBME structure $\mathcal{S}$ that encodes random messages, can learn any useful information about the deployed indices, even if she knows some of the indices and/or messages. We refer to [33] for the formal definition of the index-hiding property. Here we only recall the polynomial-based construction of perfect IHME from [33]. It is defined over $\mathcal{I} = \mathcal{M} = \mathbb{F}$ for an arbitrary finite field $\mathbb{F}$ (e.g., $\mathbb{F} = GF(p)$ as in [33]) and provides the index-hiding property in an information-theoretic sense.

iEncode($\mathcal{P}$): On input of $\mathcal{P} = \{(i_1, m_1), \ldots, (i_n, m_n)\} \subseteq \mathcal{I} \times \mathcal{M} = \mathbb{F}^2$, the encoding is defined as the list $\mathcal{S} = (c_{n-1}, \ldots, c_0)$ of coefficients of the polynomial $f = \sum_{k=0}^{n-1} c_k x^k \in \mathbb{F}[x]$ that interpolates all points in $\mathcal{P}$, i.e. $f(i_j) = m_j$ for all $(i_j, m_j) \in \mathcal{P}$. Note that this polynomial exists uniquely, i.e., the iEncode algorithm is deterministic.

iDecode($\mathcal{S}, i$): On input of $\mathcal{S} = (c_{n-1}, \ldots, c_0)$ and index $i \in \mathcal{I}$, this algorithm outputs the evaluation $m = f(i) = \sum_{k=0}^{n-1} c_k i^k$ of f at position i.

Our protocol deploys the IHME scheme in a black-box way, thus proposing another application of this recent primitive. Furthermore, in our experimental analysis, we will also take into account recent optimizations regarding the improved polynomial interpolation algorithms and implementation of the IHME scheme presented in [34].

3 Private Contact Discovery

3.1 Contact Discovery: Syntax and Correctness

A Contact Discovery Scheme CDS is defined as a tuple of four algorithms and protocols:

$\mathsf{Init}(1^\kappa)$: This algorithm is executed once by each user U. On input of a security parameter 1^κ, it initializes internal parameters $U.\mathsf{params}$ and clears U's contact revocation list, i.e., $U.\mathsf{crl} = \emptyset$. $U.\mathsf{crl}$ is authenticated by U and will be distributed to other users (i.e., all contacts of U should have access to the up-to-date $U.\mathsf{crl}$). In contrast, $U.\mathsf{params}$ is private to U.

$\mathsf{AddContact}(U \leftrightarrow V)$: This is a protocol, executed between user U and user V, who wishes to become a contact of U. User U adds identity of V to her contact list. In addition, a corresponding contact certificate $\mathsf{cc}_{U\to V}$ is output to V. Note that we model contact establishment as a unidirectional process: If U should become a contact of V as well then they additionally will execute $\mathsf{AddContact}(V \leftrightarrow U)$.

$\mathsf{RevokeContact}(U, V)$: This algorithm is executed by user U. On input identity of V, the contact revocation list of U is updated to $U.\mathsf{crl} \leftarrow U.\mathsf{crl} \cup \{V\}$.

$\mathsf{Discover}(V \leftrightarrow V')$: This is an interactive algorithm (protocol), executed between users V and V', to discover common contacts. V's private input is $(\mathsf{role}, \mathsf{CL}, \mathsf{partner})$, where $\mathsf{role} \in \{\mathsf{init}, \mathsf{resp}\}$ specifies the role of the session as initializer or responder, contact list CL is a set of pairs of the form $(U, \mathsf{cc}_{U\to V})$, for some users U, and $\mathsf{partner}$ is the name/id of the supposed protocol partner. All values $\mathsf{cc}_{U\to V}$ are contact certificates previously obtained as output of $\mathsf{AddContact}(U \leftrightarrow V)$. V''s private input is $(\mathsf{role}', \mathsf{CL}', \mathsf{partner}')$, defined analogously. Further, users keep track of the state of created $\mathsf{Discover}(\mathsf{role}, \mathsf{CL}, \mathsf{partner})$ protocol sessions π through session variables that are initialized as follows: $(\pi.\mathsf{role}, \pi.\mathsf{CL}, \pi.\mathsf{partner}) \leftarrow (\mathsf{role}, \mathsf{CL}, \mathsf{partner})$, $\pi.\mathsf{state} \leftarrow \mathsf{running}$, $\pi.\mathsf{SCL} \leftarrow \emptyset$, and $\pi.\mathsf{id}$ is set to the own identity. After the protocol completes, $\pi.\mathsf{state}$ is updated to either $\mathsf{rejected}$ or $\mathsf{accepted}$. In the latter case, *shared contact list* $\pi.\mathsf{SCL}$ holds a non-empty set of user identifiers.

Definition 3 (Correctness of CDS). *Assume that users V and V' interact in a* Discover *protocol on input* $(\mathsf{role}, \mathsf{CL}, \mathsf{partner})$ *and* $(\mathsf{role}', \mathsf{CL}', \mathsf{partner}')$, *respectively. Let π and π' denote the corresponding sessions. Let $\mathsf{CL}_\cap$ denote the set of users (contacts) U that appear in both* CL *and* CL' *with the restriction that neither* partner *nor* partner′ *are contained in the respective contact revocation lists.* CDS *scheme is* correct *if: (1) π and π' complete in the same state, which is* accepted *iff* $(\mathsf{role} \neq \mathsf{role}' \wedge \mathsf{CL}_\cap \neq \emptyset \wedge \mathsf{partner} = \pi'.\mathsf{id} \wedge \mathsf{partner}' = \pi.\mathsf{id})$, *and (2) if the sessions accept then* $\pi.\mathsf{SCL} = \pi'.\mathsf{SCL} = \mathsf{CL}_\cap$.

3.2 Protocol Specification

We now present our CDS construction and describe the instantiation of Init, AddContact, RevokeContact, and Discover algorithms. We assume that AddContact sessions among (honest) users of CDS are protected by secure channels, whereas, during Discover, the channel does not need to be confidential. Note that many applications that would use CDS, such as social networks or further group applications, already provide an authentication infrastructure for their users, e.g., they deploy a PKI or use some password-based techniques. Such an authentication infrastructure can then be used for various types of communication, including the execution of CDS protocols. With this assumption in mind, we can now focus on the core functionality of the CDS scheme, namely the private discovery of shared contacts, for which potential attacks may be mounted by other application users, i.e., from the inside.

Let $\kappa, \kappa' \in \mathbb{N}$ denote security parameters, where κ' is polynomially dependent on κ. As a building block, our construction utilizes the $\mathsf{IHME} = (\mathsf{iEncode}, \mathsf{iDecode})$ scheme from [33] (see also Section 2), defined over the finite field $\mathbb{F} = GF(p) \cong \mathbb{Z}_p$, where p is the smallest prime number satisfying $p > 2^{2\kappa'+\kappa}$. In addition, the protocol makes use of two hash functions

$$H : \{0,1\}^* \to [0, p-1] \qquad \text{and} \qquad H^* : \{0,1\}^* \to [0, p-1],$$

modeled as random oracles. For convenience, for each $N \in \mathbb{N}$ of length $2\kappa'$, we define:

$$H_N : \{0,1\}^* \to \mathbb{Z}_N;\ x \mapsto H^*(N \,\|\, x) \bmod N.$$

The four algorithms and protocols of CDS are instantiated as follows:

$\mathsf{Init}(1^\kappa)$. The setup routine run by each user U mainly consists of the generation of safe RSA parameters. Given security parameter κ', two κ'-bit safe primes $P = 2P' + 1$ and $Q = 2Q' + 1$ are picked randomly. The RSA modulus is set to $N = PQ$, and a pair $e, d \in \mathbb{Z}_{\varphi(N)}$ is chosen s.t. $ed = 1 \pmod{\varphi(N)}$. Observe that $\varphi(N) = (P-1)(Q-1) = 4P'Q'$.

The largest element order in $\mathbb{Z}_N^\times$ is $\lambda(N) = \mathrm{lcm}(P-1, Q-1) = 2P'Q' = \varphi(N)/2$. [26] and [35] show that for half of the elements $g \in \mathbb{Z}_N^\times$ it holds that $\mathrm{ord}(g) = \lambda(N)$ and $-1 \notin \langle g \rangle$, i.e., $\mathbb{Z}_N^\times \cong \langle -1 \rangle \times \langle g \rangle$. Let $\mathsf{Init}()$ algorithm find such $g \in \mathbb{Z}_N^\times$ (e.g., by random sampling and testing) and assign $U.\mathsf{params} \leftarrow (N, e, d, g)$.

Finally, the algorithm initializes U's contact revocation list by setting $U.\mathsf{crl} \leftarrow \emptyset$.

$\mathsf{AddContact}(U \leftrightarrow V)$. In this protocol user U, on input $U.\mathsf{params} = (N, e, d, g)$ and identifier id of a user V, computes contact certificate $\mathsf{cc}_{U\to V} = (N, e, g, \sigma_V)$ with $\sigma_V = (H_N(\mathsf{id}))^d \bmod N$, i.e., the Full-Domain-Hash RSA signature [4] on id, and confidentially hands it out to V.

$\mathsf{RevokeContact}(U, V)$. User U revokes given user V by inserting V into its contact revocation list: $U.\mathsf{crl} \leftarrow U.\mathsf{crl} \cup \{V\}$. It is assumed that an up-to-date version of this list is distributed authentically to all contacts of U.

$\mathsf{Discover}(V \leftrightarrow V')$. The contact discovery protocol is executed between two users V and V' with inputs ($\mathsf{role}, \mathsf{CL}, \mathsf{partner}$) and ($\mathsf{role}', \mathsf{CL}', \mathsf{partner}'$), respectively (see Section 3.1 for a description of parameters). The protocol is specified in detail in Figure 1. Each user obtains its contact list and, for each entry in it, parses the friendship certificate (lines 2–3). Next, our protocol combines Okamoto's technique [37] for RSA-based identity-based key agreement (lines 4–5 and 18) with a special padding scheme (introduced in [16]) to hide the size of deployed RSA-moduli (lines 6–7 and 17). Note that several instances of Okamoto's protocol are run in parallel — one for each contact in contact list CL — and all transferred messages are IHME-encoded into a single structure before transmission (lines 8 and 10). Upon receiving the IMHE-encoded structure, each user, for every unrevoked certificate, decodes the messages (line 16) and removes the probabilistic padding applied in lines 6–7 (line 17). As we demonstrate in Section 3.3, values r calculated in line 18 are equal for both protocol participants, when computed for the same common contact. Confirmation messages (c_0, c_1) are derived from this value (lines 19–20), IHME-encoded in lines 23–24, and verified after the last communication round (line 28). Each common contact is then added to the SCL list (line 29). Note that contact revocation is handled in lines 15 and 22. Finally, the protocol terminates with "accepted", unless SCL is empty, in which case "rejected" is returned (lines 31-34).

	V ON INPUT $(\mathsf{init}, \mathsf{CL}, \mathsf{partner})$:		V' ON INPUT $(\mathsf{resp}, \mathsf{CL}', \mathsf{partner}')$:
1	$\mathcal{P} \leftarrow \emptyset,\ \mathcal{T} \leftarrow \emptyset$		$\mathcal{P} \leftarrow \emptyset,\ \mathcal{T} \leftarrow \emptyset$
2	FOR ALL $(U, \mathsf{cc}_{U \to V}) \in \mathsf{CL}$:		FOR ALL $(U, \mathsf{cc}_{U \to V'}) \in \mathsf{CL}'$:
3	PARSE $\mathsf{cc}_{U \to V}$ AS (N, e, g, σ_V)		PARSE $\mathsf{cc}_{U \to V'}$ AS $(N, e, g, \sigma_{V'})$
4	$(b, t) \leftarrow_R \mathbb{Z}_2 \times \mathbb{Z}_{N/2}$		$(b, t) \leftarrow_R \mathbb{Z}_2 \times \mathbb{Z}_{N/2}$
5	$\vartheta \leftarrow (-1)^b g^t \sigma_V \bmod N$		$\vartheta \leftarrow (-1)^b g^t \sigma_{V'} \bmod N$
6	$k \leftarrow_R [0, \lfloor p/N \rfloor - 1]$		$k \leftarrow_R [0, \lfloor p/N \rfloor - 1]$
7	$\theta \leftarrow \vartheta + kN$		$\theta \leftarrow \vartheta + kN$
8	$\mathcal{P} \leftarrow \mathcal{P} \cup \{(N, \theta)\}$		$\mathcal{P} \leftarrow \mathcal{P} \cup \{(N, \theta)\}$
9	$\mathcal{T} \leftarrow \mathcal{T} \cup \{(U, N, e, t)\}$		$\mathcal{T} \leftarrow \mathcal{T} \cup \{(U, N, e, t)\}$
10	$\mathcal{M}_V \leftarrow \mathsf{iEncode}(\mathcal{P})$	$\xrightarrow{\mathcal{M}_V}$	$\mathcal{M}_{V'} \leftarrow \mathsf{iEncode}(\mathcal{P})$
11		$\xleftarrow{\mathcal{M}_{V'}}$	
12	$\mathsf{sid} \leftarrow \mathcal{M}_V \,\|\, \mathcal{M}_{V'}$		$\mathsf{sid} \leftarrow \mathcal{M}_V \,\|\, \mathcal{M}_{V'}$
13	$\mathcal{P}' \leftarrow \emptyset,\ \mathcal{T}' \leftarrow \emptyset$		$\mathcal{P}' \leftarrow \emptyset,\ \mathcal{T}' \leftarrow \emptyset$
14	FOR ALL $(U, N, e, t) \in \mathcal{T}$:		FOR ALL $(U, N, e, t) \in \mathcal{T}$:
15	IF $\mathsf{partner} \notin U.\mathsf{crl}$:		IF $\mathsf{partner}' \notin U.\mathsf{crl}$:
16	$\theta \leftarrow \mathsf{iDecode}(\mathcal{M}_{V'}, N)$		$\theta \leftarrow \mathsf{iDecode}(\mathcal{M}_V, N)$
17	$\vartheta \leftarrow \theta \bmod N$		$\vartheta \leftarrow \theta \bmod N$
18	$r \leftarrow (\vartheta^e / H_N(\mathsf{partner}))^{2t} \bmod N$		$r \leftarrow (\vartheta^e / H_N(\mathsf{partner}'))^{2t} \bmod N$
19	$c_0 \leftarrow H(\mathsf{sid} \,\|\, r \,\|\, 0)$		$c_0 \leftarrow H(\mathsf{sid} \,\|\, r \,\|\, 0)$
20	$c_1 \leftarrow H(\mathsf{sid} \,\|\, r \,\|\, 1)$		$c_1 \leftarrow H(\mathsf{sid} \,\|\, r \,\|\, 1)$
21	$\mathcal{T}' \leftarrow \mathcal{T}' \cup \{(U, N, c_1)\}$		$\mathcal{T}' \leftarrow \mathcal{T}' \cup \{(U, N, c_0)\}$
22	ELSE: $c_0 \leftarrow_R [0, p-1]$		ELSE: $c_1 \leftarrow_R [0, p-1]$
23	$\mathcal{P}' \leftarrow \mathcal{P}' \cup \{(N, c_0)\}$		$\mathcal{P}' \leftarrow \mathcal{P}' \cup \{(N, c_1)\}$
24	$\mathcal{M}'_V \leftarrow \mathsf{iEncode}(\mathcal{P}')$		$\mathcal{M}'_{V'} \leftarrow \mathsf{iEncode}(\mathcal{P}')$
25		$\xrightarrow{\mathcal{M}'_V}$	
26	$\mathsf{SCL} \leftarrow \emptyset$	$\xleftarrow{\mathcal{M}'_{V'}}$	$\mathsf{SCL} \leftarrow \emptyset$
27	FOR ALL $(U, N, c_1) \in \mathcal{T}'$:		FOR ALL $(U, N, c_0) \in \mathcal{T}'$:
28	IF $c_1 = \mathsf{iDecode}(\mathcal{M}'_{V'}, N)$:		IF $c_0 = \mathsf{iDecode}(\mathcal{M}'_V, N)$:
29	$\mathsf{SCL} \leftarrow \mathsf{SCL} \cup \{U\}$		$\mathsf{SCL} \leftarrow \mathsf{SCL} \cup \{U\}$
30			
31	IF $\mathsf{SCL} \neq \emptyset$ THEN		IF $\mathsf{SCL} \neq \emptyset$ THEN
32	TERMINATE WITH "ACCEPT"		TERMINATE WITH "ACCEPT"
33	ELSE		ELSE
34	TERMINATE WITH "REJECT"		TERMINATE WITH "REJECT"

Fig. 1. Specification of $\mathsf{Discover}(V \leftrightarrow V')$

3.3 Protocol Correctness

Suppose that users V, V' have valid contact certificates $\mathsf{cc}_{U\to V}, \mathsf{cc}_{U\to V'}$, respectively, for a shared contact U. Then $\mathsf{cc}_{U\to V} = (N, e, g, \sigma_V)$ and $\mathsf{cc}_{U\to V'} = (N, e, g, \sigma_{V'})$ for common parameter set N, e, g. In a $\mathsf{Discover}(V \leftrightarrow V')$ protocol session, V would receive $\vartheta = (-1)^{b'} g^{t'} \sigma_{V'}$ from V' (lines 5 and 17) and compute $r = (\vartheta^e / H_N(\mathsf{partner}))^{2t}$ (line 18). From $\sigma_{V'} = H_N(\mathsf{partner})^d \bmod N$ (see AddContact protocol) it follows that $r = g^{2ett'}$. User V' obtains the same value r by executing analogous computations (with $\mathsf{partner}'$ and t'). The protocol's correctness is now implied by IHME's correctness, and verifiable by inspection of Figure 1. The security analysis of the protocol is postponed to Section 4.3, after the specification of the security model.

3.4 Protocol Efficiency and Performance Analysis

We now discuss the efficiency of our CDS construction. We focus on the protocol Discover since Init is run only once per user, while AddContact and RevokeContact are executed only once for each added or removed contact, respectively, and can be performed off-line.

Computational Complexity and Bandwidth Requirements. The computational complexity of the Discover protocol is essentially related to the number of (relatively more expensive) exponentiations, executed for each contact in lines 5 and 18. Any user V needs to compute $2|\mathsf{CL}_V|$ modular exponentiations with modulus size $2\kappa'$, where $|\mathsf{CL}_V|$ denotes the number of contacts of V. If the polynomial-based IHME constructions from [33] or [34] are used to encode messages, the polynomial interpolations and evaluations only require inexpensive operations, such as multiplications in $\mathbb{F}$. Specifically, the number of multiplications in $\mathbb{F}$ would amount to $O(|\mathsf{CL}_V|^2)$ and $O(|\mathsf{CL}_V| \cdot |\mathsf{CL}_{V'}|)$, respectively, where V' denotes the protocol partner. Nevertheless, the corresponding workload can be considered small in practice, as discussed in [33, 34].

The CRL-based check for revocation of partner's pseudonym in line 15 can be implemented in logarithmic complexity (assuming sorted crls). Note that users need to keep revocation lists of their contacts up-to-date. In practice, this does not impose a significant overhead, as revocation lists grow incrementally and include only (short) identifiers of revoked contacts. Thus, the related communication overhead is negligible compared to that of an actual Discover session.

The overall communication complexity of the Discover protocol (including the IHME-encoded transmission) is linear in the number of contacts. More precisely, each user sends and receives in total approximately $2(|\mathsf{CL}_V| + |\mathsf{CL}_{V'}|)(2\kappa' + \kappa)$ bits. Observe that this value can be lowered to $2(|\mathsf{CL}_V| + |\mathsf{CL}_{V'}|)(\kappa' + \kappa)$ by shortening confirmation messages c_0, c_1 to κ bits, in lines 19 and 20 (see also [34]).

Experimental Analysis. In addition to our asymptotic analysis, we also measured the performance of our scheme, experimentally. To this end, we conducted several experiments involving laptops and mobile devices. Our prototypes use the recent optimizations to IHME scheme, proposed by Manulis and Poettering [34].

Following [34], for $|\mathsf{CL}| < 100$, the overall costs of running the protocol is dominated by the time consumed in the exponentiations, when related to IHME encoding. If certain IHME-related precomputations from [34] are possible, then this bound increases to $|\mathsf{CL}| < 250$. Therefore, the computational overhead of our CDS construction is, in practice, almost linear in the number of contacts.

Figure 2 presents running times of our Discover protocol, using different CPUs: a single core of an Intel XEON 2.6GHz CPU, an AMD NEO 1.6GHz processor (often found in Netbook computers), and an ARMv7 600MHz CPU (installed on many today's smartphones). All measurements were performed using the GMP library [18], thus, execution on smartphones can be even speeded up using different cryptographic libraries optimized for mobile environments.

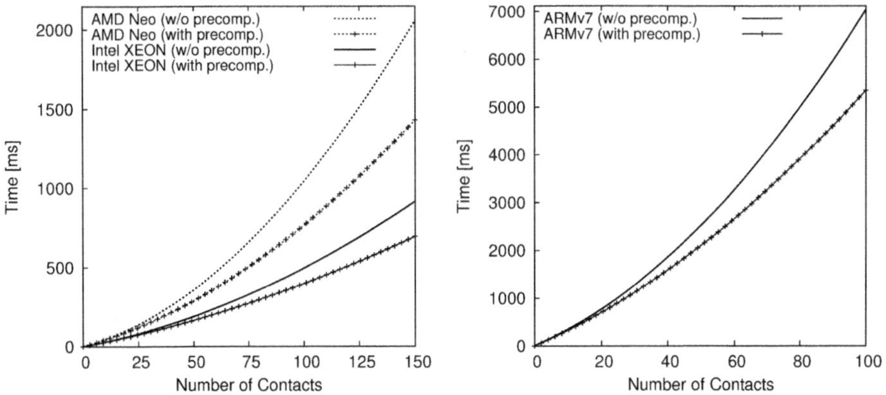

(a) AMD Neo and Intel Xeon (b) ARMv7

Fig. 2. Running times of our Discover protocol on different CPUs with an increasing number of contacts. For each CPU, we also consider session-independent (off-line) precomputations from [34]. All measurements are performed for 80-bit (symmetric) security and 1024-bit RSA moduli.

We observe that our protocol for Private Contact Discovery scales fairly well. For security level $(\kappa, 2\kappa') = (80, 1024)$, i.e., 80-bit symmetric security and 1024-bit RSA moduli, on laptops and server machines, a full protocol execution requires less than a second, even for 100 or more contacts per user. On cores with smaller footprint, e.g., on recent smartphones like Nokia's N900 (equipped with the ARMv7 600MHz processor), protocol execution with 100 contacts requires about 5 seconds, which is an acceptable overhead. Note that smartphones' CPU speeds are envisioned to increase rapidly in the near future (e.g., the iPhone 4G is already equipped with a 1GHz processor). Finally, we computed that each user sends and receives around 300 Bytes per user of his contact list, where we assume $|\mathsf{CL}_V| = |\mathsf{CL}_{V'}|$ for simplicity. That is, in the protocol execution with 100 contacts, a total of 30KB is transmitted.

We conclude that our Private Contact Discovery solution is efficient and practical enough for actual deployment, also on smartphones widely available *today.* Yet, our technique does not give up solid privacy guarantees, as we show in the next section.

4 Security Model for Contact Discovery Protocols

In this section, we introduce our security model for a Contact Discovery Scheme (CDS). We formalize this notion by describing adversarial capabilities and defining Contact-Hiding security. Finally, we analyze the properties of our scheme with respect to this model.

4.1 Adversary Model

The adversary $\mathcal{A}$ is modeled as a PPT machine interacting with protocol participants and having access to the following set of queries, where $\mathcal{U}$ denotes the set of honest users in the system.

$\mathsf{Discover}(U, \mathsf{role}, \mathsf{CL}, \mathsf{partner})$**:** This query results in initiating, on behalf of user $U \in \mathcal{U}$, a new session π of Discover. Query's input is a role identifier $\mathsf{role} \in \{\mathsf{init}, \mathsf{resp}\}$, a contact list $\mathsf{CL} \subseteq \mathcal{U}$ of users, and an identifier partner of the protocol partner. Query's output is a first protocol message M (if available).

$\mathsf{Send}(\pi, M)$**:** With this query, message M is delivered to session π. After processing M, the output (if any) is given to $\mathcal{A}$. The query is ignored if π is not waiting for input.

$\mathsf{Reveal}(\pi)$**:** This query is ignored if $\pi.\mathsf{state} = \mathsf{running}$. Otherwise, the query returns $(\pi.\mathsf{state}, \pi.\mathsf{SCL})$.

$\mathsf{RevealCC}(V, U)$**:** This query gives the adversary contact certificate $\mathsf{cc}_{U \to V}$ of user V for contact U. It models the possibility of selective contact corruptions.

$\mathsf{Revoke}(U, V)$**:** This query lets user U include user V in its contact revocation list $U.\mathsf{crl}$.

4.2 Contact-Hiding Security

Informally, the Contact-Hiding property protects users from disclosing non-matching contacts to other participants. We model CH-security with a game, following the indistinguishability approach. The goal of the adversary is to decide which of two contact lists, CL_0^* or CL_1^*, is used by some challenge session π^*. The adversary can also invoke any number of Discover sessions, and perform Reveal and RevealCC queries at will.

Definition 4 (Contact-Hiding Security). *Let* $\mathsf{CDS} = \{\mathsf{Init}, \mathsf{AddContact}, \mathsf{RevokeContact}, \mathsf{Discover}\}$, *$b$ be a randomly chosen bit, and* $\mathcal{Q} = \{\mathsf{Discover}, \mathsf{Send}, \mathsf{Reveal}, \mathsf{RevealCC}, \mathsf{Revoke}\}$ *denote the set of queries the adversary $\mathcal{A}$ has access to. We consider the following game between a challenger and the adversary $\mathcal{A}$:*

$\mathsf{Game}^{\mathsf{ch},b}_{\mathcal{A},\mathsf{CDS}}(\kappa, n)$:

- *The challenger creates n users, denoted by* $\mathcal{U} = \{U_1, \ldots, U_n\}$*. The adversary* $\mathcal{A}$ *specifies a set* $\mathcal{U}^c \subseteq \mathcal{U}$ *of initially corrupted users. Let* $\mathcal{U}^h = \mathcal{U} \setminus \mathcal{U}^c$*.* $\mathsf{Init}(1^\kappa)$ *is run for all* $U \in \mathcal{U}^h$*, and, for all combinations* $(U, V) \in \mathcal{U}^h \times \mathcal{U}^h$*, contact certificates* $\mathsf{cc}_{U \to V}$ *are created by respective user U and given to V, each time by running the* $\mathsf{AddContact}(U \leftrightarrow V)$ *protocol.*
  For all $U \in \mathcal{U}^c$*, the adversary sets up all parameters himself, including* $U.\mathsf{crl}$*. He then specifies a list* $\mathcal{L} \subseteq \mathcal{U}^h \times \mathcal{U}^c$*, and for all* $(U, V) \in \mathcal{L}$*, protocol* $\mathsf{AddContact}(U, V)$ *is run, and the respective certificate* $\mathsf{cc}_{U \to V}$ *is given to* $\mathcal{A}$*;*
- $\mathcal{A}^{\mathcal{Q}}$ *interacts with all (honest) users using the queries in* $\mathcal{Q}$*; at some point* $\mathcal{A}^{\mathcal{Q}}$ *outputs a tuple* $(U^*, \mathsf{role}^*, \mathsf{CL}^*_0, \mathsf{CL}^*_1, \mathsf{partner}^*)$ *where* $U^* \in \mathcal{U}^h$*,* $\mathsf{role}^* \in \{\mathsf{init}, \mathsf{resp}\}$*,* $\mathsf{CL}^*_0, \mathsf{CL}^*_1 \subseteq \mathcal{U}^h$ *with* $|\mathsf{CL}^*_0| = |\mathsf{CL}^*_1|$*, and* $\mathsf{partner}^*$ *is any user id (in* $\mathcal{U}$*). Set* $\mathcal{D}^* = (\mathsf{CL}^*_0 \setminus \mathsf{CL}^*_1) \cup (\mathsf{CL}^*_1 \setminus \mathsf{CL}^*_0) = (\mathsf{CL}^*_0 \cup \mathsf{CL}^*_1) \setminus (\mathsf{CL}^*_0 \cap \mathsf{CL}^*_1)$ *is called the* distinguishing set*;*
- *the challenger invokes a* $\mathsf{Discover}(U^*, \mathsf{role}^*, \mathsf{CL}^*_b, \mathsf{partner}^*)$ *session* π^* *(and provides all needed credentials);*
- $\mathcal{A}^{\mathcal{Q}}$ *continues interacting via queries (including on session* π^**) until it terminates and outputs bit* b'*;*
- *the output of the game is* b' *if all of the following hold; else the output is 0:*
  (a) *if there is a* $\mathsf{Discover}$ *session* π' *with* $\mathcal{D}^* \cap \pi'.\mathsf{CL} \neq \emptyset$ *and* $(\pi'.\mathsf{id}, \pi'.\mathsf{partner}) = (\pi^*.\mathsf{partner}, \pi^*.\mathsf{id})$ *which was in state* $\mathsf{running}$ *while* π^* *was in state* $\mathsf{running}$*, then neither* $\mathsf{Reveal}(\pi^*)$ *nor* $\mathsf{Reveal}(\pi')$ *was asked,*
  (b) *for no* $U \in \mathcal{D}^*$ *a* $\mathsf{RevealCC}(\mathsf{partner}^*, U)$ *query has been posed or a pair* $(U, \mathsf{partner}^*)$ *is contained in* $\mathcal{L}$*, i.e. the adversary did not ask for a contact certificate for* $\mathsf{partner}^*$ *issued by any user in the distinguishing set.*

We define

$$\mathsf{Adv}^{\mathsf{ch}}_{\mathcal{A},\mathsf{CDS}}(\kappa, n) := \left| \Pr\left[\mathsf{Game}^{\mathsf{ch},0}_{\mathcal{A},\mathsf{CDS}}(\kappa, n) = 1\right] - \Pr\left[\mathsf{Game}^{\mathsf{ch},1}_{\mathcal{A},\mathsf{CDS}}(\kappa, n) = 1\right] \right|$$

and denote with $\mathsf{Adv}^{\mathsf{ch}}_{\mathsf{CDS}}(\kappa, n)$ *the maximum advantage over all PPT adversaries* $\mathcal{A}$*. We say that* CDS *is* CH-secure *if this advantage is negligible in* κ *(for all n polynomially dependent on* κ*).*

Conditions (a) and (b) exclude some trivial attacks on contact hiding. Condition (a) thwarts the attack where $\mathcal{A}$ starts a $\mathsf{Discover}(U', \mathsf{role}', \mathsf{CL}', \mathsf{partner}')$ session π' with $\mathsf{CL}' \cap \mathcal{D}^* \neq \emptyset$ and $(\pi'.\mathsf{id}, \pi'.\mathsf{partner}) = (\pi^*.\mathsf{partner}, \pi^*.\mathsf{id})$, relays all messages between π^* and π', and finally asks $\mathsf{Reveal}(\pi^*)$ or $\mathsf{Reveal}(\pi')$. By protocol correctness, $\pi^*.\mathsf{SCL} = \pi'.\mathsf{SCL}$ would contain elements from $\mathcal{D}^*$, and it would be trivial to correctly decide about b. Condition (b) prevents $\mathcal{A}$ to ask for a contact certificate issued by a user $U \in \mathcal{D}^*$ for a user $V \in \mathcal{U}$, to simulate a protocol session on behalf of V, to relay all messages between that session and π^*, and to decide about bit b from the results.

Remark 2. One may also define a stronger notion of contact-hiding by requiring that distinct sessions of the Discover protocol executed by the same user remain *unlinkable*. We observe that our Discover protocol in its plain form would not guarantee such unlinkability, since credentials (i.e., certified friendships) are re-used across multiple protocol executions, which also allows an efficient realization. Although linkable protocols may yield traceability concerns (with respect to eavesdropping adversaries) and leak sensitive information about users, we address this issue by executing the protocol over secure (encrypted and authenticated) channels. Nonetheless, it is an interesting open problem to design a Discover protocol, which would preserve the linear complexity of our solution and, simultaneously, achieve such stronger property of unlinkability.

4.3 Security Analysis of Our Protocol

Following the definition in Section 4.2, we argue that our CDS protocol, described in Section 3, offers Contact-Hiding security. We refer to Appendix A for the proof.

Theorem 1 (Contact-Hiding Security). *The* CDS *protocol in Section 3.2 is CH-secure under the RSA assumption on safe moduli, in Random Oracle Model.*

5 Conclusion

This paper motivated the importance, and introduced the concept of, Private Contact Discovery. Following a cryptographic treatment of the problem, we presented an efficient and provably secure construction. During protocol design, we overcame several challenges, such as the arbitrary expansion of contact lists, by using contact certification. Our solution relies on Full-Domain-Hash RSA signatures and on the recent IHME primitive [33]. We also showed, through experimental evaluation, that our solution is practical enough to be deployed in real-world applications, including those running on mobile devices.

Private Contact Discovery provides a valuable privacy-preserving tool that can serve as building block for many collaborative applications, including popular social networks. Since this work represents an initial foray into Private Contact Discovery, much remains to be done. First, we plan to extend our techniques to privately discover *communities*: consider, for example, two smartphone users in proximity willing to find out whether or not they are member of the same social community (e.g., a Facebook group or an Awarenet community [1]), in a privacy-preserving manner. Users may receive (from a community manager) credentials for community membership, and execute our CDS protocol to discover common memberships. Then, we intend to address the (privacy-preserving) discovery of i-th grade contacts and cliques [38], which currently seems impossible without relying on some trusted third party. Another interesting direction is to consider *fairness* [5] of the contact discovery process, a property that ensures a balanced gain of knowledge of protocol participants even against insider adversaries.

Acknowledgments. We are grateful to Nokia for donating the Nokia N900 devices used in our experiments. M. Manulis and B. Poettering acknowledge partial support from the German Science Foundation (DFG) project PRIMAKE (MA 4957), DAAD project PACU (PPP 50743263), and BMBF project POC (AUS 10/046).

References

1. Ahtiainen, A., Kalliojarvi, K., Kasslin, M., Leppanen, K., Richter, A., Ruuska, P., Wijting, C.: Awareness Networking in Wireless Environments: Means of Exchanging Information. IEEE Vehicular Technology Magazine, 48–54 (2009)
2. Ateniese, G., De Cristofaro, E., Tsudik, G.: (If) Size Matters: Size-Hiding Private Set Intersection. In: Catalano, D., Fazio, N., Gennaro, R., Nicolosi, A. (eds.) PKC 2011. LNCS, vol. 6571, pp. 156–173. Springer, Heidelberg (2011)
3. Balfanz, D., Durfee, G., Shankar, N., Smetters, D.K., Staddon, J., Wong, H.-C.: Secret Handshakes from Pairing-Based Key Agreements. In: IEEE Symposium on Security and Privacy, pp. 180–196 (2003)
4. Bellare, M., Rogaway, P.: Random Oracles are Practical: A Paradigm for Designing Efficient Protocols. In: ACM CCS, pp. 62–73 (1993)
5. Boudot, F., Schoenmakers, B., Traoré, J.: A Fair and Efficient Solution to the Socialist Millionaires' Problem. Discrete Applied Mathematics 111(1-2), 23–36 (2001)
6. Bradshaw, R., Holt, J., Seamons, K.: Concealing Complex Policies with Hidden Credentials. In: ACM CCS, pp. 146–157 (2004)
7. Brands, S.: Rethinking Public Key Infrastructures and Digital Certificates: Building in Privacy. MIT Press, Cambridge (2000)
8. Camenisch, J., Casati, N., Groß, T., Shoup, V.: Credential Authenticated Identification and Key Exchange. In: Rabin, T. (ed.) CRYPTO 2010. LNCS, vol. 6223, pp. 255–276. Springer, Heidelberg (2010)
9. Camenisch, J., Lysyanskaya, A.: An Efficient System for Non-transferable Anonymous Credentials with Optional Anonymity Revocation. In: Pfitzmann, B. (ed.) EUROCRYPT 2001. LNCS, vol. 2045, pp. 93–118. Springer, Heidelberg (2001)
10. Camenisch, J., Zaverucha, G.M.: Private intersection of certified sets. In: Dingledine, R., Golle, P. (eds.) FC 2009. LNCS, vol. 5628, pp. 108–127. Springer, Heidelberg (2009)
11. Castelluccia, C., Jarecki, S., Tsudik, G.: Secret Handshakes from CA-Oblivious Encryption. In: Lee, P.J. (ed.) ASIACRYPT 2004. LNCS, vol. 3329, pp. 293–307. Springer, Heidelberg (2004)
12. Chiou, S., Chang, S., Sun, H.: Common Friends Discovery with Privacy and Authenticity. In: IAS, pp. 337–340 (2009)
13. Dachman-Soled, D., Malkin, T., Raykova, M., Yung, M.: Efficient Robust Private Set Intersection. In: Abdalla, M., Pointcheval, D., Fouque, P.-A., Vergnaud, D. (eds.) ACNS 2009. LNCS, vol. 5536, pp. 125–142. Springer, Heidelberg (2009)
14. De Cristofaro, E., Kim, J., Tsudik, G.: Linear-Complexity Private Set Intersection Protocols Secure in Malicious Model. In: Abe, M. (ed.) ASIACRYPT 2010. LNCS, vol. 6477, pp. 213–231. Springer, Heidelberg (2010)
15. De Cristofaro, E., Tsudik, G.: Practical Private Set Intersection Protocols with Linear Complexity. In: Sion, R. (ed.) FC 2010. LNCS, vol. 6052, pp. 143–159. Springer, Heidelberg (2010)
16. Desmedt, Y.: Securing Traceability of Ciphertexts - Towards a Secure Software Key Escrow System. In: Guillou, L.C., Quisquater, J.-J. (eds.) EUROCRYPT 1995. LNCS, vol. 921, pp. 147–157. Springer, Heidelberg (1995)

17. Diehl, C., Namata, G., Getoor, L.: Relationship Identification for Social Network Discovery. AAAI 22(1), 546–552 (2007)
18. Free Software Foundation. The GNU MP Bignum Library, `http://gmplib.org/`
19. Freedman, M.J., Nicolosi, A.: Efficient Private Techniques for Verifying Social Proximity. In: IPTPS (2007)
20. Freedman, M.J., Nissim, K., Pinkas, B.: Efficient Private Matching and Set Intersection. In: Cachin, C., Camenisch, J.L. (eds.) EUROCRYPT 2004. LNCS, vol. 3027, pp. 1–19. Springer, Heidelberg (2004)
21. Goldreich, O., Micali, S., Wigderson, A.: How to Play any Mental Game or A Completeness Theorem for Protocols with Honest Majority. In: 19th STOC, pp. 218–229 (1987)
22. Günther, F., Manulis, M., Strufe, T.: Cryptographic Treatment of Private User Profiles. In: Financial Cryptography Workshops. Springer, Heidelberg (2011), `http://eprint.iacr.org/2011/064`
23. Hazay, C., Lindell, Y.: Efficient Protocols for Set Intersection and Pattern Matching with Security Against Malicious and Covert Adversaries. In: Canetti, R. (ed.) TCC 2008. LNCS, vol. 4948, pp. 155–175. Springer, Heidelberg (2008)
24. Hazay, C., Nissim, K.: Efficient Set Operations in the Presence of Malicious Adversaries. In: Nguyen, P.Q., Pointcheval, D. (eds.) PKC 2010. LNCS, vol. 6056, pp. 312–331. Springer, Heidelberg (2010)
25. Huberman, B., Franklin, M., Hogg, T.: Enhancing Privacy and Trust in Electronic Communities. In: ACM Conference on Electronic Commerce, pp. 78–86 (1999)
26. Jarecki, S., Kim, J., Tsudik, G.: Beyond Secret Handshakes: Affiliation-Hiding Authenticated Key Exchange. In: Malkin, T. (ed.) CT-RSA 2008. LNCS, vol. 4964, pp. 352–369. Springer, Heidelberg (2008)
27. Jarecki, S., Liu, X.: Affiliation-Hiding Envelope and Authentication Schemes with Efficient Support for Multiple Credentials. In: Aceto, L., Damgård, I., Goldberg, L.A., Halldórsson, M.M., Ingólfsdóttir, A., Walukiewicz, I. (eds.) ICALP 2008, Part II. LNCS, vol. 5126, pp. 715–726. Springer, Heidelberg (2008)
28. Jarecki, S., Liu, X.: Private Mutual Authentication and Conditional Oblivious Transfer. In: Halevi, S. (ed.) CRYPTO 2009. LNCS, vol. 5677, pp. 90–107. Springer, Heidelberg (2009)
29. Jarecki, S., Liu, X.: Fast Secure Computation of Set Intersection. In: Garay, J.A., De Prisco, R. (eds.) SCN 2010. LNCS, vol. 6280, pp. 418–435. Springer, Heidelberg (2010)
30. Kissner, L., Song, D.X.: Privacy-Preserving Set Operations. In: Shoup, V. (ed.) CRYPTO 2005. LNCS, vol. 3621, pp. 241–257. Springer, Heidelberg (2005)
31. Korolova, A., Motwani, R., Nabar, S., Xu, Y.: Link Privacy in Social Networks. In: CIKM, pp. 289–298 (2008)
32. Lindell, Y., Pinkas, B.: An Efficient Protocol for Secure Two-Party Computation in the Presence of Malicious Adversaries. In: Naor, M. (ed.) EUROCRYPT 2007. LNCS, vol. 4515, pp. 52–78. Springer, Heidelberg (2007)
33. Manulis, M., Pinkas, B., Poettering, B.: Privacy-Preserving Group Discovery with Linear Complexity. In: Zhou, J., Yung, M. (eds.) ACNS 2010. LNCS, vol. 6123, pp. 420–437. Springer, Heidelberg (2010)
34. Manulis, M., Poettering, B.: Practical Affiliation-Hiding Authentication from Improved Polynomial Interpolation. In: ACM ASIACCS, pp. 286–295 (2011), `http://eprint.iacr.org/2010/659`
35. Manulis, M., Poettering, B., Tsudik, G.: Affiliation-Hiding Key Exchange with Untrusted Group Authorities. In: Zhou, J., Yung, M. (eds.) ACNS 2010. LNCS, vol. 6123, pp. 402–419. Springer, Heidelberg (2010)

36. Manulis, M., Poettering, B., Tsudik, G.: Taming Big Brother Ambitions: More Privacy for Secret Handshakes. In: Atallah, M.J., Hopper, N.J. (eds.) PETS 2010. LNCS, vol. 6205, pp. 149–165. Springer, Heidelberg (2010)
37. Okamoto, E.: Key Distribution Systems Based on Identification Information. In: Pomerance, C. (ed.) CRYPTO 1987. LNCS, vol. 293, pp. 194–202. Springer, Heidelberg (1988)
38. Pons, P., Latapy, M.: Computing Communities in Large Networks Using Random Walks. In: Yolum, p., Güngör, T., Gürgen, F., Özturan, C. (eds.) ISCIS 2005. LNCS, vol. 3733, pp. 284–293. Springer, Heidelberg (2005)
39. The Facebook, Inc. Facebook's statistics (2010), `http://www.facebook.com/press/info.php?statistics`
40. Von Arb, M., Bader, M., Kuhn, M., Wattenhofer, R.: Veneta: Serverless Friend-of-Friend Detection in Mobile Social Networking. In: WiMob, pp. 184–189 (2008)
41. Xu, S., Yung, M.: k-Anonymous Secret Handshakes with Reusable Credentials. In: ACM CCS, pp. 158–167 (2004)
42. Yao, A.: How to Generate and Exchange Secrets. In: 27th FOCS, pp. 162–167 (1986)
43. Yu, P.S., Han, J., Faloutsos, C.: Link Mining: Models, Algorithms, and Applications. Springer, Heidelberg (2010)
44. Zheleva, E., Getoor, L., Golbeck, J., Kuter, U.: Using Friendship Ties and Family Circles for Link Prediction. In: SNA-KDD, pp. 97–113 (2008)

A Proof of Contact-Hiding Security of CDS from Fig. 1

We now prove the Contact-Hiding (CH) security (Definition 4) of our CDS protocol (illustrated in Fig. 1), relying on the RSA assumption on safe moduli (Definition 1) and on the security of the IHME scheme from [33]. By $\mathsf{Adv}^{\mathsf{ihide}}_{\mathsf{IHME}}(\kappa)$ we denote the advantage probability for breaking the index-hiding property of the IHME scheme, which (since the utilized IHME scheme is perfect) is equal to 0. Note that our security arguments have similarities with those of the multi-credential AHA protocol in [33], which proves the "affiliation hiding" property of the proposed AHA under the same assumptions. We only sketch the proof steps that mirror those in [33], as the reader can find more details in [33] and its predecessor [26]. In contrast, we detail arguments regarding the use of the IHME scheme. We prove CH-Security of CDS by presenting a sequence of games $\mathbf{G}_0, \ldots, \mathbf{G}_5$.

Game $\mathbf{G}_0$. We start with $\mathbf{G}_0 = \mathsf{Game}^{\mathsf{ch},b}_{\mathcal{A},\mathsf{CDS}}(\kappa, n)$, in which $\mathcal{A}$ interacts with simulator $\mathcal{C}$, which answers all queries honestly according to the specification of the game (see Definition 4).

Game $\mathbf{G}_1$. Game $\mathbf{G}_1$ is like $\mathbf{G}_0$, except that the simulation is aborted (the game outputs 0) if there exists a Discover session $\pi' \neq \pi^*$ that sends out the same $\mathcal{M}$ structure as π^* (see line 10 in Figure 1).

Note that $\mathcal{M}$ sent by π^* contains for each contact in CL_b^* a specific θ value. These are almost uniformly distributed in a set of size $p \approx 2^{2\kappa'+\kappa}$. Thus, the probability of finding a collision in the $\mathcal{M}$'s is upper-bounded by $q_q/2^{2\kappa'+\kappa}$ (where q_q denotes the number of Discover queries), and thus negligible in κ.

Game $\mathbf{G}_2$. Let $R = (r_1, \ldots, r_k)$ denote the list of r-values for the contacts in $\mathsf{CL}_b^* \setminus \mathsf{CL}_{1-b}^* = \mathsf{CL}_b^* \cap \mathcal{D}^*$ of session π^*, as computed in line 18 of the protocol. Game $\mathbf{G}_2$ is like

$\mathbf{G}_1$, except that all confirmation messages c_0, c_1 of π^* (see lines 19 and 20), computed based on the values in R, are replaced by random elements in the respective range.

By the Random Oracle Model (ROM), the modification introduced in Game $\mathbf{G}_2$ can only be detected by adversaries that can compute and query the H oracle on at least one of the r-values in R. Let $r_t \in R$ be such a value (and assume that the simulator guesses t correctly). By embedding an RSA challenge into user identifiers (by programming the H_N oracle) and public user parameters (by choosing N and g appropriately), the problem of computing r_t can be reduced to the hardness of the RSA problem (see [26, Section 3] for more details). We conclude that the computational distance between $\mathbf{G}_2$ and $\mathbf{G}_1$ is polynomially dependent on $\mathsf{Succ}^{\mathsf{rsa}}(\kappa')$, and is thus negligible in κ.

Remark 3. Note that for all contacts in $\mathsf{CL}_b^* \cap \mathcal{D}^*$, the adversary $\mathcal{A}$ cannot distinguish correct confirmation messages for π^* from random ones. Therefore, the output set $\pi^*.\mathsf{SCL}$ is disjoint with $\mathcal{D}^*$.

Game $\mathbf{G}_3$. This game is like Game $\mathbf{G}_2$, except that, for session π^*, the θ-values for all contacts in $\mathsf{CL}_b^* \cap \mathcal{D}^*$ (as computed in line 7) are replaced by values uniformly random in $[0, p-1]$.

Observe from the protocol definition that the θ-values replaced in this game only affect the computation of the $r_t \in R$ (in Game $\mathbf{G}_2$), which cannot be computed and checked by the adversary anyway by a result of Game $\mathbf{G}_2$. Hence, the only detectable difference between $\mathbf{G}_2$ and $\mathbf{G}_3$ may arise from different distributions of the original and the modified θ-values. Although the original values are *not* uniformly distributed in $[0, p-1]$, their distribution is statistically indistinguishable from the uniform distribution. In [26], the corresponding statistical difference is proven to be bounded by $2^{-\kappa}$, what is negligible in κ.

Game $\mathbf{G}_4$. Game $\mathbf{G}_4$ is like $\mathbf{G}_3$, except that, for session π^*, in the IHME-encoding step in line 10, for all contacts in $\mathsf{CL}_b^* \cap \mathcal{D}^*$, the indices N are replaced by the indices N that correspond to the contacts in $\mathsf{CL}_0^* \cap D^*$. For all contacts in $\mathsf{CL}_b^* \cap \mathsf{CL}_{1-b}^*$ the indices remain unchanged.

As Games $\mathbf{G}_3$ and $\mathbf{G}_4$ are exactly the same in case $b = 0$ (as nothing has changed), in the following we assume $b = 1$. We show that, if an efficient distinguisher $\mathcal{D}^{3,4}$ that distinguishes between Game $\mathbf{G}_3$ and $\mathbf{G}_4$ exists, then we can use it to construct an adversary $\mathcal{A}^{\mathsf{ihide}}$ against index-hiding of IHME as follows. Adversary $\mathcal{A}^{\mathsf{ihide}}$ acts as a challenger for $\mathcal{D}^{3,4}$, i.e., $\mathcal{A}^{\mathsf{ihide}}$ sets up all users, and answers all protocol queries honestly (but following the rules of Game $\mathbf{G}_3$), with the only exception that the encoding step in line 10 for session π^* is performed by the IHME challenger (i.e., the latter receives (I_0, I_1, M') where I_0 and I_1 are the sets of indices N of CL_0^* and CL_1^*, respectively, and M' is the list of the θ-values honestly computed for the contacts in $\mathsf{CL}_0^* \cap \mathsf{CL}_1^*$), which returns an encoding $\mathcal{M}$ using either the indices corresponding to CL_0^* or CL_1^*. The bit output by $\mathcal{D}^{3,4}$ serves as output b' for $\mathcal{A}^{\mathsf{ihide}}$. We see that the success probability of $\mathcal{D}^{3,4}$ is bounded by $\mathsf{Adv}^{\mathsf{ihide}}_{\mathsf{IHME}}(\kappa)$, which is 0 (since the IHME construction is perfect). Hence, the computational difference between Games $\mathbf{G}_3$ and $\mathbf{G}_4$ is 0.

Game $\mathbf{G}_5$. The step between Game $\mathbf{G}_4$ and $\mathbf{G}_5$ is very similar to the previous transition: this time, it is the IHME-encoding in line 24 for which the indices N of all contacts in $\mathsf{CL}_b^* \cap D^*$ are replaced by the indices N that correspond to the contacts in $\mathsf{CL}_0^* \cap D^*$.

The difference between $\mathbf{G}_5$ and $\mathbf{G}_4$ is bounded by $\mathsf{Adv}^{\mathsf{ihide}}_{\mathsf{IHME}}(\kappa) = 0$, exactly for the same reason above.

We conclude that the computational difference between $\mathbf{G}_0$ and $\mathbf{G}_5$ is negligible in κ. Therefore, a CH-adversary cannot distinguish between $\mathsf{Game}^{\mathsf{ch},b}_{\mathcal{A},\mathsf{CDS}}$ and $\mathsf{Game}^{\mathsf{ch},0}_{\mathcal{A},\mathsf{CDS}}$ with non-negligible probability, neither by analyzing the exchanged messages, nor by interpreting the results of Reveal queries. As the latter game contains no information about bit b, it follows that $\mathsf{Adv}^{\mathsf{ch}}_{\mathsf{CDS}}(\kappa, n)$ is negligible in κ (for all N polynomially dependent on κ). $\square$

Sanitizable Signatures in XML Signature — Performance, Mixing Properties, and Revisiting the Property of Transparency*

Henrich C. Pöhls, Kai Samelin, and Joachim Posegga

Chair of IT Security, University of Passau, Germany
{hp,jp}@sec.uni-passau.de, samelin@fim.uni-passau.de

Abstract. We present the performance measures of our Java Cryptography Architecture (JCA) implementation that integrates sanitizable signature schemes into the XML Signature Specification. Our implementation shows mostly negligible performance impacts when using the *Ateniese* scheme with four different chameleon hashes and the *Miyazaki* scheme in XML Signatures. Thus, sanitizable signatures can be added to the XML Security Toolbox. Applying the new tools we show how to combine different hash algorithms over different document parts adding and removing certain properties of the sanitizable signature scheme; this mixing comes very natural in XML Signatures. Finally, we motivate that existing definitions for the property of Transparency are counterintuitive in these combinations. Our conclusion is that the document-level Transparency property is independent of the sub-document properties Weak and Strong Transparency.

Keywords: Sanitizable Signatures, Transparency, Performance, XML Signature Framework.

1 Introduction

A sanitizable signature scheme (SSS) allows a defined third party, the so-called sanitizer, to alter an already signed document without invalidating the signature and without involving the original signer again. This comes in handy for many applications, examples thereof can be found in each scheme's literature. However, sanitizable signature schemes are not part of the proposed signature methods in the XML Signature Syntax and Processing W3C Standard [7][1]. We have implemented the sanitizable signature scheme proposed in 2005 by *Ateniese* [1] in Java. The *Ateniese* scheme is build upon a chameleon hash; we implemented chameleon hashes proposed by *Krawczyk* and *Rabin* [12], *Ateniese* and *de Medeiros* [2], *Chen* et al. [6] and *Zhang* et al. [20]. Additionally, a redaction based scheme proposed by *Miyazaki* et al. [16] was implemented to allow performance comparison between both approaches.

* This research is funded by BMBF (FKZ:13N10966) and ANR as part of the ReSCUe-IT project.

[1] We will refer to this as XML Signature for brevity.

J. Lopez and G. Tsudik (Eds.): ACNS 2011, LNCS 6715, pp. 166–182, 2011.

All schemes have been integrated into the Java Cryptography Architecture (JCA), thus they are usable for XML Signatures. We present XML Signature integrations of a representative subset of the different "flavours" of sanitizable signature schemes following the *Ateniese* scheme and redaction based schemes, which have not been developed explicitly for tree-structured documents. This integration is non-trivial: Sanitizable signature schemes do not work as straight forward as the standard hash-and-sign SHA–RSA-Scheme, since some schemes follow a different process when generating the signatures for document parts. In this paper we present our solution that integrates sanitizable signature schemes into XML Signatures without invalidating the W3C Standard.

Our contribution consists of the actual implementation and integration of five different schemes and an analysis that demonstrates a practical performance penalty, compared to RSA and SHA for most of the schemes presented. Additionally, we show how XML helps to mix and allows to add resp. remove properties from the original sanitizable signature scheme. We have shown this for the properties of transparency, accountability, consecutive sanitization control, restricting to values and restricting to sanitizers. These additions are still compliant with the XML Signature Specification and do not need major modifications of the underlying sanitizable signature algorithms. Finally, we offer two observations: First, not all sanitizable signature schemes are suitable for use in XML Documents, since some do not allow overlapping references. Second, the properties of transparency and its "flavours" weak and strong transparency, as originally defined by *Ateniese* et al. [1], are actually independent properties and should not imply each other.

The rest of the paper is structured as follows: Sec. 3 gives technical details on the integration into the existing frameworks and standards: JCA and XML Signature. A short summary of our detailed performance measures is given in Sec. 4. In Sec. 5 is shown how alterations using given primitives will add, respectively remove, certain properties from the schemes. We then use these alterations to motivate a revised view on the existing property of Transparency and its definition in Sec. 6.

2 Related Work

From an implementer's point of view, *Tan* et al. integrated the *Ateniese* scheme into XML Signature using one chameleon hash developed by *Krawczyk* et al. [12] in [18]. However, they just showed that one scheme can be integrated into XML Signature without offering a performance analysis. Their implementation work is similar to that of *Ateniese* et al. in [1]: they also implemented the *Krawczyk* scheme, but in conjunction with OpenSSL and not JCA and XML Signature. Additionally, they did not discuss the possibility to alter properties of the schemes and their integration into the XML Signature Standard in more detail.

Brzuska et al. formalized the most common properties in [5], which we used for our work. However, they do not take XML Signature into account any further. They constructed a tag-based chameleon hash, which allows to add sanitizer accountability to the *Ateniese* scheme. We did not implemented this, as this chameleon hash is similar to the *Krawczyk* one.

Kundu et al. developed a sanitizable signature scheme, which addresses the specific needs of tree-structured documents. [13] These needs have been formalized by *Liu* et. al in [15]. This has also been addressed and tailored for redaction based signatures by *Bruzska* et al. in [4]. However, both approaches are not based on the hash-and-sign paradigm in the "natural way". For brevity their schemes are left out of this work. A performance analysis of the *Kundu* scheme can be found in [14].

Wu et al. also implemented a sanitizable signature scheme in [19]. However, they also do not provide a performance analysis and their scheme relies on the *Merkle*-Hash-Tree-Technique.

3 Implementation in JAVA and Integration into XML

We used the Java Cryptography Architecture (JCA) as an existing open-source framework and we implemented a new "Cryptographic Service Provider", following the design patterns given in JCA. Whenever the schemes required additional cryptographic algorithms these were added via external libraries. In particular: *GnuCrypto*[2] for EMSA-PSS [10], required by *Ateniese* [2], and a library from the *National University of Maynooth*[3] for elliptic curves and bilinear pairings, required by *Zhang* [20]. Our implementations make use of Java's `BigInteger` class.

Within the XML Signature Specification we use `<Reference>` to split the XML Document m into subdocuments $m_1, ..., m_n$. Subdocuments, or references, are actually pointers to subsets ($\subseteq$) of the whole document, e.g. a nodeset, an element or the complete document itself.[4] The union $m_1 \cup ... \cup m_n$ represents the part of the document which is covered by the final signature. Each of these references can be digested using a different hash-algorithm, allowing a maximum of flexibility for the signer. The resulting digests, along with additional information, are combined into the `SignedInfo` element. The additional information stored contains, among others, the following: Applied digest method, C14n-algorithm, reference URI, and applied transforms. The `SignedInfo` element is canonicalized, hashed, and the resulting digest is signed. The W3C specification allows to assign parameters for each message-digest, we used this to integrate the keyed-hash-functions.

3.1 JCA Implementation Details for the Five Schemes

This section will shortly introduce the schemes we implemented and evaluated.

***Ateniese* Scheme.** Standard cryptographic hashes, denoted $\mathcal{H}$, do not allow attackers to find collisions in the hash-domain ("Collision-Resistance"). Chameleon hashes, denoted $\mathcal{CH}$, however do allow to compute collisions for arbitrary input,

[2] `http://www.gnu.org/software/gnu-crypto/`
[3] `http://www.nuim.ie/`
[4] *XPath* allows even more complex expressions.

as long as a trapdoor is known; this trapdoor must be kept secret for chameleon signature schemes. [12] *Ateniese* et al. propose to use those hashes to construct a sanitizable signature scheme: [1]

$$\sigma = SIGN(d_1||...||d_n)$$

where

$$d_i = \begin{cases} \mathcal{H}(m_i) & \text{if } m_i \text{ is not sanitizable} \\ \mathcal{CH}(m_i) & \text{if } m_i \text{ is sanitizable} \end{cases}$$

$\mathcal{H}$ is a standard cryptographic hash like SHA-512, $\mathcal{CH}$ a chameleon hash, m_i subdocument i and n is the number of subdocuments.

Each sanitizer receives the trapdoor needed to find collisions, thus he is able to do arbitrary changes to the sanitizable subdocuments. For this work four different chameleon hashes have been implemented and used within the *Ateniese* scheme, as each chameleon hash can result in different properties of the resulting signature [5]:

1. *Krawczyk* as the first chameleon hash, based on the DLP assumption [12]
2. *Ateniese* as an ID-based approach [2]
3. *Zhang* as an ID-based approach without an UForge-algorithm [20]
4. *Chen* as an ID-based approach without the key-exposure-problem [6,3]

As we are concerned with the application within XML, let us share the following observation concerning message splitting: For the *Ateniese* scheme it is necessary, that in general: $\forall i, j : m_i \neq m_j$, where $0 \leq i < j \leq n$. Suppose $m_i = m_j$ yields, which means that two subdocuments are the same. If now m_i is sanitizable while m_j is not, m_j cannot be altered since $SIGN(\mathcal{H}(m_i)||\mathcal{CH}(m_j))$ must remain the same value under all circumstances. Since $\mathcal{H}$ serves as a random oracle, uniformerly distributing its digests over $dom(\mathcal{H})$, finding a collision such that $SIGN(\mathcal{H}(m_i')||\mathcal{CH}(m_j')) = SIGN(\mathcal{H}(m_i)||\mathcal{CH}(m_j))$, is infeasible by definition of $\mathcal{H}$.

The prior statement is just correct if $m_i = m_j$. Let's assume that $m_i \subsetneq m_j$. Note, m_j can be expressed as $m_j = m_k||m_i||m_l$ where $k \neq i \neq l, m_k \neq \emptyset \vee m_l \neq \emptyset$. Further assume that m_i is not sanitizable while m_j is. Now, the subsubdocument m_i which is contained within m_j cannot be altered while the complement $m_j \backslash m_i = \{m_k, m_l\}$ can. Let the signature be $\sigma = SIGN(\mathcal{H}(m_i)||\mathcal{CH}(m_j))$ while $m_i \subsetneq m_j$. If now m_i is not changed in any way $\mathcal{H}(m_i)$ is not changed either. Since a collision in $dom(\mathcal{CH})$ can be found for arbitrary $m \in \{0,1\}^*$, m_i does not have to be changed. Therefore $m_j \backslash m_i = \{m_k, m_l\}$ can be altered in any way, since $SIGN(\mathcal{H}(m_i)||\mathcal{CH}(m_j)) = SIGN(\mathcal{H}(m_i)||\mathcal{CH}(m_j')) \wedge SIGN(\mathcal{H}(m_i)||\mathcal{CH}(m_j)) = SIGN(\mathcal{H}(m_i)||\mathcal{CH}(m_k'||m_i||m_l'))$ remains true. This is important if the document to be sanitized follows a strict ordering, i.e. text-documents. Note, for $m_j \subsetneq m_i$ this does not work, since all alterations to m_j also affect m_i, which means that the resulting digest $d_i = \mathcal{H}(m_i)$ will differ. One may argue that splitting the document up into three parts, namely m_k, m_i and m_l, while making m_k and m_l sanitizable fulfills the same requirements. This is possible if the subdocument itself is just text. If additional information is signed, i.e. structure, or the node-based nature of XML documents is taken into account this may not be always possible. Consider an *XPath* expression that counts the number of elements and is signed using a standard cryptographic hash, while the nodeset used by this expression is signed using a chameleon hash. This prohibits the alteration of the number of elements contained. Since *XPath* is *Turing*-Complete [8] this construction allows the protection of all structural information, if the expression used is wisely chosen. A formal construction of these expressions is future work.

***Miyazaki* scheme.** *Miyazaki* et al. developed a redactable signature scheme which allows disclosure control in [16]. The scheme just allows deletion and of-

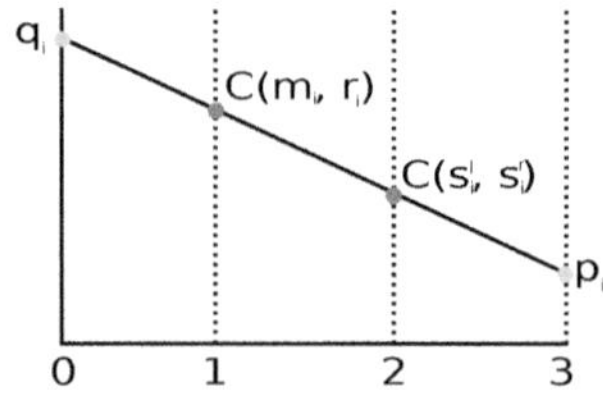

Fig. 1. Illustration of the *Miyazaki* Scheme

fers the possibility to prohibit additional sanitization for a consecutive sanitizer. Each subdocument m_i is transfered on a coordinate system. The transform is a commitment to m_i using e.g. the *Halevi*-Commitment-Scheme. [9] This commitment $(C(m_i, r_i))$[5] will then be transfered onto the point $(1, C(m_i, r_i))$. In a second step, two random numbers s_i^l, s_i^r with $|s_i^l| = |s_i^r| = |r_i|$ are committed to by calculating an additional commitment $C(s_i^l, s_i^r)$. The second commitment constructs the point $(2, C(s_i^l, s_i^r))$. Finally, the signer calculates two auxiliary points $(3, P_i)$ and $(0, Q_i)$ as illustrated in Fig. 1. The resulting signature is $\sigma_{miyazaki} = (P_1||...||P_n||Q_1||...||Q_n)$. The signature $\sigma_{miyazaki}$, all P_i, all m_i, and the corresponding (de)commitment values will be distributed with m. To sanitize a sanitizer just has to remove m_i and r_i along with its (de)commitment values; a verifier is able to reconstruct Q_i given P_i and $C(s_i^l, s_i^r)$. To prohibit sanitization an adversary has to remove s_i^l and s_i^r along with the (de)commitment values; a verifier is still able to verify the signature, calculating Q_i using $C(m_i, r_i)$ and P_i. Removing both values will result in a non-verifyable signature. *Miyazaki* requires completely disjunct subdocuments ($\forall i, j : m_i \cap m_j = \emptyset$ where $0 \leq i < j \leq n$)[6] since the scheme just allows deleting subdocuments rather then arbitrary alterations. Hence, any partial deletion will alter another subdocument which is not possible with this scheme. This limits its usefulness in XML Signatures, since *Miyazaki* does not allow overlapping references.

[5] Note, the *Halevi* scheme requires a purely random number r.

[6] They need to form a disjunct partition.

3.2 Integration into an XML Signature

In this section we give a brief introduction on the JCA code changes resp. extension made to allow XML Signatures with the sanitizable schemes. Sources will be made available upon request. However, we present all the modifications necessary to make the schemes JCA implementable.

Key Generation. Every chameleon hash needs a key-pair. Obviously this key-pair has to be generated prior to calculating the digest. A Java key-generator accepts exactly two parameters, a `SecureRandom` object, which handles the generation of random numbers and a security parameter τ used to define the key-length. However, the class `BigInteger` does not allow to generate primes which have a "bit-range". Hence, each key will be generated using $\tau_{new} = \tau - 1$ to avoid too huge primes. This reduces the security by two Bits. Trivially, the workaround is to increase the bitlength by two to get at least the same security level. For the performance evaluation this has no major impact.

Key Generation Zhang. For the *Zhang* chameleon hash no keys have been defined, since this scheme has been implemented as a TTP-Service.

DigestMethod. The chameleon hashes need a public-key to be computed. We marshalled the public key into to the reference element. For example, public keys in the *Krawczyk* scheme have four elements: $PK_{CH} = (p, q, g, y)$. [12] We add a new element containing the public key as *Base64*-encoded `BigIntegers` as illustrated in Lst. 1.1. Finally, we used new URIs to identify each digest method[7].

Listing 1.1. Marshalling of the *Krawczyk* Parameters

```
<DigestMethod Algorithm="http://www.example.org/xmldsig-more#chamhashdisc">
  <ChamHashDiscKeyValue>
    <p>Aa5Mue7ppx2YD7R8KXUqQIKSTSay6jHhWm9L0dxHpL2P</p>
    <q>1yZcO3TTjswII2j4UupUgQUkmk111GPCtNCXo7iP3Xsc=</q>
    <r>FQrJPkWb0JwiffjrAdbWAoyropQmNohMgEy6ABsvptQ=</r>
    <g>JtqJ1HONL0Is+6Y797XKQ1hbHc+HYgoGQAkvK8h+q8Y=</g>
    <y>AVwdxMlXF6HIHRH10r7Xoojb0VoB7ZBP4Dxc83BDDgxG</y>
  </ChamHashDiscKeyValue>
</DigestMethod>
```

For the *Zhang* scheme the coin consists of a point P on an elliptic curve $E(\mathbb{F}_q)$. The point is stored as its (x, y) coordinates.

For the *Miyazaki* and *Chen* scheme some workarounds were necessary due to the JCA not being able to store negative numbers. Therefore, everything must exclusively be encoded using positive numbers. For the *Chen* scheme we store a negative b, i.e. -1, as a Base64-encoded 2 and reverse it to -1 during unmarshalling. For the *Miyazaki* scheme a negative point P is not unlikely, if $\Delta = C(s_i^l, s_i^r) - C(m_i, r_i) < 0$. To prohibit this the result of $C(s_i^l, s_i^r)$ is multiplied by $2^{|p|-1}$. This ensures that $\forall C(m_i, r_i) : \Delta > 0 \Rightarrow P > 0$, where p is the prime used in the *Halevi* scheme. Note, neither of the workarounds impacts security since no Bits are omitted and the lower Bits are shifted out during unmarshalling.

[7] `http://www.example.org/xmldsig-more#chamhashdisc`, `...#miyazaki`, etc.

Our second workaround in the implementation of the *Miyazaki* scheme addresses the output of the reference digest, which is $(P_i||Q_i)$ for each reference, since the references are sequentially processed by the JCA framework. Hence, we add an additional hashing step which extends the reference "digest" to $\mathcal{H}(P_i||Q_i)$ in order to gain a fixed length output. Additionally, any deletable element has been placed in an `Signature` element not covered by the signature itself. Trivially, this must be done to ensure that the signature verification can be done correctly if any changes are done. E.g. for the *Miyazaki* scheme the (de-)commitment values are added outside of `SignedInfo` as shown in Lst. 1.2, since they can be removed by a sanitizer. Note, we link them to the corresponding references by using the reference's `URI` value as `Id` attribute.

For the chameleon hashes the random value r resp. b are also part of the `DigestMethod` element, as shown in Lst. 1.1. Just to be clear: This removes the transparency property. However, storing them here or forward referencing them has negligible impact on the performance itself. Another way to allow random coin change would be to add an additional transform which removes them prior to digesting the `Reference` element. The latter approach was not chosen since an additional transform would not have been as self-explaining as this version.

Listing 1.2. Marshalled *Miyazaki* scheme (de-)commitment values

```
<Signature>
  <SignedInfo>...</SignedInfo>
  <SignatureValue>...</SignatureValue>
  <KeyInfo>...</KeyInfo>
  <MiyazakiKeyValue Id="#xpointer(id('8492341'))">
    <sr>cjyt7T7qu3j+ieBksyE0J+yVv/0=</sr>
    <sl>S5NrovXVK0YkswUERiz0EfjR+fc=</sl>
    <r>bWTFplIzjCpiCWReRGzCL8m20yY=</r>
    <a>DKFCyGUL</a>
    <b>ASWQM3JEo0cpm77v0oc+NLwujCX+</b>
    <y>7Zo2Etyvy3Umwf8LlyRfDSCvLc4=</y>
    <a_s>CbaVZ/t+Lw==</a_s>
    <b_s>AbW5KeK0BcyQMxqpZ0oqLUIwakDU</b_s>
    <y_s>W9k9TudAZ813C4gFXbA1/VM11E8=</y_s>
    <prime>AuX1hF/7HhBr5va4Ayx9HWIK1Sfj</prime>
    <prime_s>Ao8RKY6Ezd5uRikVEpLheUqxdYVz</prime_s>
  </MiyazakiKeyValue>
</Signature>
```

Additional Hashing Steps. The hashes proposed by *Ateniese* et al., the one proposed by *Zhang* et al., and *Chen* et al. perform an additional standard cryptographic hash prior to its own calculations. To have comparable evaluations this hashing-step has also been added to the other schemes. This additional hashing step has been fixed as SHA-1 for all schemes to have comparable results. The commitment-scheme required by the *Miyazaki* scheme returns an array (y_i, a_i, b_i, p_i). For further calculations y_i, a_i, b_i have to be merged into one integer. We pad all numbers with zeroes till they have the same size. This means that $|q_{i_1}| + |y_i| = |q_{i_2}| + |a_i| = |q_{i_3}| + |b_i|$, where $q_i \geq 0$ are the padding Bits. This does not allow any modifications since the width of each part is fixed. Note, the prime p is an external value and thus not part of the commitment-value itself.

Listing 1.3. Appended Values for the *Chen* scheme

```
<Signature>
  <SignedInfo>...</SignedInfo>
  <SignatureValue>...</SignatureValue>
  <KeyInfo>...</KeyInfo>
  <r Id="#xpointer(id('8492340'))">210874650874</r>
  <b Id="#xpointer(id('8492340'))">2</b>
</Signature>
```

Signing. For the *Miyazaki* scheme the digest values need to be added prior to hashing the reference. Thus, all parameters added to the `DigestMethod` element must be known prior to the digestion procedure. This would have prohibited the calculation of commitments. Our workaround runs the signing process twice, in the first run the intermediate results, i.e. the (de-)commitment values are cached and added during the second run. The second run's results will not be used. A negative performance impact, but this approach did not require major code changes in the JCA framework itself.

Sanitization. The sanitization procedure outputs an r'_i (and b'_i for the *Chen* scheme) for a given m'_i such that $\mathcal{CH}(r_i, m_i) = \mathcal{CH}(r'_i, m'_i)$ where $m_i \neq m'_i$ and $r_i \neq r'_i$. For performance reasons we cache each subdocument during the signing process to allow an easy UForge implementation. Lst. 1.3 shows how a sanitizer adds new elements to the `Signature` element for the *Chen* scheme, which requires the two values b'_i and r'_i. For *Ateniese* and *Krawczyk* the element contains only r'_i for each sanitized reference. The *Zhang* scheme requires a point (x'_i, y'_i).

The values leading to a collision are not encoded in *Base64*; this was done for convenience. A sanitizer can add the calculated values without additional encoding, except the encoding for negative values. Note, this does not have a security impact; a maliciously changed r'_i will most likely result in a wrong digest value. Thus, the whole signature validation will fail and malicious alterations will be detected. In our performance evaluation we replaced every sanitized subdocument by the fixed string `<test id="ID">xxx</test>`.

For the *Miyazaki* scheme we have a different sanitization process: The whole subdocument has to be replaced with a new element `<sanitized id="ID> ARBITRARY STRING </sanitized>`. This allows for an easy decision which commitment value to use when calculating the auxiliary points. Additionally, the corresponding de-commitment values have to be removed as well.

Special Implementation of the *Zhang* scheme. The chameleon hash proposed by *Zhang* et al. requires a very large signature and the reconstruction of $E(\mathbb{F}_q)$ for each validation. We used a client-server approach. Every method of the chameleon hash is calculated by a TTP and transfered to the calling application. This was done to avoid reconstructing the elliptic curve for every validation and to have a smaller signature, which just contains the random point $(x_i, y_i) \in E(\mathbb{F}_q)$ and the recipient S for each reference. Note, this use of an online-TTP is not enforced by the scheme itself; the verification only needs the public key to reconstruct the curve.

4 Performance Evaluation of Implemented Schemes

For each performance measurement we performed 100 runs and calculated the median to reduce the impact of the probabilistic algorithms. Every input is digested using a standard cryptographic hash prior to hashing it with the chameleon ones. Thus, the results are independent of subdocument's length. For every identity-based scheme we used a fixed S as well.

Tests were run on a *Lenovo Thinkpad T61* with an *Intel* T8300 Dual Core @ `2.40` Ghz and 4 GiB of RAM. The operating system was *Ubuntu* Version 10.04 (64 Bit), while the Java-Framework version was `1.6.0_20-b02`.

4.1 Algorithms: Setup, Hashing and Forging

To have a similar and commonly known algorithm to compare with, the RSA key-pair-generation-algorithm was also measured. e has been fixed to 65537 in the RSA algorithm. The results are shown in Tab. 1 along with SHA-512 for hashing.

Krawczyk et al. The key pair generation time was the longest. It can basically be divided into the need to find a safe-prime $p = 2q + 1$ and a generator of order q, generating $\mathbb{Z}/p\mathbb{Z}$. Detailed analysis found the generation of the safe-primes consumes the most time (99%), while finding a generator and setting up the keys is negligible with less than 1% of the whole key generation time. Our comparison yielded that key-length has the biggest performance impact for *Krawczyk*, especially notable during key-pair generation. Currently the highest practical key-size is 512 Bit. Hashing and forging do not use much time and are almost calculated instantly. We suggest a pool of pre-generated key-pairs to avoid long wait times for new key-pairs.

Ateniese et al. The key-pair-generation of the *Ateniese* scheme is not as expensive as the *Krawczyk* scheme since no safe-primes are needed. We found the prime generation to be the limiting factor; all other operations are almost instantly calculated. The argumentation is basically the same as for the *Krawczyk*

Table 1. Median Runtime: Input: 160 Bit; Hash-Output: 512 Bit; all in μs

Algorithm	**Setup**	**Hash**	**IForge**	**UForge**
RSA/SHA-512	400,119	7	-	-
Krawczyk	11,082,481	3	4	16
Ateniese	7,856	410	424	522
Chen	68,319	1,878	248,280	7,661
Zhang 512Bit-Key	9,570,008	25,547,333	8,267,775	-
Zhang 128Bit-Key	243,040	929,648	381,215	-
Miyazaki[a]	4,700	generation: 11,400 verification: 70	0	-

[a] Miyazaki does not use hashing, times are given for the comparable operation.

scheme. The setup is fast and even huger security parameters can be used. However, hashing and forging are more complex algorithms than *Krawczyk*'s and take longer. Pre-generated primes could increase the performance. Compared to *Krawczyk*'s it will be faster if you consider that *Krawczyk* would require new keys, while the ID-based concept used by *Ateniese* just needs one "master-key-pair". In practice, the speed gained during key generation compensates the longer hashing times.

Chen et al. As the *Ateniese* scheme, the *Chen* scheme does not require safe-primes. However, the needed primes cannot be arbitrary, since $p \equiv q \equiv 3 \pmod 4$ must be true. ω has been fixed to 64 for all measurements. Notably, *Chen*'s *IForge* algorithm takes a lot of time. Further investigation showed the extraction process to consume the most time, in particular the square root calculation is very expensive, consuming $> 99\%$ of the time. However, the practical impact is reduced since *Chen* is key-exposure-free. Compared to *Ateniese*, which does not offer this property, the impact of key-exposure must not be limited. *Ateniese* et al. suggested in [2] to append a `TransactionID` to the `ID`, resulting in the extraction procedure being performed more often in *Ateniese* than in *Chen*. A practical performance increase for the TTP for *Chen* is to extract the trapdoor along with the registration or on request, as *Ateniese* proposed for his scheme [2].

***Zhang* et al.** The *Zhang* scheme relies on elliptic curves and pairings. Hence, the Key-Pair-Generation is actually constructing an elliptic curve along with its parameters. Note, the evaluation of the *Zhang* scheme includes the delays caused by our TCP-socket-connections. Additionally a delay of $1s$ had to be introduced to avoid a timeout of the signature-generation due to the TCP-transfers and calculation on the server-side. This implementation-specific overhead has been subtracted from the presented timings. Our observation showed that even after the curve has been constructed, the setup for *Zhang* still takes a long time. This is caused by adding the system parameters; however due to the library approach no further investigation was possible.

The *Zhang* scheme suffers from certain limitations; first of all the key-pair-generation is very slow. Tab. 1 shows that for a key length of 512 Bit the *Zheng* algorithms take considerately more time. The speeds are getting practical when the key length for the elliptic curve based scheme is lowered to 128 Bit, which offers comparable security. The practical impact of the slow *IForge* algorithm depends on how often sanitization is done.

Miyazaki et al. Finally, the *Miyazaki* scheme. It works completely different; instead of calculating a digest it calculates a commitment, which can be seen as a key-pair. Obviously, most time is spent calculating the commitments ($11,400\mu s$), while verifying a commitment is almost instantly ($70\mu s$). As for the other schemes, this is due to the fact that primes must be generated. Redaction, comparable to *IForge*, is just removing the element from the XML, and does not require any calculation at all.

4.2 Signature Generation and Verification

Comparing algorithms with such different approaches is like comparing apples and oranges; however, all schemes aim to provide sanitizable signatures. The preceding section presented the bare runtime of each algorithm. To see how the different schemes scale in praxis, they have been used to generate an XML signature over the document listed in Lst. 1.4. By `xpointers` we split the document into two subdocuments; namely `#xpointer(id('8492340'))` and `#xpointer(id('8492341'))` have been used, returning the subtrees `Item` and `Address`.

Listing 1.4. The XML File used

```
<?xml version="1.0" encoding="UTF-8"?>
<PurchaseOrder>
 <Item id="8492341">
  <Description>Video Game</Description>
  <Price>10.29</Price>
 </Item>
 <Buyer>
  <Name>My Name</Name>
  <Address id="8492340">
   <Street>One Network Drive</Street>
   <Town>Burlington</Town>
   <State>MA</State>
   <Country>United States</Country>
   <PostalCode>01803</PostalCode>
  </Address>
 </Buyer>
</PurchaseOrder>
```

For this evaluation each scheme will be measured with 512 Bit keys, while the key-generations, both for the underlying RSA-Signature and the chameleon hashes, have been omitted to have comparable results. This was done since the *Zhang* scheme relies on a TTP to generate the curve on start-up. SHA-512 has been measured as well to see the performance impact of a sanitizable schemes. Tab. 2 shows the results for a "real" signing and verification step. As before, to have meaningful results a $1s$ delay was subtracted for the *Zhang* scheme and the server was already running.

4.3 Summary

Surprisingly, our evaluation showed that all schemes come with similar runtime figures for generation and for validation; this finding, however, does not account

Table 2. Scheme Comparison (Generation/Validation) for 512 Bit keys in μs

	SHA-512	*Krawczyk*	*Ateniese*	*Chen*	*Zhang 128*	*Zhang 512*[a]	*Miyazaki*
Generation	2,010,624	2,219,806	2,472,972	2,424,781	3,645,070	52,434,635	2,675,284
Validation	1,373,907	1,301,410	1,366,284	1,278,295	1,908,047	50,256,292	1,339,633

[a] Zhang with 128 Bit offers a comparable security.

for the pre-generated key-pair. Taken the key-pair generation into account as well, the chameleon hash proposed by *Ateniese* et al. achieves the best performance, in particular since just one key-pair has to be generated. Note, that it suffers from the key-exposure-problem as discussed in [2]. Using transaction-based IDs helps reducing the key-exposure-problem's impact, hence the *Ateniese* scheme is not as heavily affected as the *Krawczyk* scheme. *Krawczyk* needs a new key-pair for each subdocument to prohibit non-wanted key-exposure. Thus, all ID-based chameleon hashes perform better in this respect, even if they require more complex algorithms. The notable exception is the *Zhang* scheme, which requires calculations on elliptic curves. However, cryptographic algorithms based on ECC offer a higher degree of security. Hence, smaller key-sizes are acceptable which decreases the time needed for the algorithms.

The *Miyazaki* scheme is not as suitable for XML files as the *Ateniese* scheme regardless of the chameleon hash. Since in *Miyazaki* it is not possible to protect structural information as it requires non-overlapping subdocuments. However, we showed it has a comparable runtime.

5 Property Changes of Schemes due to Mixing

The schemes introduced suffer from certain limitations, e.g. a verifier is able to figure out what parts are sanitizable in the *Ateniese* scheme by just looking at the used digest method. We propose some additions to these schemes which add resp. remove certain properties.

5.1 Properties of Sanitizable Signature Schemes

Brzuska et al. already formalized most of the properties given in [5]. Hence just the basic idea is given here.

1. *Unforgeability:* For an outsider (i.e. for anyone not being the signer or a sanitizer) it is infeasible to forge signatures.
2. *Immutability:* The sanitizer should not be able to change parts of the document which are not designated to be sanitized.
3. *Privacy:* Sanitized parts should not leak any information which may be used to reconstruct information which has been censored.
4. *Transparency:* It should be impossible for the verifier to decide whether a given document was sanitized or not. *Ateniese* et al. in [1] further distinguish between:
   - *Weak Transparency:* Weak Transparency means that the verifier is able to identify the subdocuments m_i that can potentially be sanitized.
   - *Strong Transparency:* Is the opposite of Weak Transparency; the recipient is not able to distinguish whether a subdocument m_i is sanitizable or not.

   To avoid confusion, *Brzuska* et al. formalized only *Ateniese* et al.'s notion of Weak Transparency under the term Transparency. [5]
5. *Accountability:* The responsibility of a message should not be transferable to another party, in particular the original signer should not be held responsible

for a maliciously sanitized message. *Brzuska* et. al showed that this property must be split up as well: [5]

- *Signer Accountability:* If a document was not signed by the signer even a sanitizer should not be able to accuse him.
- *Sanitizer Accountability:* If a document was signed by a signer, the signer should not be able to accuse the sanitizer if he did not alter document.

Not formalized byare the following properties:

6. *Restrict to Values:* A sanitizer is just able to replace a subdocument with certain preset values instead of ones of his own choice.
7. *Sanitization Control:* Defines if a signer has control over the parts which can be sanitized by a sanitizer.
8. *Consecutive Sanitization:* Defines if an already sanitized document can be sanitized again by another sanitizer.
9. *Consecutive Sanitization Control:* Defines if a sanitizer can prohibit further sanitization by another sanitizer.

5.2 Extensions of the *Ateniese* Scheme

This subsection will introduce some additions to the *Ateniese* scheme, using a mixture of given primitives. The proposed extensions will lead to the redefinition of transparency in Sec. 6.

Removing Transparency. The *Ateniese* scheme [1] is defined as $\sigma_{ateniese} = SIGN(d_1||...||d_n)$, where

$$d_i = \begin{cases} \mathcal{H}(m_i) & \text{if } m_i \text{ is not sanitizable} \\ \mathcal{CH}(m_i) & \text{if } m_i \text{ is sanitizable} \end{cases}$$

This allows to see which subdocuments m_i are potentially sanitizable, but not if they have actually been sanitized. We change this by adding the original r_i for each chameleon hash to allow a "m_i has been sanitized" detection. Therefore the signature is expanded to $\sigma'_{ateniese} = SIGN(s_1||...||s_n||d_1||...||d_n)$, where

$$s_i = \begin{cases} \text{The } r_i \text{ in } \mathcal{CH}(m_i, r_i) & \text{if } m_i \text{ is sanitizable} \\ \emptyset & \text{else} \end{cases}$$

Every s_i is covered the signature in cleartext. This implies that every recipient is able to decide whether a given (sub-)document has been sanitized, since he can compare the r'_i used for calculating the chameleon hashes with the s_i provided by the signature $\sigma'_{ateniese}$. This has no impact on security, because the verifier is still not able to find additional collisions since he needs the original m_i. This is the scheme implemented in this paper, as proposed by *Tan* et al. [18].

Removing Transparency II. The previous approach can be fine-tuned further. The r_i used to calculate each chameleon hash are now not directly appended to the signature, but hashed using a cryptographic hash, i.e.: $\sigma''_{ateniese} = SIGN(d_1||...||d_n||h_1||...|h_n)$ while $h_i = \mathcal{H}(s_i)$. However, this construction can be enriched with an interesting flavour; the hash $\mathcal{H}$ could also be replaced with a chameleon hash ($h_i = \mathcal{CH}_{trans}(s_i)$), therefore allowing only the sanitizer who holds an additional private key sk_{trans} to transparently sanitize a subdocument.

This extended scheme requires that every r_i is part of the respective hash; however it is possible that the property of non-transparency is just desired for certain subdocuments. Obviously, this can be achieved by removing r_i from the concatenation. This allows a signer to decide where he wants that property. However, an attacker cannot add r_i' to falsely indicate a sanitization; the verifier can use the r_i for verification which will reassure him that m_i was not sanitized. The trivial proof is omitted for brevity.

Adding Strong Transparency. *Ateniese* et al. already note in [1] that it is very simple to gain strong transparency by generating distinct key pairs for each chameleon hash, i.e.

$$d_i = \begin{cases} \mathcal{CH}_{k_i}(m_i) & \text{if } m_i \text{ is sanitizable} \\ \mathcal{CH}_{u_i}(m_i) & \text{if } m_i \text{ is not sanitizable} \end{cases}$$

We will later make use of this approach. This can also be used to improve the overall performance. Originally, each chameleon hash needed a separate key pair; However, if the standard chameleon hash is replaced with an ID-based construction, just S needs to be different for each generated hash since the private key B differs. This leads to another interesting application, a signer can define groups, similar to networking domains. This means the TTP distributes the secret keys with respect to a sanitizer's group membership.

A similar approach without using a TTP is to use the extended chameleon hash presented by *Ren* et al. which allows a multi-user hash. [17] Hence, a signer can select which sanitizers are able to sanitize while hiding the group membership.

Adding Restricting to Values. One way to restrict someone to values is the use of an chameleon hash without an *UForge* algorithm, like the one proposed by *Zhang* et al. [20]. An additional way is to use Bloom-Filters as proposed by *Klonoswki* et al. [11].

5.3 *Miyazaki* Scheme

This section will introduce some mechanisms which change the properties of the *Miyazaki* scheme.

Adding Weak Transparency. Adding weak transparency can be achieved by adding a random number of "fake-documents" between the real subdocuments. This fake-subdocuments will be redacted (sanitized) by the signer itself prior to sending it to the sanitizers. As a result the signer gives only sanitized documents to the first sanitizer. Thus, neither a sanitizer nor a verifier knows if the document has been sanitized by sanitizers, since he cannot distinguish between a sanitized original subdocument and a sanitized fake-subdocument. However, the signer now no longer distributes technically not sanitized documents, which might not be wanted in all use cases.

Adding Accountability, Consecutive Sanitization Control & Restricting to Sanitizers. A way to add both signer and sanitizer accountability is to use a similar mechanism as with the *Ateniese* scheme. The signer adds an additional chameleon hash over the whole message m to the signature using a tag-based-chameleon hashi.e. the one proposed by *Brzuska* et al. in [5].

$$\sigma_{miyazaki} = SIGN(d_{ch}||P_1||...||P_n||Q_1||...||Q_n) \text{ where } d_{ch} = \mathcal{CH}_{tag}(m).$$

If a subdocument is redacted now, the tag-based-chameleon hash needs to be adjusted as well. This allows a signer to trace back the changes and thus to find a malicious sanitizer. Note, a sanitizer is not able to do arbitrary changes to the subdocument since the signature still covers the commitments. This construction also implies that just sanitizers defined by the signer are able to alter a signed document, since only they know the trapdoor needed to calculate collisions.

To prohibit consecutive sanitization control an additional chameleon hash over the (de-)commitment values of the mask can be constructed. The argumentation is essentially the same as before.

6 The Transparency Property Revisited

Ateniese et al. define the property of transparency (T) as follows:

> Given a signed message with a valid signature, no party — except the censor and the signer — should be able to correctly guess whether the message has been sanitized. [1]

They further divide the property into "weak" (WT) and "strong transparency" (ST):

> We further distinguish among two flavors of transparency: weak and strong. Weak transparency means that the verifier knows exactly which parts of the message are potentially sanitizable and, consequently, which parts are immutable. In contrast, strong transparency guarantees that the verifier does not know which parts of the message are immutable and thus does not know which parts of a signed message could potentially be sanitizable. [1]

Essentially, T always implies exactly one of WT or ST and vice versa.

$$(T \implies (ST\dot{\vee}WT)) \wedge (ST\dot{\vee}WT)) \implies T) \equiv T \Leftrightarrow (ST\dot{\vee}WT)$$

where $\dot{\vee}$ denotes "exclusive or". Practically, a verifier either knows which m_i is potentially sanitizable or he does not. Hence, $(ST\dot{\vee}WT)$ should be a tautology, i.e. $\models (ST\dot{\vee}WT)$. Since $T \Leftrightarrow (ST\dot{\vee}WT)$ this results that T is always true. This is counterintuitive and contradicts the *Miyazaki* scheme. Also, our construction in Sec 5.2 removes transparency in general on the message level, but leaves WT defined on subdocument level untouched.

Strong transparency does not always imply transparency either, as the following construction demonstrates. Let $\sigma = SIGN(d_1||...||d_n||\mathcal{H}(m))$ where all

d_i are calculated as in the *Ateniese* scheme with strong transparency. (Sec. 5.2) Now a verifier is given the digest $\mathcal{H}(m)$ along with m' and the signature σ. m' is the potentially sanitized message, and he can use σ to verify the signature over m'. With this construction a verifier can deduce that m has been altered by comparing $\mathcal{H}(m)$ and $\mathcal{H}(m')$, but gains no knowledge about what parts m_i are potentially sanitizable. Hence, this scheme does not have the property of transparency while it maintains the property of strong transparency. On a closer look, the scheme given in 5.2 can also be modified to fulfill this requirement; the only addition is to hash the *concatenation* of $r_1, ..., r_n$ prior to signing, i.e. $\sigma = SIGN(\mathcal{H}(r_1||...||r_n)||d_1||...||d_n)$. The verifier is provided with $\mathcal{H}(r_1||...||r_n)$. He further knows each chameleon hash's actual value r_i', which can be different or equal to r_i. If the verifier compares $\mathcal{H}(r_1||...||r_n)$ from the signature with $\mathcal{H}(r_1'||...||r_n')$ he can identify that m was sanitized, but cannot deduce which m_i.

A scheme without transparency but with weak transparency can be constructed in a similar way; instead of using the *Ateniese* scheme with strong transparency the standard one is used. This allows to see what subdocuments are potentially sanitizable but only to deduce if the document as a whole has been sanitized.

We conclude that the properties of weak and strong transparency are actually independent from the property of transparency. Hence, we do not see them as "different flavours" [1] as stated by *Ateniese* et al. However, *Ateniese* et al. correctly differentiated a difference in scope linguistically. We want to emphasize this further by explicitly differentiating the scope:

Transparency makes a statement about a sanitiz***ed*** document as a whole.
Weak and Strong Transparency make statements about sanitiz***able*** ***sub***documents.

7 Conclusion

We have integrated sanitizable signature schemes into the existing XML Digital Signature Specification without any major adjustments, and by disregarding the workarounds caused by bugs in the JCA. This also covers schemes which do not rely on the standard hash-and-sign paradigm like the *Miyazaki* scheme and schemes based on a TTP. Moreover, we showed that the performance of all schemes are comparable to the standard SHA-512 - RSA signature procedure. Furthermore, we have shown that the existing schemes are extendable while still being compliant with the XML Digital Signature Specification. This also lead to the finding that the properties of Weak and Strong Transparency, covering subdocuments, are independent from the overall property of Transparency, which concerns the document as a whole.

References

1. Ateniese, G., Chou, D.H., de Medeiros, B., Tsudik, G.: Sanitizable signatures. In: di Vimercati, S.d.C., Syverson, P.F., Gollmann, D. (eds.) ESORICS 2005. LNCS, vol. 3679, pp. 159–177. Springer, Heidelberg (2005)

2. Ateniese, G., de Medeiros, B.: Identity-Based Chameleon Hash and Applications. In: Juels, A. (ed.) FC 2004. LNCS, vol. 3110, pp. 164–180. Springer, Heidelberg (2004)
3. Ateniese, G., de Medeiros, B.: On the Key Exposure Problem in Chameleon Hashes. In: Blundo, C., Cimato, S. (eds.) SCN 2004. LNCS, vol. 3352, pp. 165–179. Springer, Heidelberg (2005)
4. Brzuska, C., Busch, H., Dagdelen, O., Fischlin, M., Franz, M., Katzenbeisser, S., Manulis, M., Onete, C., Peter, A., Poettering, B., Schröder, D.: Redactable Signatures for Tree-Structured Data: Definitions and Constructions. In: Zhou, J., Yung, M. (eds.) ACNS 2010. LNCS, vol. 6123, pp. 87–104. Springer, Heidelberg (2010)
5. Brzuska, C., Fischlin, M., Freudenreich, T., Lehmann, A., Page, M., Schelbert, J., Schröder, D., Volk, F.: Security of sanitizable signatures revisited. In: Jarecki, S., Tsudik, G. (eds.) PKC 2009. LNCS, vol. 5443, pp. 317–336. Springer, Heidelberg (2009)
6. Chen, X., Tian, H., Zhang, F.: Comments and Improvements on Chameleon Hashing Without Key Exposure Based on Factoring. In: IACR Cryptology ePrint Archive, number 319 (2009)
7. Eastlake, Reagle, Solo: XML-signature syntax and processing. W3C recommendation (February 2002), www.w3.org/TR/xmldsig-core/
8. Gottlob, G., Koch, C., Pichler, R.: The complexity of XPath query evaluation. In: Proceedings of the 22nd Symposium on Principles of Database Systems, PODS, pp. 179–190. ACM, New York (2003)
9. Halevi, S., Micali, S.: Practical and Provably-Secure Commitment Schemes from Collision-Free Hashing. In: Koblitz, N. (ed.) CRYPTO 1996. LNCS, vol. 1109, pp. 201–215. Springer, Heidelberg (1996)
10. Jonsson, J., Kaliski, B.: Public-Key Cryptography Standards (PKCS) #1: RSA Cryptography Specifications Version 2.1. RFC 3447, Informational (2003)
11. Klonowski, M., Lauks, A.: Extended Sanitizable Signatures. In: Rhee, M.S., Lee, B. (eds.) ICISC 2006. LNCS, vol. 4296, pp. 343–355. Springer, Heidelberg (2006)
12. Krawczyk, H., Rabin, T.: Chameleon Hashing and Signatures. In: Symposium on Network and Distributed Systems Security, pp. 143–154 (2000)
13. Kundu, A., Bertino, E.: Structural Signatures for Tree Data Structures. In: Proc. of PVLDB 2008. ACM, New Zealand (2008)
14. Kundu, A., Bertino, E.: CERIAS Tech Report 2009-1 Leakage-Free Integrity Assurance for Tree Data Structures (2009)
15. Liu, B., Lu, J., Yip, J.: XML Data Integrity Based on Concatenated Hash Function. CoRR, abs/0906.3772 (2009)
16. Miyazaki, K., Iwamura, M., et al.: Digitally Signed Document Sanitizing Scheme with Disclosure Condition Control. IEICE Transactions (2005)
17. Ren, Q., Mu, Y., Susilo, W.: Mitigating Phishing by a New ID-based Chameleon Hash without Key Exposure. In: Proceedings of AusCERT (2007)
18. Tan, K.W., Deng, R.H.: Applying Sanitizable Signature to Web-Service-Enabled Business Processes: Going Beyond Integrity Protection. In: IEEE International Conference on Web Services, pp. 67–74 (2009)
19. Wu, Z.-Y., Hsueh, C.-W., Tsai, C.-Y., Lai, F., Lee, H.-C., Chung, Y.: Redactable Signatures for Signed CDA Documents. Journal of Medical Systems, 1–14 (December 2010)
20. Zhang, F., Safavi-naini, R., Susilo, W.: ID-Based Chameleon Hashes from Bilinear Pairings. In: IACR Cryptology ePrint Archive, number 208 (2003)

Double-Trapdoor Anonymous Tags for Traceable Signatures

Masayuki Abe[1], Sherman S.M. Chow[2,⋆], Kristiyan Haralambiev[3], and Miyako Ohkubo[4]

[1] Information Sharing Platform Laboratories NTT Corporation, Japan
[2] Department of Combinatorics and Optimization University of Waterloo, Canada
[3] Computer Science Department New York University, U.S.A.
[4] NICT, Japan

Abstract. This paper introduces a novel tool, *public-key anonymous tag system*, which is useful in building controlled privacy-protecting protocols. The double-trapdoor structure of the system not only allows the authority to create a token which can trace someone's tags without violating anonymity of the tag-issuer, but also allows the issuer to claim or deny the authorship of a tag in the stateless manner. An efficient instantiation based on simple assumptions in the standard model is presented. We then use it for a modular construction of traceable signatures. Our scheme supports a signature authorship claiming (and denial) that binds a claim to the public-key of the signer unlike that in known schemes. It is also the first scheme in the literature which features concurrent joining of users, stronger anonymity and so on without random oracles.

1 Introduction

Group signatures allow members of a group to sign on behalf of the group, without revealing their identity. A line of research efforts has been focused on incorporating a variety of anonymity revocation mechanisms for balancing between privacy and fairness. For example, the group manager (GM) can use a trapdoor to reveal the identity of a group signature's signer. In verifier-local revocation (VLR) group signatures [6], this opening mechanism is usually absent. Instead, the GM can trace or revoke a user by publishing a revocation token of this user so that any verifier can locally check if the author of the signature has been revoked. Traceable signatures [12], in addition to the opening and the user tracing, allow the signer to claim the authorship of a signature, without maintaining any state other than the signing key. Thus traceable signatures can be regarded as group signatures with additional anonymity management mechanism.

Group signatures have been extensively studied since the introduction in [8]. In [10], a semi-modular construction is done based on selective-tag weakly chosen

⋆ Part of the work was done while visiting NTT Information Sharing Platform Lab.

J. Lopez and G. Tsudik (Eds.): ACNS 2011, LNCS 6715, pp. 183–200, 2011.

ciphertext secure encryption and other building blocks in the standard model. A modular construction which achieves concurrent security (i.e., the joining of different members can be arbitrarily interleaved) is given in [3]. On the other hand, as seen in [13] for instance, the tracing and the claiming mechanisms of traceable signatures are often constructed in a hand-crafted manner. While the quasi-modular construction of traceable signatures in [9] essentially realizes the tracing via the use of pseudorandom function, this scheme does not satisfy the usual unlinkability guarantee. A systematic way to realize the anonymity management functionalities provided by traceable signatures seems to be lacking.

1.1 Public-Key Anonymous Tags with Double-Trapdoor

In this paper, we first propose a new tool, which we call *public-key anonymous tag system* (AT for short) with the double-trapdoor property. We then use it as a building block for constructing a traceable signature scheme.

AT supports tag tracing (by the authority), claiming (by the tag issuer) and denial (by non-issuers). A tag can be created with a user's individual secret-key, and cannot be associated to the user in normal circumstances. Whenever needs arise, the authority can use the master key to publish a tracing token that allows anyone to link all tags issued by the same user. *Yet the tracing token does not reveal the link to the public-key of the user.* Thus all signatures of a given user can be traced while retaining the anonymity. The same token can also be created from the user's secret-key. This double-trapdoor property makes it possible for the users to claim or deny the authorship of tags by first creating the token and proving in zero-knowledge that the committed token is linked to his/her public-key and (not) linked to the tag in question. We present a concrete construction based on simple assumptions in the standard model.

1.2 Modular Construction and CCA-Anonymity

Unlike that in the group signatures, the anonymity in the VLR framework [6] only ensures that any adversary cannot compromise the anonymity of a signer whose signatures have never been opened. Such a level of anonymity is called CPA-anonymity. In this paper, we construct traceable signatures with stronger level of anonymity, called CCA-anonymity, that allows the adversary to revoke the anonymity of adaptively chosen signatures. This better models the requirement that an honest user's signature remains anonymous even when other users' signatures have been opened.

Previous constructions, e.g., [12,10,13,9], require an explicit encryption step in signing. Indeed, it was shown that public-key encryption is necessary for the opening feature of group signatures [1]. With AT, we can construct a traceable signature scheme in a modular fashion without encryption as a building block. (The removal of encryption is conceptual since we use non-interactive zero-knowledge proof (NIZK) of knowledge, which is equivalent to encryption. It therefore does not contradict to the result in [1]. Yet it brings simplicity to the construction when stronger level of anonymity is considered.)

1.3 Real Authorship Claiming and Denial Mechanism

A more important benefit brought by the public-key nature of the tag system is the possibility of realizing a better signature authorship claiming and denial of traceable signatures. In the existing schemes, e.g., [13], the approach for claiming the signature authorship is to produce a proof of some witnesses that govern the relationship between the various parts of a signature. However, it is essentially impossible to settle once and for all the linkage between a signature and the identity of the issuer because nothing about the identity is in public. A public-key anonymous tag system, on the other hand, allows a "real" claiming that binds to the identity of the issuing user.

We remark that our approach, that uses a public certificate to bind a user's identity to a public-key, has a "drawback" as the original traceable signatures in [12]. That is, the GM (as the issuer of the certificates) can create a signature which can be linked to any given signature. Yet this is a reasonable compromise since it is of the GM's interest to identify the "bad" signatures issued by users.

Finally, note that the authorship denial functionality is uncommon in traceable signature schemes. While it is theoretically possible by proving in zero-knowledge that a signature and a public-key of a user in question are *not* in a proper relation, efficient instantiation is often difficult due to the nature of the "proof of inequality". Our concrete instantiation of AT allows very efficient denial of authorship which is of an independent interest.

2 Anonymous Tag System

2.1 Definitions

Definition 1 (Anonymous Tag System). *An anonymous tag system, AT, is a tuple of algorithms, AT.{Setup, Key, Tag, Reveal, Claim, Link}:*

- *Setup: (mtpk, mtsk) ← AT.Setup(1^λ) is a probabilistic algorithm that generates a common parameter, mtpk, and a master secret-key mtsk.*
- *Key: (utpk, utsk) ← AT.Key(mtpk) is a key generation algorithm that outputs a pair of public-key and a secret-key, (utpk, utsk), for an individual user.*
- *Tag: tag ← AT.Tag(mtpk, utsk) is a probabilistic algorithm that generates a tag, tag, from user's secret-key utsk.*
- *Reveal: tkn ← AT.Reveal(mtsk, utpk) is a deterministic algorithm that generates a link token tkn for user's public-key utpk using master secret-key mtsk.*
- *Claim: tkn ← AT.Claim(mtpk, utsk) is an algorithm that generates a link token tkn from the user secret-key utsk.*
- *Link: 1/0 ← AT.Link(mtpk, tag, tkn) is an algorithm that decides if tag is linked to the given link token tkn. It outputs 1 for input (tag, tkn) generated correctly as (utpk, utsk) ← AT.Key(mtpk), tag ← AT.Tag(mtpk, utsk) and tkn ← AT.Reveal(utsk, utpk), or tkn ← AT.Claim(mtpk, utsk).*

A secure AT provides three properties, *unlinkability*, *anonymity*, and *traceability*. While unlinkability and anonymity are considered as comparable notions in some other privacy-protecting protocols, they are incomparable in our framework. We start with unlinkability that captures the idea that it is hard for the public to see whether tags are generated by the same user or not. In the following formal definition, the adversary is given tags (through oracles) whose issuers are known as side-channel information.

Definition 2 (Unlinkability). *AT is unlinkable if, for all polynomial-time algorithm $\mathcal{A}$, the advantage of $\mathcal{A}$ in making the following experiment returns* 1, *i.e.,* $\mathsf{Adv}^{\mathcal{A}}_{\mathrm{Unlink}}(\lambda) = |\Pr[\mathsf{Exp}^{\mathcal{A}}_{\mathrm{Unlink}}(\lambda) = 1] - \frac{1}{2}|$, *is negligible in* λ.

$$\begin{aligned}
&\text{Experiment } \mathsf{Exp}^{\mathcal{A}}_{\mathrm{Unlink}}(\lambda):\\
&\quad mtpk \leftarrow \mathsf{AT.Setup}(1^\lambda),\\
&\quad (utpk_0, utsk_0) \leftarrow \mathsf{AT.Key}(mtpk), \quad (utpk_1, utsk_1) \leftarrow \mathsf{AT.Key}(mtpk), \mathrm{b} \leftarrow \{0,1\},\\
&\quad tag^* \leftarrow \mathsf{AT.Tag}(mtpk, utsk_{\mathrm{b}}).\\
&\quad \tilde{\mathrm{b}} \leftarrow \mathcal{A}^{\mathsf{AT.Tag}(mtpk, utsk_0), \mathsf{AT.Tag}(mtpk, utsk_1)}(mtpk, utpk_0, utpk_1, tag^*).\\
&\quad \text{Return 1 if } \mathrm{b} = \tilde{\mathrm{b}}. \text{ Return 0, otherwise.}
\end{aligned}$$

Anonymity captures the idea that it is hard to associate a tag to a public-key even after the link token for the public-key is published. Unlike the unlinkability, the adversary is not given any reference tags whose issuer is known in advance since otherwise the anonymity is trivially lost through the link tokens.

Definition 3 (Anonymity). *AT is anonymous if, for all polynomial-time algorithm $\mathcal{A}$, the advantage of $\mathcal{A}$ in making the following experiment returns* 1, *i.e.,* $\mathsf{Adv}^{\mathcal{A}}_{\mathrm{Anon}}(\lambda) = |\Pr[\mathsf{Exp}^{\mathcal{A}}_{\mathrm{Anon}}(\lambda) = 1] - \frac{1}{2}|$, *is negligible in* λ.

$$\begin{aligned}
&\text{Experiment } \mathsf{Exp}^{\mathcal{A}}_{\mathrm{Anon}}(\lambda):\\
&\quad mtpk \leftarrow \mathsf{AT.Setup}(1^\lambda),\\
&\quad (utpk_0, utsk_0) \leftarrow \mathsf{AT.Key}(mtpk), \; (utpk_1, utsk_1) \leftarrow \mathsf{AT.Key}(mtpk), \mathrm{b} \leftarrow \{0,1\},\\
&\quad tkn_{\mathrm{b}} \leftarrow \mathsf{AT.Reveal}(mtsk, utpk_{\mathrm{b}}), \; tkn_{1-\mathrm{b}} \leftarrow \mathsf{AT.Reveal}(mtsk, utpk_{1-\mathrm{b}}).\\
&\quad \tilde{\mathrm{b}} \leftarrow \mathcal{A}^{\mathsf{AT.Tag}(mtpk, utsk_{\mathrm{b}}), \mathsf{AT.Tag}(mtpk, utsk_{1-\mathrm{b}})}(mtpk, utpk_0, utpk_1, tkn_{\mathrm{b}}, tkn_{1-\mathrm{b}}).\\
&\quad \text{Return 1 if } \mathrm{b} = \tilde{\mathrm{b}}. \text{ Return 0, otherwise.}
\end{aligned}$$

Finally, we require tags be traceable in the sense that the link token computed by the authority associates all tags generated by the target user. We do not consider the traceability for maliciously generated public-keys.

Definition 4 (Traceability). *For* $(mtpk, mtsk) \leftarrow \mathsf{AT.Setup}(1^\lambda)$, *and two independent runs of* $\mathsf{AT.Key}$ $(utpk_0, utsk_0) \leftarrow \mathsf{AT.Key}(mtpk)$ *and* $(utpk_1, utsk_1) \leftarrow \mathsf{AT.Key}(mtpk)$, $\mathsf{AT.Reveal}(mtsk, utpk_0) = \mathsf{AT.Reveal}(mtsk, utpk_1)$ *happens only with negligible probability. Furthermore, for any tag, there exists at most one* tkn *that fulfills* $1 = \mathsf{AT.Link}(mtpk, tag, tkn)$.

2.2 Efficient Instantiation

Let $\Lambda = (p, \mathbb{G}, \mathbb{G}_T, e, g)$ be a description of groups $\mathbb{G}$ and $\mathbb{G}_T$ of prime order p equipped with an efficient bilinear map $e : \mathbb{G} \times \mathbb{G} \to \mathbb{G}_T$. It includes a random generator, $g \in \mathbb{G}$. We assume that Λ is determined according to security parameter λ. We assume that the Decision Linear Assumption (DLIN) [4] holds for Λ. Namely, given $(g_1, g_2, g_3, g_1^a, g_2^b, g_3^c)$, it is hard to decide whether $c = a + b$ or not. We also make the following novel assumption, which is in the Uber-assumption family [7] and can be easily justified in the generic (bilinear) group model [16].

Definition 5. *(Decision Reciprocity Assumption (DRA)) Given Λ and $(h_1, h_2, h_3, h_1^u, h_2^v, h_3^c)$, it is hard to decide whether $c = u/v$ or not.*

In Fig. 1, we present an efficient construction of AT based on DLIN and DRA over Λ. The underlying idea is to "encrypt" a DLIN instance by using the DRA structure. Slightly more in detail, we set a master-key, a user-key, and a link token to be in the correct form of DLIN. Namely, for a linear relation $(g_1, g_2, g_3, {g_1}^a, {g_2}^b, {g_3}^{a+b})$, we set (g_1, g_2, g_3) be the master public-key *mtpk*, and $({g_1}^a, {g_2}^b)$ be user's public-key *utpk*, and ${g_3}^{a+b}$ be the link token. A tag is formed by "encrypting" the link token as $(g^u, g^v, ({g_3}^{a+b})^{u/v})$. Once the token is published, tags can be linked by checking the relation $e(g_3^{a+b}, g^u) = e((g_3^{a+b})^{u/v}, g^v)$.

Parameter $\Lambda = (p, \mathbb{G}, \mathbb{G}_T, e, g)$ is common for all algorithms below.

- AT.Setup: Choose $x, y \leftarrow \mathbb{Z}_p^*$, $g_3 \leftarrow \mathbb{G}$ and compute $g_1 = g_3^{1/x}$, $g_2 = g_3^{1/y}$. Output $mtpk = (g_1, g_2, g_3)$ and $mtsk = (x, y)$.
- AT.Key: Given $mtpk = (g_1, g_2, g_3)$, choose $a, b \leftarrow \mathbb{Z}_p^*$, and output $utpk = (g_1^a, g_2^b)$ and $utsk = (a, b)$.
- AT.Tag: Choose $u, v \leftarrow \mathbb{Z}_p^*$, and output $tag = (g^u, g^v, (g_3^{a+b})^{u/v})$.
- AT.Reveal: Given $utpk = (g_1^a, g_2^b)$ of a user, compute $tkn = (g_1^a)^x (g_2^b)^y = g_3^{a+b}$ by using master trapdoor (x, y).
- AT.Claim: Given secret-key (a, b), publish g_3^{a+b}.
- AT.Link: Given $tag = (g^u, g^v, (g_3^{a+b})^{u/v}) \in (\mathbb{G}^*)^3$ and $tkn = g_3^{a+b} \in \mathbb{G}^*$, output 1 if
$$e(g_3^{a+b}, g^u) = e((g_3^{a+b})^{u/v}, g^v) \qquad (1)$$
holds. Output 0, otherwise.

Fig. 1. Anonymous Tags System (AT) based on DLIN and DRA

Lemma 6. *AT in Fig. 1 is anonymous if DLIN holds for Λ.*

Proof. We proceed in a sequence of games, from a game where the challenger behaves like in the above anonymity game, and end up with one in which the tokens (and the tags produced by the oracles) are computed independently from the public-keys. Let W_i be the event that the adversary $\mathcal{A}$ outputs $\tilde{\mathrm{b}}$ such that $\mathrm{b} = \tilde{\mathrm{b}}$ in Game i. We show that all the games are computationally indistinguishable.

Game 0. This is the anonymity game. $\Pr[W_0] = \frac{1}{2} + \mathsf{Adv}^{\mathcal{A}}_{\text{Anon}}(\lambda)$.

Game 1. The tags from $\mathsf{AT.Tag}(mtpk, utsk_{\mathrm{b}})$ are computed differently. For each query, choose $u, v \leftarrow \mathbb{Z}_p^*$ and compute $tag = (g^u, g^v, (tkn_{\mathrm{b}})^{u/v})$. Similarly, each tag from $\mathsf{AT.Tag}(mtpk, utsk_{1-\mathrm{b}})$ is computed as $tag = (g^u, g^v, (tkn_{1-\mathrm{b}})^{u/v})$ for randomly chosen u and v. The change is only syntactic, so $\Pr[W_1] = \Pr[W_0]$.

Game 2. The two pairs of public/secret-keys $\{(utpk_i, utsk_i)\}_{i=0}^{1}$ are computed differently. For $mtpk = (g_1, g_2, g_3)$, choose $a, b \leftarrow \mathbb{Z}_p$, and compute $A = g_1^a$, $B = g_2^b$. Then for random $\alpha_i, \beta_i, \gamma_i \leftarrow \mathbb{Z}_p$, $i = 0, 1$, set $utpk_i = (g_1^{a\gamma_i + \alpha_i}, g_2^{b\gamma_i + \beta_i})$ $= \left(A^{\gamma_i} g_1^{\alpha_i}, B^{\gamma_i} g_2^{\beta_i}\right)$ and the secret-key accordingly. Note that if $C = g_3^{a+b}$, then $tkn_i = C^{\gamma_i} g_3^{\alpha_i + \beta_i}$, for $i = 0, 1$. The public-keys follow the correct distribution and the tokens are computed correctly, so $\Pr[W_2] = \Pr[W_1]$.

Game 3. The tokens are computed as $tkn_{\mathrm{b}} = g_3^{c_{\mathrm{b}}}$ and $tkn_{1-\mathrm{b}} = g_3^{c_{1-\mathrm{b}}}$, for random $c_0, c_1 \leftarrow \mathbb{Z}_p$. Game 2 and Game 3 are indistinguishable due to DLIN. For $(g_1, g_2, g_3, g_1^a, g_2^b, g_3^c)$, tuple $(g_1, g_2, g_3, g_1^{a\gamma+\alpha}, g_2^{b\gamma+\beta}, g_3^{c\gamma+\alpha+\beta})$ is in the linear relation if $c = a + b$. On the other hand, if c is random, the tuple distributes uniformly. So, if we are given a tuple (g_1, g_2, g_3, A, B, C) by a DLIN challenger, we compute the public-keys and the tokens as in Game 2 (if the tuple is a DLIN one) or as in Game 3 (if not). Therefore, $|\Pr[W_3] - \Pr[W_2]| = \mathrm{negl}(\lambda)$.

In the last game $\Pr[W_3] = \frac{1}{2}$ as the tokens are independent of the public-keys (the role of b in $tkn_{\mathrm{b}} = g_3^{c_{\mathrm{b}}}$ is symmetric to the variable $(1 - \mathrm{b})$ and its value is never revealed). Moreover, the tags returned by the oracles are computed so that they match the corresponding token but are independent from the public-keys. Therefore, if DLIN holds, $\mathcal{A}$ has a negligible advantage of winning. ∎

Lemma 7. *AT in Fig. 1 is unlinkable if DLIN and DRA hold for* Λ.

Proof. We proceed in a sequence of games. We start from the original unlinkability game in Definition 2, and proceed like in Game 1, Game 2, and Game 3 by computing the challenge tag in different ways as if it is the output from an oracle query. This way the tags become independent from the public-keys. Then, we go through a couple more games so that the challenge tag becomes independent from the tags returned by the oracle queries. In the last game, $\mathcal{A}$ cannot do any better than guessing. Let W_i denote the event that the adversary $\mathcal{A}$ outputs $\tilde{\mathrm{b}}$ such that $\mathrm{b} = \tilde{\mathrm{b}}$ in Game i.

Game 0. This is the unlinkability game. $\Pr[W_0] = \frac{1}{2} + \mathsf{Adv}^{\mathcal{A}}_{\text{Unlink}}(\lambda)$.

Game 1. Proceed like in Game 1 in the previous proof. The change is only syntactic regarding the way the tags are computed, so $\Pr[W_1] = \Pr[W_0]$.

Game 2. Proceed like in Game 2 in the previous proof. The public-keys follow correct distribution and the tokens are computed correctly, so $\Pr[W_2] = \Pr[W_1]$.

Game 3. Proceed like in Game 3 in the previous proof. Note that the tags (including the challenge tag) are computed using fresh randomness from the corresponding tokens and do not depend on the public-keys. If DLIN holds, then $|\Pr[W_3] - \Pr[W_2]| = \mathrm{negl}(\lambda)$.

Game 4. Change again the way $\mathsf{AT.Tag}(mtpk, utsk_{\mathrm{b}})$ computes the tags. In Game 3, for a fixed random c_{b}, tags were computed as $(g^u, g^v, (g_3^{c_{\mathrm{b}}})^{u/v})$ for randomly chosen u, v. Instead, choose $h_1, h_2, h_3 \leftarrow \mathbb{G}^*$ and compute each tag returned by $\mathsf{AT.Tag}(mtpk, utsk_{\mathrm{b}})$ (and the challenge tag) as $tag = (h_1^u, h_2^v, h_3^{u/v})$ for randomly chosen $u, v \leftarrow \mathbb{Z}_p^*$. The change is only syntactic and does not affect the output distribution, so $\Pr[W_4] = \Pr[W_3]$.

Game 5. Process the oracle queries for $\mathsf{AT.Tag}(mtpk, utsk_{\mathrm{b}})$ as in the previous game but compute the challenge tag as (h_1^u, h_2^v, h_3^w) for randomly chosen $u, v, w \leftarrow \mathbb{Z}_p^*$. If DRA holds, then $\mid \Pr[W_5] - \Pr[W_4] \mid = \mathrm{negl}(\lambda)$ as any adversary distinguishing the two games with non-negligible probability breaks DRA.

Now the challenge tag, which is the only object in the experiment related to the variable b, consists of three randomly chosen group elements. The adversary's advantage is clearly no better than guessing[1]. Therefore, if DLIN and DRA hold $\mathsf{Adv}_{\mathrm{Unlink}}^{\mathcal{A}}(\lambda) = \mathrm{negl}(\lambda)$. ∎

Lemma 8. *AT in Fig. 1 is traceable.*

Proof. The probability of having the same $a + b$ in $\mathbb{Z}_p$ is negligible since a and b are chosen randomly in each run of $\mathsf{AT.Key}$. It is also clear that there is a unique token that fulfills the relation (1) for a given tag. ∎

3 A Modular Construction of Traceable Signatures

3.1 Syntax and Security Model

A traceable signature scheme, TS, consists of algorithms and protocols, used by three kinds of entities – the group manager ($\mathcal{G}$), group members ($\mathcal{P}$), and verifiers ($\mathcal{V}$), as follows.

- Setup : $(\mathsf{vk}, \mathsf{rk}, \mathsf{ik}, \mathsf{lk}) \leftarrow \mathsf{TS.Setup}(1^\lambda)$ is an algorithm that generates a group verification-key vk, opening-key rk, certificate issuing key ik, and link-key lk.
- Join : $(\mathsf{cert}, (\mathsf{cert}, \mathsf{usk})) \leftarrow \langle \mathsf{Join}_{\mathcal{G}}(\mathsf{vk}, \mathsf{ik}), \mathsf{Join}_{\mathcal{P}_i}(\mathsf{vk}) \rangle$ is a pair of interactive algorithms $\mathsf{Join}_{\mathcal{G}}$ and $\mathsf{Join}_{\mathcal{P}_i}$ run by the manager $\mathcal{G}$ and a group member $\mathcal{P}_i$, respectively. Algorithm $\mathsf{Join}_{\mathcal{G}}$ issues a certificate by using issuing-key ik. The resulting certificate will be appended to the membership list $\mathcal{L}$. Algorithm $\mathsf{Join}_{\mathcal{P}_i}$ outputs certificate cert and private signing-key usk.
- Sign : $\mathsf{sig} \leftarrow \mathsf{TS.Sign}_{\mathsf{usk}}(\mathsf{m})$ is a probabilistic algorithm that generates a signature sig for message m by using the signing-key usk.
- Vrf : $1/0 \leftarrow \mathsf{TS.Vrf}_{\mathsf{vk}}(\mathsf{sig}, \mathsf{m})$ is a deterministic algorithm that decides the correctness of signature sig on m. It outputs 1 for any input created through Setup, Join and Sign.

[1] It is possible to strengthen the notion of unlinkability by allowing $\mathcal{A}$ to query oracles $\mathsf{AT.Tag}(mtpk, utsk_{\mathrm{b}})$ and $\mathsf{AT.Tag}(mtpk, utsk_{1-\mathrm{b}})$. For the proof to go through, we can simulate these oracles as $\mathsf{AT.Tag}(mtpk, utsk_0)$ and $\mathsf{AT.Tag}(mtpk, utsk_1)$, but each tag will be changed like what we did on the challenge tag in Game 5.

- Open : $\mathsf{open}/\bot \leftarrow \mathsf{TS.Open}(\mathsf{vk}, \mathsf{rk}, \mathsf{sig})$ is an algorithm that identifies the owner of a valid signature sig by using the opening-key rk. It outputs an opening information open that identifies certificate cert owned by a member. In case of failure, the algorithm outputs $\bot$.
- Judge : $1/0 \leftarrow \mathsf{TS.Judge}(\mathsf{vk}, \mathsf{sig}, \mathsf{m}, \mathsf{open}, \mathsf{cert})$ is an algorithm that decides whether or not opening information open is correct with respect to sig and cert. It outputs 1 for any input created through Setup, Join, Sign, and Open.
- Reveal : $\mathsf{tkn} \leftarrow \mathsf{TS.Reveal}(\mathsf{lk}, \mathsf{cert})$ is an algorithm that publishes a link token tkn with respect to the member associated to cert.
- Trace : $1/0 \leftarrow \mathsf{TS.Trace}(\mathsf{vk}, \mathsf{sig}, \mathsf{tkn})$ is a deterministic algorithm that decides whether a signature sig is associated to link token tkn.
- Claim : $\mathsf{claimpf}/\bot \leftarrow \mathsf{TS.Claim}(\mathsf{vk}, \mathsf{sig}, \mathsf{m}, \mathsf{usk})$ generates a proof claimpf that associates signature sig on m to certificate cert identified by usk. In case of failure, the algorithm outputs $\bot$.
- ClaimVer : $1/0 \leftarrow \mathsf{TS.ClaimVer}(\mathsf{vk}, \mathsf{sig}, \mathsf{m}, \mathsf{claimpf}, \mathsf{cert})$ verifies the proof created by TS.Claim.
- Deny : $\mathsf{denypf}/\bot \leftarrow \mathsf{TS.Deny}(\mathsf{vk}, \mathsf{sig}, \mathsf{m}, \mathsf{usk})$ generates a proof denypf that signature sig on m is not associated with cert identified by usk. In case of failure, the algorithm outputs $\bot$.
- DenyVer : $1/0 \leftarrow \mathsf{TS.DenyVer}(\mathsf{vk}, \mathsf{sig}, \mathsf{m}, \mathsf{denypf}, \mathsf{cert})$ verifies the proof created by TS.Deny.

We assume that Setup is run by a trusted party. The resulting vk is published, and $(\mathsf{rk}, \mathsf{ik}, \mathsf{lk})$ is privately given to the group manager $\mathcal{G}$. Our model assumes a public directory service, denoted by $\mathcal{C}$, that maintains certificates issued by $\mathcal{G}$. Every certificate is registered to $\mathcal{C}$ and the consistency is verified publicly.

We borrow some notations from [13]. Let $\mathcal{M}^i_{\mathsf{signed}}$, $\mathcal{P}_{\mathsf{honest}}$ and $\mathcal{P}_{\mathsf{corrupt}}$ be initially empty lists. By $\mathcal{P}_{\mathsf{all}}$, we denote $\mathcal{P}_{\mathsf{honest}} \cup \mathcal{P}_{\mathsf{corrupt}}$. Adversary $\mathcal{A}$ is given some of the oracles listed as follows.

- $\mathcal{O}_{\mathsf{p\text{-}join}}$ is an oracle that executes $\langle \mathsf{Join}_{\mathcal{G}}(\mathsf{vk}, \mathsf{ik}), \mathsf{Join}_{\mathcal{P}_i}(\mathsf{vk})\rangle$ on receiving a request. When $\mathsf{Join}_{\mathcal{P}_i}$ outputs $(\mathsf{cert}_i, \mathsf{usk}_i)$, the oracle updates $\mathcal{P}_{\mathsf{honest}} \leftarrow \mathcal{P}_{\mathsf{honest}} \| \mathsf{cert}_i$ and $\mathcal{C} \leftarrow \mathcal{C} \| \mathsf{cert}_i$.
- $\mathcal{O}_{\mathsf{a\text{-}join}}$ is an oracle that works on behalf of $\mathcal{G}$ in Join protocol. The interaction with $\mathsf{Join}_{\mathcal{G}}$ is shown to and controlled by the adversary. On receiving a request, it executes $\langle \mathsf{Join}_{\mathcal{G}}(\mathsf{vk}, \mathsf{ik}), \mathcal{A}\rangle$. When $\mathsf{Join}_{\mathcal{G}}$ is completed, the oracle updates $\mathcal{P}_{\mathsf{corrupt}} \leftarrow \mathcal{P}_{\mathsf{corrupt}} \| \mathsf{cert}_i$ and $\mathcal{C} \leftarrow \mathcal{C} \| \mathsf{cert}_i$.
- $\mathcal{O}_{\mathsf{b\text{-}join}}$ is an oracle that works on behalf of a new member $\mathcal{P}_i$ in Join protocol. On receiving a request, it picks a new identity i and executes $\langle \mathcal{A}, \mathsf{Join}_{\mathcal{P}_i}\rangle$. When $\mathsf{Join}_{\mathcal{P}_i}$ is completed with output cert_i, the oracle updates $\mathcal{P}_{\mathsf{honest}} \leftarrow \mathcal{P}_{\mathsf{honest}} \| \mathsf{cert}_i$ and $\mathcal{C} \leftarrow \mathcal{C} \| \mathsf{cert}_i$.
- $\mathcal{O}_{\mathsf{sig}}$ is the non-anonymous signing oracle. It receives (i, m) as input and returns $\mathsf{sig} \leftarrow \mathsf{TS.Sign}_{\mathsf{usk}_i}(\mathsf{m})$. It then appends $(\mathsf{sig}, \mathsf{m})$ to $\mathcal{M}^i_{\mathsf{signed}}$. This oracle models the situation that the adversary has some side-channel information about the identity of the signer.

- $\mathcal{O}^i_{\mathsf{a\text{-}sig}}$ is the anonymous signing oracle for $\mathcal{P}_i$. It receives message m as input and returns $\mathsf{sig} \leftarrow \mathsf{TS.Sign}_{\mathsf{usk}_i}(\mathsf{m})$. It then appends $(\mathsf{sig}, \mathsf{m})$ to $\mathcal{M}^i_{\mathsf{signed}}$. Note that i will be seen as a symbol instead of a concrete value in the eyes of $\mathcal{A}$.
- $\mathcal{O}_{\mathsf{open}}$ is the opening oracle that receives $(\mathsf{sig}, \mathsf{m})$ on request, runs $\mathsf{open} \leftarrow \mathsf{TS.Open}(\mathsf{vk}, \mathsf{rk}, \mathsf{sig})$ and outputs open if $1 \leftarrow \mathsf{TS.Vrf}_{\mathsf{vk}}(\mathsf{sig}, \mathsf{m})$.
- $\mathcal{O}_{\mathsf{reveal}}$ is the revocation oracle that receives $i \in \mathcal{P}_{\mathsf{all}}$. It executes $\mathsf{tkn} \leftarrow \mathsf{TS.Reveal}(\mathsf{rk}, \mathsf{cert}_i)$ and outputs tkn.
- $\mathcal{O}_{\mathsf{claim}}$ is the claiming oracle that takes $(i, \mathsf{sig}, \mathsf{m})$. If $1 \leftarrow \mathsf{TS.Vrf}_{\mathsf{vk}}(\mathsf{sig}, \mathsf{m})$, it runs $\mathsf{claimpf} \leftarrow \mathsf{TS.Claim}(\mathsf{vk}, \mathsf{sig}, \mathsf{m}, \mathsf{usk}_i)$, and returns $\mathsf{claimpf}$. If $\mathsf{claimpf} \neq \perp$, then $(\mathsf{sig}, \mathsf{m})$ is appended to $\mathcal{S}^i_{\mathsf{claimed}}$. Note that this oracle can naturally be used to verify that if a signature belongs to $\mathcal{P}_i$ or not.
- $\mathcal{O}_{\mathsf{deny}}$ is the oracle that takes $(i, \mathsf{sig}, \mathsf{m})$. If $1 \leftarrow \mathsf{TS.Vrf}_{\mathsf{vk}}(\mathsf{sig}, \mathsf{m})$, it runs $\mathsf{denypf} \leftarrow \mathsf{TS.Deny}(\mathsf{vk}, \mathsf{sig}, \mathsf{m}, \mathsf{usk}_i)$, and returns denypf.

NON-FRAMEABILITY. A malicious GM should not be able to convince a verifier about the honest member's ownership of a signature that has never actually been created by the member. There are two cases depending on how the verifier is convinced, either by the opening protocol or the claiming protocol.

The malicious GM may create a signature that is traceable by using the tracing-key of $\mathcal{P}_i$ (and perhaps also prove that the tracing-key in question is indeed for $\mathcal{P}_i$), however, we exclude this in our definition since we want to keep the only "official" mean for a GM to reveal the identity of a signature to public is to use Open. The Trace algorithm is designed to link the signatures created by the same signer while keeping this signer anonymous, and is never meant to be a way for a GM to convince others about the true signer of a signature.

We also exclude the unavoidable attack that all but one member use TS.Deny to deny the authorship of a given signature to "reveal" its signer.

Definition 9 (Non-Frameability). *A traceable signature scheme is non-frameable if the following game returns* 1 *only with negligible probability in* λ.

Experiment **Frame** :

$(\mathsf{vk}, \mathsf{rk}, \mathsf{ik}, \mathsf{lk}) \leftarrow \mathsf{TS.Setup}(1^\lambda)$

$(i^*, \mathsf{claimpf}^*, \mathsf{open}^*, \mathsf{sig}^*, \mathsf{m}^*) \leftarrow \mathcal{A}^{\mathcal{O}_{\mathsf{b\text{-}join}}, \mathcal{O}_{\mathsf{sig}}, \mathcal{O}_{\mathsf{claim}}, \mathcal{O}_{\mathsf{deny}}}(\mathsf{vk}, \mathsf{rk}, \mathsf{ik}, \mathsf{lk})$

Return 1 if $\mathsf{cert}_{i^*} \in \mathcal{P}_{\mathsf{honest}}$ and

$$\Big((\mathsf{sig}^*, \mathsf{m}^*) \notin \mathcal{M}^{i^*}_{\mathsf{signed}} \wedge 1 \leftarrow \mathsf{TS.Judge}(\mathsf{vk}, \mathsf{sig}^*, \mathsf{m}^*, \mathsf{open}^*, \mathsf{cert}_{i^*})\Big) \vee$$

$$\Big((\mathsf{sig}^*, \mathsf{m}^*) \notin \mathcal{S}^{i^*}_{\mathsf{claimed}} \wedge 1 \leftarrow \mathsf{TS.ClaimVer}(\mathsf{vk}, \mathsf{sig}^*, \mathsf{m}^*, \mathsf{claimpf}^*, \mathsf{cert}_{i^*})\Big)$$

Return 0, otherwise.

SECURITY NOTIONS AGAINST OUTSIDERS AND MEMBERS. In the following notions, the adversary is given access to oracles $\mathcal{O}_{\mathsf{p\text{-}join}}$, $\mathcal{O}_{\mathsf{a\text{-}join}}$, $\mathcal{O}_{\mathsf{sig}}$, $\mathcal{O}_{\mathsf{open}}$, $\mathcal{O}_{\mathsf{claim}}$,

$\mathcal{O}_{\mathsf{deny}}$, $\mathcal{O}_{\mathsf{reveal}}$ which are wrapped and denoted by $\mathcal{O}_{\mathsf{atk}}$ for readability. More oracles may be given. Any restrictions on the oracles will be noted explicitly.

Definition 10 ((Bi-directional) Traceability). *A traceable signature scheme is traceable if the following game returns* 1 *only with negligible probability.*

Experiment **Trace** :

$(\mathsf{vk}, \mathsf{rk}, \mathsf{ik}, \mathsf{lk}) \leftarrow \mathsf{TS.Setup}(1^\lambda)$

$(\mathsf{sig}^*, \mathsf{m}^*) \leftarrow \mathcal{A}^{\mathcal{O}_{\mathsf{atk}}}(\Lambda, \mathsf{vk})$

Return 1 if $1 \leftarrow \mathsf{TS.Vrf}(\mathsf{vk}, \mathsf{sig}^*, \mathsf{m}^*)$ and

$^\forall\mathsf{cert} \in \mathcal{P}_{\mathsf{all}},\ 0 \leftarrow \mathsf{TS.Judge}(\mathsf{vk}, \mathsf{sig}^*, \mathsf{m}^*, \mathsf{TS.Open}(\mathsf{vk}, \mathsf{rk}, \mathsf{sig}^*), \mathsf{cert}) \vee$

$^\forall\mathsf{cert} \in \mathcal{P}_{\mathsf{all}},\ 0 \leftarrow \mathsf{TS.Trace}(\mathsf{vk}, \mathsf{sig}^*, \mathsf{TS.Reveal}(\mathsf{lk}, \mathsf{cert})) \vee$

$(^\exists\mathsf{cert}_i, \mathsf{cert}_j \in \mathcal{P}_{\mathsf{all}},\ \mathsf{cert}_i \neq \mathsf{cert}_j,$

$1 \leftarrow \mathsf{TS.Judge}(\mathsf{vk}, \mathsf{sig}^*, \mathsf{m}^*, \mathsf{TS.Open}(\mathsf{vk}, \mathsf{rk}, \mathsf{sig}^*), \mathsf{cert}_i) \wedge$

$1 \leftarrow \mathsf{TS.Trace}(\mathsf{vk}, \mathsf{sig}^*, \mathsf{TS.Reveal}(\mathsf{lk}, \mathsf{cert}_j))\ )$

Return 0, otherwise.

The notion of anonymity claims that the outsider should not be able to see which of the two users issued the target signature even when the link tokens for those users are available. It can be seen as a natural extension of the anonymity for the anonymous tag system. The adversary is given access to the oracles $\mathcal{O}_{\mathsf{sig}}$ and $\mathcal{O}_{\mathsf{reveal}}$, but with no query about the target users i_0, i_1. Nevertheless, the adversary can obtain "anonymous" signatures (via $\mathcal{O}^{i_b}_{\mathsf{a\text{-}sig}}$ and $\mathcal{O}^{i_{1-b}}_{\mathsf{a\text{-}sig}}$ oracles) and the link tokens $(\mathsf{tkn}_b, \mathsf{tkn}_{1-b})$ of the target users. The adversary can also query $\mathcal{O}_{\mathsf{open}}$, $\mathcal{O}_{\mathsf{claim}}$, and $\mathcal{O}_{\mathsf{deny}}$, but not on the anonymous signatures of i_0, i_1.

In the security model of group signatures in the literature, e.g., [10], some of the secret-keys among $(\mathsf{rk}, \mathsf{ik}, \mathsf{lk})$ can be leaked to the adversary. For instance, when considering anonymity, the adversary is given the certificate issuing key ik. Such a fine grained security was not considered in the context of traceable signatures [12,13], and we follow that model. Thus our trust model is slightly stronger than that of [10]. We will revisit this point in Section 3.2.

Definition 11 (Anonymity). *A traceable signature scheme is anonymous if the following game returns* 1 *only with negligibly better probability than* $\frac{1}{2}$. *In the following, target identities,* i_0 *and* i_1*, must not be sent to either* $\mathcal{O}_{\mathsf{sig}}$ *or* $\mathcal{O}_{\mathsf{reveal}}$. *It is also prohibited to send the signatures given from anonymous signing oracle* $\mathcal{O}_{\mathsf{a\text{-}sig}}$ *to any of* $\mathcal{O}_{\mathsf{open}}$, $\mathcal{O}_{\mathsf{claim}}$ *and* $\mathcal{O}_{\mathsf{deny}}$.

Experiment **Anonymity** :

$(\mathsf{vk}, \mathsf{rk}, \mathsf{ik}, \mathsf{lk}) \leftarrow \mathsf{TS.Setup}(1^\lambda)$

$(i_0, i_1) \leftarrow \mathcal{A}^{\mathcal{O}_{\mathsf{atk}}}(\Lambda, \mathsf{vk})$

If $\mathsf{cert}_{i_0} \notin \mathcal{P}_{\mathsf{honest}} \vee \mathsf{cert}_{i_1} \notin \mathcal{P}_{\mathsf{honest}}$, return 0.

$\mathsf{tkn}_0 \leftarrow \mathsf{TS.Reveal}(\mathsf{lk}, \mathsf{cert}_{i_0}),\ \mathsf{tkn}_1 \leftarrow \mathsf{TS.Reveal}(\mathsf{lk}, \mathsf{cert}_{i_1})$

$b \leftarrow \{0, 1\}$

$\tilde{b} \leftarrow \mathcal{A}^{\mathcal{O}_{\mathsf{atk}}, \mathcal{O}^{i_b}_{\mathsf{a\text{-}sig}}, \mathcal{O}^{i_{1-b}}_{\mathsf{a\text{-}sig}}}(\mathsf{tkn}_b, \mathsf{tkn}_{1-b})$

Return 1 if $b = \tilde{b}$. Return 0, otherwise.

Next is the unlinkability which has been named as anonymity in some literatures about group signatures. Like the ones for anonymous tag system, anonymity and unlinkability for traceable signatures are incomparable in our framework.

Definition 12 (Unlinkability). *A traceable signature scheme is unlinkable if the following game returns* 1 *only with negligibly better probability than* $\frac{1}{2}$. *In the following, target identities,* i_0 *and* i_1, *must not be sent to* $\mathcal{O}_{reveal}$. *It is also prohibited to send the target signature* sig *to* $\mathcal{O}_{open}$, $\mathcal{O}_{claim}$ *or* $\mathcal{O}_{deny}$.

Experiment **Unlinkability** :

$(\mathsf{vk}, \mathsf{rk}, \mathsf{ik}, \mathsf{lk}) \leftarrow \mathsf{TS.Setup}(1^\lambda)$

$(i_0, i_1, \mathsf{m}) \leftarrow \mathcal{A}^{\mathcal{O}_{\mathsf{atk}}}(\Lambda, \mathsf{vk})$

If $\mathsf{cert}_{i_0} \notin \mathcal{P}_{\mathsf{honest}} \vee \mathsf{cert}_{i_1} \notin \mathcal{P}_{\mathsf{honest}}$, return 0.

$b \leftarrow \{0, 1\},\ \mathsf{sig} \leftarrow \mathsf{TS.Sign}_{\mathsf{usk}_{i_b}}(\mathsf{m})$

$\tilde{b} \leftarrow \mathcal{A}^{\mathcal{O}_{\mathsf{atk}}}(\mathsf{sig})$

Return 1 if $b = \tilde{b}$. Return 0, otherwise.

We consider two more security notions: no one but the real signer can claim the ownership of a signature (Non-Snatching), and the real signer should not be able to deny the ownership of his/her signatures (Undeniability).

Definition 13 (Non-Snatching). *A traceable signature scheme is non-snatching if the following game returns* 1 *only with negligible probability. The list* $\mathcal{P}_{all}$ *is managed by the oracles.*

Experiment **Snatch** :

$(\mathsf{vk}, \mathsf{rk}, \mathsf{ik}) \leftarrow \mathsf{Setup}(1^\lambda)$

$(\mathsf{sig}, \mathsf{m}, \mathsf{claimpf}, \mathsf{cert}_i) \leftarrow \mathcal{A}^{\mathcal{O}_{\mathsf{atk}}}(\Lambda, \mathsf{vk})$

Return 1 if $\mathsf{cert}_i \in \mathcal{P}_{\mathsf{corrupt}}$ and $^\exists\mathsf{cert}_j \in \mathcal{P}_{\mathsf{honest}}$ such that

$(\mathsf{sig}, \mathsf{m}) \in \mathcal{M}^j_{\mathsf{signed}} \wedge 1 \leftarrow \mathsf{TS.ClaimVer}(\mathsf{vk}, \mathsf{sig}, \mathsf{m}, \mathsf{claimpf}, \mathsf{cert}_i)$

Return 0, otherwise.

Definition 14 (Undeniability). *A traceable signature scheme is undeniable if the following game returns* 1 *only with negligible probability. The list* $\mathcal{P}_{all}$ *is managed by the oracles.*

Experiment **Deny** :

$(\mathsf{vk}, \mathsf{rk}, \mathsf{ik}) \leftarrow \mathsf{Setup}(1^\lambda)$

$(\mathsf{sig}^*, \mathsf{m}^*, \mathsf{denypf}^*, \mathsf{cert}^*) \leftarrow \mathcal{A}^{\mathcal{O}_{\mathsf{atk}}}(\Lambda, \mathsf{vk})$

Return 1 if $\mathsf{cert}^* \in \mathcal{P}_{\mathsf{all}}$ and

$1 \leftarrow \mathsf{TS.Judge}(\mathsf{vk}, \mathsf{sig}^*, \mathsf{m}^*, \mathsf{TS.Open}(\mathsf{vk}, \mathsf{rk}, \mathsf{sig}^*), \mathsf{cert}^*) \wedge$

$1 \leftarrow \mathsf{TS.DenyVer}(\mathsf{vk}, \mathsf{sig}^*, \mathsf{m}^*, \mathsf{denypf}^*, \mathsf{cert}^*)$

Return 0, otherwise.

3.2 Construction

BUILDING BLOCKS. Other than AT from Section 2, we use the following building blocks. Efficient and inter-operable instantiations are shown in Section 3.4.

Non-interactive Zero-Knowledge (NIZK) Proof of Knowledge System: It consists of a suite of six algorithms; ZK.{Crs, Prf, Vrf, Ext, SimCrs, SimPrf}. ZK.Crs is the common reference string (CRS) generator, which also outputs a trapdoor used by the extractor algorithm, ZK.Ext. Algorithm ZK.SimCrs generates a CRS and a trapdoor used by ZK.SimPrf that simulates proofs. Algorithm ZK.Prf takes CRS generated by ZK.Crs and outputs a proof that passes the verification algorithm ZK.Vrf. Algorithm ZK.Prf also works with a simulated CRS generated by ZK.SimCrs, but ZK.Ext does not work with it. ZK.Ext outputs $\perp$ if it fails to extract a correct witness. It happens only with negligible probability for any input that passes the verification with ZK.Vrf.

Digital Signatures: We use two signature schemes that are existentially unforgeable against adaptive chosen message attacks (EUF-CMA). Let $\mathsf{S1}$ and $\mathsf{S2}$ be the signature schemes that consist of three algorithms $\{\mathsf{Key}, \mathsf{Sign}, \mathsf{Vrf}\}$ as usual. We require that one can issue a certificate in the form of a signature produced by $\mathsf{S1}$ that authorizes a public-key of $\mathsf{S2}$, and it must be further possible to prove the possession of the certificate in the NIZK manner.

Strongly Unforgeable (SUF) One-time Signatures: To make our scheme SUF, we use one-time signature scheme $\mathsf{OTS}.\{\mathsf{Key}, \mathsf{Sign}, \mathsf{Vrf}\}$ that is SUF against one-time chosen message attacks.

Design Intuition. Our construction follows the structure of "three-level" certification. In the first level, the group manager certifies the public-key (consists of the public-keys for the underlying signature scheme and anonymous tag system) of each member. In the second level, the member signs on a one-time public-key. The signatures and keys in the first and second level will be hidden by NIZK, then a tag is appended. Finally, in the third level, the message and everything else are signed with the one-time key. To claim the signature ownership, a member proves in NIZK the relation between the tag and the public tag-key. For tracing, the link token is published by the group manager so that anyone can use the token to check against the tag of a signature in question.

Our main idea of getting CCA-anonymity is similar to the technique used in convertible undeniable signatures (e.g., [15]), which uses strong unforgeability plus signature list checking. However, applying this technique for our usage requires more work since the adversary may ask for the opening of a valid signature which is signed with the key of a compromised user. To resolve this issue, we require the GM in the simulation to know the user tracing trapdoors of compromised users. We stress that it does not kill the advantages brought by our public-key anonymous tag system. The GM in the real world is still not required to securely store the tracing tokens of users. This approach has similarity to the Naor-Yung construction [14] of CCA-secure public-key encryption. Given a signature, identity of the signer can be retrieved either by applying the extractor of $\mathsf{ZK1}$ to the zero-knowledge part $zsig$, or by using the tracing mechanism to the tag part tag. The former identifies uvk and the latter identifies $utpk$. The consistency of the retrieved public-keys are guaranteed by the certificate issued by the group manager. This is the reason that the issuing key should not be leaked to the adversary in our framework.

The Scheme. Fig. 2 illustrates the scheme. Relations and detailed description of zero-knowledge (ZK) proofs are presented in Fig. 3. We make the following remarks:

- The input signature to $\mathsf{TS.Open}$, $\mathsf{TS.Judge}$ and $\mathsf{TS.Trace}$ must be verified by $\mathsf{TS.Vrf}$ beforehand.
- Relation (3) in $\mathsf{ZK1}$ and (6) in $\mathsf{ZK3}$ are for the same fact. Actually, the latter can be replaced with the former in a special case (and our instantiation from Section 2.2 fits to the case).

- ZK1, ZK3 and ZK4 can share the CRS and it can be provided by the group manager since they are conducted by group members. On the other hand, the CRS for ZK2 must be provided separately for the group manager.

3.3 Security

Theorem 15. *TS is non-frameable if ZK2 is knowledge sound, S2 is EUF-CMA, and OTS is strongly one-time EUF-CMA.*

Theorem 16. *TS is traceable if ZK1 is knowledge-sound, S1 is EUF-CMA, and AT is traceable.*

Theorem 17. *TS is anonymous if ZK1 is knowledge sound and zero-knowledge, ZK2 is zero-knowledge, ZK3 is zero-knowledge, AT is anonymous and traceable, S1 and S2 are EUF-CMA, and OTS is one-time strong EUF-CMA.*

Theorem 18. *TS is unlinkable if ZK1 is knowledge sound, ZK1, ZK2, ZK3 and ZK4 are zero-knowledge, AT is unlinkable and traceable, S1 and S2 are EUF-CMA, and OTS is one-time strong EUF-CMA.*

Theorem 19. *TS is non-snatchable if ZK3 is knowledge sound and AT is traceable.*

Theorem 20. *TS is undeniable if ZK1 and ZK4 are knowledge sound.*

The proofs for these theorems are omitted due to page limitations.

3.4 Efficient Instantiation

For zero-knowledge proofs, we use the Groth-Sahai proof system [11], GS, with a symmetric bilinear group setting Λ. The proof system has two serious constraints to provide the zero-knowledge and proof of knowledge properties. One is that the relations in question are limited to pairing product equations where the witnesses are group elements of $\mathbb{G}$. The other is that the constants in $\mathbb{G}_T$ are represented as a product of pairings of known elements in $\mathbb{G}$. Other building blocks should meet these constraints to inter-operate with GS. Fig. 4 shows concrete relations to prove in our instantiation.

For S1, we use a signature scheme from [2,3], denoted by AHO. It is EUF-CMA under the SFP Assumption [3]. The messages, public-keys and signatures of AHO are elements of $\mathbb{G}$. When signing n group elements $(m_1, \ldots, m_n) \in \mathbb{G}^n$, a public-key is $vk = (\{g_i, h_i\}_{i=1}^{n+2}, \{a_i, b_i\}_{i=1}^{4})$, and a signature is $(z_1, \ldots, z_7)$ verified by $A = (\prod_{i=1}^{n} e(g_i, m_i))\, e(g_{n+1}, z_1)\, e(g_{n+2}, z_2)\, e(z_3, z_4)$ and $B = (\prod_{i=1}^{n} e(h_i, m_i))\, e(h_{n+1}, z_1)\, e(h_{n+2}, z_5)\, e(z_6, z_7)$ where $A = e(a_1, a_2)\, e(a_3, a_4)$ and $B = e(b_1, b_2)\, e(b_3, b_4)$. Values z_2, z_3, z_4, z_5, z_6, and z_7 are randomized every time ZK1 is done (see [3] for details of the randomization).

$\mathsf{TS.Setup}(1^\lambda)$:
$(cvk, csk) \leftarrow \mathsf{S1.Key}(1^\lambda)$
$(mtpk, mtsk) \leftarrow \mathsf{AT.Setup}(1^\lambda)$
$(crs_{zs}, trap_{zs}) \leftarrow \mathsf{ZK1.Crs}(1^\lambda)$
$(crs_{zj}, trap_{zj}) \leftarrow \mathsf{ZK2.Crs}(1^\lambda)$
$\mathsf{vk} = (crs_{zs}, crs_{zj}, mtpk, cvk)$, $\mathsf{rk} = trap_{zs}$, $\mathsf{ik} = csk$, $\mathsf{lk} = mtsk$
Output $(\mathsf{vk}, \mathsf{rk}, \mathsf{ik}, \mathsf{lk})$.

$\mathsf{TS.Join}_{\mathcal{G}}(\mathsf{vk}, \mathsf{ik})$		$\mathsf{TS.Join}_{\mathcal{P}_i}(\mathsf{vk}, \mathcal{C})$
	$\xleftarrow{uvk}$	$(uvk, usk) \leftarrow \mathsf{S2.Key}(1^\lambda)$
$(utpk, utsk) \leftarrow \mathsf{AT.Key}(mtpk)$		
$asig \leftarrow \mathsf{S1.Sign}_{csk}(utpk \Vert uvk)$	$\xrightarrow{asig,\ utpk,\ (utsk)}$	$1 \stackrel{?}{=} \mathsf{S1.Vrf}_{cvk}(utpk \Vert uvk, asig)$
		$\mathsf{cert} = (asig, utpk, uvk)$
		$\mathsf{usk} = (usk, utsk, \mathsf{cert})$
		Output $(\mathsf{cert}, \mathsf{usk})$.

$\mathsf{TS.Sign}(\mathsf{usk}, \mathsf{m})$:
$(ovk, osk) \leftarrow \mathsf{OTS.Key}(1^\lambda)$
$psig \leftarrow \mathsf{S2.Sign}_{usk}(0 \Vert ovk)$
$tag \leftarrow \mathsf{AT.Tag}(mtpk, utpk)$
$zsig \leftarrow \mathsf{ZK1.Prf}_{crs_{zs}}$
$osig \leftarrow \mathsf{OTS.Sign}_{osk}(\mathsf{m} \Vert zsig \Vert tag)$
$\mathsf{sig} = (zsig, tag, osig, ovk)$
Output sig.

$\mathsf{TS.Vrf}(\mathsf{vk}, \mathsf{m}, \mathsf{sig})$:
$\mathsf{sig} \rightarrow (zsig, tag, osig, ovk)$
Output 1 if
$1 = \mathsf{OTS.Vrf}_{ovk}(\mathsf{m} \Vert zsig \Vert tag, osig)$ and
$1 = \mathsf{ZK1.Vrf}_{crs_{zs}}$.

$\mathsf{TS.Open}(\mathsf{vk}, \mathsf{rk}, \mathsf{sig})$:
$(utpk, \ldots, psig) \leftarrow \mathsf{ZK1.Ext}_{trap_{zs}}(zsig)$
$info \leftarrow \mathsf{ZK2.Prf}_{crs_{zj}}$
$\mathsf{open} = (utpk, info)$
Output (i, open).

$\mathsf{TS.Judge}(\mathsf{vk}, \mathsf{sig}, \mathsf{m}, \mathsf{open}, \mathsf{cert})$:
Output 1 if
$(utpk \in \mathsf{open}) = (utpk \in \mathsf{cert})$,
$1 = \mathsf{ZK2.Vrf}_{crs_{zj}}$.

$\mathsf{TS.Reveal}(\mathsf{lk}, \mathsf{cert})$:
$mtsk \leftarrow \mathsf{lk}$, $utpk \leftarrow \mathsf{cert}$
$tkn \leftarrow \mathsf{AT.Reveal}(mtsk, utpk)$
Output $\mathsf{tkn} = tkn$.

$\mathsf{TS.Trace}(\mathsf{vk}, \mathsf{sig}, \mathsf{tkn})$:
$mtpk \leftarrow \mathsf{vk}$, $tag \leftarrow \mathsf{sig}$, $tkn \leftarrow \mathsf{tkn}$
Output 1 if
$1 = \mathsf{AT.Link}(mtpk, tag, tkn)$.

$\mathsf{TS.Claim}(\mathsf{vk}, \mathsf{sig}, \mathsf{m}, \mathsf{usk})$:
$mtpk \leftarrow \mathsf{vk}$, $utsk \leftarrow \mathsf{usk}$, $tag \leftarrow \mathsf{sig}$
$tkn \leftarrow \mathsf{AT.Claim}(mtpk, utsk)$
If $1 \leftarrow \mathsf{AT.Link}(mtpk, tag, tkn)$, do:
  $claim \leftarrow \mathsf{ZK3.Prf}$
  $csig \leftarrow \mathsf{S2.Sign}_{usk}(1 \Vert claim \Vert \mathsf{sig} \Vert \mathsf{m})$
  Output $\mathsf{claimpf} = (claim, csig)$.
Else output $\perp$.

$\mathsf{TS.ClaimVer}(\mathsf{vk}, \mathsf{sig}, \mathsf{m}, \mathsf{claimpf}, \mathsf{cert})$:
$(claim, csig) \leftarrow \mathsf{claimpf}$, $uvk \leftarrow \mathsf{cert}$
Output 1 if
$1 = \mathsf{ZK3.Vrf}_{crs_{zs}}$ and
$1 = \mathsf{S2.Vrf}_{uvk}(1 \Vert claim \Vert \mathsf{sig} \Vert \mathsf{m}, csig)$.

$\mathsf{TS.Deny}(\mathsf{vk}, \mathsf{sig}, \mathsf{m}, \mathsf{usk})$:
$mtpk \leftarrow \mathsf{vk}$, $utsk \leftarrow \mathsf{usk}$, $tag \leftarrow \mathsf{sig}$
$tkn \leftarrow \mathsf{AT.Claim}(mtpk, utsk)$
If $0 \leftarrow \mathsf{AT.Link}(mtpk, tag, tkn)$, do:
  $deny \leftarrow \mathsf{ZK4.Prf}$
  Output $\mathsf{denypf} = deny$.
Else output $\perp$.

$\mathsf{TS.DenyVer}(\mathsf{vk}, \mathsf{sig}, \mathsf{m}, \mathsf{denypf}, \mathsf{cert})$:
$deny \leftarrow \mathsf{denypf}$, $uvk \leftarrow \mathsf{cert}$
Output 1 if
$1 = \mathsf{ZK4.Vrf}_{crs_{zs}}$

Fig. 2. Protocol specification of a traceable signature scheme TS

ZK1: ("I have public-keys, *utpk* and *uvk*, certified by the GM, and *uvk* is used to sign the one-time key *ovk*, and *utpk* is associated by *tag*.")

$$1 = \mathsf{S1.Vrf}_{cvk}(\underline{utpk}\|\underline{uvk}, \underline{asig}) \wedge 1 = \mathsf{S2.Vrf}_{\underline{uvk}}(0\|ovk, \underline{psig}) \wedge \tag{2}$$

$$tag \in \mathsf{AT.Tag}(mtpk, \underline{utsk}) \wedge (\underline{utpk}, \underline{utsk}) \in \mathsf{AT.Key}(mtpk) \tag{3}$$

ZK2: ("I know a valid signature *psig* w.r.t. message *ovk* and public-key *uvk*.")

$$1 = \mathsf{S2.Vrf}_{uvk}(0\|ovk, \underline{psig}) \tag{4}$$

ZK3: ("Key *utpk* is associated with *tag*.")

$$1 = \mathsf{AT.Link}(mtpk, tag, \underline{tkn}) \wedge \tag{5}$$

$$\underline{tkn} = \mathsf{AT.Claim}(mtpk, \underline{utsk}) \wedge (utpk, \underline{utsk}) \in \mathsf{AT.Key}(mtpk) \tag{6}$$

ZK4: ("Key *utpk* is *not* associated with *tag*.")

$$0 = \mathsf{AT.Link}(mtpk, tag, \underline{tkn}) \wedge \tag{7}$$

$$\underline{tkn} = \mathsf{AT.Claim}(mtpk, \underline{utsk}) \wedge (utpk, \underline{utsk}) \in \mathsf{AT.Key}(mtpk) \tag{8}$$

Fig. 3. Relations for Zero-Knowledge proofs in Fig. 2 (Witnesses are underlined.)

- (ZK1) For $A = e(a_1, \underline{\underline{a_2}})\ e(a_3, \underline{\underline{a_4}})$, $B = e(b_1, \underline{\underline{b_2}})\ e(b_3, \underline{\underline{b_4}})$, $\mathbb{G} = e(g, \underline{\underline{g}})$, and $H = \mathsf{Hash}(\eta)$;

$$A = e(g_1', \underline{z_1})\ e(g_2', \underline{z_2})\ e(z_3, z_4)\ e(g_3', \underline{u})\ e(g_4', \underline{v})\ e(g_5', \underline{g_1^a})\ e(g_6', \underline{g_2^b}) \wedge \tag{9}$$

$$B = e(h_1', \underline{z_1})\ e(h_2', \underline{z_5})\ e(z_6, z_7)\ e(h_3', \underline{u})\ e(h_4', \underline{v})\ e(h_5', \underline{g_1^a})\ e(h_6', \underline{g_2^b}) \wedge \tag{10}$$

$$G = e(\underline{s},\ \underline{c}\, g^H\, \underline{d}^r) \wedge \tag{11}$$

$$e(\underline{g_3^{a+b}}, g^u) = e((g_3^{a+b})^{u/v}, \underline{\underline{g^v}}) \wedge \tag{12}$$

$$e(g_3, \underline{g^a})\, e(g_3, \underline{g^b}) = e(\underline{g_3^{a+b}}, g) \wedge e(g_1, \underline{g^a})\, e(g_2, \underline{g^b}) = e(\underline{g_1^a}\, \underline{g_2^b}, g) \tag{13}$$

- (ZK2) The same as (11) except that c and d are public.
- (ZK3) The same as (12) and (13) except that g_1^a and g_2^b are public.
- (ZK4) Use (13) but replacing the rightmost pairing with $e(g_1^a\, g_2^b, \underline{\underline{g}})$ where g_1^a and g_2^b are public. Also, for $Y \neq 1$, $tag = (t_1, t_2, t_3)$;

$$e(\underline{g_3^{a+b}}, \underline{t_1^r})/e(\underline{t_3^r}, t_2) = Y \wedge e(t_3, \underline{t_1^r}) = e(\underline{t_3^r}, t_1) \tag{14}$$

Fig. 4. Concrete Relations for ZK: Underlined values are the witnesses. Double underlined values are public but committed for the sake of zero-knowledge simulation. It requires extra proof that the commitment is made correctly from the known value. (Such a proof costs 2 group elements. See [11] for details.)

Table 1. Summary of the properties among the state of the art group signature and traceable signature schemes that provide non-frameability (the signature size counts the number of group elements.)

Scheme	Construction Type	Claim & Trace	Deny Function	Anonymity w/ Trace	Anonymity Level	Concurrent Join	Sig. Size
AHO10 [3]	generic	no	no	no	CCA	yes	58
LY09 [13]	tailor-made	yes	no	no	CPA	no	83
This paper	generic	yes	yes	yes	CCA	yes	107

For S2, we use the full Boneh-Boyen signature scheme [5], which is EUF-CMA under the Strong Diffie-Hellman Assumption [4]. A public-key consists of two group elements; $vk = (c, d) \in \mathbb{G}^2$. For message $m \in \mathbb{Z}_p$, a signature is $(s, r) \in \mathbb{G} \times \mathbb{Z}_p^*$ which is verified by $e(s,\ c\, g^m\, d^r) = e(g, g)$. A long message is compressed to a value in $\mathbb{Z}_p$ with a collision resistant hash function Hash.

For OTS, we use the weak version of Boneh-Boyen scheme. A public-key consists of one group element $\eta \in \mathbb{G}$ and the signature is μ that can be verified by $e(g, g) = e(\mu,\ \eta \cdot g^m)$. (Base g can be re-used without endangering the security.)

For AT, we use the scheme in Section 2.2. Let (t_1, t_2, t_3) denote a tag. To show that a tag is not related to a public-key, as in ZK4, the prover chooses $r \in \mathbb{Z}_p^*$ and compute $Y = \left\{e(g_3^{a+b}, t_1)/e(t_3, t_2)\right\}^r$. The verifier accepts if $Y \neq 1$ and the proof that the value Y is correctly made from the tag and the public-key in question. If the tag is indeed made from the public-key, $Y = 1$ for any $r \in \mathbb{Z}_p$. The zero-knowledge simulation for ZK4 is done as follows. Choose r randomly and commit to t_1^r and t_3^r as well as the normal proof. Then commit to 1 for other witnesses, g^a, g^b, g_3^{a+b}, and g, and simulate the proof that g is committed correctly (this simulation is possible as the simulation trapdoor for GS is known to the simulator). Finally compute $Y = e(t_3^r, t_2)$. The output perfectly follows the proper distribution.

With above instantiations, our scheme yields a signature that consists of 107 elements. (Precisely, 88 in $\mathbb{G}$ and 19 in $\mathbb{Z}_p$. We assume in the following that elements in $\mathbb{G}$ and $\mathbb{Z}_p$ can be represented in the similar bit-size and only counts the total number of elements for the sake of better overview of the comparison.) More precisely, the number of group elements in the proof and the commitments for showing relation (9) and (10) for S1.Vrf is 59, (11) for S2.Vrf is 19, and (12)-(13) for AT.Tag and AT.Key is 24. Other than those, a tag costs 3, and a public-key and a signature for OTS costs 2 group elements, which adds up to 107. The size of proofs for opening (ZK2) and claiming (ZK3) are 16 and 32, respectively. A proof for denying a ownership (ZK4) costs 39 elements in the base group and 1 in the target group.

Table 1 summarizes the comparison. Compared to [13], which is the state of the art traceable signature scheme without random oracles, the increased cost of 24 group elements is the price for gaining the properties of CCA-anonymity with tracing, and concurrent join.

References

1. Abdalla, M., Warinschi, B.: On the Minimal Assumptions of Group Signature Schemes. In: López, J., Qing, S., Okamoto, E. (eds.) ICICS 2004. LNCS, vol. 3269, pp. 1–13. Springer, Heidelberg (2004)
2. Abe, M., Fuchsbauer, G., Groth, J., Haralambiev, K., Ohkubo, M.: Structure-Preserving Signatures and Commitments to Group Elements. In: Rabin, T. (ed.) CRYPTO 2010. LNCS, vol. 6223, pp. 209–237. Springer, Heidelberg (2010)
3. Abe, M., Haralambiev, K., Ohkubo, M.: Signing on Group Elements for Modular Protocol Designs. Cryptology ePrint Archive, Report 2010/133 (2010)
4. Boneh, D., Boyen, X., Shacham, H.: Short Group Signatures. In: Franklin, M. (ed.) CRYPTO 2004. LNCS, vol. 3152, pp. 41–55. Springer, Heidelberg (2004)
5. Boneh, D., Boyen, X.: Short Signatures without Random Oracles and the SDH Assumption in Bilinear Groups. J. Cryptology 21(2), 149–177 (2008)
6. Boneh, D., Shacham, H.: Group Signatures with Verifier-Local Revocation. In: CCS 2004, pp. 168–177 (2004)
7. Boyen, X.: The Uber-Assumption Family: a unified complexity framework for bilinear groups. In: Galbraith, S.D., Paterson, K.G. (eds.) Pairing 2008. LNCS, vol. 5209, pp. 39–56. Springer, Heidelberg (2008)
8. Chaum, D., van Heyst, E.: Group Signatures. In: Davies, D.W. (ed.) EUROCRYPT 1991. LNCS, vol. 547, pp. 257–265. Springer, Heidelberg (1991)
9. Chow, S.S.M.: Real Traceable Signatures. In: Jacobson Jr., M.J., Rijmen, V., Safavi-Naini, R. (eds.) SAC 2009. LNCS, vol. 5867, pp. 92–107. Springer, Heidelberg (2009)
10. Groth, J.: Fully Anonymous Group Signatures without Random Oracles. In: Kurosawa, K. (ed.) ASIACRYPT 2007. LNCS, vol. 4833, pp. 164–180. Springer, Heidelberg (2007)
11. Groth, J., Sahai, A.: Efficient Non-interactive Proof Systems for Bilinear Groups. In: Smart, N.P. (ed.) EUROCRYPT 2008. LNCS, vol. 4965, pp. 415–432. Springer, Heidelberg (2008)
12. Kiayias, A., Tsiounis, Y., Yung, M.: Traceable Signatures. In: Cachin, C., Camenisch, J.L. (eds.) EUROCRYPT 2004. LNCS, vol. 3027, pp. 571–589. Springer, Heidelberg (2004)
13. Libert, B., Yung, M.: Efficient Traceable Signatures in the Standard Model. In: Shacham, H., Waters, B. (eds.) Pairing 2009. LNCS, vol. 5671, pp. 187–205. Springer, Heidelberg (2009)
14. Naor, M., Yung, M.: Public-key Cryptosystems Provably Secure against Chosen Ciphertext Attacks. In: STOC 1990, pp. 427–437 (1990)
15. Schuldt, J.C.N., Matsuura, K.: An Efficient Convertible Undeniable Signature Scheme with Delegatable Verification. In: Kwak, J., Deng, R.H., Won, Y., Wang, G. (eds.) ISPEC 2010. LNCS, vol. 6047, pp. 276–293. Springer, Heidelberg (2010)
16. Shoup, V.: Lower Bounds for Discrete Logarithms and Related Problems. In: Fumy, W. (ed.) EUROCRYPT 1997. LNCS, vol. 1233, pp. 256–266. Springer, Heidelberg (1997)

Hierarchical Identity-Based Chameleon Hash and Its Applications

Feng Bao[1], Robert H. Deng[2], Xuhua Ding[2], Junzuo Lai[2,*], and Yunlei Zhao[3]

[1] Institute for Infocomm Research, Singapore
baofeng@i2r.a-star.edu.sg
[2] Singapore Management University, Singapore
{robertdeng,xhding,junzuolai}@smu.edu.sg
[3] Fudan University, China
ylzhao@fudan.edu.cn

Abstract. At ACNS 2008, Canard et al. introduced the notion of trapdoor sanitizable signature (TSS) based on identity-based chameleon hash (IBCH). Trapdoor sanitizable signatures allow the signer of a message to delegate, at any time, the power of sanitization to possibly several entities who can modify predetermined parts of the message and generate a new signature on the sanitized message without interacting with the original signer. In this paper, we introduce the notion of hierarchical identity-based chameleon hash (HIBCH), which is a hierarchical extension of IBCH. We show that HIBCH can be used to construct other cryptographic primitives, including hierarchical trapdoor sanitizable signature (HTSS) and key-exposure free IBCH. HTSS allows an entity who has the sanitization power for a given signed message, to further delegate its power to its descendants in a controlled manner. Finally, we propose a concrete construction of HIBCH and show that it is t-threshold collision-resistant.

Keywords: Chameleon Hash, Trapdoor Sanitizable Signature, Hierarchical Identity-Based Chameleon Hash, Hierarchical Trapdoor Sanitizable Signature.

1 Introduction

Chameleon hash was introduced by Krawczyk and Rabin [20] as a tool to construct chameleon signatures. Informally, a chameleon hash function is a trapdoor collision-resistant hash function: without knowledge of the trapdoor, the chameleon hash function is collision-resistant; however, collisions can be easily computed once the trapdoor is known. Similar to undeniable signatures [9], chameleon signatures possess the properties of non-repudiation and *non-transferability* for the signed messages; however, chameleon signatures are non-interactive protocols. In order to provide a recipient with a non-transferable signature, a signer hashes the message to be signed with a recipient's chameleon

* Corresponding author.

J. Lopez and G. Tsudik (Eds.): ACNS 2011, LNCS 6715, pp. 201–219, 2011.

hash function and signs on the resulting digest value. The recipient knows the trapdoor of the chameleon hash function and hence is able to re-use the hash value to obtain a signature on a second message. On the other hand, the signer can prove knowledge of a hash collision, since the original signed message and the claimed signed message have the same hash value. Such a collision can be seen as proof of forgery by the signature recipient, as nobody apart from the recipient has more than a negligible probability of successfully finding a collision. One limitation of the original chameleon signature scheme [20] is that signature forgery results in the signer discovering the recipient's trapdoor information. This deterrent effect of key/trapdoor exposure on forgeries threatens the claims of non-transferability provided by the scheme. In fact, a third party will likely believe claims made by the recipient, because the potential devastating damage to the recipient would result from the forgery of a signature.

Identity-based chameleon hash (IBCH) and identity-based chameleon signature (IBCS), introduced by Ateniese and Medeiros [2], partly addressed the problem of key exposure. In IBCH/IBCS, a unique transaction-specific public key, called *customized identity*, is used to compute the chameleon hash of a transaction. The customized identity is computed by the signer from special strings that describe the transaction, including the signer and recipient information as well as a nonce value or time-stamp. As a result, the trapdoor corresponding to the customized identity is transaction specific and signature forgery only results in the signer recovering the trapdoor information associated with a single transaction.

Based on chameleon hash, Ateniese et al. [1] introduced the notion of sanitizable signature and presented its generic construction. Sanitizable signatures allow a signer to partly delegate signing rights to a semi-trusted party, called a sanitizer. During generation of a signature on a message, the signer chooses a specific sanitizer who can later modify predetermined parts of the message and generate a new signature on the sanitized message without interacting with the signer. The capability of modification renders sanitizable signatures valuable for many applications, such as authenticated multicast, authenticated database outsourcing and secure routing.

At ACNS 2008, Canard et al. [8] introduced the notion of trapdoor sanitizable signature (TSS) and showed its generic construction based on IBCH. TSS allows the signer to delegate the power of sanitization for a specific signed message to possibly several entities. Different from the sanitizable signatures in [1] where the sanitizer is predetermined at the time of signature generation by the signer, the signer in TSS can choose to whom and when it will provide the trapdoor information and therefore, any entity can potentially act as a sanitizer. This property makes a crucial difference from the conventional sanitizer signatures and is essential for applications where the potential sanitizers are not known at the time of signature generation.

In this paper, we introduce the notions of hierarchical identity-based chameleon hash (HIBCH) and hierarchical trapdoor sanitizable signature (HTSS), which are the hierarchical extensions of IBCH and TSS, respectively. We present a generic construction of HTSS from HIBCH. Like TSS, HTSS allows a signer to delegate the power of sanitization for a specific signed message to any sanitizers at any time. In addition, HTSS allows a sanitizer to further delegate sanitization power to its descendants in an identity hierarchy. This distinguishing feature of cascaded delegation of sanitization powers makes HTSS especially powerful in protecting information flows in distributed settings, such as automated web-service-enabled business processes [26] and tiered multimedia distribution systems [25].

As an example of automated web-service-enabled business processes, let us consider a simple quotation response process, involving an electronic distributor (ED), a transportation company (TC) and an electronic manufacturer (EM). A business document *Quotation* in XML format is transferred between various entities with use of document-styled web services. The quotation response process begins when ED receives a request for quotation from an electronic retailer (ER). ED generates the *Quotation* by providing quotes for each item and the taxes associated and forwards the *Quotation* document to TC via a SOAP message. Upon receipt of the document, TC adds the delivery cost, delivery information and updates the total cost. TC then forwards the document to EM, which inputs additional product information based on the retailer's information and then forwards the *Quotation* to ER. Upon receipt of the document, ER informs ED that the quotation has been received. A basic security requirement of the quotation response process is integrity and authenticity of the *Quotation* document. Such a requirement can be fulfilled readily using HTSS: the document originator ED generates a signature and a trapdoor on the *Quotation* document and forwards them to TC. With the knowledge of the trapdoor, TC is able to perform predetermined modification on the document, such as adding delivery cost, without invaliding the original signature. TC then generates a new trapdoor and forwards the updated document and the trapdoor to EM, which in turn modifies the document based on its input.

Another application of HTSS is end-to-end content authentication in tiered multimedia distribution systems, where multimedia contents are distributed from a top-tier primary content provider to multiple levels of lower-tier affiliating providers each with its own user groups. An example is a multinational company that has a global headquarter, a number of regional headquarters and many country level offices worldwide. To promote a new product, the company produces a video advertisement for the product and delivers it to all the regional headquarters for processing, which then disseminate the processed video clips to the country level offices. In order to better fit local markets, regional headquarters and country level offices are entitled to derive their own local versions (e. g., adding subtitles in the local language) based on the original advertisement. In scenarios like this, higher-tier content providers may authorize lower-tier content providers performing transcoding operations on the original content, such as content downscaling, content alteration, and content insertion. Apparently, the capability of cascaded delegation of sanitization powers makes HTSS an ideal solution for authenticated content delivery in such environments.

1.1 Our Contributions

In this paper, we make the following contributions:

1. We introduce the notion of HIBCH, which is a hierarchical extension of IBCH. We also introduce the security definitions of HIBCH.
2. We introduce the notion of HTSS, which is a hierarchical extension of TSS. In an HTSS scheme, a signer can delegate at any time the power of sanitization for a specific signed message to an entity; the entity in turn can further delegate its power to its descendants *in a controlled manner* (details are given in Section 4.1). We extend the standard security definitions of TSS for the hierarchical setting and propose a generic construction of HTSS based on HIBCH.
3. We show that a key-exposure free IBCH can be obtained from a two-level HIBCH, though the latter is not key-exposure free. The construction of key-exposure free IBCH from HIBCH with key-exposure is similar to the construction of key-exposure free chameleon hash from IBCH [2,3]. In our construction, forgery only results in recovering the trapdoor information associated to a specific transaction, and therefore offering a partial answer to the key exposure problem of IBCH. We also point out in the full version of the paper a flaw in [11] which was the first full construction of a key-exposure free IBCH.
4. We present a concrete construction of HIBCH, which is resilient against compromise of a threshold number of entities in every level of the underlying hierarchy.

1.2 Related Work

Chameleon hash was introduced by Krawczyk and Rabin [20]. The original construction of chameleon hash [20] suffers from the key exposure problem. The problem was partly addressed by IBCH, which was introduced by Ateniese and Medeiros [2].

Chen et al. [10] presented the first full construction of a key-exposure free chameleon hash, which works in the setting of gap groups with bilinear pairings. Ateniese and Medeiros [3] proposed three key-exposure free chameleon hash functions, two based on RSA and one based on pairings. Other key-exposure free chameleon hash constructions [14,13,12] were proposed subsequently.

Zhang et al. [28] proposed two IBCH schemes from bilinear pairing. Recently, Chen et al. [11] considered the key exposure problem of IBCH and proposed a concrete construction of key-exposure free IBCH. However, in the Appendix, we point out a fault of the construction in [11].

The notion of sanitizable signature was introduced by Ateniese et al. [1]. Santizable signature allows a sanitizer to modify predetermined parts of a signed message and generate new signature on the sanitized message without interacting with the signer. Klonowski and Lauks [19] presented several extensions of sanitizable signature, including limitation of the set of possible modifications of a single mutable block and limitation of the number of modifications of mutable blocks.

Ateniese et al. [1] identified five security requirements of sanitizable signature schemes, including *unforgeability*, *immutability*, *privacy*, *transparency* and *accountability*. Recently, Brzuska et al. [7] revisited the security requirements for sanitizable signatures and investigated the relationship of the security requirements, showing for example that transparency implies privacy.

Miyazaki et al. [23] also used the notion of sanitizable signature in a slightly different vein. Such sanitizable signature schemes [23,17,22] allow the sanitizer to only *delete* predetermined parts of a signed message.

The notions of incremental cryptography [4] and homomorphic signatures, which encompass transitive [21], redactable [18] and context-extraction signatures [24], are also related to sanitizable signatures. We refer the reader to [1] for details.

Canard et al. [8] introduced the notion of trapdoor sanitizable signatures (TSS), in which the power of sanitization is given to possibly several entities. Based on IBCH, Canard et al. [8] proposed a generic construction of TSS. Recently, Yum et al. [27] presented a generic construction of trapdoor sanitizable signatures from ordinary signature schemes; therefore, one-way functions imply trapdoor sanitizable signatures.

1.3 Organization

The rest of the paper is organized as follows. Some preliminaries are given in Section 2. We introduce the notion and security requirements of HIBCH in Section 3. We introduce the notion of HTSS and propose a generic construction of HTSS from HIBCH in Section 4. In Section 5, we describe the generic construction of key-exposure free IBCH from HIBCH with key exposure. We describe and analysis our concrete HIBCH scheme in Section 6. Finally, we state our conclusion in Section 7.

2 Preliminaries

If L is a positive integer, then $[1, L] = \{1, 2, \ldots, L\}$. If S_1, S_2 are two sets, $S_1 \backslash S_2 = \{x \in S_1 | x \notin S_2\}$. Let $\mathbb{Z}_p$ denote the set $\{0, 1, 2, \ldots, p-1\}$ and $\mathbb{Z}_p^*$ denote $\mathbb{Z}_p \backslash \{0\}$. For a finite set S, $x \xleftarrow{\$} S$ means choosing an element $x \in S$ with a uniform distribution. If $x_1, x_2, \ldots$ are strings, then $x_1 \| x_2 \| \ldots$ denotes their concatenation. If A is a probabilistic algorithm, then $A(x, r)$ is the result of running A on input x and coins r. We denote by $A(x; \mathcal{R})$ the random variable of choosing coins r uniformly at random from $\mathcal{R}$ and outputting $A(x, r)$.

We say that a function $f(\lambda)$ is *negligible* if for every $c > 0$ there exists an λ_c such that $f(\lambda) < 1/\lambda^c$ for all $\lambda > \lambda_c$. We say that two distribution ensembles $\{X(\lambda, z)\}_{\lambda \in N, z \in \{0,1\}^*}$ and $\{Y(\lambda, z)\}_{\lambda \in N, z \in \{0,1\}^*}$ are *computationally indistinguishable*, if for any probabilistic polynomial-time (PPT) algorithm D, and for all sufficiently large λ and any $z \in \{0, 1\}^*$, it holds that $|\Pr[D(\lambda, z, X) = 1] - \Pr[D(\lambda, z, Y) = 1]|$ is negligible in λ.

2.1 Bilinear Pairings

Let $\mathbb{G}$ be a cyclic multiplicative group of prime order p and $\mathbb{G}_T$ be a cyclic multiplicative group of the same order p. A bilinear pairing is a map $e : \mathbb{G} \times \mathbb{G} \rightarrow \mathbb{G}_T$ with the following properties:

- Bilinearity: $\forall g_1, g_2 \in \mathbb{G}, \forall a, b \in \mathbb{Z}_p^*$, we have $e(g_1^a, g_2^b) = e(g_1, g_2)^{ab}$;
- Non-degeneracy: There exist $g_1, g_2 \in \mathbb{G}$ such that $e(g_1, g_2) \neq 1$;
- Computability: There exists an efficient algorithm to compute $e(g_1, g_2)$ for $\forall g_1, g_2 \in \mathbb{G}$.

2.2 Identity-Based Chameleon Hash

A identity-based chameleon hash (IBCH) scheme [2] is a tuple of algorithms described as follows:

Setup takes as input a security parameter λ. It generates a public/private key pair (pk, sk), publishes pk and keeps sk secret. This algorithm is run by a trusted party, called private key generator (PKG).

$$(pk, sk) \leftarrow \mathsf{Setup}(\lambda).$$

Extract takes as input sk and an identity ID. It outputs the trapdoor information sk_{ID} associated with the identity. This algorithm is run by PKG.

$$sk_{\mathsf{ID}} \leftarrow \mathsf{Extract}(sk, \mathsf{ID}).$$

Hash takes as input pk, an identity ID and a message m. It chooses a randomness r and outputs a hash value h.

$$h \leftarrow \mathsf{Hash}(pk, \mathsf{ID}, m, r).$$

Forge takes as input an identity ID, the trapdoor information sk_{ID} associated with ID, the hash value h on a message m with r, and a new message m'. It outputs a value r'.

$$r' \leftarrow \mathsf{Forge}(sk_{\mathsf{ID}}, \mathsf{ID}, m, r, h, m').$$

For correctness, it requires that

$$\mathsf{Hash}(pk, \mathsf{ID}, m, r) = h = \mathsf{Hash}(pk, \mathsf{ID}, m', r' = \mathsf{Forge}(sk_{\mathsf{ID}}, \mathsf{ID}, m, r, h, m')) \text{ and } m' \neq m.$$

The security of an IBCH scheme consists of two requirements: *resistance to collision forgery under active attacks* and *semantic security*. In the following and throughout the rest of the paper, we use $\mathcal{A}$ to denote an adversary which can be any probabilistic polynomial-time algorithm.

Resistance to collision forgery under active attacks: The IBCH scheme is secure against (existential) collision forgery under active attacks, if for any

PPT adversary $\mathcal{A}$, for all sufficiently large λ, any $(pk, sk) \leftarrow \mathsf{Setup}(\lambda)$, the probability that $\mathcal{A}(\lambda, pk)$ outputs $(\mathsf{ID}, m, r, m', r')$, satisfying $\mathsf{Hash}(pk, \mathsf{ID}, m, r) = \mathsf{Hash}(pk, \mathsf{ID}, m', r')$ and $m' \neq m$, is negligible. $\mathcal{A}$ is allowed to query an oracle $\mathcal{O}_{IBCH}^{\mathsf{Extract}}$ [1] on adaptively chosen identities other than ID.

Semantic security: The IBCH scheme is said to be semantically secure if, for all sufficiently large λ, any $(pk, sk) \leftarrow \mathsf{Setup}(\lambda)$, any target identity ID and all pairs of messages m and m', the distribution ensembles $\{\mathsf{Hash}(pk, \mathsf{ID}, m; \mathcal{R})\}_{\lambda, pk, \mathsf{ID}, m, m'}$ and $\{\mathsf{Hash}(pk, \mathsf{ID}, m'; \mathcal{R})\}_{\lambda, pk, \mathsf{ID}, m, m'}$ are computationally indistinguishable.

3 Hierarchical Identity-Based Chameleon Hash

Like an IBCH scheme, a hierarchical identity-based chameleon hash (HIBCH) scheme consists of four algorithms: Setup, Extract, Hash and Forge. In HIBCH, however, identities are organized into a hierarchy, where an identity at depth k of the hierarchy is represented as a vector of dimension k, and the trapdoor information for an identity is generated by its parent. Concretely, an ℓ-HIBCH scheme consists of the following algorithms:

Setup takes as input a security parameter λ and the maximum hierarchy depth ℓ that is polynomial in λ. It generates a public/private key pair (pk, sk), publishes pk and keeps sk secret. This algorithm is run by PKG.

$$(pk, sk) \leftarrow \mathsf{Setup}(\lambda, \ell).$$

Extract takes as an identity $\mathsf{ID} = (\mathsf{ID}_1, \ldots, \mathsf{ID}_k)$ at depth $k \leq \ell$, and the trapdoor information $sk_{\mathsf{ID}_{|k-1}}$ of the parent identity $\mathsf{ID}_{|k-1} = (\mathsf{ID}_1, \ldots, \mathsf{ID}_{k-1})$ at depth $k-1$. It outputs the trapdoor information sk_{ID} for identity ID.

$$sk_{\mathsf{ID}} \leftarrow \mathsf{Extract}(sk_{\mathsf{ID}_{|k-1}}, \mathsf{ID}).$$

Note that, if $k = 1$, the trapdoor information $sk_{\mathsf{ID}_{|k-1}}$ of the identity $\mathsf{ID}_{|k-1}$ is sk. Running Extract algorithm recursively, an identity $\mathsf{ID} = (\mathsf{ID}_1, \ldots, \mathsf{ID}_k)$, using its trapdoor information sk_{ID}, can generate trapdoor information for all its descendants. So, we also can denote this algorithm as

$$sk_{\mathsf{ID}'} \leftarrow \mathsf{Extract}(sk_{\mathsf{ID}}, \mathsf{ID}'),$$

where ID' is a descendant of ID.

Hash takes as input pk, an identity ID and a message m. It chooses a randomness r and outputs a hash value h.

$$h \leftarrow \mathsf{Hash}(pk, \mathsf{ID}, m, r).$$

[1] When the adversary $\mathcal{A}$ queries $\mathcal{O}_{IBCH}^{\mathsf{Extract}}$ on an identity ID', the simulator gives the trapdoor information $sk_{\mathsf{ID}'}$ associated with ID' to $\mathcal{A}$.

Forge takes as input an identity ID, the trapdoor information sk_{ID} associated with ID, the hash value h on a message m with randomness r, and a new message m'. It outputs r'.

$$r' \leftarrow \mathsf{Forge}(sk_{\mathsf{ID}}, \mathsf{ID}, m, r, h, m').$$

For correctness, it requires that

$$\mathsf{Hash}(pk, \mathsf{ID}, m, r) = h = \mathsf{Hash}(pk, \mathsf{ID}, m', r' = \mathsf{Forge}(sk_{\mathsf{ID}}, \mathsf{ID}, m, r, h, m')) \text{ and } m' \neq m.$$

We now introduce the security requirements of HIBCH, including *resistance to collision forgery under active attacks*, *semantic security* and *forgery indistinguishability*. The security requirements of *resistance to collision forgery under active attacks* and *semantic security* are extended from the security requirements of IBCH. For *forgery indistinguishability*, informally, it requires that an adversary be not able to decide whether (m, r, h) is a forgery or not.

Resistance to collision forgery under active attacks: The HIBCH scheme is secure against (existential) collision forgery under active attacks, if for any PPT adversary $\mathcal{A}$, for any sufficiently large λ, any hierarchy depth ℓ that is polynomial in λ, any $(pk, sk) \leftarrow \mathsf{Setup}(\lambda, \ell)$, the probability that $\mathcal{A}(\lambda, pk)$ outputs $(\mathsf{ID}, m, r, m', r')$, satisfying $\mathsf{Hash}(pk, \mathsf{ID}, m, r) = \mathsf{Hash}(pk, \mathsf{ID}, m', r')$ and $m' \neq m$, is negligible. $\mathcal{A}$ is allowed to query an oracle $\mathcal{O}^{\mathsf{Extract}}_{HIBCH}$ [2] on adaptively chosen identities other than ID or an ancestor of ID.

We say that an HIBCH scheme is t-threshold resistant to collision forgery under active attacks (or t-threshold collusion-resistant for simplicity.), if $\mathcal{A}$ issues at most t queries to its $\mathcal{O}^{\mathsf{Extract}}_{HIBCH}$ oracle on identities at each depth of the hierarchy.

Semantic security: The HIBCH scheme is said to be semantically secure if, for all sufficiently large λ, any hierarchy depth ℓ that is polynomial in λ, any $(pk, sk) \leftarrow \mathsf{Setup}(\lambda, \ell)$, all identities ID and all pairs of messages m and m', the distribution ensembles $\{\mathsf{Hash}(pk, \mathsf{ID}, m; \mathcal{R})\}_{\lambda, pk, \mathsf{ID}, m, m'}$ and $\{\mathsf{Hash}(pk, \mathsf{ID}, m'; \mathcal{R})\}_{\lambda, pk, \mathsf{ID}, m, m'}$ are computationally indistinguishable.

Forgery indistinguishability: The HIBCH scheme is said to be forgery-indistinguishable if, for all sufficiently large λ, any hierarchy depth ℓ that is polynomial in λ, any $(pk, sk) \leftarrow \mathsf{Setup}(\lambda, \ell)$, all identities ID and all pairs of messages m and m', the following distribution ensembles are computationally indistinguishable:

$$\begin{aligned} \mathcal{D}_{\mathsf{Forge}} &= \{(m', \hat{r}, h) | r \xleftarrow{\$} \mathcal{R}, h \leftarrow \mathsf{Hash}(pk, \mathsf{ID}, m, r), sk_{\mathsf{ID}} \leftarrow \mathsf{Extract}(sk, \mathsf{ID}), \\ &\qquad \hat{r} \leftarrow \mathsf{Forge}(sk_{\mathsf{ID}}, \mathsf{ID}, m, r, h, m')\}_{\lambda, pk, \mathsf{ID}}, \\ \mathcal{D}_{\mathsf{Hash}} &= \{(m', r', h') | r' \xleftarrow{\$} \mathcal{R}, h' \leftarrow \mathsf{Hash}(pk, \mathsf{ID}, m', r')\}_{\lambda, pk, \mathsf{ID}}. \end{aligned}$$

[2] When the adversary $\mathcal{A}$ queries $\mathcal{O}^{\mathsf{Extract}}_{HIBCH}$ on an identity ID', the simulator gives the trapdoor information $sk_{\mathsf{ID}'}$ associated with ID' to $\mathcal{A}$.

4 Hierarchical Trapdoor Sanitizable Signature and Its Construction from HIBCH

In this section, we first introduce the notion of hierarchical trapdoor sanitizable signature (HTSS), which is a hierarchical extension of TSS, and extend the standard security definitions of TSS for the hierarchical setting. Then, we propose a generic construction of HTSS from HIBCH. The construction is similar to the construction of TSS from IBCH [8].

4.1 Hierarchical Trapdoor Sanitizable Signature

Informally, in an HTSS scheme, an identity associated with an entity who has the power of sanitization for a given signed message, can delegate its rights to its descendant identities in a controlled manner. In the following, to simplify the description, we will use the terms identity and entity interchangeably.

In the sequel we assume that each signed message $m = m_1 \| \cdots \| m_L$ is partitioned into L blocks, where L is an positive integer. We define a hierarchical sanitizable description $\mathtt{ADM}$ of m using a tree. Each leaf node of the tree is labeled by a distinct block index $i \in [1, L]$, which indicates the block is sanitizable. Each node of the tree is associated with an identity. The identity of a node at depth k is a $k+1$-dimensional vector, and the first k components of the identity is inherited from its parent. The identity of the root node is computed from special strings, which may include the signer and recipient information as well as some nonce or time-stamp. We say that an identity ID matches $\mathtt{ADM}$ if an internal node of the tree is associated with the identity ID.

A pictorial depiction of a hierarchical sanitizable description $\mathtt{ADM}$ is given in Figure 1, where $L = 8$. We can obtain from the $\mathtt{ADM}$ that the set of indices $I = \{1, 4, 5, 8\}$ that are sanitizable, and identities ID_0, $(\mathsf{ID}_0, \mathsf{ID}_1^1)$, $(\mathsf{ID}_0, \mathsf{ID}_1^2)$, $(\mathsf{ID}_0, \mathsf{ID}_1^1, \mathsf{ID}_2^1)$ and $(\mathsf{ID}_0, \mathsf{ID}_1^1, \mathsf{ID}_2^2)$ match the $\mathtt{ADM}$.

Like a TSS scheme, an HTSS scheme consists of five algorithms: KeyGen, Sign, $\mathsf{Trapdoor}$, $\mathsf{Sanitize}$ and Verify. In HTSS, however, Sign algorithm takes as input a hierarchical sanitizable description $\mathtt{ADM}$, not only a set of the indices $I \subseteq [1, L]$

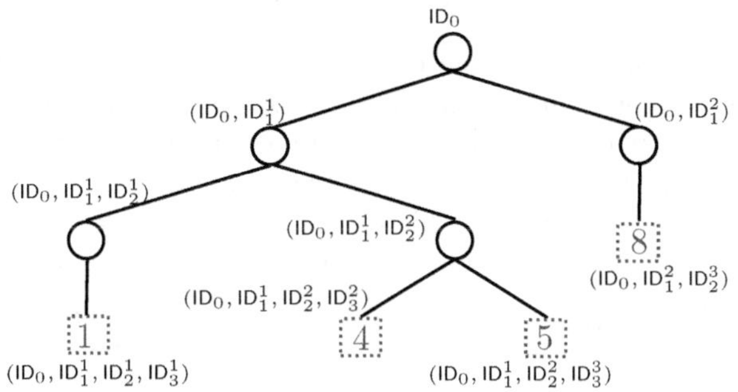

Fig. 1. An example of hierarchical sanitizable description

that are sanitizable in TSS, and the trapdoor information for an identity is generated by its parent, which runs $\mathsf{Trapdoor}$ algorithm. Concretely, an HTSS scheme consists of the following algorithms:

KeyGen takes as input a security parameter λ and the maximum hierarchy depth ℓ of hierarchical sanitizable descriptions. It generates a public/private key pair (pk, sk), publishes pk and keeps sk secret.

$$(pk, sk) \leftarrow \mathsf{KeyGen}(\lambda, \ell).$$

Sign takes as input a message $m = m_1 \| \cdots \| m_L$, a hierarchial sanitizable description $\mathtt{ADM}$ and sk. It outputs a signature σ on the message m.

$$\sigma \leftarrow \mathsf{Sign}(m, \mathtt{ADM}, sk).$$

$\mathsf{Trapdoor}$ takes as input a message m, a valid signature σ on m, a hierarchial sanitizable description $\mathtt{ADM}$, the trapdoor $sk_{\mathsf{ID}_{|k-1}}$ of the identity $\mathsf{ID}_{|k-1} = (\mathsf{ID}_1, \ldots, \mathsf{ID}_{k-1})$, and a child identity $\mathsf{ID} = (\mathsf{ID}_1, \ldots, \mathsf{ID}_k)$. It outputs a trapdoor sk_{ID} associated with ID.

$$sk_{\mathsf{ID}} \leftarrow \mathsf{Trapdoor}(m, \mathtt{ADM}, \sigma, \mathsf{ID}, sk_{\mathsf{ID}_{|k-1}}).$$

Note that, if $k = 1$, the trapdoor $sk_{\mathsf{ID}_{|k-1}}$ of the identity $\mathsf{ID}_{|k-1}$ is sk. Running $\mathsf{Trapdoor}$ algorithm recursively, an identity $\mathsf{ID} = (\mathsf{ID}_1, \ldots, \mathsf{ID}_k)$ with its trapdoor sk_{ID} can generate the trapdoors for all its descendants.

$\mathsf{Sanitize}$ takes as input pk, a message m, a valid signature σ on m, a hierarchial sanitizable description $\mathtt{ADM}$, a trapdoor sk_{ID} associated with identity ID, a new message m'. It outputs a new signature σ' on m'.

$$\sigma' \leftarrow \mathsf{Sanitize}(pk, m, \mathtt{ADM}, \sigma, m', \mathsf{ID}, sk_{\mathsf{ID}}).$$

Verify takes as input pk, a message m, a putative signature σ and a hierarchial sanitizable description $\mathtt{ADM}$. It outputs 1 if the signature σ on m is valid and 0 otherwise.

$$0/1 \leftarrow \mathsf{Verify}(pk, m, \mathtt{ADM}, \sigma).$$

For an HTSS scheme the usual correctness properties should hold, saying that genuinely signed or sanitized messages are accepted. Formally, for correctness, an HTSS scheme must satisfy the following condition. For any security parameter λ and maximum hierarchy depth ℓ of hierarchical sanitizable descriptions, any message $m = m_1 \| \cdots \| m_L$, any hierarchial sanitizable description $\mathtt{ADM}$, any identity $\mathsf{ID} = (\mathsf{ID}_1, \ldots, \mathsf{ID}_k)$ and the trapdoor $sk_{\mathsf{ID}_{|k-1}}$ of the parent identity $\mathsf{ID}_{|k-1} = (\mathsf{ID}_1, \ldots, \mathsf{ID}_{k-1})$, letting $(pk, sk) \leftarrow \mathsf{KeyGen}(\lambda, \ell)$, $\sigma \leftarrow \mathsf{Sign}(m, \mathtt{ADM}, sk)$, $sk_{\mathsf{ID}} \leftarrow \mathsf{Trapdoor}(m, \mathtt{ADM}, \sigma, \mathsf{ID}, sk_{\mathsf{ID}_{|k-1}})$, $\sigma' \leftarrow \mathsf{Sanitize}(pk, m, \mathtt{ADM}, \sigma, m', \mathsf{ID}, sk_{\mathsf{ID}})$,

1. $\mathsf{Verify}(pk, m, \mathtt{ADM}, \sigma) = 1$.
2. $\mathsf{Verify}(pk, m', \mathtt{ADM}, \sigma') = 1$.

The security requirements of an HTSS scheme include *unforgeability* and *indistinguishability*, which are extended from the security requirements of TSS. Informally, unforgeability requires that an outsider be not able to forger a signature on the original or the sanitized message, and indistinguishability requires that an outsider be not able to decide whether a message has been sanitized or not.

Unforgeability: An HTSS scheme is existential unforgeable under adaptive chosen message attacks, if for all sufficiently large λ, any hierarchy depth ℓ that is polynomial in λ, any $(pk, sk) \leftarrow \mathsf{Setup}(\lambda, \ell)$, any PPT adversary $\mathcal{A}(\lambda, pk)$, after issuing $\mathcal{O}_{HTSS}^{\mathsf{Sign}}$, $\mathcal{O}_{HTSS}^{\mathsf{Trapdoor}}$ and $\mathcal{O}_{HTSS}^{\mathsf{Sanitize}}$[3] oracle queries adaptively, with only negligible probability, can output $(m^*, \mathtt{ADM}^*, \sigma^*)$ such that:

1. $\mathsf{Verify}(pk, m^*, \mathtt{ADM}^*, \sigma^*) = 1$;
2. $\mathcal{A}$ never queries $\mathcal{O}_{HTSS}^{\mathsf{Sign}}$ oracle on $(m^*, \cdot)$;
3. (m^*, σ^*) does not come from $\mathcal{O}_{HTSS}^{\mathsf{Sanitize}}$ oracle; i. e., $\mathcal{A}$ never queries $\mathcal{O}_{HTSS}^{\mathsf{Sanitize}}$ oracle on $(m, \cdot, \sigma, m^*, \cdot)$;
4. $\mathcal{A}$ never queries $\mathcal{O}_{HTSS}^{\mathsf{Trapdoor}}$ oracle on $(m, \mathtt{ADM}, \sigma, \mathsf{ID})$ such that $m_i = m_i^*$ for all $i \notin I$, where I is extracted from $\mathtt{ADM}$ and is a set of indices $I \subseteq [1, L]$ that are sanitizable.

Indistinguishability: Indistinguishability of an HTSS scheme demands that the output distributions of Sign algorithm and $\mathsf{Sanitize}$ algorithm be computationally indistinguishable. In other words, for all sufficiently large λ, any hierarchy depth ℓ that is polynomial in λ, any $(pk, sk) \leftarrow \mathsf{Setup}(\lambda, \ell)$, any hierarchial sanitizable description $\mathtt{ADM}$, all message pairs m, m' such that $m_i = m'_i$ for all $i \notin I$, where I is extracted from $\mathtt{ADM}$ and is a set of indices $I \subseteq [1, L]$ that are sanitizable, any identity $\mathsf{ID} = (\mathsf{ID}_1, \ldots, \mathsf{ID}_k)$ that matches $\mathtt{ADM}$ and the trapdoor $sk_{\mathsf{ID}_{|k-1}}$ of the parent identity $\mathsf{ID}_{|k-1} = (\mathsf{ID}_1, \ldots, \mathsf{ID}_{k-1})$, the following distribution ensembles $\mathcal{D}_{\mathsf{Sanitize}}$ and $\mathcal{D}_{\mathsf{Sign}}$ are computationally indistinguishable:

$$\begin{aligned}\mathcal{D}_{\mathsf{Sanitize}} &= \{(m', \hat{\sigma}) | \sigma \leftarrow \mathsf{Sign}(m, \mathtt{ADM}, sk), sk_{\mathsf{ID}} \leftarrow \mathsf{Trapdoor}(m, \mathtt{ADM}, \sigma, \mathsf{ID}, sk_{\mathsf{ID}_{|k-1}}), \\ &\qquad \hat{\sigma} \leftarrow \mathsf{Sanitize}(pk, m, \mathtt{ADM}, \sigma, m', \mathsf{ID}, sk_{\mathsf{ID}})\}_{\lambda, pk, \mathsf{ID}, \mathtt{ADM}}, \\ \mathcal{D}_{\mathsf{Sign}} &= \{(m', \sigma') | \sigma' \leftarrow \mathsf{Sign}(m', \mathtt{ADM}, sk)\}_{\lambda, pk, \mathsf{ID}, \mathtt{ADM}}.\end{aligned}$$

4.2 Generic Construction of HTSS from HIBCH

In our construction, to sign a message $m = m_1 \| \cdots \| m_L$, the signer first sets $\tilde{m} = \tilde{m}_1 \| \cdots \| \tilde{m}_L$, where $\tilde{m}_i = m_i$ if $i \notin I$ and otherwise, $\tilde{m}_i = h_i = \mathsf{HIBCH.Hash}(pk, \mathsf{ID}^{(i)}, m_i, r_i)$. The set of indices $I \subseteq [1, L]$ that are sanitizable and the identity $\mathsf{ID}^{(i)}$ associated with the sanitizable block index $i \in I$ are given in the hierarchial sanitizable description $\mathtt{ADM}$. Then, the signer signs the message

[3] When the adversary $\mathcal{A}$ queries $\mathcal{O}_{HTSS}^{\mathsf{Sign}}$ on a message $(m, \mathtt{ADM})$, the simulator gives a valid signature $\sigma = \mathsf{Sign}(m, \mathtt{ADM}, sk)$ on m to $\mathcal{A}$. When the adversary $\mathcal{A}$ queries $\mathcal{O}_{HTSS}^{\mathsf{Trapdoor}}$ on $(m, \mathtt{ADM}, \sigma, \mathsf{ID})$, the simulator gives the corresponding trapdoor sk_{ID} to $\mathcal{A}$. When the adversary $\mathcal{A}$ queries $\mathcal{O}_{HTSS}^{\mathsf{Sanitize}}$ on $(m, \mathtt{ADM}, \sigma, m', \mathsf{ID})$, the simulator gives a valid signature σ' on m' to $\mathcal{A}$.

$\tilde{m}$ using a conventional signature scheme. Obviously, an entity with the trapdoor associated with $\mathsf{ID}^{(i)}$ or an ancestor of $\mathsf{ID}^{(i)}$ can modify m_i and generate a new signature on the sanitized message.

Given a conventional signature scheme $\Sigma = (\Sigma.\mathsf{KeyGen}, \Sigma.\mathsf{Sign}, \Sigma.\mathsf{Verify})$ and an HIBCH scheme $\Pi = (\Pi.\mathsf{Setup}, \Pi.\mathsf{Extract}, \Pi.\mathsf{Hash}, \Pi.\mathsf{Forge})$, we define the 5-tuple algorithms (KeyGen, Sign, Trapdoor, Sanitize, Verify) of an HTSS scheme as follows:

KeyGen Given a security parameter λ and the maximum hierarchy depth ℓ of hierarchical sanitizable descriptions, it first runs

$$(pk_\Sigma, sk_\Sigma) \leftarrow \Sigma.\mathsf{KeyGen}(\lambda), \ (pk_\Pi, sk_\Pi) \leftarrow \Pi.\mathsf{Setup}(\lambda, \ell + 1).$$

Then, it sets the public key $pk = (pk_\Sigma, pk_\Pi)$ and the private key $sk = (sk_\Sigma, sk_\Pi)$. Finally, it publishes pk and keeps sk secret.

Sign Given a message $m = m_1 \| \cdots \| m_L$, a hierarchial sanitizable description ADM and $sk = (sk_\Sigma, sk_\Pi)$, it first extracts a set of indices $I \subseteq [1, L]$ that are sanitizable from ADM. Then, it proceeds as follows.

1. For all $i \in [1, L] \backslash I$, it sets $\tilde{m}_i = m_i$.
2. For all $i \in I$, let $\mathsf{ID}^{(i)}$ be the identity of the leaf node of ADM labeled by the block index i, it chooses a randomness r_i uniformly, and computes $h_i = \Pi.\mathsf{Hash}(pk_\Pi, \mathsf{ID}^{(i)}, m_i, r_i)$ and sets $\tilde{m}_i = h_i$. Let r be the concatenation of all random values r_i, $i \in I$.
3. It sets $\tilde{m} = \tilde{m}_1 \| \cdots \| \tilde{m}_L$ and runs
$$\tilde{\sigma} \leftarrow \Sigma.\mathsf{Sign}(\tilde{m}, sk_\Sigma).$$
4. Finally, it sets $\sigma = \tilde{\sigma} \| r$ and outputs the signature σ on m.

Trapdoor Given a message m, a valid signature σ on m, a hierarchial sanitizable description ADM, an identity $\mathsf{ID} = (\mathsf{ID}_1, \ldots, \mathsf{ID}_k)$, and the trapdoor $sk_{\mathsf{ID}_{|k-1}}$ of the parent identity $\mathsf{ID}_{|k-1} = (\mathsf{ID}_1, \ldots, \mathsf{ID}_{k-1})$, it first checks whether ID matches ADM. If not, it outputs $\perp$, denoted an error. Otherwise, it runs

$$sk_{\mathsf{ID}} \leftarrow \Pi.\mathsf{Extract}(sk_{\mathsf{ID}_{|k-1}}, \mathsf{ID}),$$

and outputs the trapdoor sk_{ID} associated with ID.

Note that, if $k = 1$, the trapdoor $sk_{\mathsf{ID}_{|k-1}}$ of the identity $\mathsf{ID}_{|k-1}$ is sk_Π.

Sanitize Given $pk = (pk_\Sigma, pk_\Pi)$, a message $m = m_1 \| \cdots \| m_L$, a valid signature $\sigma = \tilde{\sigma} \| r$ on m, a hierarchial sanitizable description ADM, a trapdoor sk_{ID} associated with identity ID, a new message $m' = m'_1 \| \cdots \| m'_L$, it proceeds as follows.

1. Let $I' = \{i \in [1, L] | m_i \neq m'_i\}$. It extracts a set of indices $I \subseteq [1, L]$ that are sanitizable from ADM. Then, it checks whether $I' \subseteq I$. If not, it outputs $\perp$, denoted an error.
2. It checks whether ID matches ADM. If not, it outputs $\perp$.

3. For all $i \in I'$, let $\mathsf{ID}^{(i)}$ be the identity of the leaf node of ADM labeled by the block index i, it checks whether $\mathsf{ID}^{(i)}$ is a descendant of ID. If not, it outputs $\perp$. Otherwise, it runs
$$sk_{\mathsf{ID}^{(i)}} \leftarrow \Pi.\mathsf{Extract}(sk_{\mathsf{ID}}, \mathsf{ID}^{(i)}),$$
to obtain the trapdoor $sk_{\mathsf{ID}^{(i)}}$ associated with $\mathsf{ID}^{(i)}$.
4. It retrieves $\{r_i | i \in I\}$ from the signature $\sigma = \tilde{\sigma} \| r$.
5. For all $i \in I'$, it computes $h_i \leftarrow \Pi.\mathsf{Hash}(pk_\Pi, \mathsf{ID}^{(i)}, m_i, r_i)$ and
$$r'_i \leftarrow \Pi.\mathsf{Forge}(sk_{\mathsf{ID}^{(i)}}, \mathsf{ID}^{(i)}, m_i, r_i, h_i, m'_i).$$
6. For all $i \in I \backslash I'$, it sets $r'_i = r_i$. Let r' be the concatenation of all random values r'_i, $i \in I$.
7. It sets $\sigma' = \tilde{\sigma} \| r'$ and outputs the new signature σ' on m'.

Verify Given $pk = (pk_\Sigma, pk_\Pi)$, a message $m = m_1 \| \cdots \| m_L$, a putative signature $\sigma = \tilde{\sigma} \| r$ and a hierarchial sanitizable description ADM, it proceeds as follows.
1. It extracts a set of indices $I \subseteq [1, L]$ that are sanitizable from ADM and retrieves $\{r_i | i \in I\}$ from the signature $\sigma = \tilde{\sigma} \| r$.
2. For all $i \in [1, L] \backslash I$, it sets $\tilde{m}_i = m_i$.
3. For all $i \in I$, let $\mathsf{ID}^{(i)}$ be the identity of the leaf node of ADM labeled by the block index i, it computes $h_i = \Pi.\mathsf{Hash}(pk_\Pi, \mathsf{ID}^{(i)}, m_i, r_i)$ and sets $\tilde{m}_i = h_i$.
4. It sets $\tilde{m} = \tilde{m}_1 \| \cdots \| \tilde{m}_L$ and outputs $\Sigma.\mathsf{Verify}(pk_\Sigma, \tilde{m}, \tilde{\sigma})$.

It is obvious that the above HTSS scheme satisfies correctness. We now state the security theorems of the above HTSS scheme, including *unforgeability* and *indistinguishability*. The proofs of the security theorems are similar to those in [8] and will be given in the full version of the paper.

Theorem 1 (Unforgeability). *If the signature scheme Σ is existential unforgeable under adaptive chosen message attacks [16] and the HIBCH scheme Π is resistant to collision forgery under active attacks, the above construction of HTSS is existential unforgeable under adaptive chosen message attacks.*

Theorem 2 (Indistinguishability). *If the HIBCH scheme Π is forgery indistinguishable, the following distributions $\mathcal{D}_{Sanitize}$ and $\mathcal{D}_{Sign}$ are indistinguishable for all sufficiently large λ, any hierarchy depth ℓ that is polynomial in λ, any $(pk, sk) \leftarrow Setup(\lambda, \ell)$, any hierarchial sanitizable description ADM, messages m, m' such that $m_i = m'_i$ for all $i \notin I$, where I is extracted from ADM and is a set of indices $I \subseteq [1, L]$ that are sanitizable, any identity $ID = (ID_1, \ldots, ID_k)$ that matches ADM and the trapdoor $sk_{ID_{|k-1}}$ of the parent identity $ID_{|k-1} = (ID_1, \ldots, ID_{k-1})$:*

$$\begin{aligned}\mathcal{D}_{Sanitize} &= \{(m', \hat{\sigma}) | \sigma \leftarrow Sign(m, ADM, sk), sk_{ID} \leftarrow Trapdoor(m, ADM, \sigma, ID, sk_{ID_{|k-1}}), \\ &\qquad \hat{\sigma} \leftarrow Sanitize(pk, m, ADM, \sigma, m', ID, sk_{ID})\}_{\lambda, pk, ID, ADM}, \\ \mathcal{D}_{Sign} &= \{(m', \sigma') | \sigma' \leftarrow Sign(m', ADM, sk)\}_{\lambda, pk, ID, ADM}.\end{aligned}$$

5 Key-Exposure Free IBCH from HIBCH

As mentioned in [2,3], a key-exposure free chameleon hash scheme can be obtained from an IBCH scheme. In the construction of key-exposure free chameleon hash from IBCH, each transaction uses a different public key (corresponding to a different private key), so that a forgery only results in the user recovering the trapdoor information associated with a single transaction. The transaction-specific public key, called *customized identity*, is computed from special strings that describe the transaction. Based on the same idea, we show that a 2-HIBCH can be used to construct a key-exposure free IBCH scheme.

In this section, we first review the notion of key exposure freeness. Then, we describe the generic construction of key-exposure free IBCH from a two-level HIBCH formally.

Key Exposure Freeness: An identity-based chameleon hash scheme is key-exposure free if, for any PPT adversary $\mathcal{A}$, for all sufficiently large λ, any $(pk, sk) \leftarrow \mathsf{Setup}(\lambda)$, the probability that, $\mathcal{A}(\lambda, pk)$ outputs $(\mathsf{ID}, \mathcal{L}, m, r, m', r')$, satisfying $\mathsf{Hash}(pk, \mathsf{ID}, \mathcal{L}, m, r) = \mathsf{Hash}(pk, \mathsf{ID}, \mathcal{L}, m', r')$ and $m' \neq m$, is negligible. $\mathcal{A}$ is allowed to query $\mathcal{O}_{IBCH}^{\mathsf{Forge}}$[4] oracle on the adaptively chosen tuples $(\mathsf{ID}, \mathcal{L}_i, m_i, r_i, m_i')$, except that $\mathcal{L}_i$ must be different from the target *customized identity* $\mathcal{L}$.

Now, given an HIBCH scheme $\Pi = (\Pi.\mathsf{Setup}, \Pi.\mathsf{Extract}, \Pi.\mathsf{Hash}, \Pi.\mathsf{Forge})$, we define the 4-tuple algorithms (Setup, Extract, Hash, Forge) of an IBCH scheme as follows:

Setup Given a security parameter λ, PKG first runs

$$(pk, sk) \leftarrow \Pi.\mathsf{Setup}(\lambda, 2).$$

Then, it publishes the public key pk and keeps the private key sk secret.

Extract Given the private key sk and an identity ID, it first runs

$$sk_{\mathsf{ID}} \leftarrow \Pi.\mathsf{Extract}(sk, \mathsf{ID}).$$

Then, it outputs the trapdoor information sk_{ID} associated with the identity.

Hash Given the public key pk, an identity ID and a message m, it first computes the *customized identity* $\mathcal{L}$ for this transaction, and chooses a randomness r. Then, it sets a 2-level identity $\widetilde{\mathsf{ID}} = (\mathsf{ID}, \mathcal{L})$ and runs

$$h \leftarrow \Pi.\mathsf{Hash}(pk, \widetilde{\mathsf{ID}}, m, r).$$

Finally, it outputs the hash value h.

Forge Given an identity ID, the trapdoor information sk_{ID} association with ID, the hash value h on a message m with *customized identity* $\mathcal{L}$ and randomness r, and a new message m', it first sets a 2-level identity $\widetilde{\mathsf{ID}} = (\mathsf{ID}, \mathcal{L})$ and runs

[4] When the adversary $\mathcal{A}$ queries $\mathcal{O}_{IBCH}^{\mathsf{Forge}}$ on $(\mathsf{ID}, \mathcal{L}_i, m_i, r_i, m_i')$, the simulator gives the randomness r_i' to $\mathcal{A}$ such that $\mathsf{Hash}(pk, \mathsf{ID}, \mathcal{L}_i, m_i, r_i) = \mathsf{Hash}(pk, \mathsf{ID}, \mathcal{L}_i, m_i', r_i')$.

$$sk_{\widetilde{\mathsf{ID}}} \leftarrow \Pi.\mathsf{Extract}(sk_{\mathsf{ID}}, \widetilde{\mathsf{ID}}).$$

Then it runs

$$r' \leftarrow \Pi.\mathsf{Forge}(sk_{\widetilde{\mathsf{ID}}}, \widetilde{\mathsf{ID}}, m, r, h, m'),$$

and outputs r'.

It is obvious that, if the HIBCH scheme Π satisfies correctness, the above IBCH scheme also satisfies correctness, and if the HIBCH scheme Π is resistant to collision forgery under active attacks and semantically secure, so is the above IBCH scheme. Next, we prove that the above IBCH scheme is key-exposure free.

Theorem 3. *If the HIBCH scheme Π is resistant to collision forgery under active attacks, the above IBCH scheme is key-exposure free.*

Proof. To prove this theorem, we will show that, given pk, if a PPT adversary $\mathcal{A}$ can output $(\mathsf{ID}, \mathcal{L}, m, r, m', r')$ such that $\mathsf{Hash}(pk, \mathsf{ID}, \mathcal{L}, m, r) = \mathsf{Hash}(pk, \mathsf{ID}, \mathcal{L}, m', r')$ and $m' \neq m$, we can construct another algorithm $\mathcal{B}$, which is a forger against the HIBCH scheme Π.

Given a public key pk of the HIBCH scheme Π, using $\mathcal{A}$ as a sub-routine, $\mathcal{B}$ simulates a forger against the HIBCH scheme Π. First, $\mathcal{B}$ sends pk to $\mathcal{A}$. When $\mathcal{A}$ issues $\mathcal{O}^{\mathsf{Forge}}_{IBCH}$ oracle queries on the adaptively chosen tuples $(\mathsf{ID}, \mathcal{L}_i, m_i, r_i, m'_i)$, $\mathcal{B}$ sets $\widetilde{\mathsf{ID}}_i = (\mathsf{ID}, \mathcal{L}_i)$ and queries its $\mathcal{O}^{\mathsf{Extract}}_{HIBCH}$ oracle on $\widetilde{\mathsf{ID}}_i$ to obtain the trapdoor information $sk_{\widetilde{\mathsf{ID}}_i}$; then $\mathcal{B}$ computes $\Pi.\mathsf{Hash}(pk, \widetilde{\mathsf{ID}}_i, m_i, r_i) = h_i$ and $\Pi.\mathsf{Forge}(sk_{\widetilde{\mathsf{ID}}_i}, \widetilde{\mathsf{ID}}_i, m_i, r_i, h_i, m'_i) = r'_i$ and sends r'_i to $\mathcal{A}$. Finally, $\mathcal{A}$ outputs $(\mathsf{ID}, \mathcal{L}, m, r, m', r')$ such that $\mathsf{Hash}(pk, \mathsf{ID}, \mathcal{L}, m, r) = \mathsf{Hash}(pk, \mathsf{ID}, \mathcal{L}, m', r')$. $\mathcal{B}$ also outputs $(\widetilde{\mathsf{ID}} = (\mathsf{ID}, \mathcal{L}), m, r, m', r')$, which is a collision against the HIBCH scheme Π.

6 Construction of HIBCH

In this section, based on multivariate polynomials, we propose a concrete construction of HIBCH, which is t-threshold collusion-resistant. Blundo et al. [6] first used multivariate polynomials to construct key distribution schemes, and Gennaro et al. [15] extended their schemes for hierarchical systems.

In our construction, the private key of the HIBCH scheme is a random multivariate polynomial $f(x_1, \ldots, x_\ell)$, where the degree of x_i is a threshold parameter t. The trapdoor information of an identity $\mathsf{ID} = (\mathsf{ID}_1, \ldots, \mathsf{ID}_k)$ is $f(\mathsf{ID}_1, \ldots, \mathsf{ID}_{k-1}, \mathsf{ID}_k, x_{k+1}, \ldots, x_\ell)$, which can be derived from his parent's trapdoor information $f(\mathsf{ID}_1, \ldots, \mathsf{ID}_{k-1}, x_k, \ldots, x_\ell)$. Blundo et al. [6] proved that, if an adversary colludes with at most t entities in each depth of the hierarchy, the multivariate polynomial $f(x_1, \ldots, x_\ell)$ can still be kept secret.

In fact, if we choose a random multivariate polynomial $f(x_1, \ldots, x_\ell)$ as the private key, where the degree of x_i is t_i, our HIBCH scheme is resilient against the adversary who colludes with at most t_i entities at depth i of the hierarchy.

The scheme consists of the following algorithms:

Setup Given a security parameter λ and the maximum hierarchy depth ℓ, it first generates a bilinear map group system $\langle p, \mathbb{G}, \mathbb{G}_T, e\rangle$. Then it chooses a random polynomial (over $\mathbb{Z}_p$) $f(x_1, \ldots, x_\ell) = a_{t,t,\ldots,t}x_1^t x_2^t \cdots x_\ell^t + a_{t-1,t,\ldots,t}x_1^{t-1}x_2^t \cdots x_\ell^t + \cdots + a_{0,0,\ldots,0}$, where the degree of x_i is the threshold parameter t. Next, it chooses a generator g of $\mathbb{G}$ and an idle identity $\overline{\mathsf{ID}} \in \mathbb{Z}_p$. Finally, it chooses a cryptographic hash function $H : \{0,1\}^* \to \mathbb{G}$. The published public key is

$$pk = (p, \mathbb{G}, \mathbb{G}_T, e,\ g,\ H,\ \overline{\mathsf{ID}},\ g^{a_{t,\ldots,t}}, \ldots, g^{a_{0,\ldots,0}}),$$

and the private key is $sk = f(x_1, \ldots, x_\ell)$.

Extract Given an identity $\mathsf{ID} = (\mathsf{ID}_1, \ldots, \mathsf{ID}_k)$ of depth $k \le \ell$, and the trapdoor information $sk_{\mathsf{ID}_{|k-1}} = f(\mathsf{ID}_1, \ldots, \mathsf{ID}_{k-1}, x_k, \ldots, x_\ell)$ of the parent identity $\mathsf{ID}_{|k-1} = (\mathsf{ID}_1, \ldots, \mathsf{ID}_{k-1})$ at depth $k-1$, it first checks whether $\mathsf{ID}_i \neq \overline{\mathsf{ID}}$ for $1 \le i \le k$. If not, it outputs $\perp$, denoted an error. Otherwise, it computes

$$sk_{\mathsf{ID}} = f(\mathsf{ID}_1, \ldots, \mathsf{ID}_{k-1}, \mathsf{ID}_k, x_{k+1}, \ldots, x_\ell),$$

and outputs the trapdoor information sk_{ID} for identity ID.

Note that, if $k = 1$, the trapdoor information $sk_{\mathsf{ID}_{|k-1}}$ of the identity $\mathsf{ID}_{|k-1}$ is sk.

Hash Given pk, an identity $\mathsf{ID} = (\mathsf{ID}_1, \ldots, \mathsf{ID}_k)$ of depth $k \le \ell$, and a message $m \in \{0,1\}^*$, it first chooses a randomness $R \in \mathbb{G}$ uniformly. Then it computes

$$h = e(R, g) \cdot e(H(m), g^{f(\mathsf{ID}_1, \ldots, \mathsf{ID}_k, \overline{\mathsf{ID}}, \ldots, \overline{\mathsf{ID}})}),$$

and outputs the hash value h.

Note that, given $g^{a_{t,\ldots,t}}, \ldots, g^{a_{0,\ldots,0}}$, one can compute $g^{f(\mathsf{ID}_1, \ldots, \mathsf{ID}_k, \overline{\mathsf{ID}}, \ldots, \overline{\mathsf{ID}})}$.

Forge Given an identity $\mathsf{ID} = (\mathsf{ID}_1, \ldots, \mathsf{ID}_k)$, the trapdoor information $sk_{\mathsf{ID}} = f(\mathsf{ID}_1, \ldots, \mathsf{ID}_k, x_{k+1}, \ldots, x_\ell)$ associated with ID, the hash value h on a message $m \in \{0,1\}^*$ with randomness R, and a new message $m' \in \{0,1\}^*$, it first computes $f(\mathsf{ID}_1, \ldots, \mathsf{ID}_k, \overline{\mathsf{ID}}, \ldots, \overline{\mathsf{ID}})$ using sk_{ID}. Then it computes

$$R' = R \cdot (H(m) \cdot H(m')^{-1})^{f(\mathsf{ID}_1, \ldots, \mathsf{ID}_k, \overline{\mathsf{ID}}, \ldots, \overline{\mathsf{ID}})},$$

and outputs the randomness R'.

Note that

$$\begin{aligned}
\mathsf{Hash}(pk, \mathsf{ID}, m', R') &= e(R', g) \cdot e(H(m'), g^{f(\mathsf{ID}_1, \ldots, \mathsf{ID}_k, \overline{\mathsf{ID}}, \ldots, \overline{\mathsf{ID}})}) \\
&= e(R \cdot (H(m) \cdot H(m')^{-1})^{f(\mathsf{ID}_1, \ldots, \mathsf{ID}_k, \overline{\mathsf{ID}}, \ldots, \overline{\mathsf{ID}})}, g) \\
&\quad \cdot e(H(m'), g^{f(\mathsf{ID}_1, \ldots, \mathsf{ID}_k, \overline{\mathsf{ID}}, \ldots, \overline{\mathsf{ID}})}) \\
&= e(R, g) \cdot e(H(m), g^{f(\mathsf{ID}_1, \ldots, \mathsf{ID}_k, \overline{\mathsf{ID}}, \ldots, \overline{\mathsf{ID}})}) \\
&= \mathsf{Hash}(pk, \mathsf{ID}, m, R).
\end{aligned}$$

So, the above construction of HIBCH satisfies *correctness*. We now state the security theorems of the above HIBCH scheme. The proofs of the security theorems will be given in the full version of the paper.

Theorem 4. *In the random oracle model [5], the above construction of HIBCH is t-threshold resistant to collision forgery under active attacks.*

Theorem 5. *The above construction of HIBCH is semantically secure.*

7 Conclusions

In this paper, we introduced the notion of HIBCH, which is a hierarchical extension of IBCH. We showed that HIBCH can be used to construct other cryptographic primitives, including HTSS, which is a hierarchical extension of TSS, and key-exposure free IBCH, even the HIBCH is not key-exposure free. Finally, we proposed a concrete construction of HIBCH, which is t-threshold collision-resistant. A future direction is to find other constructions of HIBCH, which are fully collusion-resistant and key-exposure free.

Acknowledgement

We are grateful to the anonymous reviewers for their helpful comments. This research is supported in part by A*STAR SERC Grant No. 102 101 0027 in Singapore. Yunlei Zhao is partly supported by a grant from the Major State Basic Research Development (973) Program of China (No. 2007CB807901) and a grant from the National Natural Science Foundation of China NSFC (No. 61070248) and the QiMingXing Program of Shanghai.

References

1. Ateniese, G., Chou, D.H., de Medeiros, B., Tsudik, G.: Sanitizable Signatures. In: di Vimercati, S.d.C., Syverson, P.F., Gollmann, D. (eds.) ESORICS 2005. LNCS, vol. 3679, pp. 159–177. Springer, Heidelberg (2005)
2. Ateniese, G., de Medeiros, B.: Identity-based chameleon hash and applications. In: Juels, A. (ed.) FC 2004. LNCS, vol. 3110, pp. 164–180. Springer, Heidelberg (2004)
3. Ateniese, G., de Medeiros, B.: On the key exposure problem in chameleon hashes. In: Blundo, C., Cimato, S. (eds.) SCN 2004. LNCS, vol. 3352, pp. 165–179. Springer, Heidelberg (2005)
4. Bellare, M., Goldreich, O., Goldwasser, S.: Incremental cryptography: The case of hashing and signing. In: Desmedt, Y.G. (ed.) CRYPTO 1994. LNCS, vol. 839, pp. 216–233. Springer, Heidelberg (1994)
5. Bellare, M., Rogaway, P.: Random oracles are practical: A paradigm for designing efficient protocols. In: ACM Conference on Computer and Communications Security, pp. 62–73 (1993)
6. Blundo, C., De Santis, A., Herzberg, A., Kutten, S., Vaccaro, U., Yung, M.: Perfectly secure key distribution for dynamic conferences. Inf. Comput. 146(1), 1–23 (1998)
7. Brzuska, C., Fischlin, M., Freudenreich, T., Lehmann, A., Page, M., Schelbert, J., Schröder, D., Volk, F.: Security of sanitizable signatures revisited. In: Jarecki, S., Tsudik, G. (eds.) PKC 2009. LNCS, vol. 5443, pp. 317–336. Springer, Heidelberg (2009)

8. Canard, S., Laguillaumie, F., Milhau, M.: Trapdoor sanitizable signatures and their application to content protection. In: Bellovin, S.M., Gennaro, R., Keromytis, A.D., Yung, M. (eds.) ACNS 2008. LNCS, vol. 5037, pp. 258–276. Springer, Heidelberg (2008)
9. Chaum, D., van Antwerpen, H.: Undeniable signatures. In: Brassard, G. (ed.) CRYPTO 1989. LNCS, vol. 435, pp. 212–216. Springer, Heidelberg (1990)
10. Chen, X., Zhang, F., Kim, K.: Chameleon hashing without key exposure. In: Zhang, K., Zheng, Y. (eds.) ISC 2004. LNCS, vol. 3225, pp. 87–98. Springer, Heidelberg (2004)
11. Chen, X., Zhang, F., Susilo, W., Tian, H., Li, J., Kim, K.: Identity-based chameleon hash scheme without key exposure. In: Steinfeld, R., Hawkes, P. (eds.) ACISP 2010. LNCS, vol. 6168, pp. 200–215. Springer, Heidelberg (2010)
12. Chen, X., Zhang, F., Tian, H., Wei, B., Kim, K.: Key-exposure free chameleon hashing and signatures based on discrete logarithm systems. Cryptology ePrint Archive, Report 2009/035 (2009), `http://eprint.iacr.org/`
13. Gao, W., Li, F., Wang, X.: Chameleon hash without key exposure based on schnorr signature. Computer Standards & Interfaces 31(2), 282–285 (2009)
14. Gao, W., Wang, X., Xie, D.: Chameleon hashes without key exposure based on factoring. J. Comput. Sci. Technol. 22(1), 109–113 (2007)
15. Gennaro, R., Halevi, S., Krawczyk, H., Rabin, T., Reidt, S., Wolthusen, S.D.: Strongly-resilient and non-interactive hierarchical key-agreement in mANETs. In: Jajodia, S., Lopez, J. (eds.) ESORICS 2008. LNCS, vol. 5283, pp. 49–65. Springer, Heidelberg (2008)
16. Goldwasser, S., Micali, S., Rivest, R.L.: A digital signature scheme secure against adaptive chosen-message attacks. SIAM J. Comput. 17(2), 281–308 (1988)
17. Izu, T., Kanaya, N., Takenaka, M., Yoshioka, T.: PIATS: A partially sanitizable signature scheme. In: Qing, S., Mao, W., López, J., Wang, G. (eds.) ICICS 2005. LNCS, vol. 3783, pp. 72–83. Springer, Heidelberg (2005)
18. Johnson, R., Molnar, D., Song, D.X., Wagner, D.: Homomorphic signature schemes. In: Preneel, B. (ed.) CT-RSA 2002. LNCS, vol. 2271, pp. 244–262. Springer, Heidelberg (2002)
19. Klonowski, M., Lauks, A.: Extended sanitizable signatures. In: Rhee, M.S., Lee, B. (eds.) ICISC 2006. LNCS, vol. 4296, pp. 343–355. Springer, Heidelberg (2006)
20. Krawczyk, H., Rabin, T.: Chameleon signatures. In: NDSS (2000)
21. Micali, S., Rivest, R.L.: Transitive signature schemes. In: Preneel, B. (ed.) CT-RSA 2002. LNCS, vol. 2271, pp. 236–243. Springer, Heidelberg (2002)
22. Miyazaki, K., Hanaoka, G., Imai, H.: Invisibly sanitizable digital signature scheme. IEICE Transactions 91-A(1), 392–402 (2008)
23. Miyazaki, K., Iwamura, M., Matsumoto, T., Sasaki, R., Yoshiura, H., Tezuka, S., Imai, H.: Digitally signed document sanitizing scheme with disclosure condition control. IEICE Transactions 88-A(1), 239–246 (2005)
24. Steinfeld, R., Bull, L., Zheng, Y.: Content extraction signatures. In: Kim, K.-c. (ed.) ICISC 2001. LNCS, vol. 2288, pp. 285–304. Springer, Heidelberg (2002)
25. Suzuki, T., Ramzan, Z., Fujimoto, H., Gentry, C., Nakayama, T., Jain, R.: A system for end-to-end authentication of adaptive multimedia content. In: Dittmann, J., Katzenbeisser, S., Uhl, A. (eds.) CMS 2005. LNCS, vol. 3677, pp. 237–249. Springer, Heidelberg (2005)
26. Tan, K.W., Deng, R.H.: Applying santitizable signature to web-service-enabled business processes: going beyond integrity protection. In: ICWS, pp. 67–74 (2009)

27. Yum, D.H., Seo, J.W., Lee, P.J.: Trapdoor sanitizable signatures made easy. In: Zhou, J., Yung, M. (eds.) ACNS 2010. LNCS, vol. 6123, pp. 53–68. Springer, Heidelberg (2010)
28. Zhang, F., Safavi-Naini, R., Susilo, W.: Id-based chameleon hashes from bilinear pairings. Cryptology ePrint Archive, Report 2003/208 (2003), http://eprint.iacr.org/

Efficient Generic Constructions of Signcryption with Insider Security in the Multi-user Setting

Daiki Chiba[1], Takahiro Matsuda[2], Jacob C.N. Schuldt[2], and Kanta Matsuura[1]

[1] The University of Tokyo, Japan
{chi-ba,kanta}@iis.u-tokyo.ac.jp
[2] Research Center for Information Security, AIST, Japan
{t-matsuda,jacob.schuldt}@aist.go.jp

Abstract. Signcryption is a primitive which provides the combined security properties of encryption and digital signatures i.e. confidentiality and unforgeability. A number of signcryption schemes have been presented in the literature, but up until now, no scheme which simultaneously achieves the currently strongest notions of insider confidentiality and strong insider unforgeability in the multi-user setting, has been proposed, without relying on random oracles or key registration. In this paper, we propose two new generic constructions of signcryption schemes from the combination of standard primitives and simple extensions of these. From our constructions, we instantiate a number of concrete and efficient signcryption schemes which satisfy the strongest notions of insider security in the multi-user setting while still being provably secure in the standard model.

Keywords: signcryption, insider security, multi-user setting, generic construction.

1 Introduction

Public key encryption and digital signatures aim at providing very different security properties. More specifically, encryption provides confidentiality of data whereas digital signatures provides authenticity of data. However, many practical applications, such as secure e-mail, require both properties simultaneously. To address this need, Zheng [25] proposed *signcryption* which is a single primitive aiming at efficiently providing the security guarantees of both encryption and digital signatures. Although the scheme proposed in [25] was not formally proved secure, this was done in subsequent works [5,6].

Since the introduction of the primitive, many signcryption schemes have been proposed, e.g. [25,4,5,18,19,13,6,17,23,21]. However, due to the many variations of the used security models, the security level achieved by these schemes vary. The simplest security model for signcryption, which was adopted in a few of the early papers [4,13], considers the so-called two-user setting in which only a single sender and a single receiver interact. However, as pointed out by Dent [13],

J. Lopez and G. Tsudik (Eds.): ACNS 2011, LNCS 6715, pp. 220–237, 2011.

security in the two-user setting does not imply security in the multi-user setting in which several senders address the same receiver, and several receivers receive messages from the same sender. Hence, to ensure security in a more realistic environment, a multi-user security model must be adopted. Furthermore, another aspect of the signcryption security definition, is the concept of "insider" and "outsider" attacks. While some security models require the adversary to attack an uncompromised sender and receiver pair (i.e. the key material of both parties are unknown to the adversary), others allow an "insider" attack in which the adversary has access to the key material of one of the parties. Since security against an insider attack implies security against an outsider attack, the former is preferred. The currently strongest security definitions which capture insider confidentiality and strong insider unforgeability in the multi-user setting, were first defined and used in [18]. For a more detailed overview of the used security models, see [21].

Up until now, several practical signcryption schemes which achieve insider confidentiality and strong insider unforgeability in the multi-user setting have been proposed [19,17,21], but these are only shown secure in the random oracle model. While a number of signcryption schemes which are secure in the standard model, e.g. [4,23,21], these do *not* achieve the same level of security as the random oracle model schemes. For example, An et al. [4] showed how to strengthen the traditional Encrypt-then-Sign and Sign-then-Encrypt schemes to achieve security in the multi-user setting. However, Encrypt-then-Sign construction only achieves the so-called *generalized chosen ciphertext security* which is strictly weaker than ordinary chosen ciphertext security. Sign-then-Encrypt construction achieves ordinary chosen ciphertext security for confidentiality while it only achieves weak unforgeability for authenticity. While weak unforgeability might be sufficient in some scenarios, strong insider unforgeability guarantees that a ciphertext is non-malleable, even for a malicious receiver, and might be needed when the signcryption scheme is used as a building block in a higher level protocol.

Tan [23] proposed a scheme which achieves insider confidentiality in the strongest sense, but the strong insider unforgeability of his scheme is only achieved if *key registration* is used i.e. the adversary is required to reveal the private key corresponding to any public key he makes use of when trying to attack the scheme. In practice, this assumption requires that a traditional public-key infrastructure (PKI) is employed and that all parties engage in a zero-knowledge proof with the certificate authority (CA), which proves knowledge of their private key, before they obtain a certificate for their public key. However, executing these proofs places a heavy burden on the CA, and this type of registration is not used in most practical systems.

Matsuda et al. [21] showed generic constructions of signcryption schemes from existing basic primitives and their simple extensions such as tag-based encryption (TBE) [20,16]. Although their generic constructions were shown to achieve insider confidentiality in the strongest sense, their constructions only achieve either weak insider unforgeability or strong insider unforgeability under the assumption that key registration is used.

To the best our knowledge, there is no standard model signcryption scheme which achieves both insider confidentiality and strong insider unforgeability in the multi-user setting, without relying on key registration. The main motivation of this paper is to construct such signcryption schemes.

Our Contribution. In this paper, we propose two new generic constructions of simple but efficient signcryption schemes. Instantiations of our constructions are the first to achieve strong insider unforgeability and insider confidentiality without relying on random oracles or key registration.

Our first construction makes use of an IND-tag-CCA secure tag-based key encapsulation mechanism (TBKEM), an IND-CCA secure data encapsulation mechanism (DEM) which has a one-to-one property, and a sUF-CMA secure signature scheme. Note that TBKEMs are fairly easy to construct from already existing primitives, and many efficient concrete constructions are possible (see Section 2.2). Our second construction makes use of an IND-CCA secure key encapsulation mechanism (KEM), a one-time secure DEM with a one-to-one property, a one-time secure message authentication code (MAC) with a one-to-one property, and a sUF-CMA secure signature scheme. Both of our constructions are based on the ideas related to the well-known Sign-then-Encrypt approach, but they exploit the functionality of tag-based primitives and hybrid encryption (KEM/DEM approach) to overcome the limitation of the previous approaches (see Section 4 for more details).

From these two constructions, we instantiate a number of efficient concrete signcryption schemes which are insider secure in the multi-user setting (in the standard model), and compare these with the existing standard model schemes (see Section 5 for details). We emphasize that the advantage of the above constructions lies not only in the efficiency and security properties achieved by our concrete instantiations, but also in being generic constructions, which allows us to make use of any present or future instantiation of the underlying primitives and the established security results for these.

2 Preliminaries

In this section, we review the notation used throughout the paper and the definitions of the primitives that will be used in our first construction. The additional primitives needed for our second construction, which includes a KEM and a MAC, will be given in Appendix A.

2.1 Notation

In this paper, "$x \leftarrow y$" denotes that x is chosen uniformly at random from y if y is a finite set, x is output from y if y is a function or an algorithm, or y is assigned to x otherwise. "$x||y$" denotes a concatenation of x and y. "PPT" denotes probabilistic polynomial time. "κ" always denotes the security parameter. We say that a function $f(\kappa)$ is negligible in κ if $f(\kappa) \leq 1/p(\kappa)$ for any positive polynomial $p(\kappa)$ and all sufficiently large κ. In this paper, when we say a function is negligible then we always mean that it is negligible in the security parameter κ.

2.2 Tag-Based Key Encapsulation Mechanism

A tag-based key encapsulation mechanism (TBKEM) is a key encapsulation mechanism whose encapsulation and decapsulation algorithms take an arbitrary string called *tag* as an additional input. Note that a TBKEM is a KEM-analogue of tag-based encryption (TBE) [20,16], and must not be confused with a tag-KEM, which is a building block of hybrid encryption proposed by Abe et al. [2].

Formally, a TBKEM is given by the following four algorithms.

TSetup: Given input a security parameter 1^κ, this algorithm returns a set of public parameters prm (included in prm is a description of a key space $\mathcal{K}$).
TKG: Given input prm, this algorithm returns a public/private key pair (pk, sk).
TEncap: Given input prm, a public key pk and a tag tag, this algorithm returns an encapsulation and encapsulated key pair (c, K).
TDecap: Given input prm, a private key sk, a tag tag and an encapsulation c, this algorithm returns a key K or an error symbol $\perp$.

It is required for all $prm \leftarrow \mathsf{TSetup}(1^\kappa)$, all $(pk, sk) \leftarrow \mathsf{TKG}(prm)$, all tags tag, and all $(c, K) \leftarrow \mathsf{TEncap}(prm, pk, \mathsf{tag})$, that $K = \mathsf{TDecap}(prm, sk, \mathsf{tag}, c)$

IND-tag-CCA Security. For a TBKEM, indistinguishability against adaptive tag and adaptive chosen ciphertext attacks (IND-tag-CCA) is defined by the following game between an adversary $\mathcal{A}$ and an IND-tag-CCA challenger $\mathcal{CH}$.

Setup. $\mathcal{CH}$ computes $prm \leftarrow \mathsf{TSetup}(1^\kappa)$ and $(pk, sk) \leftarrow \mathsf{TKG}(prm)$, and then forwards (prm, pk) to $\mathcal{A}$ and keeps sk to itself.
Phase 1. $\mathcal{A}$ can adaptively submit decapsulation queries (tag, c) to $\mathcal{CH}$. $\mathcal{CH}$ responds to each query by returning $K \leftarrow \mathsf{TDecap}(prm, sk, \mathsf{tag}, c)$.
Challenge. $\mathcal{A}$ chooses a challenge tag tag^* and sends this to $\mathcal{CH}$. $\mathcal{CH}$ computes $(c^*, K_1^*) \leftarrow \mathsf{TEncap}(pk, \mathsf{tag}^*)$, and chooses $K_0^* \in \mathcal{K}$ uniformly at random. Then $\mathcal{CH}$ flips a fair coin $b \in \{0, 1\}$, and returns the challenge encapsulation and key (c^*, K_b^*) to $\mathcal{A}$.
Phase 2. $\mathcal{A}$ can submit decapsulation queries in the same way as in Phase 1, except that $\mathcal{A}$ is not allowed to submit the challenge pair (tag^*, c^*).
Guess. $\mathcal{A}$ outputs a bit b' as its guess for b.

We define the IND-tag-CCA advantage of $\mathcal{A}$ attacking the TBKEM TK as $\mathsf{Adv}_{TK,\mathcal{A}}^{\text{IND-tag-CCA}} = |\Pr[b' = b] - \frac{1}{2}|$.

Definition 1. *We say that a TBKEM TK is IND-tag-CCA secure if* $\mathsf{Adv}_{TK,\mathcal{A}}^{\text{IND-tag-CCA}}$ *is negligible for any PPT adversary* $\mathcal{A}$.

How to construct TBKEMs. Although one might think that a TBKEM is not a basic primitive, any IND-CCA secure public key encryption scheme can be converted into a IND-tag-CCA secure TBKEM by encrypting a random session key together with the tag [16]. Furthermore, any IND-CCA secure tag-KEM can be used as an IND-tag-CCA secure TBKEM, which allows the many practical constructions of tag-KEMs to be used (e.g. [2,1]). Lastly, Abe et al. [2] show that IND-CCA secure tag-KEMs can be generically built from any IND-CCA secure ordinary KEM and a one-time secure MAC, which implies that this technique can be used for the construction of IND-tag-CCA secure TBKEMs as well.

2.3 Data Encapsulation Mechanism

A data encapsulation mechanism (DEM) is given by the following two algorithms.

DEnc: Given input a symmetric key $K \in \mathcal{K}$ and a message m, this algorithm returns a ciphertext c. $\mathcal{K}$ is a key space.

DDec: Given input symmetric key $K \in \mathcal{K}$ and a ciphertext c, this algorithm returns a message m or an error symbol $\perp$.

It is required for all $K \in \mathcal{K}$ and all messages m that $\mathsf{DDec}(K, \mathsf{DEnc}(K, m)) = m$.

IND-CCA Security. For a DEM, indistinguishability against adaptive chosen ciphertext attacks (IND-CCA) is defined by the following game between an adversary $\mathcal{A}$ and an IND-CCA challenger $\mathcal{CH}$.

Setup. $\mathcal{CH}$ chooses $K \in \mathcal{K}$ uniformly at random, where $\mathcal{K}$ is a key space.

Phase 1. $\mathcal{A}$ can adaptively submit decryption queries c to $\mathcal{CH}$. $\mathcal{CH}$ responds to each query by returning $m \leftarrow \mathsf{DDec}(K, c)$.

Challenge. $\mathcal{A}$ chooses two plaintexts (m_0, m_1) of equal length, and sends them to $\mathcal{CH}$. $\mathcal{CH}$ flips a fair coin $b \in \{0, 1\}$, and then returns the challenge ciphertext $c^* \leftarrow \mathsf{DEnc}(K, m_b)$ to $\mathcal{A}$.

Phase 2. $\mathcal{A}$ can submit decryption queries in the same way as in Phase 1, except that $\mathcal{A}$ is not allowed to submit the challenge ciphertext c^*.

Guess. $\mathcal{A}$ outputs a bit b' as its guess for b.

Note that, in the above game, $\mathcal{A}$ is not allowed to make any encryption queries i.e. the above defined IND-CCA security for a DEM is strictly weaker, and therefore easier to achieve, than ordinary IND-CCA security for symmetric encryption.

We define the IND-CCA advantage of $\mathcal{A}$ attacking the DEM D as $\mathsf{Adv}_{D,\mathcal{A}}^{\text{IND-CCA}} = |\Pr[b' = b] - \frac{1}{2}|$.

Furthermore, we define indistinguishability against one time attacks (IND-OT) by an IND-OT game which is defined in the same way as the IND-CCA game except that the adversary is not allowed to submit any decryption queries. The advantage $\mathsf{Adv}_{D,\mathcal{A}}^{\text{IND-OT}}$ of an adversary attacking the IND-OT security of a DEM D is defined similarly to the IND-CCA advantage.

Definition 2. *We say that a DEM D is IND-CCA (resp. IND-OT) secure if $\mathsf{Adv}_{D,\mathcal{A}}^{\text{IND-CCA}}$ (resp. $\mathsf{Adv}_{D,\mathcal{A}}^{\text{IND-OT}}$) is negligible for any PPT adversary $\mathcal{A}$.*

One-to-One Property. A DEM is said to be *one-to-one* if given a key K and a plaintext m, there is at most one c such that $\mathsf{DDec}(K, c) = m$. We consider this property to be quite natural for a DEM, and all IND-CCA secure DEMs based on strong pseudorandom permutations [22] satisfy this. Furthermore, the well-known Encrypt-then-MAC [7] construction of IND-CCA secure DEMs preserves the one-to-one property of the underlying (IND-OT secure) DEM and MAC i.e. if both the used DEM and MAC are one-to-one, then so is the DEM resulting from a Encrypt-then-MAC construction (the definition of the one-to-one property of a MAC is given in Section A.2). We are not aware of any natural construction of a DEM which does not have this property.

2.4 Signature

A signature scheme is given by the following four algorithms.

SSetup: Given input a security parameter 1^κ, this algorithm returns a set of public parameters prm.
SKG: Given input prm, this algorithm returns a public/private key pair (pk, sk).
Sign: Given input prm, private key sk and a message m, this algorithm returns a signature σ.
SVer: Given input prm, a public key pk, a message m, and a signature σ, this algorithm returns $\top$ or $\bot$.

It is required for all $prm \leftarrow \mathsf{SSetup}(1^\kappa)$, all $(pk, sk) \leftarrow \mathsf{SKG}(prm)$ and all messages m, that $\mathsf{SVer}(prm, pk, m, \mathsf{Sign}(prm, sk, m)) = \top$

sUF-CMA Security. For a signature scheme, strong unforgeability against chosen message attacks (sUF-CMA) is defined by the following game between an adversary $\mathcal{A}$ and a sUF-CMA challenger $\mathcal{CH}$.

Setup. $\mathcal{CH}$ computes $prm \leftarrow \mathsf{SSetup}(1^\kappa)$ and $(pk, sk) \leftarrow \mathsf{SKG}(prm)$, and generates an empty list $\mathcal{L}$ into which message/signature pairs will be stored. $\mathcal{CH}$ then forwards (prm, pk) to $\mathcal{A}$ and keeps sk to itself.
Query. $\mathcal{A}$ can adaptively submit signature queries m to $\mathcal{CH}$. $\mathcal{CH}$ responds to each query by returning $\sigma \leftarrow \mathsf{Sign}(prm, sk, m)$. $\mathcal{CH}$ furthermore stores the message/signature pair (m, σ) in $\mathcal{L}$.
Output. $\mathcal{A}$ outputs a message/signature pair (m^*, σ^*).

We define the sUF-CMA advantage of $\mathcal{A}$ attacking the signature scheme S as follows:

$$\mathsf{Adv}_{S,\mathcal{A}}^{\text{sUF-CMA}} = \Pr[\mathsf{SVer}(prm, pk, m^*, \sigma^*) = \top \wedge (m^*, \sigma^*) \notin \mathcal{L}]$$

Definition 3. *We say that a signature S is sUF-CMA secure if $\mathsf{Adv}_{S,\mathcal{A}}^{\text{sUF-CMA}}$ is negligible for any PPT adversary $\mathcal{A}$.*

3 Signcryption

In this section we review the formal definition of a signcryption scheme.

Algorithms. A signcryption scheme is given by the following five algorithms.

Setup: Given input a security parameter 1^κ, this algorithm returns a set of public parameters prm.

$\mathsf{KeyGen_S}$: This is the sender's key generation algorithm which takes prm as input and returns a public/private sender key pair (pk_S, sk_S).

$\mathsf{KeyGen_R}$: This is the receiver's key generation algorithm which takes prm as input and returns a public/private receiver key pair (pk_R, sk_R).

SC: This is the signcryption algorithm which takes prm, a private sender key sk_S, a public receiver key pk_R, and a message m as input, and returns a ciphertext c.

USC: This is the unsigncryption algorithm which takes prm, a public sender key pk_S, a private receiver key sk_R, and a ciphertext c as input, and returns either a message m or an error symbol $\perp$.

It is required for all $prm \leftarrow \mathsf{Setup}(1^\kappa)$, all $(pk_S, sk_S) \leftarrow \mathsf{KeyGen_S}(prm)$, all $(pk_R, sk_R) \leftarrow \mathsf{KeyGen_R}(prm)$, all messages m and all $c \leftarrow \mathsf{SC}(prm, sk_S, pk_R, m)$, that $m = \mathsf{USC}(prm, pk_S, sk_R, c)$.

3.1 Security

In this subsection, we review the security model in which we will prove our signcryption constructions secure using the notations introduced in [21]. More specifically, for confidentiality, we will use the notion indistinguishability against insider chosen ciphertext attacks in the dynamic multi-user model (*dM-IND-iCCA*), and for unforgeability, we use the notion strong unforgeability against insider chosen message attacks in the dynamic multi-user model (*dM-sUF-iCMA*).

dM-IND-iCCA Security. For a signcryption scheme, dM-IND-iCCA security is defined by the following game between an adversary $\mathcal{A}$ and a dM-IND-iCCA challenger $\mathcal{CH}$.

Setup. $\mathcal{CH}$ computes $prm \leftarrow \mathsf{Setup}(1^\kappa)$ and $(pk_R, sk_R) \leftarrow \mathsf{KeyGen_R}(prm)$, forwards (prm, pk_R) to $\mathcal{A}$, and keeps sk_R to itself.

Phase 1. $\mathcal{A}$ can adaptively submit unsigncryption queries (pk_S, c) to $\mathcal{CH}$. $\mathcal{CH}$ responds to each query by returning $m \leftarrow \mathsf{USC}(prm, pk_S, sk_R, c)$ to $\mathcal{A}$.

Challenge. $\mathcal{A}$ chooses two plaintexts (m_0, m_1) of equal length and a challenge sender public/private key pair (pk_S^*, sk_S^*), and sends these to $\mathcal{CH}$. It is required that (pk_S^*, sk_S^*) is a valid key pair i.e. (pk_S^*, sk_S^*) lies in the range of $\mathsf{KeyGen_S}(prm)$. $\mathcal{CH}$ flips a fair coin $b \in \{0, 1\}$, and then returns the challenge ciphertext $c^* \leftarrow \mathsf{SC}(prm, pk_R, sk_S^*, m_b)$ to $\mathcal{A}$.

Phase 2. $\mathcal{A}$ can submit unsigncryption queries in the same way as in Phase 1, except that $\mathcal{A}$ is not allowed to submit the pair (pk_S^*, c^*).

Guess. $\mathcal{A}$ outputs a bit b' as its guess for b.

We define the dM-IND-iCCA advantage of $\mathcal{A}$ attacking signcryption SC as $\mathsf{Adv}_{SC,\mathcal{A}}^{\text{dM-IND-iCCA}} = |\Pr[b' = b] - \frac{1}{2}|$.

Definition 4. *We say that a signcryption scheme SC is dM-IND-iCCA secure if* $\mathsf{Adv}^{\text{dM-IND-iCCA}}_{SC,\mathcal{A}}$ *is negligible for any PPT adversary* $\mathcal{A}$.

Note that, in the dM-IND-iCCA security game, $\mathcal{A}$ can freely choose the sender keys submitted in the unsigncryption queries, and is not required to prove knowledge of or reveal the private keys corresponding to these (i.e. $\mathcal{A}$ is allowed to employ an attack strategy in which these private keys are unknown to $\mathcal{A}$). This will ensure security in the multi-user scenario in which a receiver decrypts ciphertexts from multiple senders. Furthermore, we would like to emphasize that the challenge sender key pair (pk_S^*, sk_S^*) is also freely chosen by $\mathcal{A}$, and in particular, $\mathcal{A}$ knows sk_S^*. This implies that confidentiality is maintained, even if the private sender key is compromised i.e. insider confidentially is guaranteed. This is currently the strongest security notion used in the signcryption literature, and is strictly stronger than the security notion used in a number of previous papers [5,4,13,6]. See [21] for a more detailed di scussion of the security models for signcryption.

dM-sUF-iCMA Security. For a signcryption scheme, dM-sUF-iCMA security is defined by the following game between an adversary $\mathcal{A}$ and a dM-sUF-iCMA challenger $\mathcal{CH}$.

Setup. $\mathcal{CH}$ computes $prm \leftarrow \mathsf{Setup}(1^\kappa)$ and $(pk_S, sk_S) \leftarrow \mathsf{KeyGen_S}(prm)$, and generates an empty list $\mathcal{L}$ into which the adversary's queries and answers will be stored. $\mathcal{CH}$ forwards (prm, pk_S) to $\mathcal{A}$ and keeps sk_S to itself.

Query. $\mathcal{A}$ can adaptively submit signcryption queries (pk_R, m) to $\mathcal{CH}$. $\mathcal{CH}$ responds to each query by returning $c \leftarrow \mathsf{SC}(prm, sk_S, pk_R, m)$. Furthermore, $\mathcal{CH}$ stores the tuple (pk_R, c, m) in $\mathcal{L}$.

Output. $\mathcal{A}$ outputs a tuple (pk_R^*, sk_R^*, c^*). It is required that (pk_R^*, sk_R^*) is a valid key pair.

We define the dM-sUF-iCMA advantage of $\mathcal{A}$ attacking signcryption SC as follows:

$$\mathsf{Adv}^{\text{dM-sUF-iCMA}}_{SC,\mathcal{A}} = \Pr[\mathsf{USC}(prm, pk_S^*, sk_R^*, c^*) = m^* \neq \perp \wedge (pk_R^*, c^*, m^*) \notin \mathcal{L}]$$

Definition 5. *We say that a signcryption scheme SC is dM-sUF-iCMA secure if* $\mathsf{Adv}^{\text{dM-sUF-iCMA}}_{SC,\mathcal{A}}$ *is negligible for any PPT adversary* $\mathcal{A}$.

Similar to the confidentiality definition, $\mathcal{A}$ can freely choose public receiver keys in the signcryption queries, and is not required to prove knowledge of or reveal the private keys corresponding to these i.e. $\mathcal{A}$ can observe signcryptions under multiple receiver keys as required for multi-user security. Furthermore, $\mathcal{A}$ can freely choose the challenge receiver key pair (pk_R^*, sk_R^*), and proving security in this model ensures insider unforgeability. Like dM-IND-iCCA, this is the strongest unforgeability notion used in the signcryption literature.

4 Proposed Generic Constructions

In this section, we show our generic constructions of signcryption schemes which achieve both insider confidentiality and strong insider unforgeability in the multi-user setting. The first construction is based on a TBKEM, a signature scheme

$\mathsf{Setup}(1^\kappa)$:
$prm_{tk} \leftarrow \mathsf{TSetup}(1^\kappa)$
$prm_{sig} \leftarrow \mathsf{SSetup}(1^\kappa)$
Output $prm \leftarrow (prm_{tk}, prm_{sig})$.

$\mathsf{SC}(prm, sk_S, pk_R, m)$:
$\mathsf{tag} \leftarrow pk_S$
$(c_1, K) \leftarrow \mathsf{TEncap}(prm_{tk}, pk_R, \mathsf{tag})$
$\sigma \leftarrow \mathsf{Sign}(prm_{sig}, sk_S, (m||c_1||pk_R))$
$c_2 \leftarrow \mathsf{DEnc}(K, (m||\sigma))$
Output $c \leftarrow (c_1, c_2)$

$\mathsf{KeyGen}_\mathsf{S}(prm)$:
Output $(pk_S, sk_S) \leftarrow \mathsf{SKG}(prm_{sig})$.
$\mathsf{KeyGen}_\mathsf{R}(prm)$:
Output $(pk_R, sk_R) \leftarrow \mathsf{TKG}(prm_{tk})$.
$\mathsf{USC}(prm, pk_S, sk_R, c)$:
Parse c as (c_1, c_2)
$\mathsf{tag} \leftarrow pk_S$
$K \leftarrow \mathsf{TDecap}(prm_{tk}, sk_R, \mathsf{tag}, c_1)$
(if $K = \perp$, then return $\perp$)
$(m||\sigma) \leftarrow \mathsf{DDec}(K, c_2)$
(if DDec outputs $\perp$, then return $\perp$)
If $\mathsf{SVer}(prm_{sig}, pk_S, (m||c_1||pk_R), \sigma) = \perp$
then return $\perp$
Output m

Fig. 1. Construction using TBKEM: SC_{tk}

and a DEM, and is given in Section 4.1. The second construction is based on a KEM, a signature scheme, a MAC and a DEM, and is given in Section 4.2.

4.1 Composition Using TBKEM

Let TK = (TSetup, TKG, TEncap, TDecap) be an IND-tag-CCA secure TBKEM, let D = (DEnc, DDec) be an IND-CCA secure DEM, and let S = (SSetup, SKG, Sign, SVer) be a sUF-CMA secure signature scheme. Then, we construct a signcryption scheme SC_{tk} as shown in Fig.1. For simplicity, we assume the symmetric key space of TK and that of D are $\{0,1\}^\kappa$.

Construction Idea. The proposed construction can be regarded as a variant of the Sign-then-Encrypt construction. However, as shown by the previous results [4,21], the ordinary Sign-then-Encrypt constructions cannot achieve strong insider unforgeability in the multi-user setting (dM-sUF-iCMA security), because an insider adversary who has a receiver's private key and receives a ciphertext from a signcryption query can decrypt the ciphertext and then re-encrypt it, which can be a successful forgery in the sense of strong unforgeability. However, in the proposed construction, we adopt the KEM/DEM approach and thus can sign the KEM ciphertext together with the message. Because of this, an insider adversary can no longer mount the above attack, and thus to break the strong unforgeability he has to modify the DEM ciphertext so that the whole ciphertext remains valid. However, such modification of the DEM ciphertext is impossible if the underlying DEM is one-to-one, which guarantees the dM-sUF-iCMA security of our construction. To achieve security in the multi-user setting, we make use of a similar idea to [21] and employ a tag-based KEM. More specifically, we sign the receiver's public key with a signature scheme, and use the sender's public key as a tag of the TBKEM. This will ensure that the ciphertext is only valid for a specific sender and receiver key pairs.

Security. The security of SC_{tk} is formally guaranteed by the following theorems.

Theorem 1. *If the TBKEM TK is IND-tag-CCA secure and the DEM D is IND-CCA secure, then the proposed signcryption scheme SC_{tk} is dM-IND-iCCA secure.*

Proof. Suppose there exists an adversary $\mathcal{A}$ that breaks the dM-IND-iCCA security of the signcryption scheme SC_{tk}. We then consider the following games. (Values used in the challenge phase are marked with an asterisk (*))

Game_0: This is the ordinary dM-IND-iCCA game regarding SC_{tk}.

Game_1: In this game, a uniformly random key $K' \in \mathcal{K}$ is chosen at the beginning of the game, and the second component c_2^* of the challenge ciphertext is computed by using this K' ($c_2^* \leftarrow \mathsf{DEnc}(K', (m||\sigma))$). Moreover, for an unsigncryption query of the form (pk_S^*, c_1^*, c_2) with $c_2 \neq c_2^*$, c_2 is decrypted using K'. The rest is unchanged from Game_0.

We define Succ_i as the event that $\mathcal{A}$, in Game i, succeeds in guessing the bit b used for generating the challenge ciphertext. Then, we have

$$\mathsf{Adv}^{\text{dM-IND-iCCA}}_{SC_{tk},\mathcal{A}} = |\Pr[\mathsf{Succ}_0] - \frac{1}{2}| \leq |\Pr[\mathsf{Succ}_0] - \Pr[\mathsf{Succ}_1]| + |\Pr[\mathsf{Succ}_1] - \frac{1}{2}| \quad (1)$$

To complete the proof, we prove the following lemmas.

Lemma 1. $|\Pr[\mathsf{Succ}_0] - \Pr[\mathsf{Succ}_1]|$ *is negligible.*

Proof. (of Lemma 1) Towards a contradiction, assume $|\Pr[\mathsf{Succ}_0] - \Pr[\mathsf{Succ}_1]|$ is non-negligible. Using $\mathcal{A}$ as a building block, we then show how to construct a PPT adversary $\mathcal{S}$ which breaks the IND-tag-CCA security of the TBKEM scheme TK, thereby proving the lemma by contradiction. $\mathcal{S}$ plays its own IND-tag-CCA game regarding the TBKEM scheme TK as follows.

Setup. Given (prm_{tk}, pk_R), $\mathcal{S}$ runs $prm_{sig} \leftarrow \mathsf{SSetup}(1^\kappa)$ and sets $prm \leftarrow (prm_{tk}, prm_{sig})$. Then $\mathcal{S}$ runs $\mathcal{A}$ with input (prm, pk_R).

Phase 1. $\mathcal{S}$ responds to $\mathcal{A}$'s unsigncryption queries $(pk_S, (c_1, c_2))$ as follow. $\mathcal{S}$ first sets $\mathsf{tag} \leftarrow pk_S$, issues (tag, c_1) to $\mathcal{CH}$ as his decapsulation query and then obtains K. Then $\mathcal{S}$ computes $(m||\sigma) \leftarrow \mathsf{DDec}(K, c_2)$ and $\mathsf{SVer}(prm_{sig}, pk_S, (m||c_1||pk_R), \sigma)$. Finally, if the value returned from SVer is $\top$, $\mathcal{S}$ returns m to $\mathcal{A}$, otherwise returns $\perp$.

Challenge. When $\mathcal{A}$ submits $(m_0, m_1, pk_S^*, sk_S^*)$ to $\mathcal{S}$, $\mathcal{S}$ regards pk_S^* as its challenge tag $\mathsf{tag}^* = pk_S^*$ and submits tag^* to $\mathcal{CH}$ and obtains $\mathcal{S}$'s challenge ciphertext/key pair (c_1^*, K_β^*), where β is the challenge bit of $\mathcal{S}$ in the IND-tag-CCA game. Then $\mathcal{S}$ flips a coin $b \in \{0,1\}$ uniformly at random, and computes $\sigma^* \leftarrow \mathsf{Sign}(prm_{sig}, sk_S^*, (m_b||c_1^*||pk_R))$ and $c_2^* \leftarrow \mathsf{DEnc}(K_\beta^*, (m_b||\sigma^*))$. $\mathcal{S}$ finally returns $c^* = (c_1^*, c_2^*)$ to $\mathcal{A}$ as $\mathcal{A}$'s challenge ciphertext.

Phase 2. $\mathcal{S}$ responds to $\mathcal{A}$'s unsigncryption queries $(pk_S, (c_1, c_2))$ as follows.

1. **If** $(pk_S, c_1) = (pk_S^*, c_1^*)$**:** $\mathcal{S}$ computes $(m||\sigma) \leftarrow \mathsf{DDec}(K_\beta^*, c_2)$ and $\mathsf{SVer}(prm_{sig}, pk_S^*, (m||c_1^*||pk_R), \sigma)$. If the value returned from SVer is $\top$, $\mathcal{S}$ returns m to $\mathcal{A}$, otherwise returns $\perp$.

2. **Otherwise:** $\mathcal{S}$ returns $m/\bot$ in the same way as the unsigncryption queries in Phase 1.

Guess. $\mathcal{A}$ outputs a bit b'. If $b' = b$, $\mathcal{S}$ outputs $\beta' = 1$ and otherwise, outputs $\beta' = 0$ as its guess.

Note that $\mathcal{S}$ never issues the prohibited decapsulation query $(\mathsf{tag}^*, c_1^*) = (pk_S^*, c_1^*)$ to $\mathcal{CH}$.

Then, the IND-tag-CCA advantage of the adversary $\mathcal{S}$ is estimated as:

$$\begin{aligned}\mathsf{Adv}_{TK,\mathcal{S}}^{\text{IND-tag-CCA}} &= |\Pr[\beta' = \beta] - \frac{1}{2}| \\ &= \frac{1}{2}|\Pr[\beta' = 0|\beta = 1] - \Pr[\beta' = 0|\beta = 0]| \\ &= \frac{1}{2}|\Pr[b' = b|\beta = 1] - \Pr[b' = b|\beta = 0]|\end{aligned}$$

Now, consider the case when $\beta = 1$, i.e. $K_\beta^* = K_1^*$ is a real symmetric key corresponding to c_1^* under the tag $\mathsf{tag} = pk_S^*$. Then, it is easy to see that $\mathcal{S}$ perfectly simulates Game_0 for $\mathcal{A}$ in which the challenge bit for $\mathcal{A}$ is b. Specifically, $\mathcal{A}$'s responses to decryption queries are perfectly answered as in Game_0, and the challenge ciphertext c^* for $\mathcal{A}$ is generated as in Game_0 (c_2^* is computed with a correct symmetric key K_1^* as in the proposed signcryption scheme SC_{tk}). Under this situation, the event $b' = b$ corresponds to the event Succ_0, i.e. $\Pr[b' = b|\beta = 1] = \Pr[\mathsf{Succ}_0]$.

When $\beta = 0$, i.e. $K_\beta^* = K_0^*$ is a uniformly random value in $\{0,1\}^\kappa$, on the other hand, $\mathcal{S}$ perfectly simulates Game_1 for $\mathcal{A}$ in which the challenge bit for $\mathcal{A}$ is b. Specifically, c_2^* in the challenge ciphertext is an encryption of m_b under the random key K_0^*, and an unsigncryption query of the form (pk_S^*, c_1^*, c_2) where $c_2 \neq c_2^*$ is answered using K_0^*. Under this situation, the event $b' = b$ corresponds to the event Succ_1, i.e. $\Pr[b' = b|\beta = 0] = \Pr[\mathsf{Succ}_1]$.

In summary, we have $\mathsf{Adv}_{TK,\mathcal{S}}^{\text{IND-tag-CCA}} = \frac{1}{2}|\Pr[\mathsf{Succ}_0] - \Pr[\mathsf{Succ}_1]|$ which is not negligible by the assumption we made at the beginning of the proof of this lemma. Since this contradicts IND-tag-CCA security of the TBKEM TK, $|\Pr[\mathsf{Succ}_0] - \Pr[\mathsf{Succ}_1]|$ must be negligible. This completes the proof of Lemma 1. □

Lemma 2. $|\Pr[\mathsf{Succ}_1] - \frac{1}{2}|$ *is negligible.*

Proof. (of Lemma 2) Assume towards a contradiction that $|\Pr[\mathsf{Succ}_1] - \frac{1}{2}|$ is not negligible. Then we show that we can construct another PPT adversary $\mathcal{S}$ that uses $\mathcal{A}$ as a subroutine and has non-negligible IND-CCA advantage against the DEM D, thereby proving the lemma by contradiction. The description of the IND-CCA adversary $\mathcal{S}$ is as follows.

Setup. $\mathcal{S}$ first computes $prm_{tk} \leftarrow \mathsf{TSetup}(1^\kappa)$ and $prm_{sig} \leftarrow \mathsf{SSetup}(1^\kappa)$, then sets $prm \leftarrow (prm_{tk}, prm_{sig})$. Then $\mathcal{S}$ computes $(pk_R, sk_R) \leftarrow \mathsf{TKG}(prm_{tk})$ and runs $\mathcal{A}$ with input (prm, pk_R).

Phase 1. $\mathcal{S}$ responds to $\mathcal{A}$'s unsigncryption queries $(pk_S, (c_1, c_2))$ by returning $m \leftarrow \mathsf{USC}(prm, pk_S, sk_R, (c_1, c_2))$. This is possible because $\mathcal{S}$ owns sk_R.

Challenge. When $\mathcal{A}$ submits $(m_0, m_1, pk_S^*, sk_S^*)$ to $\mathcal{S}$, $\mathcal{S}$ first sets $\mathsf{tag} \leftarrow pk_S^*$, computes $(K', c_1^*) \leftarrow \mathsf{TEncap}(prm_{tk}, pk_R, \mathsf{tag})$. Next $\mathcal{CH}$ computes $\sigma_0 \leftarrow \mathsf{Sign}(prm_{sig}, sk_S^*, (m_0||c_1^*||pk_R^*))$ and $\sigma_1 \leftarrow \mathsf{Sign}(prm_{sig}, sk_S^*, (m_1||c_1^*||pk_R^*))$. Then $\mathcal{S}$ submits two plaintexts $M_0 = (m_0||\sigma_0)$ and $M_1 = (m_1||\sigma_1)$ to $\mathcal{CH}$ as $\mathcal{S}$'s challenge and obtains c_2^*. $\mathcal{S}$ finally returns $c^* = (c_1^*, c_2^*)$ to $\mathcal{A}$ as $\mathcal{A}$'s challenge ciphertext.

Phase 2. $\mathcal{S}$ responds to $\mathcal{A}$'s unsigncryption queries $(pk_S, (c_1, c_2))$ as follows.
1. **If** $(pk_S, c_1) = (pk_S^*, c_1^*)$**:** $\mathcal{S}$ issues c_2 as a decryption query and obtains $(m||\sigma)$. Then $\mathcal{S}$ computes $\mathsf{SVer}(prm_{sig}, pk_S, (m||c_1||pk_R), \sigma)$. Finally, if the returned value of SVer is $\top$, $\mathcal{S}$ returns m to $\mathcal{A}$, otherwise returns $\perp$.
2. **Otherwise:** $\mathcal{S}$ returns $m/\perp$ in the same way as the unsigncryption queries in Phase 1.

Guess. $\mathcal{A}$ outputs a bit b'. $\mathcal{S}$ outputs b' as its guess.

Note that $\mathcal{S}$ never issues the prohibited decryption query c_2^* to $\mathcal{CH}$. Moreover it is easy to see that $\mathcal{S}$ perfectly simulates Game_1 for $\mathcal{A}$ in which $\mathcal{A}$'s challenge bit is that of $\mathcal{S}$'s. Therefore, $\mathcal{S}$'s IND-CCA advantage can be caluculated as $\mathsf{Adv}_{D,\mathcal{S}}^{\text{IND-CCA}} = |\Pr[\mathsf{Succ}_1] - \frac{1}{2}|$ which is not negligible according to the assumption we made at the beginning of the proof of this lemma. Since this contradicts the IND-CCA security of the DEM D, $|\Pr[\mathsf{Succ}_1] - \frac{1}{2}|$ must be negligible. This completes the proof of Lemma 2. □

According to the inequality (1) and Lemmas 1 and 2, $\mathsf{Adv}_{SC_{tk},\mathcal{A}}^{\text{dM-IND-iCCA}}$ is upper-bounded to be negligible for any PPT adversary $\mathcal{A}$. This completes the proof of Theorem 1. □

Theorem 2. *If the signature scheme S is sUF-CMA secure and the DEM D is one-to-one, then the proposed signcryption scheme SC_{tk} is dM-sUF-iCMA secure.*

Proof. Suppose there exists an adversary $\mathcal{A}$ that breaks the dM-sUF-iCMA security of the signcryption scheme SC_{tk}. Using $\mathcal{A}$ as a building block, we then show how to construct an algorithm $\mathcal{S}$ which breaks sUF-CMA security of the signature scheme S, thereby proving the theorem by contradiction. $\mathcal{S}$ plays its own sUF-CMA game regarding the signature scheme S and is defined as follows.

Setup. Given (prm_{sig}, pk_S), $\mathcal{S}$ runs $prm_{tk} \leftarrow \mathsf{TSetup}(1^\kappa)$, sets $prm \leftarrow (prm_{tk}, prm_{sig})$ and sets $\mathsf{tag} \leftarrow pk_S$. Then $\mathcal{S}$ runs $\mathcal{A}$ with input (prm, pk_S).

Query. When $\mathcal{A}$ issues a signcryption query of the form (pk_R, m), $\mathcal{S}$ first computes $(c_1, K) \leftarrow \mathsf{TEncap}(prm_{tk}, pk_R, \mathsf{tag})$. $\mathcal{S}$ then issues $(m||c_1||pk_R)$ to $\mathcal{CH}$ as a signing query and obtains σ. Then $\mathcal{S}$ computes $c_2 \leftarrow \mathsf{DEnc}(K, (m||\sigma))$, and finally returns (c_1, c_2) to $\mathcal{A}$ as an answer to the signcryption query.

Output. When $\mathcal{A}$ terminates with output a receiver key pair (pk_R^*, sk_R^*) and a ciphertext $c^* = (c_1^*, c_2^*)$, $\mathcal{S}$ firstly computes $K^* \leftarrow \mathsf{TDecap}(prm, sk_R^*, \mathsf{tag}, c_1^*)$ and $(m^*||\sigma^*) \leftarrow \mathsf{DDec}(K^*, \sigma^*)$, and then outputs $((m^*||c_1^*||pk_R^*), \sigma^*)$ as its own forgery and terminates. (If either TDecap or DDec returns $\perp$, then $\mathcal{A}$ simply gives up and aborts.)

It is straightforward to see that $\mathcal{S}$'s simulation for $\mathcal{A}$ is perfect. Specifically, the parameters $prm = (prm_{tk}, prm_{sig})$ and the key pk_S given to $\mathcal{A}$ are distributed identically to those given to $\mathcal{A}$ in the dM-sUF-iCMA game. Furthermore, $\mathcal{A}$'s signcryption queries are properly answered by using signing queries.

Hence, in order to complete the proof, all we need to show is that $\mathcal{S}$'s forgery is valid whenever $\mathcal{A}$'s forgery is valid. Let q be the number of $\mathcal{A}$'s signcryption queries. For $i \in \{1, ..., q\}$, let $(pk_R^{(i)}, m^{(i)})$ be $\mathcal{A}$'s i-th signcryption query, let $(c_1^{(i)}, c_2^{(i)})$ be $\mathcal{S}$'s response to $\mathcal{A}$'s i-th query, and let $K^{(i)}$ be the symmetric key which is the output value of $\mathsf{TEncap}(prm_{tk}, pk_R^{(i)}, \mathsf{tag})$ used to compute $c_2^{(i)}$, where $\mathsf{tag} = pk_S$. If $\mathcal{A}$ succeeds in the dM-sUF-CMA game regarding SC_{tk}, it must hold that

$$\begin{aligned} \mathsf{TDecap}(prm_{tk}, sk_R^*, \mathsf{tag}, c_1^*) &= K^* \neq \perp \\ \mathsf{DDec}(K^*, c_2^*) &= (m^* || \sigma^*) \neq \perp \\ \mathsf{SVer}(prm_{sig}, pk_S, (m^* || c_1^* || pk_R^*), \sigma^*) &= \top \\ {}^\forall i \in \{1, ..., q\} : \qquad (pk_R^*, m^*, c_1^*, c_2^*) &\neq (pk_R^{(i)}, m^{(i)}, c_1^{(i)}, c_2^{(i)}) \end{aligned}$$

where $\mathsf{tag} = pk_S$. Under the above conditions, in order for $\mathcal{S}$ to succeed in the sUF-CMA game regarding S, it must hold that there exists no $i \in \{1, ..., q\}$ such that $((m^* || c_1^* || pk_R^*), \sigma^*) = ((m^{(i)} || c_1^{(i)} || pk_R^{(i)}), \sigma^{(i)})$.

Now assume towards a contradiction that such i exists. Then, due to the correctness of TK and the fact that in this simulation $\mathcal{S}$ always uses the same tag $\mathsf{tag} = pk_S$, $pk_R^* = pk_R^{(i)}$ and $c_1^* = c_1^{(i)}$ implies $K^* = K^{(i)}$. Next, due to the one-to-one property of D, $(K^*, (m^* || \sigma^*)) = (K^{(i)}, (m^{(i)} || \sigma^{(i)}))$ implies $c_2^* = c_2^{(i)}$. Putting everything together, for this i we have $(pk_R^*, m^*, c_1^*, c_2^*) = (pk_R^{(i)}, m^{(i)}, c_1^{(i)}, c_2^{(i)})$. But this contradicts the fourth succeeding condition of $\mathcal{A}$ above, and thus such i never exists.

Therefore, $\mathcal{S}$ succeeds in the sUF-CMA game regarding S whenever $\mathcal{A}$ succeeds, and we have $\mathsf{Adv}_{S,\mathcal{S}}^{\text{sUF-CMA}} = \mathsf{Adv}_{SC_{tk},\mathcal{A}}^{\text{dM-sUF-iCMA}}$ is non-neglibible, which contradicts that S is sUF-CMA secure. Hence, $\mathsf{Adv}_{SC_{tk},\mathcal{A}}^{\text{dM-sUF-iCMA}}$ must be negligible for any PPT adversary $\mathcal{A}$. This completes the proof of Theorem 2. □

4.2 Composition Using KEM

Let KM = (KSetup, KKG, Encap, Decap) be an IND-CCA secure KEM, let D = (DEnc, DDec) be an IND-CCA secure DEM, let M = (Mac, MVer) be a sUF-OT secure MAC, and let S = (SSetup, SKG, Sign, SVer) be a sUF-CMA secure signature scheme (the definitions of a KEM and a MAC are given in Appendix A). Then, we construct a signcryption scheme SC_{kem} as shown in Fig.2. We assume symmetric key space of KM is $\{0,1\}^{2\kappa}$ and that of D and M are $\{0,1\}^{\kappa}$. (We can always achieve this by using an appropriate key derivation function and/or a pseudorandom generator.)

Construction Idea. The basic idea of SC_{kem} is almost the same as that of SC_{tk}. In SC_{kem}, we use a KEM and a MAC as building blocks, instead of a TBKEM. As

$\mathsf{Setup}(1^\kappa)$:
  $prm_{kem} \leftarrow \mathsf{KSetup}(1^\kappa)$
  $prm_{sig} \leftarrow \mathsf{SSetup}(1^\kappa)$
  Output $prm \leftarrow (prm_{kem}, prm_{sig})$.

$\mathsf{SC}(prm, sk_S, pk_R, m)$:
  $(c_1, K) \leftarrow \mathsf{Encap}(prm_{kem}, pk_R)$
  $(K_m \| K_a) \leftarrow K$
  $\sigma \leftarrow \mathsf{Sign}(prm_{sig}, sk_S, (m \| c_1 \| pk_R))$
  $c_2 \leftarrow \mathsf{DEnc}(K_m, (m \| \sigma))$
  $\tau \leftarrow \mathsf{Mac}(K_a, (pk_S \| c_1 \| c_2))$
  Output $c \leftarrow (c_1, c_2, \tau)$

$\mathsf{KeyGen_S}(prm)$:
  Output $(pk_S, sk_S) \leftarrow \mathsf{SKG}(prm_{sig})$.
$\mathsf{KeyGen_R}(prm)$:
  Output $(pk_R, sk_R) \leftarrow \mathsf{KKG}(prm_{kem})$.
$\mathsf{USC}(prm, pk_S, sk_R, c)$:
  Parse c as (c_1, c_2, τ)
  $K \leftarrow \mathsf{Decap}(prm_{kem}, sk_R, c_1)$
    (if $K = \bot$, then return $\bot$)
  $(K_m \| K_a) \leftarrow K$
  If $\mathsf{MVer}(K_a, (pk_S \| c_1 \| c_2), \tau) = \bot$
    then return $\bot$
  $(m \| \sigma) \leftarrow \mathsf{DDec}(K_m, c_2)$
    (if DDec outputs $\bot$, then return $\bot$)
  If $\mathsf{SVer}(prm_{sig}, pk_S, (m \| c_1 \| pk_R), \sigma) = \bot$
    then return $\bot$
  Output m

Fig. 2. Construction using KEM: SC_{kem}

mentioned earlier, we can construct an IND-tag-CCA secure TBKEM using an IND-CCA secure KEM and a one-time secure MAC [2], and we can also construct an IND-CCA secure DEM using a one-time secure DEM and a sUF-OT secure MAC [7]. However, we find that if we use an IND-tag-CCA secure TBKEM and an IND-CCA secure DEM constructed as above in the first construction SC_{tk}, the one-time secure MAC can be shared and thus can be slightly more efficient. SC_{kem} is constructed based on this observation.

Security. The security of SC_{kem} is formally guaranteed by the following theorems. The proofs of these theorems are similar to the proofs of Theorems 1 and 2, and will not be given here.

Theorem 3. *If the KEM KM is IND-CCA secure, the DEM D is IND-OT secure and MAC M is sUF-OT secure, then the proposed signcryption scheme SC_{kem} is dM-IND-iCCA secure.*

Theorem 4. *If the signature scheme S is sUF-CMA secure, the DEM D is one-to-one and MAC M is one-to-one, then the proposed signcryption scheme SC_{kem} is dM-sUF-iCMA secure.*

5 Comparison

In Fig. 3, we compare concrete instantiations of standard model signcryption schemes. In the figure, tBMW1 denotes the TBKEM (and the TBE scheme) which is obtained from the PKE scheme by Boyen, Mei, and Waters [11] based on the observation in [21, Sect. 7.2], and BMW2 denotes the KEM by Boyen, Mei, and Waters in [12, Sect. 4]. MMS-StTE(X, Y) denotes "Sign-then-Tag-based-Encrypt" schemes [21, Sect. 5] where the TBE scheme X and the signature

Scheme	Confidentiality /Assumption	Unforgeability /Assumption	Comp. Cost SC / USC	Ciphertext OH Elements/Bits
Tan [23]	dM-IND-iCCA /DBDH	dM-sUF-iCMA(KR) /q-SDH	[3,2;0] / [3,1;4]	$3\|\mathbb{G}_p\| + 2\|\mathbb{Z}_p\|$ / 800
MMS-StTE (tBMW1,BB[8])	dM-IND-iCCA /DBDH	dM-wUF-iCMA /q-SDH	[4,0;0] + 1W / [1,1;2]	$3\|\mathbb{G}_p\| + \|\mathbb{Z}_p\|$ / 640
MMS-SC (tBMW1,Waters[24])	dM-IND-iCCA /DBDH	dM-wUF-iCMA(KR) /co-CDH	[4,0;0] / [1,0;3] + 1W	$3\|\mathbb{G}_p\|$ / 480
MMS-SC (tBMW1,BSW[10])	dM-IND-iCCA /DBDH	dM-sUF-iCMA(KR) /co-CDH	[4,1;0] / [1,1;3] + 1W	$3\|\mathbb{G}_p\| + \|\mathbb{Z}_p\|$ / 640
Ours: SC_{tk} (tBMW1,BB[8])	dM-IND-iCCA /DBDH	dM-sUF-iCMA /q-SDH	[4,0;0] + 1W / [1,1;2]	$3\|\mathbb{G}_p\| + \|\mathbb{Z}_p\|$ / 640
Ours: SC_{kem} (BMW2,BB[8])	dM-IND-iCCA /DBDH	dM-sUF-iCMA /q-SDH	[3,1;0] / [1,1;2]	$3\|\mathbb{G}_p\| + \|\mathbb{Z}_p\|$ $+\|\mathsf{MAC}\|$ / 720

Fig. 3. Comparison of existing and proposed signcryption schemes with insider security in the dynamic multi-user model. Columns "Confidentiality" and "Unforgeability" list the achieved security notions as well as the underlying security assumptions. In these columns, the security notions followed with (KR) are security in the key registered model. Column "Comp. Cost" lists the computational overhead for signcryption (SC) and unsigncryption (USC), where $[a, b; c]$ denotes a exponentiations, b multi-exponentiations and c pairing computations, and W denotes computation of the so called Waters hash [24] (other multiplications, computation costs of hash functions and symmetric key primitives are ignored). In the overhead column, $|\mathbb{G}|$ denotes the size of a group element of an elliptic curve equipped with an asymmetric pairing, and $|\mathbb{Z}_p|$ denotes the size of an exponent (i.e. the order of the group), and $|\mathrm{MAC}|$ denotes the size of a (sUF-OT secure) MAC tag. When instantiated to achieve 80 bits of security and minimize ciphertext overhead, listed in the last column, we set $|\mathbb{G}| = |\mathbb{Z}_p| = 160$ bits, and $|\mathsf{MAC}| = 80$ bits. We assume that a DEM has no ciphertext overhead [22] (i.e. length preserving), regardless of whether it is IND-OT or IND-CCA secure. "Ciphertext OH" means the difference between the ciphertext size and the plaintext size.

scheme Y are used as concrete building blocks. MMS-SC(X, Y) denotes a signcryption scheme constructed from "signcryption composable" TBKEM X and signature scheme Y [21, Sect. 6]. SC_{tk}(X, Y) (resp. SC_{kem}(X, Y)) denotes the signcryption scheme constructed from SC_{tk} (resp. SC_{kem}) in Section 4 where the TBKEM (resp. the KEM) X and a signature scheme Y are used as concrete building blocks.

As shown in Fig. 3, our schemes are the only ones which achieve dM-IND-iCCA security and dM-sUF-iCMA security simultaneously without random oracles or key registration. SC_{tk}(tBMW1, BB) has better efficiency in all aspects than Tan and MMS-StTE(tBMW1,BB). Furthermore, we see that there exists a tradeoff in the achieved security/assumption, computational costs, and ciphertext size (overhead) among SC_{tk}(tBMW1, BB), MMS-SC(tBMW1, Waters), and MMS-CS(tBMW1, BSW). However, we would like to stress that if we admit the q-SDH assumption and if we need strong unforgeability, then SC_{tk}(tBMW1, BB) is the best choice over other schemes. We remark that SC_{kem} and Tan have constant size user (i.e. sender and receiver) public keys, and SC_{kem} is better in all aspects than Tan.

Although we have chosen tBMW1, BMW2, and BB as concrete building blocks for comparison in Fig. 3 we would like to stress that our proposed constructions SC_{tk} and SC_{kem} are generic constructions (which is also true in the constructions in [21]), and thus a number of different combinations of concrete instantiations are possible. For example, if one wants to construct an insider secure signcryption scheme from factoring-based assumptions rather than discrete-logarithm-based assumptions, one can use the recent KEM (and its TBKEM-analogue that can be obtained through the observation in [21]) by Hofheinz and Kiltz [14] whose security is proved from the factoring assumption and the recent signature scheme by Hohenberger and Waters [15] whose security is proved from the RSA assumption. Alternatively, one can also instantiate insider secure signcryption schemes based on lattice-based assumptions from recent progress regarding identity-based encryption and signature schemes in this area, such as the schemes from [3]. (Note that we can always construct CCA secure KEMs from identity-based encryption schemes due to [9])

Acknowledgement

The authors would like to thank the reviewers for their useful comments. Takahiro Matsuda and Jacob Schuldt are supported by a JSPS Fellowship for Young Scientists.

References

1. Abe, M., Cui, Y., Imai, H., Kiltz, E.: Efficient hybrid encryption from ID-based encryption. Designs, Codes and Cryptography 54(3), 205–240 (2010)
2. Abe, M., Gennaro, R., Kurosawa, K.: Tag-KEM/DEM: A new framework for hybrid encryption. J. of Cryptology 21(1), 97–130 (2008)
3. Agrawal, S., Boneh, D., Boyen, X.: Lattice basis delegation in fixed dimension and shorter-ciphertext hierarchical IBE. In: Rabin, T. (ed.) CRYPTO 2010. LNCS, vol. 6223, pp. 98–115. Springer, Heidelberg (2010)
4. An, J., Dodis, Y., Rabin, T.: On the security of joint signature and encryption. In: Knudsen, L.R. (ed.) EUROCRYPT 2002. LNCS, vol. 2332, pp. 83–107. Springer, Heidelberg (2002)
5. Baek, J., Steinfeld, R., Zheng, Y.: Formal proofs for the security of signcryption. In: Naccache, D., Paillier, P. (eds.) PKC 2002. LNCS, vol. 2274, pp. 80–98. Springer, Heidelberg (2002)
6. Baek, J., Steinfeld, R., Zheng, Y.: Formal proofs for the security of signcryption. J. of Cryptology 20(2), 203–235 (2007)
7. Bellare, M., Namprempre, C.: Authenticated encryption: Relations among notions and analysis of the generic composition paradigm. In: Okamoto, T. (ed.) ASIACRYPT 2000. LNCS, vol. 1976, pp. 531–545. Springer, Heidelberg (2000)
8. Boneh, D., Boyen, X.: Secure identity based encryption without random oracles. In: Franklin, M. (ed.) CRYPTO 2004. LNCS, vol. 3152, pp. 443–459. Springer, Heidelberg (2004)

9. Boneh, D., Canetti, R., Halevi, S., Katz, J.: Chosen-ciphertext security from identity-based encryption. SIAM J. Computing 36(5), 1301–1328 (2007)
10. Boneh, D., Shen, E., Waters, B.: Strongly unforgeable signatures based on computational diffie-hellman. In: Yung, M., Dodis, Y., Kiayias, A., Malkin, T. (eds.) PKC 2006. LNCS, vol. 3958, pp. 229–240. Springer, Heidelberg (2006)
11. Boyen, X., Mei, Q., Waters, B.: Direct chosen ciphertext security from identity-based techniques. In: Proc. of CCS 2005, pp. 320–329. ACM, New York (2005)
12. Boyen, X., Mei, Q., Waters, B.: Direct chosen ciphertext security from identity-based techniques, Updated version of [11]. Cryptology ePrint Archive: Report 2005/288 (2005), `http://eprint.iacr.org/2005/288/`
13. Dent, A.: Hybrid signcryption schemes with outsider security (extended abstract). In: Zhou, J., López, J., Deng, R.H., Bao, F. (eds.) ISC 2005. LNCS, vol. 3650, pp. 203–217. Springer, Heidelberg (2005)
14. Hofheinz, D., Kiltz, E.: Practical chosen ciphertext secure encryption from factoring. In: Joux, A. (ed.) EUROCRYPT 2009. LNCS, vol. 5479, pp. 313–332. Springer, Heidelberg (2009)
15. Hohenberger, S., Waters, B.: Short and stateless signatures from the RSA assumption. In: Halevi, S. (ed.) CRYPTO 2009. LNCS, vol. 5677, pp. 654–670. Springer, Heidelberg (2009)
16. Kiltz, E.: Chosen-ciphertext security from tag-based encryption. In: Halevi, S., Rabin, T. (eds.) TCC 2006. LNCS, vol. 3876, pp. 581–600. Springer, Heidelberg (2006)
17. Li, C., Yang, G., Wang, D., Deng, X., Chow, S.: An efficient signcryption scheme with key privacy. In: López, J., Samarati, P., Ferrer, J.L. (eds.) EuroPKI 2007. LNCS, vol. 4582, pp. 78–93. Springer, Heidelberg (2007)
18. Libert, B., Quisquater, J.-J.: Improved signcryption with key privacy from gap Diffie-Hellman groups. In: Bao, F., Deng, R., Zhou, J. (eds.) PKC 2004. LNCS, vol. 2947, pp. 187–200. Springer, Heidelberg (2004)
19. Libert, B., Quisquater, J.-J.: Improved signcryption with key privacy from gap Diffie-Hellman groups, Updated version of [18] (2004), `http://www.dice.usl.ac.be/~libert/`
20. MacKenzie, P.D., Reiter, M.K., Yang, K.: Alternatives to non-malleability: Definitions, constructions, and applications. In: Naor, M. (ed.) TCC 2004. LNCS, vol. 2951, pp. 171–190. Springer, Heidelberg (2004)
21. Matsuda, T., Matsuura, K., Schuldt, J.: Efficient constructions of signcryption schemes and signcryption composability. In: Roy, B., Sendrier, N. (eds.) INDOCRYPT 2009. LNCS, vol. 5922, pp. 321–342. Springer, Heidelberg (2009)
22. Phan, D., Pointcheval, D.: About the security of ciphers (semantic security and pseudo-random permutations). In: Handschuh, H., Hasan, M.A. (eds.) SAC 2004. LNCS, vol. 3357, pp. 182–197. Springer, Heidelberg (2004)
23. Tan, C.: Signcryption scheme in multi-user setting without random oracles. In: Matsuura, K., Fujisaki, E. (eds.) IWSEC 2008. LNCS, vol. 5312, pp. 64–82. Springer, Heidelberg (2008)
24. Waters, B.: Efficient identity-based encryption without random oracles. In: Cramer, R. (ed.) EUROCRYPT 2005. LNCS, vol. 3494, pp. 114–127. Springer, Heidelberg (2005)
25. Zheng, Y.: Digital signcryption or how to achieve cost (Signature & encryption) ¡¡ = cost(Signature) + cost(Encryption). In: Kaliski Jr., B.S. (ed.) CRYPTO 1997. LNCS, vol. 1294, pp. 165–179. Springer, Heidelberg (1997)

A Additional Definitions

A.1 Key Encapsulation Mechanism

A key encapsulation mechanism (KEM) is a special case of a TBKEM in which the tag is always given by an empty string i.e. the encapsulation and decapsulation algorithms do not support the additional input tag. We use the notation KSetup, KKG, Encap, and Decap to denote the setup, key generation, encapsulation and decapsulation algorithms, respectively. Like for a TBKEM, it is required for all $prm \leftarrow \mathsf{KSetup}(1^\kappa)$, all $(pk, sk) \leftarrow \mathsf{KKG}(prm)$, and all $(c, K) \leftarrow \mathsf{Encap}(prm, pk)$, that $K = \mathsf{Decap}(prm, sk, c)$. Lastly, IND-CCA security of a KEM is defined almost as IND-tag-CCA security of a TBKEM, except that decapsulation queries do not contain a tag, and the adversary is given a challenge encapsulation and key pair without choosing a challenge tag.

A.2 Message Authentication Code

A message authentication code (MAC) is given by the following two algorithms $(\mathsf{Mac}, \mathsf{MVer})$: Mac, which is given input a symmetric key $K \in \mathcal{K}$ and a message m, where $\mathcal{K}$ is a key space, returns a message authentication tag τ; and MVer, which given input a symmetric key $K \in \mathcal{K}$, a message m, and a message authentication tag τ, returns $\top$ or $\bot$.

It is required for all $K \in \mathcal{K}$ and all m that $\mathsf{MVer}(K, m, \mathsf{Mac}(K, m)) = \top$.

sUF-OT security. For a MAC, strong unforgeability against one time attacks (sUF-OT) is defined by the following game between an adversary $\mathcal{A}$ and a sUF-OT challenger $\mathcal{CH}$.

Setup. $\mathcal{CH}$ chooses a key $K \in \mathcal{K}$ at random.

Query. $\mathcal{A}$ can submit a single MAC query m to $\mathcal{CH}$. $\mathcal{CH}$ responds to this query by returning $\tau \leftarrow \mathsf{Mac}(K, m)$.

Output. $\mathcal{A}$ outputs a message/tag pair (m^*, τ^*).

We define the sUF-OT advantage of $\mathcal{A}$ attacking the MAC M as follows:

$$\mathsf{Adv}_{M,\mathcal{A}}^{\mathrm{sUF\text{-}OT}} = \Pr[\mathsf{MVer}(K, m^*, \tau^*) = \top \wedge (m^*, \tau^*) \neq (m, \tau)]$$

Definition 6. *We say that a MAC M is sUF-OT secure if $\mathsf{Adv}_{M,\mathcal{A}}^{\mathrm{sUF\text{-}OT}}$ is negligible for any PPT adversary $\mathcal{A}$.*

One-to-One Property. A MAC is said to be *one-to-one* if given a key $K \in \mathcal{K}$ and a message m, there is only one message authentication tag τ such that $\mathsf{MVer}(K, m, \tau) = \top$. It is satisfied by many MAC schemes such as the CMAC, whose Mac algorithm is deterministic and the MVer algorithm re-computes the MAC tag from a key and given message, and compares it with the given tag.

Quantitatively Analyzing Stealthy Communication Channels*

Patrick Butler, Kui Xu, and Danfeng (Daphne) Yao

Department of Computer Science
Virginia Tech
Blacksburg, VA 24060

Abstract. Attackers in particular botnet controllers use stealthy messaging systems to set up large-scale command and control. Understanding the capacity of such communication channels is important in detecting organized cyber crimes. We analyze the use of domain name service (DNS) as a stealthy botnet command-and-control channel, which allows multiple entities to pass messages stored in DNS records to each other. We describe and quantitatively analyze new techniques that can be used to hide malicious DNS activities both at the host and network levels.

We also present and experimentally evaluate statistical content-analysis techniques as a countermeasure, which require deep packet inspection. Our techniques are beyond the specific DNS security problem studied. We give a formal definition for the *perfect stealth* of a communication channel; point out the fundamental limits in achieving it, as well as the practical issues in the detection. We perform comprehensive statistical analysis that makes use of a two-month-long 4.6GB campus network dataset and 1 million domain names obtained from `alexa.com`.

Keywords: DNS tunneling, command and control, information theory.

1 Introduction

Botnet command and control (C&C) channel refers to the protocol used by bots and botmaster (i.e., botnet controller) to communicate to each other, e.g., for bots to receive new attack commands and updates from botmaster, or to submit stolen data. A C&C channel for a botnet needs to be reliable, redundant, non-centralized, and easily disguised as legitimate traffic. Many botnet operators used the Internet Relay Chat protocol (IRC) or HTTP servers to pass information. Botnet operators constantly explore new stealthy communication mechanisms to evade detection. HTTP-based command and control is difficult to distinguish from legitimate Web traffic. For example, detecting frequent and periodic HTTP requests was proposed for identifying botnet traffic [12]. However, this method may give high false positives as legitimate websites also automatically refresh pages. The feasibility of email as a stealthy botnet command and control protocol

* This work has been supported in part by NSF grant CAREER CNS-0953638 and CNS-0831186, and ARO grant STIR-45008.

J. Lopez and G. Tsudik (Eds.): ACNS 2011, LNCS 6715, pp. 238–254, 2011.

was studied by researchers in [22]. In this paper, we systematically investigate the sole use of DNS queries for botnet command and control.

The decentralized nature of domain name systems (DNS) with a series of redundant servers potentially provides an effective channel for covert communication of a large distributed system, including botnets. The focus of this paper is on analyzing the feasibility of a pure DNS-based command-and-control. We analyze a stealthy communication channel based on DNS updates, queries, and responses with existing infrastructure, without enlisting any Web or special-purpose servers. The DNS channel is aided by being a high traffic channel such that data can be easily hidden. As virtually anyone can create their own domain name and DNS servers, it is a system that can easily be infiltrated by hackers and botnet operators.

DNS tunneling is a technique known for transmitting arbitrary data via DNS protocol [3,4,9]. One application is to bypass firewalls, as both inbound and outbound DNS connections are usually allowed by organizational firewall rules. Because DNS is often overlooked in current security measures, it offers a command-and-control channel that is unimpeded. Because nearly all traffic requires DNS to translate domain names to IP addresses and back, simple firewall rules can not easily be created less they harm legitimate traffic. Whereas, other channels such as HTTP might be limited to well known sites. Compared to the existing studies on DNS security and botnet C&C, the novelty of our work is two-fold: *i)* presenting and quantitatively evaluating new DNS-based techniques for distributed and stealthy communication, and *ii)* more importantly, analyzing the practical limitations of information-theoretic based detection techniques. These limitations largely contribute to the current arm race between attackers and defenders.

Our analysis is useful beyond the specific DNS tunneling problem studied. There has not been a systematic study on its robustness against statistical detection methods. Understanding the capacity of botnets communication power helps identify and eliminate nefarious attacks launched from them. Therefore, our work is not yet another botnet command-and-control solution. Our techniques – including the countermeasure based on analyzing content distributions and a model for perfect stealth in content-based communication channel – are useful beyond the specific DNS problem studied.

Our contributions. While the technology of DNS tunneling for command and control has been observed [11], it was still unclear how effective and feasible to use this technique to sustain large botnets. We provide the first systematic analysis on the constraints associated with DNS-based botnet communication channels. Our technical contributions are summarized as follows.

- We give a formal description of a botnet command-and-control protocol through DNS queries and responses associated with domains under botmaster's control. We describe a new *codeword mode* of communication, which employs a shared vocabulary between bots and botmaster for stealthy dissemination of attack commands or code updates.

- We describe techniques for hiding query activities, including *i) piggybacking query strategy* – a bot blends its (outbound) DNS queries with legitimate DNS queries and *ii) exponentially distributed query strategy* – a bot probabilistically distributes DNS queries so that inter-arrival times follow an exponential distribution.
- We give the first formal definition for the perfect covert channel in terms of the distinguishability between normal traffic and attacker's traffic. We discuss how information theoretic analysis can be used to detect malicious traffic. We evaluate statistical methods for detecting anomalies in the content of DNS packets, through comparing the probability distributions of normal DNS traffic and tunneling traffic. More importantly, we point out the practical limitations (namely efficiency and scalability) associated with these information theoretic methods.

  We perform comprehensive experiments to evaluate the behaviors of proposed query strategies in terms of how quickly new commands are disseminated to a large number of bots. Our analysis utilizes a 4.6GB two-month-long wireless network trace obtained from an organization.
- We also raise an open question on how to efficiently and automatically generate short-lived domain names that resemble legitimate domain names and evade anomaly detection. We give evidences on the difficulty of the problem.

Organization. We describe the basic DNS tunneling mechanisms in Section 2. We present new strategies for improving the stealthiness of DNS-based command and control in Section 3. Our experimental evaluation results are described in Section 3.2. We describe a countermeasure that requires examining the content of DNS packets and performing statistical analysis in Section 4. We raise an open question regarding how to automatically generate practical domain names in Section 5. Related work is given in Section 6. Conclusions are given in Section 7.

2 Communication Modes

In this section, we describe protocols that pass messages over the DNS between distributed entities, and illustrate the ease of setting up large-scale command-and-control via DNS. We describe two forms of communication modes: *codeword* mode and *tunneled* mode. *Codeword communication* allows one-way communication from botmaster to a bot client, which is suitable for issuing attack commands. *Tunneled communication* allows for the transmitting of arbitrary data in both directions between bot and botmaster, which may be used for both issuing commands and collecting stolen data. The former only requires the ability to set a particular domain name response, this could be done via any free DNS service, while the latter requires setting up an authoritative domain server.

The controller of the botnet first needs to create a domain or subdomain, which is administered from a special DNS server. This DNS server waits for special name lookups, which it then translates into incoming data. The DNS server then responds with the appropriate data using the agreed-upon semantics. We assume that the botnet controller (i.e., botmaster) has access to the authoritative domain name server for some domains or sub-domains. Bots across the Internet frequently receive commands and updates from a botmaster and launch attacks accordingly, as well as submit stolen data to the botmaster. We give brief background information on DNS records.

DNS Resources Records. The DNS system allows a name server administrator to associate different types of data with either a fully qualified domain name or an IP address. To send a message to a bot, an adversary can store data in any one of these types of records.

- `A` record specifies an IP address for a given host name.
- `CNAME` and `MX` records can point to textual data representing the alias or mailing host of a particular host name.
- `TXT` records are designed to store arbitrary textual data up to 255 characters.
- `EDNS0` record allows storing up to a 1280 byte payload [9].

2.1 Codeword Mode

The *codeword mode* is a new stealthy communication mechanism. It requires a botnet operator to decide upon a set of agreed upon codewords *a priori*. Each codeword represents a specific type of commands or attacks. The codeword appears in the DNS query as an innocent hostname, for example `codeword.domain.com`. This hostname may be stored as any type of record (e.g. `A`, `MX`, `CNAME`). A request for an `A` or `CNAME` record tends to be the most common and therefore a preference should be given to these records types so that queries would appear most like legitimate traffic. The client queries `codeword.domain.com`, and waits for a particular value in the server's response. Upon receiving the query, the DNS server (controlled by the botnet operator) returns the pre-set response that contains command information. If the codeword corresponds to denial-of-service (DoS) attacks, then the response may represent a target of DoS attacks. If the codeword corresponds to update, the client may contact the IP address returned for updated code or other instructions.

It is important to note that the codeword can be chosen *arbitrarily* and does not need to correspond to a specific host or service. The codeword method allows a stealthy *one-way* commanding system. It can effectively evade detection approaches based on non-conforming packet sizes [11], i.e., DNS packets whose sizes are outside the range of [28, 300] bytes. Codewords may be arbitrarily generated, or may be common service names such as `www`, `mail`, or `ftp`. In the latter case, packet statistics cannot be performed to find anomalies.

2.2 Tunneled Mode

The purpose of tunneled mode is to allow the *two-way* transfer of arbitrary binary data between a server and a client. This mode is referred to as tunneled mode, as one can tunnel streaming data over this DNS communication method.

- *Upstream communication* is for a client to submit data to a (malicious) domain server. The client submits the data as a `CNAME` query by *i)* encoding the data using a base32 encoding, *ii)* using the encoded string to construct a host name, and *iii)* send a `CNAME` DNS query. An example is shown in Figure 1.
- *Downstream communication* is for the server to issue commands to clients. Upon receiving the above query from the client on a hostname h, the server *i)* encodes the response as base32 data, and *ii)* constructs and returns a `CNAME` record for h. An example is shown in Figure 1.

- Upstream: Ask CNAME for:
`NBSWY3DPFQQHO33SNRSA000.domain.com`
- Downstream: CNAME points to:
`NBUSYIDCN5ZXG000.domain.com`
`3600`
`CNAME`
`NBSWY3DPFQQHO33SNRSA000.domain.com`

Fig. 1. Example data packets sent to and from a server in tunneled mode: To server: "hello, world". From server: "hi, boss". In this example, the domain server for domain.com is the malicious server and the response has one hour TTL.

To prevent DNS caching from disrupting the communications, the server may set a short *time-to-live* (TTL). This tunneling method gives an operator the most options after implementation as the data stream can be arbitrary. Because of the arbitrary payload, the distribution of packet bytes may differ significantly from conventionally DNS payload. We perform more analysis in Section 4.

There are two main characteristics of DNS-based communication. First, because the DNS protocol does not allow the server to initiate a connection with the client, the client needs to continually *pull* updates from the server. Second, DNS is based on UDP, and thus does not guarantee reliable data transfer or message order. To mitigate the problem, sequence numbers have to be appended to messages for bookkeeping purposes.

Both the tunneled mode and codeword mode require clients to frequently pull updates from name servers by querying the corresponding botnet's domain. Straightforward querying patterns are easy to detect (e.g., periodically sending DNS queries) and susceptible to simple aggregate analysis, such as counting DNS queries for each unique domains and identifying domains with abnormally large query volume at the host, local area network, or internet service provider levels. We analyze several simple-yet-effective methods for bots to hide their DNS traffic in the next section.

3 Query Strategies and Quantitative Evaluation

In this section, we play the devil's advocate and describe and experimentally evaluate new techniques for hiding DNS query activities on a host, in order to defeat anomaly detection that targets abnormal temporal patterns. The proposed strategies are useful for both the tunneling and codeword modes. We quantitatively analyze the detection countermeasures in Section 4.

3.1 Exponentially Distributed Query and Piggybacking Query

We describe an exponentially distributed query strategy and a piggybacking query strategy, both can be used to hide bot activities while communicating with a botmaster in a timely fashion. In our experiments in Section 3.2, we provide an experimental evaluation on both query methods.

Exponentially distributed query strategy. The Poisson process is previously believed to be a suitable model for representing stochastic processes where arrivals are independent on each other, i.e., memoryless. In [19], client-side DNS request arrivals are modeled by Poisson processes with exponential random variables with different rates λ (e.g., 2.63 queries/hour for `www.google.com` and 0.78 queries/hour for `www.cnn.com`). In our exponentially distributed query strategy, a bot probabilistically distributes DNS queries so that their intervals follow an exponential distribution with a parameterized arrival rate λ_b. Because of the memoryless feature of the model, the bot does not need to store the previous communication history. One simple way to implement this query strategy is as follows.

1. The bot sends a DNS query;
2. It computes an interval t by drawing from an exponential distribution with parameter λ_b (hardcoded or dynamically generated);
3. The bot sleeps for t, and repeats from Step 1.

There is a trade-off between being stealthy and communication efficiency. We study a bot's strategy in finding an optimal λ_b in Section 3.2, given the host-wide DNS query rates.

Piggybacking query strategy. Many (legitimate) websites contain content from multiple independent domains due to third-party content delivery, advertisements, or content mashup. Therefore, multiple DNS queries are usually issued by a host with temporal proximity. The composition of domains is usually dynamic. The piggybacking query strategy leverages this fact. A bot passively listens on the host's DNS traffic or name-translation related function calls and sends DNS queries when legitimate DNS queries are being made. Thus, the bot's query is blended among a group of legitimate DNS queries.

In the piggybacking query strategy, a bot's communication with the controller is constrained by the host's activities. Therefore, we focus on analyzing its timeliness, in terms of the dissemination efficiency of new command and data. We define *time-to-communicate* (TTC) and *minimum TTC* as follows. Minimum TTC is a threshold aiming to prevent a bot from sending queries too frequently.

Definition 1. *Time-to-communicate (TTC) is defined as the time interval between two network connections (DNS queries in our setting) of a bot for retrieving information from or submitting data to the botmaster server.*

Definition 2. *Minimum TTC is the lower bound of time-to-communicate. A bot does not send any DNS query if the bot's last DNS query was sent within the minimum TTC.*

In this piggybacking mode, bots need to know when a legitimate query is made. Since the DNS service in a server listens on port 53 for incoming requests, an outgoing packet from host to a destination IP on port 53 is an indication of a DNS request. Therefore, the bot program may monitor the host's network traffic through functions in a packet capture library, (e.g., `pcap`), and launch its own communication DNS query upon successful detection of legitimate queries.

An alternative method for learning the host's (legitimate) DNS traffic is to watch the calls for DNS-related APIs such as `gethostbyname()` function in Linux `libbind` library. `gethostbyname` looks up all IP addresses associated with a host name and is implemented in the resolver library. One way of hooking into the API function is for the bot to register a `.so` file (shared library) to the `LD_PRELOAD` environmental variable. In the registered `.so` file, the target API function is replaced by the attacker's version which can notify the bot whenever this function is called. Similarly, in Windows a bot may replace the corresponding Winsock DLL file in order to implant a DNS-notifier function.

3.2 Experimental Evaluation

The goal of this evaluation is to understand how effective the aforementioned stealthy query strategies are. Specifically, how soon botmaster disseminates commands to all or most bots; and how soon stolen data is harvested bots by botmaster? We do not allow bots to submit DNS queries at will, in order to avoid detection. We only allow bots to either piggyback their queries with legitimate DNS queries from the victim host, or follow a special inter-query distribution.

Our implementation uses the Python Modular DNS Server (`pymds`) and a specially designed plugin to respond to DNS requests. PyMDS implements the full DNS protocol while allowing the user to implement a programmatic and dynamic backend to generate the DNS records returned. Instead of returning records from a static file, PyMDS allowed for the decoding of codewords and the creation of appropriate responses.

To evaluate the *piggyback query strategy*, our dataset is a two-month-long network trace obtained from a university and collected with the IPAudit tool. The trace covered users from three departments and several research and education centers. (All machines were connected to the Internet wirelessly, i.e., there was no wired connection.) The raw dataset is 4.6GB. We identify and analyze the DNS traffic on port 53 of remote destinations. For data preprocessing, we select the most active 200 users from the our dataset by partitioning users by their (static) MAC address and sorting users by their traffic volume. We simulate the piggyback DNS-query strategy by having a bot send outbound communication whenever a host issues a UDP datagram on remote host port 53. Figure 2 shows the percentage of packets whose TTC is above the given a minimum TTC in a 10-hour-span. Three minimum TTC values are analyzed: 1, 30, and 60 minutes.

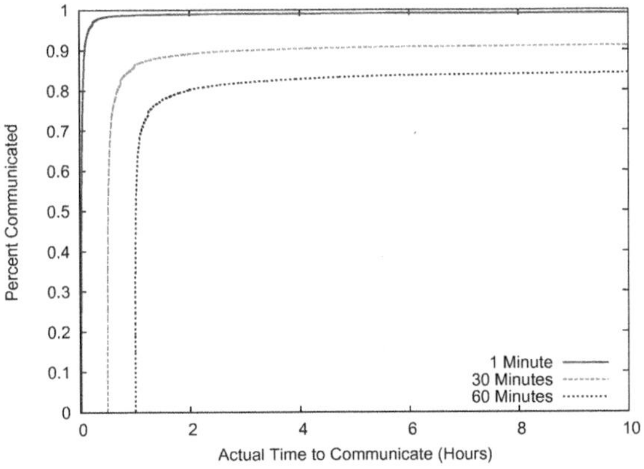

Fig. 2. Cumulative density function on the percentage of bots that have successfully sent at least one DNS query by piggybacking after a time period (X-axis). Each line corresponds to a different minimum TTC (1, 30, and 60 minutes). The figure shows a 10-hour span.

Results in Figure 2 show that the piggybacking query strategy is quite effective – at least 80% of bots are able to communicate with the botmaster within 2 hours. Clearly, there is a trade-off between minimum TTC and how soon bots communicate with the headquarter. For an active botnet where commands may change every day, minimum TTC may be set to 60 minutes.

Piggybacking case studies. We select four hosts from our dataset to simulate the piggybacking behaviors on them and evaluate the mean time-to-communicate. The four hosts are the first, 50-th, 100-th, and 200-th most active hosts according to their total traffic volume during the 2-month period. Figure 3 plots how the mean TTC changes with the minimum TTC in a piggybacking query strategy. The results show that bot's communication efficiency is higher on more active hosts. Mean time-to-communicate grows with minimum TTC and is almost always greater than minimum TTC. Their relationships for the 200 hosts studied are shown in Figure 4.

For the *exponentially distributed query strategy*, our goal is to identify an optimal range for λ_b – bot's query arrival rate on a host. We analyze the difference between two distributions: *i)* host-wide inter-arrival time for regular DNS queries with arrival rate λ, and *ii)* inter-arrival time for the bot-mixed DNS queries, i.e., new arrival rate $\lambda + \lambda_b$, where λ_b is the bot's query rate.

We use Kolmogorov-Smirnov (KS) test, which is suitable for comparing unbinned distributions that are functions of a *single* independent variable as in our case [8]. In our KS test, a higher p value ([0, 1]) represents a higher resemblance between the normal and the bot-mixed distributions. To simulate the Poisson process, we use two estimated λ values – high arrival rate of 131.5 queries/hour and low arrival rate of 39 queries/hour – based on results from [19].

Intuitively, a higher legitimate DNS query rate makes it easier for a bot to blend in its traffic. Our results in Figure 5 and Figure 6 confirm the intuition.

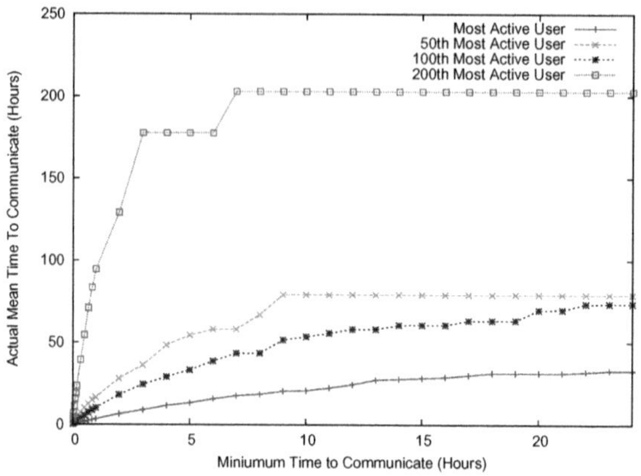

Fig. 3. Case studies on the time-to-communicate for four hosts with varying active traffic volume, given minimum TTC values shown on X-axis

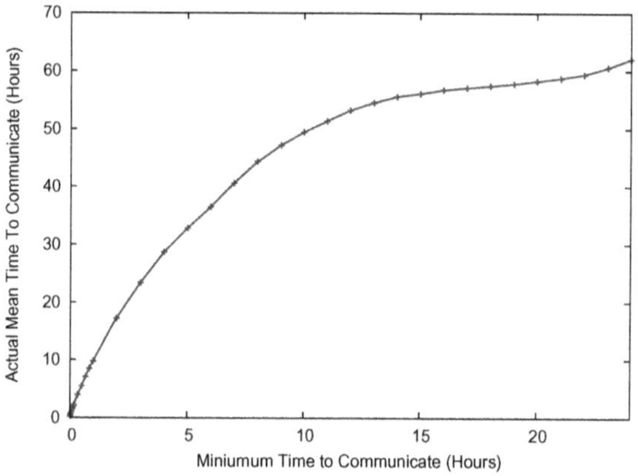

Fig. 4. Mean TTC vs. minimum TTC for 200 hosts

High rate $\lambda = 131.5$ is shown in Figure 5, and low rate $\lambda = 39$ in Figure 6, where each line represents a different amount of data collected: 10, 24, 48, and 100 hours. X-axis is the varying λ_b value. The horizontal line represents a 5% cut-off threshold that may be used for detecting anomalies.

Our results show that longer traces make it easier to discern data. Higher λ tolerates higher λ_b, allowing bots to communicate more often. Given a p value threshold, the KS test can be used to find a suitable λ_b. The experiments show that even when data is collected for long periods of time, such as 100 hours, it can be difficult to detect bots using a small λ_b. In the case of less active hosts, λ_b can be come undetectable at 4 requests per hour and with more active hosts, λ_b can be as high as 10 requests per hour.

Summary. The experiments suggest that both the piggybacking and exponentially-distributed query strategies can be effective in allowing the

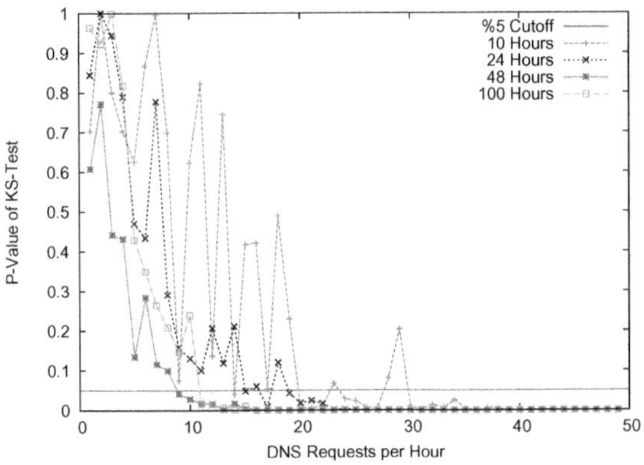

Fig. 5. KS test results between queries with the arrival rate of $\lambda = 131.5$ queries/hour and bot-mixed queries of $\lambda + \lambda_b$ (X-axis). Four runs of simulation lasting for 10, 24, 48, and 100 hours are shown.

majority of bots to communicate in a reasonable time frame without being detected. The exponentially-distributed query strategy gives the bot slightly more control over when to query. On the other hand, the optimal query rate λ_b depends on the host-wide query rate, which may change.

4 Perfect Stealth and Countermeasures

In this section, we describe and experimentally evaluate a countermeasure against DNS-based stealthy messaging systems that requires deep packet inspection and statistical analysis. Deep packet inspection examines packet payload beyond the packet header. Specifically, we quantitatively analyze the probability distributions of (bot's) DNS-packet content. We also give a formal definition for the perfect stealth in content-based communication channels in terms of the distinguishability of probability distributions between legitimate and attacker's traffic.

4.1 Perfect Stealth in Content-Based Covert Channel

Despite the existing work on constructing, measuring, and detecting general covert channels [15] and specific covert timing channels [6], there still lacks any formal definition for specifying *perfect* stealth in covert channels. We refer to content-based stealthy communication channel as where an adversary aims to hide her communication among normal payload. Such a channel differs from covert timing channel [6] where the adversary cannot modify the packet payload, but can tamper with the packet-sending schedule to reveal sensitive data. Next, we present a formal definition aiming to capture the ultimate goal of the adversary in realizing stealthy communication. We further discuss the fundamental limits and practical issues in encapsulating and detecting stealthy communication channels over the Internet.

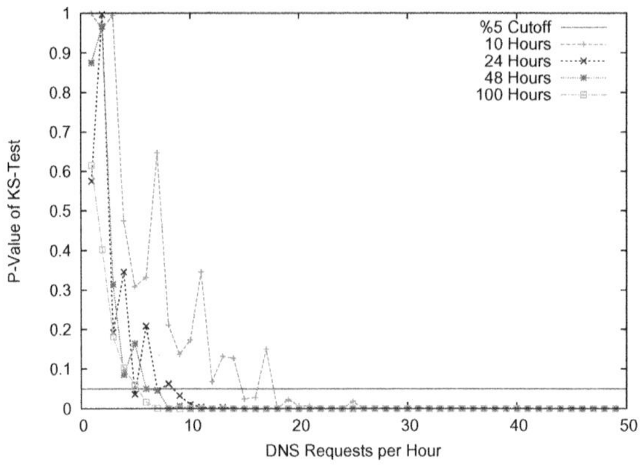

Fig. 6. KS test results between queries with the arrival rate of $\lambda = 39$ queries/hour and bot-mixed queries with $\lambda + \lambda_b$ (X-axis). Four runs of simulation lasting for 10, 24, 48, and 100 hours are shown.

Our definition for the perfect stealth in content-based covert channel is formulated following the indistinguishability in Equation 1, and is specified as a game between a challenger and a defender as follows. Specifically, our definition is given in terms of the indistinguishability between attacker's communication distribution and that of normal communication (not random distribution). Given the legitimate message space $\mathcal{M}_0$ and malicious message space $\mathcal{M}_1$ of equal size, our definition is modeled as a game between a challenger $\mathcal{C}$ – who chooses a challenge message M_b^* randomly from $\mathcal{M}_0 \cup \mathcal{M}_1$, and a defender $\mathcal{D}$ – who aims to break the perfect covert channel and guess the bit b. During the game, the defender $\mathcal{D}$ can learn some classified messages (with labels) (as the preparation) from the challenger, denoted by $\{\hat{M}\}$. The channel is perfectly covert if and only if defender's guess b' equals b with $\frac{1}{2} + \epsilon$ probability, where ϵ is negligible.

$$Pr[b' = b \mid \mathcal{C} \xrightarrow{M_b^* \notin \{\hat{M}\}} \mathcal{D}; \mathcal{D} \xrightarrow{outputs} b'] = \frac{1}{2} + \epsilon \text{ (negligible)} \quad (1)$$

We note that this definition applies to all types of communication protocols, not limited to DNS. Similar definitions can be given to capture the indistinguishability of the temporal property in legitimate traffic and adversary's traffic, e.g., probability distributions of query intervals.

The perfect stealth imposes a very strong requirement for attackers. Because botnet is designed to carry out special information and data, known botnet communication indeed has characteristics patterns different from legitimate traffic in practice – making it possible to detect. This observation implies the limitation of attackers/malware in hiding their content over the Internet.

On the other hand, as long as the malware communication has a different probability distribution from the normal traffic, then given sufficient storage and computation power, defenders can deploy deep packet inspection to detect suspicious sets of packets. We demonstrate the use of information theoretic measures, namely Jensen-Shannon (JS) divergence, for the analysis in the next section.

4.2 A Countermeasure Based on Deep Packet Inspection

We describe and evaluate a concrete countermeasure against stealthy DNS channels through statistically analyzing traffic content. To compute the byte distribution in normal and tunneling traces, we use the Jensen-Shannon (JS) Divergence D_{JS}, which is a common metric for quantifying the difference between two probability distributions P and Q, and is a commutative version of Kullback-Leibler divergence of Q from P. A lower D_{JS} value means a higher similarity in two probability distributions. The JS Divergence is particularly suited in situations where the random variable is discretized. It is computed as follows.

$$M = \frac{1}{2}(P + Q) \tag{2}$$

$$D_{KL}(P, Q) = \sum_{i=0}^{n} p_i \log \frac{p_i}{q_i} \tag{3}$$

$$D_{JS} = \frac{1}{2}(D_{KL}(P, M) + D_{KL}(Q, M)) \tag{4}$$

We experimentally compare DNS packet traces recorded on a host, specifically, on how different tunneling packets are from legitimate ones in terms of the probability distribution of content, assuming that content is not encrypted.

Such probability measures may be taken on a per-host or per subnet basis, however since a filter based on these methods must only keep an probability distribution of the bytes in a packet, no identifying information can be inferred. In this way privacy concerns can be kept at a minimum.

In the following tests, three normal DNS traces were recorded and one tunneling DNS trace via tunneled mode was recorded. Each trace corresponds to an hour-long network activities on a host. Sizes of our traces are as follows: 862KB for the tunneling trace, 823KB for normal trace 1, 699KB for normal trace 2, and 153KB for normal trace 3. In addition, the tunnel trace contained 191 `A` queries and 1433 `TXT` queries, while the normal trace 1 contained 1750 `A` queries and no `TXT` queries, and normal trace 2 contained 2417 `A` queries and no `TXT` queries. Tunneling trace contains encrypted Secure Shell (SSH) activities, i.e., SSH traffic through DNS tunneling.

When the entire packet including header is analyzed, we find that the divergence of normal traces (normal 1 and normal 3) is large (not shown). To get a more stable comparison, we drop the UDP headers and only observe the DNS payload. Figure 7 shows how the Jensen-Shannon divergence changes as more tunneling message carrying packets are mixed in. The X-axis is the ratio of tunnel trace to normal trace 1. Our results show that a divergence threshold of 0.015 can sufficiently distinguish normal traces from mixed traces containing more than 30% bot queries. These results indicate that analyzing DNS payload gives a better result than the entire DNS datagram; and the JS divergence can be used for determining anomalies in a stream of DNS packets.

Practical limitations: This countermeasure requires the access of DNS packet content and thus may not be scalable for real-time analysis, especially for

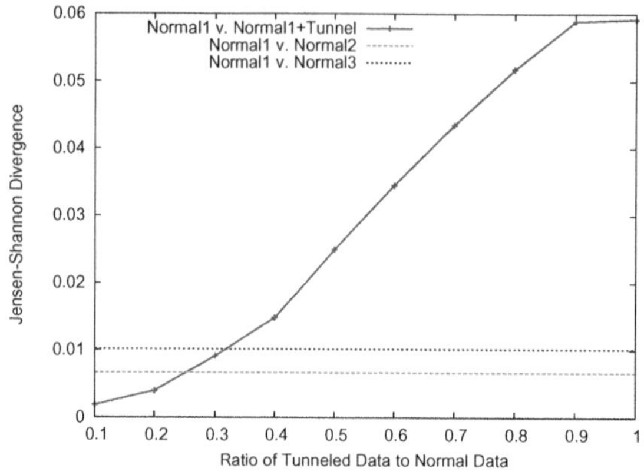

Fig. 7. Divergence computed from the payload of UDP datagrams. Horizontal lines represent the divergence of normal streams. The red line is the divergence of mixed traces.

high-bandwidth routers. Each host needs a 256-item floating point array to hold the probability distribution of DNS packets. This requirement may incur storage overhead for a large network. In addition, many legitimate applications use DNS for storing non-IP data, such as public keys in the DomainKeys protocols [5,7]. The above content-distribution based analysis may result in false positives.

There are also several practical issues and constraints when executing the detection in reality, besides the computational and storage costs. *i)* Because of traffic diversity such as in HTTP and SMTP, it is difficult to generate the standard probability distributions representing legitimate traffic in general. *ii)* The use of end-to-end encryption may prevent defenders from analyzing byte distribution of payload. *iii)* For stealthy communication as in our codeword mode described in Section 2.1, the attacker's extra payload is small and subtle without significantly affecting the overall byte distributions. The abnormal traffic is mixed with and diluted by normal traffic, making it harder to detect. Conventional signature-based detection relies on known patterns, and is not effective against new malware activities.

5 An Open Question on Domain Names

Understanding the capability of adversaries in setting up stealthy DNS-based communication channel is important. In this section, we describe an open question about how to automatically generate a large number of realistic-looking domain names for command and control purposes.

Long-lived domain names are easy to manage and cheaper to maintain, however, they are susceptible to aggregate analysis. Domain flux refers to using short-lived domain names in botnet C&C [18,24]. Domain flux typically requires bots and botnet controller to independently derive new domain names periodically. To have short-lived domains, a static approach is to have a botmaster

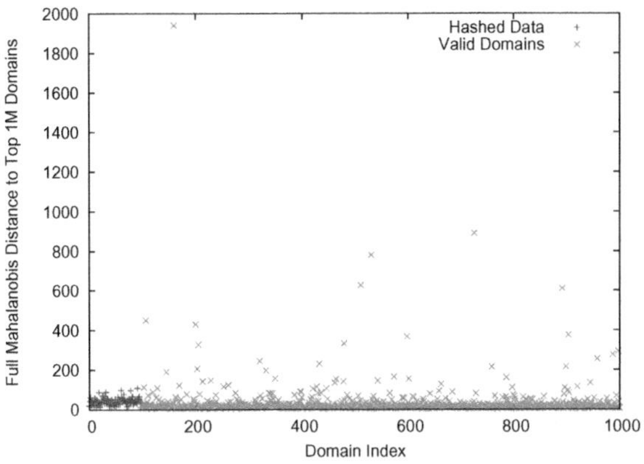

Fig. 8. Mahalanobis distances of 100 hash-based domains and 1,000 legitimate domains. X-axis is domain index with hash-based domains in 1, 100].

generate an ordered list of domain names and pack the list in malware code for bot to look up. However, there are two disadvantages for this method: large storage and high code-homogeneity – long lists of domain names shared by all bot code making the code susceptible to signature-based malware detection. An alternative is for the botmaster to send to all bots the new domain name during the current epoch. However, a communication failure may prevent the bots from learning the correct name for the next epoch.

One simple approach is for bots and their controller to independently compute the hash value of an incremental counter and a shared secret at each epoch, i.e., $H(counter \| secret)$, where H is a one-way collision-resistant hash function. However, automatically generating realistic-looking domain names by distributed parties is still an open question, which is further explained next.

Popular (and legitimate) domain names usually have semantics, i.e., meanings, whereas automatically generated domains do not. This difference (among other features) was recently used to identify anomalous domains [1]. For hash-generated domains and legitimate domains, their entropy may differ. For example, we obtained the top one million popular domains from `alexa.com` on May 25, 2010, which we used to represent legitimate domain names. For hash-generated domains, we use the first 32 bits of a hash value to generate 8 characters. The average entropy for hash-generated domains is 2.97, which is close to the maximum entropy for an 8-character string: $-\sum_{i=1}^{8} \frac{1}{8} \log_2 \frac{1}{8} = 3$. In comparison, the averaged entropy for legitimate domains is 2.17. We also evaluate the Mahalanobis distance for comparing byte distributions in `alexa.com` domain names and hash-generated domains in Figure 8. hash-generated domains give slightly higher Mahalanobis distances than legitimate domains. An interesting finding is that outlier domains from `alexa.com` that give high Mahalanobis distance tend to contain digits, e.g., `8555.com` and `c8048.com`.

These evidences prompt us to raise an interesting open question as to how to design algorithms and protocols that enable multiple parties to automatically generate synchronized domain names that resemble legitimate domains and evade statistics-based detection.

6 Related Work

Despite the fact that DNS tunneling is known for bypassing firewalls and encapsulating arbitrary data such as SSL traffic [9,3], Exploring DNS protocol as a practical command-and-control channel and identifying its limitations have not been scientifically studied. Various proof-of-concept botnet command and control systems via unconventional media exist, such as via bluetooth [21] and social networks [13]. In comparison, our work is useful beyond the specific DNS-based communication channel studied in two aspects.

- We present new quantitative techniques and evaluation regarding the detection and construction of general-purpose distributed stealthy communication systems, including temporal strategies for making stealthy communication and statistical content analysis.
- We give the first attempt to formalize the perfect stealth in content-based covert channel and point out its practical implications. We also describe an open question with a cryptographic and algorithmic flavor regarding how to automatically generate useful domain names.

For DNS-based anomaly detection, Karasaridis *et al* described the use of the Kullback-Leibler distance mentioned in Section 4 to measure byte distribution in DNS datagrams [11]. Dagon [2] proposed to quantify how anomalous the number of queries for each domain name during an hour in a day with Chebyshev's inequality and distance measures previously used for examining anomalous payloads. DNS-based anomaly detection approaches are presented in [25] for detecting botnet C&C activities. One method is to detect dynamic domain names whose query rates are abnormally high or temporally concentrated using outlier detection metrics such as Chebyshev's inequality. Our work describes stealthy DNS behaviors whose querying patterns are hard to distinguish with legitimate domains, which make the counting based detection less effective. We note that DNSSec protocol does not prevent stealthy communication via DNS.

Stone-Gross *et al* observed the use of domain flux in Torpig botnet [24], where new communication domains are generated periodically and registered by the C&C server. Torpig bots communicated with the server over HTTP, after resolving the domain name. In comparison, we investigate the feasibility of solely DNS-based command and control, without requiring any additional Web servers.

Our piggybacking DNS queries should not be confused with previously reported piggybacking methods for reducing DNS traffic. Those techniques usually take advantage of empty payload space in UDP datagrams. For example, renewal using piggyback method was proposed to piggyback cached DNS records to DNS queries to refresh expired cached records [10]. Related domains may also be piggybacked in DNS queries [20], e.g., to include `i.cnn.net` in the DNS packet for `www.cnn.com` as they are likely to be requested together by the browser.

Millen did pioneering work on covert-channel analysis [14,15], in particular in a system (host) environment. Covert channel has been heavily analyzed in the context of traffic-analysis prevention [17] and routing anonymity [16]. Our perfect covert channel definition (for one-to-one communication) can be applied to traffic matrix (for n-to-n communication) defined in [17]. Our work differs from them in that we focus on experimentally evaluating and detecting practical covert channels across the Internet.

Our work is complementary to host-based malware detection and prevention solutions, such as the cryptographic provenance verification technique by Stefan, Wu, Yao, and Xu [23].

7 Conclusions

We conducted a systematic study on the use of pure DNS queries for massive-scale stealthy communications among entities on the Internet. Our work shows that DNS – in particular the codeword mode combined with advanced querying strategies – can be used as a stealthy command-and-control channel. Because almost all computers need domain-name resolution, it is impossible to block DNS traffic. For the tunneling mode, we presented a payload-inspection based countermeasure for detecting anomalies in DNS traffic through analyzing the probability distributions of content. However, the payload inspection techniques do not apply to codeword systems.

The focus of this paper is *not* on presenting offensive techniques for attackers. Rather, we used information theoretic analysis and experiments to illustrate the need and importance of understanding the potential capabilities of adversaries. We further pointed out that although it may be difficult for attackers to achieve a perfect stealth for C&C, practical constraints may prevent detection methods such as the JS divergence test from being effective. We leave an open question on how to algorithmically generate short-lived and realistic-looking domain names.

References

1. Bilge, L., Kirda, E., Kruegel, C., Balduzzi, M.: EXPOSURE: Finding malicious domains using passive dns analysis. In: Proceedings of the Annual Network and Distributed System Security (NDSS) (February 2011)
2. Dagon, D.: Botnet detection and response, the network is the infection, 2005. In: Domain Name System Operations Analysis and Research Center Workshop (2005)
3. DeNiSe, `http://c0re.23.nu/c0de/snap/DeNiSe-snap-20021026.tar.gz`
4. DNScat, `http://tadek.pietraszek.org/projects/DNScat/`
5. Yahoo! Anti-Spam Resource Center - DomainKeys, This is an electronic document (2008), `http://antispam.yahoo.com/domainkeys` (Date retrieved: February 1, 2007)
6. Gianvecchio, S., Wang, H.: Detecting covert timing channels: an entropy-based approach. In: Ning, P., di Vimercati, S.D.C., Syverson, P.F. (eds.) ACM Conference on Computer and Communications Security, pp. 307–316. ACM, New York (2007)
7. Goodrich, M.T., Tamassia, R., Yao, D.: Accredited DomainKeys: a service architecture for improved email validation. In: Proceedings of the Conference on Email and Anti-Spam (CEAS 2005) (July 2005)

8. Hollander, M., Wolfe, D.A. (eds.): Nonparametric Statistical Methods, 2nd edn. Wiley-Interscience, Hoboken (1999)
9. Horenbeeck, M.V.: DNS tunneling, `http://www.daemon.be/maarten/dnstunnel.html`
10. Jang, B., Lee, D., Chon, K., Chul Kim, H.: DNS resolution with renewal using piggyback. Journal of Communications and Networks 11(4) (August 2009)
11. Karasaridis, A., Meier-Hellstern, K.S., Hoeflin, D.A.: Detection of DNS anomalies using flow data analysis. In: GLOBECOM. IEEE, Los Alamitos (2006)
12. Karasaridis, A., Rexroad, B., Hoeflin, D.: Wide-scale botnet detection and characterization. In: HotBots 2007: Proceedings of the First Conference on First Workshop on Hot Topics in Understanding Botnets. USENIX Association, Berkeley (2007)
13. Kartaltepe, E., Morales, J., Xu, S., Sandhu, R.: Social network-based botnet command-and-control: Emerging threats and countermeasures. In: Zhou, J., Yung, M. (eds.) ACNS 2010. LNCS, vol. 6123, pp. 511–528. Springer, Heidelberg (2010)
14. Millen, J.K.: Covert channel capacity. In: IEEE Symposium on Security and Privacy, pp. 60–66 (1987)
15. Millen, J.K.: 20 years of covert channel modeling and analysis. In: IEEE Symposium on Security and Privacy, pp. 113–114 (1999)
16. Moskowitz, I., Newman, R.E., Crepeau, D.P., Miller, A.R.: Covert channels and anonymizing networks. In: Workshop on Privacy in the Electronic Society (WPES 2003), pp. 79–88. ACM, New York (2003)
17. Newman, R.E., Moskowitz, I.S., Syverson, P.F., Serjantov, A.: Metrics for traffic analysis prevention. In: Dingledine, R. (ed.) PET 2003. LNCS, vol. 2760, pp. 48–65. Springer, Heidelberg (2003)
18. Ollmann, G.: Botnet communication topologies – understanding the intricacies of botnet command-and-control, `http://www.damballa.com/`
19. Rajab, M.A., Monrose, F., Terzis, A., Provos, N.: Peeking through the cloud: DNS-based estimation and its applications. In: Bellovin, S.M., Gennaro, R., Keromytis, A.D., Yung, M. (eds.) ACNS 2008. LNCS, vol. 5037, pp. 21–38. Springer, Heidelberg (2008)
20. Shang, H., Wills, C.E.: Piggybacking related domain names to improve DNS performance. Comput. Netw. 50(11), 1733–1748 (2006)
21. Singh, K., Sangal, S., Jain, N., Traynor, P., Lee, W.: Evaluating bluetooth as a medium for botnet command and control. In: Kreibich, C., Jahnke, M. (eds.) DIMVA 2010. LNCS, vol. 6201, pp. 61–80. Springer, Heidelberg (2010)
22. Singh, K., Srivastava, A., Giffin, J.T., Lee, W.: Evaluating email's feasibility for botnet command and control. In: DSN, pp. 376–385. IEEE Computer Society, Los Alamitos (2008)
23. Stefan, D., Wu, C., Yao, D., Xu, G.: Cryptographic provenance verification for the integrity of keystrokes and outbound network traffic. In: Proceedings of the 8th International Conference on Applied Cryptography and Network Security (ACNS) (June 2010)
24. Stone-Gross, B., Cova, M., Cavallaro, L., Gilbert, B., Szydlowski, M., Kemmerer, R., Kruegel, C., Vigna, G.: Your botnet is my botnet: Analysis of a botnet takeover. In: Proceedings of the 16th ACM Conference on Computer and Communications Security (CCS) (November 2009)
25. Villamarín-Salomón, R., Brustoloni, J.C.: Identifying botnets using anomaly detection techniques applied to DNS traffic. In: Proceedings of the 5th IEEE Consumer Communications and Networking Conference, CCNC (2008)

Fully Non-interactive Onion Routing with Forward-Secrecy

Dario Catalano[1], Mario Di Raimondo[1], Dario Fiore[2,⋆], Rosario Gennaro[3], and Orazio Puglisi[1]

[1] Dipartimento di Matematica ed Informatica – Università di Catania, Italy
{catalano,diraimondo}@dmi.unict.it, puglisi.o@gmail.com
[2] École Normale Supérieure, CNRS - INRIA, Paris, France
dario.fiore@ens.fr
[3] IBM T.J. Watson Research Center. Hawthorne, New York 10532.
rosario@us.ibm.com

Abstract. In this paper we put forward a new onion routing protocol which achieves forward secrecy in a *fully non-interactive* fashion, without requiring any communication from the router and/or the users and the service provider to update time-related keys. We compare this to TOR which requires $O(n^2)$ rounds of interaction to establish a circuit of size n. In terms of the computational effort required to the parties, our protocol is comparable to TOR, but the network latency associated with TOR's high round complexity ends up dominating the running time. Compared to other recently proposed alternative to TOR (such as the PB-OR and CL-OR protocols) our scheme still has the advantage of being non-interactive (both PB-OR and CL-OR require some interaction to update time-sensitive information), and achieves similar computational performances. We performed extensive implementation and simulation tests that confirm our theoretical analysis. Additionally, while comparing our scheme to PB-OR, we discovered a flaw in the security of that scheme which we repair in this paper.

Our solution is based on the application of forward-secure encryption. We design a forward-secure encryption scheme (of independent interest) to be used as the main encryption scheme in an onion routing protocol.

1 Introduction

As we move to use network communication in more and more aspects of everyday's life, it has become apparent that our privacy is at stake. The ability to monitor electronic communication, to store large amount of data, and to run sophisticated analytics on it, allows a sufficiently motivated party to "connect the dots" between various on-line activities of a specific user and get a pretty accurate picture of his/her private life.

These privacy concerns were recognized since the beginning of the Internet age, and *anonymous communication* was conceived as a possible approach to

⋆ Work partially done while student at University of Catania.

J. Lopez and G. Tsudik (Eds.): ACNS 2011, LNCS 6715, pp. 255–273, 2011.

their solutions. Anonymity is the user's ability to hide not only her identity but also her network information (e.g. her network address). This is of utter importance in many real life applications, where a user's identity should be decoupled from her network activities (e.g. voting, e-cash, anonymous credentials, etc.).

Chaum in 1981 [8], proposed the notion of a *anonymous channel*, realized through a *mix-net*: very informally, his idea was to route messages through a series of nodes (the mix-net). The messages are "wrapped" in several layers of encryptions and sent to the first node in the mix-net. Each node, batches a number of received ciphertexts, peels off a layer of encryption from each of them, and sends the resulting values in permuted order to the next node. The last node in the mix-net delivers the messages to the recipients. Anonymity derives from the fact that since each node permutes the messages in a randomized order before forwarding, no traffic analysis can actually link the sender to the receiver.

Goldschlag *et al.* introduced in [17] the so-called *Onion Routing* approach which is based on Chaum's idea as follows. Consider a setting defined by: a service provider, a set of users and a set of nodes (called *onion routers*). The user's goal is to establish an anonymous channel that allows him to send messages over the network without being identified. In order to do so, he selects a random set of nodes (called a *circuit*), wraps the message with several layers of encryption, one for each node, and sends it through these intermediate nodes. Because of their layered composition such wrapped messages are called *onions*. Whenever a node receives a message, it decrypts it and immediately sends the resulting value to the next node. Note that differently from Chaum's mixers, an onion router does not collect and permute a batch of messages before forwarding, but it immediately forwards what it receives. Roughly speaking, since the user chooses a random subset of the routers to forward his messages, anonymity is guaranteed by the assumption that the adversary cannot monitor the entire set of onion routers, but has only a *local* view of the network communication. This very simple and elegant idea has proven itself very popular over the Internet. Besides leading to several other constructions and implementations (e.g. [17,9,25,18,15,26,14]), it gave birth to the Onion Routing Project, later replaced by Tor [14] (the second generation onion router) which provides privacy and anonymity to a large number of users over the Internet. At the moment it counts, roughly, 1000 onion routers and hundreds of thousands of users over the world.

An important aspect of onion routing protocols is how messages travel securely through each onion router. The idea given in [17] proposes that the user encrypts a random symmetric key with the public key of each onion router: the symmetric key is used to encrypt the corresponding onion layer and the name of the next node in the circuit. This approach, unfortunately, is not robust in the face of possible server corruptions. Indeed if an adversary obtains the long-term secret key of a router O, it can then decrypt *all* the ciphertexts received by O: particularly troubling is that the adversary can decrypt communication that happened *before* the leakage of the secret key. Resistance to such attacks has been already recognized as an important security property (that is called *forward secrecy*) in other cryptographic contexts. So, even for onion routing, ideally we would like to have a protocol that is *forward-secure*, namely such that a router's corruption does not reveal anything about communication prior to the corruption.

In the context of Tor, Dingledine *et al.* [14] proposed a solution which relies on using the routers' public keys only to establish a temporary session key via an (interactive) Diffie-Hellman [12] key agreement. In order to get anonymity such interaction is made part of a specific protocol called *Tor Authentication Protocol* (TAP) which was proven secure by Goldberg [16]. The main idea of TAP is the *telescoping* technique that allows to construct the circuit and to exchange the temporary keys anonymously. However, this technique has a major drawback: its bandwidth and round complexity. In fact, in order to build a circuit of length n it is required to exchange $O(n^2)$ (symmetrically encrypted) messages. Øverlier and Syverson [24] later improved the efficiency of TAP by proposing the use of only half-certified Diffie-Hellman key-agreement, but the round complexity of telescoping is still quadratic.

A related notion of forward secrecy (sometimes called *eventual forward secrecy* in the literature) can be realized by frequently changing the long-term server keys, in order to minimize the security impact of key leakage. If the adversary learns the secret key of a server O, it may only learn the communication related to the validity period of that key. The trivial implementation of this idea (changing the servers public keys) would be very complicated in practice as it forces the routers to generate new keys with corresponding valid certificates, and the users to repeatedly obtain such certified keys.

Recently, two approaches were proposed to achieve eventual forward secrecy in a more efficient and simple way. In 2007, Kate *et al.* [20,21] proposed using identity-based encryption schemes such as [4,29] to construct an onion routing protocol called PB-OR (for pairing-based onion routing). In identity-based cryptography (introduced by Adi Shamir in [27]) the parties' public keys are their identities, and the secret keys are provided to them by a trusted Key Generation Center (KGC). PB-OR uses the original onion-routing idea to encrypt messages using the public key of the routers, except that in this case the routers' public keys are their identities together with the validity period. Therefore a router's corruption reveals only the messages encrypted during the particular period of the corruption. PB-OR has two major drawbacks: (i) the existence of a trusted KGC who can decrypt any message in the network; (ii) it requires the routers to interact with the KGC at each validity period to obtain new secret keys. While the former can be solved by using known techniques (e.g. a distributed KGC), the latter is more annoying and seems to be inherent in that construction.

These two drawbacks were addressed in a subsequent paper by Catalano *et al.* [7] who suggested using *certificateless* encryption (rather than identity-based) to construct the onions. Certificateless encryption [1] is an hybrid setting that lies between public key and identity-based cryptography: each user has an identity string ID with a matching secret key produced by a KGC and also a public/secret key pair, as in the traditional public key model but with the advantage that such key needs *not* be certified. Certificateless encryption does not suffer the problem of key escrow as the KGC cannot decrypt the messages sent to a user. The CL-OR protocol in [7] achieves eventual forward secrecy by having the routers periodically change their public keys: compared to PB-OR, CL-OR requires the users to interact with the service provider at each time period to obtain the routers' new public keys (but with the advantage of not having to manage and verify certificates).

Our Contribution. This paper presents a new onion routing protocol which outperforms TOR (and the other proposals such as PB-OR and CL-OR as well). The main improvement of our proposal is that it is *fully non-interactive.* Our main idea is to achieve eventual forward secrecy by using *forward-secure identity-based encryption (fs-IBE)* for the public keys of the routers.

Forward-secure public-key encryption (fs-PKE) was originally proposed by Anderson in [2] exactly as a way to achieve eventual forward secrecy for public-key encryption, without requiring users to continuously change their public keys. In Anderson's idea, a user U of a fs-PKE scheme publishes a static public key pk_U and senders encrypt messages using this public key and a time value t. To decrypt such ciphertexts, U uses a secret key $sk_{U,t}$. At the beginning, U holds the secret key $sk_{U,0}$, and at each time period U *updates* its secret key from $sk_{U,t}$ to $sk_{U,t+1}$ and erases $sk_{U,t}$. This process must be one-way – while it is easy to compute $sk_{U,t+1}$ from $sk_{U,t}$ the reverse must be hard – in order to achieve eventual forward secrecy: if $sk_{U,t}$ is compromised past communication remains secret. Our contribution can be described as follows:

- First, we propose a new onion routing protocol *fs-ID-OR*, which uses the "classical" onion-routing approach to construct onions by using the static public keys of the routers, except that we use an identity-based forward secure encryption (*fs-IBE*) scheme for the routers' public-keys.
- Next, we build an fs-IBE scheme by carefully applying a generic paradigm by Canetti *et al.* [6] to the Hierarchical Identity-Based Encryption of Boneh *et al.* [3]. This scheme is tailored to the onion-routing application and has other properties (discussed below) which can make it of independent interest.

The advantages of fs-ID-OR compared to PB-OR and CL-OR are substantial. Compared to PB-OR, our new scheme does not require the KGC to be involved in the generation of new secret keys for the routers: indeed in fs-ID-OR the update of the secret key at each time period is a local, non-interactive procedure performed by the router. Compared to CL-OR the public keys of the routers are fixed throughout all time periods (only the secret keys change) so the users do not need to obtain new public keys for the routers after each time period. Compared to Tor, fs-ID-OR has a completely non-interactive circuit-building protocol with linear round complexity. This makes fs-ID-OR a *truly non-interactive solution* as it requires no interaction between the routers, the KGC or the users to update time information after each time change.

It is fair to notice that in practice clients have to receive up-to-date informations about the state of the network to ensure that they create correct circuits (e.g. restrictions on the paths, status of online nodes, delays, etc.) since these informations are all security sensitive items. Therefore a truly secure solution seems to be interactive anyway, and the advantage of our proposal limited. However, we argue that the cost of exchanging and processing cryptographic information related to the protocol is likely to be orders of magnitude larger than the cost of receiving network status updates, and therefore removing interaction from the cryptographic part of the protocol is not just a theoretical exercise, but a real practical advantage.

Note that the level of protection afforded by eventual forward-security is related to the frequency of updates of the long-term keys (more frequent updates imply less past information being leaked in case of a key compromise). Because our solution is non-interactive, we have removed the major cost of key updates, thus making very frequent updates possible. Remarkably, our non-interactive solution does not come with high efficiency cost. In terms of computational load, our protocol is comparable with PB-OR, and definitely better than Tor (which is saddled by the cost of telescoping). The performance details of our protocol are discussed in Section 6 where we report an extensive implementation and simulation tests. The basic version of our protocol works in the identity-based setting: therefore we must assume a trusted KGC who has the ability to decrypt all communications, as in PB-OR. However, we stress on that it is possible to modify our protocol to work in both the classical PKI setting and in the certificateless setting so to avoid the key-escrow problem. We discuss these variations in Section 4.

NOTE. While proving the security of our scheme, we noticed a small flaw in one of the security arguments in [21]. They claim that the anonymity property can be achieved by assuming that the encryption schemes used in the Onion Routing protocol are simply semantically secure. But our proof of security shows that anonymity relies in a crucial way on assuming that the encryption schemes are secure against chosen ciphertext attack.

OTHER RELATED WORK. We refer the reader to the work in [23,5] for formal security definitions for the problem of onion routing. We discuss the relationship of our work with respect to the Camenisch and Lysyanskaya formal model [5] in Section 2. A forward-secure (hierarchical) identity-based public key encryption is presented by Yao *et al.* in [30]. Our new scheme is somewhat uncomparable to theirs: our scheme was designed to satisfy only the minimal security properties needed for the onion-routing application, while the scheme in [30] proposes additional security properties which might be useful in other contexts. As a result our scheme is simpler and more efficient (in particular it allows for constant size ciphertexts), but does not satisfy all the security properties proposed in [30].

2 Forward-Secure Identity-Based Onion Routing

In this section we introduce the notion of *Forward-Secure Identity-Based Onion Routing* (*fs-ID-OR*). As usual, an onion routing protocol is characterized by a service provider, a set of "onion routers" and users. The goal of the protocol is to provide anonymity over a network to users and the basic idea is that users route their traffic throughout an encrypted circuit of randomly chosen onion routers.

In addition, our solution considers *forward-secrecy*, a property which is in general quite important for cryptographic protocols. Informally, it guarantees that the security properties still hold even if the adversary can corrupt *all* the parties and learn their secret keys *after* a protocol session is expired. If one focuses only on adversaries that can corrupt parties *after* a specific time period τ, then we call this property *eventual* forward-secrecy. Otherwise it is called *immediate* forward-secrecy. In our work we will focus on eventual forward-secrecy since it is the strongest notion that is achievable in a non-interactive way. Our definition of fs-ID-OR follows the traditional notion of onion routing but focuses on the identity-based setting where each onion router is represented by a unique identity string OR_i (e.g. its name, address, etc.) and receives a secret key related to such string by a trusted entity, called the *Key Generation Center* (KGC).

In order to consider forward-secrecy, we assume that the time is split into time periods of the same length. A fs-ID-OR protocol consists of the following phases:

Setup and Key Generation. The service provider generates the global parameters of the system and makes them available to all users. When an onion router with identity OR_i joins the system at time t, the service provider acts as a KGC and uses its master secret to generate a private key $\mathsf{sk}_{OR_i,t}$ for OR_i. When a time period t expires, each onion router is required to update its secret key, that means that it runs a specific algorithm that on input $\mathsf{sk}_{OR_i,t}$ outputs $\mathsf{sk}_{OR_i,t+1}$ while the old key is erased.

Circuit construction. In this phase the user firstly has to obtain a list L containing the identities of all the available onion routers. Such list is maintained by the KGC and is updated at a specific interval t'. We notice that t' might also be different from the τ used for updating the keys. Next, the user chooses an ordered set of n onion routers $OR_1, \ldots, OR_n$ at random among those in L. This ordered set is called *circuit* and the number n is typically fixed and specified in the protocol parameters. In order to send messages through the circuit at time t, the user builds a special ciphertext O_1, called "onion ciphertext" such that, for all $i = 1$ to n onion router OR_i is able to partially decrypt the onion O_i (using its secret key of time t). From such partial decryption it obtains: (i) the address of the next router OR_{i+1} in the circuit (ii) and another onion ciphertext O_{i+1}. The user sends O_1 to OR_1 and whenever router OR_i receives O_i, it decrypts it and forwards O_{i+1} to OR_{i+1}. Finally, the last router of the circuit gets the message m and the address P of the actual recipient, and forwards m to P.

2.1 Security of Forward-Secure Identity-Based Onion Routing

Now we describe the security properties that a fs-ID-OR protocol should satisfy.

Integrity and Correctness. *Correctness* states that when parties follow the protocol, then the recipient should get the message that was originally sent and encrypted by the sender. Let n be a pre-specified upper bound for the number of routers in a circuit. Then we say that an onion routing protocol satisfies *integrity* if it is possible to recognize an onion ciphertext which is intended for more than n routers.

Cryptographic Unlinkability. Cryptographic unlinkability formalizes in a cryptographic way the fact that a fs-ID-OR protocol provides anonymity. Informally, this property says that an attacker should not be able to find a link between the sender and the receiver of a given communication. We point out, as explained in [21], that network-level attacks are not considered at this stage.

Consider the following game between an adversary $\mathcal{A}$ and a Challenger:

Setup. The Challenger generates the public parameters and gives them to $\mathcal{A}$.

Phase 1. In this phase the adversary is allowed to:
- corrupt onion routers and learn their secret keys (at specific time t);
- submit a tuple (OR, t, O) to get the decryption of O under OR's secret key at time t.

Challenge. At some point the adversary is allowed to choose a message m, a time period t^* and routers $OR_1, OR'_1, OR_2, OR'_2, OR_H$ such that OR_H is honest (i.e. OR_H has not been corrupted in the previous phase, or it has been corrupted at time $t > t^*$). The Challenger flips a binary coin $b \xleftarrow{\$} \{0,1\}$ and proceeds as follows. If $b = 0$ it creates:
- O_1 as the onion for the circuit (OR_1, OR_H, OR_2)
- and O'_1 as the onion for the circuit (OR'_1, OR_H, OR'_2).

Otherwise, if $b = 1$ it creates
- O_1 as the onion for the circuit (OR_1, OR_H, OR'_2)
- and O'_1 as the onion for the circuit (OR'_1, OR_H, OR_2).

Let O_H, O_2 and O'_H, O'_2 be the onion ciphertexts contained into O_1 and O'_1 respectively. Finally (O_1, O'_1) is given to the adversary.

Phase 2. $\mathcal{A}$ can proceed as in Phase 1 except that:
- OR_H cannot be corrupted at time $t \leq t^*$,
- $\mathcal{A}$ cannot submit (OR_H, t^*, O_H) and (OR_H, t^*, O'_H) to the decryption oracle (otherwise the adversary would trivially win);
- $\mathcal{A}$ can ask to the Challenger the decryption of a pair (O, O') under OR_H's secret key at time t^*. However in this case the Challenger does the following. It first decrypts O and O' and gets $(\overline{OR}, \overline{O})$ and $(\overline{OR'}, \overline{O'})$ respectively. If $\overline{OR} = OR_2$ and $\overline{OR'} = OR'_2$ then the Challenger outputs $((\overline{OR}, \overline{O}), (\overline{OR'}, \overline{O'}))$. Otherwise if $\overline{OR} = OR'_2$ or $\overline{OR'} = OR_2$ then $\mathcal{A}$ is given $((\overline{OR'}, \overline{O'}), (\overline{OR}, \overline{O}))$ (i.e. the tuple corresponding to OR_2 is always given first). We notice that such a requirement is essential, otherwise the adversary might trivially win the game.

Guess. At the end, the adversary outputs a guess b' for b and it wins if $b' = b$.

We define the advantage of an adversary $\mathcal{A}$ in this game as $\mathbf{Adv}^{anon}_{ID-OR}(\mathcal{A}) = 2\Pr[b' = b] - 1$ and we say that an onion routing protocol has cryptographic unlinkability if for any PPT adversary $\mathcal{A}$, $\mathcal{A}$'s advantage is at most negligible.

Remark. A more general definition would consider an adversary $\mathcal{A}$ that in the challenge phase can choose circuits of length n instead of length 3. However, since we assume that all but one (i.e. OR_H) routers can be corrupted, one may think to OR_1 (resp. OR_2) as the collapsed set of adversarially controlled routers before (resp. after) the single honest one (i.e. OR_H), and the same can be done for OR'_1, OR_2 and OR'_2.

Circuit Position Secrecy. This property says that it should not be possible to learn a router's position in the circuit by looking at the ciphertext it is receiving. In those constructions where the onion ciphertexts are built as re-encryptions with several keys (e.g. $\Gamma = E_{K_1}(E_{K_2}(\cdots E_{K_n}(m)\cdots)))$ this property cannot hold. Camenisch and Lysyanskaya [5] showed that it is in fact sufficient to look at the ciphertext's size to derive such information. Indeed, in every randomized encryption scheme the ciphertext space is bigger than the plaintext one (typically by a constant). However solutions to this problem have been proposed [5,21,11,19].

3 A Generic Construction of FS-ID Onion Routing

In this section we show how to construct a fs-ID-OR protocol in a black-box way from any forward-secure identity-based key encapsulation mechanism (fs-IB-KEM) and a symmetric encryption scheme.

Our construction generalizes the idea of Kate *et al.* [21] whose scheme can be seen as an instantiation of our generic construction when using the IB-KEM of Boneh and Franklin [4] and considering an interactive protocol for updating routers' keys (i.e. the KGC generates new keys every time period).

In what follows we define the primitives that are relevant for our construction.

Forward-Secure Identity-Based Key Encapsulation. A Forward-Secure Identity-Based Key Encapsulation Mechanism (fs-IB-KEM) is defined by the following algorithms:

$\mathsf{Setup}(1^k, T)$. It takes as input the security parameter k and the total number of time periods T and outputs a public key MPK and a master secret key MSK.

$\mathsf{KeyGen}(MSK, \mathsf{ID}, t)$. The key generation algorithm uses the master secret key to produce a private key $\mathsf{sk}_{\mathsf{ID},t}$ that is related to the identity ID and the time period t.

$\mathsf{KeyUpdate}(\mathsf{sk}_{\mathsf{ID},t})$. Given in input $\mathsf{sk}_{\mathsf{ID},t}$ (the secret key of identity ID for time period t) the key update algorithm outputs a new key for time period $t+1$.

$\mathsf{Encap}(MPK, \mathsf{ID}, t)$. Given the master public key, an identity ID and a time period t, the encapsulation algorithm outputs a ciphertext C and a session key K.

$\mathsf{Decap}(\mathsf{sk}_{\mathsf{ID},t}, C)$. The decapsulation algorithm uses the secret key of identity ID and time period t to recover the session key K from a ciphertext C.

Correctness requires that for all identities $\mathsf{ID} \in \{0,1\}^*$ and time periods $0 \leq t < T$:

$$\Pr\left[\begin{array}{l}(MPK, MSK) \xleftarrow{\$} \mathsf{Setup}(1^k, T); \mathsf{sk}_{\mathsf{ID},t} \xleftarrow{\$} \mathsf{KeyGen}(MSK, \mathsf{ID}, t);\\ (C, K) \xleftarrow{\$} \mathsf{Encap}(MPK, \mathsf{ID}, t); K' \leftarrow \mathsf{Decap}(\mathsf{sk}_{\mathsf{ID},t}, C) : K' = K\end{array}\right] = 1$$

We notice that the same holds when the secret key $\mathsf{sk}_{\mathsf{ID},t}$ is obtained via the key update algorithm. Below we define the notion of forward-security against chosen-ciphertext attacks (fs-ID-IND-CCA) for fs-IB-KEM schemes. Consider the following game between an adversary $\mathcal{A}$ and a Challenger:

Setup. A pair of master keys $(MPK, MSK) \xleftarrow{\$} \mathsf{Setup}(1^k, T)$ is generated and the adversary is given MPK.

Phase 1. In this phase the adversary is given access to oracles $\mathsf{breakin}(\cdot,\cdot)$ and $\mathsf{Decap}(\cdot,\cdot,\cdot)$ as follows:
- On input (ID, i) the $\mathsf{breakin}$ oracle computes the secret key $\mathsf{sk}_{\mathsf{ID},i}$ and gives it to the adversary.
- On query (ID, i, C) to the decapsulation oracle, the Challenger computes the key $\mathsf{sk}_{\mathsf{ID},i}$ and gives $K \leftarrow \mathsf{Decap}(\mathsf{sk}_{\mathsf{ID},i}, C)$ to the adversary.

Challenge. At some point the adversary is allowed to output a pair (ID^*, t^*) such that either ID^* is different from all the identities queried to $\mathsf{breakin}$ in the previous phase or $\mathsf{ID}^* = \mathsf{ID}_j$ and $t^* < t_j$ (where (ID_j, t_j) was the j-th query to $\mathsf{breakin}$). The Challenger computes $(C^*, K_0) \xleftarrow{\$} \mathsf{Encap}(MPK, \mathsf{ID}^*, t^*)$ and picks a random session key $K_1 \xleftarrow{\$} \mathcal{K}$. Then it flips a random bit $b \xleftarrow{\$} \{0,1\}$ and gives (C^*, K_b) to the adversary.

Phase 2. As Phase 1 except that the adversary is not allowed to query the decapsulation oracle on $(\mathsf{ID}^*, t^*, C^*)$ and the $\mathsf{breakin}$ oracle on (ID^*, j) with $j \leq t^*$.

Guess. At the end of the game the adversary outputs a bit b' as its guess for b.

We define the advantage of $\mathcal{A}$ in this game as

$$\mathbf{Adv}_{\mathcal{IB}}^{fs-IND-ID-CCA}(\mathcal{A}) = |2\Pr[b = b'] - 1|.$$

Definition 1 (fs-IND-CCA security). *A fs-IB-KEM is* forward-secure against chosen-ciphertext attacks *if for any PPT adversary $\mathcal{A}$ we have:* $\mathbf{Adv}_{\mathcal{IB}}^{fs-IND-ID-CCA}(\mathcal{A}) \leq \epsilon$*, where ϵ is negligible in the security parameter.*

3.1 Our Generic Construction

Let $\mathcal{IB}$ = $(\mathsf{Setup}, \mathsf{KeyGen}, \mathsf{KeyUpdate}, \mathsf{Encap}, \mathsf{Decap})$ be a fs-IB-KEM and $\mathcal{E}$ = (KG, E, D) be a symmetric encryption scheme (whose notion is quite standard, and is omitted for lack of space). We construct the following protocol:

Setup. In this phase the KGC runs the setup algorithm of $\mathcal{IB}$ to obtain a master public key MPK and a master secret key MSK. The master public key is made available to all users together with informations about the time. We assume the time to be split into time periods of length τ (e.g. τ = one hour) such that when such a period expires, each onion router updates his secret key as explained in the next phase. Moreover, the KGC maintains a list L containing the identities of all the onion routers available at a specific time. Such list is updated at a specific interval τ', which does not have to be necessarily equal to τ.

Key Generation. Whenever an onion router OR_i joins the system, at time t, the KGC generates a secret key $\mathsf{sk}_{OR_i,t} \xleftarrow{\$} \mathsf{KeyGen}(MSK, OR_i, t)$ for it. To achieve forward secrecy, when a time period t expires each onion router OR_i is required to update his secret key by running $\mathsf{sk}_{OR_i,t+1} \leftarrow \mathsf{KeyUpdate}(\mathsf{sk}_{OR_i,t})$ and erasing $\mathsf{sk}_{OR_i,t}$ from its memory.

Circuit construction. Assume that a user wants to build a circuit at time t. First he obtains the updated list L from the KGC and then he chooses an ordered sequence of n onion routers $OR_1, \ldots, OR_n$ at random among those in L. Next, for all $i = n$ to 1 it proceeds as follows. It runs $(C_i, K_i) \xleftarrow{\$} \mathsf{Encap}(MPK, OR_i, t)$ and creates "onion ciphertext" $O_i = (C_i, \Gamma_i)$ where $\Gamma_i = E_{K_i}(OR_{i+1}, O_{i+1})$. The user sends O_1 to the first onion router in the circuit. Whenever onion router OR_i gets a pair $O_i = (C_i, \Gamma_i)$ it recovers $K_i \leftarrow \mathsf{Decap}(\mathsf{sk}_{OR_i,t}, C_i)$ and then runs $(OR_{i+1}, O_{i+1}) \leftarrow D_{K_i}(\Gamma_i)$. Finally it sends O_{i+1} to OR_{i+1} (which is the next router of the circuit).

The first time a user is using a circuit, he wants to be aware that all the chosen routers are available. Therefore it sends a special message $\perp$ through the circuit (i.e. $\Gamma_n = E_{K_n}(\perp)$). When an onion router decrypts and obtains $\perp$ it learns that it is the last router of the circuit and sends back a confirmation message $E_{K_n}(Ack)$ to the previous router. Upon the receipt of a confirmation message an onion router OR_i encrypts it using K_i and sends it to the previous router. For this reason, we assume that each router keeps in memory a session state containing the two adjacent nodes and the session keys. We notice that this is also useful to prevent replay attacks. Finally, upon the receipt of a confirmation message, the user verifies its validity by decrypting it using the session keys $K_1, \ldots, K_n$.

Once the circuit has been successfully established, the user will use it to send messages over the network. In particular, he will re-use the same session keys $K_1, \ldots, K_n$ to form the onions. This allows to avoid expensive asymmetric encryption (and decryption) operations.

3.2 Security

Integrity and Correctness. Let n be the fixed upper bound for the number of routers in the circuit. We notice that an onion ciphertext containing more than n layers of encryption can be easily recognized by looking at its length. Therefore our protocol has integrity. On the other hand correctness easily follows from the construction and the correctness of the two employed encryption schemes.

Cryptographic Unlinkability. The property is proven by the following theorem whose proof is omitted for lack of space.

Theorem 1. *If $\mathcal{IB}$ is fs-ID-IND-CCA secure and $\mathcal{E}$ is IND-CCA secure, then the protocol given in Section 3 satisfies cryptographic unlinkability.*

A remark on cryptographic unlinkability. The previous theorem proves the cryptographic unlinkability of our generic construction by assuming that both the fs-IB-KEM and the symmetric encryption schemes are secure in an IND-CCA sense. We need this property because of the adversary's (realistic) ability to ask decryption of onions (e.g. he may simply send an onion to a router and look for its outgoing packets).

Cryptographic unlinkability was first defined in [21] (though in a slightly less formal way). There the authors claimed that for their construction this property is implied by the IND-CPA security of the symmetric encryption scheme. The proof of this claim is not formal and it is unclear how the proof can manage the adversary's decryption queries in the "C processing" phase.

Moreover, we notice that if only IND-CPA security is considered, then the adversary might modify only one of the two challenge ciphertexts O_H, O'_H in such a way that, after seeing their decryptions, it can recognize which of the two onions they come from. More precisely, $\mathcal{A}$ (who owns all secret keys but OR_H's) may keep O'_H the same and modify O_H such that the encrypted onion will decrypt to a random message[1]. When $\mathcal{A}$ later receives the two decrypted onions, it will be able to recognize what was the path chosen by the challenger. On the other hand, in our case assuming IND-CCA security allows to obtain a correct and formal proof of cryptographic unlinkability.

Circuit Position Secrecy. Unfortunately, our protocol does not satisfy this property as it is vulnerable to the attack showed by Camenisch and Lysyanskaya in [5] that allows to learn the circuit's position of a ciphertext's recipient. Precisely, this can be done by looking at the length of a ciphertext. However, if one is interested into this property, then it is possible to slightly modify our protocol using the technique proposed by Kate *et al.* in [21]. Its application to our protocol is straightforward and thus we can obtain a protocol with circuit position secrecy, even if this comes at the cost of having longer ciphertexts.

4 Certificateless and PKI Variants

The onion routing scheme we presented in Section 3, uses an *identity-based* forward-secure encryption for the routers. This means that the routers' identities serve as their public keys and the secret keys are provided to them by a trusted KGC. We chose this approach to minimize the size of the public information required to run the system: public keys and certificates (users need to know only the KGC's). The obvious drawback of this approach is *key escrow:* the KGC has the ability to decrypt any message. In this section we describe two simple variations to eliminate the key escrow problem from our scheme. One will yield a scheme in the classical PKI setting: each router has his own public key and certificate. The second variation will be a certificateless (CL) scheme: in this case together with the KGC's keys, routers will hold public keys which however need not be certified. Both variations pay some price compared to the identity-based scheme presented earlier: when instantiated with our scheme of Section 5,

[1] The definition of IND-CPA security does not rule out that this is possible, and indeed it can be done in many IND-CPA secure schemes.

the PKI version has to face long public keys (but no increase in computation); the CL one requires a few extra exponentiations to the user. Details follow.

A PKI VARIATION. To obtain eventual forward secrecy for an onion-routing protocol, it is sufficient to use any forward-secure encryption scheme, not necessarily an *identity-based* one. In particular, one could use our scheme where each router acts as his own KGC, and give himself different keys for each time period. If we were to follow this approach, there would be no centralized KGC and no key escrow problem. To create an onion, a user would have to do the same amount of work as in the identity-based scheme above. The only problem is that the concrete scheme we propose in Section 5 has longer public keys. Thus the amount of data to be stored at each user would be large.

A CERTIFICATELESS VARIATION. There is a generic way to transform any ID-based encryption into a CL one [1]. The receiver R, who already holds a secret key sk_R related to his identity and provided to him by the KGC, also publishes an independent public key PK and keeps the secret key SK. To encrypt a message m for R, a sender splits $m = m_1 \oplus m_2$, with m_1 random, and sends m_1 encrypted with the ID-based scheme, and m_2 encrypted under PK. As pointed out in [1] the public key PK needs not be certified to belong to R (intuitively this is because only R can decrypt m_1). The advantage is that now the KGC cannot decrypt the message m. This generic paradigm can be efficiently implemented in our case. In our protocol the user establishes a shared symmetric key k_i with the ith router in the circuit using the id-based KEM described in the previous section. The key k_i is used to encrypt the ith layer of the onion. To transform this scheme into a CL one, each router can publish a public key and the user runs another KEM to establish another key k_i' with it, and the ith layer of the onion is encrypted with $k_i \oplus k_i'$. An efficient instantiation of this KEM could be any KEM that works over the same cyclic group used for the ID-based scheme (so that no new public information must be generated), e.g. establishing a random key using ElGamal. This approach requires the user to compute n extra exponentiations to create an onion (n is again the length of the circuit).

5 The Proposed Construction

In this section we present a concrete scheme that realizes a fs-IB-KEM with $T = 2^{\ell+1} - 1$ time periods. Our solution is presented in two steps. First, we give a forward secure identity based encryption scheme (fs-IBE) that is probably secure only in an IND-CPA sense. Next, we apply a simple variant of Dent's transformation [10] in order to convert our basic fs-IBE into an IND-CCA secure fs-IB-KEM.

As for the first construction, we construct our fs-IBE as follows. Following the idea of Canetti-Halevi-Katz [6], we use a binary tree of height ℓ where each time period is associated with a node of the tree according to a pre-order traversal. If w^t is the node of the tree associated with time period t we have:

- $w^0 = \epsilon$, i.e. the root of the tree;
- if w^i is an internal node, then $w^{i+1} = w^i||0$ (where $||$ is the concatenation operator);
- if w^i is a leaf node (and $i < T - 1$) then $w^{i+1} = w'1$ where w' is the longest string such that $w'0$ is a prefix of w^i.

The proposed scheme builds upon the HIBE of Boneh, Boyen and Goh [3] and the generic construction of Canetti-Halevi-Katz [6] as follows. The first level of the hierarchy contains the identities and then each identity has below a binary tree that represents the evolving time. In this setting a user who is given $\mathsf{sk}_{\mathsf{ID},0}$ can derive the secret keys for all the nodes in its binary subtree, that is for all successive time periods. In order to achieve forward-security, users are required to update their keys every time the period expires. More precisely a user computes $\mathsf{sk}_{\mathsf{ID},t+1} \xleftarrow{\$} \mathsf{KeyUpdate}(\mathsf{sk}_{\mathsf{ID},t})$ and deletes $\mathsf{sk}_{\mathsf{ID},t}$. The construction is based on the decisional weak ℓ-Bilinear Diffie Hellman Inversion Assumption (ℓ-wBDHI* for short) that was introduced by Boneh, Boyen and Goh in [3]. The ℓ-wBDHI* problem is defined in a bilinear group G of prime order p where $g \in G$ is a generator. Given $D = (g, h, g^{\alpha}, g^{\alpha^2}, ..., g^{\alpha^{\ell}})$, for random $\alpha \in \mathbb{Z}_p^*$ and $h \in G$, we say that an algorithm $\mathcal{A}$ has advantage ϵ in solving decisional ℓ-wBDHI* in G if

$$\left| \Pr[\mathcal{A}(D, e(g,h)^{\alpha^{\ell+1}}) = 0] - \Pr[\mathcal{A}(D, e(g,g)^z) = 0] \right| \leq \epsilon$$

where the probability is taken over the random choices of $\alpha, z \in \mathbb{Z}_p^*$ and $h \in G$.

The ℓ-wBDHI* assumption holds in a bilinear group G if, for any ℓ polynomial in k, any polynomially bounded adversary $\mathcal{A}$ has at most negligible advantage.

The scheme follows:

$\mathsf{Setup}(1^k, \ell)$. Let G and G_T be two groups of prime order p equipped with a bilinear map $e : G \times G \to G_T$. Let $g \in G$ be a generator. Pick a random $\alpha \xleftarrow{\$} \mathbb{Z}_p^*$ and set $g_1 = g^{\alpha}$. Then take random elements $g_2, u, v, h_1, \ldots, h_{\ell} \xleftarrow{\$} G$, compute $z = e(g_1, g_2)$ and select an hash function $H : \{0,1\}^* \to \mathbb{Z}_p^*$. The master public key is $MPK = (p, G, G_T, g, g_1, g_2, z, u, v, h_1, \ldots, h_{\ell}, H)$ and the master secret key is $MSK = g_2^{\alpha}$.

$\mathsf{KeyGen}(MSK, \mathsf{ID}, t)$. Let $\mathsf{sk}_{\mathsf{ID},w}$ be the key of the node w where w is a binary string of length at most ℓ. A key $\mathsf{sk}_{\mathsf{ID},t}$ is organized as a stack of node keys where $\mathsf{sk}_{\mathsf{ID},w^t}$ is on top.
A node key $\mathsf{sk}_{\mathsf{ID},w^t}$ is computed as follows. Let $w^t = w_1 \cdots w_k$ (with $0 \leq k \leq \ell$) be the binary string representing the node w^t. Pick a random $r \xleftarrow{\$} \mathbb{Z}_p^*$ and compute $d_0 = g_2^{\alpha}(uv^{H(\mathsf{ID})} \prod_{i=1}^{k} h_i^{f(w_i)})^r$, $d_1 = g^r$, $b_i = h_i^r$ for $i = k+1$ to ℓ. Since $0 \notin \mathbb{Z}_p^*$ $f : \{0,1\} \to \mathbb{Z}_p^*$ is a function that maps 0 and 1 to specific values of $\mathbb{Z}_p^*$ (e.g $f(0) = 1, f(1) = 2$). Thus we have $\mathsf{sk}_{\mathsf{ID},w^t} = (d_0, d_1, b_{k+1}, \ldots, b_{\ell})$. Finally $\mathsf{sk}_{\mathsf{ID},t}$ contains all the node keys of the stack that are needed to derive the keys of successive time periods. We notice that the stack will contain at most $O(\ell)$ node keys.

$\mathsf{KeyUpdate}(\mathsf{sk}_{\mathsf{ID},t})$. First pop the first key from the stack. If w^t is a leaf node, then $\mathsf{sk}_{\mathsf{ID},w^{t+1}}$ is the next key on the stack. Otherwise, if w^t is an internal node, compute $(\mathsf{sk}_{\mathsf{ID},w^t0}, \mathsf{sk}_{\mathsf{ID},w^t1})$ as described below and push $\mathsf{sk}_{\mathsf{ID},w^t1}$ and then $\mathsf{sk}_{\mathsf{ID},w^t0}$ onto the stack. In both cases the node key $\mathsf{sk}_{\mathsf{ID},w^t}$ is erased. Given the node key $\mathsf{sk}_{\mathsf{ID},w^t} = (d_0, d_1, b_{k+1}, \ldots, b_\ell)$, a key $\mathsf{sk}_{\mathsf{ID},w^tb}$ for $b \in \{0,1\}$ is obtained as follows. Pick a random $t \xleftarrow{\$} \mathbb{Z}_p^*$ and compute

$$d_0' = d_0 (uv^{H(\mathsf{ID})} \prod_{i=1}^{k} h_i^{f(w_i)} h_{k+1}^{f(b)})^t b_{k+1}^{f(b)}, \quad d_1' = d_1 g^t, \quad b_i' = b_i h_i^t$$

for $i = k+2$ to ℓ. It is easy to notice that such key is correctly distributed for randomness $r + t$.

$\mathsf{Encrypt}(MPK, \mathsf{ID}, t, m)$. Let $w^t = w_1 \cdots w_k$ be the node of the tree associated with t. Pick a random $s \xleftarrow{\$} \mathbb{Z}_p^*$ and compute $C_0 = (uv^{H(\mathsf{ID})} \prod_{i=1}^{k} h_i^{f(w_i)})^s$, $C_1 = g^s$ and $C_2 = z^s m$. Finally output $C = (C_0, C_1, C_2)$.

$\mathsf{Decrypt}(\mathsf{sk}_{\mathsf{ID},t}, C)$. The message is recovered by computing $m = \frac{C_2 e(C_0, d_1)}{e(C_1, d_0)}$.

The security of the scheme follows from the following theorem whose proof is omitted for lack of space.

Theorem 2. *The scheme is fs-IND-ID-CPA secure if the decisional $(\ell+1)$-wBDHI* holds and H is modeled as a random oracle.*

Now we show how to convert the construction given above into a IND-CCA secure fs-IB-KEM using a very simple variant of Dent's transform [10]. Such a transform allows to convert a forward secure IBE satisfying very weak security requirements into an IND-CCA secure fs-IB-KEM. Specifically the underlying IBE is required to be only one-way forward secure.

Suppose $\Pi = (\mathsf{Setup}, \mathsf{KeyGen}, \mathsf{KeyUpdate}, \mathsf{Encrypt}, \mathsf{Decrypt})$ be a secure (in the weak sense mentioned above) fs-IBE scheme with a finite and efficiently sampleable message space $\mathcal{M}$. We assume that the $\mathsf{Encrypt}$ algorithm uses random values taken from a set R. We can write $\mathsf{Encrypt}$ as a deterministic algorithm $C \leftarrow \mathsf{Encrypt}(MPK, \mathsf{ID}, t, m; r)$ where $r \xleftarrow{\$} R$. The only difficulty in applying the method of Dent [10] is that we must re-encrypt the recovered message for integrity check. In the context of forward secure IBEs, this means one must know the time period and the identity under which the message was originally encrypted. In our setting (i.e. the specific application of onion routing) we overcome this difficulty as such information is available to routers.

We transform the fs-IBE scheme Π with a finite and efficiently sampleable message space $\mathcal{M}$ and maximum number of time periods T into a fs-IB-KEM scheme $\Pi' = (\mathsf{Setup}, \mathsf{KeyGen}, \mathsf{KeyUpdate}, \mathsf{Encap}, \mathsf{Decap})$ using two hash functions:

$$H_1 : \{0,1\}^* \times \{0,1\}^* \times \mathcal{M} \to R \qquad \text{and} \qquad H_2 : \{0,1\}^* \to \{0,1\}^k .$$

6 Efficiency and Comparisons

In this section we compare the efficiency of our proposal with those of the other known solutions: the certificateless onion routing (CL-OR) protocol of Catalano *et al.* [7], the pairing-based onion routing (PB-OR) scheme of Kate *et al.* [21]

and the actual Tor protocol (we refer to the official specifications [13]). Basically, all these solutions differ only in the way the symmetric keys are established, so we decided to analytically compare the cost of building a circuit of length n from the perspective of both a user and an onion router. All the tests are carried on considering security parameters of 80 and 128 bits: as widely suggested in [31,32], the latter should be considered in order to gain an adequate long-term security level.

In what follows we briefly describe the operations involved during the building of a circuit in the considered protocols.

Tor. The Tor protocol incrementally builds the circuit using the telescoping technique and each new key is established using a Diffie-Hellman (DH) key-exchange [12]. Its specifications require that the user sends to each onion router the DH component encrypted using an RSA key associated to the router. It follows that: a user computes 1 RSA encryption and 2 exponentiations for each of the n routers; an onion router performs 1 RSA decryption and 2 exponentiations. Even if not required by Tor's specifications [13] we consider pre-computation on the fixed base for one of the two exponentiations of the DH key-exchange. For a security level of 80 bits we need a 1024-bits RSA modulus and a 1024-bits finite field for Diffie-Hellman. The specifications given in [13] suggest to use 65537 as fixed RSA exponent and to optimize DH with exponents of 360 bits and generator 2. In order to get a 128-bits security level, the sizes of the RSA modulus as well as of the DH finite-field have to be of 3072 bits. Tor's specifications [13] require a periodic update of the onion routers' keys.

PB-OR. We consider an implementation of the pairing-based onion routing protocol over a group of points of elliptic curves using the PBC library [22]. More specifically, following the indications of the authors, a type A (in the PBC nomenclature) curve is used in order to get fast pairing operation $e : \mathbb{G} \times \mathbb{G} \rightarrow \mathbb{G}_T$. Kate *et al.* suggest that each user can pre-compute a pairing application for each onion router (as a function of some public parameter and of onion router's identity); such values have to be re-computed every time the KGC's keys change (e.g. every day). Therefore, in order to build a circuit of length n, a user has to compute n exponentiations in $\mathbb{G}$ and n exponentiations in $\mathbb{G}_T$: both the operations can be speed-up using pre-computation on the fixed base. On the other hand, each onion router has to compute one pairing but an optimized implementation can exploit a pre-computation on the pairing application that makes use of a fixed parameter (such functionality is offered by PBC library).

CL-OR. For the CL-OR protocol[2] of Catalano *et al.* we also consider an implementation over EC using the PBC library but, as suggested by the authors, using a type F curve in order to gain fast operations on smaller group elements. The user can pre-precompute some values that are function of the onion router's identities and public-keys. The on-line computation of the user requires

[2] In [7] there are two implementations available: we consider the fastest that makes use of the Strong Diffie-Hellman assumption in the Random Oracle model.

3 exponentiations on the working group: only 2 of them can be performed using pre-computation on the fixed bases. In such protocol, each onion router requires 2 exponentiations to compute the session-key: none of them can use such kind of pre-computation.

Our Protocol. For this comparison we consider our proposal of Section 5 implemented on a type A curve (like the PB-OR protocol). Observe that to compute $C_0 = (uv^{H(\mathsf{ID})} \prod_{i=1}^{k} h_i^{f(w_i)})^s$, many values can be entirely pre-computed off-line by the user: the values $uv^{H(\mathsf{ID})}$ as well as the values $h_i^{f(w_i)}$. The remaining notable tasks for the user, for each onion router in the circuit, are: two exponentiations over $\mathbb{G}$ (one for C_0 and one for $C_1 = g^s$) and one exponentiation in $\mathbb{G}_T$ to compute $C_2 = z^s m$. Notice that the latter two exponentiations can be optimized with pre-computation on the fixed bases g, z. Each onion router involved in the circuit establishment has to compute 2 pairings for decryption (but with partial pre-computation as in PB-OR) as well as a new encryption to fulfill the integrity check required by Dent's transformation.

In a fully operational implementation of an onion routing network, the key establishment phase involves the use of other minor tools: a symmetric encryption scheme (e.g. AES) to protect the passing messages using the negotiated session keys and fixed TLS channels among the connected onion routers. For sake of simplicity we ignore the computational load related to such operations since they are used by all the protocols considered in our comparisons. Moreover, in the case of AES, its time complexity is negligible if compared with the other involved cryptographic tools.

As a first step, all the operations were implemented using the PBC library (version 0.4.18) on a 2.4GHz Intel Core 2 Duo workstation running Mac OS X 10.5.6[3].

As one can see from Table 1, from a purely computational perspective our solution is not faster than previous protocols but its computational costs are definitely practical. In these comparisons it is worth noting that the Tor circuit construction is an interactive protocol that requires a quadratic number of exchanged messages. Therefore if we consider the natural network latency we obtain that, even for the shortest possible circuit (i.e. 3 nodes), our protocol is faster than Tor in constructing an entire new circuit. In fact, assuming a network latency of 50 ms, Tor requires 627 ms to complete the circuit construction while our protocol needs only 370 ms.

A LOOK AT INTERACTION. We stress that the main contribution of our work is that the resulting OR protocol is totally non-interactive solving a problem that is yet unsolved in currently known solutions. Indeed, all the other onion routing protocols require interaction in some phase of the protocol. Tor is clearly interactive in the circuit construction phase due to the use of telescoping. In PB-OR the routers have to obtain new private keys from the KGC every time the key validity period expires (a heavy workload for the KGC!). On the other hand,

[3] The computational costs of each operation are ommitted for lack of space. They are available in the full version of this work.

Table 1. Total times for building a circuit of n routers and protocol features

Times and features	Tor		PB-OR		CL-OR		our protocol	
	user	OR	user	OR	user	OR	user	OR
Time for 80 bits security (ms)	$2.3n$	6.9	$1.1n$	3.9	$2.1n$	3.4	$7.8n$	15.6
Time for 128 bits security (ms)	$16.5n$	93.3	$9.3n$	57.3	$5.1n$	8.2	$63.4n$	178.0
Number of exchanged messages	$n(n+1)$		$2n$		$2n$		$2n$	
IND-CPA security level	×		✓		✓		✓	
IND-CCA security level	×		×		✓		✓	
Non-inter. circuit construction	×		✓		✓		✓	
Non-inter. key-update by OR	×		×		✓		✓	
Non-inter. by user after OR key-update	×		✓		×		✓	
Absence of key-escrow by KGC	✓		×		✓		×	

in CL-OR the routers can generate the updated keys by themselves but the users have to obtain such new keys (e.g. by querying a directory server) every time they are changed.

We stress that in our protocol onion routers can update their keys without interacting with any party, and this process is transparent for the users who keep using always the same public keys (i.e. routers' identities). It is interesting to note that the non-interactive nature of our key update allows to reduce the security gap between eventual and immediate forward-security. We can indeed arbitrarily reduce the refresh period of routers' keys without any further network overhead: this is not true for PB-OR and CL-OR. Finally, we observe that the slightly higher computational cost of our solution is due only to the fact the best fs-IB-KEM we can achieve has to perform pairings computation. Although this is currently a limitation, we believe that the purely non-interactive nature of our protocol, more than compensate for the slight increase in computational cost.

References

1. Al-Riyami, S., Paterson, K.: Certificateless public key cryptography. In: Laih, C.-S. (ed.) ASIACRYPT 2003. LNCS, vol. 2894, pp. 452–473. Springer, Heidelberg (2003)
2. Anderson, R.: Two remarks on public key cryptology. In: ACM-CCS 1997 (1997) (invited lecture), `http://www.cl.cam.ac.uk/ftp/users/rja14/forwardsecure.pdf`
3. Boneh, D., Boyen, X., Goh, E.: Hierarchical Identity Based Encryption with Constant Size Ciphertext. In: Cramer, R. (ed.) EUROCRYPT 2005. LNCS, vol. 3494, pp. 440–456. Springer, Heidelberg (2005)
4. Boneh, D., Franklin, M.: Identity-Based Encryption from the Weil Pairing. In: Kilian, J. (ed.) CRYPTO 2001. LNCS, vol. 2139, pp. 213–615. Springer, Heidelberg (2001)
5. Camenisch, J., Lysyanskaya, A.: A Formal Treatment of Onion Routing. In: Shoup, V. (ed.) CRYPTO 2005. LNCS, vol. 3621, pp. 169–187. Springer, Heidelberg (2005)

6. Canetti, R., Halevi, S., Katz, J.: A Forward-Secure Public Key Encryption Scheme. In: Biham, E. (ed.) EUROCRYPT 2003. LNCS, vol. 2656, pp. 255–271. Springer, Heidelberg (2003)
7. Catalano, D., Fiore, D., Gennaro, R.: Certificateless Onion Routing. In: Proc. of the 16th ACM Conference on Computer and Comm. Security (CCS 2009), pp. 151–160. ACM Press, New York (2009)
8. Chaum, D.: Untraceable Electronic Mail, return address and digital pseudonyms. Communications of the ACM 24(2), 84–88 (1981)
9. Dai, W.: PipeNet 1.1, http://www.weidai.com/pipenet.txt
10. Dent, A.W.: A designer's guide to kEMs. In: Paterson, K.G. (ed.) Cryptography and Coding 2003. LNCS, vol. 2898, pp. 133–151. Springer, Heidelberg (2003)
11. Danezis, G., Goldberg, I.: Sphinx: A Compact and Provably Secure Mix Format. In: IEEE Symposium on Security and Privacy 2009, pp. 269–282 (2009)
12. Diffie, W., Hellman, M.: New Directions in Cryptography. IEEE Transactions on Information Theory 22(6), 644–654 (1976)
13. Dingledin, R., Mathewson, N.: Tor Protocol Specification (2008), http://www.torproject.org/svn/trunk/doc/spec/tor-spec.txt
14. Dingledin, R., Mathewson, N., Syverson, P.: Tor: The Second-Generation Onion Router. In: Proc. of the 13th USENIX Security Symposium, pp. 303–320 (2004)
15. Freedman, M., Morris, R.: Tarzan: A Peer-to-Peer Anonymizing Networ Layer. In: Proc. of 9th ACM Conference on Computer and Comm. Security (CCS 2002), pp. 193–206 (2002)
16. Goldberg, I.: On the Security of the Tor Authentication Protocol. In: Danezis, G., Golle, P. (eds.) PET 2006. LNCS, vol. 4258, pp. 316–331. Springer, Heidelberg (2006)
17. Goldschlag, D., Reed, M., Syverson, P.: Hiding Routing Informations. In: Anderson, R. (ed.) IH 1996. LNCS, vol. 1174, pp. 137–150. Springer, Heidelberg (1996)
18. Goldschlag, D., Reed, M., Syverson, P.: Onion Routing for Anonymous and Private Internet Connections. Communications of the ACM 42(2), 84–88 (1999)
19. Kate, A., Goldberg, I.: Using Sphinx to Improve Onion Routing Circuit Construction. In: Sion, R. (ed.) FC 2010. LNCS, vol. 6052, pp. 359–366. Springer, Heidelberg (2010)
20. Kate, A., Zaverucha, G., Goldberg, I.: Pairing-Based Onion Routing. In: Borisov, N., Golle, P. (eds.) PET 2007. LNCS, vol. 4776, pp. 95–112. Springer, Heidelberg (2007)
21. Kate, A., Zaverucha, G., Goldberg, I.: Pairing-Based Onion Routing with Improved Forward Secrecy. ACM Transactions on Information and System Security (2009)
22. Lynn, B.: PBC: The Pairing-Based Crypto Library, http://crypto.stanford.edu/pbc
23. Möller, B.: Provably Secure Public-Key Encryptionfor Length-Preserving Chaumian Mixes. In: Joye, M. (ed.) CT-RSA 2003. LNCS, vol. 2612, pp. 244–262. Springer, Heidelberg (2003)
24. Øverlier, L., Syverson, P.F.: Improving Efficiency and Simplicity of Tor Circuit Establishment and Hidden Services. In: Borisov, N., Golle, P. (eds.) PET 2007. LNCS, vol. 4776, pp. 134–152. Springer, Heidelberg (2007)
25. Reed, M., Syverson, P., Goldschlag, D.: Anonymous Connections and Onion Routing. IEEE Journal on Selected Ares in Communications 16(4), 482–494
26. Renhard, M., Plattner, B.: Introducing MorphMix: Peer-toPeer based Anonymous Internet Usage with Collusion Detection. In: The Workshop on Privacy in the Electronic Society (WPES 2002), pp. 91–102. ACM, New York (2002)

27. Shamir, A.: Identity-Based Cryptosystems and Signature Schemes. In: Blakely, G.R., Chaum, D. (eds.) CRYPTO 1984. LNCS, vol. 196, pp. 47–53. Springer, Heidelberg (1985)
28. Shoup, V., Gennaro, R.: Securing threshold cryptosystems against chosen ciphertext attack. In: Nyberg, K. (ed.) EUROCRYPT 1998. LNCS, vol. 1403, pp. 1–16. Springer, Heidelberg (1998)
29. Sakai, R., Ohgishi, K., Kasahara, M.: Cryptosystems based on pairing. In: Symposium on Cryptography and Information Security, Okinawa, Japan (2000)
30. Yao, D., Fazio, N., Dodis, Y., Lysyanskaya, A.: ID-Based Encryption for Complex Hierarchies with Applications to Forward Security and Broadcast Encryption. In: Proc. of the ACM Conference on Computer and Comm. Security, CCS 2004 (2004)
31. NIST Recommendations for Key Management Part 1: General NIST Special Publication 800-57 (August 2005), http://csrc.nist.gov/publications/nistpubs/800-57/SP800-57-Part1.pdf
32. ECRYPT Yearly Report on Algorithms and Key Sizes (2007-2008) (July 2008), http://www.ecrypt.eu.org/ecrypt1/documents/D.SPA.28-1.1.pdf

Generic Fully Simulatable Adaptive Oblivious Transfer

Kaoru Kurosawa[1], Ryo Nojima[2], and Le Trieu Phong[2]

[1] Ibaraki University, Japan
kurosawa@mx.ibaraki.ac.jp
[2] NICT, Japan
{ryo-no,phong}@nict.go.jp

Abstract. We aim at constructing adaptive oblivious transfer protocols, enjoying fully simulatable security, from various well-known assumptions such as DDH, DLIN (and more generally, d-linear), QR, DCR. To this end, we present two generic constructions of adaptive OT, one of which utilizes verifiable shuffles together with threshold decryption schemes, while the other uses permutation networks together with what we call *loosely-homomorphic* key encapsulation schemes. We then show that specific choices of the building blocks lead to concrete adaptive OT protocols with fully simulatable security in the standard model under the targeted assumptions. Our generic method can be further used to construct the first (memory) leakage-resilient adaptive OT.

Keywords: adaptive OT, fully-simulatable, verifiable shuffles, permutation networks, loose homomorphism.

1 Introduction

1.1 Background

Oblivious transfer (OT) with adaptive queries, or adaptive OT for short, was first examined by Naor and Pinkas in [19], in which there are a sender and a receiver. The sender holds n messages, and the receiver would like to retrieve k of them, one after the other, so that: (1) the sender does not know what the receiver obtains, and (2) the receiver gets nothing more beside the k messages. The key applications of this type of OT are in patent searches, oblivious search, medical databases etc.

The security notion capturing the above requirements has evolved in the literature. The notion of full simulatability was introduced by Camenisch, Neven, and Shelat in [3], following the real-world, ideal-world paradigm. In the ideal world, there exists a trusted third party (TTP), to which the sender gives all of his messages. When a receiver wants to obtain a message, he simply sends the corresponding index to the TTP. On the other hand, in the real world, there is no TTP at all, and the protocol of adaptive OT is run by the sender and the receiver. The intuition of full simulatability is that the real world is indistinguishable from the ideal world, with respect to any poly-time adversary.

J. Lopez and G. Tsudik (Eds.): ACNS 2011, LNCS 6715, pp. 274–291, 2011.

Camenisch et al. additionally provided us with some first constructions of adaptive OT which were fully simulatable, in both the random oracle model (ROM) and the standard model. In particular, they showed with a refinement that the scheme in ROM of Ogata and Kurosawa [22] achieved fully simulatable security. They furthermore gave a construction in the standard model, using $\mathfrak{q}$-based assumptions (in which $\mathfrak{q}$ depends on n) in pairing groups.

After the work of Camenisch et al., much effort has been devoted to further extending the direction. In [9], Green and Hohenberger constructed a universally-composable, so fully-simulatable, scheme under the $\mathfrak{q}$-hidden LRSW assumption. Jarecki and Liu [14] joined the research line with a scheme based on the $\mathfrak{q}$-DHI assumption yet in RSA groups.

With respect to assumptions which are not $\mathfrak{q}$-based, Kurosawa and Nojima [16] showed a simple scheme fully simulatable under the DDH assumption. However, the scheme suffered from a large communication cost of $O(n)$ in each transfer, as pointed out by Green and Hohenberger in [10], who further gave a construction under the decision 3-party DDH (3DDH) assumption in pairing groups. Concurrently, Kurosawa, Nojima, and Phong [17], using a verifiable shuffle protocol, overcome the demerit of [16], reducing the cost to $O(1)$, while still maintaining the DDH assumption for security. Specifically, they used the verifiable shuffle protocol of Neff [21] which is a 7-move honest verifier ZKIP for proving the relation between $(g, X_1, \ldots, X_n)$ and $(g^c, X^c_{\pi(1)}, \ldots, X^c_{\pi(n)})$, where π is a random permutation and c is random. Note that Neff's shuffle protocol is computationally zero-knowledge under the DDH assumption, so that it seems impossible to utilize the shuffle beyond the DDH case.

1.2 Our Contribution

We present two generic methods for constructing fully simulatable adaptive OT in the standard model. They yield numerous protocols from various assumptions, including the DDH, d-linear ($d \geq 2$), quadratic residuosity (QR), and decisional composite residuosity (DCR) assumptions. A comparison with previous works is given in Table 1, in which our DDH-based OT protocol has less number of moves in the initialization phase than that of [17]. Note that our schemes based on the QR, and DCR assumptions induce a bit higher communication cost for initialization.

Our first method can be applied to any public-key encryption scheme E which satisfies two conditions: (1) It must be a homomorphic encryption scheme such that the message space is a group of prime public order; and (2) It can be used as a 2-out-of-2 threshold decryption scheme.

The first condition allows us to use the verifiable shuffle protocol of Groth and Lu [11] which is a 3-move honest verifier ZKIP for proving the relation between the tuples $(E(m_1), \ldots, E(m_n))$ and $(E(m_{\pi(1)}), \ldots, E(m_{\pi(n)}))$, where π is a random permutation. However, we cannot obtain any adaptive OT even if we directly replace Neff's shuffle protocol by Groth-Lu's shuffle protocol into [17]. This is because the sender can compute π from $(E(m_{\pi(1)}), \ldots, E(m_{\pi(n)}))$. To overcome this problem, we use 2-out-of-2 threshold decryption. From this

Table 1. Fully simulatable adaptive OT schemes without random oracles

Scheme	Assumption	Comm. Cost (each transfer)	Init. Cost
CNS [3]	q-strong DH and q-PDDH	$O(1)$	$O(n)$
GH [9]	q-hidden LRSW (UC secure)	$O(1)$	$O(n)$
JL [14]	q-DHI (RSA group)	$O(1)$	$O(n)$
KN [16]	DDH	$O(n)$	$O(n)$
GH [10]	decision 3-party DH (3DDH)	O(1)	$O(n)$
KNP [17]	DDH	$O(1)$	$O(n)$ (more moves)
This work	DDH	$O(1)$	$O(n)$ (less moves)
	d-Linear		$O(n)$
	DCR		$O(n \log n)$
	QR		$O(n \log n)$

method, new adaptive OTs are obtained under the DDH assumption and the d-linear ($d \geq 2$) assumption, respectively.

Our second method can be applied to any key encapsulation mechanisms (KEM) satisfying what we call *loosely-homomorphic* property. We use permutation networks for this case while we do not use threshold decryption. From this method, new adaptive OTs are obtained under the QR assumption and the DCR assumption, respectively.

Our generic method can be further used to construct the first (memory) leakage resilient adaptive OT protocol as shown in Sect.5.

2 Preliminaries

2.1 Notations

Throughout the paper, $\mathrm{OT}^n_{k\times 1}$ denote the adaptive OT with n messages of the sender and k choices of the receiver. ZKPK stands for zero-knowledge proof of knowledge, while ZKPM for zero-knowledge proof of membership. WIPK means witness-indistinguishable proof of knowledge. Furthermore, $\mathsf{ZKPK}\{(x) : X = g^x\}$ means a ZKPK protocol showing the knowledge of secret x satisfying the equation; and similar notations for more complex ZKPK, ZKPM, WIPK protocols will be used. Taking an element a randomly from a set A is denoted by $a \stackrel{\$}{\leftarrow} A$. We use $a[i]$ to indicate the i-th component of a. For example, when a is a bit string, $a[i]$ is the i-th bit; when a is a tuple of elements, $a[i]$ becomes the i-th element.

2.2 Fully-Simulatable $\mathrm{OT}^n_{k\times 1}$

We use almost the same presentation as [16], and consider a weak model of universally composable (UC) framework as follows.

- At the beginning of the game, an adversary $\mathcal{A}$ can corrupt either a sender S or a receiver R, but not both of them.
- $\mathcal{A}$ can send a message, denoted by $\mathcal{A}_{\text{out}}$, to an environment $\mathcal{Z}$ after the end of the protocol. However, $\mathcal{A}$ cannot communicate with $\mathcal{Z}$ during the protocol execution. (This property makes the definitions weaker than standard UC security.)

The ideal functionality of $\text{OT}^n_{k\times 1}$ will be shown below. For a protocol $\Pi = (\mathsf{S}, \mathsf{R})$, define the advantage of $\mathcal{Z}$ as

$$\mathbf{Adv}(\mathcal{Z}) \stackrel{\text{def}}{=} \Big| \Pr(\mathcal{Z} = 1 \text{ in the real world}) - \Pr(\mathcal{Z} = 1 \text{ in the ideal world}) \Big|$$

where the real and ideal worlds are defined below.

The ideal world: there are a few parties consisting of the ideal functionality $\mathcal{F}_{\text{adapt}}$, an ideal world adversary $\mathcal{A}'$, and the environment $\mathcal{Z}$. Also we have dummy sender S' and receiver R'. The parties behave as follows.

Initialization phase

1. The environment $\mathcal{Z}$ sends $(M_1, \ldots, M_n)$ to the dummy sender S'.
2. S' sends $(M_1^*, \ldots, M_n^*)$ to $\mathcal{F}_{\text{adapt}}$, where $(M_1^*, \ldots, M_n^*) = (M_1, \ldots, M_n)$ if S' is not corrupted.

Transfer phase $i = 1, \ldots, k$

1. $\mathcal{Z}$ sends σ_i to the dummy receiver R', where $1 \le \sigma_i \le n$.
2. R' sends σ_i^* to $\mathcal{F}_{\text{adapt}}$, where $\sigma_i^* = \sigma_i$ if R' is not corrupted.
3. $\mathcal{F}_{\text{adapt}}$ sends `received` to $\mathcal{A}'$.
4. $\mathcal{A}'$ sends $b = 1$ or 0 to $\mathcal{F}_{\text{adapt}}$, where $b = 1$ if S' is not corrupted.
5. $\mathcal{F}_{\text{adapt}}$ sends E_i to R', where $E_i = M^*_{\sigma_i^*}$ if $b = 1$, and $E_i = \perp$ if $b = 0$.
6. R' sends E_i to $\mathcal{Z}$.

After the end of the protocol, $\mathcal{A}'$ sends a message $\mathcal{A}'_{\text{out}}$ to $\mathcal{Z}$. Finally $\mathcal{Z}$ outputs 1 or 0.

The real world: Simply in this world, the protocol $\Pi = (\mathsf{S}, \mathsf{R})$ is executed as specified by its construction (thus without $\mathcal{F}_{\text{adapt}}$). The environment $\mathcal{Z}$ and the real world adversary $\mathcal{A}$ behave in the same way as above.

For proving the security of our schemes, we will require that an honest receiver will not ask for the same message twice.

Definition 1 (Full simulatability). *Protocol $\Pi = (\mathsf{S}, \mathsf{R})$ is secure against the sender (resp, receiver) corruption if for any real world adversary $\mathcal{A}$ who corrupts the sender S (resp, receiver R), there exists an ideal world adversary $\mathcal{A}'$ who corrupts the dummy sender S' (resp, dummy receiver R') such that for any poly-time environment $\mathcal{Z}$, the advantage* $\mathbf{Adv}(\mathcal{Z})$ *is negligible. Moreover, protocol $\Pi = (\mathsf{S}, \mathsf{R})$ is a* fully simulatable *$OT^n_{k\times 1}$ if it is secure against the sender corruption and the receiver corruption.*

3 Generic Adaptive OT from Verifiable Shuffles

3.1 Building Blocks

Threshold PKE. We need an 2-out-of-2 threshold PKE scheme TPKE, which consists of the following algorithms.

- TGen: Two parties S and R run a protocol so that they respectively obtain (pk, sk_S) and (pk, sk_R) where pk is the agreed public key and $sk_\mathsf{S}, sk_\mathsf{R}$ are the shares of secret key. (The public key is needed for all algorithms below, and we omit writing it for clarity.)
- $\mathsf{TEnc}(M; r)$: output a ciphertext C for a plaintext M and a random coin r.
- $\mathsf{TDec}(sk_\mathsf{P}, C)$: for $\mathsf{P} \in \{\mathsf{S}, \mathsf{R}\}$, output μ_P which is the decryption share of the ciphertext C under secret key sk_P.
- $\mathsf{TComb}(C, \mu_\mathsf{S}, \mu_\mathsf{R})$: output a plaintext M by combining the input $C, \mu_\mathsf{S}, \mu_\mathsf{R}$.

We require the following properties on the TPKE scheme.

Homomorphism: Namely, $\mathsf{TEnc}(M; r) \otimes \mathsf{TEnc}(M'; r') = \mathsf{TEnc}(M \oplus M'; r \odot r')$, where $\otimes, \oplus, \odot$ are the operators on the corresponding spaces.
Semantic security: We require that for all M, the ciphertext $\mathsf{Enc}(M; r)$ for random r is (almost) uniformly distributed over the ciphertext space.

Verifiable shuffles. Consider a set of ciphertexts $C_i = \mathsf{TEnc}(M_i; r_i)$ for $1 \leq i \leq n$ of the TPKE scheme formed by S. Let I be the identity element of the message space. It is easy enough for R to choose a permutation π on $\{1, \ldots, n\}$, and random s_i to form the set of $C'_i = C_{\pi(i)} \otimes \mathsf{TEnc}(I; s_i)$ for $1 \leq i \leq n$, so that both sets of ciphertexts contain the same plaintexts. The set of $C'_i (1 \leq i \leq n)$ is called a shuffle of the original one. If the scheme TPKE is semantically secure, publishing the shuffle $C'_i (1 \leq i \leq n)$ reveals nothing on the permutation π to S. Correctness of the shuffle is verified via the following protocol

$$\mathsf{ZKPK}\big\{(\pi, s_i) : C'_i = C_{\pi(i)} \otimes \mathsf{TEnc}(I; s_i) \forall 1 \leq i \leq n\big\},$$

which has efficient implementations for homomorphic encryption schemes such as ElGamal or Paillier[1] as shown in the work of Groth and Lu [11]. More generally, the results of Groth and Lu apply for homomorphic encryption schemes with the following properties:

Proper message space: the order of the message space does not have any small prime factor (say less than 2^{80}).
Root extraction: from M, R, and $C^e = \mathsf{TEnc}(M; R)$, it is possible to efficiently extract (m, r) such that $C = \mathsf{TEnc}(m; r)$ for every e co-prime with the order of the message space.

[1] However, the Paillier encryption scheme with *threshold decryption* needs a setup assumption for the secret keys.

The protocols for verifiable shuffles given in [11] are statistical strong HVZK arguments of three rounds, and can be turned into fully zero-knowledge by standard techniques. The additional property below will be needed in proving sender security.

Computing μ_{S} without sk_{S}: Let $\mu_{\mathsf{S}} = \mathsf{TDec}(sk_{\mathsf{S}}, C')$ where $C' = C'_{\pi^{-1}(\sigma)}$ as above for some $1 \leq \sigma \leq n$. Note that $C' = C_\sigma \otimes \mathsf{TEnc}(I; s_{\pi^{-1}(\sigma)})$ for $C_\sigma = \mathsf{TEnc}(M_\sigma; r_\sigma)$. We require that μ_{S} can be alternatively expressed as a function of $pk, C_\sigma, M_\sigma, s_{\pi^{-1}(\sigma)}$, and sk_{R}. Namely there exists an efficiently-computable function f such that we have $\mu_{\mathsf{S}} = f(pk, C_\sigma, M_\sigma, s_{\pi^{-1}(\sigma)}, sk_{\mathsf{R}})$.

3.2 The OT Protocol

Initialization:

1. The sender S and the receiver R run the protocol TGen so that they obtain a common public key pk; and S gets secret key sk_{S}, R gets secret key sk_{R}.The receiver R proves in ZKPK that he knows sk_{R} corresponding to pk.
2. For $1 \leq i \leq n$, S computes and sends
$$C_i = \mathsf{TEnc}(M_i; r_i)$$
to R where r_i are randomness used by TEnc.
3. S then proves to R by ZKPK that he knows M_i for all i. (This is equivalent to proving the knowledge of r_i in our below instantiations.)
4. (**Shuffling**) R chooses a permutation π on $\{1, \ldots, n\}$ and randomness s_i for $1 \leq i \leq n$, then computes and sends to S for all i
$$C'_i = C_{\pi(i)} \otimes \mathsf{TEnc}(I; s_i),$$
where I is the unit element of the message space.
5. R proves to S in ZKPK that he knows π and $s_i (1 \leq i \leq n)$ satisfying the equation at Step 4.

The j-th transfer:

6. R obtains an index σ as input, and sends $C' = C'_{\pi^{-1}(\sigma)}$ to S.
7. S checks $C' \in \{C'_1, \ldots, C'_n\}$, then computes and sends $\mu_{\mathsf{S}} = \mathsf{TDec}(sk_{\mathsf{S}}, C')$ to R.
8. S then proves in ZKPM that he did the right decryption in the above step.
9. R lets $\mu_{\mathsf{R}} = \mathsf{TDec}(sk_{\mathsf{R}}, C')$, and then obtaining M_σ by $\mathsf{TComb}(pk, C', \mu_{\mathsf{S}}, \mu_{\mathsf{R}})$.

To prove correctness of the OT, note that

$$C' = C'_{\pi^{-1}(\sigma)} = C_\sigma \otimes \mathsf{TEnc}(I; s_{\pi^{-1}(\sigma)}) = \mathsf{TEnc}(M_\sigma, r_\sigma) \otimes \mathsf{TEnc}(I; s_{\pi^{-1}(\sigma)})$$

which means C' encrypts the plaintext M_σ thanks to the homomorphic property of the threshold PKE scheme. Now, by the correctness of the threshold PKE scheme, $\mathsf{TComb}(pk, C', \mu_{\mathsf{S}}, \mu_{\mathsf{R}})$ is exactly M_σ as required.

Theorem 1. *The generic* $\mathrm{OT}^n_{k\times 1}$ *from verifiable shuffles above is fully simulatable, if the* TPKE *scheme has semantic security.*

The proof is postponed in Appendix A.

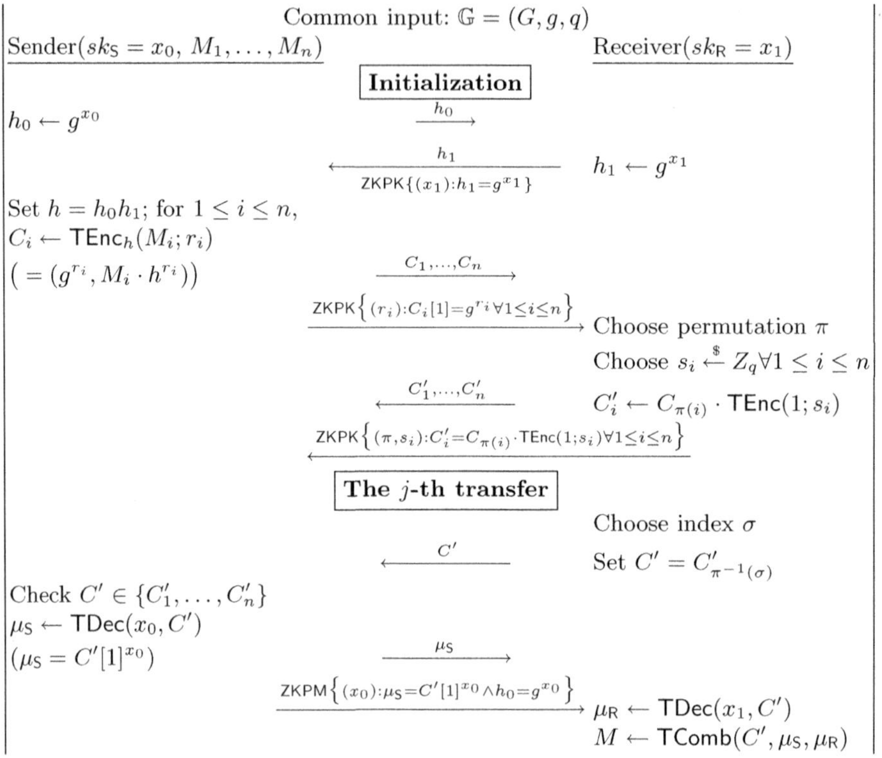

Fig. 1. The $\mathrm{OT}^n_{k\times 1}$ secure under the DDH assumption

3.3 Instantiations from DDH and Linear Assumptions

$\mathrm{OT}^n_{k\times 1}$ from the DDH assumption. We will use the threshold ElGamal encryption scheme. The scheme works on a cyclic group $\mathbb{G} = (G, g, q)$ where g is the generator of prime order q, and has semantic security under the DDH assumption on $\mathbb{G}$.

- TGen: S chooses $sk_\mathsf{S} = x_0 \xleftarrow{\$} Z_q$, computes and sends $h_0 \leftarrow g^{x_0}$ to R. Similarly, R chooses $sk_\mathsf{R} = x_1 \xleftarrow{\$} Z_q$ and sends $h_1 \leftarrow g^{x_1}$ to S. The agreed public key is then $h = h_0 h_1$.
- $\mathsf{TEnc}(M; r)$: Output $C = (C[1], C[2]) = (g^r, M \cdot h^r)$ for $r \xleftarrow{\$} Z_q$ and $M \in G$.
- $\mathsf{TDec}(sk_\mathsf{P}, C)$: Output $\mu_\mathsf{P} = C[1]^{sk_\mathsf{P}}$ for P is either S or R.
- $\mathsf{TComb}(C, \mu_\mathsf{S}, \mu_\mathsf{R})$: Output $C[2]/(\mu_\mathsf{S}\mu_\mathsf{R})$.

The TPKE scheme satisfies all requirements described in Sect.3.1. Our $\mathrm{OT}^n_{k\times 1}$ instantiation from the threshold ElGamal encryption scheme is depicted in Fig.1. In the figure, the element $\mu_\mathsf{S} = C'[1]^{x_0}$ can be alternatively expressed as[2]

[2] Let us elaborate a bit on the formula. We have $C_\sigma[2] M_\sigma^{-1} C_\sigma[1]^{-x_1} h_0^{s_{\pi^{-1}(\sigma)}} = (M_\sigma h^{r_\sigma}) M_\sigma^{-1} C_\sigma[1]^{-x_1} h_0^{s_{\pi^{-1}(\sigma)}} = h^{r_\sigma} (g^{r_\sigma})^{-x_1} h_0^{s_{\pi^{-1}(\sigma)}} = (h_0 h_1)^{r_\sigma} (h_1^{-r_\sigma}) h_0^{s_{\pi^{-1}(\sigma)}} = h_0^{r_\sigma + s_{\pi^{-1}(\sigma)}} = g^{(r_\sigma + s_{\pi^{-1}(\sigma)}) x_0} = C'[1]^{x_0} = \mu_\mathsf{S}$, as required.

$$\mu_{\mathsf{S}} = f(pk, C_\sigma, M_\sigma, s_{\pi^{-1}(\sigma)}, sk_{\mathsf{R}}) \stackrel{\text{def}}{=} C_\sigma[2] M_\sigma^{-1} C_\sigma[1]^{-x_1} h_0^{s_{\pi^{-1}(\sigma)}},$$

which is the formula needed when proving sender security.

Since the threshold ElGamal encryption scheme has semantic security under the DDH assumption, thanks to Theorem 1, the $\mathrm{OT}^n_{k\times 1}$ in Fig.1 is fully-simulatable under the same assumption.

$\mathrm{OT}^n_{k\times 1}$ from the d-linear assumptions. We also works on $\mathbb{G} = (G, g, q)$, and let us introduce some more notations. For two vectors $v = (v[1], \ldots, v[l]) \in G^{1\times l}$, $u = (u[1], \ldots, u[l]) \in Z_q^{1\times l}$ define

$$v \cdot u^{\mathsf{T}} = u \cdot v^{\mathsf{T}} = \prod_{i=1}^{l} v[i]^{u[i]} \in G.$$

Matrix-matrix and matrix-vector multiplications are defined in the same manner. Sometimes, the $\cdot$ operators are implicitly understood. Also recall that for $u, u' \in Z_q^{1\times l}$, we have $u + u' = (u[1] + u'[1], \ldots, u[l] + u'[l])$ as normal. It is easy to check that $(u + u') \cdot v^{\mathsf{T}} = (u \cdot v^{\mathsf{T}})(u' \cdot v^{\mathsf{T}}) \in G$, and $v \cdot (u + u')^{\mathsf{T}} = (v \cdot u^{\mathsf{T}})(v \cdot u'^{\mathsf{T}}) \in G$.

For $d \geq 2$, the following PKE scheme, introduced by Naor and Segev [20], has semantic security under the d-linear assumption. The algorithm Gen generates $sk \stackrel{\$}{\leftarrow} Z_q^{(d+1)\times 1}$, $\phi \stackrel{\$}{\leftarrow} G^{d\times(d+1)}$. The secret key is sk, and the public key is $pk = (\phi, \psi)$ for $\psi = \phi \cdot sk \in G^{d\times 1}$. The encryption algorithm $\mathsf{Enc}(M; R)$, on message $M \in G$ and random $R \in Z_q^{1\times d}$ as input, outputs the ciphertext $C = (R\phi, (R\psi)M) \in G^{1\times(d+1)} \times G$. The decryption algorithsm $\mathsf{Dec}(sk, C)$ outputs $C[2]/(C[1] \cdot sk)$. Note that the correctness of the PKE scheme comes from the equation $(R \cdot \phi) \cdot sk = R \cdot (\phi \cdot sk)$. The semantic security of the PKE scheme implies that, given ϕ, ψ, the pair $\mathsf{Enc}(1; R) = (R\phi, R\psi)$ is indistinguishable from random over $G^{1\times(d+1)} \times G$.

We now present the 2-out-of-2 threshold variant of the above PKE. In TGen, the parties S and R, using $\mathbb{G}$, agree on the matrix $\phi \in G^{d\times(d+1)}$. They then choose secrets sk_{S} and sk_{R} respectively in $Z_q^{(d+1)\times 1}$; S publishes $\psi_{\mathsf{S}} = \phi \cdot sk_{\mathsf{S}} \in G^{d\times 1}$ while R does the same with $\psi_{\mathsf{R}} = \phi \cdot sk_{\mathsf{R}} \in G^{d\times 1}$. The agreed common public key is ϕ, $\psi_{\mathsf{S}}, \psi_{\mathsf{R}}$ in which $\psi = \psi_{\mathsf{S}}\psi_{\mathsf{R}} = (\psi_{\mathsf{S}}[1]\psi_{\mathsf{R}}[1], \ldots, \psi_{\mathsf{S}}[d]\psi_{\mathsf{R}}[d])^{\mathsf{T}} \in G^{d\times 1}$ will be used in encryption. Note that $\psi = \phi \cdot (sk_{\mathsf{S}} + sk_{\mathsf{R}})$. The algorithm $\mathsf{TEnc}(M; R)$ outputs $C = \mathsf{Enc}(M; R) = (R\phi, (R\psi)M) \in G^{1\times(d+1)} \times G$ as above. The algorithm $\mathsf{TDec}(sk_{\mathsf{P}}, C)$ outputs $\mu_{\mathsf{P}} = C[1] \cdot sk_{\mathsf{P}} \in G$ for $\mathsf{P} \in \{\mathsf{S}, \mathsf{R}\}$. Finally, $\mathsf{TComb}(C, \mu_{\mathsf{S}}, \mu_{\mathsf{R}})$ outputs $C[2]/(\mu_{\mathsf{S}}\mu_{\mathsf{R}})$. The resulting OT scheme is given in Fig.2.

Note that the above TPKE satisfies all properties mentioned in Sect.3.1. In particular, let us check some of them. For root extraction, given $C^e = \mathsf{TEnc}(M; R) = \big(R\phi, (R\psi)M\big)$ with $(e, q) = 1$, we want to extract (m, r) satisfying $C = \mathsf{TEnc}(m; r)$. This is done by just putting $m = M^{[e^{-1} \bmod q]}$, and $r = R \cdot [e^{-1} \bmod q] = \big(R[1](e^{-1} \bmod q), \ldots, R[d](e^{-1} \bmod q)\big)$. For the property of computing μ_{S} without sk_{S}, referring to Fig.2, we show that the element $\mu_{\mathsf{S}} = \mathsf{TDec}(sk_{\mathsf{S}}, C') = C'[1] \cdot sk_{\mathsf{S}}$ can be computed by R in case the

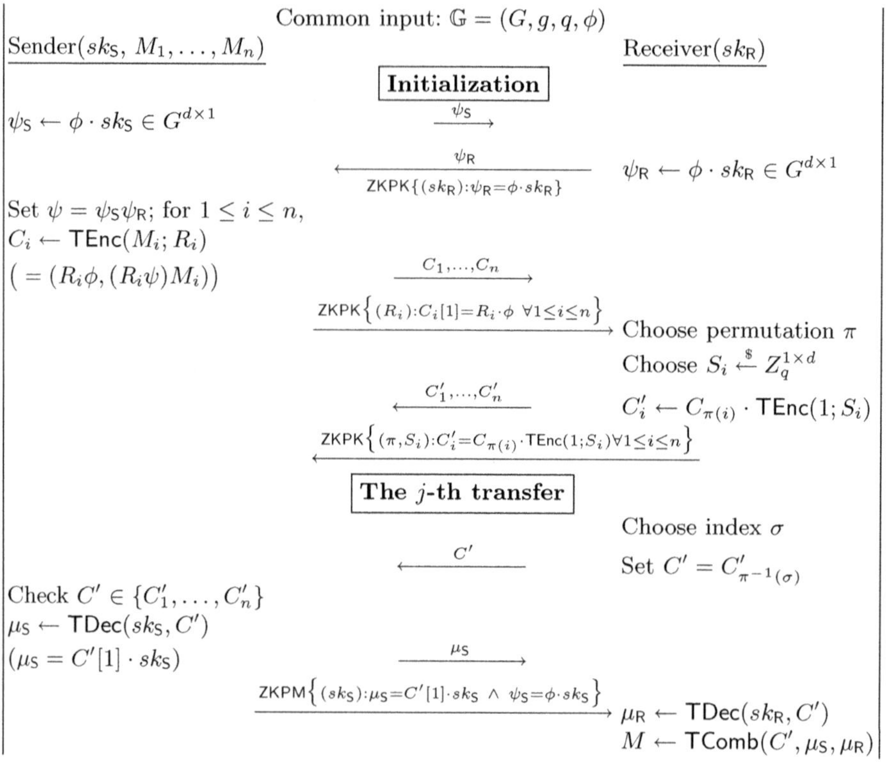

Fig. 2. The $\mathrm{OT}^n_{k\times 1}$ secure under the d-linear assumption

receiver already knew M_σ. Note that $C' = C'_{\pi^{-1}(\sigma)} = C_\sigma \cdot \mathsf{TEnc}(1; S_{\pi^{-1}(\sigma)}) = \mathsf{TEnc}(M_\sigma; R_\sigma + S_{\pi^{-1}(\sigma)})$ so that $C'[1] = (R_\sigma + S_{\pi^{-1}(\sigma)})\phi$, and hence

$$\begin{aligned}
\mu_S &= (R_\sigma + S_{\pi^{-1}(\sigma)})\phi \cdot sk_S = (R_\sigma\phi \cdot sk_S)(S_{\pi^{-1}(\sigma)}\phi \cdot sk_S) \\
&= \big(R_\sigma\phi \cdot (sk - sk_R)\big)\big(S_{\pi^{-1}(\sigma)}\phi \cdot (sk - sk_R)\big) \text{ for } sk = sk_S + sk_R \\
&= (R_\sigma\phi sk) \cdot (S_{\pi^{-1}(\sigma)}\phi sk) \cdot \big\{R_\sigma\phi(-sk_R)\big\} \cdot \big\{S_{\pi^{-1}(\sigma)}\phi(-sk_R)\big\} \\
&= (C_\sigma[2]M_\sigma^{-1}) \cdot (S_{\pi^{-1}(\sigma)}\psi) \cdot \big\{C_\sigma[1](-sk_R)\big\} \cdot \big\{S_{\pi^{-1}(\sigma)}\phi(-sk_R)\big\} \\
&\stackrel{\text{def}}{=} f(\phi, \psi, C_\sigma, M_\sigma, S_{\pi^{-1}(\sigma)}, sk_R),
\end{aligned}$$

which is a function of what R has, as desired.

4 Generic Adaptive OT from Permutation Networks

We present $\mathrm{OT}^n_{k\times 1}$ with $O(1)$ communication cost for the transfer phase, while with $O(n \log n)$ for the initialization phase. The assumptions used for security will be DCR or QR.

4.1 Loosely Homomorphic KEM

A key encapsulation mechanism KEM consists of algorithms Gen, Encap, Decap as follows: Gen produces keys (pk, sk); $\mathsf{Encap}(pk)$ outputs (ψ, K) where ψ is the encapsulation of the key K; $\mathsf{Decap}_{sk}(\psi)$ returns K as the decapsulation of ψ. We write $\mathsf{Encap}(pk; r)$ to emphasize the random coin r. We need the following conditions on KEM.

Semantic security: Suppose $\mathsf{Encap}(pk) = (\psi, K)$. Given pk, ψ, the key K is indistinguishable from random.

Loose homomorphism: Given (ψ, K) and (ψ', K'), there are efficiently computable functions f_1, f_2 such that

$$\mathsf{Decap}_{sk}(\psi \cdot \psi') = f_1(\psi, \psi', K, K') \text{ and } K' = f_2\left(\psi, \psi', K, \mathsf{Decap}_{sk}\left(\psi \cdot \psi'\right)\right).$$

The former equation is used in proving sender security, while the latter is for the OT's correctness. It is clear that a KEM is loosely homomorphic if it is homomorphic (namely, satisfying $\mathsf{Decap}_{sk}(\psi \cdot \psi') = K \oplus K'$).

Let us show some examples of loosely homomorphic KEM.

First example $\mathsf{KEM}_{\mathrm{DCR}}$: Gen generates primes p, q, setting $pk = N = pq$, and $sk = (p, q)$. Encap takes $r \xleftarrow{\$} Z_N$ and computes $(\psi, K) \in Z_N^2$ satisfying $r^N = \psi + K \cdot N \bmod N^2$. Note that $\psi = [r^N \bmod N] \in Z_N$. Using sk, $\mathsf{Decap}_{sk}(\psi)$ first computes r satisfying $r^N = \psi \bmod N$, an then outputs $K = (r^N - \psi \bmod N^2)/N$. The computation $\psi \cdot \psi'$ is normally defined over Z_N.

The semantic security of $\mathsf{KEM}_{\mathrm{DCR}}$ comes from the DCR assumption. To show that it is loosely homomorphic, consider (ψ, K) and (ψ', K') satisfying $r^N = \psi + K \cdot N \bmod N^2$, and $r'^N = \psi' + K' \cdot N \bmod N^2$. Writing $\psi\psi' = S + TN \bmod N^2$, we have

$$(rr')^N = [(\psi + KN)(\psi' + K'N) \bmod N^2] = [S + (T + K\psi' + K'\psi)N \bmod N^2],$$

so that $\hat{K} = \mathsf{Decap}_{sk}(\psi\psi' \in Z_N) = T + K\psi' + K'\psi \bmod N$, which is the function f_1. Moreover, since $(\psi + KN)(\psi' + K'N) = S + \hat{K}N \bmod N^2$, the key K' can be computed as

$$K' = \frac{[(S + \hat{K}N)(\psi + KN)^{-1} - \psi'] \bmod N^2}{N},$$

which expresses the function f_2.

Second example $\mathsf{KEM}_{\mathrm{QR}}$: To apply the recent 3-move ZKPK of Cramer and Damgård [5], we will use an expanded version of the Goldwasser-Micali encryption scheme. In particular, Gen is the same as above, except that a quadratic non-residue $y \in \mathcal{QNR}_N^{+1}$ is added to pk. The algorithm Encap takes $K \xleftarrow{\$} \{0,1\}^\ell$ and $r \xleftarrow{\$} Z_N^\ell$, returning the key K and its encapsulation

$$\psi = \left(y^{K[1]} r[1]^2 \bmod N, \ldots, y^{K[\ell]} r[\ell]^2 \bmod N\right).$$

The algorithm $\mathsf{Decap}_{sk}(\psi)$, for $1 \leq i \leq \ell$, returns $K[i] = 0$ if $\psi[i]$ is a quadratic residue modulo N; otherwise returns $K[i] = 1$. The scheme $\mathsf{KEM}_{\mathrm{QR}}$ is homomorphic, and has semantic security under the QR assumption.

In [5], the protocol $\mathsf{WIPK}\{(K, r) : \psi = \mathsf{Encap}(N; (K, r))\}$, is realized by a Σ-protocol, with soundness error $2^{-\ell}$ and communication cost $O(\ell)$ (instead of $O(\ell^2)$ via the cut-and-choose technique). Turning the Σ-protocol into a fully zero-knowledge one can be done by standard techniques (e.g., see [7]).

4.2 The OT Protocol

We show that an adaptive $\mathrm{OT}^n_{k\times 1}$ can be constructed from a loosely homomorphic $\mathsf{KEM} = (\mathsf{Gen}, \mathsf{Encap}, \mathsf{Decap})$.

Initialization Phase

1. The sender S generates $(pk, sk) \leftarrow \mathsf{Gen}$ and sends pk to R. The sender proves that pk is a valid public-key by ZKPM.
2. For $i = 1, \ldots, n$, the sender S generates $(\psi(r_i), K_i) = \mathsf{Encap}(pk; r_i)$ by choosing r_i randomly and sends to R
$$C_i = (A_i, B_i) = (\psi(r_i), K_i M_i),$$
where r_i is a random string used by Encap.
3. The sender proves by ZKPK that he knows r_i of $\psi(r_i)$ for every $1 \leq i \leq n$. Alternatively, he proves that he knows sk by ZKPK.
4. (**Permuting and Blinding**) The receiver chooses u_i randomly for $1 \leq i \leq n$, and generates
$$\mathsf{Encap}(pk; u_i) = (\varphi(u_i), K'_i).$$
He then randomly picks a permutation π on $\{1, \ldots, n\}$, computes $U_i = A_{\pi(i)} \cdot \varphi(u_i)$, and sends $U_1, \ldots, U_n$ to the sender. The receiver, equipped with secrets $(u_1, \ldots, u_n)$ and π, proves in ZKPK that
$$\left[U_1 = A_{\pi(1)} \cdot \varphi(u_1)\right] \wedge \cdots \wedge \left[U_n = A_{\pi(n)} \cdot \varphi(u_n)\right].$$
We will describe in Sect.4.3 how to perform the ZKPK with $O(n \log n)$ communication cost.

The jth Transfer Phase

5. The receiver chooses an index $1 \leq \sigma \leq n$, then sends $U = U_{\pi^{-1}(\sigma)}$.
6. The sender checks $U \in \{U_1, \ldots, U_n\}$, computes $\hat{K} = \mathsf{Decap}_{sk}(U)$ and sends $\hat{K}$ to the receiver.
7. The sender proves that $\hat{K} = \mathsf{Decap}_{sk}(U)$ by ZKPM.
8. Note that $U = A_\sigma \cdot \varphi(u_{\pi^{-1}(\sigma)})$. The receiver computes
$$K_\sigma = f_2\left(A_\sigma, \varphi(u_{\pi^{-1}(\sigma)}), K'_{\pi^{-1}(\sigma)}, \hat{K}\right),$$
and then obtains M_σ by computing $B_\sigma K_\sigma^{-1}$. (In additive groups, this becomes $B_\sigma - K_\sigma$, and all B_i as above are $K_i + M_i$.)

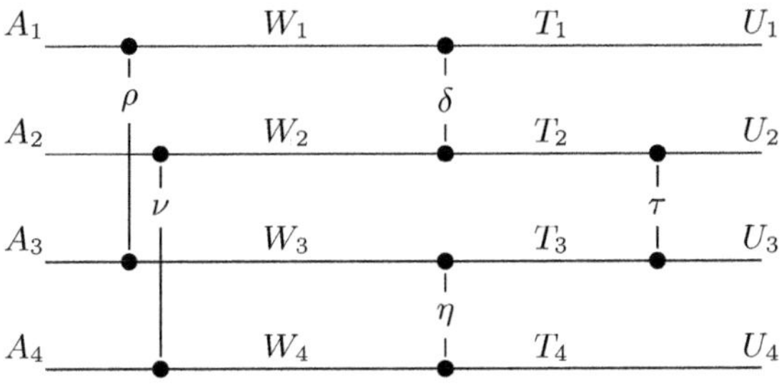

Fig. 3. From $n = 2$ to $n = 4$ with a permutation network of five switches

Theorem 2. *The generic* $\mathrm{OT}^n_{k\times 1}$ *from permutation networks above is fully simulatable, if the* KEM *scheme has semantic security.*

The proof is postponed in Appendix B.

4.3 How to Execute the ZKPK at Step 4

The case $n = 2$**:** First, let us focus on $n = 2$, proving the knowledge of u_1, u_2 such that

$$U_1 = A_{\pi(1)} \cdot \varphi(u_1) \wedge U_2 = A_{\pi(2)} \cdot \varphi(u_2),$$

for *some* permutation π on $\{1, 2\}$. The task is equivalent to proving

$$\Big(U_1 = A_1 \cdot \varphi(u_1) \wedge U_2 = A_2 \cdot \varphi(u_2)\Big) \vee \Big(U_1 = A_2 \cdot \varphi(u_1) \wedge U_2 = A_1 \cdot \varphi(u_2)\Big),$$

depending on whether $\big(\pi(1), \pi(2)\big) = (1, 2)$ or $(2, 1)$. Expanding further, what is proved becomes

$$\Big(U_1 = A_1 \cdot \varphi(u_1) \vee U_1 = A_2 \cdot \varphi(u_1)\Big) \wedge \Big(U_1 = A_1 \cdot \varphi(u_1) \vee U_2 = A_1 \cdot \varphi(u_2)\Big)$$
$$\wedge \Big(U_2 = A_2 \cdot \varphi(u_2) \vee U_1 = A_2 \cdot \varphi(u_1)\Big) \wedge \Big(U_2 = A_2 \cdot \varphi(u_2) \vee U_2 = A_1 \cdot \varphi(u_2)\Big).$$

The above are exactly four OR-proofs. If one can implement the interactive proof $\mathrm{WIPK}\{(u) : U = A \cdot \varphi(u)\}$ by a Σ-protocol, then it is well-known that one can efficiently realize the OR-proofs also with Σ-protocols. Note that if u_1, u_2 are known, then the permutation π can be extracted as well. Transforming Σ-protocols to ZKPK ones can be done by well-known techniques [6]. Therefore, the ZKPK for $n = 2$ in consideration can be implemented in four rounds, and we count its communication cost asymptotically as $O(1)$.

From 2 **to general** n: We will use the idea of n-permutation networks, which turn n inputs to n outputs, and the outputs are a permutation of the inputs. It is known that n-permutation networks can be built from 2-ones, which are called switches. There are constructions of n-permutation networks with $O(n \log^2 n)$ [4] or even $O(n \log n)$ switches [1, 8]. A comprehensive treatment on the topic can be found in [4, Chapter 28].

The idea is now we replace the switches by the WIPK protocol for $n = 2$ described above. We need $O(n \log n)$ protocols as switches, and each protocol requires $O(1)$ communication cost, so that the total communication cost becomes $O(n \log n)$.

Let us concretely illustrate how one proceeds from $n = 2$ to $n = 4$, using the permutation network depicted in Fig.3 of five switches. The elements W_i, T_i, U_i are sent to the sender by the receiver[3]. The first two switches ρ and ν prove that

$$W_1 = A_{\rho(1)} \cdot \varphi(w_1), W_3 = A_{\rho(3)} \cdot \varphi(w_3), W_2 = A_{\nu(2)} \cdot \varphi(w_2), W_4 = A_{\nu(4)} \cdot \varphi(w_4).$$

Consequently, the second two switches δ and η ensure

$$T_1 = W_{\delta(1)} \cdot \varphi(t_1), T_2 = W_{\delta(2)} \cdot \varphi(t_2), T_3 = W_{\eta(3)} \cdot \varphi(t_3), T_4 = W_{\eta(4)} \cdot \varphi(t_4).$$

The final switch τ is between T_2 and T_3, showing

$$U_2 = T_{\tau(2)} \cdot \varphi(v_2) \wedge U_3 = T_{\tau(3)} \cdot \varphi(v_3).$$

To ease the illustration, let us take concrete switches $\tau = (2\ 3)$ (namely 2 to 3 and vice versa), $\delta = (1\ 2)$, $\nu = (2\ 4)$, and the others are identity switches. Denote $U \sim A$ if there is u such that $U = A \cdot \varphi(u)$, so that

$$\begin{aligned} U_1 &\sim T_1 \sim W_2 \sim A_4 \\ U_2 &\sim T_3 \sim W_3 \sim A_3 \\ U_3 &\sim T_2 \sim W_1 \sim A_1 \\ U_4 &\sim T_4 \sim W_4 \sim A_2 \end{aligned}$$

which means (U_1, U_2, U_3, U_4) blinds and permutes (A_1, A_2, A_3, A_4) as expected.

Instantiations: As shown above, we just need to implement the atomic WIPK$\{(u) : U = A \cdot \varphi(u)\}$ by a Σ-protocol.

From the DCR assumption: Set $\varphi(u) = u^N \bmod N$ for $u \in Z_N$, so that the atomic WIPK is similar to the GQ proof [12].

From the QR assumption: Set

$$\varphi(u = (K, r)) = \left(y^{K[1]} r[1]^2 \bmod N, \ldots, y^{K[\ell]} r[\ell]^2 \bmod N \right)$$

for $u = (K, r) \in Z_2^\ell \times Z_N^\ell$. The elegant result of Cramer and Damgård [5] gives us the desired 3-move WIPK with soundness error $2^{-\ell}$.

4.4 How to Execute Other ZKPK Protocols

The ZKPM at step 7, in the case of $\mathsf{KEM}_{\mathrm{DCR}}$, is equivalent to proving $r^N = U + \hat{K}N \bmod N^2$ for some r, for which the 4-move ZK protocol can be found

[3] In general, the receiver needs to send n (= 4 in Fig.3) elements at $O(\log n)$ (= 3 in Fig.3) steps.

in [18]. For KEM_{QR}, proving $U = (y^{\hat{K}[1]}r[1]^2 \bmod N, \ldots, y^{\hat{K}[\ell]}r[\ell]^2 \bmod N)$ for some $r \in Z_N^\ell$ is needed, which can be accomplished by the 4-move ZK protocol for the knowledge of ℓ square roots in [5].

We now turn to the necessary protocols for the validity of the public key. Proving y is a quadratic non-residue can be done in 4 moves as in [5]. To prove the validity of N, namely $N = pq$ for some distinct primes p, q, we can use the protocols in [2,15].

5 Leakage-Resilient Adaptive OT

Our generic method in Sect.3 can be further used to construct the first (memory) leakage resilient adaptive OT protocol. We define an adaptive OT scheme by $(\mathsf{G}, \mathsf{S}, \mathsf{R})$, where G is a PPT algorithm, and S(ender) and R(eceiver) are interactive PPT algorithms. G outputs a pair of public-key and secret-key (pk, sk). S takes (pk, sk) and $(M_1, \ldots, M_n)$ as inputs, where $(M_1, \ldots, M_n)$ are given from outside. R takes pk as input. It is also given a choice index σ from outside in each transfer phase. Consider a scenario in which R mounts a side-channel attack against S to steal some information on sk.

The leakage-resilient encryption scheme shown by Naor and Segev at the end of Sect.5.2 in [20, Eprint] satisfies our conditions in Sect.3. Hence we can obtain a leakage-resilient adaptive OT protocol under the DDH assumption. The details will be given in the final paper.

References

1. Ajtai, M., Komlós, J., Szemerédi, E.: An $O(n \log n)$ sorting network. In: STOC, pp. 1–9. ACM, New York (1983)
2. Boyar, J., Friedl, K., Lund, C.: Practical zero-knowledge proofs: Giving hints and using deficiencies. J. Cryptology 4(3), 185–206 (1991)
3. Camenisch, J., Neven, G., Shelat, A.: Simulatable adaptive oblivious transfer. In: Naor, M. (ed.) EUROCRYPT 2007. LNCS, vol. 4515, pp. 573–590. Springer, Heidelberg (2007)
4. Cormen, T.H., Leiserson, C.E., Rivest, R.L., Stein, C.: Introduction to algorithms, 2nd edn. (2001)
5. Cramer, R., Damgård, I.: On the amortized complexity of zero-knowledge protocols. In: Halevi, S. (ed.) [13], pp. 177–191
6. Cramer, R., Damgård, I., MacKenzie, P.D.: Efficient zero-knowledge proofs of knowledge without intractability assumptions. In: Imai, H., Zheng, Y. (eds.) PKC 2000. LNCS, vol. 1751, pp. 354–373. Springer, Heidelberg (2000)
7. Damgård, I.: On Σ-protocols. In: Course notes in Cryptologic Protocol Theory (2010), `http://www.daimi.au.dk/~ivan/CPT.html`
8. Goodrich, M.T.: Randomized shellsort: A simple oblivious sorting algorithm. In: Proceedings 21st ACM-SIAM Symposium on Discrete Algorithms, SODA (2010)
9. Green, M., Hohenberger, S.: Universally composable adaptive oblivious transfer. In: Pieprzyk, J. (ed.) ASIACRYPT 2008. LNCS, vol. 5350, pp. 179–197. Springer, Heidelberg (2008)

10. Green, M., Hohenberger, S.: Practical adaptive oblivious transfer from simple assumptions. In: Ishai, Y. (ed.) TCC 2011. LNCS, vol. 6597, pp. 347–363. Springer, Heidelberg (2011)
11. Groth, J., Lu, S.: Verifiable shuffle of large size ciphertexts. In: Okamoto, T., Wang, X. (eds.) PKC 2007. LNCS, vol. 4450, pp. 377–392. Springer, Heidelberg (2007)
12. Guillou, L.C., Quisquater, J.-J.: A "Paradoxical" identity-based signature scheme resulting from zero-knowledge. In: Goldwasser, S. (ed.) CRYPTO 1988. LNCS, vol. 403, pp. 216–231. Springer, Heidelberg (1990)
13. Halevi, S. (ed.): CRYPTO 2009. LNCS, vol. 5677. Springer, Heidelberg (2009)
14. Jarecki, S., Liu, X.: Efficient oblivious pseudorandom function with applications to adaptive OT and secure computation of set intersection. In: Reingold, O. (ed.) TCC 2009. LNCS, vol. 5444, pp. 577–594. Springer, Heidelberg (2009)
15. Kurosawa, K., Katayama, Y., Ogata, W., Tsujii, S.: General public key residue cryptosystems and mental poker protocols. In: Damgård, I.B. (ed.) EUROCRYPT 1990. LNCS, vol. 473, pp. 374–388. Springer, Heidelberg (1991)
16. Kurosawa, K., Nojima, R.: Simple adaptive oblivious transfer without random oracle. In: Matsui, M. (ed.) ASIACRYPT 2009. LNCS, vol. 5912, pp. 334–346. Springer, Heidelberg (2009)
17. Kurosawa, K., Nojima, R., Phong, L.T.: Efficiency-improved fully simulatable adaptive OT under the DDH assumption. In: Garay, J.A., De Prisco, R. (eds.) SCN 2010. LNCS, vol. 6280, pp. 172–181. Springer, Heidelberg (2010)
18. Kurosawa, K., Takagi, T.: New approach for selectively convertible undeniable signature schemes. In: Lai, X., Chen, K. (eds.) ASIACRYPT 2006. LNCS, vol. 4284, pp. 428–443. Springer, Heidelberg (2006)
19. Naor, M., Pinkas, B.: Oblivious transfer with adaptive queries. In: Wiener, M. (ed.) CRYPTO 1999. LNCS, vol. 1666, pp. 573–590. Springer, Heidelberg (1999)
20. Naor, M., Segev, G.: Public-key cryptosystems resilient to key leakage. In: Halevi, S. (ed.) [13], pp. 18–35, Full version available at `http://eprint.iacr.org/2009/105.pdf`
21. Neff, C.A.: A verifiable secret shuffle and its application to e-voting. In: ACM Conference on Computer and Communications Security, pp. 116–125 (2001)
22. Ogata, W., Kurosawa, K.: Oblivious keyword search. J. Complexity 20(2-3), 356–371 (2004), `http://eprint.iacr.org/2002/182`

A Proof of Theorem 1

Lemma 3 (Receiver security). *The* $\mathrm{OT}^n_{k\times 1}$ *protocol in Sect.3 is secure against sender corruption.*

Proof. For every real-world adversary $\mathcal{A}$ who corrupts the sender, we construct an ideal-world adversary $\mathcal{A}'$ such that the advantage $\mathbf{Adv}(\mathcal{Z})$ is negligible. We will consider a sequence of games beginning from game G_0, which is the real world experiment, and proceed to the final game, which is the ideal world experiment as in Sec.2.2. For each integer i, let $\Pr(G_i) = \Pr(\mathcal{Z} = 1 \text{ in game } G_i)$, and denote $\Pr(G_i) \approx \Pr(G_j)$ when the two values are negligibly close.

Game G_0: This is the real world experiment such that the sender is controlled by the adversary $\mathcal{A}$. By definition $\Pr(G_0) = \Pr(\mathcal{Z} = 1 \text{ in the real world})$.

Game G_1: This game is the same as the previous one except the following. In the initialization phase, the game extracts $M_i^*(1 \le i \le n)$ from $\mathcal{A}$ by using the knowledge extractor of the ZKPK. If it fails, then the protocol stops. Since the failure occurs with negligible probability, we have $\Pr(G_0) \approx \Pr(G_1)$.

Game G_2: This game is the same as game G_1 except that, in the initialization phase, the game uses two zero-knowledge simulators of the ZKPKs for proving the knowledge of sk_{R}, and (π, s_i) respectively. Since ZKPK protocols are zero-knowledge, we have $\Pr(G_1) \approx \Pr(G_2)$.

Game G_3: This game is the same as the previous one, except that $C_i'(1 \le i \le n)$ are randomly chosen from the ciphertext space. We have $\Pr(G_3) \approx \Pr(G_2)$, thanks to the semantic security of the threshold encryption scheme. (Namely, $\mathsf{TEnc}(I; s_i)$ are almost random, so are $C_i' = C_{\pi(i)}\mathsf{TEnc}(1; s_i)$.)

Game G_4: This game is the same as the previous one except the following. In each transfer phases, the receiver chooses C' randomly and distinctly from the set $\{C_1', \ldots, C_n'\}$. Since the view of $\mathcal{A}$ is unchanged, we have $\Pr(G_4) = \Pr(G_3)$.

Game G_5: This game is the ideal world experiment in which an ideal-world adversary $\mathcal{A}'$ uses $\mathcal{A}$ as a black-box as follows: (1) $\mathcal{A}'$ receives $(M_1, \ldots, M_n)$ from $\mathcal{Z}$, and sends $(M_1, \ldots, M_n)$ to $\mathcal{A}$; (2) $\mathcal{A}'$ runs game G_4 with $\mathcal{A}$, where $\mathcal{A}'$ plays the role of the receiver, which can be accomplished since σ, the secret index of the receiver, is not used in game G_4; (3) $\mathcal{A}'$ sends the extracted $(M_1^*, \ldots, M_n^*)$ as in game G_1 to $\mathcal{F}_{\text{adapt}}$; (4) In each transfer phase, if $\mathcal{A}$ behaved in an acceptable way, then $\mathcal{A}'$ sends $b = 1$ to $\mathcal{F}_{\text{adapt}}$. Otherwise $\mathcal{A}'$ sends $b = 0$ to $\mathcal{F}_{\text{adapt}}$; (5) Suppose that $\mathcal{A}$ sends $\mathcal{A}_{\text{out}}$ to $\mathcal{Z}$ at the end of the game. Then $\mathcal{A}'$ sends $\mathcal{A}'_{\text{out}} = \mathcal{A}_{\text{out}}$ to $\mathcal{Z}$.

We have $\Pr(G_4) = \Pr(G_5)$, and by definition $\Pr(\mathcal{Z} = 1 \text{ in the ideal world}) = \Pr(G_5)$. Summing up all above, we have $\mathbf{Adv}(\mathcal{Z}) = |\Pr(G_0) - \Pr(G_5)|$ is negligible as required. □

Lemma 4 (Sender security). *The* $\mathrm{OT}^n_{k\times 1}$ *protocol in Sect.3 is secure against receiver corruption.*

Proof. For every real-world adversary $\mathcal{A}$ who corrupts the receiver, we construct an ideal-world adversary $\mathcal{A}'$ such that the advantage of the environment $\mathbf{Adv}(\mathcal{Z})$

is negligible. We again consider a sequence of games in which the first is the real world experiment of Sec.2.2, while the final is the ideal world experiment. Again, let $\Pr(G_i) = \Pr(\mathcal{Z} = 1 \text{ in game } G_i)$.

Game G_0: In this game the receiver is controlled by the adversary $\mathcal{A}$, and by definition $\Pr(G_0) = \Pr(\mathcal{Z} = 1 \text{ in the real world})$.

Game G_1: This game is the same as game G_0 except the following. In the initialization phase, the game extracts sk_{R}, and (π, s_i) by using the extractors of the corresponding ZKPKs respectively. Unless the extractors fail, which occurs with negligible probability, games G_1 and G_0 are identical, so that $\Pr(G_1) \approx \Pr(G_0)$.

Game G_2: In this game the index σ used in the transfer phase is extracted as follows. Since $\mathcal{A}$ sends C' such that $C' \in \{C'_1, \ldots, C'_n\}$, the sender searches the index $1 \le \rho \le n$ satisfying $C' = C'_\rho$. Recall $C' = C'_{\pi^{-1}(\sigma)}$, so $\pi^{-1}(\sigma) = \rho$, and hence $\sigma = \pi(\rho)$. Since the change is syntactic, we have $\Pr(G_2) = \Pr(G_1)$.

Game G_3: This game is the same as the previous one except the following. In each transfer phase, the game computes μ_{S} as

$$\mu_{\mathsf{S}} = f(pk, C_\sigma, M_\sigma, s_{\pi^{-1}(\sigma)}, sk_{\mathsf{R}}).$$

Since the change is syntactic, we have $\Pr(G_3) = \Pr(G_2)$.

Game G_4: This game is the same as the previous one except the following. In each transfer phase, instead of running the ZKPK proving the correct decryption of C' under sk_{S}, the zero-knowledge simulator of the ZKPK is run so that $\Pr(G_4) = \Pr(G_3)$.

Game G_5: This game is the same as the previous one except the following. In the initialization phase, each C_i is randomly chosen. It is easy to see that $\Pr(G_5) \approx \Pr(G_4)$ thanks to the semantic security of the TPKE scheme.

Game G_6: This game is the ideal world experiment in which an ideal-world adversary $\mathcal{A}'$ uses $\mathcal{A}$ as a black-box as follows: (1) $\mathcal{A}'$ runs game G_5 with $\mathcal{A}$, where $\mathcal{A}'$ plays the role of the sender; (2) In each transfer phase, $\mathcal{A}'$ sends σ which is extracted as in game G_2 to $\mathcal{F}_{\text{adapt}}$, and obtains M_σ. Then $\mathcal{A}'$ computes μ_{S} as in game G_3; (3) Suppose that $\mathcal{A}$ sends $\mathcal{A}_{\text{out}}$ to $\mathcal{Z}$ at the end of the game. Then $\mathcal{A}'$ sends $\mathcal{A}'_{\text{out}} = \mathcal{A}_{\text{out}}$ to $\mathcal{Z}$.

We have by definition $\Pr(G_6) = \Pr(\mathcal{Z} = 1 \text{ in the ideal world})$. Summing up all above, we have $\mathbf{Adv}(\mathcal{Z}) = |\Pr(G_0) - \Pr(G_6)|$ is negligible as required. □

B Proof of Theorem 2

Lemma 5 (Receiver security). *The* $\mathrm{OT}^n_{k\times 1}$ *protocol in Sect.4 is secure against sender corruption.*

Proof. For every real-world adversary $\mathcal{A}$ who corrupts the sender, we construct an ideal-world adversary $\mathcal{A}'$ such that the advantage $\mathbf{Adv}(\mathcal{Z})$ is negligible. We consider a series of games as follows. Game G_0 is exactly the real-world experiment where the sender is corrupted. Game G_1 is the same as the previous game except that it extracts the secret key sk from the corrupted sender. It is easy to see that $\Pr(G_1) \approx \Pr(G_0)$. In game G_2, the difference is that $U_1, \ldots, U_n$ is chosen randomly and sent to the sender. Then the simulator for the corresponding ZKPK (at step 4) is run. It is clear that $\Pr(G_2) \approx \Pr(G_1)$. Game G_3 is the ideal experiment in which $\mathcal{A}'$ runs game G_2 with $\mathcal{A}$. Since $\mathcal{A}'$ extracts sk from $\mathcal{A}$, it obtains M_i^* from C_i for $1 \leq i \leq n$ and sends all the messages to $\mathcal{F}_{\text{adapt}}$. In each transfer, $\mathcal{A}'$ chooses U randomly and distinctly from $\{U_1, \ldots, U_n\}$. Moreover, if the ZKPM (at step 7) passes, $\mathcal{A}'$ sends 1, otherwise sends 0 to $\mathcal{F}_{\text{adapt}}$. We thus have $\Pr(G_3) = \Pr(G_2)$ and hence $\Pr(G_3) \approx \Pr(G_0)$, meaning the ideal and real worlds are indistinguishable, so that $\mathbf{Adv}(\mathcal{Z})$ must be negligible as required. □

Lemma 6 (Sender security). *The* $\mathrm{OT}^n_{k \times 1}$ *protocol in Sect.4 is secure against receiver corruption.*

Proof. For every real-world adversary $\mathcal{A}$ who corrupts the receiver, we construct an ideal-world adversary $\mathcal{A}'$ such that the advantage of the environment $\mathbf{Adv}(\mathcal{Z})$ is negligible. We consider a series of games as follows. First, game G_0 is the real-world experiment. Game G_1 is identical to game G_0, except that it extracts the secrets $u_1, \ldots, u_n$ and π from the corrupted receiver. We have $\Pr(G_1) \approx \Pr(G_0)$. In game G_2, the index σ is extracted as follows. In the transfer phase, when the receiver sends U, the game searches for an index $1 \leq \rho \leq n$ such that $U = U_\rho$. By the construction $U = U_{\pi^{-1}(\sigma)}$ so that $\rho = \pi^{-1}(\sigma)$ and hence $\sigma = \pi(\rho)$. We have $\Pr(G_2) \approx \Pr(G_1)$. In game G_3, $\hat{K} = \mathsf{Decap}_{sk}(U_{\pi^{-1}(\sigma)}) = \mathsf{Decap}_{sk}(A_\sigma \cdot \varphi(u_{\pi^{-1}(\sigma)}))$ is alternatively computed as $\hat{K} = f_1(A_\sigma, \varphi(u_{\pi^{-1}(\sigma)}), B_\sigma M_\sigma^{-1}, K'_{\pi^{-1}(\sigma)})$. We have $\Pr(G_3) \approx \Pr(G_2)$. In game G_4, all $C_i = (A_i, B_i)$ are randomly chosen. By the semantic security of KEM, we have $\Pr(G_4) \approx \Pr(G_3)$. Game G_5 is the ideal world in which $\mathcal{A}'$ runs $\mathcal{A}$ as in game G_4. The adversary $\mathcal{A}'$ extracts σ as in game G_2, and the index is sent to $\mathcal{F}_{\text{adapt}}$ to obtain M_σ. Then the key $\hat{K}$ is computed as in game G_3. All the zero-knowledge proofs to the corrupted receiver are replaced by the simulated ones. It is clear that $\Pr(G_5) \approx \Pr(G_4)$ so that $\Pr(G_5) \approx \Pr(G_0)$, and hence $\mathbf{Adv}(\mathcal{Z})$ is negligible as required. □

Simple and Efficient Single Round almost Perfectly Secure Message Transmission Tolerating Generalized Adversary

Ashish Choudhury[1,*], Kaoru Kurosawa[2], and Arpita Patra[3,**]

[1] Applied Statistics Unit
Indian Statistical Institute Kolkata India
partho31@gmail.com, partho_31@yahoo.co.in

[2] Department of Computer and Information Sciences
Ibaraki University, Hitachi, Ibaraki Japan
kurosawa@mx.ibaraki.ac.jp

[3] Department of Computer Science
Aarhus University, Denmark
arpitapatra10@gmail.com, arpita@cs.au.dk

Abstract. Patra et al. (IJACT '09) gave a necessary and sufficient condition for the possibility of *almost perfectly secure message transmission* protocols[1] tolerating *general, non-threshold* $\mathcal{Q}^2$ adversary structure. However, their protocol requires *at least three* rounds and performs *exponential* (exponential in the size of the adversary structure) computation and communication. They have left it as an open problem to design efficient protocol for almost perfectly secure message transmission, tolerating $\mathcal{Q}^2$ adversary structure.

In this paper, we show the first *single* round almost perfectly secure message transmission protocol tolerating $\mathcal{Q}^2$ adversary structure. The computation and communication complexities of the protocol are both *polynomial* in the size of underlying *linear secret sharing scheme* (LSSS). This solves the open problem posed by Patra et al.

When we restrict our general protocol to a *threshold adversary*, we obtain a single round, *communication optimal* almost secure message transmission protocol tolerating threshold adversary, which is much more *computationally efficient* and *relatively simpler* than the previous single round, communication optimal protocol of Srinathan et al. (PODC '08).

Keywords: Information theoretic security, non-threshold adversary, Byzantine corruption, Efficiency.

* Financial Support from Department of Information Technology, Govt. of India Acknowledged.

** Financial Support from Center for Research in Foundations of Electronic Markets, Denmark Acknowledged.

[1] Patra et al. [25] called this problem as *unconditionally secure message transmission* (USMT).

J. Lopez and G. Tsudik (Eds.): ACNS 2011, LNCS 6715, pp. 292–308, 2011.

1 Introduction

Consider the following problem: there exists a sender **S** and a receiver **R**, who are part of a large distributed network and connected by n disjoint channels. There exists a *computationally unbounded adversary*, who can listen and forge communication over some of these channels in any arbitrary manner. However, neither **S**, nor **R** knows which of the channels are under the control of the adversary. **S** has a message $m^{\mathbf{S}}$, which is a sequence of ℓ elements from a finite field $\mathbb{F}$, where $\ell \geq 1$. The challenge is to design a protocol, such that after interacting with **S** as per the protocol, the following should hold at **R**'s end:

1. **Perfect Reliability:** **R** outputs $m^{\mathbf{R}} = m^{\mathbf{S}}$.
2. **Perfect Secrecy:** Adversary should not get any *extra* information about $m^{\mathbf{S}}$. In other words, $m^{\mathbf{S}}$ should be *information theoretically secure.*

This problem is called *perfectly secure message transmission* (PSMT) [12].

Motivation and Different Models for Studying PSMT. PSMT is a well known and fundamental problem in secure distributed computing. If **S** and **R** are directly connected by a secure channel, as assumed in generic *multiparty computation* (MPC) protocols [3,4,29], then PSMT is trivial. However, if **S** and **R** are not directly connected by a secure channel, then PSMT protocols help to simulate a *virtual secure channel* between **S** and **R**. The second motivation for PSMT is to achieve information theoretic security. The security of all existing public key cryptosystems is based on hardness assumptions of certain number theoretic problems and security of these schemes holds only against a computationally bounded adversary. However, with the advent of new computing paradigms like Quantum computing [33] and with the increase in computing speed, these assumptions may tend to be useless. But all these factors have no effect on PSMT protocols, as security of these protocols holds good against a computationally unbounded adversary.

Over the past two decades, PSMT problem has been studied by several researchers in different settings. Specifically, we can consider the following settings:

1. **Type of Channels:** The channels between **S** and **R** can be *bi-directional.* This setting has been considered in [12,31,16,36,1,13,19,26,24]. On the other hand, channels may be *uni-directional*, having direction associated with them [10,23,21,40,6].
2. **Adversary Capacity:** The adversary may be characterized by a *threshold*, say t, such that the adversary can control *any* t out of the n channels [12,31,36,19] or the adversary may be characterized as a more general *non-threshold* adversary, specified by an adversary structure [16,28,40,41,17,18].
3. **Adversary Behavior:** The adversary may be *static* who corrupts the same channels throughout the protocol [12,31,36,19] or the adversary may be *mobile*, who corrupts different set of channels, during different stages of the protocol [39,26,5,27].

4. **Type of Underlying Network:** The underlying network may be *synchronous*, where there is a global clock in the system and the delay in the transmission over any channel is bounded by a constant [12,31,36,38,19] or the network may be *asynchronous*, having no such global clock [30,6].

Any PSMT protocol is analyzed by the following parameters:

1. **Round Complexity**: It is the number of communication rounds taken by the protocol, where a round is a communication from **S** to **R** or vice-versa.
2. **Communication Complexity**: It is the total number of field elements sent by **S** and **R** in the protocol.
3. **Computational Complexity**: It is amount of computation which is done by **S** and **R** in the protocol.

We call a PSMT protocol against a non-threshold adversary as *efficient*, if the round complexity, communication complexity and computational complexity of the protocol is *polynomial* in n and the size of the Monotone Span Programme (MSP) for *the adversary structure* (adversary structure is presented in Sec. 1.1 and MSP is presented Sec. 2). On the other hand, a PSMT protocol against a t-active threshold adversary is called efficient, if its round, communication and computational complexity is polynomial in n and t. Irrespective of the settings in which PSMT problem is studied, the following questions are fundamental:

- **Possibility**: What are the necessary and sufficient conditions for the existence of any PSMT protocol, tolerating a given type of adversary?
- **Feasibility**: Once the possibility of a protocol is ascertained, the next obvious question is whether there exists an *efficient* protocol or not?
- **Optimality**: Given a message of some specific length, what is the lower bound on the round complexity and communication complexity of any PSMT protocol to send the message? Moreover, do we have a protocol, whose total round complexity and communication complexity matches these bounds?

Different techniques are used to answer the above questions in different settings. For details, see [7]. The issue of **Possibility, Feasibility** and **Optimality** of PSMT has been completely resolved tolerating *threshold adversary.* However, not too much is known regarding the **Feasibility** and **Optimality** of protocols against non-threshold adversary (see [7] for complete details).

1.1 Non-Threshold Adversary

Let the set of n channels be denoted by $\mathcal{W} = \{w_1, \ldots, w_n\}$. Then a *threshold adversary* is characterized by a threshold t, such that the adversary can control any t channels out of the n channels for corruption. We denote such an adversary by $\mathcal{A}_t$. On the other hand, a *non-threshold general adversary* $\mathcal{A}$ is characterized by an *adversary structure* Γ, which is a collection of subsets of channels that the adversary $\mathcal{A}$ can *potentially* corrupt. That is,

$$\Gamma = \{B \subset \mathcal{W} \mid \mathcal{A} \text{ can corrupt } B\}.$$

Moreover, we assume that if $B \in \Gamma$ and if $B' \subset B$, then $B' \in \Gamma$. It is easy to see that a threshold adversary is a special case of non-threshold adversary, such that all possible $B \subset \mathcal{W}$ with $|B| \leq t$, are present in Γ.

Definition 1 ($\mathcal{Q}^k$ Condition [15]). *We say that $\mathcal{A}$ satisfies the $\mathcal{Q}^k$ condition with respect to $\mathcal{W}$, if there exists no k sets in Γ, which adds up to the whole set $\mathcal{W}$. That is:*

$$\forall B_1, \ldots, B_k \in \Gamma : B_1 \cup \ldots \cup B_k \neq \mathcal{W}.$$

PSMT Tolerating Non-Threshold Adversary. Modeling the adversary by a threshold helps in easy characterization of protocols and it also helps in analyzing protocols. However, as mentioned in [15], modeling the (dis)trust in the network as a threshold adversary is not always appropriate because threshold protocol requires more *stringent* requirements than the reality [16]. Motivated by this, Kumar et al. [16] studied PSMT tolerating non-threshold adversary for the first time in the literature. The work of Kumar et al. is followed by [11,28,17,40,41], where the issues related to the **Possibility** and **Feasibility** of PSMT against non-threshold adversary have been studied. In short, there exists efficient PSMT protocols tolerating non-threshold adversary for bi-directional channels [41,17] as well as for uni-directional channels [41]. However, there exists another variant of PSMT, known as *almost perfectly secure message transmission* (almost-PSMT), which got relatively less attention in the context of non-threshold adversary.

1.2 Almost Perfectly Secure Message Transmission: Almost-PSMT

In PSMT, it is required that **R** should output $m^{\mathbf{R}} = m^{\mathbf{S}}$ without any error. In [14], the authors considered a variant of PSMT called almost-PSMT, where they relaxed this requirement. Specifically, a protocol is called almost-PSMT, if it satisfies the following requirements:

1. **Perfect Secrecy**: Same as in the case of PSMT.
2. **Almost Perfect Reliability**: **R** outputs $m^{\mathbf{R}} = m^{\mathbf{S}}$ with probability at least $1 - 2^{-\Omega(\kappa)}$, where κ is the error parameter and $\kappa > 0$.

In [14], the authors studied almost-PSMT tolerating threshold adversary and showed that almost-PSMT protocols require less number of channels than PSMT protocols for tolerating a threshold adversary with the same threshold. *That is, allowing a negligible error probability in protocol outcome reduces the connectivity requirement.* The work of [14] is followed by [10,37,20,35,2,9,22] where almost-PSMT tolerating threshold adversary is studied rigorously and the issues related to the **Possibility, Feasibility** and **Optimality** of almost-PSMT tolerating threshold adversary has been completely resolved. In summary, all these works show that *allowing a negligible error probability in the protocol output (without compromising the secrecy) results in significant reduction in the round complexity, communication complexity and also connectivity requirement (number of channels) of PSMT protocols.*

Remark 1. (**On the Term almost-PSMT**): In the literature, almost-PSMT protocols are also known by various other names. In [34,37], the authors called these protocols as *probabilistic PSMT* (PPSMT). On the other hand, [25,35] called these protocols as *unconditionally secure message transmission* (USMT) protocols. Finally, [7] called these protocols as *statistically secure message transmission* (SSMT) protocols. However, all the above terms stand for almost-PSMT. In this article, we prefer to use the original name, namely almost-PSMT.

1.3 Almost-PSMT Tolerating Non-Threshold Adversary: Motivation of Our Work

Unlike almost-PSMT tolerating threshold adversary, almost-PSMT against non-threshold adversary has got very less attention. In [25], Patra et al. have studied almost-PSMT tolerating non-threshold adversary. They showed that *single round as well as multi-round almost-PSMT* is possible iff $\mathcal{A}$ satisfies $\mathcal{Q}^2$ condition. This is to be compared with the results of [11] and [16], according to which *single round* and *multi-round* PSMT is possible iff $\mathcal{A}$ satisfies $\mathcal{Q}^3$ and $\mathcal{Q}^2$ condition respectively. Unfortunately, the almost-PSMT protocol tolerating non-threshold adversary presented in [25] is very inefficient and requires computation and communication complexity, which is exponential in the size of adversary structure[2]. Moreover, it requires at least three rounds. In [25], the authors have left it as an open problem to design efficient almost-PSMT protocol tolerating non-threshold adversary, satisfying $\mathcal{Q}^2$ condition. In this paper, we solve this open problem.

1.4 Our Results and Comparison with the Existing Results

In this paper, we present the first single round almost-PSMT protocol tolerating non-threshold adversary $\mathcal{A}$, specified by an adversary structure, satisfying $\mathcal{Q}^2$ condition. Our protocol is *round optimal*, requiring minimum number of rounds. Moreover, our protocol is very simple and efficient and thus significantly outperforms the almost-PSMT protocol of [25].

As a special case of our single round protocol, when we restrict it to *threshold adversary*, we get a single round *communication optimal* almost-PSMT tolerating threshold adversary. Though there exists single round, communication optimal almost-PSMT protocol tolerating threshold adversary [35], we find that our protocol is much more *computationally efficient* and *relatively simpler* than the protocol of [35]. In practical networks like sensor networks, it is desirable to have protocols which perform simple computation. In such a situation, our communication optimal protocol (tolerating threshold adversary) fits the bill more appropriately than the communication optimal protocol of [35].

In [9] the authors have designed single round almost-PSMT protocol tolerating threshold adversary, which performs simple computations. However, their protocol is *not communication optimal.* On the other hand, our protocol tolerating threshold adversary enjoys the property of *being both simple and also communication optimal.*

[2] The protocol of [25] does not use LSSS and is based on the principle of Induction.

Table 1. Comparison of our almost-PSMT protocol tolerating $\mathcal{Q}^2$ adversary structure with best known almost-PSMT protocol tolerating $\mathcal{Q}^2$ adversary structure

Reference	Number of Rounds	Efficient/Inefficient
[25]	At least three	Inefficient
This paper	One	Efficient

Table 2. Comparison of our single round almost-PSMT protocol tolerating threshold adversary with $n = 2t + 1$ with the best known single round almost-PSMT protocols tolerating threshold adversary with $n = 2t + 1$

Reference	Communication Optimal	Computational Complexity
[35]	Yes	Efficient (Polynomial in n)
[9]	**No**	More efficient than [35]
This paper	Yes	More efficient than [35]

In Table 1 and 2, we compare our protocols with the best known almost-PSMT protocols in non-threshold and threshold settings respectively.

1.5 Tools and Techniques Used in Our Protocol

To design our protocol, we use *Linear Secret Sharing Scheme* (LSSS) [8]. In addition, we also use a new method of authenticating multiple values concurrently in information theoretic sense. Together this leads to our efficient single round almost-PSMT protocol.

2 Primitives

Our protocol involves a negligible error probability of $2^{-\Omega(\kappa)}$. To bound the error probability by $2^{-\Omega(\kappa)}$, our protocol operates over a finite field $\mathbb{F}$, where $|\mathbb{F}| = 2^{\kappa}$. In our protocol, the error probability comes from the fact that adversary has to guess a value (unknown to the adversary), selected uniformly and randomly by **S** from $\mathbb{F}$. If the adversary can correctly guess the value, then the protocol output will be incorrect. However, the probability of this event is $\frac{1}{|\mathbb{F}|} = 2^{-\kappa}$. Without loss of generality, we assume that $\frac{\ell}{|\mathbb{F}|} \approx 2^{-\Omega(\kappa)}$ and hence is negligible (this is assumed in all the previous almost-PSMT protocols). We now discuss LSSS.

2.1 Linear Secret Sharing Scheme: LSSS

In a *secret sharing scheme*, a dealer D distributes a secret $s \in \mathbb{F}$, to a set of n parties $\mathcal{P} = \{P_1, \ldots, P_n\}$ in such a way that some subsets of the participants (called access sets) can reconstruct s from their shares, while the other subsets of the participants (called forbidden sets) have no information about s from their shares. The family of access sets is called *access structure*. Moreover, we assume that the access structure is monotone, which is defined as follows:

Definition 2. *An access structure Σ is monotone if $A \in \Sigma$ and $A' \supseteq A$, then $A' \in \Sigma$.*

Corresponding to the access structure Σ, we have the adversary structure $\Gamma = \Sigma^c$, where c denotes the complement. The sets in Γ are called *forbidden* sets. There exists a *computationally unbounded* adversary $\mathcal{A}$, who can control any set in Γ.

A secret sharing scheme for any monotone access structure Σ can be realized by a linear secret sharing scheme (LSSS) [8] as follows: Let $\mathcal{M}$ be a $d \times e$ matrix over $\mathbb{F}$ and $\psi : \{1, \cdots, d\} \to \{1, \cdots, n\}$ be a labeling function, where $d \geq e$ and $d \geq n$.

Sharing algorithm

1. To share a secret $s \in \mathbb{F}$, D first chooses a random vector $\boldsymbol{\rho} \in \mathbb{F}^{e-1}$ and compute a vector
$$\boldsymbol{v} = (v_1, \cdots, v_d)^T = \mathcal{M} \cdot \begin{pmatrix} s \\ \boldsymbol{\rho} \end{pmatrix}. \tag{1}$$
2. Let
$$\mathsf{LSSS}(s, \boldsymbol{\rho}) = (\mathtt{share}_1, \cdots, \mathtt{share}_n), \tag{2}$$
where $\mathtt{share}_i = \{v_j \mid \psi(j) = i\}$. The dealer gives $\mathtt{share}_i$ to P_i as a share of s for $i = 1, \cdots, n$.

Reconstruction algorithm: A set of parties $A \in \Sigma$ can reconstruct the secret s if and only if $(1, 0, \cdots, 0)$ is in the linear span of
$$\mathcal{M}_A = \{m_j \mid \psi(j) \in A\},$$
where m_j denotes the jth row of $\mathcal{M}$. If this is indeed the case then there exists a vector $\boldsymbol{\alpha}_A$ called *recombination vector*, such that $\boldsymbol{\alpha}_A \cdot \mathcal{M}_A = (1, 0, \ldots, 0)$. Let $\boldsymbol{s}_A$ denote the set of shares corresponding to the parties in A. Then the parties in A can reconstruct s by computing $s = \langle \boldsymbol{\alpha}_A, \boldsymbol{s}_A^T \rangle$, where $\langle x, y \rangle$ denotes the *dot product* of x and y and x^T denotes the transpose of x.

Definition 3 (Monotone Span Programme (MSP) [8]). *We say that the above $(\mathcal{M}, \psi)$ is a monotone span program which realizes Σ. The size of the MSP is the number of rows d in M.*

Theorem 1 ([8]). *The above algorithm constitutes a valid secret sharing scheme.*

We are now ready to present our protocol.

3 Efficient Single Round Almost-PSMT Protocol Tolerating Non-Threshold Adversary

Let $\mathcal{W} = \{w_1, \ldots, w_n\}$ be the set of n channels between **S** and **R** and let $\mathcal{A}$ be a non-threshold adversary, specified by an adversary structure Γ over $\mathcal{W}$. Moreover, let $\Sigma = \Gamma^c$ be the corresponding access structure over $\mathcal{W}$. Furthermore, let

$\mathcal{A}$ satisfies $\mathcal{Q}^2$ condition with respect to $\mathcal{W}$, which is necessary for the existence of any almost-PSMT protocol tolerating $\mathcal{A}$. During the protocol, $\mathcal{A}$ can select any set of channels $B \in \Gamma$ for corruption. However, before the beginning of the protocol, neither **S** nor **R** will know which set of channels are under the control of $\mathcal{A}$. The channels which are under the control of $\mathcal{A}$ are called *corrupted*. On the other hand, the channels not under the control of $\mathcal{A}$ are called *honest*.

Let $(\mathcal{M}, \psi)$ be the MSP realizing the access structure Σ. *Without loss of generality and for simplicity, we assume that only i^{th} row of $\mathcal{M}$ is assigned to channel w_i, for $i = 1, \ldots, n$.* Thus,

$$\mathcal{M} = \begin{pmatrix} m_1 \\ \vdots \\ m_n \end{pmatrix}$$

is an $n \times e$ matrix over $\mathbb{F}$. *However, our protocol will also work when more than one row of $\mathcal{M}$ is assigned to some w_i.* Finally we use the following notation in our protocol:

Notation 1. *Let $\mathcal{Q}$ be any subset of $\mathcal{W}$ i.e. $\mathcal{Q} \subseteq \mathcal{W}$. Then $\mathcal{M}_{\mathcal{Q}}$ denotes the matrix containing the rows of $\mathcal{M}$ corresponding to the channels in $\mathcal{Q}$. For example, if $\mathcal{Q} = \{w_1, \ldots, w_t\}$, then*

$$\mathcal{M}_{\mathcal{Q}} = \begin{pmatrix} m_1 \\ \vdots \\ m_t \end{pmatrix}.$$

3.1 Underlying Idea of the Protocol

The high level idea of the protocol is as follows: let the message $m^{\mathbf{S}}$, which is a sequence of ℓ elements from $\mathbb{F}$ be denoted by $m^{\mathbf{S}} = [m_1^{\mathbf{S}}, \ldots, m_\ell^{\mathbf{S}}]$. Now using the MSP $\mathcal{M}$, **S** generates $\mathsf{LSSS}(m_i^{\mathbf{S}}, \boldsymbol{\rho}_i) = (\mathtt{share}_{i1}^{\mathbf{S}}, \ldots, \mathtt{share}_{in}^{\mathbf{S}})$, for $i = 1, \ldots, \ell$, where $\boldsymbol{\rho}_i$'s are the randomness used by **S**.

If **S** sends the jth share of all the ℓ $m_i^{\mathbf{S}}$'s, namely $\mathtt{share}_{ij}^{\mathbf{S}}$, over w_j, for $j = 1, \ldots, n$, then the communication preserves the secrecy of $m^{\mathbf{S}}$. This is because $\mathcal{A}$ can control any one set from the adversary structure Γ and hence will get the shares of each $m_i^{\mathbf{S}}$'s, sent over those channels. However, from the properties of MSP, these shares will not reveal any information about $m_i^{\mathbf{S}}$'s to $\mathcal{A}$.

However, **S** cannot ensure that $m^{\mathbf{S}}$ will be recovered correctly by **R** by simply sending the shares. This is because $\mathcal{A}$ may corrupt the shares sent over the channels under its control and there will be no way by which **R** can detect which channels have delivered correct shares. This is because there is only one round in the protocol. So **S** also need to send some *additional* information to authenticate each share, which can assist **R** to detect the corrupted shares with very high probability. So in our protocol, **S** also sends additional authentication information, using which **R** can detect the corrupted shares with very high probability (the way this is done is explained in the next section).

Though this mechanism of sending the shares, along with their authentication information is also used in the earlier almost-PSMT protocols, we use a new way of sending the authentication information, which is relatively simpler than the earlier schemes. After removing the corrupted shares, **R** will be left with the shares, which are correctly delivered with very high probability. Among these shares, there will be a set of shares which are delivered over the *honest* channels and hence they correspond to shares of the wires that constitute an access set. So if **R** applies the reconstruction algorithm of the LSSS to the retained shares, **R** will correctly recover each $m_i^{\mathbf{S}}$ with very high probability.

3.2 Sending the Authentication Information

In our protocol, the authentication of shares is done in the following way: corresponding to the jth share of all the ℓ $m_i^{\mathbf{S}}$'s, sender **S** constructs a polynomial $p_j^{\mathbf{S}}(x)$ of degree $\ell-1$ as follows: $p_j^{\mathbf{S}}(x) = \mathtt{share}_{1j}^{\mathbf{S}} + \mathtt{share}_{2j}^{\mathbf{S}} \cdot x + \ldots + \mathtt{share}_{\ell j}^{\mathbf{S}} \cdot x^{\ell-1}$. Now **S** associates $p_j^{\mathbf{S}}(x)$ with channel w_j, for $j = 1, \ldots, n$ and sends it over w_j (by sending the coefficients of $p_j^{\mathbf{S}}(x)$ over w_j). This is same as sending all the jth shares over w_j.

Now **S** associates a *random evaluation point* $\alpha_k^{\mathbf{S}}$ with every channel w_k, for $k = 1, \ldots, n$. If **S** sends $\alpha_k^{\mathbf{S}}$ and $p_j^{\mathbf{S}}(\alpha_k^{\mathbf{S}})$, for $j = 1, \ldots, n$ over w_k, then it achieves the following: if w_j is *corrupted* and if w_k is *honest*, then w_j cannot deliver $p_j^{\mathbf{R}}(x) \neq p_j^{\mathbf{S}}(x)$ to **R** over w_j without being caught by w_k with very high probability. This is because $\mathcal{A}$ will have no information about $\alpha_k^{\mathbf{S}}$ sent over w_k and also $\alpha_k^{\mathbf{R}}$ received by **R** over w_k is same as $\alpha_k^{\mathbf{S}}$. So except with probability $\frac{\ell-1}{|\mathbb{F}|}$, $p_j^{\mathbf{R}}(\alpha_k^{\mathbf{R}}) \neq p_j^{\mathbf{S}}(\alpha_k^{\mathbf{R}})$. This is because two different polynomials of degree $\ell - 1$ can have at most $\ell - 1$ common roots and $\alpha_k^{\mathbf{S}}$ is randomly selected from $\mathbb{F}$. By appropriately selecting $\mathbb{F}$, we can ensure that $\frac{\ell-1}{|\mathbb{F}|} \approx 2^{-\Omega(\kappa)}$, which is negligible. So this can help to detect corrupted shares.

However, the above communication may breach the secrecy as follows: if P_j is *honest* and P_k is *corrupted*, then earlier adversary would have no information about $p_j^{\mathbf{S}}(x)$, as no information about $p_j^{\mathbf{S}}(x)$ would have been sent over w_k. But now, adversary will know $p_j^{\mathbf{S}}(\alpha_k^{\mathbf{S}})$, as well as $\alpha_k^{\mathbf{S}}$ through w_k, thus revealing information about $p_j^{\mathbf{S}}(x)$ and hence about j^{th} share of all $m_i^{\mathbf{S}}$'s. To avoid this situation, we use the following idea: corresponding to channel w_j, **S** selects n random *masking keys*, denoted by $key_{j1}^{\mathbf{S}}, \ldots, key_{jn}^{\mathbf{S}}$. All the n masking keys (associated with w_j) are sent over w_j. Now the authentication of $p_j^{\mathbf{S}}(x)$ corresponding to the evaluation point $\alpha_k^{\mathbf{S}}$, namely $p_j^{\mathbf{S}}(\alpha_k^{\mathbf{S}})$, is masked with the kth masking key, namely $key_{jk}^{\mathbf{S}}$ and sent over w_k. That is, over w_k, **S** sends $p_j^{\mathbf{S}}(\alpha_k^{\mathbf{S}}) + key_{jk}^{\mathbf{S}}$, instead of only $p_j^{\mathbf{S}}(\alpha_k^{\mathbf{S}})$. Notice that $key_{jk}^{\mathbf{S}}$ is not sent over w_k. So if the adversary controls w_k, then even after knowing $\alpha_k^{\mathbf{S}}$ and $p_j^{\mathbf{S}}(\alpha_k^{\mathbf{S}}) + key_{jk}^{\mathbf{S}}$, adversary will not gain any information about $p_j^{\mathbf{S}}(x)$, as he has no information about the kth masking key $key_{jk}^{\mathbf{S}}$ associated with w_j. This way, we preserve the secrecy of each $p_j^{\mathbf{S}}(x)$, sent over honest p_j's. The interesting fact is that with this communication, we can also ensure that if some $p_j^{\mathbf{S}}(x)$ is changed by the adversary over some corrupted w_j, then it will be detected with very high probability by an *honest* w_k.

Remark 2. (**Comparison with the earlier mechanisms of authenticating shares**) As mentioned earlier, all the previous almost-PSMT protocols also used the idea of sending the authentication information of the shares, along with the shares. However, these protocols perform the authentication of each individual j^{th} share separately, corresponding to each of the ℓ messages, using the idea of Check vectors [29]. On the other hand, in our scheme, a single authentication information is sent for all the jth shares of the ℓ secrets. This way, we achieve more efficiency.

We are now ready to formally present our protocol, which is given in Fig. 1.

We now proceed to prove the properties of the protocol. In the proofs, we will use the following notations (For the definition of VALID, see Fig. 1):

- HW denotes the set of channels in $\mathcal{W}$ not under the control of $\mathcal{A}$.
- CW denotes the set of channels in $\mathcal{W}$ under the control of $\mathcal{A}$.
- HVALID denotes the set of channels in VALID not under the control of $\mathcal{A}$.
- CVALID denotes the set of channels in VALID under the control of A.

Remark 3. Notice that if some channel is under the control of $\mathcal{A}$ then it is not necessary that $\mathcal{A}$ changes all the information sent over the channel. The adversary may or may not change any portion of the information sent over the channels under his control.

Lemma 1. *$\mathsf{HVALID} = \mathsf{HW}$ and hence HVALID constitutes an access set.*

PROOF: First notice that every channel in the set HW will correctly deliver all the information to $\mathbf{R}$. Specifically, $p_k^{\mathbf{R}}(x) = p_k^{\mathbf{S}}(x)$, $\alpha_k^{\mathbf{R}} = \alpha_k^{\mathbf{S}}$, $(key_{k1}^{\mathbf{R}}, \ldots, key_{kn}^{\mathbf{R}}) = (key_{k1}^{\mathbf{S}}, \ldots, key_{kn}^{\mathbf{S}})$ and $val_{jk}^{\mathbf{R}} = val_{jk}^{\mathbf{S}}$, for $j = 1, \ldots, n$, for every channel $w_k \in \mathsf{HW}$. So the condition $val_{jk}^{\mathbf{R}} = p_j^{\mathbf{R}}(\alpha_k^{\mathbf{R}}) + key_{jk}^{\mathbf{R}}$ holds for every $w_j, w_k \in \mathsf{HW}$. Moreover, HW constitutes an access set. Thus, the condition $\mathcal{W} \setminus \mathsf{SUPPORT}_j \in \Gamma$ will hold for every channel $w_j \in \mathsf{HW}$. Thus, every channel in HW will be present in VALID and hence $\mathsf{HVALID} = \mathsf{HW}$. □

Lemma 2. *Every channel $w_j \in \mathsf{VALID}$ will deliver $p_j^{\mathbf{R}}(x) = p_j^{\mathbf{S}}(x)$, except with error probability $2^{-\Omega(\kappa)}$.*

PROOF: The proof holds without any error probability if $w_j \in \mathsf{HVALID}$. So we now consider the case when $w_j \in \mathsf{CVALID}$. So let w_j be a channel in CVALID. Since $w_j \in \mathsf{CVALID}$ (and hence VALID), it implies that $\mathcal{W} \setminus \mathsf{SUPPORT}_j \in \Gamma$. This further implies that there exists at least one channel in $\mathsf{SUPPORT}_j$, say w_k, such that w_k is not under the control of the adversary. Otherwise, it implies that $\mathsf{SUPPORT}_j \in \Gamma$ and hence $\mathcal{A}$ does not satisfy $\mathcal{Q}^2$ condition with respect to $\mathcal{W}$, which is a contradiction.

Now since w_k is not under the control of $\mathcal{A}$, it implies that $\alpha_k^{\mathbf{R}} = \alpha_k^{\mathbf{S}}$ and also $val_{jk}^{\mathbf{R}} = val_{jk}^{\mathbf{S}}$. Moreover, $\mathcal{A}$ will have no information about $\alpha_k^{\mathbf{R}}$ and $val_{jk}^{\mathbf{R}}$. Now suppose adversary changes $p_j^{\mathbf{S}}(x)$, so that $p_j^{\mathbf{R}}(x) \neq p_j^{\mathbf{S}}(x)$. However, since $w_k \in \mathsf{SUPPORT}_j$, it implies that $val_{jk}^{\mathbf{R}} = p_j^{\mathbf{R}}(\alpha_k^{\mathbf{R}}) + key_{jk}^{\mathbf{R}}$. But adversary can ensure the same only if he can correctly guess $\alpha_k^{\mathbf{R}} = \alpha_k^{\mathbf{S}}$. However, adversary can do the same with probability at most $\frac{\ell-1}{|\mathbb{F}|} \approx 2^{-\Omega(\kappa)}$. □

Computation by S:

1. For $i = 1, \ldots, \ell$, **S** computes $\mathsf{LSSS}(m_i^{\mathbf{S}}, \boldsymbol{\rho}_i) = (\mathtt{share}_{i1}^{\mathbf{S}}, \ldots, \mathtt{share}_{in}^{\mathbf{S}})$.
2. For $k = 1, \ldots, n$, corresponding to channel w_k, **S** selects a random value $\alpha_k^{\mathbf{S}}$, called as k^{th} *evaluation point*.
3. For $j = 1, \ldots, n$, corresponding to the jth share of all the ℓ $m_i^{\mathbf{S}}$'s, **S** constructs a polynomial $p_j^{\mathbf{S}}(x)$ of degree $\ell - 1$ as follows:

$$p_j^{\mathbf{S}}(x) = \mathtt{share}_{1j}^{\mathbf{S}} + \mathtt{share}_{2j}^{\mathbf{S}} \cdot x + \ldots + \mathtt{share}_{\ell j}^{\mathbf{S}} \cdot x^{\ell-1}.$$

4. For $j = 1, \ldots, n$, **S** evaluates each $p_j^{\mathbf{S}}(x)$ at evaluation point $\alpha_k^{\mathbf{S}}$, for $k = 1, \ldots, n$.
5. For $j = 1, \ldots, n$, corresponding to channel w_j, **S** selects n random, non-zero values $key_{j1}^{\mathbf{S}}, \ldots, key_{jn}^{\mathbf{S}}$, called as *masking keys*.

Round I: Communication from S to R: For $k = 1, \ldots, n$, **S** sends the following to **R** over channel w_k and terminates the protocol.

1. Polynomial $p_k^{\mathbf{S}}(x)$.
2. Evaluation point $\alpha_k^{\mathbf{S}}$.
3. n masking keys $key_{k1}^{\mathbf{S}}, \ldots, key_{kn}^{\mathbf{S}}$.
4. Masked authentication values $val_{jk}^{\mathbf{S}}$, for $j = 1, \ldots, n$, where $val_{jk}^{\mathbf{S}} = p_j^{\mathbf{S}}(\alpha_k^{\mathbf{S}}) + key_{jk}^{\mathbf{S}}$.

Information Received by R: For $k = 1, \ldots, n$, let **R** receive the following from **S** over channel w_k:

1. Polynomial $p_k^{\mathbf{R}}(x)$.
2. Evaluation point $\alpha_k^{\mathbf{R}}$.
3. n masking keys $key_{k1}^{\mathbf{R}}, \ldots, key_{kn}^{\mathbf{R}}$.
4. Masked authentication values $val_{jk}^{\mathbf{R}}$, for $j = 1, \ldots, n$.

Message Recovery by R: **R** does the following computation:

1. **R** initializes a set $\mathsf{VALID} = \emptyset$.
2. For $j = 1, \ldots, n$, corresponding to channel w_j, **R** constructs a set $\mathsf{SUPPORT}_j = \emptyset$.
3. **R** adds channel w_k in $\mathsf{SUPPORT}_j$ if $val_{jk}^{\mathbf{R}} = p_j^{\mathbf{R}}(\alpha_k^{\mathbf{R}}) + key_{jk}^{\mathbf{R}}$.
4. For $j = 1, \ldots, n$, **R** adds channel w_j to VALID if $\mathcal{W} \setminus \mathsf{SUPPORT}_j \in \Gamma$ [a].
5. Without loss of generality, let $w_1, \ldots, w_t$ be the channels in VALID. Moreover, for $j = 1, \ldots, t$, let $p_j^{\mathbf{R}}(x)$ be of the form

$$p_j^{\mathbf{R}}(x) = \mathtt{share}_{1j}^{\mathbf{R}} + \mathtt{share}_{2j}^{\mathbf{R}} \cdot x + \ldots + \mathtt{share}_{\ell j}^{\mathbf{R}} \cdot x^{\ell-1}.$$

6. For $i = 1, \ldots, \ell$, **R** applies reconstruction algorithm of the LSSS to $\mathtt{share}_{i1}^{\mathbf{R}}, \mathtt{share}_{i2}^{\mathbf{R}}, \ldots, \mathtt{share}_{it}^{\mathbf{R}}$ and reconstructs $m_i^{\mathbf{R}}$.
7. Finally **R** reconstructs $m^{\mathbf{R}} = [m_1^{\mathbf{R}}, \ldots, m_\ell^{\mathbf{R}}]$ and terminates the protocol.

[a] This is can be done efficiently by checking whether the target vector $(1, 0, \ldots, 0)$ lies in the span of the rows assigned to the parties in the set $\mathcal{W} \setminus \mathsf{SUPPORT}_j$ in $\mathcal{M}$.

Fig. 1. Efficient Single Round Almost-PSMT Tolerating $\mathcal{Q}^2$ Adversary Structure

Lemma 3 (Perfect Secrecy). *The protocol in Fig. 1 satisfies perfect secrecy condition.*

Proof: If $w_k \in \mathsf{CW}$, then adversary will know the polynomial $p_k^{\mathbf{S}}(x)$ and hence the shares $\mathtt{share}_{1k}^{\mathbf{S}}, \ldots, \mathtt{share}_{\ell k}^{\mathbf{S}}$. However, even after knowing all the polynomials transmitted through the channels in CW, adversary will not know $m_1^{\mathbf{S}}, \ldots, m_\ell^{\mathbf{S}}$, as adversary will only come to know the shares of $m_1^{\mathbf{S}}, \ldots, m_\ell^{\mathbf{S}}$ sent through the channels in CW and $\mathsf{CW} \in \Gamma$. However, the adversary will also know $val_{jk}^{\mathbf{S}} = p_j^{\mathbf{S}}(\alpha_k^{\mathbf{S}}) + key_{jk}^{\mathbf{S}}$, corresponding to every $w_j \in \mathsf{HW}$, which is transmitted through every $w_k \in \mathsf{CW}$. However, such $val_{jk}^{\mathbf{S}}$'s will not reveal any extra information about $p_j^{\mathbf{S}}(x)$ (corresponding to any P_j in HW) to the adversary, as the adversary will have no information about the masking key $key_{jk}^{\mathbf{S}}$, which is only sent over w_j. Thus, $val_{jk}^{\mathbf{S}}$'s corresponding to every $w_j \in \mathsf{HW}$, which are transmitted through every $p_k \in \mathsf{CW}$ will not reveal any information about $p_j^{\mathbf{S}}(x)$'s corresponding to w_j's in HW. Thus, through the information received over the channels in CW, adversary will not get any information about $m_i^{\mathbf{S}}$'s and hence the message $m^{\mathbf{S}}$.

Lemma 4 (Almost Perfect Reliability). *The protocol in Fig. 1 satisfies almost perfect reliability condition.*

Proof: To prove the lemma, we have to show that the shares (of $m_i^{\mathbf{S}}$'s) received by $\mathbf{R}$ over the channels in VALID are correct shares, except with error probability $2^{-\Omega(\kappa)}$. This further implies that every channel $w_j \in \mathsf{VALID}$ has delivered $p_j^{\mathbf{R}}(x) = p_j^{\mathbf{S}}(x)$, except with error probability $2^{-\Omega(\kappa)}$. However, this follows from Lemma 2. □

Lemma 5 (Computation and Communication Complexity). *In the protocol of Fig. 1,* $\mathbf{S}$ *and* $\mathbf{R}$ *performs computation which is polynomial in the size of the underlying* LSSS. *In the protocol,* $\mathbf{S}$ *sends* $\mathcal{O}(\ell n + n^2)$ *field elements from* $\mathbb{F}$ *to* $\mathbf{R}$.

Proof: The computational complexity is easy to verify. We now analyze the communication complexity. Through each channel, $\mathbf{S}$ sends a polynomial of degree $\ell - 1$, one evaluation point, n masking keys and n authenticated values. This results in a total communication complexity of $\mathcal{O}(\ell n + n^2)$ field elements. □

Theorem 2. *Let* $\mathbf{S}$ *and* $\mathbf{R}$ *be connected by* n *channels and let there exists a computationally unbounded adversary* $\mathcal{A}$, *specified by an adversary structure* Γ *over the* n *channels, such that* $\mathcal{A}$ *satisfies* $\mathcal{Q}^2$ *condition. Then there exists an efficient single round almost-PSMT protocol tolerating* $\mathcal{A}$.

Proof: The proof follows from Lemma 3, Lemma 4 and Lemma 5. □

4 Simple and Computationally Efficient Single Round Almost-PSMT Tolerating Threshold Adversary with Optimum Communication Complexity

As discussed earlier, a threshold adversary $\mathcal{A}_t$ is a special type of non-threshold adversary where the adversary structure Γ consists of all possible subsets of $\mathcal{W}$ of size at most t. We now recall the following results from [25].

Theorem 3 ([25]). *Any almost-PSMT (irrespective of the number of rounds) tolerating $\mathcal{A}_t$ is possible iff* **S** *and* **R** *are connected by $n \geq 2t+1$ channels. Moreover, any single round almost-PSMT protocol tolerating $\mathcal{A}_t$ has to communicate $\Omega\left(\frac{n\ell}{n-2t}\right)$ field elements to send a message containing ℓ field elements.*

Remark 4. In any almost-PSMT protocol, $|\mathbb{F}|$ is selected as a function of the error parameter κ (normally $|\mathbb{F}| = 2^{\kappa}$) and thus each field element can be represented by a number of bits, which will be function of κ. So though κ does not figure explicitly in the expression for communication complexity in Theorem 3, it is implied implicitly if we look into the total number of bits that are actually communicated.

Any single round almost-PSMT protocol designed with $n = 2t + 1$ channels is said to have *optimal resilience.* Substituting $n = 2t+1$ in the above theorem, we find that any single round almost-PSMT protocol with optimal resilience has to communicate $\Omega(n\ell)$ field elements to send a message containing ℓ field elements. Thus any single round, optimally resilient, almost-PSMT protocol whose total communication complexity is $\mathcal{O}(n\ell)$ is said to be *communication optimal.*

In [35,25], the authors presented an efficient[3] single round, optimally resilient almost-PSMT protocol tolerating $\mathcal{A}_t$. However, the protocol performs some complex (though efficient) computations, like *extrapolation technique, extracting randomness*, etc[4] to achieve its task. In practical networks like sensor network, it is desirable to design protocols which perform computationally simple steps. Motivated by this, the authors in [9] have designed a very simple, optimally resilient, single round almost-PSMT tolerating $\mathcal{A}_t$. However, their protocol is not communication optimal. Specifically, their protocol sends $\mathcal{O}(n^2)$ field elements to send a message containing one field element.

We now show that our single round almost-PSMT protocol against non-threshold adversary when restricted to threshold adversary is a single round, optimally resilient, almost-PSMT protocol tolerating $\mathcal{A}_t$ having *optimal communication complexity.* Moreover, the protocol is efficient. Furthermore, the protocol is very simple and performs much simplers steps (by avoiding steps like extrapolation technique, extracting randomness) than the communication optimal single round almost-PSMT protocol of [35].

[3] The computation and communication complexity of the protocol are polynomial in n and ℓ.

[4] See [7] for the detailed presentation of the single round almost-PSMT protocol of [35].

The first observation is that if the adversary is specified by a threshold t and if the underlying adversary structure satisfies $\mathcal{Q}^2$ condition, then it implies that **S** and **R** are connected by $n \geq 2t+1$ channels. Moreover, it is well known that there exists a very simple MSP tolerating a threshold adversary with threshold t, such that there are exactly n rows in the MSP and one row of the MSP is assigned to each channel. The MSP is nothing but an $n \times (t+1)$ Vandermonde matrix [8]. The resultant secret sharing scheme is known as Shamir secret sharing scheme [32]. So now with these observations, if we simply execute the protocol of previous section assuming that the adversary is a threshold adversary and there are $n = 2t+1$ channels between **S** and **R**, we get a simple, efficient, optimally resilient, single round almost-PSMT protocol tolerating $\mathcal{A}_t$, which communicates $\mathcal{O}(\ell n + n^2)$ field elements to send a message containing ℓ field elements. Now if we set $\ell = n$, then we find that the protocol sends a message containing n field elements by communicating $\mathcal{O}(n^2)$ field elements. From Theorem 3, any single round optimally resilient almost-PSMT protocol has to communicate $\Omega(n^2)$ field elements to securely send a message containing n field elements. Thus our resultant protocol is communication optimal. We now state this in the following theorem:

Theorem 4. *Let* **S** *and* **R** *be connected by* $n = 2t+1$ *channels. Moreover, let* **S** *has a message containing* $\ell = n$ *field elements. Then there exists a simple, efficient, optimally resilient, communication optimal single round almost-PSMT protocol tolerating* $\mathcal{A}_t$.

5 Conclusion

In this paper, we resolved one of the open problems raised in [25] by designing an optimally resilient, single round, efficient almost-PSMT protocol tolerating non-threshold adversary. This is the first ever efficient single round almost-PSMT protocol tolerating non-threshold adversary. When restricted to threshold adversary, we get a simple, efficient, optimally resilient, single round communication optimal almost-PSMT protocol.

Acknowledgments. We would like to thank the anonymous referees of ACNS 2011 for several useful suggestions.

References

1. Agarwal, S., Cramer, R., de Haan, R.: Asymptotically optimal two-round perfectly secure message transmission. In: Dwork, C. (ed.) CRYPTO 2006. LNCS, vol. 4117, pp. 394–408. Springer, Heidelberg (2006)
2. Araki, T.: Almost secure 1-round message transmission scheme with polynomial-time message decryption. In: Safavi-Naini, R. (ed.) ICITS 2008. LNCS, vol. 5155, pp. 2–13. Springer, Heidelberg (2008)
3. Ben-Or, M., Goldwasser, S., Wigderson, A.: Completeness Theorems for Non-Cryptographic Fault-Tolerant Distributed Computation. In: Proceedings of the Twentieth Annual ACM Symposium on Theory of Computing, Chicago, Illinois, USA, May 2-4, pp. 1–10. ACM, New York (1988)

4. Chaum, D., Crépeau, C., Damgård, I.: Multiparty Unconditionally Secure Protocols (extended abstract). In: Proceedings of the Twentieth Annual ACM Symposium on Theory of Computing, Chicago, Illinois, USA, May 2-4, pp. 11–19. ACM, New York (1988)
5. Choudhary, A., Patra, A., Ashwinkumar, B.V., Srinathan, K., Rangan, C.P.: Perfectly Reliable and Secure Communication Tolerating Static and Mobile Mixed Adversary. In: Safavi-Naini, R. (ed.) ICITS 2008. LNCS, vol. 5155, pp. 137–155. Springer, Heidelberg (2008)
6. Choudhary, A., Patra, A., Ashwinkumar, B.V., Srinathan, K., Rangan, C.P.: On Minimal Connectivity Requirement for Secure Message Transmission in Asynchronous Networks. In: Garg, V., Wattenhofer, R., Kothapalli, K. (eds.) ICDCN 2009. LNCS, vol. 5408, pp. 148–162. Springer, Heidelberg (2008)
7. Choudhury, A.: Protocols for reliable and secure message transmission. Cryptology ePrint Archive, Report 2010/281 (2010)
8. Cramer, R., Damgård, I., Maurer, U.M.: General secure multi-party computation from any linear secret-sharing scheme. In: Preneel, B. (ed.) EUROCRYPT 2000. LNCS, vol. 1807, pp. 316–334. Springer, Heidelberg (2000)
9. Desmedt, Y., Erotokritou, S., Safavi-Naini, R.: Simple and communication complexity efficient almost secure and perfectly secure message transmission schemes. In: Bernstein, D.J., Lange, T. (eds.) AFRICACRYPT 2010. LNCS, vol. 6055, pp. 166–183. Springer, Heidelberg (2010)
10. Desmedt, Y., Wang, Y.: Perfectly secure message transmission revisited. In: Biham, E. (ed.) EUROCRYPT 2003. LNCS, vol. 2656, pp. 502–517. Springer, Heidelberg (2003)
11. Desmedt, Y., Wang, Y., Burmester, M.: A complete characterization of tolerable adversary structures for secure point-to-point transmissions without feedback. In: Deng, X., Du, D.-Z. (eds.) ISAAC 2005. LNCS, vol. 3827, pp. 277–287. Springer, Heidelberg (2005)
12. Dolev, D., Dwork, C., Waarts, O., Yung, M.: Perfectly secure message transmission. JACM 40(1), 17–47 (1993)
13. Fitzi, M., Franklin, M.K., Garay, J.A., Vardhan, S.H.: Towards optimal and efficient perfectly secure message transmission. In: Vadhan, S.P. (ed.) TCC 2007. LNCS, vol. 4392, pp. 311–322. Springer, Heidelberg (2007)
14. Franklin, M., Wright, R.: Secure communication in minimal connectivity models. Journal of Cryptology 13(1), 9–30 (2000)
15. Hirt, M., Maurer, U.M.: Complete Characterization of Adversaries Tolerable in Secure Multi-Party Computation. In: Proceedings of the Sixteenth Annual ACM Symposium on Principles of Distributed Computing, Santa Barbara, California, USA, August 21-24, pp. 25–34. ACM Press, New York (1997)
16. Kumar, M.V.N.A., Goundan, P.R., Srinathan, K., Pandu Rangan, C.: On perfectly secure communication over arbitrary networks. In: Proceedings of the Twenty-First Annual ACM Symposium on Principles of Distributed Computing, PODC 2002, Monterey, California, USA, July 21-24, pp. 193–202. ACM, New York (2002)
17. Kurosawa, K.: General error decodable secret sharing scheme and its application. Cryptology ePrint Archive, Report 2009/263 (2009)
18. Kurosawa, K.: Round-efficient perfectly secure message transmission scheme against general adversary. Cryptology ePrint Archive, Report 2010/450 (2010)
19. Kurosawa, K., Suzuki, K.: Truly efficient 2-round perfectly secure message transmission scheme. In: Smart, N.P. (ed.) EUROCRYPT 2008. LNCS, vol. 4965, pp. 324–340. Springer, Heidelberg (2008)

20. Kurosawa, K., Suzuki, K.: Almost secure (1-round, n-channel) message transmission scheme. IEICE Transactions 92-A(1), 105–112 (2009)
21. Patra, A., Choudhary, A., Pandu Rangan, C.: Constant phase efficient protocols for secure message transmission in directed networks. In: Gupta, I., Wattenhofer, R. (eds.) Proceedings of the Twenty-Sixth Annual ACM Symposium on Principles of Distributed Computing, PODC 2007, Portland, Oregon, USA, 2007, August 12-15, pp. 322–323. ACM, New York (2007)
22. Patra, A., Choudhary, A., Rangan, C.P.: Unconditionally reliable and secure message transmission in directed networks revisited. In: Ostrovsky, R., De Prisco, R., Visconti, I. (eds.) SCN 2008. LNCS, vol. 5229, pp. 309–326. Springer, Heidelberg (2008)
23. Patra, A., Choudhary, A., Rangan, C.P.: On communication complexity of secure message transmission in directed networks. In: Kant, K., Pemmaraju, S.V., Sivalingam, K.M., Wu, J. (eds.) ICDCN 2010. LNCS, vol. 5935, pp. 42–53. Springer, Heidelberg (2010)
24. Patra, A., Choudhary, A., Srinathan, K., Pandu Rangan, C.: Constant phase bit optimal protocols for perfectly reliable and secure message transmission. In: Barua, R., Lange, T. (eds.) INDOCRYPT 2006. LNCS, vol. 4329, pp. 221–235. Springer, Heidelberg (2006)
25. Patra, A., Choudhary, A., Srinathan, K., Pandu Rangan, C.: Unconditionally reliable and secure message transmission in undirected synchronous networks: Possibility, feasibility and optimality. International Journal of Applied Cryptography 2(2), 159–197 (2010); A preliminary version appeared in [37] (2009)
26. Patra, A., Choudhary, A., Vaidyanathan, M., Rangan, C.P.: Efficient perfectly reliable and secure message transmission tolerating mobile adversary. In: Mu, Y., Susilo, W., Seberry, J. (eds.) ACISP 2008. LNCS, vol. 5107, pp. 170–186. Springer, Heidelberg (2008)
27. Patra, A., Choudhury, A., Pandu Rangan, C.: Brief announcement: perfectly secure message transmissiontolerating mobile mixed adversary with reduced phase complexity. In: PODC, pp. 245–246 (2010)
28. Patra, A., Shankar, B., Choudhary, A., Srinathan, K., Rangan, C.P.: Perfectly secure message transmission in directed networks tolerating threshold and non threshold adversary. In: Bao, F., Ling, S., Okamoto, T., Wang, H., Xing, C. (eds.) CANS 2007. LNCS, vol. 4856, pp. 80–101. Springer, Heidelberg (2007)
29. Rabin, T., Ben-Or, M.: Verifiable secret sharing and multiparty protocols with honest majority. In: Proceedings of the 21st Annual ACM Symposium on Theory of Computing, Seattle, Washigton, USA, May 14-17, pp. 73–85. ACM, New York (1989)
30. Sayeed, H., Abu-Amara, H.: Perfectly secure message transmission in asynchronous networks. In: Proceedings of 7th IEEE Symposium on Parallel and Distributed Processing, pp. 100–105. IEEE, Los Alamitos (1995)
31. Sayeed, H., Abu-Amara, H.: Efficient perfectly secure message transmission in synchronous networks. Information and Computation 126(1), 53–61 (1996)
32. Shamir, A.: How to share a secret. Communications of the ACM 22(11), 612–613 (1979)
33. Shor, P.W.: Polynomial time algorithms for Prime factorization and Discrete Logarithms on a Quantum computer. SIAM Journal on Computing 26(5), 1484–1509 (1997)
34. Srinathan, K.: Secure distributed communication. PhD Thesis, IIT Madras (2006)

35. Srinathan, K., Choudhary, A., Patra, A., Pandu Rangan, C.: Efficient Single Phase Unconditionally Secure Message Transmission with Optimum Communication Complexity. In: Bazzi, R.A., Patt-Shamir, B. (eds.) Proceedings of the Twenty-Seventh Annual ACM Symposium on Principles of Distributed Computing, PODC 2008, Toronto, Canada, August 18-21, p. 457. ACM, New York (2008)
36. Srinathan, K., Narayanan, A., Pandu Rangan, C.: Optimal perfectly secure message transmission. In: Franklin, M. (ed.) CRYPTO 2004. LNCS, vol. 3152, pp. 545–561. Springer, Heidelberg (2004)
37. Srinathan, K., Patra, A., Choudhary, A., Rangan, C.P.: Probabilistic perfectly reliable and secure message transmission – possibility, feasibility and optimality. In: Srinathan, K., Rangan, C.P., Yung, M. (eds.) INDOCRYPT 2007. LNCS, vol. 4859, pp. 101–122. Springer, Heidelberg (2007)
38. Srinathan, K., Prasad, N.R., Pandu Rangan, C.: On the optimal communication complexity of multiphase protocols for perfect communication. In: 2007 IEEE Symposium on Security and Privacy (S&P 2007), Oakland, California, USA, May 20-23, pp. 311–320. IEEE Computer Society, Los Alamitos (2007)
39. Srinathan, K., Raghavendra, P., Rangan, C.P.: On proactive perfectly secure message transmission. In: Pieprzyk, J., Ghodosi, H., Dawson, E. (eds.) ACISP 2007. LNCS, vol. 4586, pp. 461–473. Springer, Heidelberg (2007)
40. Yang, Q., Desmedt, Y.: Cryptanalysis of secure message transmission protocols with feedback. In: Kurosawa, K. (ed.) Information Theoretic Security. LNCS, vol. 5973, pp. 159–176. Springer, Heidelberg (2010)
41. Yang, Q., Desmedt, Y.: General perfectly secure message transmission using linear codes. In: Abe, M. (ed.) ASIACRYPT 2010. LNCS, vol. 6477, pp. 448–465. Springer, Heidelberg (2010)

Relaxed Security Notions for Signatures of Knowledge

Marc Fischlin and Cristina Onete

Darmstadt University of Technology & CASED, Germany
www.minicrypt.de

Abstract. We revisit the definition of signatures of knowledge by Chase and Lysanskaya (Crypto 2006) which correspond to regular signatures but where the signer also proves knowledge of the secret key to the public key through any signature. From a more abstract point of view, the signer holds a secret witness w to a public NP statement x and any signature to a message allows to extract w given some auxiliary trapdoor information. Besides extractability, Chase and Lysanskaya also demand a strong witness-hiding property, called simulatability, akin to the zero-knowledge property of non-interactive proofs. They also show that this property ensures anonymity for delegatable credentials or for ring signatures, for example.

In this work here we discuss relaxed notions for simulatability and when they are sufficient for applications. Namely, in one notion we forgo any explicit witness-hiding notion, beyond some weak requirement that signatures should not help to produce further signatures, analogously to unforgeability of regular signature schemes. This notion suffices for example for devising regular signature schemes with some additional proof-of-possession (POP) or knowledge-of-secret-key (KOSK) property. Our stronger notion resembles the witness-indistinguishability notion of proofs of knowledge and can be used to build anonymous ring signatures. Besides formal definitions we relate all notions and discuss constructions and the aforementioned applications.

Keywords: Signature of Knowledge, Anonymity, Credential, Ring Signature.

1 Introduction

Signatures of knowledge (SoK), a term coined in [8], are widely used in cryptography (e.g., [7,6,18,22]). The intuition behind SoKs is clear: besides basic signature security, signatures of knowledge should also prove that the signer "knows" the secret key. SoKs were, however, formalized only recently by Chase and Lysyanskaya [12].

In [12], SoKs are abstractly considered, as primitives allowing users $\mathcal{S}$ to sign messages such that if the signature verifies, a verifier knows that $\mathcal{S}$ has a witness w to some NP statement x; no further information is leaked about w though. Chase and Lysyanskaya give two equivalent formalizations: a simulation-based

J. Lopez and G. Tsudik (Eds.): ACNS 2011, LNCS 6715, pp. 309–326, 2011.

definition in Canetti's Universal Composition (UC) framework [10], following approaches for regular signature schemes [11], and a game-based definition —called SimExt security— containing an extractability experiment akin to knowledge extractors [2] and a simulatability notion similar to non-interactive zero-knowledge (NIZK) proofs [4].

Since SimExt security closely resembles the security of NIZK proofs of knowledge (NIZKPoK), it is unsurprising that the construction in Chase and Lysyanskaya [12] is based on such proofs. The generality of this approach on the one hand yields quite expensive solutions, deploying general NIZKPoKs, but on the other hand also supports many applications. Two applications shown in [12] are ring signatures, where signers prove knowledge of a secret key corresponding to one of the public keys of the ring but without revealing its identity, and delegatable anonymous credentials, where zero-knowledge guarantees anonymity.

1.1 Relaxing the Notion of Signature of Knowledge

Reconsider SoK-based ring signatures. In this case simulatability as defined in [12] yields very strong anonymity: the SoK is simulatable without any witness. This is stronger than the security requirements of ring signatures, where only the actual signer should be hard to identify. We may thus consider a switch to the weaker notion of witness indistinguishability (WI) for SoKs, ensuring that one cannot deduce which (valid) witness w was used to sign. This relaxation thus allows for potentially more efficient solutions.

Consider furthermore simple digital signatures with key registration, where (some) information about the secret key is shown. Such registration steps are both common in practice [1,19], where one simply signs the public key to be registered, and often required in theory to prove security of protocols based on such signature schemes [5,17,20]. The corresponding model is called the "knowledge of secret keys" (KOSK) model and it implements some kind of proof of knowledge. Extractability of SoKs combines theory with practice, because self-signed public keys then mirror the KOSK model (though one can now extract with each signature and does not need an extra registration step). However, ordinary digital signatures usually do not require simulatability, but only unforgeability.

1.2 Our Contributions

We introduce two relaxed security notions for SoKs, following the NIZKPoK approach of [12] and thus inheriting extractability; however simulatability no longer holds. Instead, we transfer the definition of unforgeability from regular signatures and introduce UnfExt (*Unf*orgeability & *Ext*ractability) security as a minimal security level for SoKs. We augment this notion by adding witness indistinguishability, deriving the stronger WIUnfExt security. The SimExt definition of [12] is yet one step stronger, replacing witness indistinguishability by simulatability (as [12] shows, SimExt security implies unforgeability). We relate all three notions formally, showing a strict hierarchy, and also provide equivalent definitions in the UC framework. [1]

We then instantiate our notions. Using a result about the security of Waters' signature scheme [23] in the KOSK model [20], we easily get an UnfExt SoK (for a special NP relation). In fact, this scheme is trivially witness-indistinguishable, too, as witnesses are unique. We next present a general construction of WIUnfExt-secure SoKs for arbitrary NP statements based on general assumptions. This construction relies on witness-indistinguishable proofs of knowledge (a.k.a. ZAPs [13]), which are a relaxation of non-interactive zero knowledge proofs; however, our construction does in fact achieve SimExt security in the definition of Chase and Lysyanskaya [12]. Our third construction achieves UnfExt security by signing message m on behalf of an extension of the original statement-witness pair.

We finally address the aforementioned applications, especially ring signatures. We discuss that adding witness indistinguishability reflects strong anonymity of ring signatures. We use anonymity and unforgeability notions from the framework for ring signatures of Bender et al. [3].

2 Signatures of Knowledge

Signatures of knowledge (SoKs) are protocols between a *signer* $\mathcal{S}$, which signs messages $m \in \mathcal{M}$, and a *verifier* $\mathcal{V}$ checking signature validity.

We identify NP languages L with arbitrary, but fixed relations $\mathcal{R}^L$, i.e., $x \in L$ iff there exists a polynomial-size witness w such that $(x, w) \in \mathcal{R}^L$. Jumping ahead, we also require that it is hard, given some x, to compute a valid witness w (we formalize this w.r.t. an instance generator, as shown in section 3). We assume efficient (i.e., polynomial in the length $|x|$ of x) verification of $(x, w) \in \mathcal{R}^L$; denote by $\mathcal{W}^L(x)$ the possibly empty set $\{w : (x, w) \in \mathcal{R}^L\}$ of witnesses to x. Note that $\mathcal{W}^L(x)$ formally depends on $\mathcal{R}^L$, not on L. Sets $S = \mathcal{M}, L, \mathcal{R}^L, \mathcal{W}^L, \ldots$ are usually indexed by the security parameter $k \in \mathbb{N}$ and S_k denotes the strings $s \in S$ of polynomial complexity in k (for some fixed polynomial).

Definition 1 (Signature of Knowledge). *A* Signature of Knowledge (SoK) *for relation* $\mathcal{R}^L$ *is a tuple of efficient algorithms* $\mathsf{SoK} = (\mathsf{Setup}, \mathsf{Sign}, \mathsf{Vf})$ *where:*

$\mathsf{par} \leftarrow \mathsf{Setup}(1^k)$. *For a security parameter k,* Setup *outputs public parameters* par. *We assume that k is efficiently recoverable from* par.

$\sigma \leftarrow \mathsf{Sign}(m, x, w, \mathsf{par})$. *For a message $m \in \mathcal{M}_k$, statement $x \in L_k$, witness $w \in \mathcal{W}_k^L(x)$, and parameters* par *(generated for k),* Sign *outputs SoK σ.*

$b \leftarrow \mathsf{Vf}(\sigma, m, x, \mathsf{par})$. *On input an SoK σ, a message m, a statement x, and parameters* par*, the algorithm* Vf *outputs bit b indicating the validity of the SoK ($b = 1$ for valid σ).*

[1] Note that a work by Zou and Sun [15], advertising to discuss stronger anonymity for signatures of knowledge, rather shows subliminal channels in some group signature schemes through malicious signers, and is therefore not discussed further here.

We require the usual correctness property: for any $x \in L_k$, *any* $w \in \mathcal{W}_k^L(x)$, *and any* $m \in \mathcal{M}_k$, *it holds that,*

$$\text{Prob}\big[\,\mathsf{par} \leftarrow \mathsf{Setup}(1^k); \sigma \leftarrow \mathsf{Sign}(m, x, w, \mathsf{par}) : \mathsf{Vf}(\sigma, m, x, \mathsf{par}) = 1\big] \approx 1,$$

i.e., is negligibly close to 1 *(as a function of k).*

3 Security Notions for SoKs

We first briefly describe SimExt security as in [12] and then introduce *UnfExt* security as a relaxation thereof. We also explain the relation between the notions and introduce WIUnfExt security as another flavor of SoK security. We then show equivalent definitions in Canetti's universal composition framework.

3.1 Simulatability, Unforgeability, and Witness Indistinguishability

In [12], SimExt security considers auxiliary inputs given to the adversary. We omit such inputs for simplicity and instead use efficient algorithms, covering both uniform and non-uniform (with auxiliary input) computational models, as needed. Furthermore SoKs are universal in [12], using (the machine verifying) the relation $\mathcal{R}^L$ as input. Here we define SoK for specific fixed $\mathcal{R}^L$, which is handier for instantiations, e.g., for specific SoK for discrete-log based relations.

Definition 2 (SimExt Security for SoK [12]). *The SoK scheme* $\mathsf{SoK} = (\mathsf{Setup}, \mathsf{Sign}, \mathsf{Vf})$ *is* SimExt secure *for* $\mathcal{R}^L$ *iff it is:*

Simulatable. *There exists an efficient simulator* $\mathsf{Sim} = (\mathsf{SimSetup}, \mathsf{SimSign})$ *such that for all efficient adversaries* $\mathcal{A}$ *it holds that*

$$\left|\begin{array}{l}\text{Prob}\big[\,(\mathsf{par}, \tau) \leftarrow \mathsf{SimSetup}(1^k) : d \leftarrow \mathcal{A}^{\mathsf{Sim}(\mathsf{par},\tau,\cdot,\cdot,\cdot)}(\mathsf{par}) : d = 1\big] \\ \quad - \text{Prob}\big[\,\mathsf{par} \leftarrow \mathsf{Setup}(1^k); d \leftarrow \mathcal{A}^{\mathsf{Sign}(\mathsf{par},\cdot,\cdot,\cdot)}(\mathsf{par}) : d = 1\big]\end{array}\right| \approx 0,$$

where, on input $(\mathsf{par}, \tau, m, x, w)$, Sim *checks that* $\mathcal{R}^L(x, w) = 1$; *if so, it returns* $\mathsf{SimSign}(\mathsf{par}, \tau, m, x)$, *otherwise it ouputs* $\perp$.

Extractable. *There additionally exists an efficient extractor* Ext *such that for all efficient* $\mathcal{A}$,

$$\begin{aligned}\text{Prob}\ \big[&(\mathsf{par}, \tau) \leftarrow \mathsf{SimSetup}(1^k); (x, m, \sigma) \leftarrow \mathcal{A}^{\mathsf{Sim}(\mathsf{par},\tau,\cdot,\cdot,\cdot)}(\mathsf{par});\\ &w \leftarrow \mathsf{Ext}(\mathsf{par}, \tau, x, m, \sigma) :\\ &(x, w) \in \mathcal{R}_k^L \vee (m, x) \in Q \vee \mathsf{Vf}(\sigma, m, x, \mathsf{par}) = 0\big] \approx 1\end{aligned}$$

Here, Q *is the list of* (m, x) *queries that* $\mathcal{A}$ *has made to* Sim.

Note that Simulatability guarantees that using Sign or SimSign is essentially equivalent for extractability. Simulatability is a strong requirement for SoKs, resembling zero-knowledge simulation for non-interactive proofs. Some witness-"protection" is necessary, however: we cannot restrict SoK security to just correctness and extractability, as this allows for insecure SoKs. Indeed, consider an SoK scheme outputting $w||m$ for each m, and where Vf checks the validity of w. This SoK is correct and trivially extractable. However, any adversary can create a SoK on fresh m^*, either by extracting a valid w from queried SoKs, or by modifying queried SoKs such that the new SoK verifies for m^*.

Thus, a minimal security of SoKs additionally requires the basic (existential) unforgeability under adaptive chosen message attacks of common signatures. This requires that computing a witness w from a statement x is infeasible, else unforgeability cannot hold. We capture this by introducing an *instance generator* IGen outputting $(x, w) \in \mathcal{R}^L$ accordingly. Consider the example where IGen outputs a group element x and its discrete logarithm w (w.r.t. some group generator). We say that IGen is a *hard-instance generator* if, in addition, no efficient algorithm can, on input x for $(x, w) \leftarrow \mathsf{IGen}(1^k)$, output some $w^* \in \mathcal{W}^L(x)$ with non-negligible probability.

We now define UnfExt SoKs in the notation of [12]. To connect Unforgeability and Extractability we assume that the parameters in the two experiments are indistinguishable (else the notions could be perfectly independent):

Definition 3 (UnfExt Security). *The SoK* $\mathsf{SoK} = (\mathsf{Setup}, \mathsf{Sign}, \mathsf{Vf})$ *is* UnfExt secure *for* $\mathcal{R}^L$ *and* IGen *iff it is:*

Extractable. *There exists an efficient extractor* $\mathsf{Extr} = (\mathsf{ExtSetup}, \mathsf{Ext})$ *such that for any efficient* $\mathcal{A}$

$$\text{Prob}[(\mathsf{par}, \tau) \leftarrow \mathsf{ExtSetup}(1^k); (x, m, \sigma) \leftarrow \mathcal{A}(\mathsf{par}); \\ w^* \leftarrow \mathsf{Ext}(\mathsf{par}, \tau, x, m, \sigma) : (x, w^*) \in \mathcal{R}^L \vee \mathsf{Vf}(\sigma, m, x, \mathsf{par}) = 0] \approx 1.$$

Unforgeable. *For all efficient* $\mathcal{A}$,

$$\text{Prob}[(x, w) \leftarrow \mathsf{IGen}(1^k); \mathsf{par} \leftarrow \mathsf{Setup}(1^k); \\ (m, \sigma) \leftarrow \mathcal{A}^{\mathsf{Sign}(\cdot, x, w, \mathsf{par})}(x, \mathsf{par}) : m \notin Q \wedge \mathsf{Vf}(\sigma, m, x, \mathsf{par}) = 1] \approx 0.$$

Here, the list Q *contains queries* m *to* Sign *(note that the oracle is initialized with the generated* x *and* w*, thus these parameters are not part of the queries).*

Parameter Indistinguishability. *The output* par *in* $(\mathsf{par}, \tau) \leftarrow \mathsf{ExtSetup}(1^k)$ *is computationally indistinguishable from the output* $\mathsf{par} \leftarrow \mathsf{Setup}(1^k)$.

Note that the extractability notion in [12] gives the adversary access to a simulated signing oracle, allowing the adversary to see simulated signatures for arbitrary messages (however, Ext need not extract witnesses from simulated signatures). Since we do not consider simulated signatures in our definition, we drop the oracle access and require extractability for any message.

As aforesaid, unforgeability in the UnfExt notion is equivalent to regular chosen-message unforgeability for digital signatures. As a first sanity check, note that our trivial example where the SoK included w is not unforgeable, as $\mathcal{A}$ can insert *any* fresh message into a forgery so that it verifies.

We show later that UnfExt security is strictly weaker than SimExt security. Indeed, as aforementioned, UnfExt SoKs may leak some information about the witness w, but not to the extent that it allows forgeries. An intermediate security level between UnfExt and SimExt security combines UnfExt security with witness indistinguishability (WI). As we show after the definition, we still need unforgeability to exclude trivial examples (in particular, WI does not imply unforgeability). We formalize WIUnfExt security as follows:

Definition 4 (WIUnfExt Security). *The SoK* $\mathsf{SoK} = (\mathsf{Setup}, \mathsf{Sign}, \mathsf{Vf})$ *is* WIUnfExt secure *for* $\mathcal{R}^L$ *and* IGen *iff it is:*

UnfExt. *The scheme is UnfExt scheme and, in addition,*
Witness Indistinguishable. *For all* $x \in L_k$*, all* $w_0, w_1 \in \mathcal{W}_k^L(x)$*, and all efficient* $\mathcal{A}$*,*

$$\mathrm{Prob}[\mathsf{par} \leftarrow \mathsf{Setup}(1^k); b \leftarrow \{0,1\}; \\ d \leftarrow \mathcal{A}^{\mathsf{Sign}(\cdot,x,w_b,\mathsf{par})}(x, w_0, w_1, \mathsf{par}) : d = b] \approx \tfrac{1}{2}.$$

Note that we demand witness indistinguishability even if $\mathcal{A}$ knows w_0, w_1. We also show that WI does not imply unforgeability. Consider a WIUnfExt SoK and change it into SoK' such that: $\mathsf{Setup}' = \mathsf{Setup}$; on input m algorithm Sign' runs Sign on message m, then runs Sign on message $\mathbf{0}$ (the all-zero string of some fixed length), and outputs $(\mathsf{Sign}(m), \mathsf{Sign}(\mathbf{0}))$ as its signature; and finally on inputs $m, (\sigma_m, \sigma_0)$, the verifier runs Vf on inputs (m, σ_m) and then on $(\mathbf{0}, \sigma_0)$. The new SoK' is still WI and extractable, but an adversary $\mathcal{A}$ against unforgeability simply queries Sign' on input $m \neq \mathbf{0}$, receives (σ_m, σ_0), and then outputs $\mathbf{0}, (\sigma_0, \sigma_0)$ as its forgery.

An alternative SoK security definition could use witness-hiding (WH) proofs of knowledge [14] where it is infeasible to recover the *entire* witness. However, unforgeability already implies WH: if a signature of knowledge is not WH, then it is also not unforgeable (the adversary can simply re-use the recovered witness w to sign a fresh message m^*).

3.2 Relationships of Security Notions

The strict hierarchy of SimExt, UnfExt, and WIUnfExt security appears in Fig. 1. Formally we have:

Proposition 1 (Relationships of Notions). *For any* $\mathcal{R}^L$ *and any hard-instance generator* IGen*, (1) any SimExt secure SoK for* $\mathcal{R}^L$ *is also WIUnfExt secure for* $\mathcal{R}^L$ *and* IGen*, and (2) any WIUnfExt secure SoK for* $\mathcal{R}^L$ *and* IGen *is also UnfExt secure. Furthermore, (3) if there exists an UnfExt secure SoK for some* $\mathcal{R}^L$ *and* IGen*, then there exists an SoK for some* $\mathcal{R}^{L'}$ *and* IGen' *that is UnfExt but not WIUnfExt; and (4) if there exists an WIUnfExt SoK for some* $\mathcal{R}^L$ *and* IGen*, and if one-way permutations (OWP) exist, then there exists an SoK for some* $\mathcal{R}^{L'}$ *and* IGen' *which is WIUnfExt but not SimExt.*

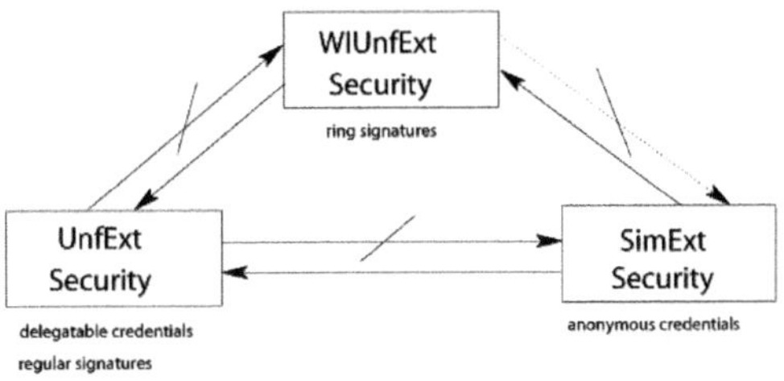

Fig. 1. Security of SoKs: arrows refer to implications, hatched arrows to separations; the dotted arrow indicates that the separation relies on an additional assumption. The figure also shows potential applications of the different notions.

Proof. We prove claim (1). Consider a SimExt secure SoK= (Setup, Sign, Vf). SimExt and UnfExt correctness are identical. Extractability follows since the definitions are almost the same; however, as aforesaid $\mathcal{A}$ uses SimSign in SimExt security, whereas in WIUnfExt security, $\mathcal{A}$ has no oracle access – though it may still simulate Sign for valid pairs (x, w). By Simulatability, $\mathcal{A}$ cannot distinguish between SimSign and Sign; thus we can interchange them with a negligible change in the success probability. Unforgeability follows as described in [12] from the fact that IGen is a hard-instance generator (despite minor technical differences). Indistinguishability of the parameters is a weaker requirement than simulatability. Finally, we prove witness indistinguishability. For this, we replace Setup by SimSetup and Sign by SimSign in the original WI game (using the trapdoor τ output by SimSetup). In the modified game, $\mathcal{A}$ cannot distinguish between SoKs for the two witnesses, as they are generated independently of the witness. By Simulability, using Sign and SimSign are indistinguishable; thus the success probability in the modified game and that of the WI game are only negligibly different. Thus the SimExt secure SoK is also WIUnfExt secure.

Statement (2) follows by definition of UnfExt and WIUnfExt security.

For claim (3) consider UnfExt secure SoK= (Setup, Sign, Vf) for $\mathcal{R}^L$ and IGen. We construct $\mathsf{SoK}^* = (\mathsf{Setup}^*, \mathsf{Sign}^*, \mathsf{Vf}^*)$ for $\mathcal{R}^{L'}$ and IGen', defined for witnesses W= $(w||b)$ (for bit b) and statements $\mathsf{X} = x$. Then $(\mathsf{X}, \mathsf{W}) \in \mathcal{R}^{L'}$ iff $(x, w) \in \mathcal{R}^L$, and IGen' samples (X, W) by running IGen and appending a random bit to w. Algorithms Setup^* and Vf^* run Setup, resp. Vf as black boxes, forwarding the output. Algorithm Sign^* on input W runs Sign as a black box, appending b from W to the output SoK. Clearly SoK^* inherits correctness, extractability, parameter indistinguishability, and unforgeability from SoK. Yet SoK^* is *not* WI, as signatures leak the added bit for witnesses $w||0$ and $w||1$. In claim (4) we assume the existence of a OWP f. Consider WIUnfExt secure SoK. We construct SoK' that is still WIUnfExt secure, but not SimExt secure. For each statement x, choose random r and set $\mathsf{X} = x||f(r)$. All $\mathsf{W} \in \mathcal{W}^L(\mathsf{X})$ also include r, i.e., $\mathsf{W} = w||r$ (this later ensures WI). It is also easy to derive IGen' from IGen as in Claim (3). Algorithm Setup' of SoK' runs Setup from SoK, forwarding the output par. Algorithm Sign' of SoK' first runs Sign from SoK to get SoK σ. The output of Sign' is $\sigma||r$. Verification by Vf' runs Vf from SoK, then checks that $f(r)$ for the r in the SoK is the one featured in X.

For the analysis, note that SoK' is still WIUnfExt secure. Completeness, extractability, and indistinguishability of the parameters are trivial. Forging SoK' also involves forging SoK. Finally, WI is preserved as r is the same for all $w \in \mathcal{W}^L(x)$ for each x. However, SoK' is *not* SimExt secure under the one-wayness of f. In particular, it is not simulatable. Assume that there exists a simulator Sim that simulates Sign' to $\mathcal{A}$. The success probability is taken over all x; if Sim is successful, then we can build an inverter $\mathcal{B}$ against f. Indeed, Sim receives for every statement x the corresponding $f(r)$ for random r. If Sim simulates Sign successfully, it outputs the correct value r (else the SoK does not verify). Algorithm $\mathcal{B}$ runs Sim, outputting r as its pre-image, and is as successful as Sim. □

3.3 Universally Composable Versions

In the Universal Composability (UC) framework due to Canetti [9], protocols are associated with ideal functionalities, describing permissible leakage of data in the protocol run. Several parties run the protocol, receiving input from a so-called environment $\mathcal{Z}$. A protocol π UC-realizes functionality F if $\mathcal{Z}$ cannot distinguish between a "real world" where parties run π around an adversary who gets inputs from, and outputs to $\mathcal{Z}$, and an "ideal world", where parties run F around a simulator Sim, also outputting to $\mathcal{Z}$. The adversary may corrupt parties, thus controlling them; these parties are marked down as corrupt. In the UC framework, π is secure if there exists a Sim such that for all $\mathcal{Z}$ and for all adversaries, $\mathcal{Z}$ cannot distinguish between the two worlds.

Chase and Lysyanskaya [12] give two equivalent definitions of SoKs. The first is UC-based, for a modification of the tweaked ideal signature functionality—[11]. For more details regarding ideal functionalities for signatures, refer to [12]. The UC definition for SoKs is equivalent to SimExt security, thus strictly stronger than UnfExt security. We show how to modify this definition to capture UnfExt security. The main difference is that we do not require simulatability for our signing algorithm. In particular, we use Sign and not SimSign for SoK generation.

Our ideal functionality (Figure 2) resembles the one in [12], but is simpler, as it is parameterized by $\mathcal{R}^L$, whereas [12] use a universal functionality and put the (code of the machine verifying the) $\mathcal{R}^L$ into session identities sid.

Note that as opposed to [12] the simulator is not among the algorithm descriptions. This follows our idea that full simulatability is not required for SoK. Note also that SoKs require some common parameter setup preceding it. As in [12], we use the CRS model and corresponding $\mathcal{F}_{\text{CRS}}^{D}$ functionality, where D is a party-chosen distribution of the parameters. If party P forwards $(\text{CRS}, \mathsf{sid})$ to $\mathcal{F}_{\text{CRS}}^{D}$, the functionality checks that no value v is associated with this sid; else, it chooses v randomly according to D and stores it, returning $(\text{CRS}, \mathsf{sid}, v)$ to both P and to the adversary.

For SoKs, D is the distribution of the parameters output by $\mathsf{Setup}(1^k)$. We run SoK= (Setup, Sign, Vf) in a hybrid CRS environment and denote the resulting protocol $\pi_{\mathsf{SoK}}(\mathcal{R}^L, \mathsf{IGen})$. During each session of π_{SoK}, every time party P receives a $(\mathsf{Setup}, \mathsf{sid})$ message from the environment $\mathcal{Z}$, sid is checked and P queries $\mathcal{F}_{\text{CRS}}^{D}$ so as to get $(\text{CRS}, \mathsf{par})$. The public par are stored by P and P also generates

$\mathcal{F}_{SOK}(\mathcal{R}^L)$: **signature of knowledge for a witness** w **with** $(x,w) \in \mathcal{R}^L$.

Setup. Upon receiving $(\mathsf{Setup}, \mathsf{sid})$ from party P, check that this is the first time a Setup request is made with parameter sid; if not, ignore, else $(\mathsf{Setup}, \mathsf{sid})$ is forwarded to the adversary, which eventually returns $(\mathsf{Algorithms}, \mathsf{sid}, x, \mathsf{Vf}, \mathsf{Sign}, \mathsf{Ext})$ to the functionality. Here Sign and Ext describe probabilistic polynomial time (PPT) algorithms (represented by PPT Turing Machines), and Vf describes a deterministic polynomial time algorithm. The algorithm descriptions and x are stored, and P receives $(\mathsf{Algorithms}, \mathsf{sid}, \mathsf{Sign}, \mathsf{Vf})$.

Signature Generation. Upon receiving $(\mathsf{Sign}, \mathsf{sid}, m)$ from P, run $\sigma \leftarrow \mathsf{Sign}(m)$ and check that $\mathsf{Vf}(\sigma, m, x) = 1$; if so, output $(\mathsf{Signature}, \mathsf{sid}, m, \sigma)$ to P and record (m, x, σ). Else, output (Completeness Error) to P and halt.

Signature Verification. Upon receiving $(\mathsf{Verify}, \mathsf{sid}, \sigma, m, x')$ from verifier V, if (m, x', σ') is stored for some σ', then output $(\mathsf{Verified}, \mathsf{sid}, \sigma, m, x', \mathsf{Vf}(\sigma, m, x'))$ to V. Else, if $x' = x$ and $\mathsf{Vf}(\sigma, m, x) = 1$ but (m, x, σ) has not been stored yet, output (Unforgeability Error) and halt. Else let $w' \leftarrow \mathsf{Ext}(m, x, \sigma)$; if $(w', x) \in \mathcal{R}^L$, output $(\mathsf{Verified}, \mathsf{sid}, \sigma, m, x, \mathsf{Vf}(\sigma, m, x))$ to V. Else, if $\mathsf{Vf}(\sigma, m, x) = 0$, output $(\mathsf{Verified}, \mathsf{sid}, \sigma, m, x, 0)$ to V. Else, output (Extraction Error) and halt.

Fig. 2. Signature of Knowledge Functionality

$(x, w) \leftarrow \mathsf{IGen}(1^k)$. Both values are then included in the descriptions returned to $\mathcal{Z}$ as $(\mathsf{Algorithms}, \mathsf{sid}, \mathsf{Sign}(\mathsf{par}, \cdot, x, w), \mathsf{Vf}(\mathsf{par}, \cdot, \cdot))$.

If $\mathcal{Z}$ sends a request $(\mathsf{Sign}, \mathsf{sid}, m)$ to party P, this party retrieves the stored par and returns $(\mathsf{Signature}, \mathsf{sid}, m, \mathsf{Sign}(\mathsf{par}, m, x, w))$. If a verifier V receives request $(\mathsf{Verify}, \mathsf{sid}, \sigma, m, x')$ from $\mathcal{Z}$, it returns $(\mathsf{Verified}, \mathsf{sid}, \sigma, m, x', \mathsf{Vf}(\mathsf{par}, \sigma, m, x'))$ to $\mathcal{Z}$.

Like in [12], we can prove the equivalence of the two definitions. In particular, we formalize the following theorem.

Proposition 2 (Equivalence of Notions). *Protocol* $\pi_{\mathsf{SoK}}(\mathcal{R}^L, \mathsf{IGen})$ *UC-realizes the functionality* $\mathcal{F}_{SOK}(\mathcal{R}^L)$ *in the* $\mathcal{F}^D_{CRS}$ *hybrid model iff* SoK *is UnfExt secure for* $\mathcal{R}^L$ *and* IGen.

The proof closely follows that of [12] and is omitted for space reasons. A UC equivalence can also be extended to the notion of WIUnfExt security.

4 SoK Instantiation

In this section we recall Waters' signature scheme [23] and show that it is UnfExt secure; it is in fact also trivially WIUnfExt, as witnesses are unique. The latter point is also discussed in [20] in the related KOSK model. We also describe a universal construction based on general assumptions and for arbitrary relations, which actually achieves the stronger SimExt security. We finally describe an UnfExt secure construction where we sign messages m for extended statement-witness pairs (x, w); an advantage here is that a single proof (ZAP) suffices for every statement-witness pair, rather than a proof for each message.

4.1 Waters' Signature Scheme

We construct UnfExt secure SoKs from Waters' unforgeable pairing-based signatures [23]. In particular, such signatures are already complete and unforgeable, becoming extractable if we add some master information about the randomness used for signature generation. We outline our construction and then consider its security. Note that we slightly abuse notation here as the key pairs now depend on the parameters, i.e., the relation depends on par; all results presented before remain valid in this setting.

Let $\mathsf{Sig}_W = (\mathsf{SKGen}_W, \mathsf{SSign}_W, \mathsf{SVf}_W)$ be the unforgeable signature scheme due to Waters. We review this construction briefly before outlining our UnfExt Secure SoK. In the schema due to Waters, the parameters (generated at Key Generation) consist of: multiplicative groups $\mathbb{G}$ and $\mathbb{G}^T$; a prime q; an element $g \in \mathbb{G}$ of prime order q; and (the description of) a bilinear mapping $\hat{e}$.

Key Generation. For security parameter k, algorithm SKGen_W runs an (external) generator $\mathcal{G}$ to generate parameters $\mathsf{par} = (\mathbb{G}, \mathbb{G}^T, q, g, \hat{e})$. The algorithm then picks a random $a \leftarrow \mathbb{Z}_q$ and computes $g_1 := g^a$. Then it chooses $g_2, u_0, \dots, u_k \leftarrow \mathbb{G}$. Finally, the algorithm outputs the public/private key-pair (pk, sk) for $pk = (\mathsf{par}, g_1, g_2, u_0, \dots, u_k)$ and $sk = g_2^a$.

Signature Generation. For message $m \in \mathcal{M}$ and private key sk, the signing algorithm SSign_W parses $m = m_1 \dots m_k$ for $m_1, \dots, m_k \in \{0,1\}$ and computes $H(m) \leftarrow u_0 \prod_{i=1}^{k} u_i^{m_i}$. It then picks a random $r \in \mathbb{Z}_q$; the signature output by SSign is $\sigma = (g_2^a \cdot H(m)^r, g^r)$.

Signature verification. For signature σ, message m, and public key pk, SVf parses σ as (σ_1, σ_2) and outputs 1 iff. $\hat{e}(g, \sigma_1) = \hat{e}(\sigma_2, H(m)) \cdot \hat{e}(g_1, g_2)$.

This signature scheme is complete and existentially unforgeable under adaptive chosen message attacks under the CDH assumption (see [23]).

We turn this construction into an UnfExt secure SoK by including $u_0, \dots, u_k$ in the public parameters, thus allowing the simulator to extract the witness if the discrete logs of $u_0, \dots, u_k$ with respect to g are in the trapdoor information generated by $\mathsf{ExtSetup}$. This idea is outlined in the following. The public parameters, besides $\mathsf{par}_W = (\mathbb{G}, \mathbb{G}^T, q, g, \hat{e})$ now also contain $u_0, \dots, u_k$. To generate x and w via IGen the values g_2 and a are chosen as by SKGen_W; g_1 is set as $g_1 = g^a$; and witness $w = sk = g_2^a$ is given to the signer. The statement is then $x = (g, g_1, g_2)$ and we define the relation $\mathcal{R}^L$ as follows: $(x, w) \in \mathcal{R}^L$ iff $\hat{e}(g_1, g_2) = \hat{e}(g, w)$. If the values are generated honestly, we note that $\hat{e}(g_1, g_2) = \hat{e}(g^a, g_2) = \hat{e}(g, g_2)^a = \hat{e}(g, g_2^a) = \hat{e}(g, w)$.

SoK Setup. For security parameter k, the algorithm Setup generates $\mathsf{par}_W = (\mathbb{G}, \mathbb{G}^T, q, g, \hat{e})$ as in Sig_W and also chooses $z_0, \dots, z_k \leftarrow \mathbb{Z}_q$ and then sets $u_i = g^{z_i}$ for $i = 0, \dots, k$. Finally, Setup outputs $\mathsf{par} = (\mathsf{par}_W, u_0, \dots, u_k)$.

SoK Generation. For message $m \in \mathcal{M}$, witness $w = g_2^a$, statement $x = (g, g^a, g_2)$, and parameters par, the signing algorithm Sign runs SSign_W and outputs signature σ.

SoK VERIFICATION. For signature σ, message m, statement x, and public parameters par, the algorithm Vf runs SVf outputting the bit b.

Theorem 1 (UnfExt Security). *The signature of knowledge scheme* SoK *defined above is UnfExt Secure under the CDH assumption.*

Proof. We have to prove correctness, extractability, unforgeability, and parameter indistinguishability. First note that correctness and unforgeability (for IGen as described above) follow from the corresponding properties of Waters' signature scheme. We now describe an extractor Extr= (ExtSetup, Ext) which, given a valid signature σ outputs witness w. The algorithm ExtSetup runs Setup and sets $\tau = (z_0, \ldots, z_k)$. In particular, the public parameters have identical distributions in both cases. For signature σ, message m, statement x, parameters par, and trapdoor information τ, the algorithm Ext parses σ as (σ_1, σ_2), calculates $d = z_0 + \sum_{i=1}^{k} z_i m_i \bmod q$ and outputs $w^* = \sigma_1 \sigma_2^{-d}$. Note that, if the signature verifies, we must have

$$\hat{e}(g, \sigma_1) = \hat{e}(\sigma_2, H(m)) \cdot \hat{e}(g_1, g_2)$$

which can be rewritten as

$$\hat{e}(g_1, g_2) = \hat{e}(g, \sigma_1) \cdot \hat{e}(\sigma_2, g^d)^{-1} = \hat{e}(g, \sigma_1 \sigma_2^{-d}) = \hat{e}(g, w^*).$$

Hence, by definition of the witness relation, it holds that $(x, w^*) \in \mathcal{R}^L$. Thus, extractability is also proved for this description of the extractor Extr. □

Note that the scheme above is actually WIUnfExt since the witness is unique, given x and par.

4.2 General Construction

By our results in Section 3.2 any SimExt signature of knowledge is WIUnfExt (if finding witnesses for instances is hard). Theoretically thus, the general construction in [12] based on simulation-sound NIZKPoK is also WIUnfExt. We show an alternative construction using the witness-indistinguishable proof systems for any NP language, also called ZAPs [13]. The other ingredients are an IND-CCA public-key encryption scheme (KGen, Enc, Dec) and a pseudorandom generator G. For space reasons we merely sketch the construction and discuss its security.

Construction. Our idea is to add into par a public key pk of the encryption scheme, a random string z of length $2k$ (whose purpose becomes clear later), and a string $\mathsf{par}_{\mathrm{ZAP}}$ for the ZAP. To sign message m with respect to witness w for x let the signer encrypt the witness w together with m to $C = \mathsf{Enc}(pk, w||m)$ and append a ZAP (with respect to $\mathsf{par}_{\mathrm{ZAP}}$) that C encrypts $w||m$ for $(x, w) \in \mathcal{R}^L$ or that z is in the range of G for inputs of length k. We remark that we formally require all witnesses w to be padded to have equal length; this is easy to implement via standard paddings as all witnesses of complexity k are polynomially bounded. The verifier simply checks the validity of the ZAP.

Extractability. The extractability of this scheme can be realized by the decryption algorithm, relying on the fact that a random z is *not* in the range of G with probability at least $1 - 2^{-k}$; thus, the validity of the ZAP implies that the encrypted witness is valid. Proving unforgeability and witness indistinguishability is more sophisticated.

Unforgeability. For unforgeability consider an adversary against the original signature algorithm, being able to produce a valid SoK for a fresh message m with non-negligible probability. As the ZAP is valid and z is not in the range of G with overwhelming probability, running the decryption algorithm on the ciphertext in the adversary's forgery yields a valid witness w with non-negligible probability. Note that we now consider an adversary's success w.r.t. successful extraction of a valid witness, not to a successful forgery. We can further condition on the adversary not outputting a forgery for a previously seen ciphertext; such ciphertexts cannot contain a fresh message.

Change the game slightly by using pseudorandom $z = G(r)$ in par. By the security of G the adversary's success cannot drop significantly. In the next game hop, the altered signing algorithm uses the preimage r to provide valid ZAPs; by the WI of the ZAPs, this negligibly increases the success probability. In the final game, instead of encrypting w in C the again modified signing process uses $0^{|w|}||m$. Note that the ZAP computation is not affected by this, and all witness have the same length. By the IND-CCA security of the encryption scheme, replacing the encryptions of $w||m$ by $0^{|w|}||m$, does not significantly change the adversary's probability of finding a forgery for a *fresh* message; this decrease is easily detected by recovering the witness and message encapsulated in the adversary's forgery attempt (this has to work by IND-CCA and by message —thus ciphertext— freshness). But now the signing oracle is independent of the actual witness; thus the adversary essentially finds a valid witness w to x without help, which is infeasible according to the security of the instance generator.

Parameter Indistinguishability. The public parameters have identical distribution in the actual scheme and the extractability experiment.

Witness Indistinguishability. Again here we can run any distinguisher on fake parameters including a pseudorandom value z, encrypting $\mathsf{Enc}(pk, 0^{|w_b|}||m)$ instead, and using the preimage of z to give a valid ZAP. As for unforgeability it follows that the behavior of the distinguisher compared to the original signature generation process (for either w_0 or w_1) cannot change significantly. But since the resulting signatures are independent of b, it also follows that the original signatures must be witness indistinguishable.

Simulatability. Actually this construction also achieves the stronger notion of Simulatability. This can be seen from the proof of Witness Indistinguishability and unforgeability, as essentially the initial game is reduced to a game where the signature is independent of the witness. The proof for simulatability would involve a SimSign procedure that does not use a valid witness at all and still outputs a verifiable signature (by using a pseudorandom value z).

4.3 Embedding Witnesses

Both the construction in section 4.2 and the one in [12] instantiate SimExt secure SoKs by encrypting the witness w and a message m, and giving a zero knowledge proof (NIZK) that the encryption is correctly formed and that w and the statement x for which the signature was created are in $\mathcal{R}^L$. This both ensures extractability for w, and it "hides" w, such that the SoK is simulatable. However, in this scheme the NIZK needs to be computed *every time a signature is generated*, as a fresh m must be encrypted every time together with w.

In this section we show how embedding the witnesses into a larger set can improve efficiency such that the proof only needs to be computed once. This, however, undermines witness indistinguishability, which does not hold in general, making the solution inapplicable e.g., to the case of ring signatures. In our construction we again use ZAPs, an IND-CCA public-key encryption scheme $(\mathsf{KGen}, \mathsf{Enc}, \mathsf{Dec})$, a pseudorandom generator, but this time also an existentially unforgeable signature scheme $\mathsf{Sig} = (\mathsf{SKGen}, \mathsf{SSign}, \mathsf{SVf})$.

Construction. We add into par the public key pk_{Enc} of the encryption scheme, a random string z of length $2k$, and a string $\mathsf{par}_{\mathrm{ZAP}}$ for the ZAP. The main idea for signature generation is that instead of signing messages m for statement-witness pairs (x, w) in the relation $\mathcal{R}^L$, we sign m with respect to an extended witness $\mathsf{W} = (w, s, r)$ and an extended statement $\mathsf{X} = (x, pk_{\mathsf{Sig}}, C, \pi)$, where π is a ZAP (with respect to $\mathsf{par}_{\mathrm{ZAP}}$) that $C = \mathsf{Enc}(pk, \mathsf{W})$ is a correct encryption of W with randomness r such that $(x, w) \in \mathcal{R}^L$, and s is the randomness which made $\mathsf{SKGen}(1^k)$ output pk_{Sig}, or that z is in the range of G for inputs of length k. The signature of knowledge is generated as $\mathsf{SSign}(sk_{\mathsf{Sig}}, m)$. Given m and x the verifier computes verifies the validity of the signature by using pk_{Sig}, and then the validity of the ZAP.

Extractability of Original Witnesses. The extractability of this scheme follows as in the previous section. Note that we extract from the proof for any X only the part of the some witness W^* which comprises a witness w and the randomness s for SKGen; the randomness r for the ciphertext would only be extractable if the encryption scheme were to support randomness recovery, i.e., the decryption algorithm could also be used to derive the randomness r.

Unforgeability. A forgery (m, σ) of a SoK must be output for a fresh message m such that σ verifies under the public key pk_{Sig}. This would straightforwardly violate the unforgeability under the signature scheme, as we can simulate the encryption and the ciphertext C and proof π without knowledge of the randomness s resp. the signing key sk_{Sig} with the same technique as in the construction in the previous section.

Parameter Indistinguishability. The parameters for the extractor and the one in the actual scheme are identically distributed.

5 Application Scenarios

SoKs allow users to sign messages on behalf of any NP statement x; in particular, if there exist more witnesses corresponding to this statement, SoKs naturally provide ring signatures. In this context, the SimExt security of SoKs due to [12] actually guarantees that signatures are simulatable without the witness. However, with our definition of Witness Indistinguishable UnfExt SoKs we ensure that witnesses are merely indistinguishable.

Below we show applications of SoKs to regular digital signatures and ring signatures.

5.1 Digital Signatures

SoKs can be easily used as simple signature schemes as described in e.g. [16] as shown in construction 2.

Construction 2. *Let* SoK= (Setup, Sign, Vf) *be a UnfExt SoK scheme for a relation* $\mathcal{R}^L$. *Define the signature scheme* Sig= (SKGen, SSign, SVf) *as follows, for some security parameter* k.

KEY GENERATION. *On input* k, *algorithm* SKGen *first runs* Setup(1^k) *and outputs* par *then it runs the instance generator* IGen *for* SoK *on input* par *to obtain a statement/witness keypair* (x, w) *such that* $(x, w) \in \mathcal{R}^L$. SKGen *outputs* $(pk = (x, \mathsf{par}), sk = (w, x, \mathsf{par}))$.

SIGNATURE GENERATION. *On input message* m *and* $sk = (w, x, \mathsf{par})$, *algorithm* SSign *runs* Sign$(m, w, x,$ par$)$ *and outputs the resulting SoK* σ *as its signature.*

SIGNATURE VERIFICATION. *On input signature* σ, *message* m, *and* $pk = (x, \mathsf{par})$, *algorithm* SVf *runs* Vf$(\sigma, m, x,$ par$)$ *and outputs the resulting bit* b.

The security of digital signatures as in [16] is defined in terms of correctness and existential unforgeability against chosen message attacks. The following holds.

Proposition 3. *If* SoK *is UnfExt secure, then construction 2 is a secure digital signature.*

Proof. Correctness is trivially inherited. Furthermore, Existential Unforgeability holds: given an efficient adversary $\mathcal{A}$ outputting forgery $s = \mathsf{SSign}_{sk}(m)$, the adversary $\mathcal{B}$ against SoK Unforgeability uses $\mathcal{A}$ as follows: whenever $\mathcal{A}$ queries SSign on m_i, adversary $\mathcal{B}$ queries Sign on the same input, forwarding the output signature σ. Finally, when $\mathcal{A}$ outputs a forgery (m, s), $\mathcal{B}$ outputs the same, together with $pk = x$. By construction, if $\mathcal{A}$ is successful, then the SoK is valid, thus $\mathcal{B}$ succeeds too. Furthermore, m must be fresh for $\mathcal{B}$, as it is fresh for $\mathcal{A}$. □

5.2 Ring Signatures

Ring signatures were formalised by Rivest, Shamir, and Tauman [21] in 2001. In this setting, a *signer* signs message m on behalf of a so-called *ring* of participants such that it is impossible to tell which ring member actually signed m. We denote ring members by U_1, U_2, ..., U_n.

Ring signatures assume the existence of a PKI where users U_i are associated with private/public key pairs (sk_i, pk_i). Message m can be signed by U_i under public keys $\{pk_i\}_{i=1}^n$ and under private key sk_i, resulting in a signature σ. Verification requires the public keys of all the users, outputting a bit. In particular, Rivest et al. [21] define ring signatures to be setup-free, i.e. any signer can dynamically select a ring just by knowing the public keys of the other ring members.

We adopt the ring signature definitions due to Bender et al. [3]. This work defines rings of n of users to be the subset of their public keys, which may be honestly generated or chosen by the adversary. In the notation of [3], we write $R = (pk_1, \ldots, pk_n)$ for the ring of users U_j with $j \in \{1, \ldots, n\}$.

Definition 5 (Ring Signatures [3]). *A* ring signature *is a tuple of efficient algorithms* $\mathsf{RSig} = (\mathsf{RSKGen}, \mathsf{RSSign}, \mathsf{RSVf})$ *such that:*

KEY GENERATION. *Run on security parameter k,* RSKGen *outputs key-pair (sk, pk).*

SIGNATURE GENERATION. *On input index i, message m, ring R of size n with n distinct elements, and sk s.t. (sk, pk_i) is a legitimate key-pair.*

SIGNATURE VERIFICATION. *On input (R, m, σ), algorithm* RSVf *returns bit b.*

We require perfect completeness, i.e., for all k, for all n key-pairs (sk_i, pk_i) for $i \in \{1, \ldots, n\}$, any $j \in \{1, \ldots, n\}$, and any message m, it holds that $\mathsf{RSVf}(R, m, \mathsf{RSSign}(j, sk_j, m, R)) = 1$ *for* $R = (pk_1, \ldots, pk_n)$.

Ring signatures have two main properties: anonymity and unforgeability. Bender et al. [3] introduce various degrees of these notions and prove strict implications between the different flavors. The adversary may query an OSign oracle with input an index j, a message m, and a ring R, and running $\mathsf{RSSign}(j, m, R, sk_j)$ to obtain the honestly generated signature σ. We reiterate only the strongest form of anonymity – anonymity against attribution attacks – and the basic-most form of unforgeability – unforgeability against fixed-rings attacks and call them anonymity, resp. unforgeability. Intuitively, this form of anonymity allows the adversary to know the secret keys of all-but-one users, but it still can't distinguish the signer of a message (i.e., if even a single signer is honest, a signature cannot be attributed to him, even by everyone else colludes together). In unforgeability against fixed-rings attacks, the signature is unforgeable if the adversary uses the same ring of users. This is equivalent, as we see below, to using the same statement x for SoKs. For further details on ring signatures, please refer to [3]. Ring signature security now follows:

ANONYMITY. In order to formalize anonymity, Bender et al. allow the adversary to learn the randomness used by RSKGenin generating the users' key pairs. Instead, we give the adversary access to all the honestly generated secret keys after they have been generated. Another subtle difference in our definition allows our adversary to be stronger: Bender et al. give the adversary access to the secret keys only *after* the adversary has chosen a challenge message m, indices i_0 and i_1, and a ring R such that pk_{i_0}, pk_{i_1} are in R. In our definition, the adversary may know the secret keys even before it has made its choice. We call the ring signature scheme anonymous iff for all integers n (depending on k), and all efficient adversaries $\mathcal{A}$, the following holds:

$$\begin{aligned}\text{Prob}\ [&(sk_i, pk_i)_{i=1}^n \leftarrow \mathsf{RSKGen}(1^k); (\mathsf{st}, i_0, i_1, m, R) \leftarrow \mathcal{A}^{\mathsf{OSign}(\cdot,\cdot,\cdot)}((sk_i, pk_i)_{i=1}^n) \\ &b \leftarrow \{0,1\}; \sigma \leftarrow \mathsf{RSSign}(i_b, m, R, sk_{i_b}); d \leftarrow \mathcal{A}^{\mathsf{OSign}(\cdot,\cdot,\cdot)}(\sigma, \mathsf{st}) : d = b] \approx \tfrac{1}{2}.\end{aligned}$$

UNFORGEABILITY. For all security parameters k, all integers n, and all efficient adversaries $\mathcal{A}$, the following holds:

$$\begin{aligned}\text{Prob}\ [&(sk_i, pk_i)_{i=1}^n \leftarrow \mathsf{RSKGen}(1^k); (m, \sigma) \leftarrow \mathcal{A}^{\mathsf{OSign}(\cdot,\cdot,R)}(\{pk_i\}_{i=1}^n) : \\ &(\cdot, m) \notin Q \wedge \mathsf{RSVf}(R, m, \sigma) = 1 \qquad] \approx 0.\end{aligned}$$

Here we denote by Q the list of queries made to the OSign oracle.

Note that, although this definition of anonymity is the strongest of the three presented by Bender et al. [3], it is not as strong as the simulatability property required by signatures of knowledge as defined by Chase and Lysyanskaya. And yet, ring signatures can be constructed from SoKs in a natural way, as long as there exists a form of witness indistinguishability. Indeed, a signature of knowledge on a message m proves that a signer who knows *a* valid witness (out of possibly many valid witnesses) to a statement has signed a statement. In fact, if we equivalate a ring to a statement, we can perceive the set of witnesses belonging to this set as each representing a user in the ring. We describe this in what follows.

We first consider a relation $\mathcal{R}^L$ with statements of the form x and witnesses w and with an efficient instance generator IGen, which, on input a security parameter k and some parameters par, outputs a statement x and a witness w with $(x, w) \in \mathcal{R}^L$. Let SoK= (Setup, Sign, Vf) be a witness indistinguishable signature of knowledge for a relation $\mathcal{R}$ with statements $R = (x_1, \ldots, x_n)$ and witnesses w such that $(w, R) \in \mathcal{R}$ iff. there exists an index $j \in \{1, \ldots, n\}$ such that $(w, x_j) \in \mathcal{R}^L$. Consider a ring signature RSig= (RSKGen, RSSign, RSVf), such that:

SETUP. Before running the ring signature scheme, the algorithm Setup is run on input k to output parameters par.

KEY GENERATION. Upon input a security parameter $K = (k, \mathsf{par})$ for an integer k and parameters par, the key generation algorithm RSKGen runs IGen, outputting the tuple (x_i, w_i).

RING SIGNATURE GENERATION. Upon input an index $i \in \{1, \ldots, n\}$, a message m, a ring $R = (x_1, \ldots, x_n)$, and a private key w_i, the signature generation algorithm RSSign runs Sign on input m, R, w, par, and returns the output signature σ.

Ring Signature verification. Upon input a ring $R = (x_1, \ldots, x_n)$, a message m, and a signature σ, the signature verification algorithm RSVf runs Vf on input $\sigma, m, R, \mathsf{par}$, and outputs the resulting bit b.

This construction is a secure ring signature in the sense of the above security definition. In particular, the completeness property follows from the correctness of the underlying signature of knowledge scheme, and unforgeability follows from the unforgeability of the SoK (but the ring is fixed, as the definition of unforgeability for SoKs fixes the statement, in this case R). To see that RSig is also anonymous in the presence of attributions, note that the witness indistinguishability definition is quantified over all witnesses, which are freely given to the adversary. Therefore the adversary knows all w_i, but cannot tell them apart anyway.

Acknowledgments

We thank the anonymous reviewers for constructive and valuable comments.

References

1. Adams, C., Farrell, S.: Internet x.509 public key infrastructure certificate. RFC 2510 (March 2009)
2. Bellare, M., Goldreich, O.: On defining proofs of knowledge. In: Brickell, E.F. (ed.) CRYPTO 1992. LNCS, vol. 740, pp. 390–420. Springer, Heidelberg (1993)
3. Bender, A., Katz, J., Morselli, R.: Ring signatures: Stronger definitions, and constructions without random oracles. In: Halevi, S., Rabin, T. (eds.) TCC 2006. LNCS, vol. 3876, pp. 60–79. Springer, Heidelberg (2006)
4. Blum, M., Feldman, P., Micali, S.: Non-interactive zero-knowledge and its applications (extended abstract). In: Proceedings of the Annual Symposium on the Theory of Computing (STOC), pp. 103–112. ACM Press, New York (1988)
5. Boldyreva, A.: Threshold signatures, multisignatures and blind signatures based on the gap-diffie-hellman-group signature scheme. In: Desmedt, Y.G. (ed.) PKC 2003. LNCS, vol. 2567, pp. 31–46. Springer, Heidelberg (2002)
6. Bresson, E., Stern, J.: Efficient revocation in group signatures. In: Kim, K.-c. (ed.) PKC 2001. LNCS, vol. 1992, pp. 190–206. Springer, Heidelberg (2001)
7. Camenisch, J.: Efficient and generalized group signatures. In: Fumy, W. (ed.) EUROCRYPT 1997. LNCS, vol. 1233, pp. 465–479. Springer, Heidelberg (1997)
8. Camenisch, J., Stadler, M.: Efficient group signature schemes for large groups (extended abstract). In: Kaliski Jr., B.S. (ed.) CRYPTO 1997. LNCS, vol. 1294, pp. 410–424. Springer, Heidelberg (1997)
9. Canetti, R.: Security and composition of multi-party cryptographic protocols. Journal of Cryptology 13, 143–202 (2000)
10. Canetti, R.: Universally composable security: a new paradigm for cryptographic protocols. In: Proceedings of the Annual Symposium on Foundations of Computer Science (FOCS), pp. 136–145. IEEE Computer Society Press, Los Alamitos (2001)
11. Canetti, R.: Universally composable security: A new paradigm for cryptographic protocol. Cryptology ePrint Archive, Report 2000/067, EPRINTURL (2005)

12. Chase, M., Lysyanskaya, A.: On signatures of knowledge. In: Dwork, C. (ed.) CRYPTO 2006. LNCS, vol. 4117, pp. 78–96. Springer, Heidelberg (2006)
13. Dwork, C., Naor, M.: Zaps and their applications. SIAM J. Comput. 36(6), 1513–1543 (2007)
14. Feige, U., Shamir, A.: Witness indistinguishable and witness hiding protocols. In: Proceedings of the Annual Symposium on the Theory of Computing, STOC (1990)
15. Guang Zou, X., Sun, S.-H.: Analysis of anonymity on the signatures of knowledge. In: IIH-MSP, pp. 621–624. IEEE Computer Society, Los Alamitos (2006)
16. Katz, J.: Digital Signatures. Springer, Heidelberg (2010)
17. Lu, S., Ostrovsky, R., Sahai, A., Shacham, H., Waters, B.: Sequential aggregate signatures and multisignatures without random oracles. In: Vaudenay, S. (ed.) EUROCRYPT 2006. LNCS, vol. 4004, pp. 465–485. Springer, Heidelberg (2006)
18. Mykletun, E., Narasimha, M., Tsudik, G.: Signature bouquets: Immutability for aggregated/condensed signatures. In: Samarati, P., Ryan, P.Y.A., Gollmann, D., Molva, R. (eds.) ESORICS 2004. LNCS, vol. 3193, pp. 160–176. Springer, Heidelberg (2004)
19. Prafullchandra, H., Schaad, J.: Diffie-Hellman proof-of-possession algorithms. RFC 2875 (July 2000)
20. Ristenpart, T., Yilek, S.: The power of proofs-of-possession: Securing multiparty signatures against rogue-key attacks. In: Naor, M. (ed.) EUROCRYPT 2007. LNCS, vol. 4515, pp. 228–245. Springer, Heidelberg (2007)
21. Rivest, R.L., Shamir, A., Tauman, Y.: How to leak a secret. In: Boyd, C. (ed.) ASIACRYPT 2001. LNCS, vol. 2248, pp. 552–565. Springer, Heidelberg (2001)
22. Shahandashti, S.F., Safavi-Naini, R.: Construction of universal designated-verifier signatures and identity-based signatures from standard signatures. In: Cramer, R. (ed.) PKC 2008. LNCS, vol. 4939, pp. 121–140. Springer, Heidelberg (2008)
23. Waters, B.: Efficient identity-based encryption without random oracles. In: Cramer, R. (ed.) EUROCRYPT 2005. LNCS, vol. 3494, pp. 114–127. Springer, Heidelberg (2005)

LBlock: A Lightweight Block Cipher

Wenling Wu and Lei Zhang

State Key Laboratory of Information Security, Institute of Software,
Chinese Academy of Sciences, Beijing 100190, P.R. China
{wwl,zhanglei}@is.iscas.ac.cn

Abstract. In this paper, we propose a new lightweight block cipher called LBlock. Similar to many other lightweight block ciphers, the block size of LBlock is 64-bit and the key size is 80-bit. Our security evaluation shows that LBlock can achieve enough security margin against known attacks, such as differential cryptanalysis, linear cryptanalysis, impossible differential cryptanalysis and related-key attacks etc. Furthermore, LBlock can be implemented efficiently not only in hardware environments but also in software platforms such as 8-bit microcontroller. Our hardware implementation of LBlock requires about 1320 GE on 0.18 μm technology with a throughput of 200 Kbps at 100 KHz. The software implementation of LBlock on 8-bit microcontroller requires about 3955 clock cycles to encrypt a plaintext block.

Keywords: Block cipher, Lightweight, Hardware efficiency, Design, Cryptanalysis.

1 Introduction

With the development of electronic and communication applications, RFID technology has been used in many aspects of life, such as access control, parking management, identification, goods tracking etc. In this kind of new cryptography environment, the applications of RFID technology and sensor networking both have similar features, such as weak computation ability, small storage space, and strict power constraints. Therefore, traditional block ciphers such as AES are not suitable for this kind of extremely constrained environment. Hence, in recent years, research on lightweight ciphers has received a lot of attention. Compared with traditional block ciphers, lightweight ciphers have the following three main properties. Firstly, applications for constrained devices are unlikely to require the encryption of large amounts of data, and hence there is no requirement of high throughput for lightweight ciphers. Secondly, in this cryptography environment, attackers are lack of data and computing ability, which means lightweight ciphers only need to achieve moderate security. Lastly, lightweight ciphers are usually implemented in hardware environment, and small part of them are also implemented on software platforms such as 8-bit microcontroller. Therefore, hardware performance will be the primary consideration for lightweight ciphers. Hardware efficiency can be measured in many different ways: the length of the critical path,

J. Lopez and G. Tsudik (Eds.): ACNS 2011, LNCS 6715, pp. 327–344, 2011.

latency, clock cycles, power consumption, throughput, area requirements, and so on. Among them area requirement is the most important parameter, since small area requirement can minimize both the cost and the power consumption efficiently. Therefore, it has become common to use the term hardware efficient as a synonym for small area requirements, and the area requirements are usually measured as gate equivalents (GE). At present, for the hardware implementation of lightweight cipher, area requirements are usually dominated by the registers storing the data state and the key, since registers typically consist of flipflops which have a rather high area and power demand. For example, when using the standard cell library it requires between 6 and 12 GE to store a single bit [26]. Therefore, in the design of lightweight block ciphers, 64-bit block size and 80-bit key size are popular parameters.

While there is a growing requirement of ciphers suited for resource-constraint applications, a series of lightweight block ciphers have been proposed recently, e.g. PRESENT[9], HIGHT[14], mCrypton[21], DESL[19], CGEN[28], MIBS[15], KATAN & KTANTAN[10], TWIS[23], SEA[30] etc. All of these ciphers are designed and targeted specifically for extremely constrained environments such as RFID tags and sensor networks. Among them, PRESENT is supposed to be very competitive, since its hardware requirement is comparable with today's leading compact stream ciphers, and it is called an ultra-lightweight block cipher. Since its publication, only a few cryptanalytic results have been proposed against PRESENT, including the related-key rectangle attack on 17-round PRESENT in [24] and the side-channel attacks described in [27,35]. HIGHT has a 32-round generalized Feistel structure. Its main feature is the compact round function which contains no S-box and all the operations are simple computations such as XOR, rotation, and addition operating on 8-bit input. In respect of cryptanalysis, a related-key attack on full-round HIGHT was presented in ICISC2010, and an impossible differential attack on 26-round HIGHT were presented in [24]. mCrypton can be considered as a miniature of the block cipher Crypton[20], and a related-key rectangle attack on 8-round mCrypton has been reported in [25]. DESL and DESXL are lightweight modified versions of the well-known DES, and they adopt only one single S-box in order to minimize the hardware implementation. CGEN employs a compact round function called *mixtable* operation, and the main design strategies include using a fixed and per-device seed key which reduces the key scheduling and the decryption operation is not needed either. MIBS is a 32-round Feistel cipher, and its round function employs SP-network with XOR operations as diffusion layer, whose hardware requirements are more expensive than the bitwise permutation used in PRESENT etc. KATAN and KTANTAN are a family of lightweight block ciphers which contain six variants altogether. The KATAN family of ciphers all employ the same components, whose design strategy exploits some features of stream cipher [11]. Meet-in-the-middle attacks to the KTANTAN family with a key of 80 bits were presented in [36]. TWIS is inspired from the existing block cipher CLEFIA [29]. However, a differential distinguisher with probability 1 for full-round TWIS was presented in [31]. SEA is a Feistel cipher with scalable block and key sizes, and its round function only consists of rotation, XOR, and a single 3-bit S-box operations. TEA [33] and XTEA [34] are lightweight block ciphers proposed several years earlier.

In this paper we propose a new lightweight block cipher called LBlock. The design of its structure and components, such as S-box layer, P permutation layer etc, all represent the trade-off between security and performance. Our security analysis shows that full-round LBlock can provide enough security margin against known cryptanalytic techniques, such as differential cryptanalysis, linear cryptanalysis, impossible differential cryptanalysis, related-key attack etc. Furthermore, the performance evaluation of LBlock shows that not only hardware efficiency but also software implementations on 8-bit/32-bit platforms are ultra lightweight. The rest of this paper is organized as follows. Sect. 2 presents the specification of LBlock. Sect. 3 introduces the design rationale briefly. Sect. 4 and Sect. 5 describe the security analysis and performance evaluation of LBlock respectively. Finally, Sect. 6 concludes the paper.

2 Specification of LBlock

The block length of LBlock is 64-bit, and the key length is 80-bit. It employs a variant Feistel structure and consists of 32 rounds. The specification of LBlock consists of three parts: encryption algorithm, decryption algorithm and key scheduling.

2.1 Notations

In the specification of LBlock, we use the following notations:

- M: 64-bit plaintext
- C: 64-bit ciphertext
- K: 80-bit master key
- K_i: 32-bit round subkey
- F: Round function
- s: 4×4 S-box
- S: S-box layer consists of eight s in parallel
- P, P_1: Permutations operate on 32-bit
- $\bigoplus$: Bitwise exclusive-OR operation
- $<<< 8$: 8-bit left cyclic shift operation
- $>>> 8$: 8-bit right cyclic shift operation
- $||$: Concatenation of two binary strings
- $[i]_2$: Binary form of an integer i

2.2 Encryption Algorithm

The encryption algorithm of LBlock consists of a 32-round iterative structure which is a variant of Feistel network. The encryption procedure is illustrated in Fig. 1. Let $M = X_1||X_0$ denote a 64-bit plaintext, and then the data processing procedure can be expressed as follows.

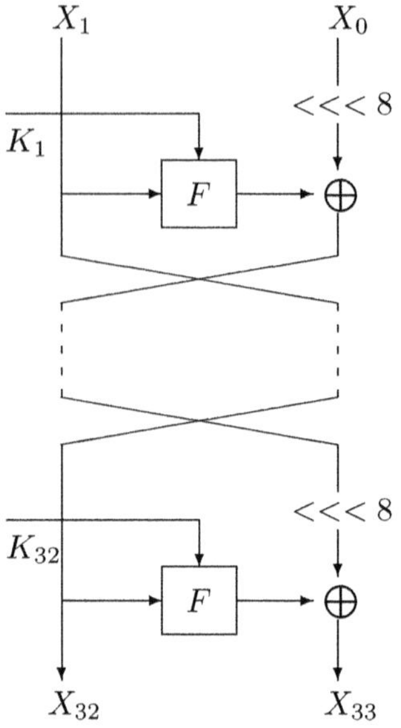

Fig. 1. Encryption procedure of LBlock

1. For $i = 2, 3, \ldots, 33$, do

$$X_i = F(X_{i-1}, K_{i-1}) \oplus (X_{i-2} <<< 8)$$

2. Output $C = X_{32}||X_{33}$ as the 64-bit ciphertext

Specifically, the components used in each round are defined as follows.

(1) Round function F

The round function F is defined as follows, where S and P denote the confusion and diffusion functions which will be defined later.

$$\begin{array}{rccc} F: & \{0,1\}^{32} \times \{0,1\}^{32} & \longrightarrow & \{0,1\}^{32} \\ & (X, K_i) & \longrightarrow & U = P(S(X \oplus K_i)) \end{array}$$

Fig. 2 illustrates the structure of round function F in detail.

(2) Confusion function S

Confusion function S denotes the non-linear layer of round function F, and it consists of eight 4-bit S-boxes s_i in parallel.

$$S: \{0,1\}^{32} \longrightarrow \{0,1\}^{32}$$
$$Y = Y_7||Y_6||Y_5||Y_4||Y_3||Y_2||Y_1||Y_0 \longrightarrow Z = Z_7||Z_6||Z_5||Z_4||Z_3||Z_2||Z_1||Z_0$$

$$\begin{array}{llll} Z_7 = s_7(Y_7), & Z_6 = s_6(Y_6), & Z_5 = s_5(Y_5), & Z_4 = s_4(Y_4), \\ Z_3 = s_3(Y_3), & Z_2 = s_2(Y_2), & Z_1 = s_1(Y_1), & Z_0 = s_0(Y_0). \end{array}$$

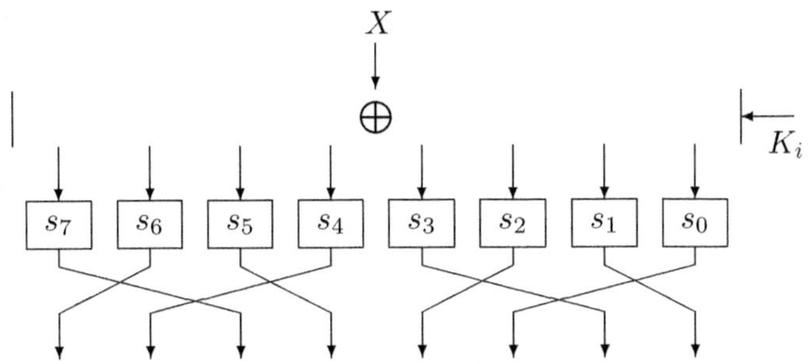

Fig. 2. Round function F

The contents of eight 4-bit S-boxes are listed in Table 1.

(3) Diffusion function P

Diffusion function P is defined as a permutation of eight 4-bit words, and it can be expressed as the following equations.

$$\begin{aligned} P : \{0,1\}^{32} &\longrightarrow \{0,1\}^{32} \\ Z = Z_7||Z_6||Z_5||Z_4||Z_3||Z_2||Z_1||Z_0 &\longrightarrow U = U_7||U_6||U_5||U_4||U_3||U_2||U_1||U_0 \end{aligned}$$

$$\begin{aligned} U_7 = Z_6,\ \ U_6 = Z_4,\ \ U_5 = Z_7,\ \ U_4 = Z_5, \\ U_3 = Z_2,\ \ U_2 = Z_0,\ \ U_1 = Z_3,\ \ U_0 = Z_1. \end{aligned}$$

2.3 Decryption Algorithm

The decryption algorithm of LBlock is the inverse of encryption procedure, and it consists of a 32-round variant Feistel structure too. Let $C = X_{32}||X_{33}$ denotes a 64-bit ciphertext, and then the decryption procedure can be expressed as follows.

1. For $j = 31, 30, \ldots, 0$, do

$$X_j = (F(X_{j+1}, K_{j+1}) \oplus X_{j+2}) >>> 8$$

2. Output $M = X_1||X_0$ as the 64-bit plaintext.

2.4 Key Scheduling

The 80-bit master key K is stored in a key register and denoted as $K = k_{79}\,k_{78}\,k_{77}\,k_{76} \ldots\ldots k_1 k_0$. Output the leftmost 32 bits of current content of register K as round subkey K_1, and then operate as follows:

1. For $i = 1, 2, \ldots, 31$, update the key register K as follows:
   (a) $K <<< 29$
   (b) $$\begin{aligned} [k_{79}\,k_{78}\,k_{77}\,k_{76}] &= s_9[k_{79}\,k_{78}\,k_{77}\,k_{76}] \\ [k_{75}\,k_{74}\,k_{73}\,k_{72}] &= s_8[k_{75}\,k_{74}\,k_{73}\,k_{72}] \end{aligned}$$
   (c) $[k_{50}k_{49}k_{48}k_{47}k_{46}] \oplus [i]_2$

(d) Output the leftmost 32 bits of current content of register K as round subkey K_{i+1}.

where s_8 and s_9 are two 4-bit S-boxes, and they are defined in Table 1.

Table 1. Contents of the S-boxes used in LBlock

s_0	14, 9, 15, 0, 13, 4, 10, 11, 1, 2, 8, 3, 7, 6, 12, 5
s_1	4, 11, 14, 9, 15, 13, 0, 10, 7, 12, 5, 6, 2, 8, 1, 3
s_2	1, 14, 7, 12, 15, 13, 0, 6, 11, 5, 9, 3, 2, 4, 8, 10
s_3	7, 6, 8, 11, 0, 15, 3, 14, 9, 10, 12, 13, 5, 2, 4, 1
s_4	14, 5, 15, 0, 7, 2, 12, 13, 1, 8, 4, 9, 11, 10, 6, 3
s_5	2, 13, 11, 12, 15, 14, 0, 9, 7, 10, 6, 3, 1, 8, 4, 5
s_6	11, 9, 4, 14, 0, 15, 10, 13, 6, 12, 5, 7, 3, 8, 1, 2
s_7	13, 10, 15, 0, 14, 4, 9, 11, 2, 1, 8, 3, 7, 5, 12, 6
s_8	14, 9, 15, 0, 13, 4, 10, 11, 1, 2, 8, 3, 7, 6, 12, 5
s_9	4, 11, 14, 9, 15, 13, 0, 10, 7, 12, 5, 6, 2, 8, 1, 3

3 Design Rationale

3.1 Structure

The structure of LBlock is a variant of Feistel network, and its design decisions contain a lot of considerations about security and efficient implementations (such as area, cost and performance etc.). In the aspect of implementation, the most important consideration is the area requirement when implemented in hardware. Therefore, we try to reduce the number of S-boxes used in each round and also minimize the size of each S-box used. Hence a Feistel-type structure seems a proper choice. Furthermore, for all kinds of generalized Feistel structures which operate less bits in each round, to achieve enough security margin they must take more rounds iteration which will affect its performance (such as speed and throughput). Therefore, in each round of LBlock, we choose only half of the data to go through round function F, and the other half applies a simple rotation operation. In the diffusion layer, we also choose to use permutation which can be implemented with no cost in hardware. However, instead of the bitwise permutation usually used, we apply a 4-bit word-wise permutation which can be implemented cheaply not only in hardware but also in software environments such as 8-bit microprocessor platforms. For example, the word-wise permutation in round function F can be combined with the S-box layer to form 8×8 table lookups. Moreover, we specifically choose the rotation offsets of right half in each round as 8 bits which can be omitted in 8-bit platform implementation. On the other hand, in the aspect of security requirement, we choose the word-wise permutation carefully so that the structure of LBlock satisfies that in both encryption and decryption directions it can achieve best diffusion [32] in 8 rounds. Furthermore, the number of differential and linear active S-boxes both increase quickly, and the following Table 2 lists the guaranteed number of active S-boxes before 20 rounds.

Table 2. Guaranteed number of active S-boxes of LBlock

Rounds	DS	LS	Rounds	DS	LS
1	0	0	**11**	22	22
2	1	1	**12**	24	24
3	2	2	**13**	27	27
4	3	3	**14**	30	30
5	4	5	**15**	32	32
6	6	6	**16**	35	35
7	8	8	**17**	36	36
8	11	11	**18**	39	39
9	14	14	**19**	41	41
10	18	18	**20**	44	44

3.2 Diffusion Layer

The diffusion permutation of LBlock consists of two parts, namely the word-wise permutation in round function which is denoted as P, and the rotation of right half data in each round which is denoted as P_1. Both of these permutations can be implemented by wiring in hardware which needs no additional area cost. For software environments such as 8-bit and 32-bit microprocessor platforms, P can be combined with the S-box layer in round function as table lookups and P_1 (8-bit rotation) can be implemented quite easily. Therefore, the diffusion permutations of LBlock can be implemented efficiently both in hardware and in software environments. Furthermore, the combination of P and P_1 can guarantee the best diffusion rounds and the least number of active S-boxes of LBlock. For example, there already exist at least 32 active S-boxes for 15-round LBlock.

3.3 S-Box Layer

On the pursuit of hardware efficiency, we use 4×4 S-boxes $s : F_2^4 \rightarrow F_2^4$ in LBlock. Compared with the regular 8×8 S-box, small S-box has much more advantage when implemented in hardware. For example, to implement the S-box of AES in hardware more than 200 GE are needed. On the other hand, for the 4×4 S-boxes used in LBlock, all of them can be implemented in hardware with only about 22 GE. Furthermore, in the aspect of security, the S-boxes used in LBlock are carefully chosen so that they all fulfill the following conditions: no fix point, completed, best non linearity, best differential probability, and good algebraic order etc.

3.4 Key Scheduling

Similar to many other lightweight block ciphers, the key scheduling of LBlock is also designed in a stream cipher way. We only apply simple rotation and non-linear operations to generate the round subkeys. First of all, the operation of 29-bit left rotation can be implemented freely in hardware, and it can also break the 4-bit word structure, which helps to improve the security of LBlock against related-key attacks. Secondly, we choose to use two 4×4 S-boxes as the non-linear operation which represents a trade-off between security and performance. Lastly, the exact values of rotation offset, constants and positions of constant addition are carefully chosen, so as to avoid weak relations between round subkeys.

4 Security Evaluation

4.1 Differential Cryptanalysis

For differential cryptanalysis, we adopt an approach to count the number of active S-boxes of differential characteristics. This is a regular method to evaluate the security against differential attack, which were adopted by many other block ciphers, such as AES [12], Camellia [1] and CLEFIA [29] etc. We found the guaranteed number of differential active S-boxes of LBlock by computer program, and the results before 20-round are listed in Table 2. Considering that there are at least 32 active S-boxes for 15-round LBlock and the best differential probabilities of s_i are all equal to 2^{-2}, then the maximum probability of differential characteristics for 15-round LBlock satisfies $DCP_{\max}^{15r} \leq 2^{32\times(-2)} = 2^{-64}$. This means there is no useful 15-round differential characteristic for LBlock, since the block length of LBlock is only 64-bit. Therefore, we believe that the full 32-round LBlock is secure against differential cryptanalysis.

4.2 Linear Cryptanalysis

We also apply the method of counting active S-boxes for the evaluation of LBlock against linear cryptanalysis. Since there are at least 32 active S-boxes for 15-round LBlock and the best linear bias of each s_i is 2^{-2}, the maximum bias of linear approximations for 15-round LBlock satisfies $LCP_{\max}^{15r} \leq 2^{32-1} \cdot 2^{32\times(-2)} = 2^{-33}$. Therefore, according to the complexity estimation of linear cryptanalysis, we can conclude that it is difficult to find useful 15-round linear-hulls which can be used to distinguish LBlock from a random permutation. As a result, we believe that the full 32-round LBlock has enough security margin against linear cryptanalysis.

4.3 Impossible Differential Cryptanalysis

Impossible differential attack [3] is one of the most powerful cryptanalytic techniques, and its applications to many block ciphers (such as Camellia and CLEFIA etc.) represent the best cryptanalytic results obtained so far. We search for the impossible differential characteristic of LBlock using the algorithm proposed by Kim et al. [16]. The best distinguisher found is the following 14-round impossible differential characteristic:

$$(00000000, 00\alpha00000) \overset{14r}{\nrightarrow} (0\beta000000, 00000000), \qquad (1)$$

where $\alpha, \beta \in \{0,1\}^4 \backslash \{0\}$ represent non-zero differences. Note that by changing the positions of α, β, we can construct other 14-round impossible differential characteristics in a similar way.

Based on the 14-round impossible differential distinguishers, we can mount a key recovery attack on 20-round LBlock. The attack procedure can be described as follows.

1. Choose a set of 2^{12} plaintexts to construct a structure, where the 4-bit words $X_{0,1}$, $X_{0,3}$ and $X_{1,2}$ take all possible values and all the other words take constants. Then each structure can generate about 2^{23} plaintext pairs satisfying the input difference $(\Delta X_1, \Delta X_0) = (00000*00, 0000*0*0)$. Choose 2^{51} different structures which can generate about 2^{74} candidate plaintext pairs.
2. For each corresponding ciphertext structure after 20-round encryption, choose the pairs satisfying the output difference $(\Delta X_{21}, \Delta X_{20}) = (**00**0*, 000*0**0)$, where $*$ denotes non-zero difference. After this test, there remains about $2^{74} \times 2^{-32} = 2^{42}$ candidate pairs.
3. For every guess of 28-bit subkey $K_{20,0}, K_{20,1}, K_{20,2}, K_{20,4}, K_{20,5}, K_{20,6}, K_{20,7}$, partially decrypt Round 20 to check if the pairs satisfying $(\Delta X_{20}, \Delta X_{19}) = (000*0**0, 00*0000*)$. After this test, there remains about $2^{42} \times 2^{-12} = 2^{30}$ pairs.
4. For every guess of the 16-bit subkey $K_{19,0}, K_{19,2}, K_{19,3}, K_{19,5}$, partially decrypt Round 19 to check if the pairs satisfying $(\Delta X_{19}, \Delta X_{18}) = (00*0000*, *0000000)$. After this test, there remains $2^{30} \times 2^{-8} = 2^{22}$ pairs.
5. For every guess of the 8-bit subkey $K_{18,1}, K_{18,7}$, partially decrypt Round 18 to check if the candidate pairs satisfying $(\Delta X_{18}, \Delta X_{17}) = (*0000000, 0*000000)$. After this test, there remains about $2^{22} \times 2^{-4} = 2^{18}$ pairs.
6. For every guess of the 4-bit subkey $K_{17,6}$, partially decrypt Round 17 to check if the candidate pairs satisfying $(\Delta X_{17}, \Delta X_{16}) = (0*000000, 00000000)$. After this test, there remains about $2^{18} \times 2^{-4} = 2^{14}$ pairs.
7. For every guess of the 8-bit subkey $K_{1,2}$, $K_{1,7}$, partially encrypt Round 1 to check if the candidate pairs satisfying $(\Delta X_2, \Delta X_1) = (00*00000, 00000*00)$. After this test, there remains about $2^{14} \times 2^{-4} = 2^{10}$ pairs.
8. For every guess of the 4-bit subkey $K_{2,5}$, partially encrypt Round 2 to check if the candidate pairs satisfying the following equation:

$$(\Delta X_3, \Delta X_2) = (00000000, 00*00000).$$

9. If there still remains a pair satisfying the impossible differential, then the 68-bit subkey guessed must be wrong. Delete it from the candidate subkey table. If the table of candidate subkey is not empty after analyzing all the remaining pairs, output the subkey remained in table as correct subkey.

For each of the candidate pair in Step 8, the probability that it satisfies the filtering condition is about 2^{-4}. Therefore, for a wrong subkey guess, the probability of its remaining after Step 8 is about $(1 - 2^{-4})^{2^{10}} \approx 2^{-95}$. Then we can expect that after all these filtering, there remains about $2^{68} \times 2^{-95} \approx 2^{-27}$ wrong subkey guess, and only the correct subkey will be output.

The data and time complexities of above attack can be estimated as follows. First of all, we choose 2^{51} structures and the data complexity is $2^{51} \times 2^{12} = 2^{63}$ chosen plaintexts. The time complexity is dominated by Step 7 to Step 8, and each step needs about 2^{78} S-box operations. Therefore, the time complexity of the attack is about $2 \times 2 \times 2^{78} \times \frac{1}{8} \times \frac{1}{20} \approx 2^{72.7}$ 20-round encryptions. According to the complexities of impossible differential attack on 20-round LBlock, we expect that the full 32-round LBlock has enough security margin against this attack.

4.4 Integral Attack

Since LBlock is a 4-bit word oriented cipher, we also consider that integral attack [18] may be one of the most powerful attacks against LBlock. The best integral characteristic found is the 15-round distinguisher. Table 3 illustrates one of the 15-round integral distinguisher in detail, where C denotes a constant word, A denotes an active word and B denotes a balanced word respectively. Note that by changing the position of C in plaintext, we can obtain similar integral distinguishers easily.

Based on the 15-round integral distinguisher, we can mount a key recovery attack up to 20-round LBlock. For simplicity, we first give the integral attack on 18-round LBlock, and the attack procedure is as follows.

1. Choose a set of 2^{60} plaintexts to construct a structure, where only 4-bit word takes a constant and all the other words take all the possible values of $\{0,1\}^{60}$. Obtain the corresponding ciphertext after 18-round encryption. Count the number of value $X_{18,6}, X_{18,4}, X_{18,1}, X_{19,6}, X_{19,0}$ occurs, and discard the values which occur even times.
2. Guess corresponding subkeys to decrypt the ciphertexts.
   (a) For every guess of the 8-bit subkey $(K_{18,1}, K_{18,4})$, partially decrypt Round 18 to compute $X_{17,4} = s_4(X_{18,4} \oplus K_{18,4}) \oplus X_{19,6}$ and $X_{17,6} = s_1(X_{18,1} \oplus K_{18,1}) \oplus X_{19,0}$.

Table 3. 15-Round integral distinguisher of LBlock

Rounds	Integral characterisitcs			
0	AAAC	AAAA	AAAA	AAAA
1	AAAC	ACAC	AAAC	AAAA
2	CCCC	AAAC	AAAC	ACAC
3	ACAC	CCCC	CCCC	AAAC
4	CCCC	ACCC	ACAC	CCCC
5	ACCC	CCCC	CCCC	ACCC
6	CCCC	CCCC	ACCC	CCCC
7	CCCC	CCAC	CCCC	CCCC
8	CCCC	CCCA	CCCC	CCAC
9	CCCC	AACC	CCCC	CCCA
10	CCCC	AAAC	CCCC	AACC
11	CCAA	ACAA	CCCC	AAAC
12	CAAB	AAAA	CCAA	ACAA
13	B?AA	BBAA	CAAB	AAAA
14	?B?B	?B?B	B?AA	BBAA
15	????	????	?B?B	?B?B

(b) For every guess of the 4-bit subkey $K_{17,4}$, partially decrypt Round 17 to compute $X_{16,4} = s_4(X_{17,4} \oplus K_{17,4}) \oplus X_{18,6}$.
(c) For every guess of the 4-bit subkey $K_{16,4}$, partially decrypt Round 16 to compute $X_{15,4} = s_4(X_{16,4} \oplus K_{16,4}) \oplus X_{17,6}$.

3. Check if the equation $\bigoplus_l X_{15,4} = 0$ is satisfied, where l is the number of plaintexts. If the equation is satisfied, then $X_{15,4}$ is a balance word. Otherwise, guess another subkey and repeat until we get the correct subkey.

The complexity of this attack can be estimated as follows. Step 1 needs about 2^{60} plaintexts which requires 2^{60} encryptions. For the five words counted in Step 1, there are at most 2^{20} values. Therefore, the time complexity of Step 1 to Step 3 are less than $2^{20} \times 2^{16}$ encryptions. For a wrong subkey guess, the probability that equation $\bigoplus_l X_{15,4} = 0$ is satisfied is about 2^{-4}. Therefore, to discard all the wrong 16-bit subkey guesses, we need about five plaintext structures. Therefore, the total data and time complexities of this attack are both 5×2^{60}.

Moreover, we can mount an integral attack on 20-round LBlock based on the 15-round integral distinguisher. The attack procedure is similar with the attack on 18-round LBlock, and we add two additional rounds in the end. Therefore, 12 subkey words need to be guessed and the data and time complexities will increase to about $13 \times 2^{60} \approx 2^{63.7}$.

4.5 Related-Key Attacks

Recently, the combination of related-key [2,17] and traditional cryptanalysis has become one of the most powerful attacks, and its application to some ciphers has improved the cryptanalytic results significantly [4,6,7,8,13]. Therefore, we have studied the possible related-key differential characteristic of LBlock so as to evaluate the security of LBlock against related-key attacks. In order to get related-key differential characteristic with high probability, we have to control the number of active S-boxes. Therefore, we first choose the output differences of 10 S-boxes (8 S-boxes in round function and 2 S-boxes in key scheduling) in Round i all have hamming weight less than 2. Then we search for the related-key differential before Round i in the decryption direction and after Round i in the encryption direction respectively, and count the total number of active S-boxes. The best related-key differential obtained so far is a 13-round distinguisher with 26 active S-boxes, and its probability is $(2^{-2})^{25} \cdot (2^{-3}) = 2^{-53}$. For the 14-round related-key differential obtained, there are 32 active S-boxes and its probability is less than $(2^{-2})^{31} \cdot (2^{-3}) = 2^{-65}$. Table 4 illustrates the propagation of 14-round related-key differential of LBlock in detail.

5 Performance Evaluation

5.1 Hardware Performance

We implemented LBlock in VHDL and synthesized it on $0.18\mu m$ CMOS technology to check for its hardware complexity. Figure 3 in Appendix III shows

Table 4. 14-Round related-key differential characteristic of LBlock

Rounds	ΔX_L	ΔRK	ΔI_S	ΔO_P	ΔX_R
1	01200101	00000000	01200101	20012100	01222121
2	02200001	00000000	02200001	20010100	01200101
3	00000001	02000000	02000001	20000100	02200001
4	00000002	00000000	00000002	00000100	00000001
5	00000000	00000008	00000008	00000200	00000002
6	00000000	00000000	00000000	00000000	00000000
7	00000000	00000000	00000000	00000000	00000000
8	00000000	00000400	00000400	00001000	00000000
9	00001000	00000000	00001000	00000010	00000000
10	00000010	00000000	00000010	00000002	00001000
11	00100002	00020000	00120002	01010100	00000010
12	01011100	00000000	01011100	21002010	00100002
13	31002210	00000000	31002210	20102012	01011100
14	21012013	04000000	25012013	41200212	31002210

Table 5. Comparison of lightweight block cipher implementations

Algorithm	Block Size	Key Size	Area #GE	Speed kbps@100KHz	Logic Process
XTEA	64	128	3490	57.1	0.13 μm
HIGHT	64	128	3048	188.2	0.25 μm
mCrypton	64	128	2500	492.3	0.13 μm
DES	64	56	2300	44.4	0.18 μm
DESXL	64	184	2168	44.4	0.18 μm
KATAN	64	80	1054	25.1	0.13 μm
KTANTAN	64	80	688	25.1	0.13 μm
PRESENT	64	80	1570	200	0.18 μm
LBlock	64	80	1320	200	0.18 μm

the datapath of an parallelization implementation of LBlock, which performs one round in one clock cycle. In this optimized implementation, we use a 64-bit width datapath and implement the eight S-boxes of round function in parallel. Then, to encrypt 64-bit plaintext with an 80-bit key occupies about 1320 GE and requires 32 clock cycles. Table 5 compares the hardware performances of LBlock with other lightweight block ciphers.

Specifically, in the above implementation the area requirement is occupied by flip-flops for storing the key and the data state. To store the 80-bit key requires about 480 GE and to store the 64-bit data state requires two 32-bit registers (denoted as *memleft* and *memright*) which are about 384 GE. For round function F, it is consisted of the following three parts. The KeyAddition is a 32-bit XOR operation which requires about 87 GE. The S-box layer consists of eight 4×4 S-boxes in parallel, which requires about $21.84 \times 8 = 174.8$ GE. The diffusion layer P can be implemented by simple wiring and costs no area. Then in the end of each round, another 32-bit XOR operation of two halves is needed which requires about 87 GE. Furthermore, another two 4×4 S-boxes and a 5-bit XOR operation are needed in key scheduling which require at most $43.7 + 13.5 \approx 57.2$ GE. Moreover, control logic and other counters require about 50 GE. Therefore, the hardware implementation of LBlock requires an estimated area of 1320 GE.

We can give a more compact implementation of LBlock with a serialization design. For example, in the key scheduling we can reuse the 32-bit register and generate each subkey by several operations. Then the area requirement of key register can be reduced to 212 GE, while additional RAM is needed. Furthermore, the data state in encryption can also reuse the 32-bit key register and the area requirements can be reduced to 192 GE. Then the control logic and other counters need about 70 GE. Therefore, this area-optimized implementation of LBlock only needs about 866.3 GE with additional RAM. Since the register is reused in both key scheduling and encryption, the generation of each round subkey will need 12 clock cycles, and the encryption procedure will need 192 clock cycles. Therefore, to encrypt 64-bit plaintext with 80-bit key needs about 576 clock cycles in total. Table 6 in Appendix II summarizes the area requirement of LBlock in detail.

5.2 Software Implementations

For some resource-constraint environments, such as smart card and sensor networking system, the embedded CPU is usually 8-bit oriented. Therefore, in the design of LBlock, we consider the implementation performance of LBlock not only in hardware environment but also in software platform such as 8-bit microcontroller. The choices of 4-bit word permutation in round function and 8-bit rotation in right half of each round are suitable for both hardware and software platforms. For example, in case of 8-bit oriented software implementation, the eight S-boxes and 4-bit word permutation P in round function can be combined together and realized as four 8-bit lookup tables. Our software implementation of LBlock on 8-bit microcontroller only requires about 3955 clock cycles to encrypt a plaintext block. Hence, LBlock can achieve competitive hardware and software performances compared with other known lightweight block ciphers.

6 Conclusion

In this paper we propose a new lightweight block cipher LBlock, whose block size is 64-bit and key size is 80-bit. Our design goal is to provide cryptography security for resource-constraint environments, e.g. RFID tags and sensor networks etc. Moreover, compared with other lightweight block ciphers, the proposal should achieve better hardware performance and also have good software efficiency on 8-bit microcontroller. Therefore, in the design of LBlock, we employ a variant Feistel structure and the encryption algorithm is 4-bit oriented which can be implemented efficiently in both hardware and software. Furthermore, the round function employs a SP-network, whose confusion layer consists of small 4×4 S-boxes and diffusion layer consists of a simple 4-bit word permutation. All of these components are designed with the consideration of both security and implementation efficiency in mind. Our hardware implementation of LBlock requires about 1320 GE on 0.18 μm technology, which satisfies the regular limitation of 2000 GE in RFID applications. Furthermore, in an area-optimized implementation, LBlock requires only 866.3 GE with additional RAM. We also evaluate the security of LBlock and our cryptanalytic results show that LBlock achieves enough security margin against known attacks. In the end, we strongly encourage the security analysis of LBlock and helpful comments.

Acknowledgments. This work is supported by the National Natural Science Foundation of China (No.60873259), and the Knowledge Innovation Project of The Chinese Academy of Sciences. Moreover, the authors are very grateful to the anonymous referees for their comments and editorial suggestions.

References

1. Aoki, K., Ichikawa, T., Kanda, M., Matsui, M., Moriai, S., Nakajima, J., Tokita, T.: *Camellia*: A 128-Bit Block Cipher Suitable for Multiple Platforms - Design and Analysis. In: Stinson, D.R., Tavares, S. (eds.) SAC 2000. LNCS, vol. 2012, pp. 39–56. Springer, Heidelberg (2001)
2. Biham, E.: New Types of Cryptanalytic Attacks Using Related Keys. Journal of Cryptology 7(4), 229–246 (1994)
3. Biham, E., Biryukov, A., Shamir, A.: Cryptanalysis of Skipjack Reduced to 31 Rounds Using Impossible Differentials. In: Cachin, C., Camenisch, J.L. (eds.) EUROCRYPT 2004. LNCS, vol. 3027, pp. 12–23. Springer, Heidelberg (2004)
4. Biham, E., Dunkelman, O., Keller, N.: A Related-Key Rectangle Attack on the Full KASUMI. In: Roy, B. (ed.) ASIACRYPT 2005. LNCS, vol. 3788, pp. 443–461. Springer, Heidelberg (2005)
5. Biham, E., Shamir, A.: Differential Cryptanalysis of the Data Encryption Standard. Springer, Berlin (1993)
6. Biryukov, A., Khovratovich, D.: Related-Key Cryptanalysis of the Full AES-192 and AES-256. In: Matsui, M. (ed.) ASIACRYPT 2009. LNCS, vol. 5912, pp. 1–18. Springer, Heidelberg (2009)
7. Biryukov, A., Khovratovich, D., Nikolić, I.: Distinguisher and related-key attack on the full AES-256. In: Halevi, S. (ed.) CRYPTO 2009. LNCS, vol. 5677, pp. 231–249. Springer, Heidelberg (2009)
8. Biryukov, A., Nikolić, I.: Automatic Search for Related-Key Differential Characteristics in Byte-Oriented Block Ciphers: Application to AES, Camellia, Khazad and Others. In: Gilbert, H. (ed.) EUROCRYPT 2010. LNCS, vol. 6110, pp. 322–344. Springer, Heidelberg (2010)
9. Bogdanov, A.A., Knudsen, L.R., Leander, G., Paar, C., Poschmann, A., Robshaw, M.J.B., Seurin, Y., Vikkelsoe, C.: PRESENT: An Ultra-Lightweight Block Cipher. In: Paillier, P., Verbauwhede, I. (eds.) CHES 2007. LNCS, vol. 4727, pp. 450–466. Springer, Heidelberg (2007)
10. De Cannière, C., Dunkelman, O., Knežević, M.: KATAN and KTANTAN — A Family of Small and Efficient Hardware-Oriented Block Ciphers. In: Clavier, C., Gaj, K. (eds.) CHES 2009. LNCS, vol. 5747, pp. 272–288. Springer, Heidelberg (2009)
11. De Canniere, C., Preneel, B.: Trivium Specifications. eSTREAM submission, `http://www.ecrypt.eu.org/stream/triviump3.html`

12. Daemen, J., Rijmen, V.: The Design of Rijndael. Springer, Berlin (2002)
13. Dunkelman, O., Keller, N., Shamir, A.: A Practical-Time Attack on the A5/3 Cryptosystem Used in Third Generation GSM Telephony. Faculty of Mathematics and Computer Science Weizmann Institute of Science P.O. Box 26, Rehovot 76100, Israel (2010)
14. Hong, D., Sung, J., Hong, S., Lim, J., Lee, S., Koo, B., Lee, C., Chang, D., Lee, J., Jeong, K., Kim, H., Kim, J., Chee, S.: HIGHT: A New Block Cipher Suitable for Low-Resource Device. In: Goubin, L., Matsui, M. (eds.) CHES 2006. LNCS, vol. 4249, pp. 46–59. Springer, Heidelberg (2006)
15. Izadi, M., Sadeghiyan, B., Sadeghian, S., Khanooki, H.: MIBS: A New Lightweight Block Cipher. In: Garay, J.A., Miyaji, A., Otsuka, A. (eds.) CANS 2009. LNCS, vol. 5888, pp. 334–348. Springer, Heidelberg (2009)
16. Kim, J.-S., Hong, S.H., Sung, J., Lee, S.-J., Lim, J.-I., Sung, S.H.: Impossible differential cryptanalysis for block cipher structures. In: Johansson, T., Maitra, S. (eds.) INDOCRYPT 2003. LNCS, vol. 2904, pp. 82–96. Springer, Heidelberg (2003)
17. Knudsen, L.R.: Cryptanalysis of LOKI91. In: Zheng, Y., Seberry, J. (eds.) AUSCRYPT 1992. LNCS, vol. 718, pp. 196–208. Springer, Heidelberg (1993)
18. Knudsen, L., Wagner, D.: Integral Cryptanalysis. In: Daemen, J., Rijmen, V. (eds.) FSE 2002. LNCS, vol. 2365, pp. 112–127. Springer, Heidelberg (2002)
19. Leander, G., Paar, C., Poschmann, A., Schramm, K.: New Lightweight DES Variants. In: Biryukov, A. (ed.) FSE 2007. LNCS, vol. 4593, pp. 196–210. Springer, Heidelberg (2007)
20. Lim, C.H.: A Revised Version of CRYPTON - CRYPTON V1.0 -. In: Knudsen, L.R. (ed.) FSE 1999. LNCS, vol. 1636, pp. 31–45. Springer, Heidelberg (1999)
21. Lim, C.H., Korkishko, T.: mCrypton – A Lightweight Block Cipher for Security of Low-Cost RFID Tags and Sensors. In: Song, J.-S., Kwon, T., Yung, M. (eds.) WISA 2005. LNCS, vol. 3786, pp. 243–258. Springer, Heidelberg (2006)
22. Matsui, M.: Linear Cryptanalysis Method for DES Cipher. In: Helleseth, T. (ed.) EUROCRYPT 1993. LNCS, vol. 765, pp. 386–397. Springer, Heidelberg (1994)
23. Ojha, S.K., Kumar, N., Jain, K., Sangeeta: TWIS – A Lightweight Block Cipher. In: Prakash, A., Sen Gupta, I. (eds.) ICISS 2009. LNCS, vol. 5905, pp. 280–291. Springer, Heidelberg (2009)
24. Özen, O., Varıcı, K., Tezcan, C., Kocair, Ç.: Lightweight Block Ciphers Revisited: Cryptanalysis of Reduced Round PRESENT and HIGHT. In: Boyd, C., González Nieto, J. (eds.) ACISP 2009. LNCS, vol. 5594, pp. 90–107. Springer, Heidelberg (2009)
25. Park, J.: Security Analysis of mCrypton Proper to Low-cost Ubiquitous Computing Devices and Applications. International Journal of Communication Systems 22(8), 959–969 (2009)
26. Parr, C., Poschmann, A., Robshaw, M.J.B.: New Designs in Lightweight Symmetric Encryption. In: Kitsos, P., Zhang, Y. (eds.) RFID Security: Techniques, Protocols and System-on-Chip Design, pp. 349–371. Springer, Heidelberg (2008)
27. Renauld, M., Standaert, F.-X.: Algebraic Side Channel Attacks. Cryptology ePrint Archive, report 2009/179, http://eprint.iacr.org/2009/279
28. Robshaw, M.J.B.: Searching for Compact Algorithms: CGEN. In: Nguyên, P.Q. (ed.) VIETCRYPT 2006. LNCS, vol. 4341, pp. 37–49. Springer, Heidelberg (2006)
29. Shirai, T., Shibutani, K., Akishita, T., Moriai, S., Iwata, T.: The 128-bit Blockcipher CLEFIA (Extended Abstract). In: Biryukov, A. (ed.) FSE 2007. LNCS, vol. 4593, pp. 181–195. Springer, Heidelberg (2007)

30. Standaert, F.-X., Piret, G., Gershenfeld, N., Quisquater, J.-J.: SEA: A Scalable Encryption Algorithm for Small Embedded Applications. In: Domingo-Ferrer, J., Posegga, J., Schreckling, D. (eds.) CARDIS 2006. LNCS, vol. 3928, pp. 222–236. Springer, Heidelberg (2006)
31. Su, B., Wu, W., Zhang, L., Li, Y.: Full-Round Differential Attack on TWIS Block Cipher. In: Chung, Y., Yung, M. (eds.) WISA 2010. LNCS, vol. 6513, pp. 234–242. Springer, Heidelberg (2011)
32. Suzaki, T., Minematsu, K.: Improving the Generalized Feistel. In: Hong, S., Iwata, T. (eds.) FSE 2010. LNCS, vol. 6147, pp. 19–39. Springer, Heidelberg (2010)
33. Wheeler, D., Needham, R.: TEA, a Tiny Encryption Algorithm. In: Preneel, B. (ed.) FSE 1994. LNCS, vol. 1008, pp. 363–366. Springer, Heidelberg (1995)
34. Wheeler, D., Needham, R.: TEA Extensions (October 1997) (Also Correction to XTEA. October 1998), `www.ftp.cl.cam.ac.uk/ftp/users/djw3/`
35. Yang, L., Wang, M., Qiao, S.: Side Channel Cube Attack on PRESENT. In: Garay, J.A., Miyaji, A., Otsuka, A. (eds.) CANS 2009. LNCS, vol. 5888, pp. 379–391. Springer, Heidelberg (2009)
36. Bogdanov, A., Rechberger, C.: Generalized Meet-in-the-Middle Attacks: Cryptanalysis of the Lightweight Block Cipher KTANTAN. In: Biryukov, A., Gong, G., Stinson, D.R. (eds.) SAC 2010. LNCS, vol. 6544, pp. 228–238. Springer, Heidelberg (2011)

Appendix I: Test Vectors

Test vectors for LBlock are shown in hexadecimal notation as follows.

Plaintext	Key	Ciphertext
00 00 00 00 00 00 00 00	00 00 00 00 00 00 00 00 00 00	c2 18 18 53 08 e7 5b cd
01 23 45 67 89 ab cd ef	01 23 45 67 89 ab cd ef fe dc	4b 71 79 d8 eb ee 0c 26

Appendix II

Table 6. Area requirement of LBlock

Module	Speed Optimized	Area Optimized
64-bit Data Register	384	192
Key Addition	87	87
S-box Layer	174.8	174.8
P Layer	0	0
32-bit XOR	87	87
80-bit Key Register	480	212
S-boxes (Key Scheule)	43.7	30
5-bit Constant XOR	13.5	13.5
Control Logic	50	70
Sum	1320 GE	866.3 GE (with RAM)

Appendix III

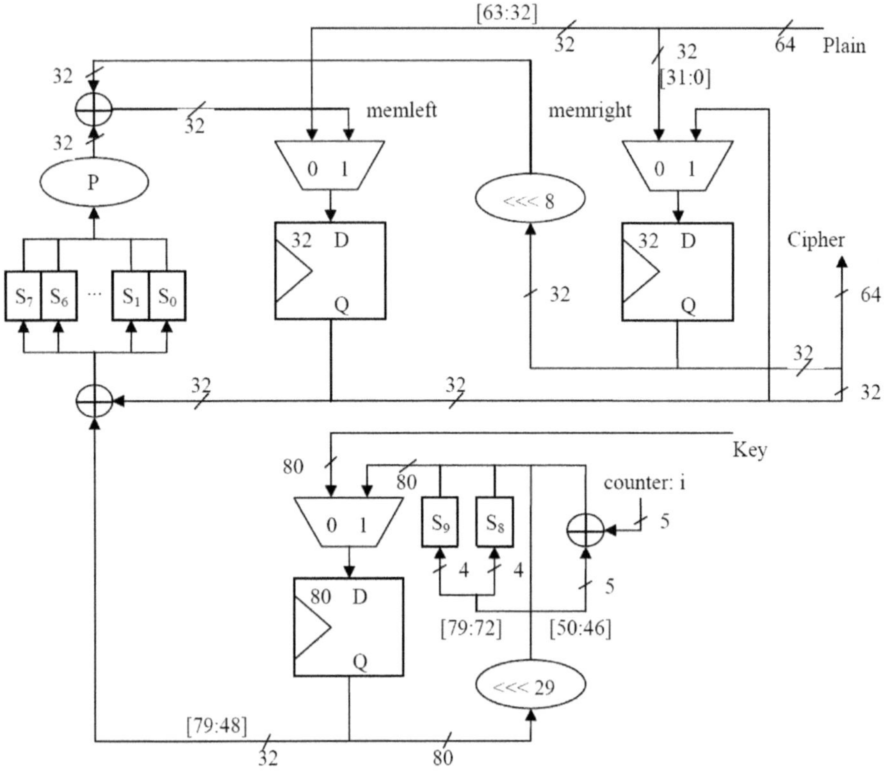

Fig. 3. The datapath of an area-optimized version of LBlock

On Hiding a Plaintext Length by Preencryption

Cihangir Tezcan and Serge Vaudenay

EPFL
CH-1015 Lausanne, Switzerland
http://lasecwww.epfl.ch

Abstract. It is a well known fact that encryption schemes cannot hide a plaintext length when it is unbounded. We thus admit that an approximation of it may leak and we focus on hiding its precise value. Some standards such as TLS or SSH offer to do it by applying some pad-then-encrypt techniques. In this study, we investigate the information leakage when these techniques are used. We define the notion of padding scheme and its associated security. We show that when a padding length is uniformly distributed, the scheme is nearly optimal. We also show that the insecurity degrades linearly with the padding length.

1 Introduction

Although an encryption process makes a plaintext unreadable to adversaries, the resulting ciphertext may still leak some information. Practically, we can always distinguish an encrypted SMS message from an encrypted HD video stream. Namely, the length of a plaintext may give some information away and it can often be deduced from the ciphertext. For instance, the lengths of a plaintext and the corresponding ciphertext are identical or differ by a small number of bits when the encryption is done by a stream or a block cipher. One way of hiding the plaintext size is to use *random padding* before the encryption which appends a padding of random length in $\{1,2,\ldots,B\}$. In this work, we investigate the information leakage when a random padding is used.

Let us consider a symmetric encryption system in which encryption under a key K is denoted by Enc_K and decryption is denoted by Dec_K. In the Shannon model [10], the plaintext and the key are defined by independent random variables X and K. Perfect secrecy is defined by the statistical independence of X and $Y = \mathsf{Enc}_K(X)$. If this property is satisfied, we can easily see that the plaintext domain must be finite: if y is a possible value for Y, then $p = \Pr[Y = y]$ is positive. For any possible value x for X, we must have $\Pr[\mathsf{Enc}_K(x) = y] = p$ due to perfect secrecy. Since $\mathsf{Enc}_K(x) = y$ implies $\mathsf{Dec}_K(y) = x$, $p \leq \Pr[\mathsf{Dec}_K(y) = x]$. By summing over all possible x's, we deduce that the number of such x is bounded by $\frac{1}{p}$, which is a finite number.[1]

[1] Actually, this proof only holds for countable sets. More generally, we should define our properties with non-discrete probabilities to be able to consider uncountably infinite sets. In theory, we *could* achieve perfect secrecy over uncountably infinite sets. However, the encryption algorithm will no longer be polynomially bounded on classical computational models. So, we only consider countable sets in the present paper. We could probably reopen this case when considering encryption over a space of quantum states. (See [9] for more discussions.)

J. Lopez and G. Tsudik (Eds.): ACNS 2011, LNCS 6715, pp. 345–358, 2011.

This impossibility result extends to weaker security notions. In [3,4], Chor and Kushilevitz consider α-weak security, given $\alpha \geq 1$, which states that for all possible x_1, x_2, y, we have

$$\frac{1}{\alpha}\Pr[Y = y|X = x_2] \leq \Pr[Y = y|X = x_1] \leq \alpha \Pr[Y = y|X = x_2]$$

Perfect secrecy corresponds to the $\alpha = 1$ case. Encryption over a countably infinite domain cannot be α-weak secure for any α: if x_2 and y are possible simultaneous values for X and Y, $p = \frac{1}{\alpha}\Pr[Y = y|X = x_2]$ is positive and we have $\Pr[Y = y|X = x_1] \geq p$ for all possible x_1. So, $\Pr[\mathsf{Dec}_K(y) = x_1] \geq p$ and the number of possible plaintexts is bounded by $\frac{1}{p}$.

In [9], Phan and Vaudenay consider ε-statistically extended *indistinguishability under one-time encryption* (extended IND-OTE game), given $\varepsilon < 1$, which means

$$\frac{1}{2}\sum_y |\Pr[Y = y|X = x_1] - \Pr[Y = y|X = x_2]| \leq \varepsilon$$

for all possible plaintexts x_1 and x_2. Again, secure (in this sense) encryption over an infinite (countable) domain is impossible.

A public-key cryptosystem is nothing but an encryption scheme in which Enc_K can be described by using public values. So, the above impossibility results also apply to public-key cryptography.

A standard security notion for encryption is the IND-CPA security (*indistinguishability under chosen plaintext attacks*) in which an adversary can make some chosen plaintext encryptions and tries to get an advantage for distinguishing the encryption of either x_1 or x_2, two plaintexts *of same length* selected by himself. For public-key encryption, the adversary makes the encryption himself by using the public key so IND-CPA and IND-OTE notions are equivalent. For symmetric encryption, he must be provided access to an encryption oracle. In the IND-OTE game, there is no such access so there may be a gap between IND-CPA and IND-OTE notions. Still, these notions impose that x_1 and x_2 have the same length so they offer no guarantee about keeping the plaintext length secret. We call *extended* IND-OTE game (E-IND-OTE) the notion where the restriction that x_1 and x_2 have the same length is relaxed.

The Phan-Vaudenay result says that if all adversaries in the E-IND-OTE game have their advantage bounded by a given $\varepsilon < 1$, then the encryption domain must be finite. Ideally, we would like to design secure encryption schemes over an infinite set. Practically, we could live with encryption domains which are finite but large enough. Indeed, we can assume that the set of bitstrings of length bounded by a few petabytes is a *virtually infinite* set. So, we could design an encryption scheme over this domain with a pretty good security. However, for efficiency reasons, we would not like that the encryption of a very small plaintext (say a few kilobytes) would lead to a ciphertext of one petabyte. Therefore, we should consider encryption schemes which are *somehow* length-preserving but also length hiding. To make it possible, we relax the E-IND-OTE security notion and consider the Δ-IND-OTE game in which the submitted plaintexts have a length difference bounded by Δ. The IND-OTE (resp. E-IND-OTE) notions correspond to $\Delta = 0$ (resp. $\Delta = +\infty$).

In the sequel we consider encryption schemes defined by

$$\mathsf{Enc}(x) = \mathsf{Enc}_0(x \| \mathsf{pad}(x))$$

where Enc_0 is a length-preserving IND-OTE-secure scheme and pad is a probabilistic padding scheme. That is, pad generates a postfix-free random string which can be extracted after decryption. Typically, $\mathsf{pad}(x)$ is a random bitstring whose length N is a random variable. This kind of construction is proposed e.g. in TLS [6] or SSH [12]. For instance, here is a quote from [6]:

> Padding that is added to force the length of the plaintext to be an integral multiple of the block cipher's block length. The padding MAY be any length up to 255 bytes, as long as it results in the TLSCiphertext.length being an integral multiple of the block length. Lengths longer than necessary might be desirable to frustrate attacks on a protocol that are based on analysis of the lengths of exchanged messages.

This suggests that we could arbitrarily pad up to $B = 32$ (resp. $B = 16$) blocks of data to hide the exact length of a plaintext, when the block cipher uses blocks of 64 bits (resp. 128 bits).

More generally, we consider preencryption schemes which are not necessarily of form $x \| \mathsf{pad}(x)$. We may consider several assumptions:

- (uniformity) the distribution of the length overhead between x and $\mathsf{Enc}(x)$ is fixed (it does not depend on x)
- (almost length-preserving property) the length overhead is bounded by B

Given B and Δ, our aim is to find the best distribution N to achieve optimal Δ-IND-OTE security.

Related work. Padding often serves another purpose. Namely, it is used to fill incomplete blocks to encrypt a plaintext using a block cipher. Our notion of preencryption scheme is similar to the notion of *encoding* by Paterson and Watson [8] who consider several practical schemes. They analyze the security of the pad-then-encrypt scheme in a practical case where the original encryption scheme is a block cipher in CTR mode. This follows some other work in which they identified a terrible interaction between the padding scheme and the decryption algorithm in CBC mode [1]. Some other padding schemes leading to decryption attacks have been identified (see e.g. [2,5,7,11]).

Our results. We first formalize in Section 2 the notion of preencryption scheme and its associated Δ-IND security notion. We formalize the notion of preencryption by padding (or the pad-then-encrypt technique). When Enc_0 is length-preserving, we show that Δ-IND-security is necessary and sufficient to make Enc Δ-IND-OTE secure.

Then, we show in Section 3 that there is always an adversary with advantage nearly $\frac{\Delta}{B}$. That is, the insecurity degrades linearly with the padding length B. This main result happens to have a very simple proof by using a diagonal argument.

We observe that a padding scheme making padding lengths uniformly distributed makes the above adversary nearly the best one. So, this preencryption scheme is nearly optimal.

In Section 4, we further precisely study the optimal padding scheme in the uniform case for $\Delta = 2$.

2 Preliminaries

In what follows we consider an alphabet Z. This can be a Boolean alphabet, or the set of bytes, or a set of blocks. We denote by Z^* the set of finite sequences of elements taken from Z. The length of an element $x \in Z^*$ is denoted by $|x|$. For $x, x' \in Z^*$, we denote by $x \| x'$ the concatenation of x and x'.

In this paper we adopt exact security notions. We can easily translate to asymptotic security by introducing security parameters in the definition of encryption schemes.

2.1 Encryption Scheme

Definition 1. *An* encryption scheme *is defined by*

- *a plaintext domain which is a subset of* Z^*
- *an algorithm to generate a key* K
- *a (probabilistic) encryption algorithm* Enc *taking a key and a plaintext as input and producing a ciphertext*
- *a (deterministic) decryption algorithm* Dec *taking a key and a ciphertext as input and producing a plaintext*

The correctness property of an encryption scheme states that if we generate a key K by the key generation algorithm, if we take a plaintext x in the plaintext domain, and if we compute $\mathsf{Dec}_K(\mathsf{Enc}_K(x))$ then we obtain x with probability 1.

*We say that an encryption scheme is B-*almost length preserving *if*

$$||\mathsf{Enc}_K(x)| - |x|| \leq B$$

with probability 1 for all x. It is length-preserving if it is 0-almost length preserving.

*We say that an encryption scheme t-*fully leaks the plaintext length *if there exists an algorithm f such that for all x in the plaintext domain, $f(\mathsf{Enc}_K(x)) = |x|$ with probability 1 within a complexity at most t.*

For instance, a length-preserving encryption scheme fully leaks the plaintext length by $f(y) = |y|$.

For symmetric encryption, the key generation algorithm simply picks a key in a given key space, following the uniform distribution. For public-key encryption, the key can be split in a public part and a private part. The encryption algorithm only use the public part. What follows applies to both cases.

We define the Δ-IND-OTE security notion as follows:

Definition 2. *We consider the following game between an adversary $\mathcal{A}$ and a challenger. Firstly, the challenger generates a key K using the key generation algorithm. In the case of a public key cryptosystem, it reveals the public part K_p of K to the* *adversary. The adversary can do some computations and then submits two plaintexts* $\mathcal{A}(K_p;\rho) = (x_0, x_1)$ *in the plaintext domain such that*

$$||x_0| - |x_1|| \leq \Delta$$

by using some random coins ρ*. The challenger flips a fair coin* b*, computes* $Y = \mathsf{Enc}_K(x_b)$ *and reveals* Y*. The adversary can then do some extra computations and yields a guess* $\mathcal{A}(K_p, Y;\rho) = b'$*. The adversary succeeds if* $b = b'$*. His advantage is* $\Pr[b = b'] - \frac{1}{2}$*.* $\mathcal{A}$ *has a complexity bounded by* t *if for any* K_p*,* Y*, and* ρ*, the total running time of* $\mathcal{A}(K_p;\rho)$ *and* $\mathcal{A}(K_p, Y;\rho)$ *is bounded by* t*. We say that the encryption scheme is* Δ-IND-OTE(t,ε)*-secure if for all adversary with time complexity limited by* t*, the advantage is at most* ε*.*

Δ-IND-OTE ***Game:***
1: *Challenger generates* K *and discloses its public part* K_p
2: *Adversary selects plaintexts* x_0 *and* x_1 *where* $||x_0| - |x_1|| \leq \Delta$
3: *Challenger flips a coin* b*, computes* $\mathsf{Enc}_K(x_b) = Y$ *and gives* Y *to the adversary*
4: *Adversary guesses* b' *and wins if* $b' = b$

IND-OTE security corresponds to the $\Delta = 0$ case. We also consider E-IND-OTE security defined by the $\Delta = +\infty$ case.

2.2 Preencryption Schemes

Definition 3. *Given two plaintext domains* $\mathcal{X}$ *and* $\mathcal{X}^0$*, a preencryption scheme from* $\mathcal{X}$ *to* $\mathcal{X}^0$ *is a pair of algorithms*

- *a (probabilistic) algorithm* pre *such that for all* $x \in \mathcal{X}$*,* $\mathsf{pre}(x) \in \mathcal{X}^0$ *with probability 1*
- *a (deterministic) algorithm* Extract

The correctness property of a preencryption scheme states that for all $x \in \mathcal{X}$*,*

$$\mathsf{Extract}(\mathsf{pre}(x)) = x$$

with probability 1.

We say that a preencryption scheme is B-almost length preserving *if*

$$||\mathsf{pre}(x)| - |x|| \leq B$$

with probability 1 for all x*. We say that a preencryption scheme is* length-increasing *if* $|\mathsf{pre}(x)| \geq |x|$ *with probability 1 for all* x*. We say that a preencryption scheme is* strictly length-increasing *if* $|\mathsf{pre}(x)| > |x|$ *with probability 1 for all* x*.*

Definition 4. *A preencryption scheme is* Δ-IND (t,ε)*-secure if for all adversary* $\mathcal{A}$ *with time complexity limited by* t*, the advantage in the following game is at most* ε*. The advantage is defined as* $\Pr[b = b'] - \frac{1}{2}$*.*

Δ-IND ***Game:***

1: Adversary selects plaintexts x_0 and x_1 where $||x_0| - |x_1|| \leq \Delta$
2: Challenger flips a coin b, computes $|\mathsf{pre}(x_b)| = L$ and gives L to the adversary
3: Adversary guesses b' and wins if $b' = b$
$\mathcal{A}$ has a complexity bounded by t if for any L and ρ, the total running time of $\mathcal{A}(;\rho)$ and $\mathcal{A}(L;\rho)$ is bounded by t.

Given a set of integers A, x_0 and x_1, we define a Δ-IND adversary $D_A(x_0,x_1)$ as the one selecting x_0 and x_1 then yielding $b' = 1$ if and only if $L \in A$. We define $\mathsf{Adv}_A(x_0,x_1)$ as the advantage of this adversary.

Lemma 5. *For any x_0 and x_1 we have*

$$\begin{aligned}\mathsf{Adv}_A(x_0,x_1) &= \frac{1}{2}\Pr[|\mathsf{pre}(x_1)| \in A] - \frac{1}{2}\Pr[|\mathsf{pre}(x_0)| \in A] \\ &= \frac{1}{2}\sum_{\ell \in A}(\Pr[|\mathsf{pre}(x_1)| = \ell] - \Pr[|\mathsf{pre}(x_0)| = \ell])\end{aligned}$$

Proof. We have

$$\begin{aligned}\mathsf{Adv}_A(x_0,x_1) &= \Pr[b = b'] - \frac{1}{2} \\ &= \frac{1}{2}\Pr[b' = 1|b = 1] + \frac{1}{2}\Pr[b' = 0|b = 0] - \frac{1}{2} \\ &= \frac{1}{2}\Pr[b' = 1|b = 1] - \frac{1}{2}\Pr[b' = 1|b = 0] \\ &= \frac{1}{2}\Pr[|\mathsf{pre}(x_1)| \in A] - \frac{1}{2}\Pr[|\mathsf{pre}(x_0)| \in A]\end{aligned}$$

and the other expression follows by separating the $\ell \in A$ cases. □

We define $\mathsf{Adv}(x_0,x_1)$ as the maximal advantage for (computationally unbounded) adversaries selecting x_0 and x_1.

Lemma 6. *For any x_0 and x_1 we have* $\mathsf{Adv}(x_0,x_1) = \mathsf{Adv}_A(x_0,x_1)$ *where*

$$A = \{\ell; \Pr[|\mathsf{pre}(x_1)| = \ell] > \Pr[|\mathsf{pre}(x_0)| = \ell]\}$$

Actually, $\mathsf{Adv}(x_0,x_1)$ is the statistical distance between the length of $\mathsf{pre}(x_0)$ and the length of $\mathsf{pre}(x_1)$.

Proof. Since we consider unbounded adversaries, an optimal one using x_0 and x_1 can be assumed to be of form $D_{A'}(x_0,x_1)$ without loss of generality. By Lemma 5 we clearly have $\mathsf{Adv}_{A'}(x_0,x_1) \leq \mathsf{Adv}_A(x_0,x_1)$. So, $A' = A$ maximizes $\mathsf{Adv}_{A'}(x_0,x_1)$ and we obtain $\mathsf{Adv}(x_0,x_1) = \mathsf{Adv}_A(x_0,x_1)$. □

Given an encryption scheme

$$C^0 = (\mathcal{X}^0, \mathsf{Gen}^0, \mathsf{Enc}^0, \mathsf{Dec}^0)$$

and a preencryption scheme $P = (\mathsf{pre}, \mathsf{Extract})$ from $\mathcal{X}$ to $\mathcal{X}^0$ we define the encryption scheme

$$C = (\mathcal{X}, \mathsf{Gen}, \mathsf{Enc}, \mathsf{Dec})$$

by $\mathsf{Gen} = \mathsf{Gen}^0$,

$$\mathsf{Enc}_K(x) = \mathsf{Enc}^0_K(\mathsf{pre}(x))$$

and

$$\mathsf{Dec}_K(y) = \mathsf{Extract}(\mathsf{Dec}^0_K(y))$$

Clearly, this defines an encryption scheme. If the preencryption scheme is B-almost length-preserving and the encryption scheme C^0 is length-preserving, then the encryption scheme C is B-almost length preserving.

Theorem 7. *We assume there exists a constant t_S and a sampling algorithm $S(1^L)$ to pick a random element of $\mathcal{X}^0$ of length L with complexity at most t_S for any $L \in \{|x|; x \in \mathcal{X}^0\}$. There exists a (small) constant c such that for any C, C_0, t, t_P, if C^0 is a $\mathsf{IND\text{-}OTE}(t+t_S+t_P+c,\varepsilon^0)$-secure encryption scheme and if P is a $\Delta\text{-}\mathsf{IND}(t+t_S+t_P+c,\varepsilon^1)$-secure preencryption scheme and* pre *can be computed within a complexity bounded by t_P, then C is a $\Delta\text{-}\mathsf{IND\text{-}OTE}(t, 2\varepsilon^0+\varepsilon^1)$-secure encryption scheme.*

When C^0 t_0-fully leaks the plaintext length, if C is $\Delta\text{-}\mathsf{IND\text{-}OTE}(t+t_0+c,\varepsilon)$-secure then P is $\Delta\text{-}\mathsf{IND}(t,\varepsilon)$-secure.

So, for an IND-OTE-secure encryption C^0 which fully leaks the plaintext length, the Δ-IND security of P is necessary and sufficient to have C Δ-IND-OTE-secure.

Proof. Let $\mathcal{A}$ be a Δ-IND-OTE adversary for C which has a time complexity bounded by t. We want to prove that its advantage is less than $2\varepsilon^0+\varepsilon^1$.

We define the following adversary $\mathcal{A}'$.

1: receive (public) key material
2: simulate $\mathcal{A}$ to get x_0 and x_1
3: flip a fair coin b
4: compute $x'_0 = \mathsf{pre}(x_b)$
5: pick a random $x'_1 = S(1^{|x'_0|})$ in $\mathcal{X}^0$
6: submit x'_0 and x'_1 and receive Y
7: continue the simulation of $\mathcal{A}$ with Y to get b'
8: output 1 if $b = b'$ and 0 otherwise

The complexity of this adversary is bounded by $t+t_S+t_P+c$ where c is the small overhead complexity beside the simulation of $\mathcal{A}$, the sampling of S, and the computation of $\mathsf{pre}(x_b)$.

Let Γ be the experiment corresponding to the IND-OTE game of $\mathcal{A}'$ against C^0 when x'_0 is selected by the challenger. So, Γ yields 1 if and only if $\mathcal{A}$ yields $b' = b$ on input $Y = \mathsf{Enc}(\mathsf{pre}(x_0))$. That is, the advantage of $\mathcal{A}$ is $\Pr[\Gamma \to 1] - \frac{1}{2}$. Therefore, to bound the advantage of $\mathcal{A}$, we just need to prove that $\Pr[\Gamma \to 1] \leq \frac{1}{2} + 2\varepsilon^0 + \varepsilon^1$.

Let Γ' be the experiment corresponding to the IND-OTE game of $\mathcal{A}'$ against C^0 when x'_1 is selected by the challenger. $\mathcal{A}'$ is an IND-OTE adversary for C^0 with advantage $\frac{1}{2}(\Pr[\Gamma' \to 1] - \Pr[\Gamma \to 1])$. Due to the IND-OTE-security of C^0, we have $|\Pr[\Gamma \to 1] - \Pr[\Gamma' \to 1]| \leq 2\varepsilon^0$.

Clearly, Γ' is equivalent to the following:

1: generate a key
2: simulate $\mathcal{A}$ to get x_0 and x_1

3: flip a fair coin b
4: compute $L = |\mathsf{pre}(x_b)|$
5: pick $X = S(1^L)$
6: compute $Y = \mathsf{Enc}(X)$
7: continue the simulation of $\mathcal{A}$ with Y to get b'
8: output 1 if $b = b'$ and 0 otherwise

This defines a Δ-IND adversary for P. So, $\Pr[\Gamma' \to 1] \leq \frac{1}{2} + \varepsilon^1$.

We deduce that $\Pr[\Gamma \to 1] \leq \frac{1}{2} + 2\varepsilon^0 + \varepsilon^1$.

For the second part of the theorem, we now let $\mathcal{A}$ be a Δ-IND adversary for P of complexity bounded by t and we want to bound its advantage. Since C^0 fully leaks the plaintext length, there is a function f to compute the plaintext length from the ciphertext. We define the following adversary:

1: get key material from a challenger
2: simulate $\mathcal{A}$ to get x_0 and x_1
3: submit x_0 and x_1 to the challenger and get ciphertext Y
4: compute $L = f(Y)$
5: continue the simulation of $\mathcal{A}$ with L to get b'
6: yield b'

Clearly, this is a Δ-IND-OTE adversary for C whose advantage is exactly the advantage of $\mathcal{A}$. Assume that its complexity is bounded by $t + t_0 + c$. Since C is Δ-IND-OTE $(t + t_0 + c, \varepsilon)$-secure, this advantage is bounded by ε. □

2.3 Pad-then-Encrypt Scheme

Definition 8. *A C subset of Z^* is postfix-free if*

$$\forall s \in Z^* \quad \forall x, y \in C \quad s \| x = y \Longrightarrow x = y$$

We observe that if the empty string belongs to C then no other string is in C. Furthermore, there exists a function Extract such that for all $s \in \mathcal{X}$ and for all $x \in \mathcal{X}$, we have

$$\mathsf{Extract}(s \| \mathsf{pad}(x)) = s$$

with probability 1. In what follows we consider a postfix-free set such that this function can be efficiently implemented.

Definition 9. *Given $\mathcal{X}^0 \subseteq Z^*$ and a postfix-free set C, a C-padding scheme on $\mathcal{X}^0$ is a probabilistic algorithm taking an element x of $\mathcal{X}^0$ as input and producing an element* $\mathsf{pad}(x)$ *of C as an output. We say that the padding scheme is* uniform *if the distribution of* $\mathsf{pad}(x)$ *does not depend on x.*

A padding scheme defines the preencryption scheme

$$\mathsf{pre}(x) = x \| \mathsf{pad}(x)$$

We note that preencryption schemes made out from a padding scheme are all length-increasing. Except in the constant 0-padding case, they are even *strictly* length increasing.

Example 10. We consider the padding scheme defined by the parameter B as follows: given x, we simply pick a sequence $100\cdots0$ of length N which is uniformly distributed in $\{1,\ldots,B\}$. This padding scheme is B-almost length preserving, strictly length-increasing, and uniform. By Lemma 5 and 6, we obtain that $\mathsf{Adv}(x_0,x_1) = \frac{||x_1|-|x_0||}{B}$. So, this preencryption scheme is Δ-$\mathsf{IND}\left(t,\frac{\Delta}{B}\right)$-secure for all Δ and any t.

In what follows we show that this scheme is nearly optimal.

To make a pad-then-encrypt construction secure with Δ large, we shall find a secure padding scheme for this Δ. A trivial solution consists of making sure that $x\|\mathsf{pad}(x)$ has a constant length no matter the plaintext x. To make it possible, this length must be at least the maximal length of a plaintext. This solution is clearly impractical. We shall rather concentrate on Δ small. So, we do not fully hide the length of plaintexts but rather their exact value.

3 Maximal Security of the Pad-then-Encrypt Scheme

In this section we consider lower bounds for the best advantage of an adversary against a preencryption scheme. We consider the case where the plaintext space is large and dense enough so that we can make sequences of plaintexts such that the length of two consecutive ones differ by Δ.

Definition 11. *We say that a sequence $(x_0,\ldots,x_n)$ of Z^* elements is a* Δ-chain *if for every $i=0,\ldots,n-1$, we have $|x_{i+1}|-|x_i|=\Delta$. We say that this sequence* represents *a length ℓ if $|x_0|\leq\ell\leq|x_n|$. We say that a subset X of Z^* is* Δ-dense *if for any $x,y\in X$, there exists a Δ-chain in X which represents $|x|$ and $|y|$. We say that X is* B-large *if there exists $x,y\in X$ such that $|x|-|y|\geq B$.*

Theorem 12. *Let P be a B-almost length-preserving preencryption scheme and Δ be an integer. We assume that the input domain of P is Δ-dense and $(2B+\Delta)$-large. Then, there exists an adversary in the Δ-IND game with advantage at least $1/\left(\left\lfloor\frac{2B}{\Delta}\right\rfloor+1\right)$.*

If P is length-increasing and B-almost length-preserving over a domain which is Δ-dense and $(B+\Delta)$-large, then there exists an adversary with advantage at least $1/\left(\left\lfloor\frac{B}{\Delta}\right\rfloor+1\right)$.

Proof. Let $n=\left\lfloor\frac{cB}{\Delta}\right\rfloor+1$ with $c=1$ for length-increasing preencryption schemes and $c=2$ otherwise. Since the domain is $(cB+\Delta)$-large and Δ-dense, we can construct a Δ-chain of $n+1$ elements $x_0,x_1,\ldots,x_n$. We have $|x_{i+1}|=|x_i|+\Delta$ for $i=0,1,\ldots,n-1$. So, $|x_i|=|x_0|+i\Delta$ for $i=0,1,\ldots,n$. Let

$$s_i=\Pr[|\mathsf{pre}(x_i)|\leq B+|x_0|]$$

which is the probability that the preencrypted version of x_i has an overhead length bounded by $B+|x_0|-|x_i|=B-i\Delta$. Clearly, $s_0=1$ since P is B-almost length preserving, and $s_n=0$ since $B-n\Delta<(1-c)B$.

So, $\sum_{i=0}^{n-1}(s_i-s_{i+1})=1$. Hence, there must exist some i such that $s_i-s_{i+1}\geq\frac{1}{n}$. Let A be the set of all integers up to $B+|x_0|$. We have $\Pr[|\mathsf{pre}(x_i)|\in A]=s_i$. We deduce that $\mathsf{Adv}_A(x_i,x_{i+1})\geq\frac{1}{n}$: there is an adversary with an advantage larger than $\frac{1}{n}$. □

Remark 13. Example 10 shows a simple B-almost length-preserving scheme which is Δ-IND $(t, \frac{\Delta}{B})$-secure. So, the optimal security which is achievable is between $\frac{\Delta}{B}$ and $\frac{1}{\lceil \frac{B}{\Delta} \rceil}$. In particular, when Δ divides B, the scheme in Example 10 is optimal.

Theorem 12 can be generalized to preencryption schemes which are unbounded, but with finite expected overhead length. In practice, we would like to have a guarantee that a preencryption overhead is not too long on average, so this is a pretty reasonable assumption.

Theorem 14. *Let P be a length-increasing preencryption scheme and Δ be an integer. We assume that the input domain of P is Δ-dense and $(2B)$-large. We assume that for all x, we have $|E(|\mathsf{pre}(x)|) - |x|| \leq B$. There exists an adversary in the Δ-IND game with advantage at least $1/\left(2\left\lceil \frac{2B}{\Delta} \right\rceil\right)$.*

Proof. We apply the same proof method as in Theorem 12. We define $n = \left\lceil \frac{\alpha B}{\Delta} \right\rceil$ and

$$s_i = \Pr[|\mathsf{pre}(x_i)| < \alpha B + |x_0|] = \Pr[|\mathsf{pre}(x_i)| - |x_i| < \alpha B - i\Delta]$$

for α such as the scheme is (αB)-large. We have $s_0 \geq 1 - \frac{1}{\alpha}$ since $E(|\mathsf{pre}(x_0)|) - |x_0| \leq B$ and $s_n = 0$ since the scheme is length-increasing. So, there is some i leading us to $\mathsf{Adv}_A(x_i, x_{i+1}) \geq \frac{1}{n}\left(1 - \frac{1}{\alpha}\right)$. We can just take $\alpha = 2$ and conclude. □

4 Uniform Padding Schemes

In this section, we consider a uniform padding scheme. We let N be a random variable following the distribution of $|\mathsf{pad}(x)|$. We assume that $\Pr[N = 0] = 0$: the padding scheme is strictly length-increasing. Since the scheme is uniform, the distribution does not depend on x. In notations, we further replace plaintexts x_0 and x_1 by their lengths a and b where $b \geq a$ without loss of generality.

Lemma 15. *We have $\Pr[N \leq b - a] \leq \mathsf{Adv}(a,b)$ and equality holds if and only if $\Pr[N = x + b - a] \leq \Pr[N = x]$ for all $x > 0$.*

Proof. Let $\varepsilon = \mathsf{Adv}(a,b)$. Due to Lemma 5 and 6, we have

$$\begin{aligned}
\varepsilon &= \sum_{\ell:\Pr[N=\ell-a]\geq\Pr[N=\ell-b]} (\Pr[N = \ell - a] - \Pr[N = \ell - b]) \\
&\geq \sum_{\ell:\ell\leq b} \Pr[N = \ell - a] \\
&= \Pr[N \leq b - a]
\end{aligned}$$

and equality holds if and only if $\Pr[N = x + b - a] \leq \Pr[N = x]$ for all $x > 0$. □

Theorem 16. *Consider a uniform strictly length-increasing padding scheme with the above notations. We assume that it is B-almost length-preserving. If $b - a = \Delta$ and B is divisible by Δ, then $\mathsf{Adv}(a,b) \geq \frac{\Delta}{B}$ and equality holds if and only if $\Pr[N \leq b - a] = \frac{\Delta}{B}$ and $\Pr[N = i]$ is periodic over $[1, \ldots, B]$ with period Δ.*

Proof. Let $\varepsilon = \mathsf{Adv}(a,b)$.

Case 1: Assume $\Pr[N \leq \Delta] > \frac{\Delta}{B}$. Due to Lemma 15, we have $\varepsilon \geq \Pr[N \leq \Delta] > \frac{\Delta}{B}$.

Case 2: Assume $\Pr[N \leq \Delta] = \frac{\Delta}{B}$. If there exists an integer $j > a+\Delta$ with $\Pr[N = j-a] > \Pr[N = j-b]$, then $A = \{a+1, a+2, \ldots, a+\Delta, j\}$ makes

$$\varepsilon \geq \mathsf{Adv}_A(a,b) = \Pr[N \leq \Delta] + \Pr[N = j-a] - \Pr[N = j-b] > \frac{\Delta}{B}$$

If no such j exists, then we have $\Pr[N = x+\Delta] \leq \Pr[N = x]$ for all $x > 0$. By Lemma 15, we obtain $\varepsilon = \frac{\Delta}{B}$. Furthermore, we get $\Pr[j\Delta < N \leq (j+1)\Delta] \leq \frac{\Delta}{B}$ for all $j \geq 0$. Therefore, we have

$$1 = \sum_{i=1}^{B} \Pr[N = i] \leq \left\lceil \frac{B}{\Delta} \right\rceil \cdot \frac{\Delta}{B}$$

Since B is divisible by Δ, this inequality is in fact an equality. Thus, we cannot have $\Pr[N = x+\Delta] < \Pr[N = x]$ for any x. Hence, $\Pr[N = x+\Delta] = \Pr[N = x]$ for all $x \in [1, B-\Delta]$, and $\Pr[N = i]$ becomes periodic over $[1, \ldots, B]$ with period Δ.

Case 3: Assume $\Pr[N \leq \Delta] < \frac{\Delta}{B}$. Then $\frac{\Delta}{B} - \Pr[N \leq \Delta] = \delta$ for some $\delta > 0$. Since

$$\sum_{j=0}^{\lceil \frac{B}{\Delta} \rceil - 1} \Pr[j\Delta < N \leq (j+1)\Delta] = 1$$

$\Pr[0 < N \leq \Delta] = \frac{\Delta}{B} - \delta$, and Δ divides B, there must exist an integer $j > 0$ such that $\Pr[j\Delta < N \leq (j+1)\Delta] > \frac{\Delta}{B}$. Thus, if we set $A = \{a+1, a+2, \ldots, a+(j+1)\Delta\}$, we obtain

$$\varepsilon \geq \mathsf{Adv}_A(a,b) = \Pr[N \leq (j+1)\Delta] - \Pr[N \leq j\Delta] > \frac{\Delta}{B}$$

Thus in all cases $\varepsilon \geq \frac{\Delta}{B}$ and equality holds if and only if $\Pr[N \leq \Delta] = \frac{\Delta}{B}$ and $\Pr[N = i]$ is periodic over $[1, \ldots, B]$ with period Δ. □

The following example shows that when B is not divisible by Δ, then ε can be less than $\frac{\Delta}{B}$.

Example 17. Let $b - a = \Delta = 2$ and $B = 5$. We define N as follows:

$$\Pr[N = 1] = \Pr[N = 3] = \Pr[N = 5] = 0.22 \qquad \Pr[N = 2] = \Pr[N = 4] = 0.17$$

Thus, the best advantage with a length difference of $\Delta = 2$ is $\varepsilon_2 = \Pr[N = 1] + \Pr[N = 2] = 0.39$ which is less than $\frac{2}{5}$. However, for $\Delta = 1$, the best advantage is $\varepsilon_1 = \Pr[N = 1] + \Pr[N = 3] + \Pr[N = 5] - \Pr[N = 2] - \Pr[N = 4] = 0.32$.

For $\Delta = 1$, B is divisible by Δ so Example 10 gives an optimal padding scheme. For $\Delta = 2$ and B even, it is the same. For $\Delta = 2$ and B odd, the optimal case is characterized as follows.

Table 1. Results of the Theorem 12 and 18 when $\Delta = 2$ and B is odd

B	Upper bound (Ex. 10)	Best Achievable (Th. 18)	Lower Bound (Th. 12)
3	0.666666666666667	0.6	0.5
5	0.4	0.384615384615385	0.333333333333333
7	0.285714285714286	0.28	0.25
9	0.222222222222222	0.219512195121951	0.2
11	0.181818181818182	0.180327868852459	0.166666666666667
13	0.153846153846154	0.152941176470588	0.142857142857143
15	0.133333333333333	0.132743362831858	0.125
17	0.117647058823529	0.117241379310345	0.111111111111111
19	0.105263157894737	0.104972375690608	0.1
21	0.0952380952380952	0.0950226244343891	0.0909090909090909
23	0.0869565217391304	0.0867924528301887	0.0833333333333333
25	0.08	0.0798722044728434	0.0769230769230769
27	0.0740740740740741	0.073972602739726	0.0714285714285714
29	0.0689655172413793	0.0688836104513064	0.0666666666666667
31	0.0645161290322581	0.0644490644490645	0.0625
33	0.0606060606060606	0.0605504587155963	0.0588235294117647
35	0.0571428571428571	0.0570962479608483	0.0555555555555556
37	0.0540540540540541	0.054014598540146	0.0526315789473684
39	0.0512820512820513	0.0512483574244415	0.05
41	0.0487804878048781	0.0487514863258026	0.0476190476190476
43	0.0465116279069767	0.0464864864864865	0.0454545454545455
45	0.0444444444444444	0.0444225074037512	0.0434782608695652
47	0.0425531914893617	0.0425339366515837	0.0416666666666667
49	0.0408163265306122	0.0407993338884263	0.04
51	0.0392156862745098	0.0392006149116065	0.0384615384615385
53	0.0377358490566038	0.0377224199288256	0.037037037037037
55	0.0363636363636364	0.0363516192994052	0.0357142857142857
57	0.0350877192982456	0.0350769230769231	0.0344827586206897
59	0.0338983050847458	0.0338885697874785	0.0333333333333333
61	0.0327868852459016	0.0327780763030629	0.032258064516129
63	0.0317460317460317	0.0317380352644836	0.03125
65	0.0307692307692308	0.0307619498343587	0.0303030303030303
67	0.0298507462686567	0.0298440979955457	0.0294117647058824
69	0.0289855072463768	0.0289794204115918	0.0285714285714286
71	0.028169014084507	0.028163427211424	0.0277777777777778
73	0.0273972602739726	0.0273921200750469	0.027027027027027
75	0.0266666666666667	0.0266619267685745	0.0263157894736842
77	0.025974025974026	0.0259696458684654	0.0256410256410256
79	0.0253164556962025	0.025312399871836	0.025

Theorem 18. *Consider a uniform strictly length-increasing padding scheme with the above notations. We assume that it is B-almost length-preserving. If B is odd, then*

$$\max_{b-a\leq 2} \mathsf{Adv}(a,b) \geq \frac{2B}{B^2+1}$$

and an equality can be reached by a distribution taking alternate values on every length.

Proof. We first note that $\frac{2B}{B^2+1} < \frac{2}{B}$ so we must find a better distribution than the uniform one from Example 10. We further note that $\frac{1}{\lfloor \frac{B}{2} \rfloor + 1} = \frac{2}{B+1} \leq \frac{2B}{B^2+1}$ so the bound to be proven is consistent with the one from Theorem 12. Let

$$\varepsilon = \max_{b-a\leq 2} \mathsf{Adv}(a,b)$$
$$\varepsilon_1 = \max_{b-a=1} \mathsf{Adv}(a,b)$$
$$\varepsilon_2 = \max_{b-a=2} \mathsf{Adv}(a,b)$$

We have $\varepsilon = \max(\varepsilon_1, \varepsilon_2)$.

We let $\alpha = \frac{B-1}{B(B^2+1)}$, $\beta = \frac{B+1}{B(B^2+1)}$. We note that $\frac{2}{B} + \alpha - \beta = \frac{2B}{B^2+1}$. Furthermore, $\frac{B+1}{2}\alpha - \frac{B-1}{2}\beta = 0$. Let N_0 be a random variable defined by the distribution $\Pr[N_0 = i] = \frac{1}{B} + \alpha$ for i odd and $\Pr[N_0 = i] = \frac{1}{B} - \beta$ for i even. That is, the distribution of N_0 takes alternate values on every length. For $N = N_0$, by using Lemma 5 and Lemma 6, we obtain $\varepsilon_1 = \frac{1}{B} + \alpha + \frac{B-1}{2}(\alpha+\beta) = \frac{2B}{B^2+1}$ with the optimal set $A = \{a+1, a+3, \ldots, a+B\}$ and $\varepsilon_2 = \frac{2}{B} + \alpha - \beta = \frac{2B}{B^2+1}$ with the optimal set $A = \{a+1, a+2\}$. So, $\varepsilon = \frac{2B}{B^2+1}$. We now want to prove that there is no distribution for N achieving a lower ε.

Let us assume that there is some $0 \leq i < B-1$ such that $\Pr[N \in \{i+1, i+2\}] > \Pr[N_0 \in \{i+1, i+2\}] = \frac{2B}{B^2+1}$. We take $A = \{a+1, a+2, \ldots, a+i+2\}$ and we obtain

$$\varepsilon \geq \mathsf{Adv}_A(a, a+2) = \Pr[N \leq i+2] - \Pr[N \leq i] > \frac{2B}{B^2+1}$$

which is not better than our above distribution. Hence, we now assume that $\Pr[N \in \{i+1, i+2\}] \leq \Pr[N_0 \in \{i+1, i+2\}]$ for $i = 0, \ldots, B-2$.

Let i be an odd integer. Since $\Pr[N \in \{u, u+1\}] \leq \Pr[N_0 \in \{u, u+1\}]$ for $u = 1, 3, \ldots, i-2, i+1, \ldots, B-3, B-1$, by summing all inequalities, we obtain $\Pr[N \neq i] \leq \Pr[N_0 \neq i]$. So,

$$\Pr[N = i] = 1 - \Pr[N \neq i] \geq 1 - \Pr[N_0 \neq i] = \Pr[N_0 = i]$$

for any odd i. Thus, $\Pr[N \text{ odd}] \geq \Pr[N_0 \text{ odd}]$.

Let now i be even. We have

$$\begin{aligned}\Pr[N = i] &= \Pr[N \in \{i, i+1\}] - \Pr[N = i+1] \\ &\leq \Pr[N_0 \in \{i, i+1\}] - \Pr[N_0 = i+1] \\ &= \Pr[N_0 = i]\end{aligned}$$

Thus, $\Pr[N \text{ even}] \leq \Pr[N_0 \text{ even}]$.

Finally, let $A = \{a+1, a+3, \ldots, B\}$. We have

$$\begin{aligned}\varepsilon \geq \mathsf{Adv}_A(a, a+1) &= \Pr[N \text{ odd}] - \Pr[N \text{ even}] \\ &\geq \Pr[N_0 \text{ odd}] - \Pr[N_0 \text{ even}] \\ &= \frac{2B}{B^2+1}\end{aligned}$$

Therefore, we cannot have ε lower than $\frac{2B}{B^2+1}$. □

Theorem 18 shows that when $b - a \leq 2$ and B is odd, the lower bound $\frac{1}{\lfloor \frac{B}{2} \rfloor + 1} = \frac{2}{B+1}$ for the maximum advantage is not achievable. Results of Theorem 12 and 18 for the case when $\Delta = 2$ and B is odd are provided in Table 1 for small values of B.

5 Conclusion

We have shown that a padding scheme adding strings with uniformly distributed length is nearly optimal and that its security is roughly $\frac{\Delta}{B}$. The optimal scheme can be slightly better but still close to this bound. This shows that the price to pay for making ε-indistinguishable two plaintexts with a single bit of length difference (i.e. 1-IND-OTE(t,ε)-security) is to append a padding of length ε^{-1}, which is impractical for the usual security levels we target for encryption (e.g. $\varepsilon = 2^{-80}$).

Acknowledgements. The authors would like to thank one of the reviewers for his/her lengthy and valuable comments.

References

1. Albrecht, M.R., Watson, G.J., Paterson, K.G.: Plaintext Recovery Attacks Against SSH. In: IEEE Symposium on Security and Privacy, Berkeley, CA, USA, pp. 16–26. IEEE, Los Alamitos (2009)
2. Canvel, B., Hiltgen, A.P., Vaudenay, S., Vuagnoux, M.: Password interception in a SSL/TLS channel. In: Boneh, D. (ed.) CRYPTO 2003. LNCS, vol. 2729, pp. 583–599. Springer, Heidelberg (2003)
3. Chor, B., Kushilevitz, E.: Secret sharing over infinite domains (Extended Abstract). In: Brassard, G. (ed.) CRYPTO 1989. LNCS, vol. 435, pp. 299–306. Springer, Heidelberg (1990)
4. Chor, B., Kushilevitz, E.: Secret Sharing over Infinite Domains. Journal of Cryptology 6, 87–95 (1993)
5. Degabriele, J.-P., Paterson, K.G.: Attacking the IPsec Standards in Encryption-only Configurations. In: IEEE Symposium on Security and Privacy, Berkeley, CA, USA, pp. 335–349. IEEE, Los Alamitos (2007)
6. Dierks, T., Rescola, C.: The TLS Protocol Version 1.2. RFC 5246, standard tracks, the Internet Society (2008)
7. Paterson, K.G., Yau, A.K.L.: Cryptography in theory and practice: The case of encryption in iPsec. In: Vaudenay, S. (ed.) EUROCRYPT 2006. LNCS, vol. 4004, pp. 12–29. Springer, Heidelberg (2006)
8. Paterson, K.G., Watson, G.J.: Plaintext-dependent decryption: A formal security treatment of SSH-CTR. In: Gilbert, H. (ed.) EUROCRYPT 2010. LNCS, vol. 6110, pp. 345–361. Springer, Heidelberg (2010)
9. Phan, R.C.-W., Vaudenay, S.: On the impossibility of strong encryption over $\aleph_0$. In: Chee, Y.M., Li, C., Ling, S., Wang, H., Xing, C. (eds.) IWCC 2009. LNCS, vol. 5557, pp. 202–218. Springer, Heidelberg (2009)
10. Shannon, C.E.: Communication Theory of Secrecy Systems. Bell System Technical Journal 28, 656–715 (1949)
11. Vaudenay, S.: Security flaws induced by CBC padding – applications to SSL, IPSEC, WTLS.. In: Knudsen, L.R. (ed.) EUROCRYPT 2002. LNCS, vol. 2332, pp. 534–545. Springer, Heidelberg (2002)
12. Ylonen, T.: The Secure Shell (SSH) Transport Layer Protocol. RFC 4253, standard tracks, the Internet Society (2006)

Fighting Pirates 2.0

Paolo D'Arco[1] and Angel L. Perez del Pozo[2]

[1] Dipartimento di Informatica
Università degli Studi di Salerno, 84084, Fisciano (SA), Italy
paodar@dia.unisa.it
[2] Universidad Rey Juan Carlos, 28933, Móstoles, Madrid, Spain
angel.perez@urjc.es

Abstract. In this paper we propose methods to cope with the Pirates 2.0 attack strategy against tracing and revoking schemes presented at Eurocrypt 2009. In the Pirates 2.0 attack model traitors collaborate in *public* and *partially* share their secret information with a *certified* guarantee of anonymity. Several classes of tracing and revoking schemes are subject to such a new threat. We focus our attention on the tree-based class of schemes. We start by discussing some simple techniques which can partially help to deal with the attack, and point out their limits. Then, we describe a new hybrid scheme which can be used to face up the Pirates 2.0 attack strategy.

Keywords: Broadcast encryption, user revocation, traitor tracing.

1 Introduction

Secure Content Distribution. A central research issue within the cryptographic community, starting from the 90's, has been providing methods to enable a center to deliver encrypted data to a large set of users, in such a way that only a privileged subset of them can decrypt the data. Applications for these schemes range from pay-tv systems to systems for delivering sensitive information stored on media like CDs, DVDs or even available through web-based services.

The research efforts have been basically directed toward solving two problems: the *access* problem, i.e., privileged users decrypt the content, unauthorized do not; the *tracing* problem, i.e., the ability to trace and, hence, discourage dishonest users, to illegally help unauthorized users to set up a decoder to gain access to the content. Specifically, in a *broadcast encryption* scheme (and its variants), during a set-up phase, every user receives a set of predefined keys. Then, at the beginning of each data transmission, the center sends a broadcast message enabling privileged users to compute a session key, by means of which, the encrypted data, that will be delivered later on, can be decrypted. The most challenging, useful and well-studied setting is the one were the users are equipped with *stateless decoders*. In such a setting *a)* from time to time, the subset of privileged users *changes*, but *b)* the sets of predefined keys held by the users *stay the same* for the lifetime of the scheme. A broadcast encryption scheme is required

J. Lopez and G. Tsudik (Eds.): ACNS 2011, LNCS 6715, pp. 359–376, 2011.

to be *collusion resistant*, in the sense that even if all the revoked users share their secret information, they cannot compute a session key they are not entitled to.

On the other hand, a *traitor tracing* scheme is designed to get the same functionality but, the key material embedded in the decoders (i.e., given to the users) in the set up phase, is diversified on *a user basis*, in such a way that when a pirate decoder, built through the contribution of legal users which collude, is found, at least the identity of one of them can be caught.

At the state of current knowledge, efficient broadcast encryption schemes, in terms of user storage and bandwidth (i.e., size of the broadcast message sent) requirements, are known (e.g. [25,19,15,3,17,12]). Similarly, tracing schemes very efficient in terms of bandwidth requirements have been proposed in the literature (e.g., [4,6]). Unfortunately, such tracing schemes are not so efficient in terms of user memory storage, when the number of traitors grows. Moreover, several schemes proposed in the literature exhibit both functionalities, e.g., some broadcast encryption schemes (and variants) support efficient tracing procedures.

A look back. Berkovits, in [2], addressed for the first time the issue of how to broadcast a secret to privileged users. Later on, Fiat and Naor, in [14], formalized the *broadcast encryption paradigm*. Since then, it has become a major topic in Cryptography, due to the large number of possible applications, and it has evolved in several directions (e.g., *multicasting schemes* [9], *traitor tracing schemes* [8], *revoking schemes*, also referred to as *user exclusion* schemes or *blacklisting* schemes, [24,23,1]).

Among these, *tracing and revoking schemes*, are schemes which combine tracing and revocation. [26] started studying efficient schemes exhibiting both functionalities. A seminal paper along this line is [25], where a framework for stateless tracing and revoking schemes enabling efficient user revocation, referred to as *Subset-Cover*, was proposed. The most efficient construction described in [25], the SD scheme, was later enhanced in [19], which proposed the LSD scheme. Related works are [11,15,18,20,17,22].

The Pirates 2.0 attack model. Historically, traitors have been considered users who *privately* share their secret information to enable unauthorized users to gain access to protected contents. In the Pirates 2.0 attack model [7], traitors collaborate in *public* and *partially* share their secret information with a *certified* guarantee of anonymity. It has been shown that several classes of tracing and revoking schemes, like tree-based revocation schemes (e.g., CS, SD, LSD...) and code-based tracing schemes (e.g., tracing schemes based on collusion secure codes, IPP codes...) are subject to such a new threat.

Our Contribution. We propose methods to cope with the new Pirates 2.0 attack strategy against tree-based tracing and revoking schemes. First, we discuss some simple techniques which can partially help to deal with the attack, and point out their limits. Then, looking through the literature, we recover some ideas, which can be used to strengthen revocation and tracing schemes. We analyze the trade-off that can be obtained by applying these ideas to the schemes. Finally, we describe a new hybrid scheme, obtained by mixing two previous constructions, which can be used to face up the Pirates 2.0 attack strategy.

2 Pirates 2.0: A New Attack Scenario

Let us briefly recall the attack model introduced in [7].

Basic Features. The main characteristics of Pirates 2.0 attacks are the following:

- *Anonymity Guarantee.* Traitors that participate are provided with guarantee, through the exhibition of a mathematical proof, that they cannot be traced by the authority.
- *Partial Contribution.* They never need to reveal their whole set of secret keys.
- *Public Collusion.* Traitors operate in a public environment: they publish secret data from their decoders.
- *Large Coalitions.* They take part in unusual large coalitions.
- *Dynamic Coalitions.* Traitors can come into action only when necessary.
- *Imperfect Decoders.* The pirate decoders are useful even if they decrypt only a certain percentage of ciphertexts.

The motivation for a Pirates 2.0 attack might be for example the need to get rid of a protection system to which a large number of users are hostile. In such an attack model the traitors contribute information at their discretion, which can be collected through a centralized system or a distributed system.

Setting. The N users in the system belong to three different groups: honest users, traitors, and pirates. *Honest users* are legitimate users who keep secret their secret information. *Traitors* are legitimate but dishonest users who (partially) disclose their secret information. *Pirates* are not legitimate users, not entitled to secret information, but able to collect relevant secret information from the public environment, in order to produce a pirate decoder.

The attack set up by traitors is modeled through the following concepts: the *contributed information*, denoted with $\mathcal{C}$, represents the sum of information released to the public domain by the traitors. Initially $\mathcal{C} = \emptyset$. Traitors, to provide information to pirates, implement and run a public available probabilistic algorithm, $Contribute()$. Such an algorithm takes as input the secret information sk of the traitor, information already published by traitors, denoted by I, and the history H of the traitor's contribution to the public. When $Contribute(sk, I, H)$ is executed, the contributed information $\mathcal{C}$ becomes $\mathcal{C} = \mathcal{C} \cup Contribute(sk, I, H)$. Moreover, the term *Public Information*, denoted by $\mathcal{P}$, refers to all public data available from the broadcaster, along with the contributed information. The *anonymity level* of a traitor is quantified through a publicly available procedure, $Anonymity()$. It takes as input the secret information sk of the traitor, information the traitor released, denoted with S, and the public information $\mathcal{P}$. $Anonymity(sk, S, \mathcal{P})$ outputs an integer $\ell \in \{1, \ldots, N\}$. The value $\ell = 1$ means the traitor is known, while the value $\ell = N$ means the traitor is indistinguishable from any other user.

Finally, a *pirate decoder* is the output of an algorithm, $Pirate()$. Such an algorithm takes as input $\mathcal{P}$ and, if the amount of public information is enough, it produces a decrypting device (pirate decoder). Otherwise, it outputs 'failed'.

Definition 1. *A traitor tracing scheme is α-secure against the Pirates 2.0 attack if it prevents the construction of pirate decoders from information published by traitors with an anonymity level greater than α.*

Anonymity Treatment. Traitors are free to contribute some pieces of secret data as long as plenty of users of the system *could have contributed* exactly the same information following the same public strategy. More formally, user anonymity is defined and estimated as follows.

Definition 2. Extraction function. *An extraction function is an efficiently computable function f that outputs information about the secret key.*

Definition 3. Masked traitor. *A traitor t is said to be masked by a user u for an extraction function f if $f(sk_t) = f(sk_u)$.*

Definition 4. Anonymity level. *The level of anonymity of a traitor t after a contribution $\cup_{1 \leq i \leq n} f_i(sk_t)$ is defined as the number α of users masking t for each of the n extraction functions f_i simultaneously:*

$$\alpha = \#\{u | \forall i, f_i(sk_t) = f_i(sk_u)\}.$$

In the following the class of extraction functions considered are projection functions, i.e., the secret information is a vector of secret keys, and f_i is a projection function, which returns the i-th coordinate (key) of the vector. Hence, projection functions model partial public release of key-material.

3 The Subset-Cover Framework: CS, SD, and LSD

In a seminal paper [25] a framework for tracing and revoking schemes enabling efficient user revocation[1], referred to as *Subset-Cover*, was proposed. We briefly recall it here and refer the reader to [25] for details.

Let $\mathcal{N}$ be the set of all users. Every user $u \in \mathcal{N}$ gets, at the beginning, some secret information I_u. Let $\mathcal{R} \subset \mathcal{N}$ be a subset of r users which should be revoked. The center sends a message M containing a new session key K such that all users $u \in \mathcal{N} \setminus \mathcal{R}$ can decrypt the message correctly, while even a coalition consisting of all members of $\mathcal{R}$ cannot decrypt it.

An algorithm defines a collection of subsets $S_1, \ldots, S_\omega \subseteq \mathcal{N}$. Each subset S_j is assigned (perhaps implicitly) a key L_j; each member $u \in S_j$ can compute L_j from its secret information I_u. Given a set of revoked users, the remaining users $\mathcal{N} \setminus \mathcal{R}$ are partitioned into disjoint subsets $S_{i_1}, \ldots, S_{i_m}$ so that $\mathcal{N} \setminus \mathcal{R} = \bigcup_{j=1}^{m} S_{i_j}$ and the session key K is encrypted m times with $L_{i_1}, \ldots, L_{i_m}$, i.e., denoting by E and F two different encryption functions (see [25] for a discussion about the properties of the functions), the broadcast message has the following structure:

$$[i_1, \ldots, i_m, E_{L_{i_1}}(K), \ldots, E_{L_{i_m}}(K), F_K(M)].$$

[1] The join operation is trivially performed by constructing at the beginning an oversized tree structure, capable of accommodating new users. Hence, the focus is only posed on efficient ways to revoke users.

A particular implementation of such a scheme is specified by *a)* the collection of subsets $S_1, \ldots, S_\omega$, *b)* the key assignment to each subset in the collection, *c)* a method to cover the non-revoked users $\mathcal{N} \setminus \mathcal{R}$ by disjoint subsets from this collection, and *d)* a method that allows each user u to find its cover-set S and compute its key L_S from I_u.

The schemes of [25] we consider in the following are all based on trees, i.e., receivers are associated to the leaves of a full binary tree.

Complete Subtree Method (CS, for short). The collection of subsets $\{S_1, \ldots, S_\omega\}$ in the CS scheme corresponds to all subtrees in the full binary tree with n leaves. For any node v_i in the full binary tree (either an internal node or a leaf, $2n-1$ altogether) let the subset S_i be the collection of receivers u that correspond to the leaves of the subtree rooted at node v_i. In other words, $u \in S_i$ iff v_i is an ancestor of u. The key assignment method is simple: assign an independent and random key L_i to every node v_i in the complete tree. Provide every receiver u with the $\log n + 1$ keys associated with nodes along the path from the root to leaf u. For a given set $\mathcal{R}$ of revoked receivers, let $\{u_1, \ldots, u_r\}$ be the leaves corresponding to the elements in $\mathcal{R}$. The method to cover $\mathcal{N} \setminus \mathcal{R}$ with disjoint subsets essentially consists in selecting the minimal number of subtrees which do not contain any of $\{u_1, \ldots, u_r\}$ (see [25] for details). It can be shown that the size of the broadcast message is $O(r \log \frac{N}{r})$.

Subset Difference Method (SD, for short). The collection of subsets in the SD scheme corresponds to subsets of the form "a group of receivers G_1 minus another group G_2." More precisely, a valid subset $S_{i,j}$ is represented by two vertices (v_i, v_j) such that v_i is an ancestor of v_j. A leaf u is in $S_{i,j}$ iff it is in the subtree rooted at v_i but not in the subtree rooted at v_j (see Figure 1).

The key assignment method associates to each *internal node* v_i of the full binary tree a random and independent value $LABEL_i$ (kept secret). This value induces the keys for all legitimate subsets of the form $S_{i,j}$. More precisely, let G be a cryptographic pseudorandom generator that triples the input, and denote

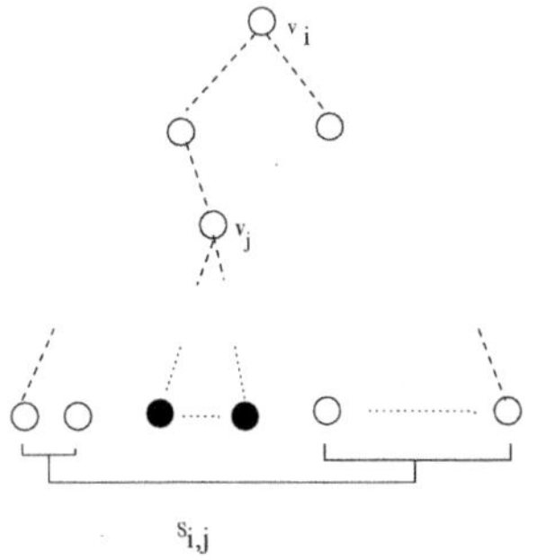

Fig. 1. $S_{i,j}$: set of leaves descending from v_i but not from v_j

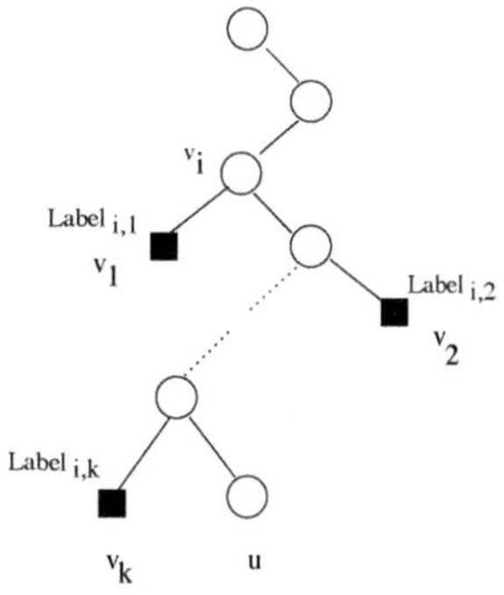

Fig. 2. Labels of nodes hanging off the path from v_i to user u

by $G_L(S), G_R(S)$, and $G_M(S)$ the left, right and middle parts, respectively. Consider a subtree T_i rooted at v_i. To the root node is assigned $LABEL_i$. From this label are computed recursively labels for all subtrees associated to subsets $S_{i,j}$: given that a parent node was labeled S, its two children are labeled $G_L(S)$ and $G_R(S)$. Let $LABEL_{i,j}$ be the label of node v_j derived in the subtree T_i from $LABEL_i$. Following such a labeling, the key $L_{i,j}$ assigned to $S_{i,j}$ is $G_M(LABEL_{i,j})$. The information I_u that each receiver u gets consists in a set of values $LABEL_{i,j}$ defined as follows: *for each* subtree T_i such that u is a leaf of T_i, consider the path from v_i to u and let $v_1, \ldots, v_k$ be the nodes adjacent to the path but not ancestor of u (see Figure 2). Then, u receives the labels $LABEL_{i,1}, \ldots, LABEL_{i,k}$. Notice that each node v_j that is not an ancestor of u is a descendant of one of the nodes $v_1, \ldots, v_k$. Therefore, u can compute the labels $LABEL_{i,j}$ for *any* j such that v_j is not an ancestor of u. It is possible to show that in SD each user stores $O((\log n)^2)$ labels and the size of the broadcast message is $O(r)$ (see [25] for details and how to construct the covering).

Layered Subset Difference Method (LSD, for short). The LSD scheme [19] shows that a small subcollection of subsets $S_{i,j}$ from the collection used by SD suffices to represent any set P as the union of $O(r)$ of the remaining subsets, with a slightly larger constant hidden within the asymptotic notation.

Security. The security of the CS scheme and of the SD scheme was shown in [25]. CS is unconditionally secure w.r.t. coalitions of users of any size (see Claim 1 of [25]). SD is computationally secure, according to an indistinguishability definition, and based on some assumptions on the encryption scheme and the generator used in the scheme (see Theorem 16 of [25]). The security of LSD follows from the security of SD.

Pirates 2.0 attack. CS, SD and LSD are subject to the new attack. All the traitors do is to publish the keys (labels) associated to nodes of the *first levels* of the tree i.e., up in the tree. The idea is that keys up in the tree are stored by many users (all leaves of the corresponding subtrees) who could have published them.

More precisely, in [7], the following theorem was proved:

Theorem 1. *On average, a randomly chosen group of $\rho \log \rho$ (operating isolated) users is able to mount a Pirates 2.0 attack against a complete subtree scheme in which the center wants to ensure a ciphertext rate of at most $\frac{\rho(N-r)}{N}$. Moreover, each traitor is guaranteed an anonymity level of N/ρ.*

In other words, traitors, in CS, by publishing keys associated to nodes belonging to levels λ such that $\lambda \leq \lfloor \log \rho \rfloor$, are guaranteed an anonymity level of N/ρ. Hence, the center, to avoid the attack has to use keys at lower levels. Unfortunately, this means that the broadcast message size from $O(r \log(N/r))$ moves to $O(\frac{\rho(N-r)}{N} + r \log(N/r))$. Similarly, in SD, by publishing direct labels (i.e., labels $LABEL_{i,j}$ were j is either the left child or the right child of i) associated to nodes belonging to levels λ such that $\lambda \leq \lfloor \log \frac{\rho}{2} \rfloor$, traitors are guaranteed an anonymity level of $N/2\rho$.

4 Partial Measures against Pirates 2.0

In this section we discuss some measures which can partially cope with the Pirates 2.0 attack. The key observation is that the attack is successful because in CS, SD and LSD, the users *know the levels* of their keys. Hence, a first attempt to reduce the impact of the attack, could be to look for a strategy which *reduces such a knowledge* but, at the same time, keeps the functionality of the scheme. The measures we discuss are intended for the CS scheme.

Permuting labels. Let us focus on the CS scheme. Let $\mathcal{V} = \{v_i\}_{i \in I}$ be the set of nodes of the complete tree and $\mathcal{W} = \{w_i\}_{i \in I}$ be a set of labels (w.l.o.g., we can assume, for example, that $\mathcal{W} = \{1, \ldots, |I|\}$). We modify the CS scheme (see Table 1) by associating *random labels to nodes*. The labels do not provide any information about the levels of the nodes.

Table 1. Permutation-based Construction

Permutation-based Construction.

Setup phase. The Broadcasting Center

- Chooses uniformly at random a *secret* bijection $\pi : \mathcal{V} \longrightarrow \mathcal{W}$.
- Associates to each node v_i a key K_i, chosen u.a.r.
- For each complete subtree S_i rooted in v_i, sends to each user corresponding to a leaf of S_i, the pair $(\pi(v_i), K_i)$.

Broadcast phase. At the beginning of a new session, to send message M, the broadcasting center

- Computes a cover $\{S_{i_1}, \ldots S_{i_m}\}$ for non-revoked users in $\mathcal{N} \setminus \mathcal{R}$.
- Chooses u.a.r a session key K, computes $F_K(M)$ and, for each subtree S_i in the cover, computes $E_{K_i}(K)$.
- Sends
$$[(\pi(v_{i_1}), E_{K_{i_1}}(K)), \ldots, (\pi(v_{i_m}), E_{K_{i_m}}(K)), F_K(M)].$$

Decryption phase. Upon receiving a broadcast message, a non-revoked user u

- Looks for a matching label contained both in the header of the message and in his set of labeled long-term keys. Let w_l be such a label.
- Computes $K = D_{K_{w_l}}(E_{K_{w_l}}(K))$ and $M = F_K^{-1}(F_K(M))$.

The purpose of this modification is to make more difficult a Pirates 2.0 type attack. In the attack proposed in [7] a traitor publishes all his keys corresponding to complete subtrees rooted at a level of the complete tree lower than a certain threshold λ. In this way, a traitor is guaranteed an anonymity level of $N/2^\lambda$, where N is the total number of users. The proposed modification thwarts in some way this attack, as a user *does not* have a priori information about the correspondence between the keys in his possession and their levels in the complete tree.

Unfortunately, notice that users could gain some information about this correspondence after seeing and decrypting broadcasted messages. Just to exemplify, pirates could get information about the level of their keys by *publicly collaborating*. For example, the users could use a shared table, in which in the first column are reported the labels $\pi(v_i)$, associated to the keys. In the second, each user puts a cross in correspondence of the label associated to the key he used to decrypt. In such a way, they can estimate the level of the key associated to a certain label. However, the important point is that pirates *lose the anonymity guarantee* they get in CS (no certificate is available any more). They can get a partial guarantee but they need *to trust* the other pirates and they have to assume that nobody else adds crosses in the table to falsify the results.

Adding some randomness. Another approach which seems to be of some help consists in looking for a strategy which *reduces the level of anonymity* of the user. To this aim, we modify the CS scheme (see Table 2), following an idea from [16], where the family of "OR"-protocols was introduced. If at each node are associated more keys and a user which descends from the node gets one of the keys chosen *uniformly at random*, then the level of anonymity depends on the level of the keys but also on the number of keys associated to that node. Of course there is a trade-off: if the node is used, the broadcast message has to be encrypted with all keys associated to the node.

Let $h = max\{h_\lambda : \lambda = 0, \ldots, \log N\}$. The parameters of the modified scheme (compared to CS) are:

	CS	modified CS
# keys per user	$\log N + 1$	$\log N + 1$
total # of keys	$2N - 1$	$\sum_{\lambda=0}^{\log N} 2^\lambda h_\lambda \leq h(2N-1)$
header length	$O(r\log(N/r))$	$O(hr\log(N/r))$

The modifications are useful to fight against Pirates 2.0 type attacks. In the modified scheme, a traitor following the strategy of making public his unique key with label equal to a certain level λ is covered, on average, by $N/2^\lambda h_\lambda$ users. Moreover, a traitor following the strategy of making public *all* his keys with label lower than a certain level λ is covered, on average, by $\frac{N}{2^\lambda}\prod_{l=0}^{\lambda}\left(\frac{1}{h_l}\right)$ users.

Notice that an interesting variant of the scheme could be the following: for each node, instead of associating to users of the subtree a key of the pool chosen *uniformly at random*, the keys could be associated according to a *non-uniform* probability distribution (kept secret by the center) e.g., a degenerate-like probability distribution which associates a sort of *trap* key to a single or few users. In such a case, if the user publishes such a key, the level of anonymity is much lower than he is expecting. Of course, also this measure is useless if the traitor try to estimate how many of them have decrypted a broadcast message by using a certain key. But, again, traitors need to trust each other and no anonymity guarantee is available anymore.

Table 2. Or-based Construction

Or-based Construction.

Setup phase. W.l.o.g., assume that $\log N$ is an integer value.

1. For each $\lambda \in \{0, \dots, \log N\}$ choose an integer $h_\lambda > 0$.
2. For each subtree S_i, rooted at node v_i at level λ in the tree, generate a set
$$\mathcal{K}_i = \{K_{i,1}, \dots, K_{i,h_\lambda}\}$$
of h_λ keys.
3. For each subtree S_i, send to each user covered by S_i a single pair $((i,j), K_{i,j})$, where $K_{i,j}$ is chosen u.a.r. from $\mathcal{K}_i$, independently for each user.

Broadcast phase. At the beginning of a new session, to send message M, the broadcasting center

- Computes a cover $\{S_{i_1}, \dots S_{i_m}\}$ for non-revoked users in $\mathcal{N} \setminus \mathcal{R}$.
- Chooses u.a.r a session key K, computes $F_K(M)$ and, for each subtree S_i in the cover, computes $E_{K_{i,1}}(K), \dots, E_{K_{i,h_\lambda}}(K)$.
- Sends
$$[i_1, E_{K_{i_1,1}}(K), \dots, E_{K_{i_1,h_\lambda}}(K), \dots, i_m, E_{K_{i_m,1}}(K), \dots, E_{K_{i_m,h_\lambda}}(K), F_K(M)].$$

Decryption phase. Upon receiving a broadcast message, a non-revoked user u

- Identifies the subtree S_i he belongs to, and looks up the key $K_{i,j}$ he holds for S_i
- Computes $K = D_{K_{i,j}}(E_{K_{i,j}}(K))$ and $M = F_K^{-1}(F_K(M))$.

5 A New Scheme

All measures described before are partial solutions against the attack. The problem is that the secret information is not bound to the user identity. We need a method through which, if a user publishes part of his secret information, then he is immediately compromised. Of course, we would like to keep the scheme as efficient as possible, comparable to the version which does not deal with the Pirates 2.0 attack. To this aim, we combine two tracing and revoking schemes previously proposed in the literature. The first one [26] in a certain sense exposes the identities of the users which publicly contribute to foil the scheme, but can be used to revoke a fixed (and a-priori known) number of users. The second, the CS scheme [25], as we have seen, has no bound on the number of users which can be revoked, but the traitors have a high level of anonymity. The hybrid inherits the nice properties of both (due to lack of space the proof of security is provided in Appendix A).

5.1 Adding Secret Sharing Schemes to CS

The hybrid scheme we describe is a modification of the CS scheme, obtained by combining it with the polynomial-based broadcast scheme described in [26] (independently introduced in [1]), which are based on a former idea of [13]. Notice that also [21] proposed a polynomial-based traitor tracing scheme, but the use of the polynomial in [21] is a bit different compared to the use done in [1,26]. Moreover, the authors of [27] showed that the scheme in [21] is not collusion resistant since two users can generate a decryption key which cannot be traced back to one of them.

Consider a cyclic group G of prime order q such that the DDH problem is believed to be hard in G. Let g be a generator of G. Each user u receives a *secret* identifier $I_u \in \mathbb{Z}_q$, chosen uniformly at random, in such a way that different users get different identifiers. Choose t different values $I_1, \ldots, I_t \in \mathbb{Z}_q$, also different from 0 and from all the previous identifiers. The scheme is given in the box in the next page. Notice that the total number of keys and the number of keys per user *remain the same* as in the basic CS scheme. For the length of the broadcasted message, $t+1$ additional elements of G are added to each fragment of the header, but, asymptotically, it remains $O(r\text{log}N/r)$.

Pirates 2.0*: No anonymity for traitors.* The scheme completely prevents a Pirates 2.0 attack. If any user u makes public just one of his secret keys, say $(I_u, P_i(I_u))$, then the authority, immediately determines the identity of the user associated with $(I_u, P_i(I_u))$ through a look up operation in its database. Actually, we can do more: we can apply a slightly modified version of the self-enforcement technique used in [26] (see paragraph 3.1) in order to discourage users to become traitors. Briefly, a public file contains, for each user u, a row where some sensitive information S_u of user u (e.g., social security number, credit card number, etc...) is encrypted by using as a key $y_{i,u} = P_i(I_u)$, for all $logN+1$ values $P_i(I_u)$ stored by user u.

The ElGamal encryption scheme, semantically secure under the DDH assumption, is used (any CPA-secure encryption scheme is fine, but by using ElGamal, the security of the scheme follows only from the DDH assumption): the system administrator, in set up phase, for each encryption for user u, chooses a random $s \in Z_q$, computes g^s and $g^{sy_{i,u}}$, computes $C = H(g^{sy_{i,u}} || I_u) \oplus S_u$, where H is a pairwise independent hash function, and publishes $(ind_{i,u}, g^s, C)$, where $ind_{i,u}$ is a prefix of $g^{sy_{i,u}}$ used for indexing the elements of the table. If user u discloses $(I_u, P_i(I_u))$, then everybody else gain access to his sensitive information.

5.2 Subset Difference and Layered Subset Difference

The technique above described does not immediately apply to SD and LSD. Indeed, a simple generalization might consist in the use of more polynomials. More precisely, in the former construction, we associate $P_i(x)$ to subset S_i. Since in SD we have subsets defined by two indices, we could associate $P_{i,j}(x)$ to $S_{i,j}(x)$ and define, at the beginning of each session, the encryption key for subset $S_{i,j}(x)$ as $L_{i,j} = g^{rP_{i,j}(0)}$. Each user u receives a point $P_{i,j}(I_u)$, for each subset $S_{i,j}(x)$ he

Hybrid CS Construction.

Setup phase. For each complete subtree S_i

1. Choose u.a.r. a secret t-degree polynomial $P_i(x) = a_0^i + a_1^i x + \ldots + a_t^i x^t \in \mathbb{Z}_q[x]$.
2. Send to each user u covered by S_i the pair $(i, P_i(I_u))$.

Broadcast phase. At the beginning of a new session, to send message M, the broadcasting center

1. Computes a cover $\{S_{i_1}, \ldots S_{i_m}\}$ for non-revoked users in $\mathcal{N} \setminus \mathcal{R}$. Then, chooses u.a.r. a session key K, computes $F_K(M)$ and, for each $S_i \in \{S_{i_1}, \ldots S_{i_m}\}$
   (a) Choose a random $r_i \in \mathbb{Z}_q$.
   (b) Computes $\{\delta_{i,j} = g^{r_i P_i(I_j)}\}_{j=1,\ldots,t}$, $K_i = g^{r_i P_i(0)}$, and $E_{K_i}(K)$.
2. Sends
$$[(i_l, g^{r_{i_l}}, \delta_{i_l,1}, \ldots, \delta_{i_l,t}, E_{K_{i_l}}(K))_{1 \le l \le m}, F_K(M)].$$

Decryption phase. Upon receiving a broadcast message, a non-revoked user u

1. Identifies the subtree S_i he belongs to, and looks up in the header of the message the corresponding tuple $(i, \gamma, \delta_1, \ldots, \delta_t, c)$.
2. Computes $\lambda_u = \prod_{k=1}^{t} \frac{I_k}{I_k - I_u}$ and $\lambda_j = \frac{I_u}{I_u - I_j} \prod_{k=1\ (k \ne j)}^{t} \frac{I_k}{I_k - I_u}$ for $j = 1, \ldots, t$.
3. Since, from Lagrange's interpolation formula,
$$P_i(0) = \lambda_u P_i(I_u) + \sum_{j=1}^{t} \lambda_j P_i(I_j)$$
   then, u computes
$$K_i = g^{r_i P_i(0)} = g^{r_i \lambda_u P_i(I_u)} \prod_{j-1}^{t} g^{r_i \lambda_j P_i(I_j)} = \gamma^{\lambda_u P_i(I_u)} \prod_{j-1}^{t} \delta^{\lambda_j}.$$
4. Finally u computes the session key $K = D_{K_i}(c)$ and $M = F_K^{-1}(F_K(M))$.

belongs to which, along with $(I_k, g^r, g^{r P_{i,j}(I_k)})$ for $k = 1, \ldots, t$, enables computing $L_{i,j}$. The drawback of this solution is that each user has to *explicitly* store a point for each subset in which he belongs to. In other words, the advantages of the pseudorandom generation of the labels (and keys) which is crucial in SD in order to store $O((\log n)^2)$ labels is lost[2]. Actually, we can do better. We could *compose* the two schemes as follows:

It is easy to check that the scheme is correct and that each user stores $O((\log n)^2)$ labels plus $\log N + 1$ points. Hence, the storage requirements stay the same of SD. Note that the same construction can be applied to the LSD scheme. Moreover, the same observations we have done for the CS case apply.

[2] A similar problem is faced in [11] when trying to generalize the SD scheme to the public key setting.

Hybrid SD Construction.

Setup phase. For each complete subtree S_i

1. Choose u.a.r. a secret t-degree polynomial $P_i(x) = a_0^i + a_1^i x + \ldots + a_t^i x^t \in \mathbb{Z}_q[x]$.
2. Generate an instance of the SD scheme with $\mathbb{Z}_q$ as set for keys $L_{i,j}$.
3. Send to each user u covered by $S_{i,*}$ the pair $(i, P_i(I_u))$ and the labels that SD assigns to him.

Broadcast phase. At the beginning of a new session, to send message M, the broadcasting center

1. Computes a cover $\{S_{i_1,j_1}, \ldots S_{i_m,j_m}\}$ for non-revoked users in $\mathcal{N}\backslash\mathcal{R}$. Then, chooses u.a.r a session key K, computes $F_K(M)$ and, for each $S_{i,j} \in \{S_{i_1,j_1}, \ldots S_{i_m,j_m}\}$
   (a) Choose a random $r_i \in \mathbb{Z}_q$.
   (b) Computes $L_{i,j}, \{g^{r_i P_i(I_k) L_{i,j}}\}_{k=1,\ldots,t}, K_{i,j} = g^{r_i P_i(0) L_{i,j}}$, and $E_{K_{i,j}}(K)$.
2. Sends

$$[(i_\ell, j_\ell, g^{r_{i_\ell}}, g^{r_{i_\ell} P_{i_\ell}(I_1) L_{i_\ell,j_\ell}}, \ldots, g^{r_{i_\ell} P_{i_\ell}(I_t) L_{i_\ell,j_\ell}}, E_{K_{i_\ell,j_\ell}}(K))]_{\ell=1,\ldots,m}.$$

Decryption phase. Upon receiving a broadcast message, a non-revoked user u

1. Identifies the subtree $S_{i,j}$ he belongs to, and looks up in the header of the message the corresponding tuple $((i,j), \gamma, \delta_1, \ldots, \delta_t, c)$.
2. Computes λ_u, λ_k for $k = 1, \ldots, t$, $L_{i,j}$ and

$$K_{i,j} = g^{r_i P_i(0) L_{i,j}} = \left(g^{r_i \lambda_u P_i(I_u)} \prod_{k=1}^{t} g^{r_i \lambda_k P_i(I_k)} \right)^{L_{i,j}} = \gamma^{\lambda_u P_i(I_u) L_{i,j}} \prod_{k=1}^{t} \delta^{\lambda_k}.$$

3. Finally u computes the session key $K = D_{K_{i,j}}(c)$ and $M = F_K^{-1}(F_K(M))$.

As reported in Theorem 1, the authors of [7] showed that in CS and LSD, traitors, by publishing keys associated to nodes belonging to levels λ such that $\lambda \leq \lfloor \log \rho \rfloor$, are guaranteed an anonymity level of N/ρ and $N/2\rho$, respectively. It is immediate to check that in the above hybrid versions, the anonymity levels drop to 1. More precisely, referring to Definition 1, we get:

Theorem 2. *The Hybrid-CS and Hybrid-SD traitor tracing schemes are 1-secure against the Pirates 2.0 attack, implemented by using projection functions and are, in terms of key and ciphertext sizes, as efficient as the former CS and SD.*

5.3 Anonymity against Private Collusion

Let us look again at the Hybrid CS Construction. Notice that, as long as traitors set up a pirate decoder by giving away privately *some pieces* of their private keys (i.e., points of the polynomials associated to nodes along the path a user belongs to), our hybrid construction inherits exactly the same (efficient) black-box tracing procedure that uses the CS scheme. Pirate decoders are either disabled or at least a traitor is found. The scheme presented in [26] has a black-box tracing procedure which requires exponential time.

However, our scheme is subject to another form of *private* collusion, which does not permit to identify traitors. A coalition consisting on $t + 1$ traitors $u_0, \ldots, u_t$, collaborating in secret, is able to recover $P_i(x)$ from their secret values $(I_{u_j}, P_i(I_{u_j}))$. Then they can learn and distribute the value $P_i(0)$ which allows a pirate decoder to compute the key $K_i = (g^r)^{P_i(0)}$ and therefore to decrypt the broadcasted content. Any $t + 1$ users covered by the same subtree S_i are able to do this. In such a case, the tracing procedure can still make the decoder useless, but the anonymity of the traitors among the subset S_i is preserved. Moreover, along the same line, the scheme enables the coalition to interpolate all polynomials for which they have $t + 1$ points. However, notice that, if the self-enforcement mechanism we have described before is used, *it does not* mean that the coalition *gain access* to the sensitive information of all other users who share the same polynomials: indeed, since the value I_v of user v is secret, they have no idea of which values of the polynomials are used by the users. This is also the reason for which, compared to [26], we have slightly modified the encryption in the self-enforcement technique by including I_u in the computation of the encryption $C = H(g^{sy_{i,u}} || I_u) \oplus S_u$.

Notice that this problem is also present in [26], where a coalition of $t+1$ users can interpolate the t-degree polynomial: unfortunately, therein the $t+1$ colluders gain access to the sensitive information of all revoked users, whose values are broadcasted. In our scheme, the points in broadcast are $I_1, \ldots, I_t$ which are not associated to any user. We emphasize, however, that the scheme provided in [26] is secure as long as the coalition of revoked users is not greater in size than the fixed a-priori threshold t. In our scheme, due to the CS revocation strategy, security holds w.r.t. coalitions of *any size*. What is lost if the coalition of traitors is greater than t is the protection against *anonymous* collaboration to the set up of a pirate decoder. Moreover, we could further strengthen the scheme in order to reduce the anonymity guarantee of the traitors, by using the technique employed within the Or-based Construction from [16]: associate to each node v_i a set of λ different polynomials, and assign to each user u of the subset S_i the value $P_{i,\ell}(I_u)$, for an $\ell \in \{1, \ldots, \lambda\}$, chosen uniformly at random. If both measures are used the total length of the broadcast message in CS moves from $O(r \log \frac{N}{r})$ to $O(\lambda t r \log \frac{N}{r})$ and in SD from $O(r)$ to $O(\lambda t r)$.

Nevertheless, notice that one of the motivations for introducing the Pirates 2.0 attack was that public collusion is easier than private collusion. Indeed, the first one enables large coalitions, since users consider themselves protected by anonymity; on the other hand, the second does not provide any guarantee to traitors, they need to trust each other and *share* their sensitive information, which means that large coalitions should be difficult to set up.

Using Cryptography against Cryptography. Notice that the above hybrid schemes could also be attacked by a group of traitors who "publicly" collaborate but still *preserve* the privacy of their inputs. Indeed, it is not difficult to see that the computation of $P_i(0)$ can be casted as an instance of the general multi-party computation problem, where each traitor has an input, the traitors jointly want to compute the output of the function, but each of them wants to keep private his own input, i.e., without disclosing any information about it, apart what can be inferred from the output of the function. However, such an attack requires a set up phase and, if the computations are publicly carried out, some other forms of protection for the traitors (e.g., anonymous communication channels) have to be considered. We do not go into details but emphasize that such a strategy could be pursued. Our scheme is not robust against it. We are only protected by the threshold on the size of the coalition of traitors. It seems an interesting open problem to study the resistance of known broadcast encryption schemes against this type of attacks.

6 Conclusions and Open Problems

In this paper we started studying measures to deal with the Pirates 2.0 attack strategy. Among these, we have proposed an hybrid scheme, which combines a scheme which enables efficient tracing and revoking with a scheme which exposes traitors who publicly disclose secret information. Several open problems need to be addressed: first of all, it is necessary to clearly conceptualize and formally define a *security model* which incorporates the Pirates 2.0 attack strategy and all forms of public collaborations. Then, a scheme which fulfills all requirements should be designed and rigorously proven secure in the model.

Acknowledgments. This work was partially done while the second author was hosted at the Dipartimento di Infomatica, University of Salerno. The authors are supported by Project MTM2010-15167 from the Spanish Ministry of Science and Technology.

References

1. Anzai, J., Matsuzaki, N., Matsumoto, T.: A quick group key distribution scheme with "Entity revocation". In: Lam, K.-Y., Okamoto, E., Xing, C. (eds.) ASIACRYPT 1999. LNCS, vol. 1716, pp. 333–347. Springer, Heidelberg (1999)
2. Berkovits, S.: How to broadcast a secret. In: Davies, D.W. (ed.) EUROCRYPT 1991. LNCS, vol. 547, pp. 535–541. Springer, Heidelberg (1991)
3. Boneh, D., Gentry, C., Waters, B.: Collusion resistant broadcast encryption with short ciphertexts and private keys. In: Shoup, V. (ed.) CRYPTO 2005. LNCS, vol. 3621, pp. 258–275. Springer, Heidelberg (2005)
4. Boneh, D., Naor, M.: Traitor tracing with constant size ciphertext. In: CCS 2008: Proceedings of the 15th ACM Conference on Computer and Communications Security, pp. 501–510. ACM, New York (2008)
5. Boneh, D.: The decision diffie-hellman problem. In: Buhler, J.P. (ed.) ANTS 1998. LNCS, vol. 1423, pp. 48–63. Springer, Heidelberg (1998)

6. Billet, O., Phan, D.H.: Efficient traitor tracing from collusion secure codes. In: Safavi-Naini, R. (ed.) ICITS 2008. LNCS, vol. 5155, pp. 171–182. Springer, Heidelberg (2008)
7. Billet, O., Phan, D.H.: Traitors collaborating in public: Pirates 2.0. In: Joux, A. (ed.) EUROCRYPT 2009. LNCS, vol. 5479, pp. 189–205. Springer, Heidelberg (2009)
8. Chor, B., Fiat, A., Naor, M.: Tracing traitors. In: Desmedt, Y.G. (ed.) CRYPTO 1994. LNCS, vol. 839, pp. 257–270. Springer, Heidelberg (1994)
9. Canetti, R., Garay, J., Itkis, G., Micciancio, D., Naor, M., Pinkas, B.: Multicast security: A taxonomy and some efficient constructions. In: Proceedings of INFOCOMM 1999, pp. 708–716 (1999)
10. D'Arco, P., Perez del Pozo, A.L.: Fighting Pirates 2.0 (full version), `http://www.dia.unisa.it/~paodar`
11. Dodis, Y., Fazio, N.: Public key broadcast encryption for stateless receivers. In: Feigenbaum, J. (ed.) DRM 2002. LNCS, vol. 2696, pp. 61–80. Springer, Heidelberg (2003)
12. Delerablée, C., Paillier, P., Pointcheval, D.: Fully collusion secure dynamic broadcast encryption with constant-size ciphertexts or decryption keys. In: Takagi, T., Okamoto, T., Okamoto, E., Okamoto, T. (eds.) Pairing 2007. LNCS, vol. 4575, pp. 39–59. Springer, Heidelberg (2007)
13. Feldman, P.: A practical scheme for non-interactive verifiable secret sharing. In: Proceedings of the 28th IEEE Symp. on Foundations of Computer Science, pp. 427–437 (1987)
14. Fiat, A., Naor, M.: Broadcast encryption. In: Stinson, D.R. (ed.) CRYPTO 1993. LNCS, vol. 773, pp. 480–491. Springer, Heidelberg (1994)
15. Goodrich, M.T., Sun, J.Z., Tamassia, R.: Efficient tree-based revocation in groups of low-state devices. In: Franklin, M. (ed.) CRYPTO 2004. LNCS, vol. 3152, pp. 511–527. Springer, Heidelberg (2004)
16. Gafni, E., Staddon, J., Yin, Y.L.: Efficient methods for integrating traceability and broadcast encryption. In: Wiener, M. (ed.) CRYPTO 1999. LNCS, vol. 1666, pp. 372–387. Springer, Heidelberg (1999)
17. Hwang, Y.H., Lee, P.J.: Efficient broadcast encryption scheme with log-key storage. In: Di Crescenzo, G., Rubin, A. (eds.) FC 2006. LNCS, vol. 4107, pp. 281–295. Springer, Heidelberg (2006)
18. Hwang, J.Y., Lee, D.H., Lim, J.: Generic transformation for scalable broadcast encryption schemes. In: Shoup, V. (ed.) CRYPTO 2005. LNCS, vol. 3621, pp. 276–292. Springer, Heidelberg (2005)
19. Halevy, D., Shamir, A.: The LSD broadcast encryption scheme. In: Yung, M. (ed.) CRYPTO 2002. LNCS, vol. 2442, pp. 47–60. Springer, Heidelberg (2002)
20. Jho, N., Hwang, J.Y., Cheon, J.H., Kim, M., Lee, D.H., Yoo, E.S.: One-way chain based broadcast encryption schemes. In: Cramer, R. (ed.) EUROCRYPT 2005. LNCS, vol. 3494, pp. 559–574. Springer, Heidelberg (2005)
21. Kurosawa, K., Desmedt, Y.: Optimum traitor tracing and asymmetric schemes. In: Nyberg, K. (ed.) EUROCRYPT 1998. LNCS, vol. 1403, pp. 145–157. Springer, Heidelberg (1998)
22. Kiayias, A., Pehlivanoglu, S.: Pirate evolution: How to make the most of your traitor keys. In: Menezes, A. (ed.) CRYPTO 2007. LNCS, vol. 4622, pp. 448–465. Springer, Heidelberg (2007)
23. Kumar, R., Rajagopalan, S., Sahai, A.: Coding constructions for blacklisting problems without computational assumptions. In: Wiener, M. (ed.) CRYPTO 1999. LNCS, vol. 1666, pp. 609–623. Springer, Heidelberg (1999)

24. Luby, M., Staddon, J.: Combinatorial bounds for broadcast encryption. In: Nyberg, K. (ed.) EUROCRYPT 1998. LNCS, vol. 1403, pp. 512–526. Springer, Heidelberg (1998)
25. Naor, D., Naor, M., Lotspiech, J.: Revocation and tracing schemes for stateless receivers. In: Kilian, J. (ed.) CRYPTO 2001. LNCS, vol. 2139, pp. 41–62. Springer, Heidelberg (2001), Full version available at `http://www.wisdom.weizmann.ac.il/~naor/`
26. Naor, M., Pinkas, B.: Efficient trace and revoke schemes. In: Frankel, Y. (ed.) FC 2000. LNCS, vol. 1962, pp. 1–20. Springer, Heidelberg (2001)
27. Stinson, D.R., Wei, R.: Key preassigned traceability schemes for broadcast encryption. In: Tavares, S., Meijer, H. (eds.) SAC 1998. LNCS, vol. 1556, pp. 144–156. Springer, Heidelberg (1999)

A Security of the Schemes

Tools. The schemes we have proposed, the Or-based construction, the Hybrid CS construction and the Hybrid SD Construction, still belong to the Subset-Cover framework. All we have changed are the key-assignment methods. Hence, in order to prove that the schemes are robust against coalitions of revoked users of any size, we can use the general results for the framework shown in [25]. Specifically, we need to show that our key-assignment methods satisfy the *key-indistinguishability* property. Let us start by recalling some definitions given in [25].

Definition 5. *(*Key-Indistinguishability*)*
Let $\mathcal{A}$ be a Subset-Cover revocation algorithm that defines a collection of subsets $S_1, \ldots, S_\omega$. Consider a feasible adversary $\mathcal{B}$ that: a) selects i, $1 \leq i \leq \omega$; b) receives the I_u's (secret information that u receives) for all $u \in \mathcal{N} \setminus S_i$. We say that $\mathcal{A}$ satisfies the key-indistinguishability property if the probability that $\mathcal{B}$ distinguishes a key L_i from a random string R_{L_i} of the same length is negligible.

The adversary we deal with is characterized as follows:

Definition 6. *(*Adversarial Model*)*
Consider an adversary $\mathcal{B}$ that:

1. *Selects adaptively a set $\mathcal{R}$ of receivers and obtains I_u for all $u \in \mathcal{R}$. By adaptively we mean that $\mathcal{B}$ may select messages $M_1, M_2, \ldots,$ and revocation sets $\mathcal{R}_1, \mathcal{R}_2, \ldots$ (the revocation sets need not correspond to the actual corrupted users) and see the encryption of M_i when the revoked sets is $\mathcal{R}_i$. Also $\mathcal{B}$ can create a ciphertext and see how any (non-corrupted) user decrypts it. It then asks to corrupt a receiver u and obtains I_u. This steps is repeated $|\mathcal{R}|$ times (for any $u \in \mathcal{R}$).*
2. *Choose a message M as the challenge plaintext and a set $\mathcal{R}$ of revoked users that must include all the ones it corrupted (but may contain more).*

$\mathcal{B}$ then receives an encrypted message M' with revoked set $\mathcal{R}$. It has to guess whether $M' = M$ or $M' = R_M$, where R_M is a random message of the same length. We say that a revocation scheme is secure if, for any (probabilistic polynomial time) adversary $\mathcal{B}$ as above, the probability that $\mathcal{B}$ distinguishes between the two cases is negligible.

The main result shown in [25] for the subset-cover framework is:

Theorem 3. *Let $\mathcal{A}$ be a subset-cover revocation algorithm where the key assignment satisfies the key-indistinguishability property and where E and F are two encryption schemes $CCA-I$ secure and CPA-secure, respectively. Then $\mathcal{A}$ is secure in the sense of Definition 6 with security parameter $\delta < \epsilon_1 + 2m\omega(\epsilon_2 + 4\omega\epsilon_3)$, where ω is the total number of subsets in the scheme, m is the maximum size of a cover, and ϵ_1, ϵ_2 and ϵ_3 are the probabilities associated to the key-assignment, E and F.*

It is immediate to see that the key-assignment method of the Or-based construction trivially satisfies the key-indistinguishability definition, since all keys are chosen independently and uniformly at random. In the following, we prove that also the key-assignment for the hybrid CS scheme satisfies the key-indistinguishability property. The proof for the SD hybrid scheme, which is obtained by using similar techniques, is included in the full version [10].

Hybrid CS Construction. The key-assignment method for the CS Hybrid Construction satisfies the key-indistinguishability property, assuming that the DDH assumption in the group G holds. The DDH assumption involves a cyclic group G and a generator g. Loosely speaking, it states that no efficient algorithm can distinguish between the two distributions $< g^a, g^b, g^{ab} >$ and $< g^a, g^b, g^c >$, where a, b, c are randomly chosen in $\{1, \ldots, |G|\}$. The reader is referred to [5] for details about the assumption.

More precisely, by using a similar technique to the one employed in [26] we show that if an adversary (i.e., a PPT algorithm) distinguishes a session key from a random value, then such an adversary can be turned into an efficient procedure which solves the DDH problem. We prove the following result:

Theorem 4. *If the DDH assumption holds, then any probabilistic polynomial time adversary $\mathcal{B}$, which operates according to Definition 5, does not distinguish a key K_i associated to subset S_i from a random value.*

Proof. By contradiction. Let us assume that there exists an efficient adversary $\mathcal{B}$, which has received the private information I_u of all users but users in S_i, and is able, with non negligible probability, after a sequence of a poly number m of executions, to distinguish a key K_i associated to S_i from a random value R_i. At the beginning of each session, $\mathcal{B}$ sees vectors of the form $< g^r, I_1, \ldots, I_t, g^{rP_i(I_1)}, \ldots, g^{rP_i(I_t)} >$ where r is a new random value, sent to enable users in S_i to compute the new key K_i. Let us denote by $<$ *history* $>$ the sequence of *all* vectors sent at the beginning of sessions $1, 2, \ldots, m-1$ to enable *all* authorized subsets to compute the new session key, and let us assume that $(< \mathit{history} >, < g^r, I_1, \ldots, I_t, g^{rP_i(I_1)}, \ldots, g^{rP_i(I_t)} >, C)$, where $< g^r, I_1, \ldots, I_t, g^{rP_i(I_1)}, \ldots, g^{rP_i(I_t)} >$ is the vector for S_i for the m-th session, and C is either the key $K_i = g^{rP_i(0)}$ or a random element $R_i \in G$, is the input of $\mathcal{B}$. On such an input, $\mathcal{B}$, independently of the specific revocation strategy, guesses with non-negligible probability what the challenge C is.

Then, we can turn $\mathcal{B}$ into an efficient algorithm $\mathcal{B}'$ which solves the DDH problem, i.e., $\mathcal{B}'$ takes in input a triple $< \alpha, \beta, \gamma >$ and outputs with non-negligible probability whether $< \alpha, \beta, \gamma > = < g^a, g^b, g^{ab} >$ or $< \alpha, \beta, \gamma > = < g^a, g^b, g^c >$, where a, b, and c are chosen u.a.r in Z_q. Such an algorithm $\mathcal{B}'$ works as follows:

- By using the public system parameters, g, q, p, $\mathcal{B}'$ generates an instance of the hybrid scheme. In particular, it generates secret keys for all users but the users in S_i. Chooses u.a.r values $(I_1, \ldots, I_t, P_i(I_1), \ldots, P_i(I_t))$, and associates it to subset S_i.
- For session $\ell = 1, \ldots, m-1$, $\mathcal{B}'$ constructs $< \mathit{history} >$ by choosing at random a revocation strategy and revoking all users but users in S_i. Since $\mathcal{B}'$ holds all secret keys but the ones held by users in S_i, then he can construct all encryptions (vectors) he needs for subsets different from S_i. Moreover, any time $\mathcal{B}'$ needs to encrypt the key for users in S_i, it chooses u.a.r a value r' and constructs the vector $< \alpha^{r'}, I_1, \ldots, I_t, \alpha^{r'P_i(I_1)}, \ldots, \alpha^{r'P_i(I_t)} >$.
- Run $\mathcal{B}$ on input $< \mathit{history} >, < \alpha, I_1, \ldots, I_t, \alpha^{P_i(I_1)}, \ldots, \alpha^{P_i(I_t)} >, \gamma$.
- If $\mathcal{B}$ outputs *key*, then output *DH triple*. Otherwise, output *random triple*.

The idea is to embed the challenge C in the m-th session, i.e., by implicitly imposing that $r = a, P_i(0) = b$. In such a way $g^r = g^a, g^b = g^{P_i(0)}$ and $g^{ab} = g^{rP_i(0)}$. It is important to note that $\mathcal{B}$ does not distinguish a *real* $< \mathit{history} >$ from the $< \mathit{history} >$ generated by $\mathcal{B}'$ by using the triple $< \alpha, \beta, \gamma >$. They are *identically distributed*. Indeed:

- the vectors for subsets different from S_i are distributed exactly as in a real execution of the protocol, i.e., we are perfectly simulating the protocol.
- the vectors for S_i use the challenge $< \alpha, \beta, \gamma >$, a different value r at each session, and the random values $P_i(I_1), \ldots, P_i(I_t)$. Hence, it follows that they are "consistent with" a *unique* t-degree polynomial P_i, chosen uniformly at random, implicitly defined by the pairs $(0, b)$, $(I_1, P_i(I_1)), \ldots, (I_t, P_i(I_t))$.

Moreover, $\mathcal{B}'$ runs in probabilistic polynomial time and distinguishes a DH triple from a random triple exactly with the *same advantage* with which $\mathcal{B}$ distinguishes a true key from a random value. Hence, if the DDH assumption in G holds, then $\mathcal{B}'$ (and thus $\mathcal{B}$) does not exist, and the key-assignment method satisfies the key-indistinguishability property. □

Security Notions for Broadcast Encryption

Duong Hieu Phan[1,2], David Pointcheval[2], and Mario Strefler[2]

[1]LAGA, University of Paris 8
[2]ENS / CNRS / INRIA

Abstract. This paper clarifies the relationships between security notions for broadcast encryption. In the past, each new scheme came with its own definition of security, which makes them hard to compare. We thus define a set of notions, as done for signature and encryption, for which we prove implications and separations, and relate the existing notions to the ones in our framework. We find some interesting relationships between the various notions, especially in the way they define the receiver set of the challenge message. In addition, we define a security notion that is stronger than all previous ones, and give an example of a scheme that fulfills this notion.

Keywords: Broadcast Encryption, Adaptive Security, Security Models.

1 Introduction

Broadcast encryption (BE) is a cryptographic primitive that was first described by Fiat and Naor in [FN94]. It provides a content holder the ability to publish the content to a specific subset of the registered users. This is used in practice by copyright protection mechanisms for digital media such as DVDs, so that if the keys for a series of DVD players become known, this series will not be able to play DVDs produced after the series is revoked [NNL01]. But while work on the related topic of multi-cast encryption progressed, BE did not receive much attention until the last decade, when Naor, Naor, and Lotspiech presented their (symmetric-key) subset-cover framework along with a security model and a security analysis [NNL01]. Since then, many BE schemes have been proposed, but for each scheme the security proof was done in a new security model. Because of these various and often *ad-hoc* security models, it is hard to compare the merits of these schemes, as it is not always clear how the security notions relate to each other.

Gentry and Waters [GW09], for example, defined a security notion they call "adaptive", because the adversary can corrupt users adaptively before the challenge phase. But there is a notion that is "even more" adaptive, where the adversary can still corrupt users *after* the challenge phase. The goal of this paper is thus to provide a better picture of the meaningful security models for BE, and to compare them. In particular, we investigate whether the various adaptive notions of corruption coincide or not.

J. Lopez and G. Tsudik (Eds.): ACNS 2011, LNCS 6715, pp. 377–394, 2011.

Related Work. The first scheme to come with a security argument was the subset-cover framework introduced by Naor, Naor, and Lotspiech [NNL01]. The framework uses symmetric keys, where the sender and the receivers share some secrets, so the security proof relies on assumptions about the symmetric primitives (one-way functions and block ciphers). Dodis and Fazio [DF03] presented the first CCA2-secure public-key Trace and Revoke (TR) scheme along with a security model covering CCA2 and generalized CCA2. When one considers possible corruption after the target ciphertext has been sent, one has to deal with forward-secrecy. This was done by Yao, Fazio, Dodis, and Lysyanskaya [YFDL04] who first considered forward-secrecy for HIBE and then by extension for BE. Boneh, Gentry, and Waters [BGW05] designed a fully collusion-resistant BE scheme and proposed a security model for it, where the adversary can corrupt all the users, except the target users. Thereafter, Boneh and Waters [BW06] presented a fully collusion-resistant TR scheme secure against adaptive attacks. Delerablée, Paillier, and Pointcheval [DPP07] also presented a fully collusion-secure dynamic BE scheme (DBE) and presented a new matching security model. More recently, Gentry and Waters [GW09] defined two additional security notions they call "semi-static" and "adaptive", as well as a generic transformation from a semi-static secure scheme into an adaptively secure scheme, and then a semi-static secure scheme to which they later apply the transformation.

Contribution. As shown above, many security notions were proposed in the literature. In this paper, we define a more systematic security model for broadcast encryption schemes, and construct a generic security framework for BE. We take into account, as usual in the "provable security framework", oracles to model the means available to the adversary, such as the possibility to join new users, to corrupt users, and to decrypt messages. It is worth noting that small details can have a high impact. For example, the choice of the set of users to which the challenge message is sent also plays a role in how the models relate to each other. We investigate the relationships between the different notions, and find that in some cases, two notions are equivalent or separated depending on the availability of some oracles or the collusion-resistance of a BE scheme. After describing the relationships between notions in our framework, we have a closer look at the security models and the schemes proposed in the literature, and discuss where they are in our framework, which then helps to compare them.

Our results are relevant for several existing BE schemes. For example, from the proof found in [GW09], it is clear that the two-key transformation actually achieves the stronger 2-adaptive-security level. We also examine the proof found in [DPP07], and see that it can fulfill a stronger security notion where the adversary can choose the target set, and then the scheme meets the requirements from [GW09]. This means that it can be made 2-adaptively secure using the generic transformation (although not efficiently).

Organization. In section 2 we provide a formal definition of broadcast encryption, or more precisely key encapsulation, and specify some terminology. In section 3 we define our security framework. Section 4 relates the different security notions to each other. In section 5 we embed the existing security models from the literature into our framework. In section 6, we describe which security notions have been achieved by existing protocols and describe an (inefficient) protocol that achieves the strongest notion.

2 Definitions

Broadcast encryption (BE) schemes enable the sender of a message to specify a subset of the registered users (the *target set* or *privileged set*), who will be able to decrypt the ciphertext sent to all users via a broadcast channel. The complement of the target set (in the set of the registered users) is called the *revoked set*. To accomplish user revocation when sending a message, a BE generally generates three parts: the *Id Header*, that is a bit-string that unambiguously identifies the target set/revoked set; the *Key Header*, that encapsulates a session key for the privileged users; and the *Message Body*, that contains the payload encrypted under the session key.

Since for all the schemes, the *Id Header* and the *Message Body* are similar, in this paper, we will focus on the *Key Header* part only, which can be seen as a key encapsulation mechanism (KEM). Furthermore, when no more information is given, we will consider a public-key key encapsulation system with possibly stateful decoders: encryption key is public, the decryption keys of the users can evolve, but the updates will be global and sent on a public channel, and ephemeral keys are distributed to be used together with symmetric encryption (DEM: Data Encapsulation Mechanism). We will nevertheless sometimes make remarks about alternative cases.

Definition 1 (Dynamic Broadcast Encapsulation). *A dynamic broadcast encapsulation scheme is a tuple of algorithms* $\mathcal{DBE} = (\mathsf{Setup}, \mathsf{Join}, \mathsf{Encaps}, \mathsf{Decaps})$*:*

- $\mathsf{Setup}(1^k)$*, where* k *is the security parameter, generates the global parameters* param *of the system (omitted in the following); and returns a master secret key* MSK *and an encryption key* EK*. It also initiates an empty list Reg. If the scheme is asymmetric,* EK *is public, otherwise it can be seen as a part of the* MSK*.*
- $\mathsf{Join}(\mathsf{MSK}, Reg, \mathsf{id})$ *takes as input the master secret key, the list Reg, and a user identifier* id*. If* $\mathsf{id} \in \mathcal{UI}$ *(where* $\mathcal{UI}$ *is the set of valid user identifiers, usually* $\mathbb{N}$*) and* $\mathsf{id} \notin Reg$*, outputs a user secret key* $\mathsf{usk}_{\mathsf{id}}$ *and a public user tag* $\mathsf{upk}_{\mathsf{id}}$*. The pair* $(\mathsf{id}, \mathsf{upk}_{\mathsf{id}})$ *is appended to Reg. Else, outputs* $\perp$*.*
- $\mathsf{Encaps}(\mathsf{EK}, Reg, S)$ *takes as input the encryption key, the list Reg, and a target set* S *and outputs a key header* H *and a session key* $K \in \{0,1\}^k$*.*
- $\mathsf{Decaps}(\mathsf{usk}_{\mathsf{id}}, S, H)$ *takes as input a user secret key, the target set* S*, and the key header* H*. If* $\mathsf{id} \in S$*, outputs the session key* K*.*

The correctness requirement is that for any (polynomial size) set of joined users $U \subset \mathcal{UI}$, any target set $S \subset U$ and for any $\mathsf{id} \in \mathcal{UI}$, if $\mathsf{id} \in S$ then the decapsulation algorithm gives back the ephemeral session key. See figure 1.

$$(\mathsf{MSK}, \mathsf{EK}, Reg) \leftarrow \mathsf{Setup}(1^k);$$
$$\texttt{for all}\ \ \mathsf{id} \in S:\ (\mathsf{usk}_{\mathsf{id}}, \mathsf{upk}_{\mathsf{id}}, Reg) \leftarrow \mathsf{Join}(\mathsf{MSK}, Reg, \mathsf{id});$$
$$(H, K) \leftarrow \mathsf{Encaps}(\mathsf{EK}, Reg, S).$$

$$i \in S \Rightarrow \mathsf{Decaps}(\mathsf{usk}_i, S, H) = K$$

Fig. 1. $\mathcal{DBE}$: Correctness

2.1 Terminologies and Various Types of Schemes

Join Algorithms. When the Join algorithm can be run at the setup phase only, with no later evolution of the group, we say the scheme is *static* (instead of *dynamic*). For a dynamic scheme, several kinds of Join functionalities are possible:

Passive, no input (except a counter i); it generates a public tag upk_i to identify the user;
Active, the input is id; it generates a public tag $\mathsf{upk}_{\mathsf{id}}$ to identify the user;
Identity-Based, the input is id, and the public tag $\mathsf{upk}_{\mathsf{id}}$ is simply id.

We stress that the default case in this paper (when no other version is specified) is that Join is passive.

Target Set. A broadcast encryption scheme is called *inclusive* when the target set is specified by the list of authorized users, and *exclusive* when the target set is specified by its complement $\mathcal{R}$, the set of revoked users.

Key Encapsulation Mechanisms. We described above a key encapsulation mechanism (KEM) where only a key is generated. The payload is then encrypted with a symmetric mechanism to get a full encryption scheme. All the broadcast encryption schemes known to the authors can be written as KEMs, e. g. the bilinear BE schemes from [BGW05, GW09] generate a random group element which is then multiplied to the message. This random group element can be considered as the symmetric key, and group multiplication as the symmetric encryption. To achieve CCA2-security for the full broadcast encryption, given a CCA2-secure key encapsulation, we additionally need to bind all the components of the ciphertext together.

Encryption and Decryption Keys. The encryption key can be either public (asymmetric) or private (symmetric), in the former case, we talk about *public-key* broadcast encryption, in the latter we say this is a *private-key* broadcast encryption. The decryption keys can either be defined and sent to the users at the join phase and never modified again, or be updated each time another user joins the system. In the former case, the decoders are said to be *stateless* since there is no state to evolve. In the latter case, the decoders are called *stateful* because they have to keep their state up-to-date. They thus have to be always on-line to receive the update information.

Default. As already mentioned, in this paper, we focus on public-key key encapsulation system with possibly stateful decoders.

3 Security Notions

Besides the various properties that a broadcast encryption scheme can satisfy, many security notions have been defined to take all the threats into consideration. We will thus review them, and try to give a cleaner view. As usual, security notions are defined by the goal the adversary want to achieve, and by the means that are available. We first define our standard security notions, and then compare them with some alternatives defined in the literature.

3.1 Standard Security Notions

Since we are studying a KEM [CS03], the goal of the adversary is to distinguish two keys in a key encapsulation, noted IND for *key indistinguishability*: after having received the public parameters, in the first phase (the FIND phase) the adversary outputs a target set S; then the challenger runs the key encapsulation algorithm, on this set S, that outputs the ephemeral K and the encapsulation H. It then chooses a random key K' and a random bit b and sets $K_b = K$ and $K_{1-b} = K'$. Upon receiving (H, K_0, K_1), the adversary runs the second phase (the GUESS) during which it has to decide whether H encapsulates K_0 or K_1, which means it has to guess the bit b.

Oracles can be available at different periods of time (Setup, FIND-phase, or GUESS-phase) which defines several kinds of attacks. Figure 2 shows the experiment $\mathrm{Exp}^{\mathsf{ind-d}x\mathsf{a}y\mathsf{cca}z}_{\mathcal{DBE},\mathcal{A}}(k)$, where the oracles OJoin_1, $\mathsf{OCorrupt}_1$ and $\mathsf{ODecaps}_1$ are available during the FIND-phase, and the oracles OJoin_2, $\mathsf{OCorrupt}_2$ and $\mathsf{ODecaps}_2$ are available during the GUESS-phase. According to the exact definition of these oracles, we have an IND-Dyn**x**-Ad**y**-CCA**z** security game, for x-Dynamic (Join), y-Adaptive (Corrupt) and CCA-z (Decaps). If not otherwise specified, use of the variables x, y, z means that they can be replaced by any level defined below.

The Join *Oracle.* It can be available at the Setup-time only. In this case, the adversary can make a number of non-adaptive Join-queries, where he receives the results only at the end of the Setup-phase, together with the parameters and MSK, EK. As said above, we then talk about a **static scheme**, and the attack is *S-Dynamic* (or Dyn**S**), and both the oracles OJoin_1 and OJoin_2 output $\perp$. The Join-oracle can be available during the first phase only, then $\mathsf{OJoin}_1 = \mathsf{Join}$ but the OJoin_2 oracle outputs $\perp$, and the attack is *1-Dynamic* (or Dyn**1**); it can be available always, then $\mathsf{OJoin}_1 = \mathsf{OJoin}_2 = \mathsf{Join}$, and the attack is *2-Dynamic* (or Dyn**2**).

The Corrupt *Oracle.* Corruptions can be more or less adaptive. Again, the adversary may have to decide before the Setup-time which users will be corrupted. This is a **selective attack** or *S-Adaptive* (also denoted Ad**S**), which is meaningful for static schemes (Dyn**S**) only (otherwise there are no users to corrupt during the Setup-phase), and then both the oracles $\mathsf{OCorrupt}_1$ and $\mathsf{OCorrupt}_2$ output $\perp$. It can be available during the first phase only, then $\mathsf{OCorrupt}_1 = \mathsf{Corrupt}$ but the

$\mathrm{Exp}_{\mathcal{DBE},\mathcal{A}}^{\mathsf{ind-d}x\mathsf{a}y\mathsf{cca}z-b}(k)$
$(\mathsf{MSK}, \mathsf{EK}) \leftarrow \mathsf{Setup}(1^k);$
$\mathcal{Q}_C \leftarrow \emptyset;\ \mathcal{Q}_D \leftarrow \emptyset;$
$(st, S) \leftarrow \mathcal{A}^{\mathsf{OJoin}_1(\cdot),\mathsf{OCorrupt}_1(\cdot),\mathsf{ODecaps}_1(\cdot,\cdot,\cdot)}(\mathsf{param});$
$(H, K) \leftarrow \mathsf{Encaps}(\mathsf{EK}, Reg, S);\ K_b \leftarrow K;\ K_{1-b} \stackrel{\$}{\leftarrow} \mathcal{K};$
$b' \leftarrow \mathcal{A}^{\mathsf{OJoin}_2(\cdot),\mathsf{OCorrupt}_2(\cdot),\mathsf{ODecaps}_2(\cdot,\cdot,\cdot)}(st; S, H, K_0, K_1);$
`if` $\exists i \in S, (i, S, H) \in \mathcal{Q}_D$ `or` $i \in \mathcal{Q}_C$
`then return` $0;$
`else return` $b';$

where $x \in \{s, 1, 2\}$, $y \in \{0, s, 1, 2\}$, $z \in \{0, 1, 2\}$.

$\mathsf{OJoin}(i)$
$(\mathsf{usk}_i, \mathsf{upk}_i) \leftarrow \mathsf{Join}(\mathsf{msk}, i);$
`return` $\mathsf{upk}_i;$

$\mathsf{OCorrupt}(i)$
$\mathcal{Q}_C \leftarrow \mathcal{Q}_C \cup \{i\};$
`return` $\mathsf{usk}_i;$

$\mathsf{ODecaps}(i, S, H)$
$\mathcal{Q}_D \leftarrow \mathcal{Q}_D \cup \{(i, S, H)\}$
$K \leftarrow \mathsf{Decaps}(\mathsf{usk}_i, S, H);$
`return` $K;$

Fig. 2. $\mathcal{DBE}$: Key Privacy (IND-Dynx-Ady-CCAz)

$\mathsf{OCorrupt}_2$ oracle outputs $\perp$, and the attack is *1-Adaptive* (or Ad**1**). It can be available during the full security game, then $\mathsf{OCorrupt}_1 = \mathsf{OCorrupt}_2 = \mathsf{Corrupt}$, and the attack is *2-Adaptive* (or Ad**2**). Eventually, the adversary can have no access at all to the Corrupt oracle: we say the attack is *0-Adaptive* (or Ad**0**).

The Decaps *Oracle.* As usual for chosen-ciphertext security, the Decaps-oracle can be available or not. It can never be available in the CPA (or CCA**0**) scenario, and both the oracles $\mathsf{ODecaps}_1$ and $\mathsf{ODecaps}_2$ output $\perp$; it can be available during the first phase only, then $\mathsf{ODecaps}_1 = \mathsf{Decaps}$ but the $\mathsf{ODecaps}_2$ oracle outputs $\perp$, and the attack is CCA**1**; it can be available during the full security game, then $\mathsf{ODecaps}_1 = \mathsf{ODecaps}_2 = \mathsf{Decaps}$, and the attack is CCA**2**.

For the IND-goal, the natural restriction for the adversary is not to ask for the decapsulation of the challenge header H nor corrupt any user in the target set S.

Remark 2. For private-key schemes, the adversary is granted access to the encapsulation oracle instead of the encryption key. In the rest of the paper, we will focus on the public-key setting for dynamic broadcast encryption schemes (noted PKDBE).

Definition 3. *A public-key DBE scheme $\mathcal{DBE}$ is said to be $(t, N, q_C, q_D, \varepsilon)$-IND-*Dynx-Ady-CCAz*-secure if in the security game presented in figure 2, the advantage, denoted $\mathrm{Adv}_{\mathcal{DBE}}^{\mathsf{ind-d}x\mathsf{a}y\mathsf{cca}z}(k, t, N, q_C, q_D)$, of any t-time adversary $\mathcal{A}$ registering at most N users (*OJoin *oracle), corrupting at most q_C of them (*OCorrupt *oracle), and asking for at most q_D decapsulation queries (*ODecaps *oracle), is bounded by ε:*

$$\mathrm{Adv}_{\mathcal{DBE}}^{\mathsf{ind-d}x\mathsf{a}y\mathsf{cca}z}(k, t, N, q_C, q_D) =$$
$$\max_{\mathcal{A}}\{\Pr[\mathrm{Exp}_{\mathcal{DBE},\mathcal{A}}^{\mathsf{ind-d}x\mathsf{a}y\mathsf{cca}z-1}(k) = 1] - \Pr[\mathrm{Exp}_{\mathcal{DBE},\mathcal{A}}^{\mathsf{ind-d}x\mathsf{a}y\mathsf{cca}z-0}(k) = 1]\} .$$

3.2 Alternatives and Variants

Forward-Secrecy. For dynamic exclusive schemes (the target set is defined by the list of revoked users), new users are by definition included in the target sets of the message headers, even if they did not exist at the time the header was sent. Furthermore, since new users are included in the challenge set S, the adversary is not allowed to corrupt them. This means the encryption does not provide forward-secrecy. To model forward-secrecy, we can allow corruption of joined users, and in this case the encryption key EK must evolve when a new user joins the system.

For dynamic inclusive schemes (the target set is defined by the list of authorized users), the $\mathsf{Ad}\mathbf{2}$ notion provides *forward-secrecy* since any user not in the target set can be corrupted in the second phase.

Target Set. In the default security game, the adversary chooses the target set S at the end of the first phase, the FIND phase which consists in finding the best S for winning the game. But some papers in the literature restrict this choice:

- The adversary announces the target set before the setup phase [BGW05]. We call this *selective security*, denoted $\mathsf{Targ}\mathbf{S}$. This can only happen in static schemes, because the adversary needs to know the set of users to choose the target set from.
- The target set is automatically set to all uncorrupted users at the end of the first phase [DF03]. We call this *fixed-target-set security*, denoted $\mathsf{Targ}\mathbf{F}$.

When needed, the default case (the adversary chooses the target set S at the end of the FIND-phase) is denoted $\mathsf{Targ}\mathbf{C}$.

Security Models in the Literature. We can now characterize all the security models defined in the literature into our formalism: These notions are summarized in table 1, when S is the target set and C the corrupted users set.

- In [YFDL04], the authors defined the full access to the $\mathsf{Corrupt}$ oracle, but for a static scheme (no Join oracle). In order to accommodate the forward-secrecy, they included time slots. Disregarding the latter, the security model is similar to IND-$\mathsf{Dyn}\mathbf{S}$-$\mathsf{Ad}\mathbf{2}$-$\mathsf{CCA}\mathbf{2}$-$\mathsf{Targ}\mathbf{F}$. Essentially the adversary is restricted to corrupting only users from a time slot later than the one the challenge message was sent in. In our model, IND-$\mathsf{Dyn}\mathbf{x}$-$\mathsf{Ad}\mathbf{2}$-$\mathsf{CCA}\mathbf{z}$-$\mathsf{Targ}\mathbf{F}$-security does only make sense for $x = 2$, as otherwise no users can be corrupted in the GUESS-phase (because the target set is fixed to $U \setminus C$ and the adversary cannot join new users after the challenge phase).
- In [Del08] the authors define a security model for IBBE they call IND-sID-CCA (selective ID CCA-security), which is IND-$\mathsf{Dyn}\mathbf{S}$-$\mathsf{Ad}\mathbf{2}$-$\mathsf{CCA}\mathbf{2}$-$\mathsf{Targ}\mathbf{S}$-security in our notation.

Table 1. Adversarial Capabilities

Security	before setup	FIND-Phase	Challenge-Phase	GUESS-Phase
Ad**2**		Corrupt	S	Corrupt
Ad**2**Targ**F**		Corrupt		Corrupt
Ad**1**		Corrupt	S	
Ad**1**Targ**F**		Corrupt		
semi-static	C		S	
static	C			

- The partial access to the Corrupt oracle has been used in [BW06] and [GW09]. In our notation, the authors used IND-Dyn**S**-Ad**1**-CCA**0** security, since no decapsulation queries were available.
- As noted above, the the fixed-target-set security was introduced in [DF03], but no Corrupt queries were allowed in the second phase, and the system was static (no Join query). In our formalism, this is IND-Dyn**S**-Ad**1**-CCA**z**-Targ**F**, according to the Decaps-oracle access.
- *Semi-static* security has been introduced in [GW09] in order to build a generic conversion into Adaptive-1. In this setting, the adversary must announce the set of corrupted users before the setup phase, as we defined as *selective attack*. In our notation, this is IND-Dyn**S**-Ad**S**-CCA**0** security.[1]
- In the *static* model due to [BGW05], the adversary also has to announce its target set before the setup phase (selective attack). In our notation, this is IND-Dyn**S**-Ad**S**-CCA**2**-Targ**F** security with fixed target set. The authors also define a CPA version.[2]

Collusion Resistance. We can also distinguish between two types of collusion-resistance: full collusion-resistance, where there is no limit on the number of Corrupt-queries, and t-collusion-resistance, where the number of queries is bounded by t (which can depend on the number of users N). With our parameters, we implicitly consider all the cases.

[1] More precisely, in the *semi-static* version of the experiment, the adversary must commit to a set $\tilde{S}$ before the setup phase. He is allowed to corrupt any user not in $\tilde{S}$ after the setup phase, and must choose a challenge set $S \subseteq \tilde{S}$. An equivalent formulation is that the adversary chooses the set C of users to corrupt before the setup phase (because he can corrupt all users not in $\tilde{S}$), but chooses S at the challenge phase. This formulation is only equivalent for fully collusion resilient schemes, but it is for these schemes that the notions were designed.

[2] In the *static* version of the experiment [BGW05], the adversary has to announce the set S of users he wants to attack before the setup phase. He then receives the private keys of all users not in S after the setup phase. An equivalent definition is that he chooses the set C of corrupted users, and the S is fixed to be all the users except C. To allow the adversary to choose the target set, the adversary announces both C and S before the setup phase. This definition where the adversary chooses both C and S can also be used in not fully collusion-secure schemes and is the one considered in this section.

4 Relationship between the Security Notions

In this section, we shed light on the relationships between the security notions we defined in the last section. We start in section 4.1 with the hierarchy of Decaps-oracles, where we expect no surprises. In section 4.2, we explore the Join-oracle, of which we defined three different versions: For the passive version, which takes no input, all notions are equivalent; For the active version, which takes input and outputs a user tag, we can separate all notions. For the IBBE version, which takes an arbitrary string as input, but does not output a user tag (the upk is the identity of the user), we can show equivalences and separations based on the availability of a Corrupt-oracle. In section 4.3, we address the Corrupt-queries, and gaps appear according to the number of such queries, and thus the level of collusion-resistance. In section 4.4, we examine the various ways in which the target set can be chosen. Due to space limitations, many proofs have been removed from this version, but are given in the full version [PPS11].

4.1 Separating CPA and CCA

We remember the well-known separation between CPA (CCA**0**), CCA**1**, and CCA**2**-security for PKE from [BDPR98]. The same separation applies in the case of broadcast encryption, because if we set $\mathsf{KeyGen}(1^k)$ to
$(\mathsf{MSK}, \mathsf{EK}) \leftarrow \mathsf{Setup}(1^k); (\mathsf{usk}_1, \mathsf{upk}_1) \leftarrow \mathsf{Join}(\mathsf{MSK}, 1); \mathsf{dk} \stackrel{\mathsf{def}}{=} \mathsf{usk}_1, \mathsf{ek} \stackrel{\mathsf{def}}{=} \mathsf{EK}||\mathsf{upk}_1$,
we obtain a single-user KEM scheme.

Theorem 4. *The following implications are strict:*

$$\mathit{IND}\text{-}\mathsf{Dynx}\text{-}\mathsf{Ady}\text{-}\mathsf{CCA}\mathbf{2} \Rightarrow \mathit{IND}\text{-}\mathsf{Dynx}\text{-}\mathsf{Ady}\text{-}\mathsf{CCA}\mathbf{1} \Rightarrow \mathit{IND}\text{-}\mathsf{Dynx}\text{-}\mathsf{Ady}\text{-}\mathsf{CCA}\mathbf{0}.$$

4.2 Separating Notions of Dynamicity

In this section, in order to compare the Join-oracle access, we also have to consider the three versions of the Join-algorithm, as defined in section 2.1: *passive*-Join, if it takes no input; *active*-Join, if it takes an input; *ID-based*-Join, if the output tag upk is the input identity.

Easy Implications. As above, there is a clear hierarchy on the Join oracle access: at the setup time only, in the first phase, or at anytime.

Theorem 5. *The following implications hold for all versions of the* Join oracle: IND-Dyn**2**-Ady-CCA**z** $\Rightarrow$ IND-Dyn**1**-Ady-CCA**z** $\Rightarrow$ IND-Dyn**S**-Ady-CCA**z**.

Passive Join. This is a standard definition in the literature. Interestingly enough, in this context all the notions are equivalent, since the adversary cannot influence the output.[3]

[3] It is interesting to note that the equivalence is for our above notions only: for passive-Join, a query in the first phase is strictly more useful than a query in the second phase. As a consequence, if we consider in details the number of queries in each phase, as done in [PP04] for the encryption and decryption oracles, we can show that $\mathsf{Dyn}(N_1 + N_2, 0) \rightarrow \mathsf{Dyn}(N_1, N_2) \rightarrow \mathsf{Dyn}(0, N_1 + N_2)$, and these implications are strict. However, in the above theorem, we do not fix the number of queries.

Theorem 6. *If* Join *takes no input,we have the following equivalences*

$$\mathit{IND}\text{-Dyn2-Ady-CCAz} \Leftrightarrow \mathit{IND}\text{-Dyn1-Ady-CCAz} \Leftrightarrow \mathit{IND}\text{-DynS-Ady-CCAz}.$$

Proof. Because of the trivial implications, it remains to show that DynS $\Rightarrow$ Dyn2. Given a successful Dyn2-adversary $\mathcal{A}^d$ that makes N_1 queries to the Join-oracle in phase 1, and N_2 queries to the Join-oracle in phase 2, we construct a successful DynS-adversary $\mathcal{A}^s$ that joins $N = N_1 + N_2$ users before the setup phase. Because the Join-oracle takes no input, its behavior is exactly the same in phase 1 and phase 2. Therefore $\mathcal{A}^s$ can store the results and then answer all Join-queries made by $\mathcal{A}^d$ later.

Active Join with Large Input. If the Join-algorithm is interactive or takes input from the adversary (that can be sufficiently large, i. e. $|\mathcal{UI}|$ is superpolynomial), the adversary can influence the Join-process:

Theorem 7. *If* Join takes input and outputs a public tag, the following implications are strict

$$\text{IND-Dyn2-Ady-CCAz} \Rightarrow \text{IND-Dyn1-Ady-CCAz} \Rightarrow \text{IND-DynS-Ady-CCAz}.$$

Identity-Based. In this case, the OJoin-oracle only outputs a user secret key $\mathsf{usk}_{\mathsf{id}}$ (because $\mathsf{upk}_{\mathsf{id}} = \mathsf{id}$). This means that in order to gain anything from the output, the adversary must also be able to corrupt users.

Theorem 8. *For ID-Based Broadcast Encryption, the following implications are strict*

$$\begin{array}{lllll}
\mathit{IND}\text{-Dyn2-Ad2-CCAz} & \Rightarrow & \mathit{IND}\text{-Dyn1-Ad2-CCAz} & \Rightarrow & \mathit{IND}\text{-DynS-Ad2-CCAz}\\
\mathit{IND}\text{-Dyn2-Ad1-CCAz} & \Leftrightarrow & \mathit{IND}\text{-Dyn1-Ad1-CCAz} & \Rightarrow & \mathit{IND}\text{-DynS-Ad1-CCAz}\\
\mathit{IND}\text{-Dyn2-AdS-CCAz} & \Leftrightarrow & \mathit{IND}\text{-Dyn1-AdS-CCAz} & \Leftrightarrow & \mathit{IND}\text{-DynS-AdS-CCAz}\\
\mathit{IND}\text{-Dyn2-Ad0-CCAz} & \Leftrightarrow & \mathit{IND}\text{-Dyn1-Ad0-CCAz} & \Leftrightarrow & \mathit{IND}\text{-DynS-Ad0-CCAz}
\end{array}$$

4.3 Separating Forms of Corruption

Theorem 9.

$$\begin{aligned}
&\mathit{IND}\text{-Dynx-Ad2-CCAz} \Rightarrow \mathit{IND}\text{-Dynx-Ad1-CCAz}\\
&\quad \Rightarrow \mathit{IND}\text{-DynS-AdS-CCAz} \Rightarrow \mathit{IND}\text{-Dynx-Ad0-CCAz},
\end{aligned}$$

and for BE schemes that are not fully collusion-secure all implications are strict.

Proof. The implications are clear, since having access to an oracle never makes an adversary weaker. The separations follow from lemmas 10, 11, 12, and 14.

Separation of no Corruption from Selective Corruption. Recall that for AdS, the only version of Dyn that makes sense is DynS (section 3.1).

Lemma 10. *IND*-Dynx-Ad**0**-CCAz $\not\Rightarrow$ *IND*-Dyn**S**-Ad**S**-CCAz.

Separation of Selective Corruption from 1-Adaptive Corruption. In a model with selective corruption, the adversary must announce the set C of corrupted users before seeing the encryption key. To make a difference, we would have to give some information on the subset of the users to corrupt in the encryption key: we thus embed such information using a secret sharing scheme to make sure all of the identified users (special users) have to be corrupted.

We need to make sure that the subset is hard to guess by chance: this is the case for IBBE, where the size of the set $\mathcal{UI}$ is exponential and any user is hard to guess. In case $\mathcal{UI}$ is of polynomial size, the size of the subset must not be too small, otherwise all of them will be corrupted even by a selective-corruption adversary with significant probability. If t is the number of special users, there are $\binom{N}{t}$ ways of choosing them, where N is polynomial in the security parameter. To make the binomial be super-polynomial for a polynomial N, we need t to be non-constant.

How can we be sure the adversary corrupts at most t users? First, it can be set by definition, using the t-collusion secure level. For IBBE, $\binom{|\mathcal{UI}|}{1}$ is already exponential in the security parameter. However, without any additional constraint, for a basic broadcast encryption scheme, if $N-t$ is constant, then $|S|$ must also be constant, and we are actually dealing with a simple multi-encryption scenario. Multi-cast security of encryption schemes has been considered in [BPS00]. The authors proved that standard IND-CPA encryption schemes remain secure even if the same message is sent to different users in parallel. This makes the case where the adversary is always sending only to a constant number of users less interesting to us. It thus seems reasonable to exclude these cases from BE. In the following, we thus focus on t-collusion secure schemes, where t must be less than the total number of users minus a non-constant number.

Lemma 11. *For a t-collusion secure scheme (for t and $N-t$ non-constant numbers),*

IND-Dyn**S**-Ad**S**-CCAz $\not\Rightarrow$ *IND*-Dyn**S**-Ad**1**-CCAz.

Separation of 1-Adaptive Corruption from 2-Adaptive Corruption.

Lemma 12. *For a t-collusion secure scheme (for t and $N-t$ non-constant numbers),*

IND-Dynx-Ad**1**-CCAz $\not\Rightarrow$ *IND*-Dynx-Ad**2**-CCAz *for* $z \in \{0,1\}$.

As noted, the proof requires t and $N-t$ to be non-constant. But we can also note that it does not work in the CCA**2**-setting, because on the one hand the scheme is malleable, and on the other hand the adversary could simply query the H_i's to the Decaps-oracle.

For the following lemma, we need the notion of a homomorphic OWF.

Definition 13 (Homomorphic One-Way Function). *Let $(G,+)$ and $(H,*)$ be two groups with $2^{k-1} \le |G| \approx |H| \le 2^k$. A PPT-computable function $f : G \to H$ is*

– one-way *if* $\forall \mathcal{A} : \Pr[x \stackrel{\$}{\leftarrow} G; y \leftarrow \mathcal{A}(1^k, f(x)); f(y) = f(x)]$ *is negligible.*
– homomorphic *if* $f(x+y) = f(x) * f(y)$.

An example of a homomorphic OWF is discrete exponentiation (assuming DLOG is hard). The security of MAC and symmetric encryption is defined as usual.

Lemma 14. *For a t-collusion secure scheme (for t and N − t non-constant numbers),if strongly-UF-CMA-secure MAC, IND-CCA2-secure symmetric encryption and homomorphic OWF exist,*

$$\textsf{\textit{IND}-Dynx-Ad1-CCA2} \not\Rightarrow \textsf{\textit{IND}-Dynx-Ad2-CCA2}.$$

4.4 Choice of the Target Set

Selective Security

Lemma 15. *The following implication is strict:*

$$\textsf{\textit{IND}-Dynx-Ady-CCAz-TargC} \Rightarrow \textsf{\textit{IND}-DynS-Ady-CCAz-TargS}.$$

Fixed Target Sets. In our definition the adversary chooses the target set S of the challenge. In the DPP security model [DPP07], S is automatically the set of all non-compromised users. The same situation appears in [BGW05], where the adversary outputs S before the setup and receives the secret keys for all users in $U \setminus S$. A similar definition is given for the BGW model. We could reformulate the BGW model so that the adversary outputs the set C of the keys he wants to know, and S is set to $U_1 \setminus C_1$. This formulation is obviously equivalent. We want to investigate the relationship between these two notions.

Note that under the "fixed" definition, the notions IND-Dynx-Ad1-CCAz and IND-Dynx-Ad2-CCAz for $x \in \{s, 1\}$ are equivalent since in any case the adversary cannot corrupt users after the challenge phase (all the non-corrupted users at the end of the first phases are in the target set and cannot be corrupted).

Theorem 16. *All the following implications are strict*

$$\begin{aligned}
\textsf{\textit{IND}-DynS-AdS-CCAz-TargC} &\Rightarrow \textsf{\textit{IND}-DynS-AdS-CCAz-TargS} \\
&\Leftrightarrow \textsf{\textit{IND}-DynS-AdS-CCAz-TargF} \\
\textsf{\textit{IND}-Dynx-Ad0-CCAz-TargC} &\Rightarrow \textsf{\textit{IND}-DynS-Ad0-CCAz-TargS} \\
&\Rightarrow \textsf{\textit{IND}-Dynx-Ad0-CCAz-TargF}
\end{aligned}$$

The theorem follows from lemmas 15, 17, 18, and 19.

Lemma 17. *IND*-DynS-Ady-CCAz-TargS $\Rightarrow$ *IND*-Dynx-Ady-CCAz-TargF *for* $y \in \{0, s\}$.

Proof. From an adversary $\mathcal{A}^f$ against the IND-Dynx-Ady-CCAz-TargF-security of a BE scheme, we build an adversary $\mathcal{A}^S$ against the IND-DynS-Ady-CCAz-TargS-security. If the model has no corruption or static corruption, $\mathcal{A}^S$ runs $\mathcal{A}^f$, who outputs C, chooses the same C and sets his target set $S = U \setminus C$.

Lemma 18. IND-Dyn**S**-Ad**S**-CCA**z**-Targ**F** $\Rightarrow$ IND-Dyn**S**-Ad**S**-CCA**z**-Targ**S**.

Proof. Given a successful adversary $\mathcal{A}^S$, we construct an adversary $\mathcal{A}^f$ as follows. $\mathcal{A}^S$ outputs his target set S and the set of users to corrupt C before the Setup phase. $\mathcal{A}^f$ chooses $C' = U \setminus S$.

Lemma 19. IND-Dyn**x**-Ad**0**-CCA**z**-Targ**F** $\not\Rightarrow$ IND-Dyn**S**-Ad**0**-CCA**z**-Targ**S**.

Proof. In the IND-Dyn**x**-Ad**0**-CCA**z**-Targ**F**-experiment, the target set is always fixed to $S = U$. Given a IND-Dyn**x**-Ad**0**-CCA**z**-Targ**F**-secure scheme Π, we modify it into a scheme Π' that is still IND-Dyn**x**-Ad**0**-CCA**z**-Targ**F**-secure, but not IND-Dyn**x**-Ad**0**-CCA**z**-Targ**S**. The only change is that if $|S| = 1$, Π'.Encaps sets $K = 0$ (or determines the key in a deterministic way by fixing all random coins e. g. to 0).

Theorem 20. *For fully collusion-resilient BE schemes, the following implications are strict*

$$\mathit{IND}\text{-Dyn}\mathbf{x}\text{-Ad}\mathbf{y}\text{-CCA}\mathbf{z}\text{-Targ}\mathbf{C} \Leftrightarrow \mathit{IND}\text{-Dyn}\mathbf{x}\text{-Ad}\mathbf{y}\text{-CCA}\mathbf{z}\text{-Targ}\mathbf{F}$$
$$\Rightarrow \mathit{IND}\text{-Dyn}\mathbf{S}\text{-Ad}\mathbf{y}\text{-CCA}\mathbf{z}\text{-Targ}\mathbf{S} \quad (y \in \{1,2\})$$

The theorem follows from lemmas 15 and 21. It seem curious at first that the relationship between fixed target set and selective security is inverted for models with no corruption, but in this case the fixed target set means that it is always set to U, while the selective security allows some freedom of the adversary to choose.

Lemma 21. *For fully collusion-resistant BE schemes*

$$\mathit{IND}\text{-Dyn}\mathbf{x}\text{-Ad}\mathbf{y}\text{-CCA}\mathbf{z}\text{-Targ}\mathbf{C} \Leftrightarrow \mathit{IND}\text{-Dyn}\mathbf{x}\text{-Ad}\mathbf{y}\text{-CCA}\mathbf{z}\text{-Targ}\mathbf{F} \quad (y \in \{1,2\}).$$

Proof. It is clear that if the adversary can choose S freely, he can set it to $U \setminus C$. Let $\mathcal{A}^{choice}$ be a successful adversary against a BE scheme that can choose his target set S. Then we construct $\mathcal{A}^{fixed}$ as follows: $\mathcal{A}^{fixed}$ faithfully forwards all queries. When $\mathcal{A}^{choice}$ outputs his challenge target set S, $\mathcal{A}^{fixed}$ first issues corrupt queries so that $U \setminus C = S$, then asks for the challenge and forwards it to $\mathcal{A}^{choice}$. He forwards the guess bit b and wins with the same probability as $\mathcal{A}^{choice}$.

Note that $\mathcal{A}^{fixed}$ corrupts more users, which could reduce the tightness of a security proof, and causes the proof to fail in a t-resilient setting where $t < N-1$ (if $t = N-1$, the scheme is fully collusion-resistant).

In the following, we denote by $\not\Leftrightarrow$ the fact that $\not\Rightarrow$ in both directions.

Theorem 22. *For BE schemes where the adversary must leave at least two users uncorrupted, the following implications are strict:*

$$\mathit{IND}\text{-Dyn}\mathbf{x}\text{-Ad}\mathbf{y}\text{-CCA}\mathbf{z}\text{-Targ}\mathbf{C} \Rightarrow \mathit{IND}\text{-Dyn}\mathbf{x}\text{-Ad}\mathbf{y}\text{-CCA}\mathbf{z}\text{-Targ}\mathbf{F}$$
$$\not\Leftrightarrow \mathit{IND}\text{-Dyn}\mathbf{S}\text{-Ad}\mathbf{y}\text{-CCA}\mathbf{z}\text{-Targ}\mathbf{S}$$

and IND-Dyn**x**-Ad**y**-CCA**z**-Targ**C** $\Rightarrow$ IND-Dyn**S**-Ad**y**-CCA**z**-Targ**S** $(y \in \{1,2\})$

The theorem follows from lemmas 15, 23, 24, and 25.

Lemma 23. *If the adversary is restricted to leaving at least 2 users uncorrupted, the following implication is strict*

$$\textsf{\textit{IND}-Dynx-Ady-CCAz-Targ}\mathbf{C} \Rightarrow \textsf{\textit{IND}-Dynx-Ady-CCAz-Targ}\mathbf{F} \quad (y \in \{1,2\}).$$

Lemma 24. *If the adversary is restricted to leaving at least 2 users uncorrupted,*

$$\textsf{\textit{IND}-Dynx-Ady-CCAz-Targ}\mathbf{F} \not\Rightarrow \textsf{\textit{IND}-Dyn}\mathbf{S}\textsf{-Ady-CCAz-Targ}\mathbf{S}.$$

We can easily see that the adversary does not get weaker if he can choose the target set freely from the set of uncorrupted users $U \setminus C$, because he can choose $S = U \setminus C$ as in the fixed case.

Lemma 25.

$$\textsf{\textit{IND}-Dyn}\mathbf{S}\textsf{-Ady-CCAz-Targ}\mathbf{S} \not\Rightarrow \textsf{\textit{IND}-Dyn}\mathbf{S}\textsf{-Ady-CCAz-Targ}\mathbf{F} \textit{ for } y \in \{1,2\}.$$

5 Relationships between Notions from the Literature

A security notion that our model does not cover is defined in [DPP07]. In this model, the adversary accesses a JoinCorrupted oracle instead of the Corrupt oracle. That means he must decide whether to corrupt a user before the user is joined, but the choice can depend on information gained previously. The model defined in [DPP07] is Dyn**1**, as the adversary has access to a Join oracle before the challenge phase, CCA**0** and Targ**F**, as the challenge set is *fixed* to $S = U \setminus C$, so it is rather similar to IND-Dyn1-Ad1-CCA0-Targ**F**-model in our framework, except that the Corrupt oracle is replaced with JoinCorrupted. We call it the *partially adaptive* model. As in the previous section, we also denote Targ**C** the default case where the adversary can choose S as any subset of $U \setminus C$.

Theorem 26. *We have the following implications*

$$\textsf{\textit{IND}-Dyn1-Ad1-CCAz-Targ}\mathbf{F} \Rightarrow \textit{partially adaptive} - \textsf{CCAz} - \textsf{Targ}\mathbf{C}$$
$$\Rightarrow \textit{partially adaptive} - \textsf{CCAz} - \textsf{Targ}\mathbf{F} \Rightarrow \textsf{\textit{IND}-Dyn}\mathbf{S}\textsf{-Ad1-CCAz-Targ}\mathbf{S}$$

that are all strict (the first one only if t-collusion secure with t and $N - t$ non-constant).

The proof can be found in the full version of this paper [PPS11].

We now have almost all the results we need to establish the relationship between the security notions that can be found in the existing literature to fill the picture on figure 3. We now complete it.

Theorem 27. *The following implication is strict*

$$\textit{Partially adaptive} - \textsf{CCAz} - \textsf{Targ}\mathbf{C} \Rightarrow \textsf{\textit{IND}-Dyn}\mathbf{S}\textsf{-Ad}\mathbf{S}\textsf{-CCA0-Targ}\mathbf{C}.$$

Proof. From any semi-static adversary $\mathcal{A}^S$ we construct a partially adaptive adversary as follows. $\mathcal{A}^S$ announces N and C before the setup phase. $\mathcal{A}^{pa}$ asks for JoinCorrupted on all users in C and simply joins all users in $U \setminus C$. The separation is analogous to the one in lemma 11.

We relate semi-static security to the version of static security with 1-adaptive corruption defined in [GW09].

Theorem 28. *The following implication is strict*

$$\textsf{\textit{IND}-DynS-AdS-CCA0-TargC} \Rightarrow \textsf{\textit{IND}-DynS-Ad1-CCA0-TargS}.$$

Proof. From any selectively 1-adaptive adversary $\mathcal{A}^a$ we construct a semi-static adversary $\mathcal{A}^s$. $\mathcal{A}^a$ announces N and S before the setup phase. $\mathcal{A}^s$ forwards N and sets $C = U \setminus S$. He now has enough information to answer all Corrupt-queries. The separation is analogous to the one in lemma 15.

Theorem 29. *Partially adaptive*-CCAz-TargF $\not\Leftrightarrow$ *IND*-DynS-AdS-CCAz-TargC.

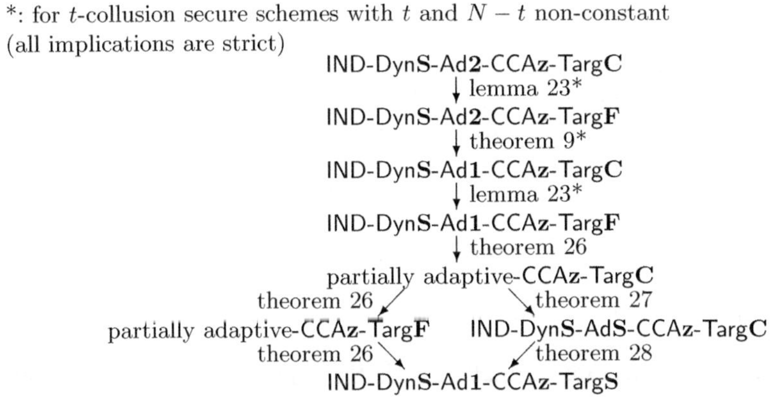

Fig. 3. Relations between Security Notions from the Literature

Remark 30. To conclude this section on the security notions found in the literature, we place some of the existing schemes more precisely. First, one can show that the DPP dynamic BE scheme [DPP07] fulfills the "choice"-definition without changing the security proof. Using theorem 28, we see that this notion implies semi-static security, so the two-key transformation from [GW09] can be used to achieve 1-adaptive security. The transformation is not efficient in this case, because the length of the ciphertext depends on the number of r revoked users, and to use the transformation, $N+r$ of $2N$ users have to be revoked. Looking at the proof in [GW09], we see that the after applying the two-key transformation, a scheme can be proved 2-adaptively secure using the same proof, because the simulator has the secret keys of each user and can answer Corrupt-queries in the GUESS-phase as easily as in the FIND-phase.

Table 2. Comparison between schemes

	DF03	BGW05	DPP07	Del08	GW09a	GW09b	Naive
Dyn	Dyn**S**	Dyn**S**	Dyn**S**	Dyn**1**	Dyn**S**	Dyn**S**	Dyn**2**
Ad	Ad**1**	Partially adaptive	Ad**2**	Ad**2**	Ad**2**	Ad**2**	Ad**2**
CCA	CCA**2**	CCA**2**	CCA**2**	CCA**0**	CCA**2**	CCA**2**	CCA**2**
Targ	Targ**S**	Targ**S**	Targ**S**	Targ**F**	Targ**C**	Targ**C**	Targ**C**

6 Previous Schemes

Let us now discuss on the previous schemes in order to compare them. Table 2 sums up the security levels for each of them.

DF03. Dodis and Fazio[DF03] proposed the first scheme that is secure against adaptive adversaries. However, their scheme is in the Targ**F** model. Consequently, the scheme can only be Ad**1**-secure, because any corrupted user in the second phase is implicitly included in the target set and can thus decrypt. One of the main disadvantages of the DF03 scheme is that the bound of maximum revoked user r_{max} should be fixed before the setup and as soon as there are more than r_{max} corrupted users, the scheme can be totally broken. The DF03 scheme can be shown to be Ad**2**-secure when the target set is adversarially chosen with the size of the revoked set bounded by r_{max} and the total number of corrupted users in both first and second phases is also bounded by r_{max}.

BGW05. In [BGW05], Bonch, Gentry, and Waters presented new methods for achieving fully collusion-resistant systems with short ciphertexts. However, the scheme is only proved secure in the static model (Dyn**S**). As discussed in [GW09], the BGW proof of security requires an "exact cancellation" and there is not an obvious way to prove BGW05 to be semi-statically secure.

DPP07. In [DPP07], Delerablée, Paillier, and Pointcheval proposed a dynamic scheme that is partially adaptive secure. As pointed out in remark 30, we can show that the adversary can be allowed to choose the target set, which implies that this scheme is semi-statically secure. Therefore, by using Gentry-Waters transformation, one can obtain a (very inefficient) adaptive secure scheme.

Del08. The identity-based broadcast encryption in [Del08] deals with 2-adaptive corruption and enjoys CCA security with constant ciphertext and private key sizes. However, the adversary has to announce its target set before the setup phase which corresponds to our selective security model.

GW09. In [GW09], the authors aim to construct efficient schemes that are adaptively secure and that resist to full collusion. The adaptive security mentioned in the paper correspond to our Ad**1** model. However, their schemes can be easily proved secure in Ad**2** model. They introduced a two-key transformation that convert a semi-static system of $2N$ users into a adaptive secure system of N users. Their schemes are not dynamic.

A Secure Broadcast Encryption Scheme

Let us now propose a simple scheme that is IND-Dyn**2**-Ad**2**-CCA**2**-secure to show that it is possible. The naive BE scheme where the center shares a key with every user is not IND-Dyn**2**-Ad**2**-CCA**2**-secure, but adding a MAC makes it secure.

Definition 31. *Let $\mathcal{PKE}$ be an IND-CCA2 secure public-key encryption scheme with key length κ, $\mathcal{MAC}$ a strongly-UF-CMA MAC. We build a BE scheme Π in the following way.*

- $\mathsf{Setup}(1^k)$ $\mathit{MSK} \stackrel{\text{def}}{=} \emptyset; \mathit{EK} \stackrel{\text{def}}{=} \emptyset; Reg \stackrel{\text{def}}{=} \emptyset$
- $\mathsf{Join}(\mathsf{MSK}, i)$ $(pk_i, sk_i) \leftarrow \mathcal{PKE}.\mathsf{KeyGen}(1^k)$. `return` (sk_i, pk_i).
- $\mathsf{Encaps}(\mathsf{EK}, S)$: $K, \mathcal{K}_m \stackrel{\$}{\leftarrow} \{0,1\}^k$;
  `for all` $i \in S : c_i \leftarrow \mathcal{PKE}.\mathsf{Encrypt}(\mathsf{pk}_i, K||\mathcal{K}_m)$;
  $\sigma \leftarrow \mathcal{MAC}_{\mathcal{K}_m}(c_1||\ldots||c_{|S|})$;
  $H \stackrel{\text{def}}{=} c_1||\ldots||c_{|S|}||\sigma$
- $\mathsf{Decrypt}(\mathsf{sk}_i, S, H)$: $K||\mathcal{K}_m = \mathcal{PKE}.\mathsf{Decrypt}(\mathsf{sk}_i, c_i)$
  `if` $\mathcal{MAC}.\mathsf{Verify}(\mathcal{K}_m, \sigma, c_1||\ldots||c_{|S|})$ `return` K,
  `else return` $\perp$

Theorem 32. *The above BE scheme is IND-*Dyn**2**-Ad**2**-CCA**2***-secure.*

The proof can be found in the full version of this paper [PPS11].

Acknowledgments

This work was supported in part by the French ANR-09-VERS-016 BEST Project.

References

[BDPR98] Bellare, M., Desai, A., Pointcheval, D., Rogaway, P.: Relations among notions of security for public-key encryption schemes. In: Krawczyk, H. (ed.) CRYPTO 1998. LNCS, vol. 1462, pp. 26–45. Springer, Heidelberg (1998), Full version available at `http://www.di.ens.fr/~pointche/pub.php`

[BGW05] Boneh, D., Gentry, C., Waters, B.: Collusion resistant broadcast encryption with short ciphertexts and private keys. In: Shoup, V. (ed.) CRYPTO 2005. LNCS, vol. 3621, pp. 258–275. Springer, Heidelberg (2005)

[BPS00] Baudron, O., Pointcheval, D., Stern, J.: Extended notions of security for multicast public key cryptosystems. In: Welzl, E., Montanari, U., Rolim, J.D.P. (eds.) ICALP 2000. LNCS, vol. 1853, pp. 499–511. Springer, Heidelberg (2000)

[BW06] Boneh, D., Waters, B.: A fully collusion resistant broadcast, trace, and revoke system. In: ACM CCS, pp. 211–220. ACM, New York (2006), Full version available at Cryptology ePrint Archive `http://eprint.iacr.org/2006/298`

[CS03] Cramer, R., Shoup, V.: Design and analysis of practical public-key encryption schemes secure against adaptive chosen ciphertext attack. SIAM Journal on Computing 33(1), 167–226 (2003)

[Del08] Delerablée, C.: Identity-based broadcast encryption with constant size ciphertexts and private keys. In: Kurosawa, K. (ed.) ASIACRYPT 2007. LNCS, vol. 4833, pp. 200–215. Springer, Heidelberg (2007)

[DF03] Dodis, Y., Fazio, N.: Public key trace and revoke scheme secure against adaptive chosen ciphertext attack. In: Desmedt, Y.G. (ed.) PKC 2003. LNCS, vol. 2567, pp. 100–115. Springer, Heidelberg (2002), Full version available at Cryptology ePrint Archive `http://eprint.iacr.org/2003/095`

[DPP07] Delerablée, C., Paillier, P., Pointcheval, D.: Fully collusion secure dynamic broadcast encryption with constant-size ciphertexts or decryption keys. In: Takagi, T., Okamoto, T., Okamoto, E., Okamoto, T. (eds.) Pairing 2007. LNCS, vol. 4575, pp. 39–59. Springer, Heidelberg (2007)

[FN94] Fiat, A., Naor, M.: Broadcast encryption. In: Stinson, D.R. (ed.) CRYPTO 1993. LNCS, vol. 773, pp. 480–491. Springer, Heidelberg (1994)

[GW09] Gentry, C., Waters, B.: Adaptive security in broadcast encryption systems (with short ciphertexts). In: Joux, A. (ed.) EUROCRYPT 2009. LNCS, vol. 5479, pp. 171–188. Springer, Heidelberg (2009), Full version available at Cryptology ePrint Archive `http://eprint.iacr.org/2008/268`

[NNL01] Naor, D., Naor, M., Lotspiech, J.: Revocation and tracing schemes for stateless receivers. In: Kilian, J. (ed.) CRYPTO 2001. LNCS, vol. 2139, pp. 41–62. Springer, Heidelberg (2001), Full version available at Cryptology ePrint Archive `http://eprint.iacr.org/2001/059`

[PP04] Phan, D.H., Pointcheval, D.: On the security notions for public-key encryption schemes. In: Blundo, C., Cimato, S. (eds.) SCN 2004. LNCS, vol. 3352, pp. 33–46. Springer, Heidelberg (2005)

[PPS11] Phan, D.H., Pointcheval, D., Strefler, M.: Security notions for broadcast encryption. In: Lopez, J., Tsudik, G. (eds.) ACNS 2011. LNCS. Springer, Heidelberg (2011); Full version available on the web page of the authors

[YFDL04] Yao, D., Fazio, N., Dodis, Y., Lysyanskaya, A.: Id-based encryption for complex hierarchies with applications to forward security and broadcast encryption. In: ACM CCS 2004. ACM, New York (2004), Full version from `http://www.cs.brown.edu/~anna/research.html`

Towards User-Friendly Credential Transfer on Open Credential Platforms

Kari Kostiainen[1], N. Asokan[1], and Alexandra Afanasyeva[2]

[1] Nokia Research Center, Helsinki
kari.ti.kostiainen@nokia.com, n.asokan@nokia.com
[2] Saint-Petersburg State University of Aerospace Instrumentation
alra@vu.spb.ru

Abstract. Hardware-based "trusted execution environments" (TrEEs) are becoming widely available and open credentials platforms allow any service provider to issue credentials to such TrEEs. Credential transfer in an open system poses usability challenges: Certain credential issuers may disallow direct credential migration and require explicit credential re-provisioning, and each credential provisioning typically requires separate user authentication. Additionally, the lack of secure user input mechanisms on existing TrEEs makes the required user identity binding to TrEEs challenging. In this paper we present a practical credential transfer protocol that can be implemented using devices available today. Our protocol makes credential transfer user-friendly with delegated, automatic re-provisioning, and can be integrated to a typical device initialization process.

Keywords: security, credential transfer, trusted computing.

1 Introduction

Traditional credentials have well-known drawbacks. User memorizable passwords suffer from bad usability and are vulnerable to phishing and dictionary attacks. Software-based credentials, such as secret keys stored on the device, may provide better usability and security, but can be compromised by exploiting security vulnerabilities inherently present in most modern operating systems. Hardware-based credentials, like dedicated security tokens, provide higher level of security, but are too expensive for most service providers, and due to their typical single-purpose nature force users to carry multiple tokens with them.

During the past decade hardware-based "trusted execution environments" (TrEEs) have started to become widely available in various commodity devices. Many current mobile phones support integrated security architectures like M-Shield [24] and TrustZone [1] that augment the mobile device central processing unit with secure storage and isolated execution. Additionally, many mobile devices are equipped with fixed or removable security elements, such as security enhanced memory cards (see e.g. [17]) and UICCs (Universal Integrated Circuit Card, used for cellular authentication), that provide similar hardware-based isolation. Many personal computers are equipped with a Trusted Platform Module

J. Lopez and G. Tsudik (Eds.): ACNS 2011, LNCS 6715, pp. 395–412, 2011.

(TPM) [25] which is a standardized security element that provides secure storage and predefined cryptographic operations. When used with suitable central processing unit TPMs can also provide isolated execution for arbitrary security-sensitive code [16]. Using such TrEEs it is possible to implement credentials that combine good usability of software-based credentials with the higher level of security traditionally only provided by dedicated security tokens.

Some TrEE-based "credential platforms" are closed systems. For example installing new credentials to an UICC typically requires approval from the TrEE owner, i.e. the mobile network operator. On-board Credentials [14] and Trusted Execution Module [6] are examples of *open* credential platforms in which *any* service provider can develop new credential types and install them to devices without prior agreement with the TrEE issuer or manufacturer.

Users should be able to transfer credentials from one compliant TrEE to another. The common use case is device replacement. Assume that Alice has several credentials provisioned to her mobile phone. When Alice buys a new phone, the already provisioned credentials should be made available to the new device as easily as possible. When TrEEs are equipped with a certified key pair, a straightforward credential transfer approach would be to first validate the target device certificate within the source TrEE and then encrypt all the TrEE-resident credentials using the target device public key. In practice, credential transfer is more challenging due to following reasons.

First, while some issuers may allow their credentials to be copied from one compliant TrEE to another, we take a practical stand and assume that others may require explicit credential re-provisioning to the target device from the original credential issuer. In a closed credential system such re-provisioning is easy. Typically, the user has to authenticate himself only towards the TrEE owner which controls all credential provisioning. In an open credential platform, such credential re-provisioning poses usability challenges, since each credential provisioning operation typically requires user authentication with respect to that issuers's domain, and thus having to re-provision credentials from multiple different issuers forces user to perform multiple user authentication operations, e.g. when a new device is taken into use. Such an user experience is clearly not optimal.

Second, validating the target device TrEE certificate only guarantees that copyable credentials are transfered from one compliant TrEE to another. To ensure that the credentials are transfered to a TrEE belonging to the *correct user*, the device certificate, or the credentials themselves, must be bound to the user identity. In case of commodity devices such user identity binding must be done by the user himself, since user identities are not known at the time of device manufacturing and can change during device lifetime. TrEEs on existing commodity devices typically lack secure (user) input mechanisms, i.e. data input to the TrEE is typically possible only via the possibly compromised device operating system, which makes secure user binding challenging.

Third, the user may not have both the source and the target device in his possession at the same time. The user may, e.g. during device replacement, have

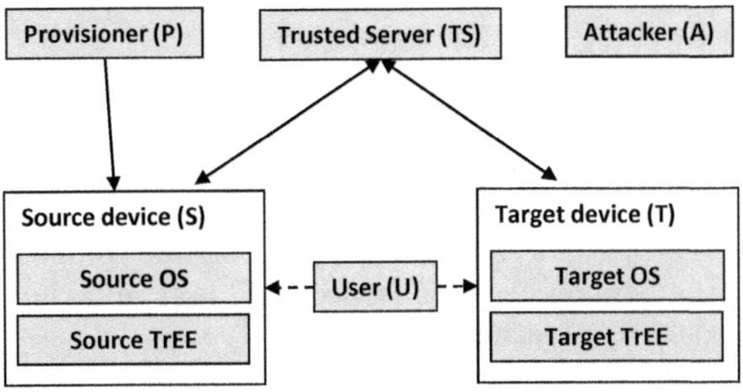

Fig. 1. Credential transfer system model

to give away his old device before he gets the new device, and thus the target device public key cannot be imported to the source device easily.

In this paper we address the problem of user-friendly credential transfer in open credential platforms. We focus on embedded TrEEs on mobile devices, although most of the discussion applies to other embedded TrEE types as well. We present a credential transfer protocol that (1) utilizes delegated, automatic re-provisioning for non-transferable credentials, and (2) handles user identity binding by relying on the "trust on first use" principle to mitigate the lack of secure user input mechanisms. The main contribution of this paper is not so much in the technical novelty of the proposed protocol, but rather in identifying the problem and presenting a practical solution that can be (a) implemented using commodity devices available today and (b) integrated into a typical mobile device initialization process for optimal user experience. We have formally validated a relevant subset of our protocol using AVISPA [26] model-checking security protocol validation tool.

2 Assumptions and Requirements

Assumptions. The entities involved in credential transfer are shown in Fig. 1. Provisioner P issues credentials to a source device S from which the credentials can be transfered to a target device T with the help of a trusted server TS.

We assume that devices S and T have an embedded TrEE with a (statistically) unique asymmetric key pair PK/SK. The key pair has been issued a certificate Cert by a trusted authority (e.g., the device manufacturer). Additionally, the TrEEs are equipped with a symmetric key K that can be used for local encryption, or sealing. TrEEs do not have persistent secure storage in addition to these fixed keys. We assume that only trusted code (e.g., code signed by the device manufacturer) can be executed within the TrEE and can access TrEE-resident keys. A trust root, such as a hash of the trusted authority public key, has been installed within the TrEEs of S and T during manufacturing.

We assume that devices S and T have an operating system level security framework using which access to the TrEE on the devices can be limited to trusted operating system level software components only (e.g., processes with a required permission). Integrity of the operating system security framework itself may be enforced with TrEE-based secure boot.

A direct and secure communication path between the user and the TrEE is often denoted "trusted user interface" (see e.g. [9] for more information). Despite of promising research prototypes [23] TrEEs on existing commodity mobile devices typically do not support trusted user interfaces or any other data input mechanisms that are not controlled by the possible compromised OS. Thus we assume that the communication between the user and the TrEE is always mediated by the device operating system.

We assume that the mobile devices S and T have a device initialization process that is automatically triggered when the device is booted for the first time (or after device reset operation). As a part of this device initialization the user is asked to log in (e.g., with username-password) to services provided by the mobile platform provider (e.g., Google services for Android devices or Ovi services for Nokia devices).

We assume that the trusted server TS also has an asymmetric key pair $\mathsf{PK_{TS}}/\mathsf{SK_{TS}}$ and a certificate $\mathsf{Cert_{TS}}$ issued by the same trusted authority. A similar trust root (e.g., hash of trusted authority public key) is fixed in TS as well.

We assume an open credential provisioning model in which credentials can be provisioned from any service provider to the TrEEs. The actual provisioning protocol can be provisioner specific, but we assume that each credential provisioning operation always requires both *device* and *user* authentication. We assume that the device authentication is based on the above mentioned TrEE certificates. Each credential issuer may have it's own user authentication mechanism, but we assume that user authentication always requires some user input, such as entering a "provisioning password" to the device. Due to previously mentioned lack of trusted user interface, the provisioning user authentication is always handled by operating system level software component.

We address two possible credential migration policies: (1) *copyable* credentials can be transfered directly from one compliant TrEE to another and (2) *non-transferable* credentials must be re-provisioning from the original issuer with user authentication.

Attacker model. We assume that the attacker A can read and modify any network traffic between P, S, TS and T based on the Dolev-Yao model [8]. The attacker cannot read or modify keys stored on, or read or modify any program execution that takes place within the TrEEs of S and T or on TS. We assume that the attacker may be able to compromise the operating system of S or T at runtime, and thus read and modify any program execution that takes place on the operating system of S or T. The attacker may have one or more devices with a compliant TrEE.

Requirements. We define single functional requirement for credential transfer:

- *No additional user interaction.* Transferring all credentials from S to T should require no additional user interaction besides the normal mobile device initialization process.

We define two security requirements:

- *Credential secrecy.* The attacker must not be able to read or modify any credentials during transfer.
- *Credential fidelity.* Credentials should be transfered only to a device belonging to the same user that the credentials were originally provisioned to.

The credential transfer protocol should guarantee "forward fidelity", i.e. the attacker should not be able to transfer securely provisioned credentials to a device that belongs to a different user if the OS gets compromised after the provisioning. (This requirement is similar to "forward secrecy" in key agreement protocols [7], i.e. the attacker should not be able to determine previously established session keys if the long-term keys used in the key agreement get compromised later.)

Since we assume that the provisioning user authentication is always handled by operating system level component, due to lack of trusted user interface on existing commodity devices, and that the attacker may be able to compromise the device OS, the credential provisioning phase is vulnerable to a man-in-the-middle attacker that can cause a credential to be provisioned to a wrong device, and thus the credential transfer solution cannot credential fidelity for credentials that are provisioned after the OS compromise.

3 Credential Transfer Protocol

The rationale behind the credential transfer protocol is as follows. User identity installation is done by using the "trust on first use" principle. During device first boot, or after device reset, when the device operating system is in an uncompromised state, the password from the normal mobile device initialization process is installed to the TrEE of the device. The installed password is bound to the credentials inside the device TrEE during each credential provisioning. The actual credential transfer happens in three phases. During credential backup, all copyable credentials are copied securely from S to TS. For each non-transferable credential, S creates a *delegation token.* During credential recovery, TS verifies that the same user identity has been installed to T, and if this is the case, all

Table 1. Notations used in credential transfer protocol

c ← Enc(PK, d)	Ciphertext c from encryption on data d using public key PK.
d ← Dec(SK, c)	Plaintext d from decryption on ciphertext c using private key SK.
s ← Sign(SK, d)	Signature s on data d using private key SK.
c ← Seal(d)	Ciphertext c from authenticated encryption on data d with key K.
d ← Unseal(c)	Plaintext d from authenticated decryption on ciphertext c with key K.
m ← MAC(k, d)	Keyed message authentication code m over data d using key k.

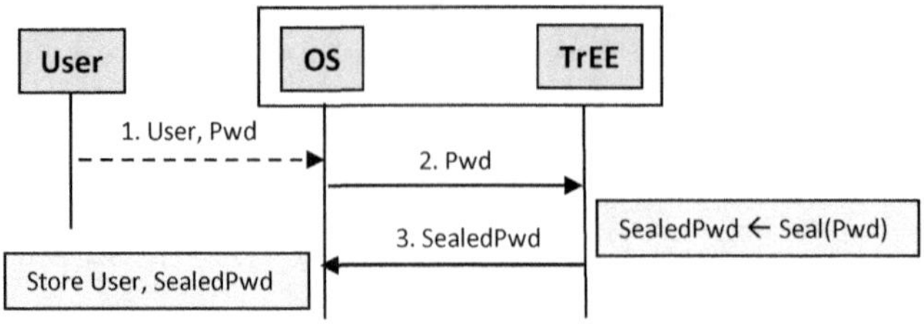

Fig. 2. User identity installation

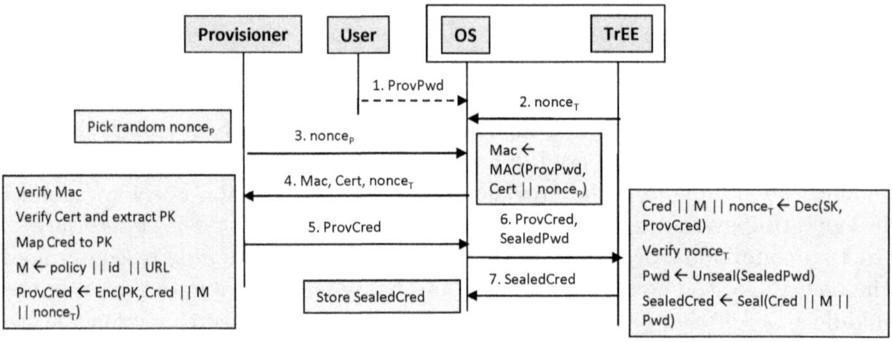

Fig. 3. Credential provisioning

copyable credentials and delegation tokens are transfered from TS to T. In re-provisioning phase, using the delegation tokens, T can automatically fetch, or re-provision, all the non-transferable credentials from the original provisioners without explicit user authentication towards each provisioner.

Table 1 lists the algorithmic notations used in the protocol description. The rest of the notations used are explained in the following text.

Identity installation. When the user starts his device for the first time, user identity should be installed to the TrEE of the device (see Fig. 2).

1. The user is asked to enter a username User and a password Pwd.
2. An operating system level software component forwards Pwd to the TrEE. Within the TrEE the password is locally sealed.
3. The resulting encryption SealedPwd is stored on the operating system side together with User.

Operating system level security mechanism, e.g. permission based access control, is used to enforce that only a trusted software component can perform this initialization only when the device is taken into use for the first time.

Provisioning. The credential provisioning protocol details may vary between different credential issuers, but we assume that the basic principles are always

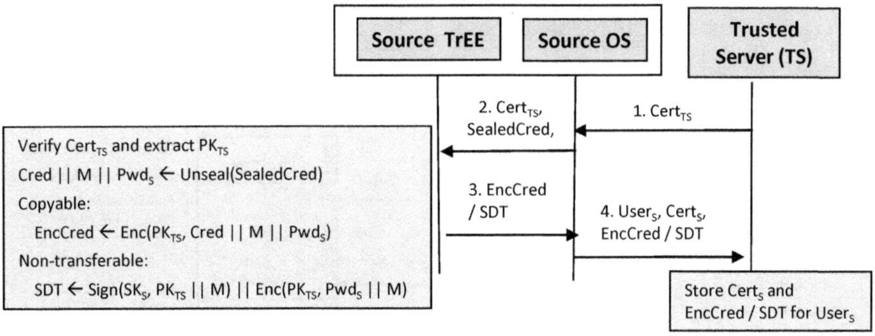

Fig. 4. Credential backup (transfer from the source device S to the trusted server TS)

the same: each credential provisioning requires both device and user authentication. Example provisioning is shown in Fig. 3. We assume that the user and the provisioner share provisioning password ProvPwd from prior out-of-band communication, such as service registration email.

1. The mobile device requests ProvPwd from the user.
2. The TrEE picks a random nonce_T and returns that to the operating system component on the mobile device.
3. Provisioner picks a random nonce_P and sends this to the device. An operating system level software component on the device calculates keyed message authentication code Mac using ProvPwd. nonce_P and the device certificate Cert is included to the message authentication code to bind the user authentication to the device authentication.
4. The device sends Mac, Cert and nonce_T to the provisioner. The provisioner verifies Mac using previously picked nonce_P and ProvPwd. The provisioner also verifies Cert to make sure that the credential is provisioned to a compliant TrEE. The provisioner stores a mapping between the provisioned credential Cred and PK. This mapping is later needed for automated re-provisioning. The provisioner encrypts the provisioned credential Cred, credential metadata M and nonce_T using PK. The credential metadata M includes migration policy, credential identifier and re-provisioning URL for non-transferable credentials.
5. The resulting encryption ProvCred is sent to the device.
6. ProvCred is loaded to the TrEE together with SealedPwd. Inside the TrEE, both of these encryptions can be recovered and Cred is sealed for local storage. The user identity password Pwd is included to the seal to bind the installed credential to current user identity. The freshness of nonce_T is also verified.
7. The resulting encryption SealedCred is stored on the operating system side.

Credential backup. Credential transfer from the source device S to the trusted server TS is described in Fig. 4.

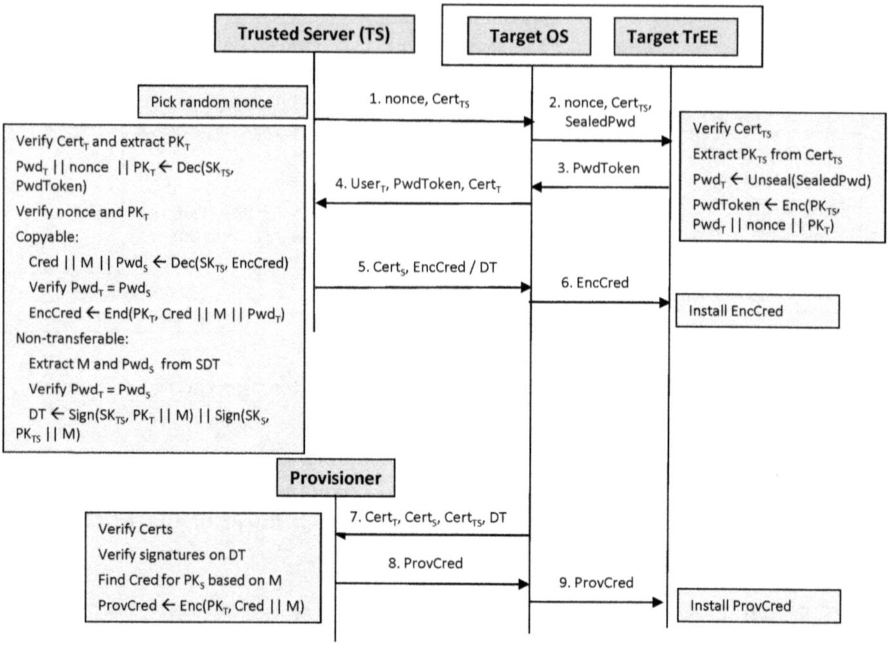

Fig. 5. Credential recovery (transfer from trusted server TS to target device T)

1. The trusted server sends its certificate $\mathsf{Cert_{TS}}$ to the source device.
2. The source device loads $\mathsf{Cert_{TS}}$ and each SealedCred to the TrEE. The source device verifies $\mathsf{Cert_{TS}}$ inside the TrEE with respect to the pre-installed trust root to make sure that credentials are transfered only via the trusted server. The source device unseals each SealedCred and in case of copyable credential the source device encrypts it using the public key of the trusted server $\mathsf{PK_{TS}}$ which can be extracted from $\mathsf{Cert_{TS}}$. The resulting encrypted credential EncCred includes the user identity password of the source device $\mathsf{Pwd_S}$. If the credential is non-transferable a server delegation token SDT is created. This token contains $\mathsf{PK_{TS}}$ and M signed with the private key of the source device $\mathsf{SK_S}$ concatenated with source device user identity password $\mathsf{Pwd_S}$ and M encrypted for the trusted server.
3. Each encrypted credential EncCred and SDT for each non-transferable credential are returned to the operating system of the source device.
4. The operating system sends these to the trusted server together with username $\mathsf{User_S}$ and $\mathsf{Cert_S}$ of the source device. The trusted server stores all received data items.

Credential recovery. Credential transfer from the trusted server TS to the target device T is described in Fig. 5. Prior to this, the user identity installation must be done to T as described before.

1. The trusted server picks a random nonce and sends this together with the server certificate $\mathsf{Cert_{TS}}$ to the target device.
2. The target device loads these to the TrEE with the local SealedPwd. Inside the target device TrEE $\mathsf{Cert_{TS}}$ is verified using the pre-installed trust root and $\mathsf{PK_{TS}}$ is extracted from the certificate. The target device creates a password token PwdToken by encrypting the user identity password of the target device $\mathsf{Pwd_T}$, nonce and $\mathsf{PK_T}$ using the public key of the server $\mathsf{PK_{TS}}$.
3. PwdToken is returned to the target device operating system.
4. The target device sends PwdToken, $\mathsf{User_T}$ and $\mathsf{Cert_T}$ to the trusted server. The trusted server verifies $\mathsf{Cert_T}$ to make sure that the credentials are transfered only to a compliant TrEE. The server decrypts PwdToken and verifies that the received nonce matches the one picked earlier and that the public key $\mathsf{PK_T}$ inside the password token matches the one in $\mathsf{Cert_T}$. For each copyable credential the trusted server does as follows: EncCred received from source device is decrypted using the private key of the server. The server verifies that the user identity password $\mathsf{Pwd_T}$ recovered from PwdToken matches $\mathsf{Pwd_S}$ recovered from EncCred. If this is the case, the target device belongs to the same user as the source device and the server encrypts the credential Cred using the public key of the target device $\mathsf{PK_T}$. The resulting encryption is denoted EncCred. For each non-transferable credential the server decrypts $\mathsf{Pwd_S}$ and M from SDT. The server verifies that $\mathsf{Pwd_T}$ received from the target device in PwdToken matches $\mathsf{Pwd_S}$ and if this is the case the server creates a new delegation token DT for the target device. This token includes a signature over the target device public key $\mathsf{PK_T}$ and M using the private key of the server $\mathsf{SK_{TS}}$ and the original signature made by the source device from SDT.
5. The server sends EncCred for each copyable credential, DT for each non-transferable credential and $\mathsf{Cert_S}$ to the target device.
6. The target device can install (decrypt and locally seal) each EncCred inside it's TrEE.
7. Using each received delegation token DT the target device can fetch non-transferable credentials from the original provisioners. The target device sends the certificates of all three devices ($\mathsf{Cert_S}$, $\mathsf{Cert_{TS}}$ and $\mathsf{Cert_T}$) and a delegation token DT to a provisioner of a non-transferable credential. The URL of the provisioning server can be extracted from the credential metadata M. The provisioner verifies these three certificates and checks that DT has correct signature chain, i.e. the source device has delegated the trusted server and that the trusted server in turn has delegated the target device. The delegation token signatures include credential metadata M that includes credential identifiers. The provisioner checks that these credential identifiers match $\mathsf{PK_S}$ (can be extracted from $\mathsf{Cert_S}$). If all the above mentioned conditions hold true, then the original credential Cred can be provisioned to the target device without explicit provisioning user authentication and the provisioner constructs ProvCred packages for the target device.
8. ProvCred is sent to the target device.
9. The target device can install ProvCred as described before.

4 Requirement Analysis

Based on our requirements the credential transfer should require no additional user interaction besides the normal mobile device initialization process. Credential backup can be automated so that credentials are copied from the source device to the trusted server automatically whenever there is a change in one of the credentials. Credential recovery can be run automatically after normal device login operation. Thus, no further user interaction besides the normal login operation is needed.

The attacker must not be able to read or modify credentials during transfer (credential secrecy). This is enforced using common public key infrastructure mechanisms. The credentials are only encrypted to a trusted server or to a compliant TrEE, and the encryption keys are validated with device certificates and using trust roots fixed to device TrEEs during manufacturing. Only trusted code is executed on the trusted server and compliant TrEEs, and trusted code will not leak credentials outside these trusted environments.

The credentials should be transfered only to a device belonging to the same user that the credentials were originally provisioned to (credential fidelity). This is enforced with a password, and thus meeting this requirement relies on the assumption that the attacker will not learn (or be able to guess) the correct password, even if he manages to compromise the operating system on the source or target device after device initialization process. User chosen and memorizable password have well known security issues, most notably vulnerability to offline dictionary attacks and phishing. In our protocol the password enforcement is done on the trusted on-line server, and thus common dictionary attack prevention techniques, such as throttling, can be applied. Our approach relies on the assumption that the user enters the password only to a trustworthy device initialization software on the device when the device is booted for the first time. If the user enters the password to any other (possible malicious) software on the device the attacker can learn the password.

Operating system level enforcement mechanisms is used to ensure that a malicious application will not replace the already installed user identity. If the attacker manages to compromise the source device operating system, he may replace currently installed user identity password with a freely chosen one and all the credentials provisioned *after* the operating system compromise will be bound to the attacker chosen password. As a result the attacker may transfer such credentials to his device. This limitation of the solution is acceptable, since an attacker that is able compromise the operating system can act as a man-in-the-middle in original provisioning protocol and direct credential provisioning to a compliant attacker device anyway.

5 Protocol Validation

We have validated (the core subset of) our credential transfer protocol using AVISPA [26]. AVISPA is an automated security protocol validation tool. Security protocols are modeled using High-Level Protocol Specification Language

(HLPSL) and the tool automatically translates such protocol models into an equivalent infinite state transition system that is then input to validation back-ends. The validation back-ends search the transition system for states that represent attacks, i.e. states and properties defined in terms of HLPSL. All the back-ends analyze protocols by considering the Dolev-Yao model [8] of an active adversary that controls the network but cannot break cryptography.

We have modeled the credential transfer protocol described in the this paper with the following two restrictions. First, our current model only addresses copyable credentials. Second, our current model does not cover operations that involve user interaction, i.e. user identity installation and provisioning. Instead we make the simplifying assumptions that (1) the user identity password has already been securely installed to the TrEE of the source device (assuming trust on first use), (2) the credential has already been provisioned to the TrEE of the source device, and (3) the same user identity password has securely been installed to the TrEE of the target device (again, trust on first use). Appendix A shows listing of the protocol validation model in HSPSL. We see that extending the model to cover the migration policy of non-transferable credentials is fairly straightforward. Modeling user identity installation and provisioning, i.e. operations that involve user interaction, pose more challenges, especially under our assumptions and threat model.

In our protocol model each entity is described as a separate role. Each role has a set of input parameters (e.g., other roles to communicate with, and keys and other data used within that role). The roles communicate over channels that the adversary can fully control. The roles are combined into a session construct, and sessions can be instantiated with constant parameters that e.g. define the keys that each role will use in that session.

HLPSL offers two mechanisms for modeling security requirements. Declaration `secret(d, id, r)` defines a security goal with identifier `id` that defines that data `d` should remain secret between set of roles `r`. Typically, authentication security requirements are modeled with `witness` and `request` declarations. For our simplified credential transfer model, both security requirements (SR1 and SR2) can be defined in one security goal `sr`. Declaration `secret(Cred, sr, {SourceTree, Server, TargetTree})` at the end of `role_SourceTree` defines that the transfered credential `Cred` should remain secret between the TrEE of the source device, the trusted server and the TrEE of the target device. Since the credential is not revealed to other parties, SR1 is met. Since the credential is not transferred to TrEE of any other compliant device, SR2 is met. We model our attacker model so that the adversary plays role of `role_TargetTree` with the knowledge of matching certified key pair. The correct user identity password `pwd` is not given to target device TrEE played by the adversary.

We have successfully validated our protocol model with three ASVISPA back-ends (OFMC, CL-AtSe and SATMC). The validation of the model can be easily re-produced from the AVISPA tool web interface.[1]

---

[1] `http://www.avispa-project.org/web-interface/`

6 Discussion

In our credential transfer protocol the user authentication is based on a password that is installed after the first boot ("trust on first use" approach). This solution meets our security requirements, provides nice user experience (assuming integration to normal device initialization process), and can be implemented with commodity devices available today. The drawbacks of this approach are vulnerability to phishing and the fact that the server has to be fully trusted.

Alternatively, the user authentication could be based on a full-length key that would is installed to the TrEE after first boot and copied e.g. to a removable memory card. First part of the key would be used for encrypting the credentials for the server and second part would be user for user authentication. Such an approach would not have the problem of phishing and would not require fully trusted server. The drawbacks of this approach would be inconvenient user interaction model, since the memory card would have to be removed from the device after first boot to prevent the secret leaking to the OS that may get compromised later. In practice the user would need a dedicated memory card for credential transfer.

Another approach would be to rely on availability of "trusted user interface" [9]. Such an approach would maintain the user interaction model of our solution. Additionally provisioning user authentication could be TrEE-based, and since the TrEE could reliably distinguish between user and application provided input, a malicious application could not replace the already installed password, and the attacker could not fool the system into transferring credentials provisioned after the device OS compromise into an unauthorized device. Trusted user interface would make phishing more difficult as well, although studies have shown that users tend to ignore security most indicators [20]. In practice, despite of promising research prototypes [23], TrEEs on existing mobiles do not support such trusted user interfaces.

Another alternative would be to use a "trusted external interface". Near Field Communication (NFC) enabled mobile devices are currently becoming increasingly popular [11] and due to many security-related NFC use cases device manufacturers are integrating NFC chips to device TrEEs. If such an interface would be available, credential transfer could be based on demonstrative identification [2], i.e. the user would touch and the target device with the source device, and the source device could transfer (and delegate) credentials directly to the target device. In practice, such devices are not yet widely available.

Finally, on mobile devices, a natural approach would be to use presence of the correct UICC for credential transfer authentication. During first boot, UICC identifier could be installed to the TrEE and that could be bound to all provisioned credentials. The target device could prove the presence of the same UICC to the trusted server e.g. using Generic Bootstrapping Architecture [12]. Such an approach would require no additional user interaction besides normal device switch operations. The drawback would be that such an approach would no meet our security model. Since UICCs are not securely integrated to TrEEs, a compromised OS could direct credential transfer to an unauthorized device.

Table 2. Comparison of user authentication alternatives for credential transfer

user authentication mechanism	vulnerability to OS compromise	vulnerability to phishing	extra user interaction	requires fully trusted server	device available today
trust on first use	no *	vulnerable	no	yes	yes
trust on first use (memory card)	no *	no	yes	no	yes
trusted user interface	no	less vulnerable	no	yes	no
trusted external interface	no	no	yes (touch)	no	no
UICC presence	yes	no	no	yes	yes

*) Subject to the assumption of OS not being compromised during first use.

We summarize key properties of above discussed user authentication alternatives in Table 2.

In this paper we have addressed two credential migration policies: copyable and non-transferable credentials. A third natural migration policy would be "movable credentials", i.e. credentials that are allowed to exist in one (or predefined number of) compliant TrEEs at the same time. The notable difference to copyable credentials is that credential deletion or disabling from the source device should be possible. Deleting or disabling credentials from a TrEE requires replay-protection support from the TrEE hardware, e.g. in the form of monotonic secure counter or non-volatile secure memory. Many existing TrEEs on mobile device do not provide such support (see e.g. [21] for rationale), and thus enforcement of movable credential migration policies must be done on the infrastructure back-end. Additionally, in many cases the credential issuers want to control credential migration even tough the end user devices would support replay protection.

Defining movable credentials as non-transferable, and thus mandating explicit re-provisioning, is one way to allow the credential provisioner to control the number of the devices to which the credential can be moved. Alternatively, such migration control could be implemented on the infrastructure back-end by verifying device identity during on-line credential usage. The drawback of such approach is that it requires changes to the actual credential usage protocols. Our approach allows the credential issuers to control credential replication by the means of explicit re-provisioning, which requires changes only to credential provisioning but not the credential usage protocols.

7 Related Work

Credential transfer between TrEEs has been studied primarily in the context of Trusted Platform Modules (TPMs). The TPM specifications define key migration commands [25]. When a new TPM based key is created, the calling application may define a migration password. The key may be migrated only by applications that know this password. The TPM specifications also define concepts of Migration Authority (MA) and Migration Selection Authority (MSA)

[25] that in principle can control to which devices TPM credentials are migrated to. The specification do not, however, define how user binding is done for such credential migration.

Several extensions and improvements have been proposed to TPM migration commands. [10] presents a protocol for migrating of TPM-based web authentication credentials using existing TPM commands. [15] proposes extensions to TPM commands. In [5] and [19] credential migration is studied in the context of digital rights management. [3] describes a mechanism for TPM virtualization and how such virtual TPMs could be migrated from one device to another. Property-based TPM virtualization and migration is described in [18]. Migration of Mobile Trusted Modules (MTM) is described in [22]. Temporary disabling of TrEE-based credentials on mobile devices has been studied in [13]. However, none of these works address the problem we are interested in, i.e. user-friendly re-provisioning of non-transferable credentials and secure user binding to prevent credentials from being transfered to a compliant but incorrect TrEE.

Our approach requires a fully trusted server. Boyen has proposed a scheme to protect credentials on a server with a single user memorizable password that is used for both access authentication and encryption in a way that the password is not revealed to the server during access authentication [4]. The untrusted server cannot recover password-encrypted credentials assuming the the credentials do not have recognizable structure, and thus brute force attacks against password-encrypted credentials are not possible. This assumption would not, however, hold true for the credential platforms we are addressing [14,6], since the credentials do have recognizable structure needed e.g. for TrEE-internal access control enforcement (and TrEE-external credential metadata).

8 Summary

In this paper we have addressed the problem of credential transfer between TrEEs on mobile devices. Credential transfer is challenging assuming an open credential provisioning model in which each credential issuer has its own user authentication domain and migration policies. Additionally, the lack of secure user input mechanisms in existing TrEEs make credential transfer challenging. The contribution of this paper is a novel and practical credential transfer protocol that requires minimal user interaction and can be implemented using commodity devices available today.

References

1. ARM. Trustzone-enabled processor, `http://www.arm.com/products/processors/technologies/trustzone.php`
2. Balfanz, D., Smetters, D.K., Stewart, P., Wong, H.C.: Talking to strangers: Authentication in ad-hoc wireless networks. In: Proc. Network and Distributed System Security Symposium (NDSS 2002) (2002)

3. Berger, S., Caceres, R., Goldman, K., Perez, R., Sailer, R., van Doorn, L.: vTPM - virtualizing the trusted platform module. In: Proc. 15th Usenix Security Symposium (2006)
4. Boyen, X.: Hidden credential retrieval from a reusable password. In: Proc. 4th International Symposium on Information, Computer, and Communications Security, ASIACCS 2009 (2009)
5. Cooper, A., Martin, A.: Towards an open, trusted digital rights management platform. In: Proc. ACM Workshop on Digital Rights Management, DRM 2006 (2006)
6. Costan, V., Sarmenta, L.F.G., van Dijk, M., Devadas, S.: The trusted execution module: Commodity general-purpose trusted computing. In: Grimaud, G., Standaert, F.-X. (eds.) CARDIS 2008. LNCS, vol. 5189, pp. 133–148. Springer, Heidelberg (2008)
7. Diffie, W., Van Oorschot, P.C., Wiener, M.J.: Authentication and authenticated key exchanges. Des. Codes Cryptography 2 (1992)
8. Dolev, D., Yao, A.C.: On the security of public key protocols. Technical report, Stanford, CA, USA (1981)
9. Fischer, T., Sadeghi, A.-R., Winandy, M.: A pattern for secure graphical user interface systems. In: Bhowmick, S.S., Küng, J., Wagner, R. (eds.) DEXA 2009. LNCS, vol. 5690. Springer, Heidelberg (2009)
10. Gajek, S., Löhr, H., Sadeghi, A.-R., Winandy, M.: Truwallet: trustworthy and migratable wallet-based web authentication. In: Proc. ACM Workshop on Scalable Trusted Computing, STC 2009 (2009)
11. Harrop, P., Das, R.: Nfc-enabled phones and contactless smart cards 2010-2020. Technical report, IDTechEx (2010), http://www.idtechex.com/research/
12. Holtmanns, S., Niemi, V., Ginzboorg, P., Laitinen, P., Asokan, N.: Cellular Authentication for Mobile and Internet Services. Wiley, Chichester (2008)
13. Kostiainen, K., Asokan, N., Ekberg, J.-E.: Credential disabling from trusted execution environments. In: Proc. of Nordic Conference in Secure IT Systems, Nordsec 2010 (2010)
14. Kostiainen, K., Ekberg, J.-E., Asokan, N., Rantala, A.: On-board credentials with open provisioning. In: Proc. ACM Symposium on Information, Computer & Communications Security, ASIACCS 2009 (2009)
15. Kühn, U., Kursawe, K., Lucks, S., Sadeghi, A.-R., Stüble, C.: Secure data management in trusted computing. In: Rao, J.R., Sunar, B. (eds.) CHES 2005. LNCS, vol. 3659, pp. 324–338. Springer, Heidelberg (2005)
16. McCune, J.M., Parno, B., Perrig, A., Reiter, M.K., Seshadri, A.: Minimal TCB Code Execution (Extended Abstract). In: Proc. IEEE Symposium on Security and Privacy (May 2007)
17. Poitner, M.: Mobile security becomes reality – the mobile security card (2008), http://www.ctst.com/CTST08/pdf/Poitner.pdf
18. Sadeghi, A.-R., Stüble, C., Winandy, M.: Property-based TPM virtualization. In: Wu, T.-C., Lei, C.-L., Rijmen, V., Lee, D.-T. (eds.) ISC 2008. LNCS, vol. 5222, pp. 1–16. Springer, Heidelberg (2008)
19. Sadeghi, A.-R., Wolf, M., Stüble, C., Asokan, N., Ekberg, J.-E.: Enabling fairer digital rights management with trusted computing. In: Garay, J.A., Lenstra, A.K., Mambo, M., Peralta, R. (eds.) ISC 2007. LNCS, vol. 4779, pp. 53–70. Springer, Heidelberg (2007)
20. Schechter, S.E., Dhamija, R., Ozment, A., Fischer, I.: The emperor's new security indicators. In: Proc. IEEE Symposium on Security and Privacy, SP 2007 (2007)

21. Schellekens, D., Tuyls, P., Preneel, B.: Embedded trusted computing with authenticated non-volatile memory. In: Lipp, P., Sadeghi, A.-R., Koch, K.-M. (eds.) Trust 2008. LNCS, vol. 4968, pp. 60–74. Springer, Heidelberg (2008)
22. Schmidt, A., Kuntze, N., Kasper, M.: On the deployment of mobile trusted modules. In: Proc. Wireless Communications and Networking Conference, WCNC 2008 (2008)
23. Selhorst, M., Stüble, C., Feldmann, F., Gnaida, U.: Towards a trusted mobile desktop. In: Acquisti, A., Smith, S.W., Sadeghi, A.-R. (eds.) TRUST 2010. LNCS, vol. 6101, pp. 78–94. Springer, Heidelberg (2010)
24. Srage, J., Azema, J.: M-Shield mobile security technology (2005), TI White paper, `http://focus.ti.com/pdfs/wtbu/ti_mshield_whitepaper.pdf`
25. TCG. TPM Specifications (July 2007), `http://www.trustedcomputinggroup.org/resources/tpm_main_specification`
26. Viganò, L.: Automated security protocol analysis with the avispa tool. Electronic Notes in Theoretical Computer Science 155, 61–86 (2006)

A HLPSL Protocol Model

```
role role_SourceTree(SourceTree, SourceOs, Server, TargetTree : agent,
                     SourceSealKey, Pwd : symmetric_key,
                     PKSource, PKServerCA : public_key,
                     SND, RCV : channel(dy))
    played_by SourceTree def=
    local
        State : nat,
        SealedPwd, SealedCred : message,
        Cred : text,
        PKServer : public_key
    init
        State := 0
    transition
    1.  State=0 /\ RCV(start) =|>
        State':=1 /\ Cred' := new() /\
        SND(SourceTree.SourceOs.{Cred'.Pwd}_SourceSealKey)
    2.  State=1 /\ RCV(SourceOs.SourceTree.{PKServer'}_inv(PKServerCA).
      {Cred'.Pwd'}_SourceSealKey) =|>
        State':=2 /\ SND(SourceTree.SourceOs.{Cred'.Pwd'}_PKServer') /\
        secret(Cred, sr, {SourceTree, Server, TargetTree})
end role

role role_SourceOs(SourceOs, SourceTree, Server : agent,
                   PKSource, PKDevCA : public_key,
                   SND, RCV : channel(dy))
    played_by SourceOs def=
    local
        State : nat,
        SealedCred, ServerCert, SourceCert, EncCred : message
    init
        State := 0
    transition
    1.  State=0 /\ RCV(SourceTree.SourceOs.SealedCred') =|>
        State':=1
    2.  State=1 /\ RCV(Server.SourceOs.ServerCert') =|>
        State':=2 /\ SND(SourceOs.SourceTree.ServerCert'.SealedCred)
    3.  State=2 /\ RCV(SourceTree.SourceOs.EncCred') =|>
        State':=3 /\ SND(SourceOs.Server.EncCred'.SourceCert)
end role

role role_Server(Server, SourceOs, TargetOs : agent,
                 PKServer, PKDevCA, PKServerCA : public_key,
                 SND, RCV : channel(dy))
```

```
    played_by Server def=
    local
        State : nat,
        Cred, Nonce : text,
        Pwd : symmetric_key,
        PKSource, PKTarget : public_key,
        EncCred : message
    init
        State := 0
    transition
    1.  State=0 /\ RCV(start) =|>
        State':=1 /\ SND(Server.SourceOs.{PKServer}_inv(PKServerCA))
    2.  State=1 /\ RCV(Server.SourceOs.{Cred'.Pwd'}_PKServer.
      {PKSource'}_inv(PKDevCA)) =|>
        State':=2 /\ Nonce':= new() /\
        SND(Server.TargetOs.Nonce'.{PKServer}_inv(PKServerCA))
    3.  State=2 /\ RCV(TargetOs.Server.{Pwd.Nonce.PKTarget'}_PKServer.
      {PKTarget'}_inv(PKDevCA) ) =|>
        State':=3 /\ EncCred' := {Cred.Pwd}_PKTarget' /\
        SND(Server.TargetOs.EncCred')
end role

role role_TargetOs(TargetOs, TargetTree, Server : agent,
                   PKTarget, PKDevCA : public_key,
                   SND, RCV : channel(dy))
    played_by TargetOs def=
    local
        State : nat,
        Nonce : text,
        SealedPwd, ServerCert, PwdToken, EncCred : message
    init
        State := 0
    transition
    1.  State=0 /\ RCV(TargetTree.TargetOs.SealedPwd') =|>
        State':=1
    2.  State=1 /\ RCV(Server.TargetOs.Nonce'.ServerCert') =|>
        State':=2 /\ SND(TargetOs.TargetTree.Nonce'.ServerCert'.SealedPwd)
    3.  State=2 /\ RCV(TargetTree.TargetOs.PwdToken') =|>
        State':=3 /\ SND(TargetOs.Server.PwdToken'.{PKTarget}_inv(PKDevCA))
    4.  State=3 /\ RCV(Server.TargetOs.EncCred') =|>
        State':=4 /\ SND(TargetOs.TargetTree.EncCred')
end role

role role_TargetTree(TargetTree, TargetOs, SourceTree : agent,
                     TargetSealKey, Pwd : symmetric_key,
                     PKTarget, PKServerCA : public_key,
                     SND,RCV : channel(dy))
    played_by TargetTree def=
    local
        State : nat,
        Nonce, Cred : text,
        PKServer : public_key
    init
        State := 0
    transition
    1.  State=0 /\ RCV(start) =|>
        State':=1 /\ SND(TargetTree.TargetOs.{Pwd}_TargetSealKey)
    2.  State=1 /\ RCV(TargetOs.TargetTree.Nonce'.{PKServer'}_inv(PKServerCA).
      {Pwd'}_TargetSealKey) =|>
        State':=2 /\ SND(TargetTree.TargetOs.{Pwd'.Nonce'.PKTarget}_PKServer')
    3.  State=2 /\ RCV(TargetOs.TargetTree.{Cred'.Pwd'}_PKTarget) =|>
        State':=3
end role

role session(Provisioner, SourceOs, SourceTree, Server : agent,
             TargetOs, TargetTree : agent,
             PKSource, PKTarget, PKServer : public_key,
             PKDevCA, PKServerCA : public_key,
```

```
                  SourceSealKey, TargetSealKey : symmetric_key,
                  SourcePwd, TargetPwd : symmetric_key)
    def=
    local
        SND1, RCV1, SND2, RCV2, SND3, RCV3,
        SND4, RCV4, SND5, RCV5 : channel(dy)
    composition
        role_SourceTree(SourceTree, SourceOs, Server,
            TargetTree, SourceSealKey, SourcePwd, PKSource,
            PKServerCA, SND1, RCV1) /\
        role_SourceOs(SourceOs, SourceTree, Server,
            PKSource, PKDevCA, SND2, RCV2) /\
        role_Server(Server, SourceOs, TargetOs, PKServer,
          PKDevCA, PKServerCA, SND3, RCV3) /\
        role_TargetOs(TargetOs, TargetTree, Server,
          PKTarget, PKDevCA, SND4, RCV4) /\
        role_TargetTree(TargetTree, TargetOs, SourceTree,
          TargetSealKey, TargetPwd, PKTarget, PKServerCA, SND5, RCV5)
end role

role environment()
    def=
    const
        provisioner, sourceos, sourcetree, server, targetos, targettree : agent,
        pksource, pktarget, pki, pkserver, pkdevca, pkserverca : public_key,
    sourcesealkey, targetsealkey, isealkey, pwd, pwdi : symmetric_key,
        sr : protocol_id

    intruder_knowledge = {provisioner, sourceos, sourcetree, server,
                          targetos, targettree, pksource, pktarget,
                          pki, inv(pki), pkdevca, pkserverca}
    composition
        session(provisioner, sourceos, sourcetree, server, targetos, i,
                pksource, pki, pkserver, pkdevca, pkserverca,
                sourcesealkey, isealkey, pwd, pwi)
end role

goal
    secrecy_of sr
end goal

environment()
```

Non-transferable User Certification Secure against Authority Information Leaks and Impersonation Attacks

Jacob C.N. Schuldt and Goichiro Hanaoka

Research Center for Information Security, AIST, Japan
{jacob.schuldt,hanaoka-goichiro}@aist.go.jp

Abstract. While standard signatures provide an efficient mechanism for information certification, the lack of privacy protecting measures makes them unsuitable if sensitive or confidential information is being certified. In this paper, we revisit nominative signatures, first introduced by Kim, Park and Won, which provides the functionality and security guarantees required to implement a certification system allowing the user (and not the authority) to control the verifiability of an obtained certificate. Unlike systems based on related primitives, the use of nominative signatures protects the user against authority information leaks and impersonation attacks based on these. We refine the security model of nominative signatures, and propose a new efficient scheme which is provably secure based on the computational Diffie-Hellman problem and the decisional linear problem. To the best of our knowledge, this is the first nominative signature scheme which is provably secure in the standard model. Furthermore, unlike the previous schemes, the proposed scheme provides signatures which hide both the signer and user identity. Hence, through our nominative signature scheme, we achieve an efficient non-transferable user certification scheme with strong security guarantees.

Keywords: user certification, nominative signatures, standard model.

1 Introduction

Information certification plays an important role in many practical information systems. While basic certification is easily achieved by the use of a standard signature scheme, this solution might not be desirable in many scenarios due to the lack of any privacy protecting measures. More specifically, if the information being certified is of a sensitive or confidential nature, the user obtaining the certificate will often be interested in restricting the verifiability of the certificate to ensure that only the intended verifiers will be able to verify the authenticity of the information in question. Due to the public verifiability of ordinary signatures, this property is not achieved when using a standard signature scheme.

However, a number of different signature primitives which provide control of verifiability have been proposed in the literature. The first type of such schemes was introduced by Chaum and Van Antwerpen with their proposal of undeniable

J. Lopez and G. Tsudik (Eds.): ACNS 2011, LNCS 6715, pp. 413–430, 2011.

signatures [7]. In an undeniable signature scheme, a verifier will have to interact with the signer to confirm the validity or invalidity of a signature, which allows the signer to control who can verify his signatures. While the functionality of undeniable signatures is not suited for certification, Kim, Park and Won [11] later introduced nominative signatures which they refer to as a "dual" to undeniable signatures. In this type of scheme, the recipient of a signature, and not the signer, is able to show validity or invalidity to a third party verifier. More specifically, a nominative signature scheme allows a signer (referred to as the *nominator*)[1] and a recipient (referred to as the *nominee*) to jointly create a signature σ which can only be verified by the nominee.[2] Furthermore, the nominee can interact with any verifier to show either validity or invalidity of a signature, but without allowing the verifier to transfer this knowledge to a third party.

Nominative signatures are closely related to designated confirmer signatures (DCS) introduced by Chaum [6], universal designated verifier signatures (UDVS) introduced by Steinfeld *et al.* [17], and the universal designated verifier signature proofs (UDVSP) by Baek *et al.* [1]. However, unlike these schemes, nominative signatures guarantee that (1) a signer cannot construct a signature associated with a nominee without the consent of the nominee, and (2) no information leaked from the signer will reveal the validity or invalidity of a signature. These properties translate into strong security guarantees for the nominee against information leaks or impersonation attacks based on these, and make nominative signatures well suited for certification of confidential information (this application was also the main motivation of UDVS [17]). To illustrate the advantages of nominative signatures for this purpose, consider the following scenario:

A patient at a hospital is diagnosed with a serious form of illness. The patient would like to keep the information about his illness private, but will at the same time have to prove that he has been diagnosed with his illness to gain access to various kinds of treatments or to be able to buy certain types of medical drugs. To address this problem, it is possible to make use of a DCS scheme which will allow an entity called the *confirmer*, to prove validity or invalidity of a signature on behalf of the signer. Hence, in this scenario, the hospital could act as a signer and issue a designated confirmer signature on the patient's diagnose, using the patient as a confirmer. This would allow the patient to prove to any third party that his diagnose is authentic. However, this approach requires the patient to place a large amount of trust in the hospital. Specifically, the patient cannot be guaranteed that rogue elements within the hospital will not, unintentionally or maliciously, leak information which will make the designated confirmer signature verifiable by unintended verifiers.

Alternatively, a UDVS or a UDVSP scheme might be used. In these schemes, the signer produces an ordinary (publicly verifiable) signature σ which is given to the recipient. In the former type of scheme, the recipient will use the public key of a verifier to convert σ to a designated signature which is only verifiable by the chosen verifier, and in the latter type of scheme, the recipient will be able to

[1] To avoid confusion, we will keep using the term *signer*.

[2] In this paper, we focus on schemes providing non-interactive signature generation.

interact with a verifier and prove that he holds σ and that σ is valid. However, applying these types of schemes in the above scenario will still leave the patient with concerns about information leakage from the hospital. In particular, σ might be leaked. Since both patient and hospital hold σ, a leak cannot be traced to either party, making it difficult to place responsibility (note that a malicious patient might try to frame the hospital to obtain compensation). Furthermore, since the conversion of a signature in a UDVS scheme and the proof of possession in a UDVSP scheme are not tied to the recipient to which σ is given, it might be a concern that rogue elements within the hospital will impersonate the patient to outside verifiers i.e. a verifier cannot tell whether the entity proving validity of a converted signature or possession of σ was in fact the intended recipient of σ. Lastly, we note that neither UDVS nor UDVSP schemes leave the verifier with any evidence of a transaction. For example, it might be questioned whether a drug supplier has been supplying drugs without confirmation of patients' needs for the drug. In this case, it cannot be confirmed that any converted signature held by the supplier is valid or not if a UDVS scheme is used, or that any transcripts of the interaction between the patient and the supplier were generated by the supplier himself, if a UDVSP is used. This issue does not occur if a DCS scheme (or nominative signature scheme) is used, since the supplier would obtain a designated confirmer signature which the patient (i.e. the confirmer) would be able to prove either valid or invalid to any third party.

The security properties of nominative signatures will eliminate the above mentioned security concerns regarding information leak, impersonation and deniability which might arise when using a DCS, UDVS or UDVSP scheme.

A different approach, which is not mentioned above, is to make use of an anonymous credential system [5]. This will allow the hospital to issue a credential to the patient for the given illness, and the patient will then have the ability to prove possession of this credential to a verifier. Furthermore, even if the verifier and the hospital collude, they will not be able to identify the patient when he proves possession of his credential. While this provides a stronger privacy guarantee than the above mentioned schemes (full anonymity as opposed to non-transferability), there might be concerns somewhat similar to the ones present when a UDVSP scheme is used i.e. a verifier might not be able to check whether a credential was issued to a legitimate patient or created by rogue elements within the hospital, and verifiers furthermore do not obtain any evidence of a transaction. When using a nominative signature scheme, only the patient can complete an execution of the confirm (or disavow) protocol involving his own public key, and if the system relies on public key certification, even an impersonation attack involving a malicious certificate authority generating a certificate binding a patient's identity to a public key of another malicious party can be detected by the presence of either two public key certificates for a legitimate patient or a public key certificate for a non-existing patient. In other words, while nominative signatures are non-transferable and not fully anonymous, they provide undeniable certificates and a high level of security against impersonation attacks, even when the system authorities are implicitly or explicitly involved.

Related Work. As mentioned above, nominative signatures were first proposed by Kim, Park and Won [11]. However, Huang and Wang [10] showed that the scheme presented in [11] is vulnerable to a malicious signer trying to determine the validity of a signature and to prove validity to a third party without the consent of the nominee. While a new scheme, which additionally supports conversion of nominative signatures into ordinary signatures, is proposed in [10], similar types of vulnerabilities were identified in the new scheme (see [18,19,12]).

The first provably secure scheme was proposed by Liu *et al.* [14] and was soon followed by another scheme by Liu *et al.* [13]. However, the invisibility definition used by [14] and [13] does not model the property that the signer will take part in the generation of all valid signatures (in fact, it is fairly easy to see that [14] allows the signer to recognize all signatures which he has been involved in generating). Furthermore, the signature space in which valid signatures cannot be distinguished from invalid ones, cannot be sampled by third party verifiers, which results in a very limited non-transferability guarantee (see Section 4 for discussion of the implications of this).

Huang *et al.* [9] introduced a stronger invisibility definition addressing the issue regarding modeling of the signer, and proposed a new provably secure scheme. Zhao *et al.* [21] adopted a similar security model, and proposed a more efficient and convertible scheme. The invisibility of both schemes is shown with respect to a signature space which can be sampled without knowledge of any secrets, but is still restricted compared to the full signature space of the schemes. As a consequence, the schemes do not provide signature anonymity i.e. given a valid signature, the identity of the signer and nominee do not remain hidden. Lastly we note that despite the claims made in [21], the scheme in [21] does not provide basic invisibility[3], which leaves [9] as the only nominative signature scheme which provably satisfies a reasonable level of security.

Our Contribution. In this paper, we present a somewhat stronger and arguably more simple security model for nominative signatures allowing conversion, and then propose a new efficient scheme. To the best of our knowledge, our proposed scheme is the first to be provably secure in the standard model, and unlike the previous schemes, our scheme provides signature which hides the identity of both signer and nominee. The security of our scheme rests on the computational Diffie-Hellman problem and the decisional linear problem, and, like the random-oracle model scheme in [9], our signatures consist of four group elements.

[3] The public/private key pairs of the signer and nominee in the scheme in [21] are given by $(pk_S, sk_S) = (g^{x_S}, x_S)$ and $(pk_N, sk_N) = (g^{x_N}, x_N)$, respectively, and a nominative signature on a message m is given by $(\sigma_1, \sigma_2) = (H(m||pk_S||pk_N)^{x_S x_N^2}, g^{x_S x_N})$. Hence, an adversary with the knowledge of a valid signature (σ_1, σ_2) on a random message m, will be able to determine if a challenge signature (σ_1^*, σ_2^*) on a message m^* is valid or not by verifying that $e(g, \sigma_2^*) = e(pk_S, pk_N)$ and $e(H(m||pk_S||pk_N), \sigma_1^*) = e(\sigma_1, H(m^*||pk_S||pk_N))$, where e is the bilinear map. The first equation ensures that $\sigma_2^* = g^{x_S x_N}$, and since the adversary knows that $\sigma_1 = H(m||pk_S||pk_N)^{x_S x_N^2}$, the second equation ensures that $\sigma_1^* = H(m^*||pk_S||pk_N)^{x_S x_N^2}$.

2 Preliminaries

Notation. We use the notation $m_1||m_2$ to mean the concatenation of the strings m_1 and m_2, and assume that m_1 and m_2 can be uniquely recovered from $m_1||m_2$. For a value p, we use $|p|_2$ to denote the bit-length of p. For a variable x and a value y, we write $x \leftarrow y$ to mean that x is assigned the value y, and for a set Y, we write $x \leftarrow Y$ to mean that x is assign an element of Y chosen uniformly at random. By $x \leftarrow \mathcal{A}^{\mathcal{O}}(y)$ we mean that the algorithm $\mathcal{A}$ is executed on input y while being allowed to make queries to the oracle $\mathcal{O}$, and that the output of $\mathcal{A}$ is assigned to x. For a pair of interactive algorithms, $\mathcal{A}$ and V, we write $z \leftarrow_2 \{\mathcal{A}(x_1) \leftrightarrow V(x_2)\}(y)$ to mean that $\mathcal{A}$ and V interact with common input y and private inputs x_1 and x_2 to $\mathcal{A}$ and V, respectively, and that the output of V, upon the completion of the interaction, is assigned to the variable z. Lastly, we will use the abbreviation PPT algorithm to mean a probabilistic polynomial time algorithm.

Negligible function. A function $\epsilon : \mathbb{N} \to [0,1]$ is said to be *negligible* if for all $c > 0$ there exists an n_c such that for all $n > n_c$ $\epsilon(n) < 1/n^c$.

Computational assumptions. In the security proof of our concrete construction of a nominative signature scheme, we will make use of the discrete logarithm problem and the decisional linear problem, which will be defined below.

For a group $\mathbb{G}$ of prime order p, we define the advantage of an algorithm $\mathcal{A}$ in solving the discrete logarithm problem as

$$\texttt{Adv}^{\texttt{DL}}_{\mathbb{G},\mathcal{A}} = \Pr[g, h \leftarrow \mathbb{G}; x \leftarrow \mathcal{A}(\mathbb{G}, g, h) : g^x = h]$$

Definition 1. *The discrete logarithm problem in $\mathbb{G}$ is said to be hard if all PPT algorithms $\mathcal{A}$ will have advantage $\texttt{Adv}^{DL}_{\mathbb{G},\mathcal{A}}$ which is negligible in $|p|_2$.*

For a group $\mathbb{G}$ of prime order p, we define the advantage of an algorithm $\mathcal{A}$ in solving the decisional linear problem as

$$\texttt{Adv}^{\texttt{DLIN}}_{\mathbb{G},\mathcal{A}} = |\Pr[\mathcal{A}(g, x_1, x_2, x_1^{y_1}, x_2^{y_2}, g^{y_1+y_2}) = 1] - \Pr[\mathcal{A}(g, x_1, x_2, x_1^{y_1}, x_2^{y_2}, g^z) = 1]|$$

where $g, x_1, x_2 \leftarrow \mathbb{G}$ and $y_1, y_2, z \leftarrow \mathbb{Z}_p$.

Definition 2. *The decisional linear problem in $\mathbb{G}$ is said to be hard if all PPT algorithms $\mathcal{A}$ will have advantage $\texttt{Adv}^{DLIN}_{\mathbb{G},\mathcal{A}}$ which is negligible in $|p|_2$.*

Signatures. A standard signature scheme is given by the following four algorithms: `Setup` which takes as input a security parameter 1^k, and returns a set of public parameters *par*; `KG` which takes as input *par*, and returns a public/private key pair (pk, sk); `Sign` which takes as input *par*, a private key *sk* and a message m, and returns a signature σ; and `Ver` which takes as input *par*, a public key *pk*, a message m and a signature σ, and returns $\top$ if σ is a valid signature on m under the public key *pk* or $\bot$ otherwise. It is required for all $par \leftarrow \texttt{Setup}(1^k)$, all $(pk, sk) \leftarrow \texttt{KG}(par)$, and all messages m, that if $\sigma \leftarrow \texttt{Sign}(par, sk, m)$, then $\texttt{Ver}(par, pk, m, \sigma) = \top$.

The standard security requirement for signatures is weak unforgeability against a chosen message attack (`wuf-cma`), which is defined as follows: Let the advantage of a `wuf-cma` adversary $\mathcal{A}$ against a signature scheme S be given by

$$\begin{aligned}\mathsf{Adv}^{\texttt{wuf-cma}}_{S,\mathcal{A}} = \Pr[&par \leftarrow \texttt{Setup}(1^k); (pk, sk) \leftarrow \texttt{KG}(par);\\ &(m^*, \sigma^*) \leftarrow \mathcal{A}^{\mathcal{O}_{sig}}(par, pk) : m^* \notin \{m_1, \ldots, m_q\} \wedge\\ &\qquad\qquad \texttt{Ver}(par, pk, m^*, \sigma^*) = \top]\end{aligned}$$

where $\mathcal{O}_{sig}$ is a signing oracle which given m_i returns $\sigma_i \leftarrow \texttt{Sign}(par, sk, m_i)$.

Definition 3. *A signature scheme S is said to be `wuf-cma` secure if all PPT adversaries $\mathcal{A}$ will have advantage $\mathsf{Adv}^{\texttt{wuf-cma}}_{S,\mathcal{A}}$ negligible in k.*

We now recall the signature scheme by Waters [20]. This scheme will play an important role in both the construction and proof of security of our concrete nominative signature scheme.

- `Setup`: Pick a group $\mathbb{G}$ of primer order p and equipped with a bilinear map $e : \mathbb{G} \times \mathbb{G} \rightarrow \mathbb{G}_T$. Furthermore, pick generator g of $\mathbb{G}$ and return the parameters $par \leftarrow (\mathbb{G}, p, g, e)$.
- `KG` : Given par, pick $\alpha \leftarrow \mathbb{Z}_p$ and $h \leftarrow \mathbb{G}$, and set $g' \leftarrow g^\alpha$. Furthermore, pick $u_0, \ldots, u_n \leftarrow \mathbb{G}$, and for a message $m \in \{0,1\}^n$, define $F(m) = u_0 \prod_{i=1}^n u_i^{m_i}$ where m_i is the ith bit of m. Finally set the public key to $pk \leftarrow (g', h, u_0, \ldots, u_n)$ and the private key to $sk \leftarrow \alpha$. Return (pk, sk).
- `Sign` : Given input (par, sk, m), where $sk = \alpha$, pick $r \leftarrow \mathbb{Z}_p$, compute $\sigma_1 \leftarrow g^r$ and $\sigma_2 \leftarrow h^\alpha F(m)^r$, and return the signature $\sigma = (\sigma_1, \sigma_2)$.
- `Ver` : Given par, a public key $pk = (g', h, u_0, \ldots, u_n)$, a message m and a signature $\sigma = (\sigma_1, \sigma_2)$, return `accept` if $e(g, \sigma_2) = e(g', h)e(\sigma_1, F(m))$.

In [20], the above signature scheme is shown to be `wuf-cma` secure given that the computational Diffie-Hellman problem is hard in $\mathbb{G}$.

3 Nominative Signatures

A nominative signature scheme involves a signer S and a nominee N, and is given by the algorithms described below. We consider a *non-interactive* scheme in which signature generation only involves the signer sending the nominee a single signature generation message. This allows the signer to be used off-line which is preferable when a high level of security is required. Furthermore, like [10,21], our description allows a nominee to selectively convert a nominative signature into a publicly verifiable signature by releasing a verification token.

- $\texttt{Setup}$: given a security parameter 1^k, this algorithm returns a set of public parameters par.
- $\texttt{KeyGen}_S$, $\texttt{KeyGen}_N$: given par, these algorithms return a public/private signer and nominee key pair, (pk_S, sk_S) and (pk_N, sk_N), respectively.
- $\texttt{Sign}$: given par, pk_N, a message m, and sk_S, this algorithm returns a signature generation message δ.
- $\texttt{Receive}$: given par, pk_S, m, a signature generation message δ, and sk_N, this algorithm returns a nominative signature σ on m.
- $\texttt{Convert}$: given par, pk_S, m, σ, and sk_N, this algorithms returns a verification token tk_σ.
- $\texttt{TkVerify}$: given par, pk_S, pk_N, m, σ, and tk_σ, this algorithm returns either $\texttt{accept}$ or $\texttt{reject}$.
- $(\texttt{Confirm}, \texttt{V}_C)$: this is a pair of interactive algorithms with common input $(par, pk_S, pk_N, m, \sigma)$. The algorithm $\texttt{Confirm}$ is furthermore given sk_N as private input. At the end of the interaction, $\texttt{V}_C$ will either output $\texttt{accept}$ or $\texttt{reject}$.
- $(\texttt{Disavow}, \texttt{V}_D)$: like in the confirm protocol, this is a pair of interactive algorithms with the common input $(par, pk_S, pk_N, m, \sigma)$, and $\texttt{Disavow}$ is given sk_N as private input. At the end of the interaction, $\texttt{V}_D$ will either output $\texttt{accept}$ or $\texttt{reject}$.

Using the above defined algorithms, a nominee can check the validity of any signature by firstly creating a verification token using $\texttt{Convert}$, and then verify the validity using $\texttt{TkVerify}$. To simplify notation, we introduce an algorithm $\texttt{Valid}$ which performs these two steps:

- $\texttt{Valid}$: given $(par, pk_S, pk_N, m, \sigma, sk_N)$, this algorithm computes the verification token $tk_\sigma \leftarrow \texttt{Convert}(par, pk_S, m, \sigma, sk_N)$ and returns the output of $\texttt{TkVerify}(par, pk_S, pk_N, m, \sigma, tk_\sigma)$.

Correctness. A nominative signature scheme is required to be *correct* i.e. for all public parameters $par \leftarrow \texttt{Setup}(1^k)$, all signer and nominee key pairs $(pk_S, sk_S) \leftarrow \texttt{KeyGen}_S(par)$ and $(pk_N, sk_N) \leftarrow \texttt{KeyGen}_N(par)$, all messages m and all signatures $\sigma \leftarrow \texttt{Receive}(par, pk_S, m, \texttt{Sign}(par, pk_N, m, sk_S), sk_N)$, it is required that $\texttt{Valid}(par, pk_S, pk_N, m, \sigma, sk_N) = \texttt{accept}$ and that the interaction $z \leftarrow_2 \{\texttt{Confirm}(sk_N) \leftrightarrow \texttt{V}_C\}(par, pk_S, pk_N, m, \sigma)$ yields $z = \texttt{accept}$. Furthermore, for all message/signature pairs (m', σ') such that $\texttt{Valid}(par, pk_S, pk_N, m, \sigma, sk_N) = \texttt{reject}$, we require that $z' \leftarrow_2 \{\texttt{Disavow}(sk_N) \leftrightarrow \texttt{V}_D\}(par, pk_S, pk_N, m, \sigma)$ yields $z' = \texttt{accept}$.

4 Security Model

For a nominative signature scheme to be secure, we require that the scheme is *unforgeable*, provides *security against malicious signers* and is *invisible*. Additionally, we require the confirm and disavow protocols to be zero-knowledge proofs. In the following, we will formally define these security requirements. Lastly, we will introduce key registration, which is required in our proof of security of our concrete construction of a nominative signature scheme.

$\texttt{Exp}^{\texttt{uf-cma}}_{NS,\mathcal{A}}(1^k)$
$L_S \leftarrow \{\}$
$par \leftarrow \texttt{Setup}(1^k)$
$(pk_S, sk_S) \leftarrow \texttt{KeyGen}_S(par)$
$(pk^*_N, m^*, \sigma^*, tk^*_\sigma, st) \leftarrow \mathcal{A}^{\mathcal{O}}(par, pk_S)$
$z \leftarrow_2 \{\mathcal{A}(st) \leftrightarrow \texttt{V}_C(par, pk_S, pk^*_N, m^*, \sigma^*)\}$
$z' \leftarrow \texttt{TkVerify}(par, pk^*_S, pk_N, m^*, \sigma^*, tk^*_\sigma)$
if $(pk^*_N, m^*) \notin L_S \wedge (z = \texttt{accept} \vee z' = \texttt{accept})$
output 1
else output 0

Fig. 1. Unforgeability security experiment

4.1 Unforgeability

Informally, unforgeability requires that valid signatures can only be obtained by interacting with the signer i.e. a malicious nominee should not be able to produce a signature on a message m, and then convince a verifier that the signature is valid, either by running the confirm protocol or by presenting a verification token, without having requested a signature on m from the signer. The definition given below is slightly stronger than the definitions from [9,21] in that the adversary (and not the challenger) generates the challenge public nominee key and is only required to convince an honest verifier about the validity of a forgery as opposed to produce a forgery which will be accepted as valid by someone holding the private nominee key. Note also that the inability of a malicious nominee to produce an accepting signature and verification token pair for a message m which was not submitted to the signer in a signature generation query, is not captured by the security models in [21] ([9] does not consider conversion). Lastly note that while [9,21] consider malicious behavior of the signer as part of their unforgeability definition, we cover these aspects in our definition of security against malicious signers.

Formally, we define unforgeability against a chosen message attack (uf-cma) of a nominative signature scheme NS via the experiment $\texttt{Exp}^{\texttt{uf-cma}}_{NS,\mathcal{A}}$ shown in Figure 1. In the experiment, $\mathcal{A}$ has access to the oracle $\mathcal{O} = \{\mathcal{O}_{Sign}\}$ defined as follows:

- $\mathcal{O}_{Sign}$: given (pk_N, m), this oracle computes $\delta \leftarrow \texttt{Sign}(par, pk_N, m, sk_S)$, adds (pk_N, m) to L_S, and returns the signature generation message δ to $\mathcal{A}$.

Definition 4. *A nominative signature scheme NS is said to be uf-cma secure if all PPT adversaries $\mathcal{A}$ have advantage $\texttt{Adv}^{uf\text{-}cma}_{NS,\mathcal{A}} = \Pr[\texttt{Exp}^{uf\text{-}cma}_{NS,\mathcal{A}}(1^k) = 1]$ which is negligible in k.*

4.2 Security against Malicious Signers

Our definition of security against malicious signers is relatively strong, and requires that not even an adversary with the knowledge of the private signer key

$$
\begin{array}{l}
\mathtt{Exp}^{\mathtt{mal\text{-}sig}}_{NS,\mathcal{A}}(1^k)\\
\quad L_R \leftarrow \{\};\ L_C \leftarrow \{\}\\
\quad par \leftarrow \mathtt{Setup}(1^k)\\
\quad (pk_N, sk_N) \leftarrow \mathtt{KeyGen}_N(par)\\
\quad (pk^*_S, m^*, \sigma^*, tk^*_\sigma, st) \leftarrow \mathcal{A}^{\mathcal{O}}(par, pk_N)\\
\quad z \leftarrow \mathtt{Valid}(par, pk^*_S, pk_N, m^*, \sigma^*, sk_N)\\
\quad z' \leftarrow \mathtt{TkVerify}(par, pk^*_S, pk_N, m^*, \sigma^*, tk^*_\sigma)\\
\quad \text{if } z = \mathtt{accept}\\
\qquad z'' \leftarrow_2 \{\mathcal{A}(st) \leftrightarrow \mathtt{V}_C(par, pk^*_S, pk_N, m^*, \sigma^*)\}\\
\quad \text{else}\\
\qquad z'' \leftarrow_2 \{\mathcal{A}(st) \leftrightarrow \mathtt{V}_D(par, pk^*_S, pk_N, m^*, \sigma^*)\}\\
\quad \text{if } (z = \mathtt{accept} \wedge (pk^*_S, m^*, \sigma^*) \notin L_R) \vee\\
\qquad (z' = \mathtt{accept} \wedge (pk^*_S, m^*, \sigma^*) \notin L_C) \vee\\
\qquad (z'' = \mathtt{accept})\\
\qquad \text{output } 1\\
\quad \text{else output } 0
\end{array}
$$

Fig. 2. Malicious signer security experiment

can (1) produce a new valid nominative signature associated to the nominee, (2) convince a verifier about the validity or invalidity of a signature through the confirm or disavow protocols, regardless of the signature being valid or not, or (3) produce an accepting verification token for a signature he has not previously seen a verification token for. Our definition implies security against a malicious signer as defined as part of the unforgeability and non-impersonation definitions of [9,21], as well as the token non-impersonation definition of [21]. Note that unlike [9,21], we do not restrict the adversary's access to the confirm and disavow oracles i.e. the adversary is allowed to query all tuples (pk_S, m, σ) to the convert, confirm and disavow oracles, including the challenge tuple (pk^*_S, m^*, σ^*).

Security against malicious signers of a nominative signature scheme NS is defined via the experiment $\mathtt{Exp}^{\mathtt{mal\text{-}sig}}_{NS,\mathcal{A}}$ shown in Figure 2

In the experiment, $\mathcal{A}$ has access to the oracles $\mathcal{O} = \{\mathcal{O}_{Receive}, \mathcal{O}_{Convert}, \mathcal{O}_{Con}, \mathcal{O}_{Dis}\}$ defined as follows:

- $\mathcal{O}_{Receive}$: given (pk_S, m, δ), this oracle computes $\sigma \leftarrow \mathtt{Receive}(par, pk_S, m, \delta, sk_N)$, adds the tuple (pk_S, m, σ) to the list L_R, and returns σ to $\mathcal{A}$.
- $\mathcal{O}_{Convert}$: given (pk_S, m, σ), this oracle adds the tuple (pk_S, m, σ) to L_C and returns the verification token $tk_\sigma \leftarrow \mathtt{Convert}(par, pk_S, m, \sigma, sk_N)$.
- $\mathcal{O}_{Con}$: given (pk_S, m, σ), this oracle interacts with $\mathcal{A}$ by running `Confirm` with the common input $(par, pk_S, pk_N, m, \sigma)$ and the private input sk_N.
- $\mathcal{O}_{Dis}$: given (pk_S, m, σ), this oracle interacts with $\mathcal{A}$ by running `Disavow` with the common input $(par, pk_S, pk_N, m, \sigma)$ and private input sk_N.

Definition 5. *A nominative signature scheme NS is said to be* `mal-sig` *secure if all PPT adversaries $\mathcal{A}$ have advantage $\mathtt{Adv}^{\mathtt{mal\text{-}sig}}_{NS,\mathcal{A}} = \Pr[\mathtt{Exp}^{\mathtt{mal\text{-}sig}}_{NS,\mathcal{A}}(1^k) = 1]$ which is negligible in k.*

$\mathtt{Exp}^{\mathtt{inv\text{-}cma}}_{NS,\mathcal{A}}(1^k)$
  $par \leftarrow \mathtt{Setup}(1^k)$
  $(pk_N^*, sk_N^*) \leftarrow \mathtt{KeyGen}_N(par)$
  $(pk_S^*, m^*, \delta^*, st) \leftarrow \mathcal{A}^{\mathcal{O}}(par, pk_N^*)$
  $b \leftarrow \{0,1\}$
  if $b = 0$
    $\sigma^* \leftarrow \mathtt{Receive}(par, pk_S^*, m^*, \delta^*, sk_N)$
  else $(b = 1)$
    $(pk_N, sk_N) \leftarrow \mathtt{KeyGen}_N(par)$
    $(pk_S, sk_S) \leftarrow \mathtt{KeyGen}_S(par)$
    $m \leftarrow \mathcal{M}(par)$
    $\delta \leftarrow \mathtt{Sign}(par, pk_N, m, sk_S)$
    $\sigma^* \leftarrow \mathtt{Receive}(par, pk_S, m, \delta, sk_N)$
  $b' \leftarrow \mathcal{A}^{\mathcal{O}}(st, \sigma^*)$
  if $b' = b$ output 1
  else output 0

Fig. 3. Invisibility security experiment

4.3 Invisibility

To ensure that no information leaked from the signer will reveal the validity of a signature, invisibility requires that an adversary with the knowledge of the private signer key, cannot distinguish between a valid signature, and a random element of the signature space. This is required to hold, even when the adversary is given access to all random choices made by the signer when the signature is generated. We insist that the invisibility definition should make use of the *full* signature space of the scheme. Note that, unlike the definitions used in [9,21], this will ensure signer and nominee anonymity. Lastly, note that our definition does not restrict the adversary's access to the sign oracle, and he can obtain signatures on the challenge message. In comparison, the definitions in [9,21] do not allow this, and hence, invisibility is only guaranteed for nominative signatures on messages which have not previously been signed. This could potentially be a problem for applications in which only a small message space is used.

Formally, we define invisibility under a chosen message attack (inv-cma) of a nominative signature scheme NS via the experiment $\mathtt{Exp}^{\mathtt{inv\text{-}cma}}_{NS,\mathcal{A}}$ shown in Figure 3 where $\mathcal{M}(par)$ is the message space defined by par. In the experiment, $\mathcal{A}$ has access to the oracles $\mathcal{O} = \{\mathcal{O}_{Receive}, \mathcal{O}_{Convert}, \mathcal{O}_{Con}, \mathcal{O}_{Dis}\}$ defined as above.

Definition 6. *A nominative signature scheme NS is said to be* inv-cma *secure if all PPT adversaries $\mathcal{A}$ have advantage $\mathtt{Adv}^{inv\text{-}cma}_{NS,\mathcal{A}} = |\Pr[\mathtt{Exp}^{inv\text{-}cma}_{NS,\mathcal{A}}(1^k) = 1] - 1/2|$ which is negligible in k.*

4.4 Protocol Security

Lastly, we require the confirm and disavow protocols to be zero-knowledge proofs. More specifically, consider the languages $L(par)$ and $\overline{L}(par)$ parameterized by par and defined by

$$L(par) = \{(pk_S, pk_N, m, \sigma) : \exists sk_N \, s.t.$$
$$(pk_N, sk_N) \in \{\mathtt{KeyGen}_N(par)\} \wedge$$
$$\mathtt{Valid}(par, pk_S, m, \sigma, sk_N) = \mathtt{accept}\}$$
$$\overline{L}(par) = \{(pk_S, pk_N, m, \sigma) : \exists sk_N \, s.t.$$
$$(pk_N, sk_N) \in \{\mathtt{KeyGen}_N(par)\} \wedge$$
$$\mathtt{Valid}(par, pk_S, m, \sigma, sk_N) = \mathtt{reject}\}$$

The confirm protocols is required to be a zero-knowledge proof of membership for L, whereas the disavow protocol is required to be a zero-knowledge proof of membership for $\overline{L}$.

The zero-knowledge property of the protocols will, in combination with invisibility, ensure that the nominative signature scheme is *non-transferable*. More specifically, non-transferability requires that a verifier can "fake" any evidence of receiving a valid signature from the nominee and confirming the validity through the confirm protocol. This ensures that the verifier cannot convince a third party that he received a valid signature on a given message from the nominee (a similar requirement is made for invalid signatures and the disavow protocol). Non-transferability follows directly from the properties above: invisibility implies that a randomly chosen signature chosen by the verifier will be indistinguishable from a valid one, and the zero-knowledge property of the confirm protocol implies that a verifier can simulate the transcript of an accepting interaction. Hence, a verifier can, by himself, produce a "fake" signature and confirmation transcript pair which is indistinguishable from a honestly generated one. Note that this type of simulation requires that the verifier is able to sample the signature space used in the invisibility definition. As mentioned above, this is not the case for the schemes in [15,13].

4.5 Key Registration

In the above security experiments, the adversary can freely choose the public keys submitted to the sign, receive, confirm and disavow oracles. However, in systems based on a traditional PKI, users are required to obtain a certificate by registering their public key at a certificate authority before the public key can be used in interaction with other users. This allows additional security measures such as requiring that a user prove knowledge of the secret key corresponding to the public key he is registering. To model security in this scenario, we give the adversary access to a *key registration oracle* in addition to the normal oracles. The key registration oracle maintains a list L_{PK} of registered key pairs and interacts with $\mathcal{A}$ as follows:

- $\mathcal{O}_{Reg}$: Given a (signer or nominee) key pair (pk, sk), the oracle checks if (pk, sk) is a valid key pair. If not, the oracle returns $\perp$. Otherwise, it adds (pk, sk) to L_{PK}, and returns $\top$.

When key registration is used, it is assumed that all public keys which is part of an oracle query or a set of challenge values, have been previously submitted to the key registration oracle. We will append the postfix (KR) to a security notion to indicate that key registration is used e.g. uf-cma(KR). Note that key registration is also used in [9,21].

5 Concrete Scheme

Our concrete scheme is inspired by the signature scheme by Waters [20]. More specifically, a signature generation message δ for a message m, consists of an ordinary Waters signature $(g^r, h_S^{\alpha_S} F_S(m)^r)$ on m, and, using the aggregate techniques by Lu *et al.* [16], a nominee will construct a compact "double" signature $(g^r, h_S^{\alpha_S} F_S(m)^r h_N^{\alpha_N} F_N(m)^r)$ on m. To ensure invisibility of the scheme, the first component g^r is split into two parts, $y_1^{r_1}$ and $y_2^{r_2}$ where $r = r_1 + r_2$, and (y_1, y_2) is part of the public nominee key, which essentially makes determining validity of a signature as hard a the decisional linear problem. Lastly, to avoid re-randomization, which is required by security against malicious signers, the nominee will furthermore apply the technique by Boneh *et al.* [2] for obtaining strong unforgeability . In the description of our scheme, we use the notation $ZKPK\{(w) : R(x, w)\}$ to denote a zero-knowledge proof of knowledge of a witness w such that the relation R holds for the common input x and w. After presenting the scheme, we will sketch how these proofs can be implemented.

- Setup: given 1^k, choose a bilinear map $e : \mathbb{G}_1 \times \mathbb{G}_1 \rightarrow \mathbb{G}_T$ where $|\mathbb{G}_1| = p$, and a generator $\langle g \rangle = \mathbb{G}_1$. Lastly pick a collision resistant hash function $H : \{0,1\}^* \rightarrow \mathbb{Z}_p$ and return $par \leftarrow (e, p, g, H)$.
- KeyGen$_S$: given par, pick $\alpha_S, v_0, \ldots, v_n \leftarrow \mathbb{Z}_p$ and $h_S \leftarrow \mathbb{G}_1$, and compute $g_S \leftarrow g^{\alpha_S}$ and $u_i \leftarrow g^{v_i}$ for $1 \leq i \leq n$. Furthermore, for $m \in \{0,1\}^n$, define $F_S(m) = u_0 \prod_{i=1}^n u_i^{m_i}$ where m_i is the i-th bit of m, and finally set $pk_S \leftarrow (g_S, h_S, u_0, \ldots, u_n)$ and $sk_S \leftarrow \alpha_S$.
- KeyGen$_N$: given par, pick $\alpha_N, y_1, y_2, v_0, \ldots, v_n \leftarrow \mathbb{Z}_p$ and $h_N, k \leftarrow \mathbb{G}_1$, and compute $g_N \leftarrow g^{\alpha_N}$. Furthermore, for $0 \leq i \leq n$, set $u_i \leftarrow g^{v_i}$, and for $m \in \{0,1\}^n$ define $F_N(m) = u_0, \prod_{i=1}^n u_i^{m_i}$ where m_i is the i-th bit of m. Lastly compute $x_1 \leftarrow g^{y_1^{-1}}$ and $x_2 \leftarrow g^{y_2^{-1}}$, and return the public key $pk_N \leftarrow (g_N, h_N, k, u_0, \ldots, u_n, x_1, x_2)$ and the private key $sk_N \leftarrow (\alpha_N, v_0, \ldots, v_n, y_1, y_2)$.
- Sign: given par, pk_N, m, and sk_S, pick $r \leftarrow \mathbb{Z}_p$, and compute $\delta_1 \leftarrow g^r$ and $\delta_2 \leftarrow h_S^{\alpha_S} F_S(pk_N || m)^r$. Lastly return $\delta \leftarrow (\delta_1, \delta_2)$.
- Receive: given par, $pk_S = (g_S, h_S, u_0, \ldots, u_n)$, m, $\delta = (\delta_1, \delta_2)$ and $sk_N = (\alpha_N, v_0, \ldots, v_n, y_1, y_2)$, firstly check that $e(g_S, h_S)e(\delta_1, F_S(pk_N || m)) = e(g, \delta_2)$ holds. If this is not the case, return $\perp$. Otherwise, pick $r, r', s \leftarrow \mathbb{Z}_p$ and re-randomize δ by computing $\delta_1' \leftarrow \delta_1 g^{r'}$ and $\delta_2' \leftarrow \delta_2 F_S(pk_N || m)^{r'}$. Then set $\sigma_1 \leftarrow (\delta_1'/g^r)^{y_1^{-1}}$ and $\sigma_2 \leftarrow (g^r)^{y_2^{-1}}$, and compute $t \leftarrow H(\sigma_1 || \sigma_2 || pk_S || m)$ and $M \leftarrow g^t k^s$. Finally set $\sigma_3 \leftarrow \delta_2' h_N^{\alpha_N} (\delta_1')^{v_0 + \sum_{i=1}^n v_i M_i}$, where M_i is the i-th bit of M, and return the signature $\sigma \leftarrow (\sigma_1, \sigma_2, \sigma_3, s)$.
- Convert: given par, $pk_S = (g_S, h_S, u_0, \ldots, u_n)$, m, $\sigma = (\sigma_1, \sigma_2, \sigma_3, s)$, and $sk_N = (\alpha_N, v_0, \ldots, v_n, y_1, y_2)$, first check that the equation

$$e(g, \sigma_3) = e(g_S, h_S)e(g_N, h_N)e(\sigma_1^{y_1}\sigma_2^{y_2}, F_S(pk_N || m)F_N(M))$$

hold, where $M = g^t k^s$ and $t = H(pk_S||\sigma_1||\sigma_2||m)$, and if this is not the case, output $\perp$. Otherwise, return the verification token $tk_\sigma \leftarrow (\sigma_1^{y_1}, \sigma_2^{y_2})$.

- `TkVerify`: given par, $pk_S = (g_S, h_S, u_0, \ldots, u_n)$, $pk_N = (g_N, h_N, u'_0, \ldots, u'_n, x_1, x_2)$, the message m, $\sigma = (\sigma_1, \sigma_2, \sigma_3, s)$, and $tk_\sigma = (tk_1, tk_2)$, output `accept` if the equations $e(\sigma_1, g) = e(tk_1, x_1)$, $e(\sigma_2, g) = e(tk_2, x_2)$, and

$$e(g, \sigma_3) = e(g_S, h_S)e(g_N, h_N)e(tk_1 tk_2, F_S(pk_N||m)F_N(M))$$

hold, where $M = g^t k^s$ and $t = H(pk_S||\sigma_1||\sigma_2||m)$. Otherwise, output `reject`.
- $(\mathtt{Confirm}, \mathtt{V}_C)$: given $(par, pk_S, pk_N, m, \sigma)$ as common input and sk_N as the private input to the nominee running `Confirm`, let $e_1 = e(g, \sigma_3)$, $e_2 = e(g_S, h_S)e(g_N, h_N)$, $e_3 = e(\sigma_1, F_S(pk_N||m)F_N(M))$ and finally let $e_4 = e(\sigma_2, F_S(pk_N||m)F_N(M))$ where $M = g^t k^s$ and $t = H(pk_S||\sigma_1||\sigma_2||m)$. Then the nominee and the verifier interacts in the protocol

$$ZKPK\{(y_1, y_2) : x_1^{y_1} = g \wedge x_2^{y_2} = g \wedge e_1 = e_2 e_3^{y_1} e_4^{y_2}\}$$

- $(\mathtt{Disavow}, \mathtt{V}_D)$: Let input be as in $(\mathtt{Confirm}, \mathtt{V}_C)$. Then the nominee and the verifier interacts in the protocol

$$ZKPK\{(y_1, y_2) : x_1^{y_1} = g \wedge x_2^{y_2} = g \wedge e_1 \neq e_2 e_3^{y_1} e_4^{y_2}\}$$

We note that it is possible to construct sigma protocols for the confirm and disavow protocols by using a combination of well known sigma protocols for proving knowledge of a discrete logarithm, equality of discrete logarithms, and inequality of discrete logarithms (see [4,3]). The zero-knowledge proofs of knowledge in the above scheme can then be obtained by using the transformation proposed by Cramer *et al.* [8] which converts a sigma protocol into a perfect zero-knowledge proof of knowledge. The resulting zero-knowledge proofs are efficient 4-move protocols, and no additional hardness assumptions are required in the transformation.

5.1 Security

The soundness and zero-knowledge properties of the confirm and disavow protocols in the above defined nominative signature scheme NS follow directly from the results by Cramer *et al.* [8], and we refer the reader to [8] for the details. In the following, we state the theorems showing that the NS satisfies the remaining security notions defined in Section 4. While the proofs of Theorem 7 and Theorem 9 can be found in Appendix A and Appendix B, respectively, the proof of Theorem 8 is not given here due to space limitation.

Theorem 7. *Assume that Waters' signature scheme is* `wuf-cma` *secure. Then* NS *is* `uf-cma(KR)` *secure.*

Theorem 8. *Assume the decisional linear problem and the discrete logarithm problem are hard in* $\mathbb{G}_1$*, that* H *is collision resistant, and that Waters signature scheme is* `wuf-cma` *secure. Then* NS *is* `mal-sig(KR)` *secure.*

Theorem 9. *Assume that* NS *is* `mal-sig(KR)` *secure and that the decisional linear problem is hard in* $\mathbb{G}_1$*. Then* NS *is* `inv-cma(KR)` *secure.*

6 Conclusion

We have presented a refined security model for nominative signatures as well as proposed a new efficient scheme. Unlike the the previous schemes, our scheme is provably secure in the standard model, and the security rests on standard assumptions. Lastly, we note that our scheme provides signatures of the same length as the random oracle model scheme [9] which is the only other scheme provide a comparable level of security. Through our proposed scheme, we obtain a highly secure and efficient non-transferable user certification scheme.

Acknowledgement

The first author is supported by a JSPS Fellowship for Young Scientists.

References

1. Baek, J., Safavi-Naini, R., Susilo, W.: Universal designated verifier signature proof (or how to efficiently prove knowledge of a signature). In: Roy, B. (ed.) ASIACRYPT 2005. LNCS, vol. 3788, pp. 644–661. Springer, Heidelberg (2005)
2. Boneh, D., Shen, E., Waters, B.: Strongly unforgeable signatures based on computational Diffie-Hellman. In: Yung, M., Dodis, Y., Kiayias, A., Malkin, T. (eds.) PKC 2006. LNCS, vol. 3958, pp. 229–240. Springer, Heidelberg (2006)
3. Camenisch, J., Shoup, V.: Practical verifiable encryption and decryption of discrete logarithms. In: Boneh, D. (ed.) CRYPTO 2003. LNCS, vol. 2729, pp. 126–144. Springer, Heidelberg (2003)
4. Camenisch, J., Stadler, M.: Proof systems for general statements about discrete logarithms. Technical Report 260, Institute for Theoretical Computer Science, ETH Zurich (March 1997)
5. Chaum, D.: Security Without Identification: Transaction Systems to Make Big Brother Obsolete. Commun. ACM, 1030–1044 (1985)
6. Chaum, D.: Designated confirmer signatures. In: De Santis, A. (ed.) EUROCRYPT 1994. LNCS, vol. 950, pp. 86–91. Springer, Heidelberg (1995)
7. Chaum, D., Van Antwerpen, H.: Undeniable signatures. In: Brassard, G. (ed.) CRYPTO 1989. LNCS, vol. 435, pp. 212–216. Springer, Heidelberg (1990)
8. Cramer, R., Damgård, I., MacKenzie, P.: Efficient zero-knowledge proofs of knowledge without intractability assumptions. In: Imai, H., Zheng, Y. (eds.) PKC 2000. LNCS, vol. 1751, pp. 354–372. Springer, Heidelberg (2000)
9. Huang, Q., Liu, D.Y.W., Wong, D.S.: An efficient one-move nominative signature scheme. IJACT 1(2), 133–143 (2008)
10. Huang, Z., Wang, Y.: Convertible nominative signatures. In: Wang, H., Pieprzyk, J., Varadharajan, V. (eds.) ACISP 2004. LNCS, vol. 3108, pp. 348–357. Springer, Heidelberg (2004)
11. Kim, S., Park, S., Won, D.: Zero-knowledge nominative signatures. In: PRAGOCRYPT, pp. 380–392. CTU Publishing House (1996)
12. Lin, H.-C., Yen, S.-M., Huang, Y.-H.: Security reconsideration of the Huang-Wang nominative signature. Inf. Sci. 178(5), 1407–1417 (2008)
13. Liu, D.Y.W., Chang, S., Wong, D.S., Mu, Y.: Nominative signature from ring signature. In: Miyaji, A., Kikuchi, H., Rannenberg, K. (eds.) IWSEC 2007. LNCS, vol. 4752, pp. 396–411. Springer, Heidelberg (2007)

14. Liu, D.Y.W., Wong, D.S., Huang, X., Wang, G., Huang, Q., Mu, Y., Susilo, W.: Formal definition and construction of nominative signature. In: Qing, S., Imai, H., Wang, G. (eds.) ICICS 2007. LNCS, vol. 4861, pp. 57–68. Springer, Heidelberg (2007)
15. Liu, D.Y.W., Wong, D.S., Huang, X., Wang, G., Huang, Q., Mu, Y., Susilo, W.: Nominative signature: Application, security model and construction. Cryptology ePrint Archive, Report 2007/069 (2007), `http://eprint.iacr.org/`
16. Lu, S., Ostrovsky, R., Sahai, A., Shacham, H., Waters, B.: Sequential aggregate signatures and multisignatures without random oracles. In: Vaudenay, S. (ed.) EUROCRYPT 2006. LNCS, vol. 4004, pp. 465–485. Springer, Heidelberg (2006)
17. Steinfeld, R., Bull, L., Wang, H., Pieprzyk, J.: Universal designated-verifier signatures. In: Laih, C.-S. (ed.) ASIACRYPT 2003. LNCS, vol. 2894, pp. 523–542. Springer, Heidelberg (2003)
18. Susilo, W., Mu, Y.: On the security of nominative signatures. In: Boyd, C., González Nieto, J.M. (eds.) ACISP 2005. LNCS, vol. 3574, pp. 329–335. Springer, Heidelberg (2005)
19. Wang, G., Bao, F.: Security remarks on a convertible nominative signature scheme. In: New Approaches for Security, Privacy and Trust in Complex Environments. IFIP, vol. 232, pp. 265–275. Springer, Boston (2007)
20. Waters, B.: Efficient identity-based encryption without random oracles. In: Cramer, R. (ed.) EUROCRYPT 2005. LNCS, vol. 3494, pp. 114–127. Springer, Heidelberg (2005)
21. Zhao, W., Lin, C., Ye, D.: Provably secure convertible nominative signature scheme. In: Yung, M., Liu, P., Lin, D. (eds.) Inscrypt 2008. LNCS, vol. 5487, pp. 23–40. Springer, Heidelberg (2009)

A Proof of Theorem 7

Proof. Let $Succ_1$ be the event that $\mathcal{A}$ succeeds by successfully completing the confirm protocol in the unforgeability experiment (i.e. $z = \texttt{accept}$), and let $Succ_2$ be the event that $\mathcal{A}$ succeeds by producing a valid verification token (i.e. $z' = \texttt{accept}$). Furthermore, let $Valid$ be the event that the element defined by pk_S and (pk_N^*, m^*, σ^*) output by $\mathcal{A}$ belongs to the language defined by the zero-knowledge protocol implementing the confirm protocol (i.e. there exist $y_1, y_2 \in \mathbb{Z}_p$ such that $x_1^{y_1} = g$, $x_2^{y_2} = g$ and $e(g, \sigma_3) = e(g_S, h_S)e(g_N, h_N)e(\sigma_1^{y_1} \sigma_2^{y_2}, F_S(pk_N||m^*)F_N(M))$ where $M = g^t k^s$ and $t = H(pk_S||\sigma_1||\sigma_2||m^*)$). Then we have

$$\begin{aligned}\texttt{Adv}^{\texttt{uf-cma}}_{NS,\mathcal{A}} &= \Pr[Succ_1 \vee Succ_2] \\ &\leq \Pr[Succ_1 \wedge Valid] + \Pr[Succ_1 \wedge \overline{Valid}] + \Pr[Succ_2]\end{aligned}$$

Note that the zero-knowledge proof implementing the confirm protocol, which is obtained by applying the transformation in [8] to a sigma protocol for proving knowledge of (y_1, y_2) such that the required equations hold, will have negligible soundness error i.e. a prover trying to complete the protocol for values $(par, pk_S, pk_N, m, \sigma)$ for which no such (y_1, y_2) exists, will have negligible probability of making a verifier accept. This follows from the ability to use the extraction techniques from [8] to extract the values (y_1, y_2) with non-negligible probability from a prover which makes a verifier accept with non-negligible probability, which will contradict that no (y_1, y_2) exist. Hence, this implies that $\Pr[Succ_1 \wedge \overline{Valid}]$ must be negligible. To complete the proof, we show the following claims.

Claim. $\Pr[Succ_1 \wedge Valid]$ is negligible

Proof. Let $\mathcal{A}$ be an adversary such that $\epsilon_{\mathcal{A}} = \Pr[Succ_1 \wedge Valid]$ is non-negligible. Using $\mathcal{A}$, we will construct a simulator $\mathcal{B}$ breaking the unforgeability of Waters' signature scheme with probability $\epsilon_{\mathcal{A}}$. $\mathcal{B}$ is constructed as follows:

Initially, $\mathcal{B}$ receives parameters $par' = (\mathbb{G}_1, p, g, e)$ and a public signer key $pk_S = (g_S, h_S, u_0, \ldots, u_n)$. Then $\mathcal{B}$ picks a collision resistant hash function $H : \{0,1\}^* \to \mathbb{Z}_p$, sets $par \leftarrow (\mathbb{G}_1, p, g, e, H)$, and runs $\mathcal{A}$ with input (par, pk_S). While running, $\mathcal{A}$ can ask sign queries (pk_N, m) which $\mathcal{B}$ responds to by forwarding $pk_N||m$ to his own signing oracle to obtain δ, and then returning δ to $\mathcal{A}$. At some point, $\mathcal{A}$ outputs challenge values $(pk_N^*, m^*, \sigma^*, tk_\sigma^*)$ where $pk_N^* = (g_N, h_N, u_0, \ldots, u_n, x_1, x_2)$ and $\sigma^* = (\sigma_1^*, \sigma_2^*, \sigma_3^*, s^*)$. $\mathcal{B}$ interacts with $\mathcal{A}$ in the confirm protocol, and upon completion, $\mathcal{B}$ retrieves the private key $sk_N^* = (\alpha_N, v_0, \ldots, v_n, y_1, y_2)$ corresponding to pk_N^* from the list L_{PK}. If $Valid$ occurs, there must exist values (y_1', y_2') such that

$$e(g, \sigma_3^*) = e(g_S, h_S)e(g_N, h_N)e((\sigma_1^*)^{y_1}(\sigma_2^*)^{y_2}, F_S(pk_N^*||m^*)F_N(M))$$

where $M = g^t k^s$ and $t = H(pk_S||\sigma_1^*||\sigma_2^*||m^*)$. Furthermore, since it is required that (pk_N^*, sk_N^*) is a valid key pair, (y_1', y_2') must correspond to (y_1, y_2) from sk_N^*. Hence, by setting $\sigma_1' \leftarrow (\sigma_1^*)^{y_1}(\sigma_2^*)^{y_2}$ and $\sigma_2' \leftarrow \sigma_3^*/(h_N^{\alpha_N}(\sigma_1')^{v_0+\sum_{i=1}^{n} v_i M_i})$, $\mathcal{B}$ obtains a signature $\sigma' \leftarrow (\sigma_1', \sigma_2')$ such that $e(g, \sigma_2') = e(g_S, h_S)e(\sigma_1', F_S(pk_N^*||m^*))$. Furthermore, since it is required that $\mathcal{A}$ did not submit (pk_N^*, m^*) to $\mathcal{O}_{Sign}$, $\mathcal{B}$ will not have submitted $pk_N^*||m^*$ to his own sign oracle i.e. σ' is a valid forgery on $pk_N^*||m^*$ under pk_S, and $\mathcal{B}$ terminates with output σ'. □

Claim. $\Pr[Succ_2]$ is negligible

Proof. Observe that if $\texttt{TkVerify}(par, pk_S^*, pk_N, m^*, \sigma^*, tk_\sigma^*) = \texttt{accept}$, then σ^* must be a valid signature. This can be seen as follows: let $pk_S^* = (g_S, h_S, u_0, \ldots, u_n)$, $pk_N = (g_N, h_N, k, u_0', \ldots, u_n', x_1, x_2)$, $\sigma^* = (\sigma_1, \sigma_2, \sigma_3, s)$ and $tk_\sigma^* = (tk_1, tk_2)$. Since both x_1 and x_2 generates $\mathbb{G}_1$, there must exist exponents $r_1, r_2 \in \mathbb{Z}_p$ such that $\sigma_1 = x_1^{r_1}$ and $\sigma_2 = x_2^{r_2}$. Furthermore, if $\texttt{TkVerify}$ outputs $\texttt{accept}$, the equations $e(\sigma_1, g) = e(tk_1, x_1)$ and $e(\sigma_2, g) = e(tk_2, x_2)$ must hold, and hence, due to the properties of the pairing, we must have $tk_1 = g^{r_1}$ and $tk_2 = g^{r_2}$. Lastly, since the equation

$$e(g, \sigma_3) = e(g_S, h_S)e(g_N, h_N)e(tk_1 tk_2, F_S(pk_N||m)F_N(M)), \qquad (1)$$

where $M = g^t k^s$ and $t = H(pk_S^*||\sigma_1||\sigma_2||m^*)$, is also required to hold, we have that σ_3 is of the form $\sigma_3 = h_S^{\alpha_S} F_S(pk_N||m^*)^{r_1+r_2} h_N^{\alpha_N} F_N(M)^{r_1+r_2}$ where $\alpha_S = \log_g g_S$ and $\alpha_n = \log_g g_N$ i.e. σ^* is a valid signature on m^* under pk_S^* and pk_N.

While this property is slightly stronger than needed for the proof of this claim (we only need the implication of Equation 1), it will be useful in the following proofs. Returning to the proof of the above claim, let $\mathcal{A}$ be an adversary such that $\epsilon_{\mathcal{A}} = \Pr[Succ_1 \wedge Valid]$ is non-negligible. Like in the proof of the claim above, we will use $\mathcal{A}$ to construct a simulator $\mathcal{B}$ that breaks the unforgeability of Waters' signature scheme with probability $\epsilon_{\mathcal{A}}$.

$\mathcal{B}$ will setup the parameters and public signer key, and respond to sign queries in the same way as in the proof of the above claim. At some point, $\mathcal{A}$ outputs $(pk_N^*, m^*, \sigma^*, tk_\sigma^*)$, where $pk_N^* = (g_N, h_N, u_0, \ldots, u_n, x_1, x_2)$, $tk^* = (tk_1, tk_2)$ and $\sigma^* = (\sigma_1^*, \sigma_2^*, \sigma_3^*, s^*)$. $\mathcal{B}$ retrieves the private nominee key $sk_N^* = (\alpha_N, v_0, \ldots, v_n, y_1, y_2)$ corresponding to pk_N^* from the list L_{PK}, and constructs the signature (σ_1', σ_2') where $\sigma_1 \leftarrow tk_1 tk_2$ and $\sigma_3^*/(h_N^{\alpha_N}(\sigma_1')^{v_0+\sum_{i=1}^n v_i M_i})$. If $\mathtt{TkVerify}(par, pk_S, pk_N^*, m^*, \sigma^*, tk^*)$ outputs $\mathtt{accept}$, (σ_1', σ_2') must satisfy $e(g, \sigma_2') = e(g_S, h_S)\, e(\sigma_1', F_S(pk_N^* || m^*))$ due to Equation 1 i.e. (σ_1', σ_2') must be a valid signature on $pk_N^*||m^*$ under pk_S. Furthermore, since a successful $\mathcal{A}$ is not allowed to have queried (pk_N^*, m^*) to $\mathcal{O}_{Sign}$, $\mathcal{B}$ will not have queried $pk_N^*||m^*$ to his own signing oracle, in which case $\mathcal{B}$ has obtained a valid forgery (σ_1', σ_2'). □

The theorem follows by the combination of the above two claims. □

B Proof of Theorem 9

Proof. Let $Succ$ denote the event that $\mathcal{A}$ successfully guesses the challenge bit, and let $Forge$ denote the event that $\mathcal{A}$ submits a query (pk_S, m, σ) to $\mathcal{O}_{Convert}$, $\mathcal{O}_{Con}$ or $\mathcal{O}_{Dis}$ such that σ is valid but was not obtained by submitting (pk_S, m, δ) to $\mathcal{O}_{Receive}$ for any δ. The advantage of $\mathcal{A}$ can be expressed as follows:

$$\mathtt{Adv}_{NS,\mathcal{A}}^{\mathtt{inv\text{-}cma}} = |\Pr[Succ] - 1/2| \leq \Pr[Forge] + |\Pr[Succ|\overline{Forge}] - 1/2|$$

Claim. $\Pr[Forge]$ is negligible

It is fairly easy to see that an adversary which makes $Forge$ happen with non-negligible probability, can be reduced to an adversary attacking the security against malicious signers i.e. an adversary who constructs a new signature associated to the nominee. We leave out the details here.

Claim. $|\Pr[Succ_3|\overline{Forge}] - 1/2|$ is negligible

Proof. Let $\mathcal{A}$ be an adversary such that $\epsilon_{\mathcal{A}} = |\Pr[Succ|\overline{Forge}] - 1/2|$ is non-negligible. Using $\mathcal{A}$, we construct a simulator $\mathcal{B}$ that break the decisional linear problem in $\mathbb{G}_1$ with probability $\epsilon_{\mathcal{A}}$. $\mathcal{B}$ is constructed as follows:

Initially, $\mathcal{B}$ receives a group description $(\mathbb{G}_1, p, g, e)$ and group elements $x_1, x_2, x_1^a, x_2^b, g^c \in \mathbb{G}_1$ and will have to decide whether $c = a + b$ or a random element in $\mathbb{Z}_p$. Firstly, $\mathcal{B}$ picks a collision resistant hash $H : \{0,1\}^* \rightarrow \mathbb{Z}_p$ and sets $par \leftarrow (\mathbb{G}_1, p, g, e, H)$. Then $\mathcal{B}$ picks $\alpha_N, v_0, \ldots, v_n \leftarrow \mathbb{Z}_p$ and $h_N, k \leftarrow \mathbb{G}_1$, computes $g_N \leftarrow g^{\alpha_N}$ and $u_i \leftarrow g^{u_i}$ for $1 \leq i \leq n$, sets $pk_N \leftarrow (g_N, h_N, u_0, \ldots, u_n, x_1, x_2)$, and stores the partial private nominee key $sk_N' = (\alpha_N, v_0, \ldots, v_n)$ (note that this partial key sk'_N is sufficient to execute Receive). Lastly, $\mathcal{B}$ runs $\mathcal{A}$ with input (par, pk_N). While running, $\mathcal{A}$ can ask receive, confirm, disavow and convert queries, which $\mathcal{B}$ responds to as follows:

- $\mathcal{O}_{Receive}$: given (pk_S, m, δ), where $pk_S = (g_S, h_S, u_0, \dots, u_n)$ and $\delta = (\delta_1, \delta_2)$, $\mathcal{B}$ first checks if $e(g, \delta_2) = e(g_S, h_S)e(\delta_1, F_S(pk_N||m))$, and if this is not the case, $\mathcal{B}$ returns $\perp$. Otherwise, $\mathcal{B}$ retrieves $sk_S = (\alpha_S, v'_0, \dots, v'_n)$ corresponding to pk_S from the list L_{PK}, and constructs a new signature $(\sigma'_1, \sigma'_2) = (g^r, h_S^{\alpha_S} F_S(pk_N||m)^r)$ (note that (σ'_1, σ'_2) corresponds to a re-randomization of δ). Then $\mathcal{B}$ picks $r', s \leftarrow \mathbb{Z}_p$ and computes $\sigma_1 \leftarrow x_1^{r-r'}$, $\sigma_2 \leftarrow x_2^{r'}$, $t \leftarrow H(pk_S||\sigma_1||\sigma_2||m)$, $M \leftarrow g^t k^s$ and $\sigma_3 \leftarrow \sigma'_2 h_N^{\alpha_N} F_N(M)^r$. Lastly, $\mathcal{B}$ returns $\sigma \leftarrow (\sigma_1, \sigma_2, \sigma_3, s)$ to $\mathcal{A}$ and stores $(pk_S, m, \sigma, r - r', r')$ for later use.
- $\mathcal{O}_{Convert}$: given (pk_S, m, σ), B returns $\perp$ if σ was not returned in a response to a receive query on (pk_S, m, δ) for some δ (note that if *Forge* does not occur, this is the correct response). Otherwise, $\mathcal{B}$ recalls the values $(pk_S, m, \sigma, r - r', r')$, and returns the token $tk_\sigma = (tk_1, tk_2) = (g^{r-r'}, g^{r'})$. This corresponds to the correct token, since $tk_1 = g^{r-r'} = (x_1^{y_1})^{r-r'} = \sigma_1^{y_1}$ and $tk_2 = g^{r'} = (x_1^{y_2})^{r'} = \sigma_2^{y_2}$ where $y_1 = \log_{x_1} g$ and $y_2 = \log_{x_2} g$ (note that (y_1, y_2) are unknown to $\mathcal{B}$).
- $\mathcal{O}_{Con}$: given (pk_S, m, σ), B returns $\perp$ if σ was not returned in a response to a receive query on (pk_S, m, δ) for some δ. Otherwise, $\mathcal{B}$ exploits the zero-knowledge property of the confirm protocol to simulates the protocol to $\mathcal{A}$ by using standard rewind and replay techniques. Note that the zero-knowledge protocol obtained by using the conversion by Cramer *et al.* provides perfect zero-knowledge proof and thereby allows $\mathcal{B}$ to provide a perfect simulation.
- $\mathcal{O}_{Dis}$: given (pk_S, m, σ), $\mathcal{B}$ returns $\perp$ if σ was returned as a response to (pk_S, m, δ) for some δ. Otherwise, $\mathcal{B}$ simulates the disavow protocol in similar manner to the confirm protocol.

At some point, $\mathcal{A}$ will output (pk_S^*, m^*, δ^*). To construct a challenge signature, $\mathcal{B}$ first retrieves $sk_S^* = (\alpha_S, v'_0, \dots, v'_n)$ corresponding to pk_S^* from L_{PK}. Then $\mathcal{B}$ picks $s \leftarrow \mathbb{Z}_p$ and sets $\sigma_1^* \leftarrow x_1^a$, $\sigma_2^* \leftarrow x_2^b$, $t \leftarrow H(pk_S^*||\sigma_1^*||\sigma_2^*||m)$, $M^* \leftarrow g^t k^s$ and $\sigma_3 \leftarrow h_S^{\alpha_S}(g^c)^{v'_0 + \sum_{i=1}^n v'_i (pk_N||m)_i} h_N^{\alpha_N}(g^c)^{v_0 + \sum_{i-1}^n v_i M_i^*}$. Finally, $\mathcal{B}$ returns the challenge signature $\sigma^* \leftarrow (\sigma_1^*, \sigma_2^*, \sigma_3^*, s)$ to $\mathcal{A}$. Note that σ^* will be a random element of the signature space if c is random in $\mathbb{Z}_p$, whereas if $c = a + b$, σ^* will correspond to a valid signature. After receiving σ^*, $\mathcal{A}$ can ask additional queries which $\mathcal{B}$ responds to as above. Lastly, $\mathcal{A}$ will terminate with output b' which $\mathcal{B}$ forwards as his own solution to the decisional linear problem. It should be clear from the above that $\mathcal{B}$ will correctly solve the given decisional linear problem whenever $\mathcal{A}$ correctly solves the invisibility challenge. □

The theorem follows by combining the above two claims. □

Composable Security Analysis of OS Services*

Ran Canetti[2,**], Suresh Chari[1], Shai Halevi[1], Birgit Pfitzmann[1], Arnab Roy[1], Michael Steiner[1], and Wietse Venema[1]

[1] IBM T.J. Watson Research Center
[2] Tel-Aviv University

Abstract. We provide an analytical framework for basic integrity properties of file systems, namely the binding of files to filenames and writing capabilities. A salient feature of our modeling and analysis is that it is *composable:* In spite of the fact that we analyze the filesystem in isolation, security is guaranteed even when the file system operates as a component within an arbitrary, and potentially adversarial system.

Our results are obtained by adapting the *Universally Composable* (UC) security framework to the analysis of software systems. Originally developed for cryptographic protocols, the UC framework allows the analysis of simple components in isolation, and provides assurance that these components maintain their behavior when combined in a large system, potentially under adversarial conditions.

1 Introduction

Contemporary software systems are complex, consisting of many millions of lines of code, spread across a myriad of components and sub-components. A natural approach for analyzing such large systems is by analyzing each component separately, and "hoping" to use the component-wise analysis to analyze the entire system. Unfortunately, applying this approach to security analysis is problematic. Even if a component is simple enough to analyze separately, its interaction with other components can yield unexpected results. Often, a component will be used in environments different from what its designers initially had in mind, alongside other components that perhaps did not even exist when the original component was analyzed, potentially violating some assumptions that were made in the analysis.

Ideally, we would like to analyze the behavior of a component in isolation, and have the assurance that this behavior remains intact *even when that component is embedded in a new environment.* Within the realm of cryptography, the frameworks of Reactive Simulatability [19,1] and Universal Composability (UC) [4,5] ensure just that. These frameworks are aimed at capturing the security of cryptographic primitives and protocols, ranging from authentication and key exchange,

* This work is supported in part by the Department of Homeland Security under the grant FA8750-08-2-0091.

** Supported by the Check Point Institute for Information Security. Part of this work was done when visiting IBM T.J. Watson Research Center.

J. Lopez and G. Tsudik (Eds.): ACNS 2011, LNCS 6715, pp. 431–448, 2011.

to public-key encryption and signatures, zero-knowledge, and more (see [5] for many examples.) However, many of the features of these frameworks appear at first to be specific to the realm of cryptographic protocols. A natural question is whether the "composable security" approach sketched above can be carried out in a meaningful way even outside the limited domain of cryptography. A positive answer could significantly reduce the overhead in analyzing the security of large systems, while at the same time provide better overall security guarantees.

In this work we demonstrate that this can indeed be done, in the context of guaranteeing some basic integrity properties of filesystems. For this purpose we adapt the UC framework to software systems by establishing new conventions for modeling process management and scheduling. The current work is one of just a few attempts to apply the UC formalism to a large and complex software system, and we believe that it will enable further application of the UC formalism to other software systems.

Analysis in the UC framework proceeds by defining an idealized specification model and an implementation, and then proving that the implementation realizes the idealized specification. Our main contribution is a very simple filesystem specification model, called SimpFS, that captures many integrity concerns in contemporary filesystems, together with an implementation over existing POSIX filesystems [20] and a proof that the implementation realizes the specification model.

> *The composability properties of our analysis imply that software systems that use our implementation over POSIX behave essentially the same as if they were using the simple, idealized specification system* SimpFS.

This is a very strong security guarantee. In particular, it allows analyzing software systems without worrying about how the filesystem is implemented, and without worrying about potential bad interactions between the analyzed system and the filesystem implementation.

Our filesystem model is geared toward ensuring integrity of files and their names, and in particular preventing filename manipulation attacks. In such attacks, a victim program expects a particular filename to have certain semantics. (E.g., a mail program may expect the file `/var/mail/root` to be the mail file for the super-user.) In the attack, the adversary creates a link by the same name in the filesystem, pointing to another file (e.g., `/var/mail/root -> /etc/passwd`), thereby "tricking" the victim program into accessing an unexpected file. (In the mail example, such a link may cause a naive mail program to write incoming email into the system's password file.) Such attacks were quite common in UNIX systems of old.

Our implementation builds on the ideas presented in Chari et al. [7], who address the problem of privilege escalation attacks via filename manipulation. To counter these attacks, Chari et al. present a "safe" name resolution procedure, and deploy this system-wide on popular POSIX systems. The SimpFS interfaces are designed to tightly bind files with their names: files can be accessed *only* via the names they were created with, which means that filename manipulation attacks are impossible in our model. Our proof — showing that implementation based on [7] realizes the model — implies in particular that it indeed eliminates these filename manipulation attacks.

SimpFS offers a simple interface that captures enough filesystem primitives for application developers to build meaningful applications. The simplicity of SimpFS is due to its very narrow interface (only four commands) and the fact that *it does not have directories.* We argue that the murky relation between files and their names in plain POSIX systems stem to a large extent from the fact that pathnames consist of many directories, each with its own permissions, which are combined in a non-obvious manner to yield the effective permissions for the entire name. In contrast, a filename in SimpFS is just a single entity with explicitly specified permissions. Thus SimpFS provide applications with radically simplified semantics, making it easier to use the filesystem without falling into traps. At the same time, we argue that the vast majority of contemporary applications in POSIX systems *do not really need directories*, and can be implemented over the simple SimpFS interface without loss of functionality.

1.1 Related Work

Triggered by Joshi and Holzmann's mini-challenge [13], there is a lot of recent work on formalization and verifications of file systems. Most notably, Freitas et al [10] specify and prove a POSIX file store in Z/Eves. This body of work focuses mostly on the correctness aspects and does not address in depth the security and access control aspects of filesystems. In the broader perspective of (secure) operating systems, there is a long history of formalization and verification, from PSOS [16] to the recent seL4 [14]. While they make considerable progress toward high-assurance OS, these works are not based on frameworks that allow easy composition of components to form larger systems. Additionally, the focus in many of these works is on mandatory access control whereas we cover a discretionary control. (We stress that although our model addresses integrity concerns, these are very different from the Biba integrity model [3].)

An abstract model of another large standard systems, the browser, suitable for proofs of cryptographic protocols exists in [12]. It includes a model of information-flow properties under attack. However, the federated identity protocols built on top of it have only been proven secure with respect to specific security properties, not in a real-world / ideal-world setting [12].

Protocol Composition Logic (PCL) [8] is a comparable general approach on reasoning about (cryptographic) network protocols in a composable fashion. Recently, PCL was applied to analyze systems [9], more specifically integrity properties provided by TPM. The symbolic and axiomatic nature of PCL leads to a more axiomatic specification of security rather than the declarative form in UC. Furthermore, the composition theorems in PCL are weaker than in the UC framework.

A noteworthy contribution to secure composition of large systems is the CHATS project [17], that identifies architectural principles to guide the structuring and decomposition of trustworthy systems. That work is largely orthogonal to ours, as it does not focus on formal modeling or proofs.

There have been many more attempts to leverage well-established formalisms such as logic, typing or process calculi to model composability of certain system security properties, e.g., McLean [15] for non-interference properties or Bengtson et al [2] for cryptographic protocols and access control mechanisms. Many of them provide tool support; but they do not provide the same composition guarantees as in the UC framework.

Organization. Due to lack of space, we did not include in this writeup the proof of security in the UC framework. A full version that includes the proof is available online [6]. In Section 2 we outline the UC framework, then in Section 3 we describe our SIMPFS model and in Section 4 we show how it is implemented over POSIX.

2 The Universal Composability Framework

We briefly describe the relevant aspects of the framework of universally composable (UC) security. The reader is referred to [4] for more details. The framework describes two probabilistic games: The *real world* that captures the protocol flows and the capabilities of an attacker, and the *ideal world* that captures what we think of as a secure system. The notion of security asserts that these two worlds are essentially equivalent.

THE REAL-WORLD MODEL. The players in the real-world model are all the entities of interest in the system (e.g., the nodes in a network, the processes in a software system, etc.), as well as *the adversary* A and *the environment* $\mathcal{Z}$. All these players are modeled as efficient, probabilistic, message-driven programs (formally, they are all interactive Turing machines).

The actions in this game should capture all the interfaces that the various participants can utilize in an actual deployment of this component in the real world. In particular, the capabilities of A should capture all the interfaces that a real-life attacker can utilize in an attack on the system. (For example, A can typically see and modify network traffic.) The environment $\mathcal{Z}$ is responsible for providing all the inputs to the players and getting all the outputs back from them. Also, $\mathcal{Z}$ is in general allowed to communicate with the adversary A. (This captures potential interactions where higher-level protocols may leak things to the adversary, etc.)

THE IDEAL-WORLD MODEL. Security in the UC framework is specified via an "ideal functionality" (usually denoted $\mathcal{F}$), which is thought of as a piece of code to be run by a completely trusted entity in the ideal world. The specification of $\mathcal{F}$ codifies the security properties of the component at hand. Formally, the ideal-world model has the same environment as the real-world model, but we pretend that there is a completely trusted party (called "the functionality"), which is performing all the tasks that are required of the protocol. In the ideal world, participants just give their inputs to the functionality $\mathcal{F}$, which produces the correct outputs (based on the specification) and hands them back to the participants. $\mathcal{F}$ may interact with an adversary, but only to the extent that the intended security allows. (E.g., it can "leak" to the adversary things that should be publicly available, such as public keys.) Specifying the code of $\mathcal{F}$ is typically a non-trivial task. It is important that $\mathcal{F}$ satisfies all the desired security properties, but also that $\mathcal{F}$ does not impose unnecessary constraints: It is only too easy to write a functionality that describes "what we intuitively want", but is not realizable by any implementation. Another crucial concern is the *simplicity* of the functionality $\mathcal{F}$, since we want $\mathcal{F}$ to capture the important security concerns, not the mundane implementation details.

UC-Security and the Composition Theorem. An implementation π **securely realizes** an ideal functionality $\mathcal{F}$ if no external environment can distinguish between running the protocol π in the real world and interacting with the trusted entity running the ideal functionality $\mathcal{F}$ in the ideal world. That is, for every adversary A in the real world, there should exist an adversary A' in the ideal world, such that no environment $\mathcal{Z}$ can distinguish between interacting with A and π in the real world and interacting with A' and $\mathcal{F}$ in the ideal world.

The striking feature of the UC framework is its ability to handle composition. Specifically, the composition theorem from [4] asserts the following: Let ρ be an arbitrary system that runs in the ideal world and uses (perhaps multiple copies of) the functionality $\mathcal{F}$. Next, consider the system ρ' in the real world, that is the same as ρ except that in ρ' each call to the ideal functionality $\mathcal{F}$ is replaced by executing the implementation π. Then, if π securely realizes $\mathcal{F}$ it is guaranteed that system ρ' behaves essentially the same as system ρ. In particular, all the security properties of ρ are inherited by protocol ρ'. This guarantee is the basis for the composable security guarantees provided by the UC framework.

2.1 Conventions for Software Systems

We briefly describe some technicalities that must be resolved when attempting to apply the UC framework to software system, and the conventions that we use to address them. The "entities of interest" in our work are processes, which differ somewhat from the interactive Turing machines (ITMs) in common cryptographic models. One aspect relates to side-channels: whereas an ITM can only influence other ITMs by sending messages, a process shares some physical resources with other processes on the same machine, so it could influence them via side channels such as timing and concurrency. In this work we ignore that aspect, i.e. we do not have any side channels in our formal model. (This does not matter for our current SimpFS model, since we do not model any secrecy requirements.) We thus just let the adversary learn "whatever it needs," so it has no use for side channels.

A more important difference is preemptive multitasking: common crypto models postulate a sequential scheduling model, where an active ITM keeps the control until it sends a message, at which point the recipient becomes active. On the other hand, processes in contemporary OSes can be made to yield control involuntarily. Resolving this discrepancy is not as hard as it may seem, since (side-channels aside) an active entity has no effect on its surroundings until it sends a message, which means that influencing the surroundings only comes with losing the control. We use the standard sequential scheduling of the UC framework, but ensure that the adversary gets the control after every message is sent, and can decide when this message will be delivered. (This is somewhat similar to the "buffer scheduler" from [1].) Hence the adversary in our formal model is able to simulate the actions that would have happened in the actual deployed system, delay delivery messages until the simulation arrives at the point where they were delivered. We thus argue that the formal adversary in our model is able to induce any behavior that can happen in the actual deployed system.

Another difference is that some processing in real systems is done not by the processes themselves, but by the kernel on their behalf. Hence also in our model we postulate the existence of a "kernel component" that can do things on behalf of processes. In our filesystem example, this kernel component is only responsible for maintaining the process privileges: Whenever a process calls a filesystem function, the kernel adds the process-id and roles of the calling process to the list of arguments, and forwards everything to the filesystem. (The kernel component gets these roles from the environment.) We note that although we do not use it in our filesystem example, in general we could have several such "kernel components" in a system, representing several physical machines.

3 SimpFS: A Simple Idealized File-System

This section describes SimpFS, our simple filesystem model. SimpFS has a minimalistic interface with simple semantics, having only basic primitives to create, read, write and delete files. Still, we believe that the this file-system functionality is sufficient for most applications. (Other aspects — such as locking — can be implemented on top of our interface.) The SimpFS model includes file write permissions, hence capturing properties of *filesystem integrity*. We currently do not model read permissions, but we expect that this work can be extended to include read permissions without too much change.

An important feature of SimpFS is that *it does not have any directories, only files and their names.* As we mention in the introduction, we believe that directories have "inherently cumbersome semantics", hence decided to do away with them in order to keep the semantics as simple as possible. We stress that the model supports names that include '/' (so applications can still store their temporary data in files with names that begin with "`/tmp/`"). But a name such as "`/a/b/foo`" is viewed as just one entity, and its existence does not imply the existence of an object with name "`/a/b`." Of course, our *implementation* over POSIX still interprets '/' as a directory separator, and name creation induces the right associations between names and paths, in spite of symlinks, adversarial write permissions etc. While directories are a useful and convenient way to manage and organize systems, we argue that directory permissions are very rarely needed in applications (if ever), and most applications can therefore directly use the SimpFS interface.

A key security property of SimpFS is that it rules out filename manipulation attacks. Our focus on this property is motivated by the large number of privilege escalation attacks due to unsafe pathname resolution that were discovered in

POSIX systems over the years. A classical example of this type of attacks is local mail delivery, where `/var/mail` may be world-writable, allowing an adversary to create a link from `/var/mail/root` to (say) `/etc/passwd`, thereby "tricking" a naive mail-delivery program (running as `root`) to write the content of incoming mail into `/etc/passwd`. Such attacks arise due to the opaque mapping of names to files in POSIX. SIMPFS features a very tight binding between files and their names: a file can be manipulated *only* with the names it was created with.

We describe an implementation of SIMPFS over contemporary POSIX filesystems and *rigorously prove* that this implementation realizes SIMPFS, using the UC framework. The proof implies that processes that use our implementation will be protected against pathname manipulation attacks such as above even if adversarial processes use the same POSIX filesystem in arbitrary ways.

3.1 A Formal Model of SimpFS

SIMPFS consists of files and their names. A newly created file is given some names, and thereafter the file can be accessed by any of these names. Existing names can be deleted, but one cannot add names to existing files. When deleting names, a file can end up with zero names, in which case it is not reachable anymore so we can consider it as deleted. We associate permissions with both the file names and the files themselves:

- Every file has a list of roles that can write in it, called the *Writers* list. A process can write to a file if it holds a role in the Writers list of the file.
- File names have a set of *Manipulators*, listing all the roles that have permission to delete that name.

In the current version we do not have read permissions, which means that SIMPFS allows every process to read every file.

In more details, our ideal SIMPFS maintains an array of files and an associative array of names: `files[]` is an array of files (indexed by integers). Each entry is a file, consisting of an array of bytes (i.e., a data blob) and a list of roles (specifying the Writers of this file). `names[]` is an associative array (indexed by strings). We refer to the index of an entry as a file-name, and each entry consists of a pointer to a file (i.e., an integer) and a list of roles (specifying the Manipulators of this name). The interface below constrains the Manipulator lists, making sure that all the names of the same file have the same set of Manipulators. (This choice is not very important, it is done mostly to simplify the presentation.)

In the initial state, the file-system is empty, with no files and no names (i.e., both arrays are empty). There are only four operations that are supported in SIMPFS: `CreateFile` creates a new file with some names, `DeleteName` deletes an existing name, `Read` reads data from a file (specified by some name), and `Write` writes data to a file (specified by some name).

The semantics of these operations is described by the pseudo-code in Figure 1. As is the case with every formal UC functionality, the pseudo-code includes not only the intended functionality as seen by the legitimate users of the system, but also all the interfaces that an adversary can utilize to attack it. This is codified

```
CreateFile(Writers, Manipulators, Names, pid, Roles)
{
  // Allow the adversary to fail the operation and decide the error code
  var retCode = AdversaryAction("CreateFile",Writers,Manipulators,Names,pid,Roles);
  if (retCode != OKAY) return retCode;

  var codes[] = empty;    // a local list of return codes, one per name
  var f = index of next available entry in the files[] array;
  files[f].data=empty, files[f].Writers=Writers;

  // Allow the adversary to decide whether to create each name
  for each fName in Names {
    var code = AdversaryAction("CreateOneName", fName);
    if (code!=OKAY) codes[i]=code;
    else {
      if (names[fName] already exists) codes[i] = FILE_EXISTS;
      else {
        names[fName].file=f, names[fName].Manipulators=Manipulators;
        codes[i]=OKAY;
  } } }
  call AdversaryAction("Done CreateFile") and then return codes;
}

DeleteName(fName, pid, Roles)
{
  // Allow the adversary to fail the operation and decide the error code
  var retCode = AdversaryAction("DeleteName",fName,pid,Roles);
  if (retCode != OKAY) return retCode;

  if (names[fName] does not exist) return FILE_DOESNT_EXIST;
  if (Roles intersect names[fName].Manipulators = emptyset) return NO_PERMISSION;

  delete names[fName]; // Note: no point deleting the file, even if not reachable
  call AdversaryAction("Done DeleteName") and then return OKAY;
}

Write(fName, atAddr, data, pid, Roles)
{
  // Allow the adversary to fail the operation
  var retCode = AdversaryAction("OpenWrite",fName,pid,Roles);
  if (retCode != OKAY) return retCode;

  if (names[fName] does not exist)  return FILE_DOESNT_EXIST;
  var f = names[fName].file;        // f serves as a "handle" to the file
  if (Roles intersect files[f].Writers = emptyset) return NO_PERMISSION;

  var numBytes = AdversaryAction("Write",fName,atAddr,data,pid,Roles);
  if (numBytes < length(data)) truncate data to numBytes bytes;   // only partial write

  var nBytes = length(data);
  if (atAddr < 0) atAddr = length(files[f].data);    // append
  else if (atAddr > length(files[f].data)) {
    prepend (atAddr-length(files[f].data)) zero bytes to data;
    atAddr = length(files[f].data);
  }
  write data to files[f].data starting at position atAddr;
  call AdversaryAction("Done Write") and then return [OKAY,nBytes];
}

Read(fName, fromAddr, nBytes, pid, Roles)
{
  // Allow the adversary to fail the operation or read less bytes
  var [retCode,numBytes] = AdversaryAction("Read",fName,fromAddr,nBytes,pid,Roles);
  if (retCode != OKAY)  return retCode;
  if (numBytes < nBytes) nBytes = numBytes;

  if (names[fName] does not exist) return FILE_DOESNT_EXIST;
  var f = names[fName].file;

  if (fromAddr < 0) fromAddr = 0;
  else if (fromAddr > length(files[f].data)) {
    fromAddr = length(files[f].data);
    nBytes = 0;
  }
  if (nBytes < 0)   // read to end-of-file
    nBytes = length(files[f].data) - fromAddr;

  data = content of files[f].data from fromAddr for nBytes;

  call AdversaryAction("Done Read") and then return [OKAY,nBytes,data];
}
```

Fig. 1. The SimpFS commands

by an `AdversaryAction` call, in which SIMPFS "leaks" to the adversary the details of its operation, and also lets the adversary influence these operations.

A key feature of SIMPFS is that a file can be accessed *only* using one of the names that were specified when the file was created, thus eliminating filename-manipulation attacks such as described above. Hence proving that an implementation realizes SIMPFS implies in particular that such attacks cannot be successfully mounted against the implementation.

We make no liveness guarantees in SIMPFS, so at the beginning of every operation the adversary is given the option to abort the operation and determine the error code. (This does not mean that an implementation of SIMPFS cannot ensure some liveness properties, but it means that a proof that an implementation realizes SIMPFS carries no such guarantees within itself.)

The pseudo-code includes with every call also the process-id and permissions (Roles) of the caller, which in our system model are filled by the kernel component. (Formally there is also an implicit "invocation id" for each call of one of the four main operations, allowing SIMPFS to handle messages received from the ideal-world adversary for different invocations.) Note also that the `AdversaryAction` at the beginning and end of every operation comply with our convention that the adversary gets the control before any message is delivered. Finally, we note that all the variables in the code in Figure 1 are local to that invocation, except for the global `files[]` and `names[]`.

PROCESS CORRUPTION. Following the standard conventions of the UC framework, SIMPFS has a special procedure to handle the case where the adversary corrupts a process. For our purposes it is more convenient to let the environment decide when a process is corrupted (as opposed to the adversary, which is the more common convention in UC-model works). When the environment corrupts a process, this process makes a call `IamCorrupted(pid,Roles)`, to inform SIMPFS that "it belongs to the adversary" now. SIMPFS informs the adversary of this call, and it remembers that this process and all its roles are now bad. Thereafter, the adversary is allowed to make all the usual calls to SIMPFS (`CreateFile`,`DeleteName`,`Read`, `Write`) on behalf of that process. SIMPFS will process these calls just as if it was the corrupted process that made the call, but will return the result to the adversary rather than to the environment.

Every call from the corrupted process (not via the adversary) will be routed directly to the adversary, and the adversary can always instruct SIMPFS to send anything to the corrupted process (which will then be forwarded to the environment). Also, if the roles of the corrupted players change then the kernel component will notify SIMPFS of this change. SIMPFS will add any new role that a corrupted process acquires to its list of bad roles, but *it will not remove any roles from that list*, even if the corrupted process loses some of its roles. (This last aspect represents the fact that the corrupted process may already have used this role to introduce artifacts into the filesystem, that will remain even after the process no longer has this role.)

ATOMICITY OF THE SIMPFS OPERATIONS. The operations `DeleteName` and `Read` are atomic, whereas `CreateFile` and `Write` are not: In `DeleteName` and

Read, once the adversary allows the operation to go through (by returning OKAY), SIMPFS holds onto the control-flow throughout the name lookup and the operation itself, and only then it yields control back to the adversary.

In Write, on the other hand, the control is returned to the adversary after the file lookup (via the call AdversaryAction("Write", ...)), and only then is the operation carried out. Similarly in CreateFile, the adversary gets the control before the creation of any name. This choice was made so that we would be able to realize SIMPFS over the POSIX interface that requires to open the file and then write in it. The real-world read can be made atomic by checking after the fact that the file did not change since it was opened, but for write such a check is meaningless since the file was already written.

MAPPING UNIX PERMISSIONS TO ROLES. The interfaces of SIMPFS above are defined with "generic roles" that encode permissions, with access control being a simple role inclusion. Our implementation over POSIX, of course, uses userids and groups, which are particular types of roles. The mapping is quite straightforward, roughly there is a different role for each userid and group in the system, and a process gets the role corresponding to its effective-uid and all the roles corresponding to its groups. There is also one role for "others", that every process has. Some care must be taken since POSIX permissions do not exactly follow role inclusion. (For example, if a file is not owner-readable then the owner cannot read it, even if the file is readable by "others".) Adjusting the mapping to this technicality is quite straightforward, and is omitted here.

4 Implementing SimpFS over POSIX

We describe simpfs, which is a concrete implementation of the SIMPFS functionality over the POSIX filesystem interface [20]. The presentation below focuses on a user-space implementation, where each simpfs operation runs with the effective uid of its caller, but we point out that the same procedures can also be implemented in the kernel.

Our implementation relies on the "safe pathname resolution" procedure of Chari et al. [7], that protects processes from opening adversarial links. While resolving paths this procedure ensures that an adversary can not manipulate the resolution to result in opening unintended components. In simpfs, very roughly speaking, each operation consists of first using that procedure to open the corresponding file and then performing the actual operation.

Before describing this implementation, we first introduce concepts that are used in the rest of the paper and describe some assumptions that we make on the POSIX filesystems underlying our implementation. Then in Section 4.2 we describe the safeDirOpen procedure, which is the heart of our implementation and builds on [7], and then in Section 4.3 we describe the rest of the implementation.

4.1 Concepts and Properties of POSIX

We assume that the reader is familiar with basic concepts of POSIX such as directories, pathnames, users and groups, hardlinks and symlinks, etc.

Definition 1 (Pathname Manipulators). *Let `/dir1/.../dirn/foo` be an absolute pathname. The* manipulators *of this pathname are all the roles (users and groups) that own, or have write permissions in, any directory visited during the resolution of this pathname.*

Note that the definition applies even when a pathname does not resolve, and that `root` is a manipulator of every pathname.

Definition 2 (Safe Names). *A pathname is* system safe *if its only manipulator is `root`. A pathname is* safe for *U (where U is a user-id) if its only manipulators are `root` and U. Otherwise, the pathname is unsafe for U.*

For example, in a typical UNIX system the pathname `/etc/passwd` is system safe, the pathname `/home/joe/mbox` is safe for user `joe`, and the pathname `/var/spool/mail/jane` is unsafe for everyone (as `/var/spool/mail` may be world- or group-writable).

Definition 3 (Simple Pathnames). *A pathname is* simple *if it is an absolute path that resolves to a regular file, its elements are only hard links (i.e., not symbolic links), no elements are named '`.`' or '`..`', and the pathname contains no repeated slashes '`//`'.*

ASSUMPTIONS. We now list some properties that we assume on the underlying POSIX system, and use in our proof of security. Most of these assumptions are justified either by the fact that they are part of the POSIX specification itself, or by the fact that many contemporary POSIX filesystems seem to satisfy them.

Assumption 1. *The underlying filesystem does not contain multiple mount points to the same filesystem, and each directory has only one parent (i.e., one hard link with a name other than '`.`' or '`..`').*

Justification. Assumption 1 is justified by the fact that nearly all contemporary POSIX implementations either do not allow processes to create additional hard links to directories (e.g., FreeBSD, Linux) or restrict this operation to the super-user (e.g., Solaris, HP-UX). A notable exception is MacOS.

We observe that given Assumption 1, for every reachable hard link to a regular file there is a unique simple name that ends with that hard link. Moreover a resolution of any absolute name that ends with that hard link will visit all the directories in this unique simple pathname.

Assumption 2 (Permissions) *1. If an operation by a process affects the content of a file, then the process must have write permission for that file. 2. Let P be an absolute pathname. If an operation by a process affects the resolution of P or changes the permissions or ownership of any of the directories visited during its resolution, then that process must have a role which is a manipulator of P.*

Justification. The only operations that affect pathname resolution are creating, removing, or renaming pathname components, and they all require write permission in the containing directory. Also, note that only the owner of a directory (or `root`) can change the permissions of that directory, and in most systems only `root` can change ownership.

Corollary 1. *Let P be some pathname, denote by $\mathcal{M}(P)$ the set of manipulators for P (user-ids and groups), and let $\mathcal{B}$ be a set of roles such that $\mathcal{M}(P) \cap \mathcal{B} \neq \emptyset$. Then changing the manipulator set for P so that $\mathcal{M}(P) \cap \mathcal{B} = \emptyset$ requires an operation by a process with some role outside of $\mathcal{B}$.*

This corollary follows from the fact that no role can perform an operation to "remove itself" from the manipulators set of P (which follows from Assumption 2). Details are deferred to the full version.

Assumption 3. *The hardlink to a directory in its parent directory can only be removed when the child directory is empty. Moreover, after the hardlink is removed from the parent directory, no further entries can be created in the child directory, even if some process still holds a handle to it.*

Justification. The last part of Assumption 3 is justified by the fact that `rmdir` implementations remove the entries '`.`' and '`..`' from the child directory before removing the hard link in the parent directory, and no new entries can be created in directories without '`.`' and '`..`'.

Corollary 2. *If a system call for creating an entry in a directory returns successfully, then the hard link for this directory in its parent directory could not have been removed before that system call, or removed after the call but before the newly-created entry is removed.*

4.2 The safeDirOpen Procedure

Underlying our simpfs implementation is a procedure for *safe name resolution*, which is adapted from the work of Chari et al. [7]. Our safeDirOpen procedure takes an absolute pathname, resolves it "in a safe manner" and returns a handle to the directory containing the final hard link to the actual file, the name of that hard link, and additional information as discussed below. The top-level operations of simpfs first call safeDirOpen and then perform the requested operation on the final hard link.

safeDirOpen resolves a pathname one atom at a time, each time opening the next atom (or reading it, if it is a symlink), while keeping track of the owners and writers of the visited directories. (Below we identify the time that a directory was visited as the time when it was opened, and the time that a symlink was visited with the time that it was read.)

The procedure can be in one of three states: system-safe, safe-for-uid, or unsafe. When invoked (by a process with effective uid U), the procedure begins in a system-safe state, switching to safe-for-uid state upon visiting a directory where U is an owner or writer, and switching to unsafe state upon visiting a directory with any writer or owner other than `root` or U. Once in unsafe state it stays in that state for the duration of the current name resolution. Likewise, there is no transition from safe-for-uid to the system-safe state.

When safeDirOpen enters the unsafe state, it does not follow symlinks for the remainder of the current name resolution. Also, for technical reasons the procedure never accepts pathnames that contain multiple slashes '//' or have components named '.' or '..', and it refuses to visit any directory whose name begins with the special prefix `_SimpFS_ephemeral_`. In any of these cases, the procedure returns an error code.

Once safeDirOpen arrives at the final atom (and verifies that it is indeed the final atom and not a symlink), it ends successfully, returning a handle to the directory containing this last hard link, as well as the name of the hard link. In addition, safeDirOpen returns its current state (system-safe, safe-for-uid, or unsafe), the set of owners and writers of the directories that it visited, and an array of (handle,name) pairs, containing handles to all visited directories, and the names that were looked-up in those directories. (These names could belong to either a directory, a symlink, or the final hard link.)

Upon failure, safeDirOpen returns an error code, a handle to the last directory pathname component that was successfully resolved, the state (system-safe, etc.) and manipulators of that directory, and the unresolved remainder of the pathname. For example, when called to resolve `/a/b/c`, if it encountered an error after visiting `/a` but before visiting `/a/b`, then it will return a handle to directory `/a`, the state and manipulators of `/a`, and the remainder of the pathname argument "`b/c`". (Note that this will be the return value even if `/a/b` happens to be a symlink and the procedure visited more directories after `/a`, but could not completely resolve `/a/b`.)

4.3 Implementing the simpfs Commands

createFile(`Writers,Manipulators,Names`). When called by a process with effective-uid U, the procedure begins by checking that U belongs to the set of manipulators specified by the `Manipulators` parameter. Then it creates a new file with an ephemeral name that begins with the special prefix `_SimpFS_ephemeral_`. This ephemeral name is created so that it is safe for U, thus ensuring that no other users can remove or rename it.[1]

Now createFile attempts to set the write permissions of the new file as specified in the `Writers` parameter. If this is successful, it proceeds to create the names, one at a time, by calling the subroutine createOneName for each name in `Names`. After all the calls to createOneName, the procedure createFile removes the ephemeral name that it created for the new file, and returns the vector of return codes that it received from all the calls to createOneName.

The subroutine createOneName(`fName`) begins by checking that the new name is an absolute name, and that it does not contain '//' or elements named '.' or '..', or elements that begin with `_SimpFS_ephemeral_`. Then it calls safeDirOpen(`fName`) thus obtaining a handle to the last successfully resolved directory on this pathname and the corresponding set of manipulators. If all the directories were resolved successfully, then createOneName checks that the set of manipulators equals the `Manipulators` parameter, and aborts if they differ.

If some directories were not resolved, createOneName verifies that the manipulator set of the prefix is not too large (i.e., it must be contained in the

[1] See Section 4.5 for a short discussion of this point.

Manipulators parameter), aborting otherwise. Then createOneName attempts to create the remaining directories, one at the time, initially creating each one so that it is only writable by owner U with an ephemeral name that begins with _SimpFS_ephemeral_. Upon success, it tries to set the write permissions of the last directory so that the resulting set of manipulators will match the Manipulators parameter. Then it goes over all the newly created directories, top to bottom, renaming each one to the name that it is supposed to have according to fName.

Once all the directories exist and have the right set of manipulators and the right names, the procedure createOneName makes a linkat system call to create a hard link in the last directory, pointing to the new file. createOneName then returns whatever code was returned from the linkat system call.

If any operation fails, then createOneName attempts to clean-up after itself, trying to remove all the directories that still have names that begin with _SimpFS_ephemeral_. However, after a directory was renamed to its "permanent name", createOneName does not remove it.

In the proof of security in the full version we rely on the following properties of our implementation of createFile:

- The initial ephemeral name for the new file is safe for the effective-uid of the calling process.
- The procedure never creates symlinks, only directories and hard links.
- The procedure only changes permissions and/or removes pathname components if these components begin with the special prefix _SimpFS_ephemeral_.
- A name fName is created if and only if the linkat system call at the end of the subroutine createOneName(fName) is successful.

deleteName(fName). When called with effective-uid U, deleteName calls safeDirOpen(fName) and aborts if that function fails. Else deleteName has an array of pairs (handle,name), and the state with which safeDirOpen arrived at the final directory (system-safe, safe-for-uid, or unsafe). If the state is not system-safe, then deleteName checks that the final directory is either world-writable, or owner-writable and owned by U, and it aborts otherwise.[2] Also, if the state is unsafe then deleteName checks that the file that the hard link points to has only a single hard link, aborting otherwise.

Then deleteName attempts to delete the final hard link, followed by attempts to delete the directories higher-up on the path. deleteName returns when any system call to remove a name fails, or when any of these names resolves to a symlink, or when it is done deleting all the names in the array. The return code from deleteName is whatever was returned from the first unlink system call (i.e., the one that deleted the hard link at the end of fName).

We note that barring a race condition, this implementation of deleteName does not delete symlinks. In the proof in the full version we show that the only

[2] This check is intended to protect against privilege-escalation attacks on setgid programs, cf. Section 4.5.

cases where these race conditions are possible are when the adversary already has permissions to delete these symlinks by itself.

read(fName,...). When called with effective-uid U, read calls safeDirOpen(fName) to get a handle for the final directory, the name of the hard link pointing to the actual file, and the state at which it arrived in this last directory: system-safe, safe-for-uid, or unsafe. Then read uses openat, lstatat and fstat to open the file and verify that it is still the same file (and not a symlink). In addition, if the state is not system-safe, then read checks that the file is either world-readable, or owner-readable and owned by U, and it aborts otherwise. Also, if the state is unsafe then read checks that the file has only a single hard link, aborting otherwise.

Then the procedure uses the read system call to read the file, and before closing the file it makes yet another lstatat system call to check that the hard link still points to the same inode as it did when it was opened. If all these checks pass, then read returns the result from the read system call.

write(fName,...). The procedure write is almost identical to read except that it adds a write-permission check on the actual file, and it does not do the final check after writing to verify that the hard link still points to the same inode. (Indeed, such check is useless since the file was already written to.)

4.4 Consistency Properties of the Implementation

In the proof of security in the full version, it is important to consider what changes may happen in the filesystem between the time that the safeDirOpen pathname resolver visits some directory and the time that the procedure that called safeDirOpen returns. An important technical observation is that if the procedure that called safeDirOpen was successful then none of those visited directories could have been removed during this time.

Lemma 4. *Consider an execution of one of the procedures createOneName, deleteName, read, or write on argument fName, and assume that the procedure succeeds (i.e., does not return an error code). Assume further that no symlink that was read during name resolution was later deleted or renamed during the execution of this procedure, and no directory was renamed after it was opened by this procedure. Then also none of these directories was deleted after it was opened and before the time that the procedure issued the system call (respectively, linkat, unlinkat or openat) for the final hard link in fName.*

Moreover, for the procedures createOneName, read, and write, as long as no symlinks are deleted or renamed, no directories are renamed, and the final hard link in fName exists in its original containing directory, then also none of these directories is deleted even after the operation returns.

The proof follows from Assumption 3, and is provided in the full version. □

Jumping ahead, we use Lemma 4 in the proof by noting that our SimpFS implementation never renames or removes symlinks, or renames directories, and hence no uncorrupted process will do any of these things. If in addition we know that no corrupted process has write permissions in any of the directories visited then also corrupted processes could not rename or remove symlinks or rename directories. Thus, we can apply Lemma 4 and conclude that all the directories stay put throughout the execution of createOneName, deleteName, read, or write.

4.5 Rationale and Discussion

Before proceeding to the formal proof of security, we discuss here some of the rationale for our implementation, including some specific attacks that the implementation was designed to foil.

Privilege-escalation attacks on `setgid` programs. Our implementation of safeDirOpen only considers the effective-uid for the purpose of determining the safety of a directory, and thus we must consider the possibility of privilege-escalation attacks between processes with the same effective-uid. In contemporary UNIX systems, two processes with the same effective-uid can have different filesystem privileges only if one of them has a group-privilege that the other does not,[3] as would happen when one of these processes runs a `setgid` program.

To see the problem, consider two processes running with effective-uid of `joe`, one having the additional group privilege of `mail` while the other is compromised by an attacker (e.g., due to a buffer-overflow vulnerability). Ideally, we would like to argue that files which have read/write permissions for the `mail` group (but not user `joe`) are still protected against the compromised process.

Assume that the non-compromised process with `mail` group privileges needs to delete a file `/home/joe/dir/foo`. The compromised process can create a symlink `/home/joe/dir -> /var/mail`, "tricking" the other process into deleting `/var/mail/foo` (assuming that `/var/mail/` is writable by group `mail`). Embedding this attack in our formal model, we have a name `/var/mail/foo` for which `joe` is *not a manipulator*, and a good process that attempts to delete an unrelated name `/home/joe/dir/foo`, and yet by some action of a compromised process with `joe` privileges, this results in the deletion of `/var/mail/foo`.

We fix this problem by adding a check to the operations deleteName, read, and write, aborting if the name is not system-safe and group privileges are needed to perform the operation. Very roughly, this defense works because it prevents the use of group privileges after following symlinks that were created by non-`root` processes. (We note that we do not need this extra precaution in createOneName. This is because the SIMPFS functionality restricts deletion of existing names, but puts no restrictions on the creation of names that do not exist.)

Another reason for the check of the final hard link after a `read` system call in read is a potential attack on open-then-read programs, which is described in the full version.

Our treatment of symbolic links. Our proof of security in relies in places on the assumption that good processes do not create symlinks. This is consistent with our simpfs implementation (that indeed does not create symlinks), but it begs the question why we allow safeDirOpen to follow symlinks at all.

[3] We ignore the `fsuid` of Linux here.

The reason is that the implementation of simpfs is useful also in situations where the filesystem includes non-adversarial symlinks. A close inspection of our proof shows that the arguments remain valid also in the presence of non-adversarial symlinks, as long as the files that have non-adversarial symlinks in their names remain static (i.e., they are not deleted, removed, or moved). It is even possible to modify the semantics of SIMPFS to accommodate non-adversarial symlinks in a dynamic filesystem, but the new semantics will not be as simple anymore.

Using the sticky bit. Recall that the initial ephemeral name for a new file must be safe for the effective-uid of the calling process (denoted U). Such a name can perhaps be created in U's home directory, but not all uid's have one. A simple way of achieving the same result in contemporary UNIX systems is creating this ephemeral name in `/tmp`, relying on the fact that `/tmp` is owned by `root` and has the sticky bit on. This does not quite fit into our definition of "safe for U" (since `/tmp` is world-writable), but it suffices for the purpose of our proof of security. Specifically, what we need is to ensure that as long as the calling process holds a handle to the new file, only U or `root` can change the resolution of the ephemeral name.

5 Conclusion

In this work we adapted the Universal Composability (UC) framework to the modeling of large software systems. Focusing on filesystem interfaces, we described SIMPFS, which is a simple filesystem abstraction intended to capture *filesystem integrity* concerns. We describe an implementation of this abstraction over real POSIX filesystems and prove that the implementation realizes the SIMPFS abstraction in the UC sense. SIMPFS is a simple but useful interface and with a few small enhancements is sufficient to build real applications.

Our work demonstrates that formal security frameworks such as Universal Composability can be used also beyond the niche of cryptographic protocols. Our modeling of POSIX-based file systems is the first example of this scale.

References

1. Backes, M., Pfitzmann, B., Waidner, M.: The reactive simulatability (RSIM) framework for asynchronous systems. Inf. Comput. 205(12), 1685–1720 (2007)
2. Bengtson, J., Bhargavan, K., Fournet, C., Gordon, A., Maffeis, S.: Refinement types for secure implementations. In: 21st IEEE Computer Security Foundations Symposium (CSFS), pp. 17–32 (2008)
3. Biba, K.: Integrity considerations for secure computer systems, MITRE TR-3153, Bedford, MA (1977)
4. Canetti, R.: Universally Composable Security: A New Paradigm for Cryptographic Protocols. In: FOCS, pp. 136–145 (2001)
5. Canetti, R.: Security and Composition of Cryptographic Protocols. SIGACT News 37(3&4) (2006)

6. Canetti, R., Chari, S., Halevi, S., Pfitzmann, B., Roy, A., Steiner, M., Venema, W.: Composable Security Analysis of OS Services, Cryptology ePrint Archive, Report 2010/213, http://eprint.iacr.org/
7. Chari, S., Halevi, S., Venema, W.: Where Do You Want to Go Today? Escalating Privileges by Pathname Manipulation. In: Proc. Symposium on Network and Distributed Systems Security, NDSS (2010)
8. Datta, A., Derek, A., Mitchell, J.C., Roy, A.: Protocol composition logic (PCL). Electronic Notes in Theoretical Computer Science (2007)
9. Datta, A., Franklin, J., Garg, D., Kaynar, D.: A logic of secure systems and its application to trusted computing. In: Proc. of the IEEE Symp. on Research in Security & Privacy, pp. 221–236 (2009)
10. Freitas, L., Woodcock, J., Fu, Z.: POSIX file store in Z/Eves. Science of Computer Programming 74(4), 238–257 (2009)
11. Goldreich, O.: Foundations of Cryptography, vol. 1, 2. Cambridge Press, Cambridge (2004)
12. Gross, T.R., Pfitzmann, B., Sadeghi, A.-R.: Browser model for security analysis of browser-based protocols. In: di Vimercati, S.d.C., Syverson, P.F., Gollmann, D. (eds.) ESORICS 2005. LNCS, vol. 3679, pp. 489–508. Springer, Heidelberg (2005)
13. Joshi, R., Holzmann, G.J.: A mini challenge: Build a verifiable filesystem. Formal Aspects of Computing 19(2), 269–272 (2007)
14. Klein, G., Elphinstone, K., Heiser, G., Andronick, J., Cock, D., Derrin, P., Elkaduwe, D., Engelhardt, K., Kolanski, R., Norrish, M., Sewell, T., Tuch, H., Winwood, S.: seL4: Formal verification of an OS kernel. In: SOSP 2009, pp. 207–220 (2009)
15. McLean, J.: A general theory of composition for a class of "possibilistic" properties. IEEE Transactions on Software Engineering 22(1), 53–67 (1996)
16. Neumann, P., Feiertag, R.: PSOS revisited. In: Omondi, A.R., Sedukhin, S.G. (eds.) ACSAC 2003. LNCS, vol. 2823, pp. 208–216. Springer, Heidelberg (2003)
17. Neumann, P.: Principled assuredly trustworthy composable architectures, DARPA Project CHATS, Final Rep. (2004), http://www.csl.sri.com/users/neumann/chats.html
18. Pfitzmann, B., Waidner, M.: A General Framework for Formal Notions of "Secure" Systems. Institut fur Informatik, Hildesheim University (April 1994)
19. Pfitzmann, B., Waidner, M.: Composition and Integrity Preservation of Secure Reactive Systems. In: Proc. CCS, pp. 245–254 (2000)
20. The Open Group Base Spec. Issue 7, IEEE Std 1003.1-2008, http://www.opengroup.org/onlinepubs/9699919799/

Practical Attacks on the Maelstrom-0 Compression Function

Stefan Kölbl and Florian Mendel

Graz University of Technology, A-8010 Graz, Austria
stefan.koelbl@student.tugraz.at

Abstract. In this paper we present attacks on the compression function of Maelstrom-0. It is based on the Whirlpool hash function standardized by ISO and was designed to be a faster and more robust enhancement. We analyze the compression function and use differential cryptanalysis to construct collisions for reduced variants of the Maelstrom-0 compression function. The attacks presented in this paper are of practical complexity and show significant weaknesses in the construction compared to its predecessor. The methods used are based on recent results in the analysis of AES-based hash functions.

Keywords: hash functions, cryptanalysis, collisions, near-collisions.

1 Introduction

Cryptographic hash functions are a fundamental part of modern cryptography. They are used in many practical applications e.g., verification of message integrity, message authentication or secure storage of passwords. Typically a hash function is used as a digital fingerprint of the information that needs authentication.

A cryptographic hash function takes as input a string of arbitrary finite length and produce a fixed sized output. Usually the input domain is larger than the output domain, therefore this functions are many-to-one. As a result the existence of collisions is unavoidable.

Hash functions have to be both fast and secure. The security can be discussed by the following properties:

– Preimage Resistance: For a given output y it should be computationally infeasible to find an input x' such that $y = f(x')$.
– Second Preimage Resistance: For given $x, y = f(x)$ find $x' \neq x$ such that $f(x') = y$.
– Collision Resistance: Find two distinct inputs x, x' such that $f(x) = f(x')$.

A hash function with n-bit output is secure if finding a (second) preimage takes at least 2^n and finding a collision $2^{n/2}$ (birthday attack) queries [1].

The most commonly used hash functions at the moment are SHA-1, SHA-256 and SHA-512 certified by NIST. They are part of several standards and

J. Lopez and G. Tsudik (Eds.): ACNS 2011, LNCS 6715, pp. 449–461, 2011.

based on MD4 and MD5. In the last few years cryptanalysis made a huge leap forward and weaknesses have been found for these functions. There are practical collisions for MD4 [2], MD5 [3] and SHA-0 [4]. The computational effort to construct collisions for SHA-1 is still impracticable, but the security bound is much lower than expected [5] and attacks on reduced rounds are possible [6]. Therefore, there is a strong interest in new hash functions.

Maelstrom-0 is based on the Whirlpool hash function which has been adopted in the ISO 10118-3:2004 standard [7]. Maelstrom-0 is designed to be a faster and more robust enhancement of Whirlpool but we will show that the new lightweight key schedule significantly weakens this hash function and allows us to construct collisions for reduced rounds of the Maelstrom-0 compression function with practical complexity. In detail, we show how to construct a collision for 6 out of 10 rounds and give a colliding message pair. Furthermore, we show attacks for 8 and 10 rounds in a weaker attack scenario (near-collision and semi-free-start near-collision) and a theoretical collision attack on 7 rounds.

The paper is structured as follows: First there will be a description of the Maelstrom-0 hash function followed by an overview of the attack in Section 2. We continue with a detailed section on how to construct a differential path for 6-rounds and how to obtain the colliding message pair in Section 3. Afterwards possible extensions on more rounds are discussed followed by the conclusion.

2 Description of Maelstrom-0

Maelstrom-0 is an iterative hash function designed by Filho, Barreto and Rijmen [8]. Maelstrom-0 processes 1024-bit message blocks and produces a 512-bit hash value. It uses the Davies-Meyer construction (see Figure 1) and 3CM chaining mode which is based on 3C [9]. If we have a message $m = M_1||M_2||\ldots||M_k$ we can compute the hash value h in the following way

$$H_0 = IV \tag{1}$$
$$H_i = E(H_{i-1}, M_i) \oplus H_{i-1}, \forall i : 0 < i \leq k \tag{2}$$
$$h = E(H_k, s_k||t_k) \tag{3}$$

where s_k is the output of the second and t_k the output of the third chain. The second chain is the XOR accumulation of all the indermediate compression function outputs and in the third chain a LFSR is involved in the accumulation process.

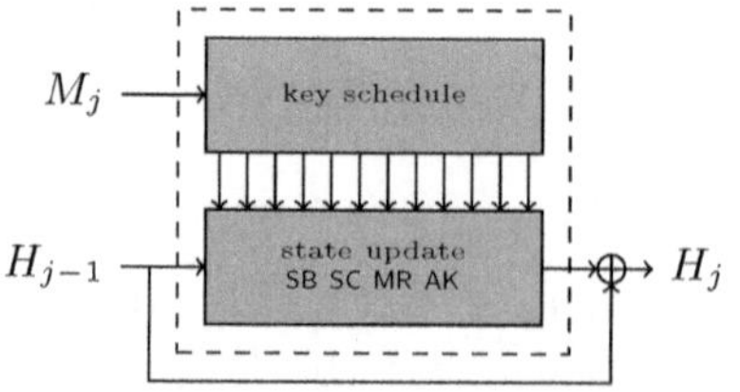

Fig. 1. Maelstrom-0 compression function

2.1 Block Cipher E

The block cipher E is based on the one used in Whirlpool. The only difference between the block cipher used for Maelstrom-0 is that it uses a different key schedule. The state update operates on a 8×8 states of 64 bytes therefore the 512-bit input is bytewise transformed. The state is updated through 10 identical rounds and one key addition at the beginning. One round consists of the application of the four transformations SubBytes, ShiftColumns, MixRows and AddRoundKey similar to AES. We only give a very brief description and a more detailed one can be found in [8].

SubBytes (SB)

The SubBytes step applies a nonlinear S-box on each byte using an 8-bit S-box. The S-box is the same as in Whirlpool. For the definition of the S-box we refer to [7].

ShiftColumns (SC)

The ShiftColumns step cyclically shifts each column $j = 0, \ldots, 7$ by j steps downwards.

MixRows (MR)

MixRows is a linear mapping based on a MDS code multiplying each row by a 8×8 matrix over $\mathbb{F}_{2^8}$. The values of the matrix are choosen such that the branch number of MixRows is 9. Therefore the sum of active bytes (byte with difference) at input and output is always at least 9.

AddRoundKey (AK)

The key addition uses bitwise xor to add the round key.

Key Schedule (KS)

The key schedule takes as input the 1024-bit message block $M = (v_0, \ldots, v_{1023})$ to generate the round keys $K^0, ..., K^{10}$. The message block M is mapped to a column vector $(K^{-2} = v_0, \ldots, v_{511}, K^{-1} = v_{512}, \ldots, v_{1023})$ and in each step two new round keys are computed in the following way

$$\begin{pmatrix} K^{2i} \\ K^{2i+1} \end{pmatrix} = \alpha \begin{pmatrix} K^{2i-2} \\ K^{2i-1} \end{pmatrix} + C_i \qquad \forall i = 0, \ldots, 5 \tag{4}$$

where

$$\alpha = \begin{pmatrix} 1 & 1 \\ x^8 & x^8 + 1 \end{pmatrix} \tag{5}$$

and C_i is some round dependent constant. For the actual values we refer to [8]. There are only 11 keys needed so the last key is dropped. Multiplication is done over $\mathbb{F}_{2^{512}}[x]/p(x)$ with $p(x) = x^{512} + x^8 + x^5 + x^2 + 1$. The 512-bit round keys are transformed to a 8×8 state. For the actual round keys Maelstrom-0 uses a key extraction function which applies the SubBytes and the MixRows step to row 3 and 7 to obtain the round key. This is the only nonlinear transformation used in the key schedule. Note that inverse key schedule looks almost the same we only need to determine the inverse matrix of α which is given by

$$\alpha^{-1} = \begin{pmatrix} x^8 + 1 & 1 \\ x^8 & 1 \end{pmatrix} \tag{6}$$

3 Outline of the Attack

For our attack on Maelstrom-0 we use differential cryptanalysis. Differential cryptanalysis observes how the difference between a pair of inputs affect the resulting output difference [10]. It was originally devised in the analysis of block ciphers but is also used for stream ciphers and hash functions. Usual differential cryptanalysis is a chosen plaintext attack. The basic method considers a pair of messages (M, M') and the xor difference $\Delta M = M \oplus M'$.

For our analysis we use *truncated differentials* [11]. This means we do not consider the full difference between two inputs, we only determine for single bytes whether there is a difference or not.

The structure of Maelstrom-0 allows us to predict how differences propagate through the key schedule and the round transformations to find a *good* differential path. For our goal to mount a collision attack we are looking for an input pair (M, M') with output difference zero. The attack can be divided into two individual steps. The first part is to find a differential path which holds with high probability. We use differences in the key input (message input) such that the resulting pattern can be fulfilled with a high probability. The second step is to find a message following this differential path.

Similiar attacks have been applied to Whirlpool in [12] but they did not use any difference in the key input due to the strong key schedule used in Whirlpool. Our attack is similiar to the attacks on the AES hash mode in [13] using local collisions to cancel out differences, but we use other techniques for finding the confirming message pair [13]. In the following we define our notation and analyze the differential properties of the round transformations and the key schedule, before we describe the attack in detail.

Notation

We denote the state after round k after the transformation $R = \{SB, SC, MR, AK\}$ by $R^k_{i,j}$ where i, j are the row and column indices. Indices are used modulo 8.

3.1 Differential Properties of the Round Transformations

SB - SubBytes. For SubBytes we consider pairs of input/output differences $\Delta a, \Delta b \in \{0,1\}^8$. Counting over all 2^{16} possible differentials the number of solutions for

$$SB(X) \oplus SB(X \oplus \Delta a) = \Delta b \tag{7}$$

can only be $0, 2, 4, 6, 8, 256$. We are interested if a given input difference can propagate to a given output difference. Counting over all possible inputs the probability for Δa to propage to Δb through SubBytes is equal to 0.395 respectively there are about 101 valid transitions on average from Δa to another difference. This can be computed by creating a difference distribution table (DDT) of size 256×256 for all possible values.

SC - ShiftColumns. This steps moves differences to different rows but the values are not changed. The 8 bytes of a full active row are moved to 8 different rows.

MR - MixRows. MixRows is a linear step, hence xor differences propagate in a deterministic way. For truncated differences we only got the position of the difference and the propagation through MixRows is probabilistic. Since the branch number of MixRows is 9, one active byte will propagate to 8 active bytes with probability 1. The probability for a transition from a to b active bytes with $1 \leq a, b \leq 8$ and $a + b \geq 9$ is in general $2^{(b-8)\cdot 8}$.

AK - AddRoundKey. AddRoundKey uses simple xor to add the round key hence difference propagate deterministic through this operation.

3.2 Differential Properties of the Key Schedule

The key schedule uses two 512-bit keys K^{-2}, K^{-1} to compute the next two round keys. Apart from the key extraction function the key schedule is linear. First we can simplify the key schedule and ignore the addition of the round constant and look at the two new keys separately.

$$K^0 = K^{-2} + K^{-1}$$
$$K^1 = x^8 K^{-2} + (x^8 + 1)K^{-1}$$

The difference propagation to K^0 is trivial because it simply xors the two input keys. For the second key we have to look how multiplication over $\mathbb{F}_{2^{512}}[x]/p(x)$ influences the differences. Multiplication by x^8 equals shifting bytewise and adding the irreducible polynomial $p(x)$ depending on the values of the first byte ($K_{0,0}$) of each round key. If we avoid any differences in the first byte the differences will just be shifted bytewise (see Figure 2). If we have a difference in the first byte it will only affect the last two bytes due to the structure of $p(x)$.

Key Extraction Function - ψ

The key extraction function applies SubBytes and MixRows to row 3 and 7 and copies all other rows. Note that ψ is only applied on the round keys and does not influence the state of the key schedule. For our attack we avoid difference in this rows so we can ignore it for our further analysis.

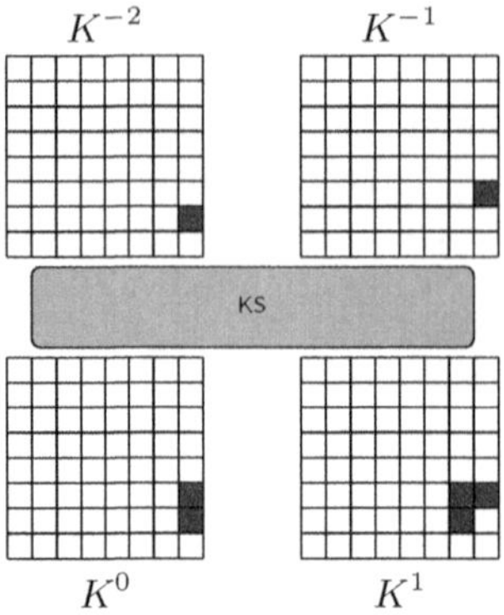

Fig. 2. Example difference propagation through the keyschedule with differences in $K^{-2}_{6,7}$ and $K^{-1}_{5,7}$. Colored bytes denote differences.

3.3 Constructing the Differential Path

Constructing the differential path is rather simple due to the structure of the round transformations. Using truncated differentials we can construct a good path by hand.

For the 6-round differential path (see Figure 3) we start with a difference in $K^{1}_{6,7}$. We can ignore K^{-2} and K^{-1} because we can apply the inverse key schedule to (K^0, K^1) to obtain the initial key. The difference introduced by K^1 will end up in a full active row in AK^2 due to the properties of MixRows. After SC^3 there will be an active byte in every row and therefore we will have a full active state in AK^3. We keep this full active state for AK^4 and use the properties of MixRows to obtain the pattern in AK^5. Note that in row 6 we want the 8 active bytes to propagate to 4 active bytes so that they are canceled out with the differences in K^5. We choose this pattern to obtain a single active row after applying ShiftColumns so that we can use the last MixRows to cancel out the differences in K^6. The number of active bytes for the rounds are:

$$0 - 1 - 9 - 64 - 64 - 8 - 0 \tag{8}$$

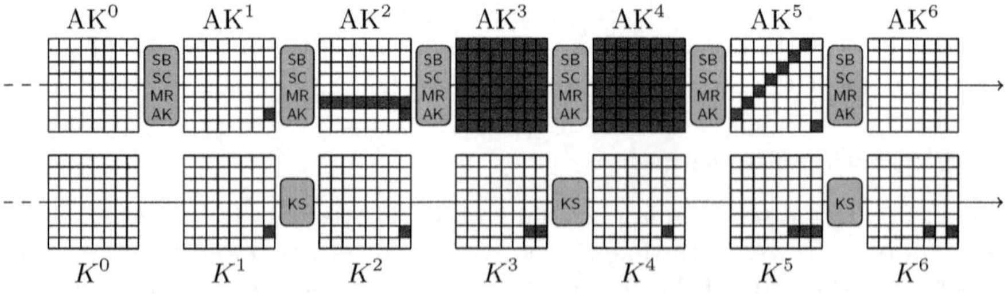

Fig. 3. 6 round differential path for Maelstrom-0

In the next step, we determine the xor differences of the differential path. For this we use the same approach that has been used in the rebound attack on Whirlpool [14]. So far we only considered truncated differentials, now we use xor differences.

Backward direction

1. Start with an arbitrary difference in $K^6_{6,5}, K^6_{6,7}$. We want to cancel this difference out so we also fix the difference in $MR^5_{6,5}, MR^5_{6,7}$.
2. The next step is to compute the MixRows step backwards to obtain the differences in SC^5 respectively SB^5. To compute backwards the SubBytes step we choose an arbitrary possible input difference to the given output difference using the difference distribution table (see Section 3.1).
3. Now we repeat this steps to propagate the differences backward through MR^4, SC^4. The differences after SB^4 are now fixed and in the following steps we show how we can propagate the difference from the start such that they match this fixed difference.

Forward direction

1. Start with the difference in $AK^1_{6,7}$ and propagate it forward. We get a full active row in MR^1. For the next SubBytes step we choose arbitrary possible output differences using the DDT to the 9 given input differences and we end up with a full active state in MR^2 and AK^2.
2. Now we start the matching process for each row $r = 0, \ldots, 7$ individual. Select the bytes $AK^2_{(8-r),i}$ for $i = 0, \ldots, 7$ and it follows that this bytes form a single row after ShiftColumns. Propagate the bytes through SB^3, SC^3, MR^3, AK^3 step. For the SubBytes step we use a random valid output difference.
   (a) Check for all bytes in row r of AK^3 if there is a valid transition from the input difference to the output difference in SB^4. For any given input difference we get up to 114 possible output difference. On average the probability that the differences in one byte transition is valid is 0.395 and the total probability is $\approx 2^{-10.72}$ that a full row matches. In practice we can improve this by using favorable differences. If we do not get a match for the row, we just choose another output difference in the previous step. For every byte we can choose from at least 89 possible output differences, therefore we can easily generate enough solutions to find a match.
   (b) Repeat the steps until all conditions are fulfilled. The expected costs are $8 \cdot 2^{10.72} = 2^{13.72}$

After finishing this steps the differential path is fully determined.

3.4 Finding the Message Pair

To find the colliding message pair we need to find a message that follows the previous differential path. We need to determine the message $M = K^{-2}||K^{-1}$. This can be done by random trials which obviously leads to a high complexity. A

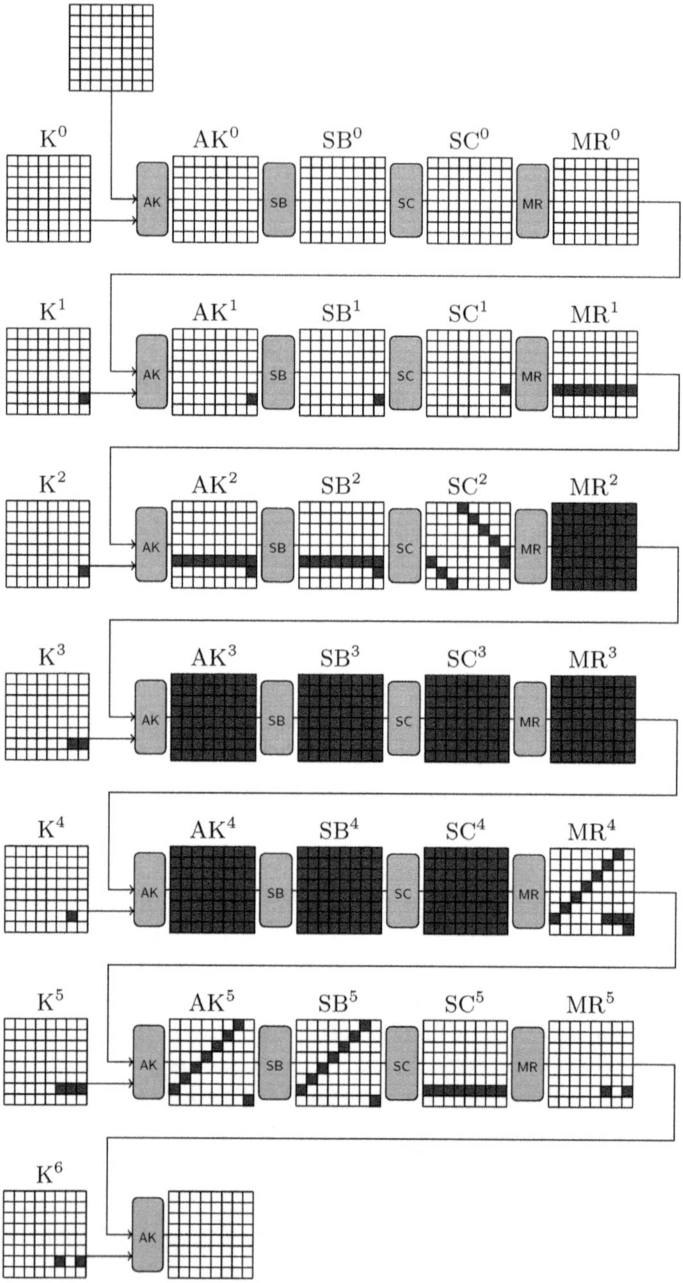

Fig. 4. Full differential path for 6 rounds using truncated differentials

more efficient method is to use the triangulation algorithm invented by Khovratovich et al. in the cryptanalysis of AES in hash mode [13]. Due to the structure of Maelstrom-0 the best result we achieved had a complexity of 2^{184} which is in fact lower than the theoretical bound but we used the following approach which turns out to be very efficient for Maelstrom-0.

1. Start at SB^4 and determine the correct values for the given differences. We need to fulfill the conditions on all 64 active bytes, but there are no restrictions yet so we can choose the right pair of values.
2. For SB^5 we got 8 conditions but we can use the according bytes in K^5 to get the correct input values to this SubBytes layer. The values for the rows affected by the key extraction function can be computed by inverting the MixRows and SubBytes step to obtain the desired value.
3. Now we need to satisfy the conditions for the fourth SubBytes layer. We can compute the values from the difference for SB^3 and apply ShiftColumns, MixRows and use K^4 to correct them and get the right input to the fourth SubBytes layer. After this step all values of K^4 and 8 bytes of K^5 are fixed.
4. Compute K^2, K^3 by applying the inverse KeySchedule. From the inverse KeySchedule it follows that K^5 is xored to K^4 in both cases. We can use the remaining 54 free bytes now to influence K^3 and therefore the values we need to satisfy the 9 byte conditions of the third SubBytes layer. By changing single bytes in K^3 we can influence the value of each active byte in SB^2. The probability that we get the right value is 2^{-8}. There are still 7 bytes that are not fixed in every row of K^5 so we can create 2^{56} possible values and are guarenteed to find a solution. Each row in K^5 only affects one respectively two active bytes in SB^2 therefore we can find the right values for each row individually. With the naive approach by trying out different values for each row we get a total complexity of 2^{16}.
5. With only one byte condition left in SB^1 to fulfill we can simply bruteforce the last S-box and repeat the previous steps. This results in a semi-free start collision with a complexity of 2^{24}.

A colliding message pair for 6 rounds is given in the Appendix A.

3.5 Extension to More Rounds

In this section, we show how the attack can be extended to more rounds. It is possible to construct collisions for 7 rounds. This could be achieved by using a different path with another state between the two full active ones. This leads to

$$0 - 1 - 9 - 64 - 8 - 64 - 8 - 0 \quad (9)$$

active bytes. The attack can be constructed similar to the attack on 6 rounds, but we have more conditions which can not be fulfilled using the message input. A possible approach would be:

1. Fix the differences in AK^4 and propagate them backward to SB^3. Fix the differences in SB^2 and propagate them forward to AK^3. We try to find a match with the same method used in the 6-round attack previously.

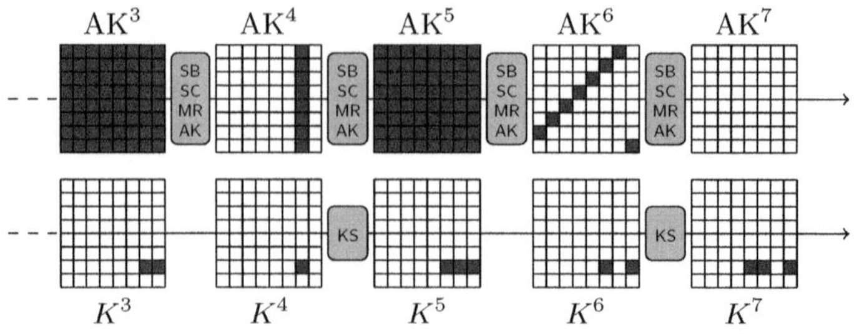

Fig. 5. 7 round differential path. The first three rounds are omitted and are the same as for the 6 round path.

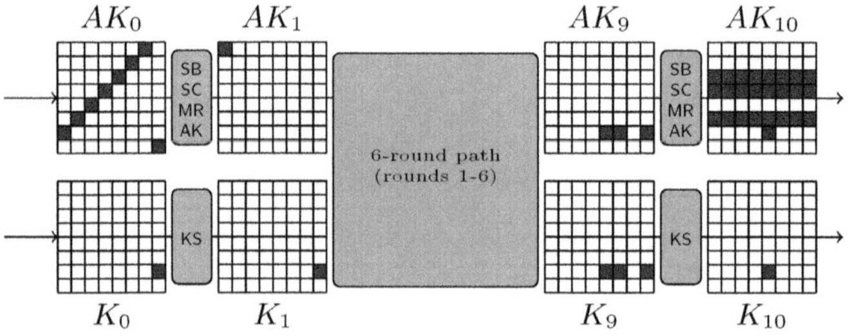

Fig. 6. Extending the differential path at the end to get a near-collisions for 8 rounds with 25 active bytes. Furthermore, adding 2 rounds at the beginning for full 10-rounds free-start near-collisions.

2. Choose a valid transition for the propagation from AK^4 to SB^4. Compute the differences forward to AK^5. Now choose a difference in AK^6 and propagate it backward to SB^5 and try to find a match again.

Finding the message can now be done in the following way:

1. Determine the values at AK^3 resp. SB^3. We got 8 byte conditions at SB^4 which can be fulfilled using the corresponding column in K^4. The 64 byte conditions for SB^5 can be fulfilled using K^5. Following onward this direction we need one transition of 8 to 3 active bytes in MR^6 which costs 2^{40} and we need them to cancel out with the last 3 bytes of K^7. So the total complexity for the forward direction is 2^{64}.
2. In backward direction this looks very similar. We got one 8 to 1 transition in MR^1 which costs 2^{56} and two conditions for the bytes $\mathrm{K}^1_{6,7}$ and $\mathrm{K}^2_{6,7}$ to cancel out. Therefore the complexity in forward direction is 2^{72}.
3. The total complexity is 2^{136} in this case.

By choosing a weaker attack scenario the previous collision attack on 6 rounds can be extended to more rounds. If we allow differences in the output we get near-collisions and additional differences at the input lead to semi-free-start near-collisions.

Table 1. Summary of attacks

Attack	rounds	complexity	generic attack
semi-free-start collision	6	2^{24}	2^{256}
semi-free-start collision	7	2^{136}	2^{256}
semi-free-start near-collision	8	2^{24}	2^{156}
free-start near-collision	10	2^{24}	2^{124}

Adding two more rounds after the 6-round attack we get near-collisions for 8 rounds with the same complexity of 2^{24}. The key addition after round 6 leads to 3 active bytes which propagate to 3 active rows. The last key adds another active byte. This gives us a near-collision with 25 active bytes (see Figure 6).

Furthermore it is also possible to additionally prepend two rounds to construct a free-start near-collisions for the full 10 rounds of Maelstrom-0.

4 Conclusion

In this paper, we have shown how to construct collisions for the Maelstrom-0 compression function. Maelstrom-0 was designed to be faster and more robust, however the new lightweight key schedule significantly weakens the compression function and allows efficient attacks. The linear key schedule allows to construct good differential paths and the key extraction function can be easily avoided. We can construct collisions for 6 and 7 rounds. Furthermore, the attack scenario can be extended to full Maelstrom-0, resulting in a free-start near-collision. In Table 1 we summarize our results for the Maelstrom-0 compression function.

Acknowledgments

We would like to thank the designers of Maelstrom-0 for providing a reference implementation and Matthias Schlaipfer for providing a implementation of the rebound attack for Whirlpool which helped us to verify our results. The work in this paper has been supported in part by the Austrian Science Fund (FWF), project P21936-N23 and by the European Commission under contract ICT-2007-216646 (ECRYPT II).

References

1. Menezes, A., van Oorschot, P.C., Vanstone, S.A.: Handbook of Applied Cryptography. CRC Press, Boca Raton (1996)
2. Wang, X., Lai, X., Feng, D., Chen, H., Yu, X.: Cryptanalysis of the Hash Functions MD4 and RIPEMD. In: Cramer, R. (ed.) EUROCRYPT 2005. LNCS, vol. 3494, pp. 1–18. Springer, Heidelberg (2005)
3. Wang, X., Yu, H.: How to Break MD5 and Other Hash Functions. In: Cramer, R. (ed.) EUROCRYPT 2005. LNCS, vol. 3494, pp. 19–35. Springer, Heidelberg (2005)

4. Wang, X., Yu, H., Yin, Y.L.: Efficient Collision Search Attacks on SHA-0. In: Shoup, V. (ed.) CRYPTO 2005. LNCS, vol. 3621, pp. 1–16. Springer, Heidelberg (2005)
5. Wang, X., Yin, Y.L., Yu, H.: Finding collisions in the full sha-1. In: Shoup, V. (ed.) CRYPTO 2005. LNCS, vol. 3621, pp. 17–36. Springer, Heidelberg (2005)
6. De Cannière, C., Mendel, F., Rechberger, C.: Collisions for 70-Step SHA-1: On the Full Cost of Collision Search. In: Adams, C.M., Miri, A., Wiener, M.J. (eds.) SAC 2007. LNCS, vol. 4876, pp. 56–73. Springer, Heidelberg (2007)
7. Barreto, P.S.L.M., Rijmen, V.: The Whirlpool Hashing Function. Submitted to NESSIE (September 2000), `http://www.larc.usp.br/~pbarreto/WhirlpoolPage.html` (revised May 2003)
8. Filho, D.G., Barreto, P.S., Rijmen, V.: The maelstrom-0 hash function. In: SBSeg 2006 (2006)
9. Gauravaram, P., Millan, W., Dawson, E., Viswanathan, K.: Constructing Secure Hash Functions by Enhancing Merkle-Damgård Construction. In: Batten, L.M., Safavi-Naini, R. (eds.) ACISP 2006. LNCS, vol. 4058, pp. 407–420. Springer, Heidelberg (2006)
10. Biham, E., Shamir, A.: Differential cryptanalysis of des-like cryptosystems. J. Cryptology 4(1), 3–72 (1991)
11. Knudsen, L.R.: Truncated and higher order differentials. In: Preneel, B. (ed.) FSE 1994. LNCS, vol. 1008, pp. 196–211. Springer, Heidelberg (1995)
12. Lamberger, M., Mendel, F., Rechberger, C., Rijmen, V., Schläffer, M.: Rebound distinguishers: Results on the full whirlpool compression function. In: Matsui, M. (ed.) ASIACRYPT 2009. LNCS, vol. 5912, pp. 126–143. Springer, Heidelberg (2009)
13. Khovratovich, D., Biryukov, A., Nikolic, I.: Speeding up collision search for byte-oriented hash functions. In: Fischlin, M. (ed.) CT-RSA 2009. LNCS, vol. 5473, pp. 164–181. Springer, Heidelberg (2009)
14. Mendel, F., Rechberger, C., Schläffer, M., Thomsen, S.S.: The rebound attack: Cryptanalysis of reduced whirlpool and grøstl. In: Dunkelman, O. (ed.) FSE 2009. LNCS, vol. 5665, pp. 260–276. Springer, Heidelberg (2009)

A Colliding Message Pair

Here a colliding message pair (M, M') and the chaining value are given. The message pair has been found by using the 6-round path and the difference in the messages are $\Delta M^1_{6,7} = \Delta M^2_{6,7} = 01$. Values are given in hex notation.

Chaining Value	After 6 rounds (AK^6)
62 c4 11 cf 0e 4e dd eb	6d 85 84 15 32 bd fc 98
7e 1f 07 7c d7 84 ae 56	b6 db 17 12 ed c5 fe 73
a4 81 51 b2 1e 91 d3 fe	f5 85 8e a7 93 ea b0 87
23 08 cd 4a b8 d4 82 b9	ac 8e da b0 e1 20 82 d8
67 89 16 74 f0 e6 7d 58	15 32 a8 61 d5 3f bc 93
76 e0 fa f9 b6 8b 01 9c	ba dd 0a 2b bb 20 87 1f
83 d8 d8 36 e3 9e 54 f2	32 45 86 6a c2 41 73 df
43 0c 85 58 a0 9b 30 38	34 81 63 4e 4a 10 18 a7

M	M'
25 fe e7 fa 16 6f 30 2b	25 fe e7 fa 16 6f 30 2b
c3 03 8e d9 79 3a d6 06	c3 03 8e d9 79 3a d6 06
8e 53 d3 da 9b 41 33 e0	8e 53 d3 da 9b 41 33 e0
66 e6 da 6 5c 9b f1 f2	66 e6 da 06 5c 9b f1 f2
31 1a ff 5c a1 ac 25 cd	31 1a ff 5c a1 ac 25 cd
2f 6e 63 a9 84 0e d5 40	2f 6e 63 a9 84 0e d5 40
00 c0 d9 9f 24 ab 7c 20	00 c0 d9 9f 24 ab 7c 21
1f 2f d8 2f bc d2 04 2a	1f 2f d8 2f bc d2 04 2a
34 8c 53 c5 17 b4 87 35	34 8c 53 c5 17 b4 87 35
e1 9c 2c e8 1d fb df 80	e1 9c 2c e8 1d fb df 80
97 3d 46 0f ee 1d 5d 4b	97 3d 46 0f ee 1d 5d 4b
63 55 37 c3 de 04 88 8e	63 55 37 c3 de 04 88 8e
b8 13 92 12 2c d2 8d 8e	b8 13 92 12 2c d2 8d 8e
ef 3b fc 5a b3 44 6b 7b	ef 3b fc 5a b3 44 6b 7b
ef f6 80 42 49 9a 5d de	ef f6 80 42 49 9a 5d df
9f 1b d8 e9 88 7f c4 73	9f 1b d8 e9 88 7f c4 73

Linear Analysis of Reduced-Round CubeHash

Tomer Ashur[1] and Orr Dunkelman[1,2]

[1] Faculty of Mathematics and Computer Science
Weizmann Institute of Science
P.O. Box 26, Rehovot 76100, Israel
tomerashur@gmail.com
[2] Computer Science Department
University of Haifa
Haifa 31905, Israel
orrd@cs.haifa.ac.il

Abstract. Recent developments in the field of cryptanalysis of hash functions has inspired NIST to announce a competition for selecting a new cryptographic hash function to join the SHA family of standards. One of the 14 second-round candidates was CubeHash designed by Daniel J. Bernstein. CubeHash is a unique hash function in the sense that it does not iterate a common compression function, and offers a structure which resembles a sponge function, even though it is not exactly a sponge function.

In this paper we analyze reduced-round variants of CubeHash where the adversary controls the full 1024-bit input to reduced-round CubeHash and can observe its full output. We show that linear approximations with high biases exist in reduced-round variants. For example, we present an 11-round linear approximation with bias of 2^{-235}, which allows distinguishing 11-round CubeHash using about 2^{470} queries. We also discuss the extension of this distinguisher to 12 rounds using message modification techniques. Finally, we present a linear distinguisher for 14-round CubeHash which uses about 2^{812} queries.

Keywords: CubeHash SHA-3 competition, Linear cryptanalysis.

1 Introduction

Recent developments in the field of hash function cryptanalysis [1,18,19,20] along with new results targeted against commonly used hash functions [6, 11, 26, 27] has urged the National Institute of Standards and Technology to announce a competition for the development of a new hash standard, SHA-3 [25].

The National Institute of Standards and Technology has received 64 hash function proposals for the competition, out of which 51 met the submission criteria and were accepted to the first round of the competition. Following the first round of analysis, in which the security and performance claims of the submitters were challenged, 14 candidates were selected to the second round of the SHA-3 competition. One of these 14 candidates was CubeHash designed by Daniel J. Bernstein [4]. Although CubeHash did not pass to the third round of the SHA-3 competition, it is still of interest to challenge its security.

J. Lopez and G. Tsudik (Eds.): ACNS 2011, LNCS 6715, pp. 462–478, 2011.

CubeHash is a family of cryptographic hash functions, parameterized by the performance and security required. CubeHash has an internal state of 1024 bits, which are processed by calling a transformation named T, a tweakable number of times r, between introductions of new b-byte message blocks (b is also a tunable parameter). At the end, after a final permutation, namely, T repeated $10r$ times, h bits of the state are used as an output. By selecting different values of h, b, and r, different security/performance tradeoffs are provided. Currently, several sets of parameters are suggested, where the "normal" security values are $r = 16, b = 32$ (for $h \in \{224, 256, 384, 512\}$) [5].[1]

In this paper we analyze the security of several variants of CubeHash against linear cryptanalysis. Our analysis found a linear approximation for 11-round CubeHash[2] with bias of $\frac{1}{4} \cdot \frac{1}{2}^{233} = 2^{-235}$. We limited the analysis to biases of no less than 2^{-256}, as we felt that a hash function offering a 512-bit security (in its strongest variant), should not be assessed with attacks taking more than 2^{512} queries. One can also extend the 11-round linear approximation into a 12-round distinguisher using simple message modification techniques [27] (or a chosen-plaintext linear cryptanalysis [22]).

We note that when removing this restriction, one can find 14-round linear approximations with bias of 2^{-406}. Exploiting this approximation requires querying T^{14} about 2^{812} times, which is outside the security model. At the same time, if T or CubeHash are ever used in different settings, this may provide some indication concerning its security.

This paper is organized as follows: In Section 2 we describe CubeHash's compression function. In Section 3 we describe the linear approximations found for CubeHash. In Section 4 we describe how bit fixing can be used to distinguish more rounds than in the approximation. In Section 5 we quickly cover a possible application of our results. Finally, Section 6 concludes this paper.

2 A Brief Description of CubeHash

As mentioned before, CubeHash is a tweakable hash function, where the shared part of all its variants is the internal state (of 1024 bits), and the use of the same round function T.

To initialize the hash function, h (the digest size), r the number of times T is iterated between message blocks, and b the size of the message blocks (in bytes), are loaded into the state. Then, the state is updated using $10r$ applications of T. At this point, the following procedure is repeated with any new message block: the b-byte block is XORed into the 128-byte state, and the state is updated by applying T^r (r times applying T) to the state. After processing the padded

[1] We note that there is a "formal" variant of CubeHash for which $r = 16, b = 1$ and $h \in \{384, 512\}$.

[2] We note that CubeHash is a full hash function which is not easily defined in the common settings. Hence, 11-round CubeHash stands for iterating 11 times the transformation T. We remind the reader that our analysis usually assumes the adversary can choose the full 1024-bit input to T and observe the full 1024-bit output from T.

message, the state is XORed with the constant 1, and is processed by applying T^{10r}. The output is composed of the first $h/8$ bytes of the state.

The 1024 bits of the internal state are viewed as a sequence of 32 4-byte words $x_{00000}, x_{00001}, \ldots, x_{11111}$ each of which is interpreted in a little-endian form as a 32-bit unsigned integer. The round function T of CubeHash is based on the following ten operations:

1. Add (modulo 2^{32}) x_{0jklm} into x_{1jklm}, for all (j, k, l, m).
2. Rotate x_{0jklm} left by 7 bits, for all (j, k, l, m).
3. Swap x_{00klm} with x_{01klm}, for all (k, l, m).
4. XOR x_{1jklm} into x_{0jklm}, for all (j, k, l, m).
5. Swap x_{1jk0m} with x_{1jk1m}, for all (j, k, m).
6. Add (modulo 2^{32}) x_{0jklm} into x_{1jklm}, for all (j, k, l, m).
7. Rotate x_{0jklm} left by 11 bits, for all (j, k, l, m).
8. Swap x_{0j0lm} with x_{0j1lm}, for all (j, l, m).
9. XOR x_{1jklm} into x_{0jklm}, for all (j, k, l, m).
10. Swap x_{1jkl0} with x_{1jkl1}, for all (j, k, l).

The structure is represented in a little endian form, i.e., x_{00000} is composed of the four least significant bytes of the state and x_{11111} is composed of the most significant four. We note that the only nonlinear operations with respect to GF(2) are the modular additions.

2.1 Previous Results on CubeHash

Following its simple structure, CubeHash has received a lot of cryptanalytic attention. Some of the attacks, such as the ones of [7, 21], can be applied to CubeHash, independent of the actual T (as long as it is invertible). These attacks target the preimage resistance of CubeHash, and exploit the fact that as all components are invertible, and as the adversary can control b bytes of the internal state directly, it is possible to find a preimage in about 2^{512-4b} CubeHash computations.

The second type of results, tried to analyze reduced-round variants of CubeHash for collisions. In [2], a collision for CubeHash2/120-512 is given. Collisions for CubeHash1/45 and 2/89 are given in [14], and for CubeHash4/48 and CubeHash4/64 are produced by [9, 10]. A more general methodology to obtain such collisions is described in [8], where variants up to CubeHash7/64 are successfully analyzed.

A third type of attacks/observations concerning CubeHash deal with the symmetric structure of T. For example, if at the input to T all x_{0jklm} words are equal, and all x_{1jklm} words are equal (not necessarily equal to the value of the x_{0jklm}), then the same property holds in the output as well. The first analysis of this type of properties is given in the original submission document [4]. In [3], several additional classes of "symmetric" states are observed, and their use is analyzed. Recently, these classes were expanded to include a larger number of states (and structures) in [16].

Despite all the above-mentioned work, CubeHash is still considered secure, as no attack comes close to offer complexity which is significantly better than generic attacks.[3] To the best of our knowledge this work is the first one that succeeds to offer some non-trivial property of more than 10 rounds of T.

3 Linear Approximation of CubeHash

Linear cryptanalysis [24] is a useful cryptanalytic tool in the world of block cipher cryptanalysis. When using linear cryptanalysis, the adversary tries to find a linear expression that approximates a non-linear function with probability different than $\frac{1}{2}$.

Once a good approximation is found for the relation between the plaintext, the ciphertext and the key (by an expression that holds with some bias) the adversary gains information concerning the key, by observing sufficient number of plaintext/ciphertext pairs until finding enough of them that satisfies the approximation.

Since a good cryptosystem is expected to behave as a random function the linear approximation is satisfied with probability $\frac{1}{2}$ regardles of the approximation correctness. Therefore, the bias induced by the approximation is of the form $\frac{1}{2} + \epsilon$.

As we expect two rounds of a good hash function to be independent of each other we can use the piling up lemma that states that the overall bias for two approximations with biases $\frac{1}{2} + \epsilon_1$ and $\frac{1}{2} + \epsilon_2$ is $\frac{1}{2} + 2\epsilon_1\epsilon_2$ making the total number of queries required to satisfy the approximation $2^{2\epsilon_1\epsilon_2}$.

In the context of hash functions, linear cryptanalysis has received very little attention, unlike differential cryptanalysis. The reason for that seems that while differential cryptanalysis can be directly used to offer collisions or preimages, linear cryptanalysis seems to be restricted to very rare cases (i.e., where the bias is extremely high).

At the same time, the use of linear approximation to assess the security of a hash function can shed some light on whether the underlying components offer the required security. Even though hash functions are unkeyed cryptosystems, the use of linear cryptanalysis supplies some information about the the uniformity of the output distributions. Moreover, linear approximations of the compression function might be useful when discussing MACs built on top of the hash function (suggesting a detectable linear bias in the output).

3.1 Linear Approximation of Addition Modulo 2^{32}

CubeHash uses a mixture of XORs, rotations, and additions. While the first two can be easily handled in the linear cryptanalysis framework, the approximation of the modular addition possess several problems, mostly due to the carry chains.

[3] We note that while the preimage attacks of [21, 7] may offer a small speed-up with respect to generic attacks, their memoryless variants are not much faster than exhaustive search. Moreover, as the submission document lists this as a known issue, this flaw is not considered too harmful by many.

One of the papers studying the cryptographic properties of modular addition is [12] which studies the carry effects on linear approximations. In the paper, Cho and Pieperzyk show that approximating two consecutive bits can overcome some of the inherent problems of carry chains. Namely, if λ is a mask of two consecutive bits (in any position) then: $\lambda \cdot (x + y) = \lambda(x \oplus y)$ with probability 3/4 (i.e., a bias of 1/4).

We analyzed several cases where λ contains pairs of consecutive bits, e.g., two pairs of consecutive pairs, and even when these pairs appear immediately after each other (i.e., λ is composed of four consecutive bits set to 1). Our analysis shows that with respect to linear cryptanalysis, these pairs can be treated as two separate independent instances. For example, the probability that $\lambda \cdot (x + y) = \lambda(x \oplus y)$ for λ whose four most significant bits are 1, while the rest are 0, is 10/16 (suggesting the expected bias of $2 \cdot (1/4)^2 = 1/8$).

3.2 The Linear Approximation of the Round Function of CubeHash

Our first attempt in understanding the security of CubeHash against linear cryptanalysis was a very simple experiment. We looked at all possible masks which had only one pair of two consecutive bits active, and tried to extend this mask as many rounds as possible in the forward direction. At some point, the resulting mask had a divided pair of bits, i.e., a pair of bits that due to the rotations used in CubeHash were sent one to the LSB of a word, and one to the MSB of the same word. Such a mask does no longer fall under the type of masks considered in [12], and our experiments show that such a mask has a very low bias when considering addition.

After performing the search in the forward direction, we repeated the experiment, this time running the light mask in the backward direction (i.e., through T^{-1}) as many rounds as possible. The results obtained in these experiments are shown in Tables 1 and 2, which present the number of possible linear approximations of that form in the forward and the backward directions (along with the associated bias). The longest of which covers 10 rounds in any direction.

Following the surprisingly long approximations, we decided to explore pairs of pairs (i.e., four active bits in the starting mask), repeating the process of analyzing the forward direction as well as the backward direction. These results are summarized in Tables 3 and 4.

We also combined the forward and the backward approximations to form a series of approximations for as many rounds as could, using the combination of this type of approximations. In Table 5 we offer input/output masks of the best approximations we found.

Following the fact that CubeHash aims to offer at most a 2^{512} security, we decided to concentrate at approximations of bias up to 2^{-256} (as detecting smaller biases requires more than 2^{512} queries). The longest possible approximation which adheres to this restriction is of 11 rounds and has a bias of 2^{-235} which is fully described in Table 6.

Table 1. Number of Linear Approximations Following the Consecutive Masks Approach (Starting from a Mask with One Consecutive Pair in the Forward Direction)

Rounds	Bias	Number of Approximations
1	$\frac{1}{4}\cdot\frac{1}{2}^{0}=2^{-2}$	480
1	$\frac{1}{4}\cdot\frac{1}{2}^{1}=2^{-3}$	16
1	$\frac{1}{4}\cdot\frac{1}{2}^{2}=2^{-4}$	480
1	$\frac{1}{4}\cdot\frac{1}{2}^{3}=2^{-5}$	16
2	$\frac{1}{4}\cdot\frac{1}{2}^{11}=2^{-13}$	432
2	$\frac{1}{4}\cdot\frac{1}{2}^{12}=2^{-14}$	16
2	$\frac{1}{4}\cdot\frac{1}{2}^{15}=2^{-17}$	16
2	$\frac{1}{4}\cdot\frac{1}{2}^{16}=2^{-18}$	416
2	$\frac{1}{4}\cdot\frac{1}{2}^{17}=2^{-19}$	16
2	$\frac{1}{4}\cdot\frac{1}{2}^{20}=2^{-22}$	16
3	$\frac{1}{4}\cdot\frac{1}{2}^{29}=2^{-31}$	384
3	$\frac{1}{4}\cdot\frac{1}{2}^{30}=2^{-32}$	16
3	$\frac{1}{4}\cdot\frac{1}{2}^{33}=2^{-35}$	16
3	$\frac{1}{4}\cdot\frac{1}{2}^{35}=2^{-37}$	352
3	$\frac{1}{4}\cdot\frac{1}{2}^{36}=2^{-38}$	16
3	$\frac{1}{4}\cdot\frac{1}{2}^{39}=2^{-41}$	16
4	$\frac{1}{4}\cdot\frac{1}{2}^{66}=2^{-68}$	336
4	$\frac{1}{4}\cdot\frac{1}{2}^{67}=2^{-69}$	16
4	$\frac{1}{4}\cdot\frac{1}{2}^{70}=2^{-72}$	16
4	$\frac{1}{4}\cdot\frac{1}{2}^{74}=2^{-76}$	288
4	$\frac{1}{4}\cdot\frac{1}{2}^{75}=2^{-77}$	16
4	$\frac{1}{4}\cdot\frac{1}{2}^{78}=2^{-80}$	16
5	$\frac{1}{4}\cdot\frac{1}{2}^{113}=2^{-115}$	272
5	$\frac{1}{4}\cdot\frac{1}{2}^{114}=2^{-116}$	240
5	$\frac{1}{4}\cdot\frac{1}{2}^{115}=2^{-117}$	16
5	$\frac{1}{4}\cdot\frac{1}{2}^{117}=2^{-119}$	16
5	$\frac{1}{4}\cdot\frac{1}{2}^{118}=2^{-120}$	16
6	$\frac{1}{4}\cdot\frac{1}{2}^{169}=2^{-171}$	208
6	$\frac{1}{4}\cdot\frac{1}{2}^{170}=2^{-172}$	16
6	$\frac{1}{4}\cdot\frac{1}{2}^{171}=2^{-173}$	160
6	$\frac{1}{4}\cdot\frac{1}{2}^{172}=2^{-174}$	16
6	$\frac{1}{4}\cdot\frac{1}{2}^{173}=2^{-175}$	16
6	$\frac{1}{4}\cdot\frac{1}{2}^{175}=2^{-177}$	16
7	$\frac{1}{4}\cdot\frac{1}{2}^{236}=2^{-238}$	96
7	$\frac{1}{4}\cdot\frac{1}{2}^{237}=2^{-239}$	16
7	$\frac{1}{4}\cdot\frac{1}{2}^{238}=2^{-240}$	144
7	$\frac{1}{4}\cdot\frac{1}{2}^{239}=2^{-241}$	16
7	$\frac{1}{4}\cdot\frac{1}{2}^{240}=2^{-242}$	16
7	$\frac{1}{4}\cdot\frac{1}{2}^{242}=2^{-244}$	16
8	$\frac{1}{4}\cdot\frac{1}{2}^{346}=2^{-348}$	32
8	$\frac{1}{4}\cdot\frac{1}{2}^{347}=2^{-349}$	16
8	$\frac{1}{4}\cdot\frac{1}{2}^{350}=2^{-352}$	16
8	$\frac{1}{4}\cdot\frac{1}{2}^{353}=2^{-355}$	80
8	$\frac{1}{4}\cdot\frac{1}{2}^{354}=2^{-356}$	16
8	$\frac{1}{4}\cdot\frac{1}{2}^{357}=2^{-359}$	16
9	$\frac{1}{4}\cdot\frac{1}{2}^{445}=2^{-447}$	16
9	$\frac{1}{4}\cdot\frac{1}{2}^{481}=2^{-483}$	32
9	$\frac{1}{4}\cdot\frac{1}{2}^{485}=2^{-487}$	16
10	$\frac{1}{4}\cdot\frac{1}{2}^{550}=2^{-552}$	16

Table 2. Number of Linear Approximations Following the Consecutive Masks Approach (Starting from a Mask with One Consecutive Pair in the Backward Direction)

Rounds	Bias	Number of Approximations
1	$\frac{1}{4} \cdot \frac{1}{2}^{2} = 2^{-4}$	496
1	$\frac{1}{4} \cdot \frac{1}{2}^{3} = 2^{-5}$	480
1	$\frac{1}{4} \cdot \frac{1}{2}^{4} = 2^{-6}$	16
2	$\frac{1}{4} \cdot \frac{1}{2}^{12} = 2^{-14}$	448
2	$\frac{1}{4} \cdot \frac{1}{2}^{13} = 2^{-15}$	16
2	$\frac{1}{4} \cdot \frac{1}{2}^{14} = 2^{-16}$	16
2	$\frac{1}{4} \cdot \frac{1}{2}^{18} = 2^{-20}$	416
2	$\frac{1}{4} \cdot \frac{1}{2}^{19} = 2^{-21}$	32
3	$\frac{1}{4} \cdot \frac{1}{2}^{29} = 2^{-31}$	368
3	$\frac{1}{4} \cdot \frac{1}{2}^{30} = 2^{-32}$	32
3	$\frac{1}{4} \cdot \frac{1}{2}^{31} = 2^{-33}$	16
3	$\frac{1}{4} \cdot \frac{1}{2}^{41} = 2^{-43}$	336
3	$\frac{1}{4} \cdot \frac{1}{2}^{42} = 2^{-44}$	48
4	$\frac{1}{4} \cdot \frac{1}{2}^{60} = 2^{-62}$	304
4	$\frac{1}{4} \cdot \frac{1}{2}^{61} = 2^{-63}$	16
4	$\frac{1}{4} \cdot \frac{1}{2}^{62} = 2^{-64}$	32
4	$\frac{1}{4} \cdot \frac{1}{2}^{85} = 2^{-87}$	256
4	$\frac{1}{4} \cdot \frac{1}{2}^{86} = 2^{-88}$	48
5	$\frac{1}{4} \cdot \frac{1}{2}^{102} = 2^{-104}$	240
5	$\frac{1}{4} \cdot \frac{1}{2}^{103} = 2^{-105}$	32
5	$\frac{1}{4} \cdot \frac{1}{2}^{104} = 2^{-106}$	16
5	$\frac{1}{4} \cdot \frac{1}{2}^{134} = 2^{-136}$	224
5	$\frac{1}{4} \cdot \frac{1}{2}^{135} = 2^{-137}$	16
6	$\frac{1}{4} \cdot \frac{1}{2}^{149} = 2^{-151}$	192
6	$\frac{1}{4} \cdot \frac{1}{2}^{150} = 2^{-152}$	16
6	$\frac{1}{4} \cdot \frac{1}{2}^{151} = 2^{-153}$	16
6	$\frac{1}{4} \cdot \frac{1}{2}^{197} = 2^{-199}$	144
6	$\frac{1}{4} \cdot \frac{1}{2}^{198} = 2^{-200}$	48
7	$\frac{1}{4} \cdot \frac{1}{2}^{212} = 2^{-214}$	112
7	$\frac{1}{4} \cdot \frac{1}{2}^{213} = 2^{-215}$	32
7	$\frac{1}{4} \cdot \frac{1}{2}^{214} = 2^{-216}$	16
7	$\frac{1}{4} \cdot \frac{1}{2}^{277} = 2^{-279}$	80
7	$\frac{1}{4} \cdot \frac{1}{2}^{278} = 2^{-280}$	32
8	$\frac{1}{4} \cdot \frac{1}{2}^{308} = 2^{-310}$	48
8	$\frac{1}{4} \cdot \frac{1}{2}^{309} = 2^{-311}$	16
8	$\frac{1}{4} \cdot \frac{1}{2}^{310} = 2^{-312}$	32
8	$\frac{1}{4} \cdot \frac{1}{2}^{407} = 2^{-409}$	48
8	$\frac{1}{4} \cdot \frac{1}{2}^{409} = 2^{-411}$	16
9	$\frac{1}{4} \cdot \frac{1}{2}^{418} = 2^{-420}$	32
10	$\frac{1}{4} \cdot \frac{1}{2}^{477} = 2^{-479}$	16

Table 3. Number of Approximations with a Given Bias Starting from a Pair of Pair of Active Bits (Forward Direction)

Rounds	Bias	Number of Approximations
1	$\frac{1}{4}\cdot\frac{1}{2}^{1}=2^{-3}$	115472
1	$\frac{1}{4}\cdot\frac{1}{2}^{3}=2^{-5}$	228128
1	$\frac{1}{4}\cdot\frac{1}{2}^{5}=2^{-7}$	113152
2	$\frac{1}{4}\cdot\frac{1}{2}^{7}=2^{-9}$	232
2	$\frac{1}{4}\cdot\frac{1}{2}^{8}=2^{-10}$	448
2	$\frac{1}{4}\cdot\frac{1}{2}^{14}=2^{-16}$	848
2	$\frac{1}{4}\cdot\frac{1}{2}^{15}=2^{-17}$ to $\frac{1}{4}\cdot\frac{1}{2}^{33}=2^{-35}$	301480
3	$\frac{1}{4}\cdot\frac{1}{2}^{23}=2^{-25}$	208
3	$\frac{1}{4}\cdot\frac{1}{2}^{25}=2^{-27}$	384
3	$\frac{1}{4}\cdot\frac{1}{2}^{35}=2^{-37}$	352
3	$\frac{1}{4}\cdot\frac{1}{2}^{37}=2^{-39}$ to $\frac{1}{4}\cdot\frac{1}{2}^{71}=2^{-73}$	188144
4	$\frac{1}{4}\cdot\frac{1}{2}^{45}=2^{-47}$	184
4	$\frac{1}{4}\cdot\frac{1}{2}^{53}=2^{-55}$	320
4	$\frac{1}{4}\cdot\frac{1}{2}^{73}=2^{-75}$	304
4	$\frac{1}{4}\cdot\frac{1}{2}^{77}=2^{-79}$ to $\frac{1}{4}\cdot\frac{1}{2}^{149}=2^{-151}$	98288
5	$\frac{1}{4}\cdot\frac{1}{2}^{87}=2^{-89}$	160
5	$\frac{1}{4}\cdot\frac{1}{2}^{94}=2^{-96}$	256
5	$\frac{1}{4}\cdot\frac{1}{2}^{121}=2^{-123}$	128
5	$\frac{1}{4}\cdot\frac{1}{2}^{122}=2^{-124}$ to $\frac{1}{4}\cdot\frac{1}{2}^{229}=2^{-231}$	61056
6	$\frac{1}{4}\cdot\frac{1}{2}^{123}=2^{-125}$	128
6	$\frac{1}{4}\cdot\frac{1}{2}^{139}=2^{-141}$	192
6	$\frac{1}{4}\cdot\frac{1}{2}^{179}=2^{-181}$	272
6	$\frac{1}{4}\cdot\frac{1}{2}^{185}=2^{-187}$ to $\frac{1}{4}\cdot\frac{1}{2}^{343}=2^{-345}$	33632
7	$\frac{1}{4}\cdot\frac{1}{2}^{181}=2^{-183}$	96
7	$\frac{1}{4}\cdot\frac{1}{2}^{201}=2^{-203}$	128
7	$\frac{1}{4}\cdot\frac{1}{2}^{249}=2^{-251}$	64
7	$\frac{1}{4}\cdot\frac{1}{2}^{257}=2^{-259}$ to $\frac{1}{4}\cdot\frac{1}{2}^{477}=2^{-479}$	14256
8	$\frac{1}{4}\cdot\frac{1}{2}^{251}=2^{-253}$	64
8	$\frac{1}{4}\cdot\frac{1}{2}^{288}=2^{-290}$	64
8	$\frac{1}{4}\cdot\frac{1}{2}^{368}=2^{-370}$	48
8	$\frac{1}{4}\cdot\frac{1}{2}^{369}=2^{-371}$ to $\frac{1}{4}\cdot\frac{1}{2}^{693}=2^{-695}$	3120
9	$\frac{1}{4}\cdot\frac{1}{2}^{371}=2^{-373}$	32
9	$\frac{1}{4}\cdot\frac{1}{2}^{395}=2^{-397}$	16
9	$\frac{1}{4}\cdot\frac{1}{2}^{423}=2^{-425}$	16
9	$\frac{1}{4}\cdot\frac{1}{2}^{481}=2^{-483}$ to $\frac{1}{4}\cdot\frac{1}{2}^{859}=2^{-861}$	336
10	$\frac{1}{4}\cdot\frac{1}{2}^{425}=2^{-427}$	16
10	$\frac{1}{4}\cdot\frac{1}{2}^{571}=2^{-573}$	32
10	$\frac{1}{4}\cdot\frac{1}{2}^{597}=2^{-599}$	16
10	$\frac{1}{4}\cdot\frac{1}{2}^{697}=2^{-699}$ to $\frac{1}{4}\cdot\frac{1}{2}^{993}=2^{-995}$	48
11	$\frac{1}{4}\cdot\frac{1}{2}^{620}=2^{-622}$	32
11	$\frac{1}{4}\cdot\frac{1}{2}^{663}=2^{-665}$	16
12	$\frac{1}{4}\cdot\frac{1}{2}^{681}=2^{-683}$	32
13	$\frac{1}{4}\cdot\frac{1}{2}^{737}=2^{-739}$	32
14	$\frac{1}{4}\cdot\frac{1}{2}^{786}=2^{-788}$	32
15	$\frac{1}{4}\cdot\frac{1}{2}^{855}=2^{-857}$	32
16	$\frac{1}{4}\cdot\frac{1}{2}^{983}=2^{-985}$	32

Table 4. Number of Approximations with a Given Bias Starting from a Pair of Pair of Active Bits (Backward Direction)

Rounds	Bias	Number of Approximations
1	$\frac{1}{4}\cdot\frac{1}{2}^{0}=2^{-2}$	464
1	$\frac{1}{4}\cdot\frac{1}{2}^{1}=2^{-3}$	240
1	$\frac{1}{4}\cdot\frac{1}{2}^{2}=2^{-4}$	448
1	$\frac{1}{4}\cdot\frac{1}{2}^{3}=2^{-5}$ to $\frac{1}{4}\cdot\frac{1}{2}^{7}=2^{-9}$	411040
2	$\frac{1}{4}\cdot\frac{1}{2}^{5}=2^{-7}$	216
2	$\frac{1}{4}\cdot\frac{1}{2}^{11}=2^{-13}$	400
2	$\frac{1}{4}\cdot\frac{1}{2}^{13}=2^{-15}$	368
2	$\frac{1}{4}\cdot\frac{1}{2}^{17}=2^{-19}$ to $\frac{1}{4}\cdot\frac{1}{2}^{37}=2^{-39}$	250224
3	$\frac{1}{4}\cdot\frac{1}{2}^{19}=2^{-21}$	184
3	$\frac{1}{4}\cdot\frac{1}{2}^{29}=2^{-31}$	352
3	$\frac{1}{4}\cdot\frac{1}{2}^{31}=2^{-33}$	304
3	$\frac{1}{4}\cdot\frac{1}{2}^{35}=2^{-37}$	152
3	$\frac{1}{4}\cdot\frac{1}{2}^{39}=2^{-41}$ to $\frac{1}{4}\cdot\frac{1}{2}^{83}=2^{-85}$	136544
4	$\frac{1}{4}\cdot\frac{1}{2}^{37}=2^{-39}$	152
4	$\frac{1}{4}\cdot\frac{1}{2}^{66}=2^{-68}$	528
4	$\frac{1}{4}\cdot\frac{1}{2}^{75}=2^{-77}$	120
4	$\frac{1}{4}\cdot\frac{1}{2}^{77}=2^{-79}$	144
4	$\frac{1}{4}\cdot\frac{1}{2}^{80}=2^{-82}$ to $\frac{1}{4}\cdot\frac{1}{2}^{172}=2^{-174}$	69664
5	$\frac{1}{4}\cdot\frac{1}{2}^{77}=2^{-79}$	120
5	$\frac{1}{4}\cdot\frac{1}{2}^{109}=2^{-111}$	96
5	$\frac{1}{4}\cdot\frac{1}{2}^{111}=2^{-113}$	192
5	$\frac{1}{4}\cdot\frac{1}{2}^{113}=2^{-115}$	240
5	$\frac{1}{4}\cdot\frac{1}{2}^{125}=2^{-127}$ to $\frac{1}{4}\cdot\frac{1}{2}^{269}=2^{-271}$	43344
6	$\frac{1}{4}\cdot\frac{1}{2}^{111}=2^{-113}$	96
6	$\frac{1}{4}\cdot\frac{1}{2}^{163}=2^{-165}$	168
6	$\frac{1}{4}\cdot\frac{1}{2}^{169}=2^{-171}$	176
6	$\frac{1}{4}\cdot\frac{1}{2}^{179}=2^{-181}$	112
6	$\frac{1}{4}\cdot\frac{1}{2}^{187}=2^{-189}$ to $\frac{1}{4}\cdot\frac{1}{2}^{387}=2^{-389}$	18672
7	$\frac{1}{4}\cdot\frac{1}{2}^{165}=2^{-167}$	56
7	$\frac{1}{4}\cdot\frac{1}{2}^{223}=2^{-225}$	24
7	$\frac{1}{4}\cdot\frac{1}{2}^{228}=2^{-230}$	64
7	$\frac{1}{4}\cdot\frac{1}{2}^{238}=2^{-240}$	112
7	$\frac{1}{4}\cdot\frac{1}{2}^{258}=2^{-260}$ to $\frac{1}{4}\cdot\frac{1}{2}^{539}=2^{-541}$	5904
8	$\frac{1}{4}\cdot\frac{1}{2}^{225}=2^{-227}$	24
8	$\frac{1}{4}\cdot\frac{1}{2}^{353}=2^{-355}$	32
8	$\frac{1}{4}\cdot\frac{1}{2}^{381}=2^{-383}$	16
8	$\frac{1}{4}\cdot\frac{1}{2}^{413}=2^{-415}$ to $\frac{1}{4}\cdot\frac{1}{2}^{617}=2^{-619}$	272
9	$\frac{1}{4}\cdot\frac{1}{2}^{481}=2^{-483}$	32
9	$\frac{1}{4}\cdot\frac{1}{2}^{527}=2^{-529}$	16
9	$\frac{1}{4}\cdot\frac{1}{2}^{679}=2^{-681}$	32
10	$\frac{1}{4}\cdot\frac{1}{2}^{550}=2^{-552}$	32
10	$\frac{1}{4}\cdot\frac{1}{2}^{773}=2^{-775}$	16
11	$\frac{1}{4}\cdot\frac{1}{2}^{599}=2^{-601}$	16
11	$\frac{1}{4}\cdot\frac{1}{2}^{863}=2^{-865}$	16
12	$\frac{1}{4}\cdot\frac{1}{2}^{953}=2^{-955}$	16

Table 5. A trade-off of biases and rounds. Each line shows the best bias in this setting

Rounds	Input mask	Output mask	Bias
7	x_{00001} = 0600 1806, x_{00011} = 0600 1806,	x_{00000} = 0018 0606, x_{00010} = 0018 0606,	$\frac{1}{4} \cdot \frac{1}{2}^{81} = 2^{-83}$
	x_{00101} = 00c0 3030, x_{00111} = 00c0 3030,	x_{01101} = 0000 0060, x_{01111} = 0000 0060,	
	x_{01001} = 000c 0303, x_{01011} = 000c 0303,	x_{10001} = 0018 0606, x_{10011} = 0018 0606,	
	x_{10100} = 0000 0030, x_{10110} = 0000 0030,	x_{10101} = c0c0 0300, x_{10111} = c0c0 0300,	
	x_{11001} = 000c 0303, x_{11011} = 000c 0303	x_{11101} = 6001 8060, x_{11111} = 6001 8060	
8	x_{00000} = 0600 1806, x_{00010} = 0600 1806,	x_{00000} = 0018 0606, x_{00010} = 0018 0606,	$\frac{1}{4} \cdot \frac{1}{2}^{121} = 2^{-123}$
	x_{01101} = 6660 0060, x_{01111} = 6660 0060,	x_{01101} = 0000 0060, x_{01111} = 0000 0060,	
	x_{10001} = 0600 1806, x_{10011} = 0600 1806,	x_{10001} = 0018 0606, x_{10011} = 0018 0606,	
	x_{10101} = 00c0 c003, x_{10111} = 00c0 c003,	x_{10101} = c0c0 0300, x_{10111} = c0c0 0300,	
	x_{11101} = 6060 0180, x_{11111} = 6060 0180	x_{11101} = 6001 8060, x_{11111} = 6001 8060	
9	x_{00001} = 0018 1998, x_{00011} = 0018 1998,	x_{00000} = 0018 0606, x_{00010} = 0018 0606,	$\frac{1}{4} \cdot \frac{1}{2}^{155} = 2^{-157}$
	x_{00101} = c0cc c000, x_{00111} = c0cc c000,	x_{01101} = 0000 0060, x_{01111} = 0000 0060,	
	x_{01001} = 0c0c cc00, x_{01011} = 0c0c cc00,	x_{10001} = 0018 0606, x_{10011} = 0018 0606,	
	x_{10100} = 00c0 c003, x_{10110} = 00c0 c003,	x_{10101} = c0c0 0300, x_{10111} = c0c0 0300,	
	x_{11001} = 0c0c cc00, x_{11011} = 0c0c cc00	x_{11101} = 6001 8060, x_{11111} = 6001 8060	
10	x_{00001} = 0018 1998, x_{00011} = 0018 1998,	x_{00001} = 0018 0606, x_{00011} = 0018 0606,	$\frac{1}{4} \cdot \frac{1}{2}^{197} = 2^{-199}$
	x_{00101} = c0cc c000, x_{00111} = c0cc c000,	x_{00101} = c030 3000, x_{00111} = c030 3000,	
	x_{01001} = 0c0c cc00, x_{01011} = 0c0c cc00,	x_{01001} = 0c03 0300, x_{01011} = 0c03 0300,	
	x_{10100} = 00c0 c003, x_{10110} = 00c0 c003	x_{10100} = 0030 3330, x_{10110} = 0030 3330,	
	x_{11001} = 0c0c cc00, x_{11011} = 0c0c cc00	x_{11001} = 0c03 0300, x_{11011} = 0c03 0300	
11	x_{00001} = 0018 1998, x_{00011} = 0018 1998,	x_{00000} = 8199 8001, x_{00010} = 8199 8001,	$\frac{1}{4} \cdot \frac{1}{2}^{233} = 2^{-235}$
	x_{00101} = c0cc c000, x_{00111} = c0cc c000,	x_{01101} = 1818 0060, x_{01111} = 1818 0060,	
	x_{01001} = 0c0c cc00, x_{01011} = 0c0c cc00,	x_{10001} = 8199 8001, x_{10011} = 8199 8001,	
	x_{10100} = 00c0 c003, x_{10110} = 00c0 c003,	x_{10101} = 0030 3330, x_{10111} = 0030 3330,	
	x_{11001} = 0c0c cc00, x_{11011} = 0c0c cc00	x_{11101} = 1819 9800, x_{11111} = 1819 9800	
12	x_{00000} = 1819 9800, x_{00010} = 1819 9800,	x_{00000} = 9980 0181, x_{00010} = 9980 0181,	$\frac{1}{4} \cdot \frac{1}{2}^{287} = 2^{-289}$
	x_{01101} = e799 9f81, x_{01111} = e799 9f81,	x_{01101} = 1800 6018, x_{01111} = 1800 6018,	
	x_{10001} = 1819 9800, x_{10011} = 1819 9800,	x_{10001} = 9980 0181, x_{10011} = 9980 0181,	
	x_{10101} = 0003 0333, x_{10111} = 0003 0333,	x_{10101} = 3033 3000, x_{10111} = 3033 3000,	
	x_{11101} = 0181 9980, x_{11111} = 0181 9980	x_{11101} = 1998 0018, x_{11111} = 1998 0018	
13	x_{00000} = 0666 0006, x_{00010} = 0666 0006,	x_{00001} = 6000 6066, x_{00011} = 6000 6066,	$\frac{1}{4} \cdot \frac{1}{2}^{345} = 2^{-347}$
	x_{01101} = e667 e079, x_{01111} = e667 e079,	x_{00101} = 0303 3300, x_{00111} = 0303 3300,	
	x_{10001} = 0666 0006, x_{10011} = 0666 0006,	x_{01001} = 0030 3330, x_{01011} = 0030 3330,	
	x_{10101} = 00c0 ccc0, x_{10111} = 00c0 ccc0,	x_{10100} = 03cf 333f, x_{10110} = 03cf 333f,	
	x_{11101} = 6066 6000, x_{11111} = 6066 6000	x_{11001} = 0030 3330, x_{11011} = 0030 3330	
14	x_{00001} = 3ccc fc0f, x_{00011} = 3ccc fc0f,	x_{00001} = 3033 3000, x_{00011} = 3033 3000,	$\frac{1}{4} \cdot \frac{1}{2}^{405} = 2^{-407}$
	x_{00101} = 67e0 79e6, x_{00111} = 67e0 79e6,	x_{00101} = 9980 0181, x_{00111} = 9980 0181,	
	x_{01001} = 667e 079e, x_{01011} = 667e 079e,	x_{01001} = 1998 0018, x_{01011} = 1998 0018,	
	x_{10100} = 6660 0060, x_{10110} = 6660 0060,	x_{10100} = 999f 81e7, x_{10110} = 999f 81e7,	
	x_{11001} = 667e 079e, x_{11011} = 667e 079e	x_{11001} = 1998 0018, x_{11011} = 1998 0018	

Table 6. The 11-round linear approximation with bias $\frac{1}{4} \cdot \frac{1}{2}^{233} = 2^{-235}$

Round	Mask (before the round)	Bias	Hamming Weight
Input	$x_{00001} = 0018\ 1998$, $x_{00011} = 0018\ 1998$ $x_{00101} = c0cc\ c000$, $x_{00111} = c0cc\ c000$, $x_{01001} = 0c0c\ cc00$, $x_{01011} = 0c0c\ cc00$, $x_{10100} = 00c0\ c003$, $x_{10110} = 00c0\ c003$, $x_{11001} = 0c0c\ cc00$, $x_{11011} = 0c0c\ cc00$	$\frac{1}{4} \cdot \frac{1}{2}^{33} = 2^{-35}$	76
1	$x_{00000} = 0600\ 1806$, $x_{00010} = 0600\ 1806$, $x_{01101} = 6660\ 0060$, $x_{01111} = 6660\ 0060$, $x_{10001} = 0600\ 1806$, $x_{10011} = 0600\ 1806$, $x_{10101} = 00c0\ c003$, $x_{10111} = 00c0\ c003$, $x_{11101} = 6060\ 0180$, $x_{11111} = 6060\ 0180$	$\frac{1}{4} \cdot \frac{1}{2}^{39} = 2^{-41}$	64
2	$x_{00001} = 0600\ 1806$, $x_{00011} = 0600\ 1806$, $x_{00101} = 00c0\ 3030$, $x_{00111} = 00c0\ 3030$, $x_{01001} = 000c\ 0303$, $x_{01011} = 000c\ 0303$, $x_{10100} = 0000\ 0030$, $x_{10110} = 0000\ 0030$, $x_{11001} = 000c\ 0303$, $x_{11011} = 000c\ 0303$	$\frac{1}{4} \cdot \frac{1}{2}^{17} = 2^{-19}$	52
3	$x_{00000} = 0001\ 8000$, $x_{00010} = 0001\ 8000$, $x_{01101} = 6018\ 1800$, $x_{01111} = 6018\ 1800$, $x_{10001} = 0001\ 8000$, $x_{10011} = 0001\ 8000$, $x_{10101} = 0000\ 0030$, $x_{10111} = 0000\ 0030$, $x_{11101} = 0000\ 1800$, $x_{11111} = 0000\ 1800$	$\frac{1}{4} \cdot \frac{1}{2}^{13} = 2^{-15}$	28
4	$x_{00001} = 0001\ 8000$, $x_{00011} = 0001\ 8000$, $x_{00101} = 0c00\ 0000$, $x_{00111} = 0c00\ 0000$, $x_{01001} = 00c0\ 0000$, $x_{01011} = 00c0\ 0000$, $x_{11001} = 00c0\ 0000$, $x_{11011} = 00c0\ 0000$	$\frac{1}{4} \cdot \frac{1}{2}^{3} = 2^{-5}$	16
5	$x_{01101} = 0000\ 0006$, $x_{01111} = 0000\ 0006$	$\frac{1}{4} \cdot \frac{1}{2}^{1} = 2^{-3}$	4
6	$x_{10100} = 0000\ 0300$, $x_{10110} = 0000\ 0300$	$\frac{1}{4} \cdot \frac{1}{2}^{5} = 2^{-7}$	4
7	$x_{00000} = 0018\ 0000$, $x_{00010} = 0018\ 0000$, $x_{10001} = 0018\ 0000$, $x_{10011} = 0018\ 0000$, $x_{10101} = 0000\ 0300$, $x_{10111} = 0000\ 0300$, $x_{11101} = 0001\ 8000$, $x_{11111} = 0001\ 8000$	$\frac{1}{4} \cdot \frac{1}{2}^{15} = 2^{-17}$	16
8	$x_{00001} = 0018\ 0000$, $x_{00011} = 0018\ 0000$, $x_{00101} = c000\ 0000$, $x_{00111} = c000\ 0000$, $x_{01001} = 0c00\ 0000$, $x_{01011} = 0c00\ 0000$, $x_{10100} = c0c0\ 0300$, $x_{10110} = c0c0\ 0300$, $x_{11001} = 0c00\ 0000$, $x_{11011} = 0c00\ 0000$	$\frac{1}{4} \cdot \frac{1}{2}^{21} = 2^{-23}$	28
9	$x_{00000} = 0018\ 0606$, $x_{00010} = 0018\ 0606$, $x_{01101} = 0000\ 0060$, $x_{01111} = 0000\ 0060$, $x_{10001} = 0018\ 0606$, $x_{10011} = 0018\ 0606$, $x_{10101} = c0c0\ 0300$, $x_{10111} = c0c0\ 0300$, $x_{11101} = 6001\ 8060$, $x_{11111} = 6001\ 8060$	$\frac{1}{4} \cdot \frac{1}{2}^{41} = 2^{-43}$	52
10	$x_{00001} = 0018\ 0606$, $x_{00011} = 0018\ 0606$, $x_{00101} = c030\ 3000$, $x_{00111} = c030\ 3000$, $x_{01001} = 0c03\ 0300$, $x_{01011} = 0c03\ 0300$, $x_{10100} = 0030\ 3330$, $x_{10110} = 0030\ 3330$, $x_{11001} = 0c03\ 0300$, $x_{11011} = 0c03\ 0300$	$\frac{1}{4} \cdot \frac{1}{2}^{35} = 2^{-37}$	64
11	$x_{00000} = 8199\ 8001$, $x_{00010} = 8199\ 8001$, $x_{01101} = 1818\ 0060$, $x_{01111} = 1818\ 0060$, $x_{10001} = 8199\ 8001$, $x_{10011} = 8199\ 8001$, $x_{10101} = 0030\ 3330$, $x_{10111} = 0030\ 3330$, $x_{11101} = 1819\ 9800$, $x_{11111} = 1819\ 9800$		76

Table 7. The 14-round linear approximation with bias $\frac{1}{4} \cdot \frac{1}{2}^{405} = 2^{-407}$ rounds 1-9

Round	Mask (before the round)	Bias	Hamming Weight
input	$x_{00001} = 3ccc\ fc0f$, $x_{00011} = 3ccc\ fc0f$, $x_{00101} = 67e0\ 79e6$, $x_{00111} = 67e0\ 79e6$, $x_{01001} = 667e\ 079e$, $x_{01011} = 667e\ 079e$, $x_{10100} = 6660\ 0060$, $x_{10110} = 6660\ 0060$, $x_{11001} = 667e\ 079e$, $x_{11011} = 667e\ 079e$	$\frac{1}{4} \cdot \frac{1}{2}^{60} = 2^{-62}$	160
1	$x_{00000} = 0003\ 0333$, $x_{00010} = 0003\ 0333$, $x_{01101} = f03c\ f333$, $x_{01111} = f03c\ f333$, $x_{10001} = 0003\ 0333$, $x_{10011} = 0003\ 0333$, $x_{10101} = 6660\ 0060$, $x_{10111} = 6660\ 0060$, $x_{11101} = 3000\ 3033$, $x_{11111} = 3000\ 3033$	$\frac{1}{4} \cdot \frac{1}{2}^{54} = 2^{-56}$	100
2	$x_{00001} = 0003\ 0333$, $x_{00011} = 0003\ 0333$, $x_{00101} = 1819\ 9800$, $x_{00111} = 1819\ 9800$, $x_{01001} = 0181\ 9980$, $x_{01011} = 0181\ 9980$, $x_{10100} = 6018\ 1800$, $x_{10110} = 6018\ 1800$, $x_{11001} = 0181\ 9980$, $x_{11011} = 0181\ 9980$	$\frac{1}{4} \cdot \frac{1}{2}^{34} = 2^{-36}$	76
3	$x_{00000} = c0c0\ 0300$, $x_{00010} = c0c0\ 0300$, $x_{01101} = 0ccc\ 000c$, $x_{01111} = 0ccc\ 000c$, $x_{10001} = c0c0\ 0300$, $x_{10011} = c0c0\ 0300$, $x_{10101} = 6018\ 1800$, $x_{10111} = 6018\ 1800$, $x_{11101} = 0c0c\ 0030$, $x_{11111} = 0c0c\ 0030$	$\frac{1}{4} \cdot \frac{1}{2}^{40} = 2^{-42}$	64
4	$x_{00001} = c0c0\ 0300$, $x_{00011} = c0c0\ 0300$, $x_{00101} = 0018\ 0606$, $x_{00111} = 0018\ 0606$, $x_{01001} = 6001\ 8060$, $x_{01011} = 6001\ 8060$, $x_{10100} = 0000\ 0006$, $x_{10110} = 0000\ 0006$, $x_{11001} = 6001\ 8060$, $x_{11011} = 6001\ 8060$	$\frac{1}{4} \cdot \frac{1}{2}^{18} = 2^{-20}$	52
5	$x_{00000} = 0000\ 3000$, $x_{00010} = 0000\ 3000$, $x_{01101} = 0c03\ 0300$, $x_{01111} = 0c03\ 0300$, $x_{10001} = 0000\ 3000$, $x_{10011} = 0000\ 3000$, $x_{10101} = 0000\ 0006$, $x_{10111} = 0000\ 0006$, $x_{11101} = 0000\ 0300$, $x_{11111} = 0000\ 0300$	$\frac{1}{4} \cdot \frac{1}{2}^{14} = 2^{-16}$	28
6	$x_{00001} = 0000\ 3000$, $x_{00011} = 0000\ 3000$, $x_{00101} = 0180\ 0000$, $x_{00111} = 0180\ 0000$, $x_{01001} = 0018\ 0000$, $x_{01011} = 0018\ 0000$, $x_{11001} = 0018\ 0000$, $x_{11011} = 0018\ 0000$	$\frac{1}{4} \cdot \frac{1}{2}^{4} = 2^{-6}$	16
7	$x_{01101} = c000\ 0000$, $x_{01111} = c000\ 0000$	$\frac{1}{4} \cdot \frac{1}{2}^{2} = 2^{-4}$	4
8	$x_{10100} = 0000\ 0060$, $x_{10110} = 0000\ 0060$	$\frac{1}{4} \cdot \frac{1}{2}^{6} = 2^{-8}$	4
9	$x_{00000} = 0003\ 0000$, $x_{00010} = 0003\ 0000$, $x_{10001} = 0003\ 0000$, $x_{10011} = 0003\ 0000$, $x_{10101} = 0000\ 0060$, $x_{10111} = 0000\ 0060$, $x_{11101} = 0000\ 3000$, $x_{11111} = 0000\ 3000$	$\frac{1}{4} \cdot \frac{1}{2}^{16} = 2^{-18}$	16

For those interested in assessing the full security that might be offered by the 1024-bit transformation T, we note that there also exists a 14-round linear approximation with a bias of 2^{-406}. We outline the full 14-round approximation in Tables 7 and 8.

Table 8. The 14-round linear approximation with bias $\frac{1}{4} \cdot \frac{1}{2}^{405} = 2^{-407}$ rounds 10-14

Round	Mask (before the round)	Bias	Hamming Weight
10	$x_{00001} = 0003\ 0000$, $x_{00011} = 0003\ 0000$, $x_{00101} = 1800\ 0000$, $x_{00111} = 1800\ 0000$, $x_{01001} = 0180\ 0000$, $x_{01011} = 0180\ 0000$, $x_{10100} = 1818\ 0060$, $x_{10110} = 1818\ 0060$, $x_{11001} = 0180\ 0000$, $x_{11011} = 0180\ 0000$	$\frac{1}{4} \cdot \frac{1}{2}^{22} = 2^{-24}$	28
11	$x_{00000} = c003\ 00c0$, $x_{00010} = c003\ 00c0$, $x_{01101} = 0000\ 000c$, $x_{01111} = 0000\ 000c$, $x_{10001} = c003\ 00c0$, $x_{10011} = c003\ 00c0$, $x_{10101} = 1818\ 0060$, $x_{10111} = 1818\ 0060$, $x_{11101} = 0c00\ 300c$, $x_{11111} = 0c00\ 300c$	$\frac{1}{4} \cdot \frac{1}{2}^{42} = 2^{-44}$	52
12	$x_{00001} = c003\ 00c0$, $x_{00011} = c003\ 00c0$, $x_{00101} = 1806\ 0600$, $x_{00111} = 1806\ 0600$, $x_{01001} = 0180\ 6060$, $x_{01011} = 0180\ 6060$, $x_{10100} = 0006\ 0666$, $x_{10110} = 0006\ 0666$, $x_{11001} = 0180\ 6060$, $x_{11011} = 0180\ 6060$	$\frac{1}{4} \cdot \frac{1}{2}^{36} = 2^{-38}$	64
13	$x_{00000} = 3033\ 3000$, $x_{00010} = 3033\ 3000$, $x_{01101} = 0303\ 000c$, $x_{01111} = 0303\ 000c$, $x_{10001} = 3033\ 3000$, $x_{10011} = 3033\ 3000$, $x_{10101} = 0006\ 0666$, $x_{10111} = 0006\ 0666$, $x_{11101} = 0303\ 3300$, $x_{11111} = 0303\ 3300$	$\frac{1}{4} \cdot \frac{1}{2}^{58} = 2^{-60}$	76
14	$x_{00001} = 3033\ 3000$, $x_{00011} = 3033\ 3000$, $x_{00101} = 9980\ 0181$, $x_{00111} = 9980\ 0181$, $x_{01001} = 1998\ 0018$, $x_{01011} = 1998\ 0018$, $x_{10100} = 999f\ 81e7$, $x_{10110} = 999f\ 81e7$, $x_{11001} = 1998\ 0018$, $x_{11011} = 1998\ 0018$		100

4 Message Modification Techniques — A Chosen-Plaintext Linear Approximations

Linear cryptanalysis relies on collecting a large number of input/output pairs, and verifying whether they satisfy the approximation or not. In [22] Knudsen and Mathiassen show that there are cases in which one can "help" the linear approximation to be satisfied by properly selecting the inputs.

In the case of modular addition, the linear approximation which we use is satisfied whenever one of the LSBs of the approximated bits is 0. This allows preselecting inputs for which the approximation holds with probability 1. This technique can be used either to increase the bias of our 11-round approximation or to extend it to a 12-round approximation.

In order to fix the bits entering the first layer of addition in CubeHash it is enough to fix the LSB of each pair of approximated bits. However, fixing the bits for the next layer is a bit more tricky, as it requires to fix some internal state bit (after an XOR or addition) to 0. This task is a bit harder due to carry issues. More precisely, to fix bit i of x_{1jklm} after the first five operations of T, it is required that bit i of $x_{1jk\bar{l}m}$ is 0 after the first operation of T. This specific bit depends on the corresponding carry chain.

Table 9. The round that extends the 11-round approximation to 12 rounds (and the bits to fix

Round	Input mask	Input bits fixed to 0
-1	$x_{00000} = 0018\ 1998$, $x_{00010} = 0018\ 1998$ $x_{01101} = 81e7\ 999f$, $x_{01111} = 81e7\ 999f$ $x_{10001} = 0018\ 1998$, $x_{10011} = 0018\ 1998$ $x_{10101} = 3300\ 0303$, $x_{10111} = 3300\ 0303$ $x_{11101} = 8001\ 8199$, $x_{11111} = 8001\ 8199$	$x_{10001} = 0008\ 0888_x$, $x_{10011} = 0008\ 0888_x$ $x_{10101} = 1100\ 0101_x$, $x_{10111} = 1100\ 0101_x$ $x_{11101} = ffff\ ffff_x$, $x_{11111} = ffff\ ffff_x$
-0.5	$x_{00101} = f30c\ 0300_x$, $x_{00111} = f30c\ 0300_x$ $x_{01000} = 0c0c\ cc00_x$, $x_{01001} = 0c0c\ cc00_x$ $x_{01010} = 0c0c\ cc00_x$, $x_{01011} = 0c0c\ cc00_x$ $x_{01101} = 8001\ 8199_x$, $x_{01111} = 8001\ 8199_x$ $x_{10001} = 0018\ 1998_x$, $x_{10011} = 0018\ 1998_x$ $x_{10101} = c00c\ 0003_x$, $x_{10111} = c00c\ 0003_x$ $x_{11000} = 0c0c\ cc00_x$, $x_{11001} = 0c0c\ cc00_x$ $x_{11010} = 0c0c\ cc00_x$, $x_{11011} = 0c0c\ cc00_x$	$x_{00001} = 000c\ 0ccc_x$, $x_{00011} = 000c\ 0ccc_x$ $x_{00101} = 6006\ 0001_x$, $x_{00111} = 6006\ 0001_x$ $x_{01000} = 0606\ 6600_x$, $x_{01001} = 0606\ 6600_x$ $x_{01010} = 0606\ 6600_x$, $x_{01011} = 0606\ 6600_x$ $x_{10001} = 000c\ 0ccc_x$, $x_{10011} = 000c\ 0ccc_x$ $x_{10101} = 6006\ 0001_x$, $x_{10111} = 6006\ 0001_x$ $x_{11010} = 0606\ 6600_x$, $x_{11011} = 0606\ 6600_x$ $x_{10001} = 0006\ 0666_x$, $x_{10011} = 0006\ 0666_x$
0	$x_{00001} = 0018\ 1998$, $x_{00011} = 0018\ 1998$ $x_{00101} = c0cc\ c000$, $x_{00111} = c0cc\ c000$ $x_{01001} = 0c0c\ cc00$, $x_{01011} = 0c0c\ cc00$ $x_{10100} = 00c0\ c003$, $x_{10110} = 00c0\ c003$ $x_{11001} = 0c0c\ cc00$, $x_{11011} = 0c0c\ cc00$	

−0.5 stands for the mask that enters the second addition of the additional round.

A simple solution would be to fix one of the words $x_{0jk\bar{l}m}$ or $x_{1jk\bar{l}m}$ to zero, ensuring no carrys are produced during the addition $x_{1jk\bar{l}m} \leftarrow x_{0jk\bar{l}m} + x_{1jk\bar{l}m}$. By additionally fixing bit i of both $x_{0jk\bar{l}m}$ and $x_{1jk\bar{l}m}$ to zero, we can guarantee that the bit that enters the second layer of additions is indeed zero.

As the above approach sets many bits to zero we offer a more efficient approach. One can fix only bits $i-1, i$ in $x_{0jk\bar{l}m}$ and $i-1, i$ in $x_{1jk\bar{l}m}$ to zero. Even if there is a carry entering bit $i-1$, it does not produce carry that affects the i'th bit, and we are assured that bit i after the addition is indeed 0.

Consider our 11-round approximation suggested in table 6. To increase its bias we can fix the bits masked by $x_{10100} = 0040\ 4001_x, x_{10110} = 0040\ 4001_x, x_{11001} = 0404\ 4400_x, x_{11011} = 0404\ 4400_x$ to 0 to ensure that our approximation holds for the first layer of addition with probability 1. To ensure the approximation holds for the second layer of the first round as well we fix the bits masked by $000c\ 0ccc_x$ in $x_{00101}, x_{00111}, x_{10101}$ and x_{10111}, the bits masked by $0060\ 6001_x$ in x_{00000}, x_{00010}, x_{10000} and x_{10010} and the bits masked by $6066\ 6000_x$ in $x_{00101}, x_{00011}, x_{10001}$ and x_{10011} to 0. Fixing these 56 bits ensures that the first round of the approximation holds with probability 1 hence increasing the bias by a factor of 2^{34} making the total bias 2^{-201}.

When considering an extension of the linear approximation shown in Table 6 by calculating it one round backward as described in Table 9, we can fix 80 input bits to zero, thus ensuring that the approximation holds for the first layer of additions with probability 1. These 80 bits are the ones masked by $x_{10001} = 0008\ 0888_x, x_{10011} = 0008\ 0888_x, x_{10101} = 1100\ 0101_x, x_{10111} = 1100\ 0101_x$ and the whole words x_{11101} and x_{11111}. We note that one can pick other sets of bits (where any fixed bit from x_{0jklm} can be exchanged for a bit in x_{1jklm}).

To ensure that all the appropriate bits in x_{1jklm} are zero one needs to set the mask bits masked by $000c\ 0ccc_x$ of $x_{00001}, x_{00011}, x_{10001}$, and x_{10011}, the bits masked by $c00c\ 0001_x$ of $x_{00101}, x_{00111}, x_{10101}$, and x_{10111}, and those masked by $0c0c\ cc00$ in $x_{01000}, x_{01001}, x_{01010}, x_{01011}, x_{11000}, x_{11001}, x_{11010}$, and x_{11011} to zero. Fixing these 116 bits (10 of which are shared with the previous 80), assures that all the additions in the first round of the 12-round approximation follow the approximation, i.e., "saving" their "contribution" to the bias, and resulting in a bias of $\frac{1}{4} \cdot \frac{1}{2}^{233} = 2^{-235}$.

We note that the number of bits set to 0 is 186, leaving 838 bits to be randomly selected. This is sufficient to generate the 2^{470} possible inputs to T^{12}, needed for using this chosen-plaintext linear approximation successfully, in a distinguishing attack on 12-round CubeHash.

5 Distinguishing Reduced-Round Variants of the Compression Function of CubeHash

Given the linear approximations presented in the previous sections, it is possible to distinguish a black box which contains up to 12-round CubeHash from a random permutation. Of course, for any unkeyed primitive this distinguishing can be done by just comparing the input/output of a few queries to the black box with the input/output produced by the publicly available algorithm. If we want to offer some cryptographic settings in which distinguishing attacks make sense, we either need to consider keyed variants (either of the round function T or of the hash function, e.g., in MACs) or to discuss known-key distinguishers [23].

Such possible "application" is an a Even-Mansour [15] variant of 11-round T (or any other number of rounds), i.e., $EM\text{-}T^{11}_{k_1,k_2}(P) = T^{11}(P \oplus k_1) \oplus k_2$. If 11-round T is indeed good as a source of nonlinearity (for a linear T, the entire security of CubeHash collapses), then XORing an unknown key before and after these 11 rounds, should result in a good pseudo-random permutation. Using our linear approximations, one can distinguish this construction from a random permutation.

We emphasize that as our results are linear in nature, they require that the adversary has access both to the input to the nonlinear function as well as its output. To the best of our knowledge, there is no way to use this directly in a hash function setting.

6 Conclusions

In this paper we presented a series of approximations for the SHA-3 former candidate CubeHash. The analysis challenges the strength of CubeHash's round function, T, and shows that (from linear cryptanalysis point of view), offers adequate security. At the same time, the security margins offered by 16 iterations of T seems to be on the smaller side, as future works on CubeHash may find better linear approximations.

Acknowledgement

The authors wish to thank Prof. Adi Shamir for his guidance and assistance analyzing CubeHash, Nathan Keller for providing core ideas in this paper, Daniel J. Bernstein for his insightful and mind-provoking comments on previous versions of this article. Finally, we wish to thank Michael Klots for his technical assistance, which was crucial for finding our results.

References

1. Andreeva, E., Bouillaguet, C., Fouque, P.-A., Hoch, J.J., Kelsey, J., Shamir, A., Zimmer, S.: Second Preimage Attacks on Dithered Hash Functions. In: Smart, N.P. (ed.) EUROCRYPT 2008. LNCS, vol. 4965, pp. 270–288. Springer, Heidelberg (2008)
2. Aumasson, J.P.: Collision for CubeHash2/120-512. NIST mailing list (2008), http://ehash.iaik.tugraz.at/uploads/a/a9/Cubehash.txt
3. Aumasson, J.-P., Brier, E., Meier, W., Naya-Plasencia, M., Peyrin, T.: Inside the Hypercube. In: Boyd, C., González Nieto, J. (eds.) ACISP 2009. LNCS, vol. 5594, pp. 202–213. Springer, Heidelberg (2009)
4. Bernstein, D.J.: CubeHash specification (2.B.1). Submission to NIST (2008)
5. Bernstein, D.J.: CubeHash specification (2.B.1). Submission to NIST (2009)
6. Biham, E., Chen, R.: Near-Collisions of SHA-0. In: [17], pp. 290–305
7. Bloom, B., Kaminsky, A.: Single Block Attacks and Statistical Tests on CubeHash. IACR ePrint Archive, Report 2009/407 (2009)
8. Brier, E., Khazaei, S., Meier, W., Peyrin, T.: Linearization framework for collision attacks: Application to cubehash and md6. In: Matsui, M. (ed.) ASIACRYPT 2009. LNCS, vol. 5912, pp. 560–577. Springer, Heidelberg (2009)
9. Brier, E., Khazaei, S., Meier, W., Peyrin, T.: Real Collisions for CubeHash-4/48. NIST mailing list (2009), http://ehash.iaik.tugraz.at/uploads/5/50/Bkmp_ch448.txt
10. Brier, E., Khazaei, S., Meier, W., Peyrin, T.: Real Collisions for CubeHash-4/64. NIST mailing list (2009), http://ehash.iaik.tugraz.at/uploads/9/93/Bkmp_ch464.txt
11. De Cannière, C., Rechberger, C.: Finding SHA-1 Characteristics: General Results and Applications. In: Lai, X., Chen, K. (eds.) ASIACRYPT 2006. LNCS, vol. 4284, pp. 1–20. Springer, Heidelberg (2006)
12. Cho, J.Y., Pieprzyk, J.: Multiple Modular Additions and Crossword Puzzle Attack on NLSv2. In: Garay, J.A., Lenstra, A.K., Mambo, M., Peralta, R. (eds.) ISC 2007. LNCS, vol. 4779, pp. 230–248. Springer, Heidelberg (2007)
13. Cramer, R. (ed.): EUROCRYPT 2005. LNCS, vol. 3494. Springer, Heidelberg (2005)
14. Dai, W.: Collisions for CubeHash1/45 and CubeHash2/89 (2008), http://www.cryptopp.com/sha3/cubehash.pdf
15. Even, S., Mansour, Y.: A Construction of a Cipher from a Single Pseudorandom Permutation. J. Cryptology 10(3), 151–162 (1997)
16. Ferguson, N., Lucks, S., McKay, K.A.: Symmetric States and their Structure: Improved Analysis of CubeHash. IACR ePrint Archive, Report 2010/273 (2010) Presented at the SHA-3 Second Workshop, Santa Barbara, USA, August 23-24 (2010)

17. Franklin, M. (ed.): CRYPTO 2004. LNCS, vol. 3152. Springer, Heidelberg (2004)
18. Joux, A.: Multicollisions in Iterated Hash Functions. Application to Cascaded Constructions. In: [17], pp. 306–316
19. Kelsey, J., Kohno, T.: Herding Hash Functions and the Nostradamus Attack. In: Vaudenay, S. (ed.) EUROCRYPT 2006. LNCS, vol. 4004, pp. 183–200. Springer, Heidelberg (2006)
20. Kelsey, J., Schneier, B.: Second Preimages on n-Bit Hash Functions for Much Less than 2^n Work. In: [13], pp. 474–490
21. Khovratovich, D., Nikolic', I., Weinmann, R.P.: Preimage attack on CubeHash512-r/4 and CubeHash512-r/8 (2008), `http://ehash.iaik.tugraz.at/uploads/6/6c/Cubehash.pdf`
22. Knudsen, L.R., Mathiassen, J.E.: A Chosen-Plaintext Linear Attack on DES. In: Schneier, B. (ed.) FSE 2000. LNCS, vol. 1978, pp. 262–272. Springer, Heidelberg (2001)
23. Knudsen, L.R., Rijmen, V.: Known-Key Distinguishers for Some Block Ciphers. In: Kurosawa, K. (ed.) ASIACRYPT 2007. LNCS, vol. 4833, pp. 315–324. Springer, Heidelberg (2007)
24. Matsui, M.: Linear Cryptanalysis Method for DES Cipher. In: Helleseth, T. (ed.) EUROCRYPT 1993. LNCS, vol. 765, pp. 386–397. Springer, Heidelberg (1994)
25. National Institute of Standards and Technology: Cryptographic Hash Algorithm Competition (2008), `http://www.nist.gov/hash-competition`
26. Stevens, M., Lenstra, A.K., de Weger, B.: Chosen-Prefix Collisions for MD5 and Colliding X.509 Certificates for Different Identities. In: Naor, M. (ed.) EUROCRYPT 2007. LNCS, vol. 4515, pp. 1–22. Springer, Heidelberg (2007)
27. Wang, X., Yu, H.: How to Break MD5 and Other Hash Functions. In: [13], pp. 19–35

On the Indifferentiability of Fugue and Luffa

Rishiraj Bhattacharyya[1] and Avradip Mandal[2]

[1] Cryptology Research Group, Applied Statistics Unit, Indian Statistical Institute, Kolkata
rishi_r@isical.ac.in
[2] Université du Luxembourg, Luxembourg
avradip.mandal@uni.lu

Abstract. Indifferentiability is currently considered to be an important security notion for a cryptographic hash function to instantiate Random Oracles in different security proofs. In this paper, we prove indifferentiability of Fugue and Luffa, two SHA3 second round candidates. We also analyze the indifferentiability of a modified Luffa mode replacing multiple small permutations by a single large permutation.

Our technique is quite general and can be applicable to any sponge based design which uses affine function for message insertion. To the best of our knowledge, our result for Luffa is the first indifferentiability analysis of a mode of operation based on variable (more than two) number of small permutations.

Keywords: Hash function, Indifferentiability, Fugue, Luffa.

1 Introduction

Design of cryptographic hash functions typically involves two steps. One constructs a fixed input length primitive (like permutation or compression function) $f : \{0,1\}^m \rightarrow \{0,1\}^n$ and applies a *domain extension* technique C^f to build the hash function. Merkle-Damgård technique along with its variants are the most popular choice for domain extension.

In the last decade, cryptographic hash functions have gained immense importance to instantiate a truly *Random Function* in cryptographic protocols. Although, previous results prove that no hash function can accurately instantiate a random function [6], one can hope to gain some confidence by using a structurally rigid construction in order to resist any generic attack. Indifferentiability of hash functions, introduced by Coron et. al. in [10] extending the result of [13], is the strongest and appropriate criteria to establish generic rigidity of the mode of a hash function. Informally, under the notion of indifferentiability, to prove rigidity of a domain extension technique C (assuming the underlying primitive f to be ideal), one has to design a simulator S that can simulate the underlying ideal primitive and still remain Consistent to a Random Oracle $\mathcal{R}$ with respect to concerned mode of operation. If no distinguisher can distinguish between the output distribution of (C^f, f) from that of $(\mathcal{R}, S^{\mathcal{R}})$ with probability ϵ, the hash function C^f is said to be indifferentiable from a Random Oracle (RO) with advantage ϵ.

J. Lopez and G. Tsudik (Eds.): ACNS 2011, LNCS 6715, pp. 479–497, 2011.

In [10], Coron et. al. proved that Merkle-Damgård mode is insecure under indifferentiability notion. However many of its variants, like strengthened-MD, chop-MD, MD with HAIFA padding was proven to be secure in [8,9,10]. In [4], Bertoni et. al. proved the indifferentiability of sponge mode of operation. In [1,5], domain extension technique of Grøstl and JH was proven to be indifferentiable. For a comparative discussion on the indifferentiability of SHA3 second round candidates, we refer the reader to [2]. In this work we obtain indifferentiability security bound for two NIST second round candidates Fugue and Luffa.

Our Results. In this paper we analyze indifferentiability of domain extension of Fugue and Luffa. Indifferentiability analysis of the mode of both the hash function was open [2]. Fugue can be viewed as a variant of the sponge construction with a post-processor and a fixed output length. However, due to the difference in message insertion algorithm and application of a different post-processor in Fugue, one cannot directly plug in the bounds from [4]. Our technique to prove indifferentiability of Fugue is based on the technique used in [5], where one gives an upper bound for the simulated world interpolation probability and lower bound for the real world interpolation probability. In Section 4 we prove that under the assumption that two permutations in Fugue are independent random permutations over $\{0,1\}^{nt}$, Fugue mode of operation is indifferentiable from a random oracle with advantage $\frac{\mathcal{O}(\sigma^2)}{2^{(t-1)n}}$ where σ is the total number of message blocks queried by the distinguisher. Recently Aumasson and Phan [3] have the final round transformation in Fg do not really behave like a random permutation by showing a distinguisher After wards Halevi et al [12], actually showed Fugue mode of operation behaves like random oracle assuming some weak ideal functionality of the underlying permutations. However in their analysis the attacker is restricted in the sense, that she can not make inverse queries to the permutation.

Domain extension technique of Luffa is also similar to sponge; but message injection algorithm of Luffa uses all the chaining bits (as opposed to at most half the chaining bits of Sponge or Fugue) and instead of one large permutation, Luffa uses t small permutations. Hence one cannot readily use previous techniques. First we consider Luffa mode of operation with a single large permutations and message block of n bits. We show that if the underlying permutation is assumed to be one fixed random permutation over $\{0,1\}^{tn}$, the modified Luffa mode is indifferentiable from a Random Oracle with advantage $\frac{\mathcal{O}(\sigma^2)}{2^{(t-2)n}}$ where σ is the total number of message blocks queried by the distinguisher.

Finally, in Section 6, we consider the actual Luffa mode of operation. We prove that under the assumption that underlying permutations are independent random n bit permutations, Luffa mode of operation is indifferentiable from a Random Oracle with advantage $\frac{\mathcal{O}(q^4)}{2^n}$, where q is the maximum number of queries the distinguisher makes.

Our result is the first indifferentiability analysis of Luffa. Although we achieve much less than the birthday bound, one can view this as the security-efficiency trade-off of Luffa due to its multiple small permutations to handle a large chaining value.

2 Preliminaries

Modes of Operation. Informally speaking, a mode of operation is an algorithm to construct a hash function from a compression function.

Definition 1. *A mode of operation C with oracle access to compression function $f\{0,1\}^m \to \{0,1\}^n$ is an algorithm which defines a function $C^f : \{0,1\}^* \to \{0,1\}^n$.*

Below we describe the well known Merkle-Damgård or MD mode of operation.

Definition 2. *Let $IV \in \{0,1\}^n$ be a fixed initial value. Given a compression function $f : \{0,1\}^m \to \{0,1\}^n$, the well known Merkle-Damgård mode of operation is defined as*

$$\mathsf{MD}^f_{IV}(m_1\|m_2\|\ldots\|m_l) = f(f(\ldots f(f(IV\|m_1)\|m_2)\ldots)\|m_l)$$

where $m_1, m_2, \ldots m_l \in \{0,1\}^{m-n}$.

There is a subtle difference between a hash function and a mode of operation. The mode of operation is actually a domain extension algorithm. If we supply a particular compression function f to the mode of operation algorithm we get a particular hash function. So when we think about a hash function, the compression function is fixed. Sometimes we will drop the subscript IV from MD^f_{IV} and write it as MD^f when the IV value is clear from the context (which is either IV_{Fg} or IV_{Lf}).

Luffa Mode of Operation . The compression function of Luffa, $f^{\Pi} : \{0,1\}^{(t+1)n} \to \{0,1\}^{tn}$ (Fig. 1(b)) is defined as, $f^{\Pi} \equiv P^{\Pi} \circ \mathsf{MI}_{\mathsf{Lf}}$, where $\mathsf{MI}_{\mathsf{Lf}} : \{0,1\}^{(t+1)n} \to \{0,1\}^{tn}$ is a fixed linear transformation and $P^{\Pi}(x^1\|\cdots\|x^t) = \pi_1(x^1)\|\cdots\|\pi_t(x^t)$, $\Pi = (\pi_1, \cdots, \pi_t)$ being a tuple of t-independent random permutations on $\{0,1\}^n$.

To process arbitrary length messages the following padding rule $\mathsf{Pad}_{\mathsf{Lf}} : \{0,1\}^* \to (\{0,1\}^n)^+$ is used in Luffa. $\mathsf{Pad}_{\mathsf{Lf}}(M) = M\|10^k$, where k is the smallest non-negative integer such that $|M| + k + 1 \equiv 0 \mod n$.

The Luffa-mode of operation is nothing but MD-mode of operation based on the compression function f^{Π} of the padded message, followed by a *finalization round.* If the required digest size n_h is same as n (the size of the permutations $\pi_1, \cdots, \pi_t$) then we pass the output of the $\mathsf{MD}^{f^{\Pi}}$, through a blank round of f^{Π} with message 0^n and process the Xor[1] of t-many permutation outputs as the final digest. If $n_h < n$, then we just Chop[2] the Xor value to give a n_h bit output. If $n < n_h \leq 2n$, then we follow the idea behind *sponge* construction, i.e. use two blank rounds f^{Π} with message 0^n each time, then concatenate the Xor of the two blank round outputs and Chop it to the desired digest size. Formally, the Luffa-mode of operation is defined as follows.

---

[1] $\mathsf{Xor} : \{0,1\}^{tn} \to \{0,1\}^n$ is defined as $\mathsf{Xor}(x^1\|\cdots\|x^t) = x^1 \oplus \cdots \oplus x^t$.

[2] $\mathsf{Chop}_s : \{0,1\}^l \to \{0,1\}^{l-s}$ is defined as $\mathsf{Chop}_s(m\|m') = m$ where $|m'| = s$.

– If $n_h \le n$,

$$\mathsf{Luffa}^{\Pi}(M) = \mathsf{Chop}_{n-n_h}(\mathsf{Xor}(\mathsf{MD}^{f^{\Pi}}(\mathsf{Pad}_{\mathsf{Lf}}(M)\|0^n))).$$

– If $n < n_h \le 2n$,

$$\mathsf{Luffa}^{\Pi}(M)$$
$$= \mathsf{Chop}_{2n-n_h}\Big(\mathsf{Xor}\Big(\mathsf{MD}^{f^{\Pi}}(\mathsf{Pad}_{\mathsf{Lf}}(M)\|0^n)\Big)\|\mathsf{Xor}\Big(\mathsf{MD}^{f^{\Pi}}(\mathsf{Pad}_{\mathsf{Lf}}(M)\|0^n\|0^n)\Big)\Big).$$

The Luffa-mode of operation uses a fixed initial value $IV_{\mathsf{Luffa}} = IV_1\|\cdots\|IV_t$ to use with the MD-mode of operation. In Luffa specification [7], the size of the permutations are always $n = 256$ bits. For $n_h = 224$ and $n_h = 256$ bit digest size, $t = 3$ permutations are used. For $n_h = 384$ and $n_h = 512$ bit digest size, there are $t = 4$ and $t = 5$ many permutations respectively.

The Message Injection function $\mathsf{MI}_{\mathsf{Lf}}$. The message injection functions can be represented by a matrix over a ring $GF(2^8)^{n/8}$ (Note, the input length of each permutation is n bits or 8 blocks of $n/8$-bits). The definition polynomial of the field is given by $\phi(x) = x^8 + x^4 + x^3 + x + 1$. The map from an 8-block value $(a_0, \cdots, a_7)$ to an element of the ring can be defined by $(\sum_{0\le k<8} a_{k,\ell}x^k)_{0\le \ell < n/8}$. The message injection function $\mathsf{MI}_{\mathsf{Lf}} : \{0,1\}^{(t+1)n} \to \{0,1\}^{tn}$ is defined as,

$$\mathsf{MI}_{\mathsf{Lf}}(h^1, \cdots, h^t, m) = [\mathsf{MI}_{\mathsf{Lf}}]_{t\times(t+1)} \cdot (h^1 \cdots h^t\ m)^T.$$

We will write $[\mathsf{MI}_{\mathsf{Lf}}]$ as $[TA]$, where $[T]$ is a $t \times t$ square matrix and A is a t-element column vector. In the specification of Luffa, some particular matrices $[\mathsf{MI}_{\mathsf{Lf}}]_{3\times4}$, $[\mathsf{MI}_{\mathsf{Lf}}]_{4\times5}$ and $[\mathsf{MI}_{\mathsf{Lf}}]_{5\times6}$ are defined. However our analysis holds whenever T is full rank (invertible) and each element of the column vector A has an inverse. For a detailed description of Luffa, the readers are referred to [7]. For simplicity, we will only consider the case $n = n_h$ in our security proofs. For other cases the same security bound can be derived in a similar but more involved manner.

LuffaS-mode of Operation. We also define a simpler version of Luffa-mode of operation, LuffaS or Luffa based on a single permutation. Here the function P is modeled as a single nt-bit random permutation π. The compression function of LuffaS, $f^{\pi} : \{0,1\}^{(t+1)n} \to \{0,1\}^{tn}$ is defined as, $f^{\pi} \equiv \pi \circ MI_{\mathsf{Luffa}}$ and in case of n bit digest we have

$$\mathsf{LuffaS}^{\pi}(M) = \mathsf{Xor}(\mathsf{MD}^{f^{\pi}}(\mathsf{Pad}_{\mathsf{Lf}}(M)\|0^n)).$$

Fugue Mode of Operation. The Fugue-mode of operation depends on $\Pi = (\pi_1, \pi_2)$ a pair of random permutations on $\{0,1\}^{nt}$. The compression function of Fugue, $g^{\pi_1} : \{0,1\}^{(n+1)t} \to \{0,1\}^{tn}$ is defined as, $g^{\pi_1} \equiv \pi_1 \circ \mathsf{MI}_{\mathsf{Fg}}$, where $\mathsf{MI}_{\mathsf{Fg}} : \{0,1\}^{(t+1)n} \to \{0,1\}^{tn}$ is a fixed linear transformation.

To process arbitrary length messages a suffix free padding rule $\mathsf{Pad}_{\mathsf{Fg}}$ is used. For our analysis it is sufficient assume that the padded message is multiple of n

bits and it is suffix free. The Fugue-mode of operation is nothing but MD-mode of operation based on the compression function g^{π_1} followed by applying the random permutation π_2 and finally Choping the output to the desired digest size. Formally, for n_h bit digest the Fugue-mode of operation is defined as,

$$\mathsf{Fugue}^{\Pi}(M) = \mathsf{Chop}_{tn-n_h}\Big(\pi_2\Big(\mathsf{MD}^{g^{\pi_1}}(\mathsf{Pad}_{\mathsf{Fg}}(M))\Big)\Big).$$

The Fugue-mode of operation uses a fixed initial value $IV_{\mathsf{Fugue}} = IV'_1\|\cdots\|IV'_t$ to use with the MD-mode of operation. Depending on parameters there are 3 different linear transformations $\mathsf{MI}_{\mathsf{Fg}}$ (TIX in Fugue specification [11]). In our security proofs we only handle the following one,

$$\mathsf{MI}_{\mathsf{Fg}}(h^1,\cdots,h^t,m) = (m, h^2 \oplus h^{t-5}, h^3, h^4, \cdots, \\ h^8, h^9 \oplus m, h^{10}, h^{11} \oplus h^1, h^{12}, h^{13}, \cdots, h^t). \quad (1)$$

For other cases, a similar security analysis holds.

Indifferentiability. The notion of indifferentiability, introduced by Maurer et. al. in [13], is a generalization of classical notion of indistinguishability. Loosely speaking, if an ideal primitive $\mathcal{G}$ is indifferentiable with a construction C based on another ideal primitive $\mathcal{F}$, then $\mathcal{G}$ can be safely replaced by $C^{\mathcal{F}}$ in any cryptographic construction. In other terms if a cryptographic construction is secure in $\mathcal{G}$ model then it is secure in $\mathcal{F}$ model.

Definition 3. Advantage:
Let F_i, G_i be probabilistic oracle algorithms. We define advantage *of the adversary $\mathcal{A}$ at distinguishing (F_1, F_2) from (G_1, G_2) as*

$$\mathsf{Adv}_{\mathcal{A}}((F_1,F_2),(G_1,G_2)) = |\Pr[\mathcal{A}^{F_1,F_2} = 1] - \Pr[\mathcal{A}^{G_1,G_2} = 1]|.$$

Definition 4. Indifferentiability [13]:
A Turing machine C with oracle access to an ideal primitive $\mathcal{F}$ is said to be $(t, q_C, q_{\mathcal{F}}, \varepsilon)$ indifferentiable from an ideal primitive $\mathcal{G}$ if there exists a simulator S with an oracle access to $\mathcal{G}$ and running time at most t, such that for any adversary $\mathcal{A}$, it holds that

$$\mathsf{Adv}_D((C^{\mathcal{F}},\mathcal{F}),(\mathcal{G},S^{\mathcal{G}})) < \varepsilon.$$

The adversary makes at most q_C queries to C or $\mathcal{G}$ and at most $q_{\mathcal{F}}$ queries to $\mathcal{F}$ or S. Similarly, $C^{\mathcal{F}}$ is said to be (computationally) indifferentiable from $\mathcal{G}$ if running time of $\mathcal{A}$ is bounded above by some polynomial in the security parameter k and ε is a negligible function of k.

We stress that in the above definition $\mathcal{G}$ and $\mathcal{F}$ can be two completely different primitives. As shown in Fig 1(a) the role of the simulator is to not only simulate the behavior of $\mathcal{F}$ but also remain consistent with the behavior of $\mathcal{G}$. Note that,

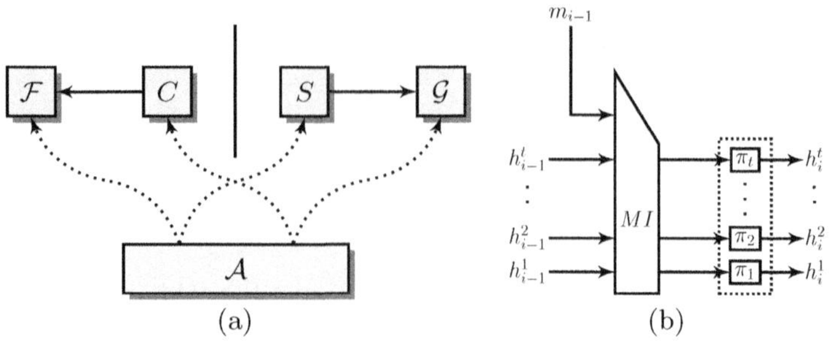

Fig. 1. (a) The indifferentiability notion and (b) The Luffa compression function

the simulator does not know the queries made directly to $\mathcal{G}$, although it can query $\mathcal{G}$ whenever it needs. In this paper $\mathcal{G}$ is a variable input length Random oracle and $\mathcal{F}$ is a random permutation. As the objective of any adversary is to build a distinguisher , we use the term adversary and distinguisher interchangeably for the rest of the paper.

3 Main Tools for Bounding Distinguisher's Advantage

We follow a similar approach to [5,8,9] for proving indifferentiability of Fugue and Luffa. In the discussion below C-mode of operation usually refers to either Fugue or Luffa mode of operation based on $\Pi = (\pi_1, \cdots, \pi_t)$ a tuple of t random permutations. In other words, $C \in \{\text{Fugue}, \text{Luffa}\}$. In case of Fugue, we have $t = 2$. Below we recall some notations from [5].

Definition 5. Consistent Oracle:
A (small domain) probabilistic oracle algorithm G_2 is said to be Consistent to a (big domain) probabilistic oracle algorithm G_1 with respect to MO-mode of operation if for any point x (from the big domain), we have

$$\Pr[G_1(x) = MO^{G_2}(x)] = 1.$$

Note, Π is always Consistent to C^{Π}-mode of operation.

There might be some point x for which the value of $MO^{G_2}(x)$ gets fixed by the relations $G_2(x_1) = y_1, \cdots, G_2(x_q) = y_q$. Such x's are called evaluatable by the relations $G_2(x_1) = y_1, \cdots, G_2(x_q) = y_q$. Formally,

Definition 6. Evaluatable Queries:
A point $x \in Domain(MO^{G_2})$ is called evaluatable *with respect to MO-mode of operation (based on G_2) by the relations $G_2(x_1) = y_1, \cdots, G_2(x_q) = y_q$, if there exist a deterministic algorithm $\mathcal{B}$ such that,*

$$\Pr[MO^{G_2}(x) = \mathcal{B}(x, (x_1, y_1), \cdots, (x_q, y_q)) | G_2(x_1) = y_1, \cdots, G_2(x_q) = y_q] = 1.$$

In this paper the adversary is modeled as a deterministic, computationally unbounded [3] adversary $\mathcal{A}$ which has access to two oracles $\mathcal{O}_1$ and $\mathcal{O}_2$. Recall that $\mathcal{A}$ tries to distinguish the output distribution of (C^{Π}, Π) from that of (R, S^R). We say $\mathcal{A}$ queries $\mathcal{O}_1$ when it queries the oracle C^{π} or R and queries $\mathcal{O}_2$ when it queries the oracle Π or S^R. As we model Π as a tuple of t random permutations, the adversary has to mention the index of the random permutation it wants to query and whether she wants to make forward or inverse query. The forward and inverse query to the ith random permutation are denoted by $\mathcal{O}_2(+i,\cdot)$ and $\mathcal{O}_2(-i,\cdot)$ respectively.

Definition 7. Distinguisher View:
The view $\mathcal{V}$ of the distinguisher is the query-response tuple

$$((M_1,h_1),\cdots,(M_{q_0},h_{q_0}),(x_1^1,y_1^1),\cdots,(x^1_{q^1+q^{-1}},y^1_{q^1+q^{-1}}),\cdots,(x^t_{q^t+q^{-t}},y^t_{q^t+q^{-t}})), \tag{2}$$

where

- $\mathcal{O}_1(M_1)=h_1,\cdots,\mathcal{O}_1(M_{q_0})=h_{q_0}$
- $\mathcal{O}_2(+1,x_1^1)=y_1^1,\cdots,\mathcal{O}_2(+1,x^1_{q^1})=y^1_{q^1}$
- $\mathcal{O}_2(-1,y^1_{q^1+1})=x^1_{q^1+1},\cdots,\mathcal{O}_2(-1,y^1_{q^1+q^{-1}})=x^1_{q^1+q^{-1}}$

$\vdots$

- $\mathcal{O}_2(+t,x_1^t)=y_1^t,\cdots,\mathcal{O}_2(+t,x^t_{q^t})=y^t_{q^t}$
- $\mathcal{O}_2(-t,y^t_{q^t+1})=x^t_{q^t+1},\cdots,\mathcal{O}_2(-t,y^t_{q^t+q^{-t}})=x^t_{q^t+q^{-t}}$.

Definition 8. Input Output View:
For any view $\mathcal{V}$ as in (2), we define Input View $\mathcal{IV}$ and Output View $\mathcal{OV}$ as follows,

$$\mathcal{IV}=(M_1,\cdots,M_{q_0},x_1^1,\cdots,x^1_{q^1},y^1_{q^1+1},\cdots,y^1_{q^1+q^{-1}},\cdots,\\ x_1^t,\cdots,x^t_{q^t},y^t_{q^t+1},\cdots,y^t_{q^t+q^{-t}}),$$
$$\mathcal{OV}=(h_1,\cdots,h_{q_0},y_1^1,\cdots,y^1_{q^1},x^1_{q^1+1},\cdots,x^1_{q^1+q^{-1}},\cdots,\\ y_1^t,\cdots,y^t_{q^t},x^t_{q^t+1},\cdots,x^t_{q^t+q^{-t}}).$$

We recall the observations made in [5].

1. $\mathcal{V}$, $\mathcal{IV}$ and $\mathcal{OV}$ are actually ordered tuples. That means, the position of any element inside the tuple actually denotes the corresponding query number. So, in general $\mathcal{O}_1(\cdot)$, $\mathcal{O}_2(+i,\cdot)$ and $\mathcal{O}_2(-i,\cdot)$ queries should not be grouped together. But we write it like this to avoid further notational complexity.
2. For any deterministic *non-adaptive* attacker $\mathcal{IV}$ is always fixed.
3. For any deterministic *adaptive* attacker $\mathcal{IV}$ is actually determined by $\mathcal{OV}$ [14].

[3] Any deterministic adversary with unlimited resources is as powerful as a randomized adversary [14].

4. For any deterministic attacker (adaptive or non-adaptive) $\mathcal{V}$ is actually determined by $\mathcal{OV}$.

$\mathcal{OV}^{\mathcal{A}}_{\mathcal{O}_1,\mathcal{O}_2}$ be the random variable corresponding to the output view of attacker $\mathcal{A}$, obtained after interacting with $\mathcal{O}_1$, $\mathcal{O}_2$.

Definition 9. Consistent Output View:
For an attacker $\mathcal{A}$ an output view $\mathcal{OV}$ is said to be Consistent *with respect to $(\mathcal{O}_1, \mathcal{O}_2)$ if $\Pr[\mathcal{OV}^{\mathcal{A}}_{\mathcal{O}_1,\mathcal{O}_2} = \mathcal{OV}] > 0$. If an output view is not Consistent, then it is an* Inconsistent *output view.*

Definition 10. Irreducible (Output) View:

$$\mathcal{V} = ((M_1, h_1), \cdots, (M_{q_0}, h_{q_0}), (x^1_1, y^1_1), \cdots, (x^1_{q^1+q^{-1}}, y^1_{q^1+q^{-1}}), \cdots, \\ (x^t_1, y^t_1), \cdots, (x^t_{q^t+q^{-t}}, y^t_{q^t+q^{-t}})) \quad (3)$$

*is called C-*Irreducible *view if,*

1. *$M_1, \cdots, M_{q_0}$ are distinct,*
2. *All the elements of $\mathcal{X} = \{x^i_1, \cdots, x^i_{q^1+q^{-1}}\}$ are distinct and all the elements of $\mathcal{Y} = \{y^1_1, \cdots, y^1_{q^1+q^{-1}}\}$ are distinct.*
3. *$M_1, \cdots, M_{q_0}$ are not evaluatable by the relations,*

$$\pi_1(x^1_1) = y^1_1, \cdots, \pi_1(x^1_{q^1+q^{-1}}) = y^1_{q^1+q^{-1}}, \cdots, \\ \pi_1(x^t_1) = y^t_1, \cdots, \pi_1(x^t_{q^t+q^{-t}}) = y^t_{q^t+q^{-t}} \quad (4)$$

with respect to C-mode of operation.

For an attacker $\mathcal{A}$, an output view $\mathcal{OV}$ is called Irreducible if the corresponding view $\mathcal{V}$ is Irreducible.

Theorem 1. *[5] If there exists a simulator S^R (aborting with a probability at most ε') Consistent to a random oracle R, with respect to C-mode of operation, such that for any attacker $\mathcal{A}$ making at most q queries , the relation*

$$\Pr[\mathcal{OV}^{\mathcal{A}}_{C^{\Pi},\Pi} = \mathcal{OV}] \geq (1-\varepsilon)\Pr[\mathcal{OV}^{\mathcal{A}}_{R,S^R} = \mathcal{OV}],$$

holds for all possible Consistent (with respect to the oracles C^{Π}, Π) and Irreducible output views $\mathcal{OV}$; then for any attacker $\mathcal{A}$ making at most q queries we have

$$\mathsf{Adv}_{\mathcal{A}}((C^{\Pi}, \Pi), (R, S^R)) \leq \varepsilon + \varepsilon'.$$

4 Indifferentiability Security Analysis of Fugue

In this section we show the Fugue-mode of operation is indifferentiable from a random function R. If one could show the compression function of Fugue, g^{π_1} is indifferentiable from a random oracle, then he can try to apply the results of

[10] to show Fugue is indifferntiable from a random oracle. However, π_1 being a random permutation attacker has access to inverse queries and g^{π_1} is not really indifferentiable from a random oracle. Hence, to prove indifferentiability security of Fugue we adopt the direct approach as outlined in Section 3. Our main result of this section is the following.

Theorem 2. *Let $\Pi = (\pi_1, \pi_2)$ be a pair of independent random permutations over $\{0,1\}^{nt}$. Let Fugue^{Π} be the* Fugue *mode of operation with n_h bit digest. There exists a simulator $S^{\mathcal{R}}$ such that for any adversary $\mathcal{A}$ which makes at most q queries*

$$\mathsf{Adv}_{\mathcal{A}}((\mathsf{Fugue}^{\Pi}, \Pi), (\mathcal{R}, S^{\mathcal{R}})) \leq \frac{\mathcal{O}(q^2 + q\sigma + \sigma^2)}{2^{n(t-1)}} + \frac{\mathcal{O}(q^2)}{2^{nt-n_h}}$$

where σ is the total number of blocks in the queries to R or Fugue^{Π}.

To start with, we build a simulator S^R Consistent to the random oracle R, with respect to Fugue-mode of operation.

Simulator of Fugue. The simulator maintains two partial permutations $\pi_1^*, \pi_2^* : \{0,1\}^{nt} \to \{0,1\}^{nt}$ initially empty and a (weighted) directed graph (in fact a tree) $G = (V, E)$. Each element of the vertex set V is an nt-bit element. V is initialized to IV_{Fugue} and the edge set E is initially empty. Whenever the simulator answers $S^R(+i, x)$ query as y or $S^R(-i, y)$ query as x it updates the partial permutation π_i^*, as $\pi_i^*(x) = y$. There exists an edge v_1 to v_2 with weight m in the edge set (or $v_1 \xrightarrow{m} v_2 \in E$) if and only if

$$g^{\pi_1^*}(v_1 \| m) = v_2.$$

While, answering a query the simulator also need to keep track of already defined input/output points. At any instance, $\mathcal{X}(\pi_i^*)$ and $\mathcal{Y}(\pi_i^*)$ be the set of already defined input and output points of the partial permutation π_i^* respectively. The simulator always answers such a way such that π_i^* behaves as a random (partial) permutation and G is a tree with root node as IV_{Fugue}. In fact, if there is a path in G from IV_{Fugue} to v with weights (messages) $m_1, \cdots, m_k$ that would imply

$$\mathsf{MD}^{g^{\pi_1^*}}(m_1 \| \cdots \| m_k) = v.$$

We recall that Fugue-mode of operation is nothing but processing the output of $\mathsf{MD}^{g^{\pi_1}}(\mathsf{Pad}_{\mathsf{Fg}}(\cdot))$ through π_2 and chopping $nt - n_h$ many bits of π_2 output to generate n_h bit digest size. Hence to remain Consistent with R, if $m_1 \| \cdots \| m_k$ is an appropriate padded message (corresponding to message M) the simulator should make sure

$$\mathsf{Chop}_{nt-n_h}(\pi_2^*(v)) = R(M).$$

The simulator needs to be careful about two things. Firstly, the tree structure of the graph is always maintained and secondly, whenever any new node v gets added to the tree the partial permutation π_2^* should not be already defined at point v. When the simulator fails to do so, it outputs $\perp$ to abort. The simulator only allows the creation of a new node during $S^R(+1, x)$ queries. Note, in this case whether a new node would be created only depends on x the query input. This can be checked easily, for each $v \in V$ find out whether there exists a message m such that

$$\mathsf{MI}_{\mathsf{Fg}}(v \| m) = x.$$

Also, a single $S^R(+1, x)$ query should not be able to create more than one new node.

$\boldsymbol{S^R(+1, x)}$ **query:** When $x \notin \mathcal{X}(\pi_1^*)$ the simulator samples y from $\{0,1\}^{tn} \setminus \mathcal{Y}(\pi_1^*)$ uniformly. The simulator aborts, when either of the following conditions get violated. Otherwise it updates the partial permutation π_1^* and returns y, if y is a new node it also updates the graph G accordingly.

- $\mathbf{Bad}_{\mathsf{Fg}}^{+11}$: y is a new node and there exist $m_1, m_2 \in \{0,1\}^n$, $v \in V$ such that $\mathsf{MI}_{\mathsf{Fg}}(y \| m_1) = \mathsf{MI}_{\mathsf{Fg}}(v \| m_2)$.
- $\mathbf{Bad}_{\mathsf{Fg}}^{+12}$: y is a new node and there exist $m \in \{0,1\}^n$, $z \in \mathcal{X}(\pi_1^*)$ such that $\mathsf{MI}_{\mathsf{Fg}}(y \| m) = z$.
- $\mathbf{Bad}_{\mathsf{Fg}}^{+13}$: y is a new node, the path from IV_{Fugue} leading to y is a valid padded message and $y \in \mathcal{X}(\pi_2^*)$.

$\boldsymbol{S^R(-1, y)}$ **query:** When $y \notin \mathcal{Y}(\pi_1^*)$ the simulator samples x from $\{0,1\}^{tn} \setminus \mathcal{X}(\pi_1^*)$ uniformly. The simulator aborts when the following condition gets violated. Otherwise it updates the partial permutation π_1^* and returns x.

- $\mathbf{Bad}_{\mathsf{Fg}}^{-11}$: There exist $m \in \{0,1\}^n$ and $v \in V$ such that $\mathsf{MI}_{\mathsf{Fg}}(v \| m) = x$.

$\boldsymbol{S^R(+2, x)}$ **query:** When $x \notin \mathcal{X}(\pi_2^*)$, at first the simulator checks whether $x \in V$ or not. If $x \notin V$ or the path from IV_{Fugue} leading to x is not a valid padded message the simulator returns y sampled uniformly from $\{0,1\}^{tn} \setminus \mathcal{Y}(\pi_2^*)$ and updates π_2^* accordingly. In case $\mathsf{Pad}_{\mathsf{Fg}}(M)$, a valid padded message, is the path from IV_{Fugue} to x then the simulator aborts if the following condition holds. Otherwise, it returns $R(M)$ and updates π_2^* accordingly.

- $\mathbf{Bad}_{\mathsf{Fugue}}^{+21}$: $R(M) \in \mathcal{Y}(\pi_2^*)$.

$\boldsymbol{S^R(-2, y)}$ **query:** When $y \notin \mathcal{Y}(\pi_2^*)$, the simulator samples $x \in \{0,1\}^{tn} \setminus \mathcal{X}(\pi_2^*)$ uniformly. It aborts if the following condition gets violated, otherwise it returns x and updates the partial permutation π_2^*.

- $\mathbf{Bad}_{\mathsf{Fg}}^{-21}$: There exists $v \in V$ such that $x = v$.

Upper bound on abort probability. Let q_1 be the upper bound on number of S^R queries. We have also seen the number of nodes in the graph G can increase by at most 1 for every S^R query. Hence, before answering any S^R query the graph G can contain only q_1 many nodes. Below we make a few observation about the function $\mathsf{MI}_{\mathsf{Fg}}$ as in (1).

1. $\mathsf{MI}_{\mathsf{Fg}}(v_1\|m_1) = \mathsf{MI}_{\mathsf{Fg}}(v_2\|m_2)$ implies $m_1 = m_2$.
2. For any $v \in \{0,1\}^{nt}$, there exist at most 2^n many $v' \in \{0,1\}^{nt}$ such that the relation $\mathsf{MI}_{\mathsf{Fg}}(v\|m) = \mathsf{MI}_{\mathsf{Fg}}(v'\|m)$, holds for some $m \in \{0,1\}^n$.
3. Observations 1 and 2 actually imply for any $v \in \{0,1\}^{nt}$ there exist at most 2^n many $v' \in \{0,1\}^{nt}$ such that the relation $\mathsf{MI}_{\mathsf{Fg}}(v\|m) = \mathsf{MI}_{\mathsf{Fg}}(v'\|m')$, holds for some $m, m' \in \{0,1\}^n$.
4. For any $z \in \{0,1\}^{nt}$ there exist at most 2^n many $v \in \{0,1\}^{nt}$ such that the relation $\mathsf{MI}_{\mathsf{Fg}}(v\|m) = z$, holds for some $m \in \{0,1\}^n$.

Utilizing the above observations one can deduce the following upper bounds on $\Pr[\mathbf{Bad}_{\mathsf{Fg}}^{\pm ij}]$, during a single S^R query.
$\Pr[\mathbf{Bad}_{\mathsf{Fg}}^{+11}] \le \frac{2^n q}{2^{nt}-q}$ $\Pr[\mathbf{Bad}_{\mathsf{Fg}}^{+12}] \le \frac{2^n q_1}{2^{nt}-q}$ $\Pr[\mathbf{Bad}_{\mathsf{Fg}}^{+13}] \le \frac{q}{2^{nt}-q}$
$\Pr[\mathbf{Bad}_{\mathsf{Fg}}^{-11}] \le \frac{2^n q_1}{2^{nt}-q}$ $\Pr[\mathbf{Bad}_{\mathsf{Fg}}^{+21}] \le \frac{q}{2^{nt}}$ $\Pr[\mathbf{Bad}_{\mathsf{Fg}}^{-21}] \le \frac{q}{2^{nt}-q}$
Note, the upper bound on $\Pr[\mathbf{Bad}_{\mathsf{Fg}}^{+21}]$ is derived assuming there is no chopping. When there is no chopping the simulator aborts when $R(M) \in \mathcal{Y}(\pi_2^*)$, otherwise it returns $R(M)$. Hence, the probability of abort is exactly $\frac{|\mathcal{Y}(\pi_2^*)|}{2^{nt}}$. In case $n - n_h$ bits are chopped, the abort probability is even less. Hence, if Abort is the event that the simulator aborts in any one of the q queries made by the attacker, assuming $q \le 2^{nt-1}$ we have

$$\Pr[\mathsf{Abort}] \le \frac{6q^2}{2^{n(t-1)}}.$$

Interpolation Probability of $\mathcal{OV}^{\mathcal{A}}_{(R,S^R)}$. There is always an one to one mapping between any (output) view and the simulators internal state (that is the graph G and the partial permutations π_1^*, π_2^*). Hence any Irreducible output view $\mathcal{OV}$, either implies that the simulator should abort, in which case $\Pr[\mathcal{OV}^{\mathcal{A}}_{(R,S^R)} = \mathcal{OV}] = 0$, or the simulator do not abort, in which case we have

$$\Pr[\mathcal{OV}^{\mathcal{A}}_{(R,S^R)} = \mathcal{OV}] = \frac{1}{2^{n_h q_0}} \prod_{i=1}^{q^1+q^{-1}} \prod_{j=1}^{q^2+q^{-2}} \frac{1}{(2^{nt}-i+1)(2^{nt}-j+1)}.$$

Hence, for any Consistent output view $\mathcal{OV}$ (with respect to Fugue^{Π}, Π oracles) we have

$$\Pr[\mathcal{OV}^{\mathcal{A}}_{(R,S^R)} = \mathcal{OV}] \le \frac{1}{2^{n_h q_0}} \prod_{i=1}^{q^1+q^{-1}} \prod_{j=1}^{q^2+q^{-2}} \frac{1}{(2^{nt}-i+1)(2^{nt}-j+1)}.$$

Also for any, Inconsistent output view $\mathcal{OV}$ (with respect to Fugue^{Π}, Π oracles) we have $\Pr[\mathcal{OV}^{\mathcal{A}}_{(R,S^R)} = \mathcal{OV}] = 0$.

Interpolation Probability of $\mathcal{OV}^{\mathcal{A}}_{(\mathsf{Fugue}^{\Pi},\Pi)}$. Here we aim to give a lower bound on $\Pr[\mathcal{OV}^{\mathcal{A}}_{(\mathsf{Fugue}^{\Pi},\Pi)} = \mathcal{OV}]$ for any Consistent Irreducible output view $\mathcal{OV}$.

For that we will consider the notion of *Good* NCFugue-Irreducible views, which is similar to the notion of MD-Irreducible views in [5,9]. Informally NCFugue-Irreducible view is an Irreducible attacker view where the attacker has access to $\mathsf{NCFugue}^{\Pi}(\cdot) \equiv \pi_2(\mathsf{MD}^{g^{\pi_1}}(\mathsf{Pad}_{\mathsf{Fg}}(\cdot)))$ oracle instead of $\mathsf{Fugue}^{\Pi}(\cdot) \equiv \mathsf{Chop}_{nt-n_h}(\pi_2(\mathsf{MD}^{g^{\pi_1}}(\mathsf{Pad}_{\mathsf{Fg}}(\cdot))))$ oracle (So the attacker essentially receives more information. However, the attacker does not use this extra information to decide its future queries). We will obtain the lower bound in two steps.

- For any Consistent *Good* NCFugue-Irreducible output view $\mathcal{OV}_{\mathsf{NC}}$ obtain a lower bound of $\Pr[\mathcal{OV}^{\mathcal{A}'}_{(\mathsf{NCFugue}^{\Pi},\Pi)} = \mathcal{OV}_{\mathsf{NC}}]$ as p (Here, $\mathcal{A}'$ is the modified attacker which has access to $\mathsf{NCFugue}^{\Pi}$ oracle instead of Fugue^{Π} oracle.)
- For any Consistent Fugue-Irreducible output view $\mathcal{OV}$ we give a lower bound (as N) on number of possible Good NCFugue-Irreducible output view $\mathcal{OV}_{\mathsf{NC}}$ such that,

$$\Pr[\mathcal{OV}^{\mathcal{A}}_{(\mathsf{NCFugue}^{\Pi},\Pi)} = \mathcal{OV} | \mathcal{OV}^{\mathcal{A}'}_{(\mathsf{MDFugue}^{\Pi},\Pi)} = \mathcal{OV}_{\mathsf{NC}}] = 1 \qquad (5)$$

This would imply, $\Pr[\mathcal{OV}^{\mathcal{A}}_{(\mathsf{Fugue}^{\Pi},\Pi)} = \mathcal{OV}] \geq pN$. Before going further we define the notion Good NCFugue-Irreducible output view.

Definition 11. *For an attacker $\mathcal{A}'$, interacting with $(\mathsf{NCFugue}^{\Pi}, \Pi)$ the view*

$$\mathcal{V}_{\mathsf{NC}} = ((M_1, r_1), \cdots, (M_{q_0}, r_{q_0}), (x^1_1, y^1_1), \cdots, (x^1_{q^1+q^{-1}}, y^1_{q^1+q^{-1}}), \cdots, \\ (x^2_1, y^2_1), \cdots, (x^2_{q^2+q^{-2}}, y^2_{q^2+q^{-2}})) \qquad (6)$$

is a Good NCFugue-*Irreducible view if $\mathcal{V}_{\mathsf{NC}}$ is an* NCFugue-*Irreducible view, $r_1, \cdots, r_{q_0}$ are all different and they are different from $y^2_1, \cdots, y^2_{q^2+q^{-2}}$. An output view $\mathcal{OV}_{\mathsf{NC}}$ is called Good* NCFugue-*Irreducible output view, if the corresponding view $\mathcal{V}_{\mathsf{NC}}$ is Good* NCFugue-*Irreducible.*

By the following theorem, we get the lower bound on $\Pr[\mathcal{OV}^{\mathcal{A}'}_{(\mathsf{NCFugue}^{\Pi},\Pi)} = \mathcal{OV}_{\mathsf{NC}}]$ or estimate of p.

Theorem 3. *For any Consistent Good* NCFugue-*Irreducible output view $\mathcal{OV}_{\mathsf{NC}}$ ($\mathcal{V}_{\mathsf{NC}}$ as in (6) being the corresponding view),*

$$\Pr[\mathcal{OV}^{\mathcal{A}'}_{(\mathsf{NCFugue}^{\Pi},\Pi)} = \mathcal{OV}_{\mathsf{NC}}] \\ \geq (1 - \frac{2\sigma(q+\sigma)}{2^{(t-1)n}}) \frac{1}{2^{ntq_0}} \prod_{i=1}^{q^1+q^{-1}} \prod_{j=1}^{q^2+q^{-2}} \frac{1}{(2^{nt}-i+1)(2^{nt}-j+1)},$$

where σ is the total number of message blocks present in NCFugue *queries, q is the total number of* NCFugue *and Π queries such that $q + \sigma < 2^{nt-1}$.*

In the following theorem we get an estimate of N.

Theorem 4. *For any Consistent* Fugue-*Irreducible view* $\mathcal{V}$ *as in* (3) *(with* $t = 2$*), there exist at least*

$$2^{(nt-n_h)q_0}\Big(1 - \frac{q^2}{2^{nt-n_h}}\Big)$$

many Consistent Good NCFugue-*Irreducible view* $\mathcal{V}_{\mathsf{NC}}$ *as in* (6) *such that,* $\mathsf{Chop}_{nt-n_h}(r_1) = h_1, \cdots, \mathsf{Chop}_{nt-n_h}(r_{q_0}) = h_{q_0}$*, where* q *is the total number of* Fugue^{Π} *and* Π *queries.*

All together Theorem 3, Theorem 4 and the upper bound from Section 4 would imply for any Consistent output view $\mathcal{OV}$ (with respect to $\mathsf{Fugue}^{\Pi}, \Pi$ oracles)

$$\begin{aligned}\Pr[\mathcal{OV}^{\mathcal{A}}_{(\mathsf{Fugue}^{\Pi},\Pi)} = \mathcal{OV}] &\geq \Big(1 - \frac{2\sigma(q+\sigma)}{2^{(t-1)n}}\Big)\Big(1 - \frac{q^2}{2^{nt-n_h}}\Big) \times \frac{1}{2^{n_h q_0}} \\ &\quad \times \prod_{i=1}^{q^1+q^{-1}} \prod_{j=1}^{q^2+q^{-2}} \frac{1}{(2^{nt}-i+1)(2^{nt}-j+1)} \\ &\geq \Big(1 - \frac{2\sigma(q+\sigma)}{2^{(t-1)n}}\Big)\Big(1 - \frac{q^2}{2^{nt-n_h}}\Big) \Pr[\mathcal{OV}^{\mathcal{A}}_{(R,S^R)} = \mathcal{OV}]\end{aligned}$$

Applying the above inequality together with Pr[Abort] to Theorem 1 we prove Theorem 2.

5 Indifferentiability Security Analysis of LuffaS

We recall that Luffa with single permutation or LuffaS is based on a random permutation $\pi : \{0,1\}^{tn} \to \{0,1\}^{tn}$. We view the the Message Insertion transformation $\mathsf{MI}_{\mathsf{Lf}}(Y\|m)$ as $TY + Am$ where $Y \in (\{0,1\}^n)^t, m \in \{0,1\}^n$. We view the chaining value Y as a vector of t many n bit elements. By "forward query input" we mean the input of forward query or the output of inverse query made to the simulator. Our main result of this section is the following.

Theorem 5. *Let* π *be an independent random permutations over* $\{0,1\}^{nt}$*. Let* LuffaS^{Π} *be the* LuffaS *mode of operation with* n *bit message blocks. There exists a simulator* $S^{\mathcal{R}}$ *such that for any adversary* $\mathcal{A}$ *which makes at most* q *queries*

$$\mathsf{Adv}_{\mathcal{A}}((\mathsf{LuffaS}^{\pi}, \pi), (\mathcal{R}, S^{\mathcal{R}})) \leq \frac{\mathcal{O}(\sigma(q+\sigma))}{2^{nt}} + \frac{\mathcal{O}(q^2)}{2^{n(t-2)}}$$

where σ *is the total number of blocks in the queries to* R *or* LuffaS^{π}*.*

As in Section 4 we start by describing a Consistent Simulator.

Simulator of LuffaS. The simulator maintains a partial permutation $\pi^* : \{0,1\}^{nt} \to \{0,1\}^{nt}$ to keep track of the all the received queries and responses it made so far. By $\mathcal{X}(\pi^*)$ and $\mathcal{Y}(\pi^*)$ we denote the set of forward query input and the set of forward query output so far. It also maintains a directed graph G whose vertex set $V = \{0,1\}^{tn}$. These vertices are actually the output of the simulator. An edge from vertex Y to vertex Y', marked by m, implies that $(TY + Am, Y') \in \pi^*$. Initially π^* is empty and $V = \{IV_{\mathsf{Luffa}}\}$. The objective of the simulator is to maintain following properties in the graph

- **No cycle:** the graph G is actually a tree, hence there is no cycle in the graph.
- **No collision:** For any three nodes Y_1, Y_2 and Y_3 in the graph, both (Y_1, Y_3) and (Y_2, Y_3) are not edges in G.

In addition to the properties of the tree, to guard against the *length-extension attack*, the simulator also tries to ensure that the final input of any path from IV_{Luffa} is not in $\mathcal{X}(\pi^*)$ yet.

Handling forward query $S^{\mathcal{R}}(+, X)$. If $X \notin \mathcal{X}(\pi^*)$,simulator checks whether for any existing node Y' and a message m, $\mathsf{MI}_{\mathsf{Lf}}(Y', m) = X$. If such a node exist for $m = 0^n$, then the simulator, retrieves M from labels of the edges of path from IV_{Luffa} to Y' and queries $\mathcal{R}(M)$. Then queries the simulator samples $Y = (y_1, \cdots, y_t)$ uniformly at random from $\{0,1\}^{nt} \setminus \mathcal{Y}(\pi^*)$ conditioned on $y_1 + \cdots + y_t = R(M)$. If $m \neq O^n$, the simulator samples $Y \leftarrow \{0,1\}^{tn} \setminus \mathcal{Y}(\pi^*)$ uniformly at random. In both the cases, a node Y is created and an edge (Y', Y) with label m is added. If no such Y' is found, simulator just the simulator samples $Y \leftarrow \{0,1\}^{tn} \setminus \mathcal{Y}(\pi^*)$ uniformly at random. If Y sets one of the following **Bad** events true, the simulator aborts. Otherwise, the simulator updates the partial permutation π^* inserting (X, Y) and returns Y as response.

We define the following events as **Bad** events for the simulator.

- $\mathbf{Bad}_{\mathsf{Lf}}^{+1}$ (Possible Collision): Let Y be the new node and Y_2 be any other node of the graph. $T(Y + Y_2) = Am$ has a solution for m.
- $\mathbf{Bad}_{\mathsf{Lf}}^{+2}$: Let Y be the new node. If, for some m, $TY + Am \in \mathcal{X}(\pi^*)$.

$\mathbf{Bad}_{\mathsf{Lf}}^{+2}$ ensures that there is no cycle in the graph and the final input (input to the finalization) for any properly padded message M is not yet in $\mathcal{X}(\pi^*)$.

Handling inverse query $S^{\mathcal{R}}(-, Y)$. While handling inverse queries, the simulator has no control over the input. The adversary might choose the inverse query Y in such a way that for some node Y' in the graph and some m, m', $TY + Am = TY' + Am'$. However, as long as the output of the inverse query do not make a new chain from IV, and the simulator remains Consistent with $\mathcal{R}$ with respect to the mode. Hence, while answering an inverse query the simulator needs to make sure that for no existing node Y' and any message m, $TY' + Am$ matches with the output. This condition together with $\mathbf{Bad}_{\mathsf{Lf}}^{+1}$ and $\mathbf{Bad}_{\mathsf{Lf}}^{+2}$ keeps the simulator Consistent with $\mathcal{R}$. So, if $Y \notin \mathcal{Y}(\pi^*)$, the simulator samples $X \leftarrow \{0,1\}^{tn} \setminus \mathcal{X}(\pi^*)$ and check for the following **Bad** event.

- $\mathbf{Bad}_{\mathsf{Lf}}^{-1}$: For some vertex Y' in the graph and some message $m \in \{0,1\}^n$, $\mathsf{MI}_{\mathsf{Lf}}(Y', m) = X$.

If the **Bad** is true then the simulator aborts. Otherwise, it adds (X, Y) to π^* and returns X.

Lemma 1. *Let $S^{\mathcal{R}}$ be the simulator for* Luffa *mode of operation with single permutation. For any attacker $\mathcal{A}$ making at most $q \leq 2^n$ queries*

$$\Pr[\mathsf{Abort}] \text{ or } \Pr[S^{\mathcal{R}} \rightarrow \perp] \leq \frac{\mathcal{O}(q^2)}{2^{(t-2)n}}.$$

Finally, similar to Section 4 for any Consistent output view $\mathcal{OV}$ (with respect to $\mathsf{LuffaS}^{\pi}, \pi$ oracles) we have

$$\Pr[\mathcal{OV}^{\mathcal{A}}_{(R,S^R)} = \mathcal{OV}] \le \frac{1}{2^{nq_0}} \prod_{i=1}^{q^1+q^{-1}} \frac{1}{(2^{nt} - i + 1)}.$$

Interpolation Probability of $\mathcal{OV}^{\mathcal{A}}_{(\mathsf{LuffaS}^{\pi},\pi)}$. In this section we aim obtain a lower bound on $\Pr[\mathcal{OV}^{\mathcal{A}}_{(\mathsf{LuffaS}^{\pi},\pi)} = \mathcal{OV}]$ for any Irreducible Consistent output view $\mathcal{OV}$. Similar to Section 4 we consider the notion of *Good* NXLuffaS-Irreducible views. Informally, an NXLuffaS-Irreducible view is an Irreducible view where the attacker has access to $\mathsf{NXLuffaS}^{\pi}(\cdot) \equiv \mathsf{MD}^{f^{\pi}}(\mathsf{Pad}_{\mathsf{Lf}}(\cdot)\|0^n)$ oracle instead of $\mathsf{LuffaS}^{\pi}(\cdot) = \mathsf{Xor}(\mathsf{MD}^{f^{\pi}}(\cdot)\|0^n)$ oracle. As before, $\mathcal{A}'$ is the modified attacker which has access to $\mathsf{NXLuffaS}^{\pi}$ oracle instead of LuffaS^{π} oracle. Our strategy is same as before, i.e. for any Consistent Good NXLuffaS-Irreducible output view $\mathcal{OV}_{\mathsf{NX}}$ obtain a lower bound on $\Pr[\mathcal{OV}^{\mathcal{A}}_{(\mathsf{NXLuffaS}^{\pi},\pi)} = \mathcal{OV}_{\mathsf{NX}}]$ and for any Consistent LuffaS-Irreducible view $\mathcal{V}$ obtain a lower bound on number of corresponding Good Consistent NXLuffaS-Irreducible view $\mathcal{V}_{\mathsf{NX}}$.

Definition 12. *For an attacker $\mathcal{A}'$, interacting with* $(\mathsf{NXLuffaS}^{\pi}, \pi)$, *the view*

$$\mathcal{V}_{\mathsf{NX}} = ((M_1, g_1), \cdots, (M_{q_0}, g_{q_0}), (X_1, Y_1), \cdots, (X_{q^1+q^{-1}}, Y_{q^1+q^{-1}})) \qquad (7)$$

is a Good NXLuffaS-*Irreducible view iff $\mathcal{V}_{\mathsf{NX}}$ is a* NXLuffaS-*Irreducible view and*

- $g_1, \cdots, g_{q_0}$ *are distinct and they do not collide with* $\{Y_1, \cdots, Y_{q^1+q^{-1}}\}$
- *For any $X \in \{IV, X_1, \cdots, X_{q^1+q^{-1}}\}$ there does not exist any $M \in \{0,1\}^n$ such that,*
  $\mathsf{MI}_{\mathsf{Lf}}(g_i, M) = X$ *for any* $i \in \{1, \cdots, q_0\}$.

The output view $\mathcal{OV}_{\mathsf{NX}}$ is called Good NXLuffaS-*Irreducible output view, if the corresponding view $\mathcal{V}_{\mathsf{NX}}$ is Good* NXLuffaS-*Irreducible.*

To obtain a lower bound on $\Pr[\mathcal{OV}^{\mathcal{A}}_{(\mathsf{NXLuffaS}^{\pi},\pi)} = \mathcal{OV}_{\mathsf{NX}}]$ we follow an approach similar to proof of Theorem 3. However, for any $z \in \{0,1\}^{nt}$, $m \in \{0,1\}^n$ there exist an unique $y \in \{0,1\}^{nt}$ such that, $\mathsf{MI}_{\mathsf{Lf}}(y\|m) = z$.[4] As a result while assigning the output value of $\mathsf{MD}^{f^{\pi}}$ one can have at most $2(q+\sigma)$ many forbidden values, where q is the total number of attacker queries and σ is the total number of message blocks present in LuffaS queries. As before, we can obtain the following theorem.

Theorem 6. *For any Consistent Good* NXLuffaS-*Irreducible output view $\mathcal{OV}_{\mathsf{NX}}$ ($\mathcal{V}_{\mathsf{NX}}$ as in* (7) *being the corresponding the view),*

$$\Pr[\mathcal{OV}^{\mathcal{A}'}_{(\mathsf{NXLuffaS}^{\pi},\pi)} = \mathcal{OV}_{\mathsf{NX}}] \ge \Big(1 - \frac{2\sigma(\sigma+q)}{2^{nt}}\Big)\frac{1}{2^{ntq_0}} \prod_{i=1}^{q^1+q^{-1}} \frac{1}{(2^{nt} - i + 1)},$$

[4] Compared to 2^n many possible $y \in \{0,1\}^{nt}$ in case $\mathsf{MI}_{\mathsf{Fg}}$.

where σ is the total number of message blocks present in NXLuffaS *queries, q is the total number* NXLuffaS *and π queries and $(q+\sigma) < 2^{nt-1}$.*

With an approach, similar to proof of Theorem 4 and using the previous observation regarding $\mathsf{MI}_{\mathsf{Lf}}$ one can also have the following theorem.

Theorem 7. *For any Consistent* LuffaS*-Irreducible view $\mathcal{V}$ as in (3) (with $t=1$), there exist at least*

$$2^{n(t-1)q_0}\left(1-\frac{q^2}{2^{n(t-2)}}\right)$$

many Consistent Good NXLuffaS*-Irreducible view $\mathcal{V}_{\mathsf{NX}}$ as in (7) such that,* $\mathsf{Xor}(g_1)=h_1,\cdots,$
$\mathsf{Xor}(g_{q_0})=h_{q_0}$*, where q is the total number of* LuffaS *and π queries.*

Following the final discussion of Section 4 we get the following. For any Consistent output view $\mathcal{OV}$(with respect to $\mathsf{LuffaS}^{\pi},\pi$ oracles) we have,

$$\Pr[\mathcal{OV}^{\mathcal{A}}_{(\mathsf{LuffaS}^{\pi},\pi)}=\mathcal{OV}] \geq \left(1-\frac{2\sigma(\sigma+q)}{2^{nt}}\right)\left(1-\frac{q^2}{2^{n(t-2)}}\right)\Pr[\mathcal{OV}^{\mathcal{A}}_{(\mathcal{R},S^{\mathcal{R}})}=\mathcal{OV}].$$

Applying the above inequality together with $\Pr[\mathsf{Abort}]$ to Theorem 1 we prove Theorem 5.

6 Indifferentiability Security Analysis of Luffa

Theorem 8. *Let $\Pi=(\pi_1,\pi_2,\cdots,\pi_t)$ be a collection of independent random permutations over $\{0,1\}^n$. Let Luffa^{Π} be the* Luffa *mode of operation with n bit message blocks and n bit digest. There exists a simulator $S^{\mathcal{R}}$ such that for any adversary $\mathcal{A}$ which makes at most q queries*

$$\mathsf{Adv}_{\mathcal{A}}((\mathsf{Luffa}^{\Pi},\Pi),(\mathcal{R},S^{\mathcal{R}})) \leq \frac{\mathcal{O}(\sigma^2+q^2)}{2^n}+\frac{\mathcal{O}(q^4)}{2^n}$$

where σ is the total number of blocks in the queries to R or Luffa^{Π}.

Simulator for Luffa. In this section we describe the simulator for Luffa mode of operation. As in Section 5, we view the $MI_{\mathsf{Lf}}(Y,m)$ transformation as an affine function $TY+Am$ where T is a $t\times t$ square matrix, A is a coefficient vector of length t and the chaining value Y is a vector of length t. For each permutation $\pi_j;\ j=1,2,\cdots,t$ the simulator keeps a partial permutation π_j^* of input-output relations derived so far. By $\mathcal{X}(\pi_j^*)$ and $\mathcal{Y}(\pi_j^*)$ we denote the set of forward query input and the set of forward query output for partial permutation π_j^*, derived so far. Simulator also maintains a directed graph G whose vertex set $V\subseteq\{0,1\}^{tn}$. Initially $V=\{(IV_1,\cdots,IV_t)\}$ These vertices are actually tuple of the outputs of the simulator that are part a chain starting from $IV_{\mathsf{Luffa}}=(IV_1,\cdots,IV_t)$. An edge from vertex Y to vertex Y', marked by m, implies that for each $j\in\{1,\cdots,t\}$, $(T_jY+A_jm,Y_j')\in\pi_j^*$. T_j represents the jth row of matrix T.

While answering a new forward query x for permutation i, the simulator first checks whether x can be an input of the finalization stage of any chain in the graph. To find it, simulator searches for each existing node Y whether $x = T_iY$ (Recall that for finalization, $m = 0^n$) and for $t-2$ many $j \neq i$, $T_jY \in \mathcal{X}(\pi_j^*)$ ($S(+, i, x)$ is the last but one query of the finalization). If simulator can find such a unique Y, it queries $R(M)$ and tries to set the output of all other permutation maintaining consistency with R. For example, when $t = 3$, the simulator checks whether there exists a state Y and an $x_j \in \mathcal{X}(\pi_j^*)$ for some $j \neq i$. If such a node exists, simulator finds a valid padded message M traversing the path from IV to Y, get $R(M)$ and set the output $y_i = \pi_i^*(x)$ and $y_k = \pi_k^*(T_kY)$ $(k \neq i, j)$ so that $y_i \oplus y_j \oplus y_k = R(M)$ $((x_j, y_j) \in \pi_j^*)$. Simulator also checks whether this new (y_1, y_2, y_3) state maintains some properties of the tree.

If, x is not the penultimate query of a finalization stage, simulator checks whether output of x can make a new node in a chain in the graph; i. e. whether there is a node Y in the graph such that for some m, $x = T_jY + A_jm$ and for all other $i \neq j, T_iY + A_im \in \mathcal{X}(\pi_j^*)$. If no such node exists, the simulator outputs randomly only maintaining the permutation property. If such a node exists, the simulator tries to sample the output so that the new node Y' maintains the following properties of the tree:

- **No cycle.** Let Y' be the new node. For any other node $Y = (y_1, \cdots, y_t)$, in the graph, for any $m \in \{0,1\}^n$, $\exists j \in \{1, \cdots, t\}$, $y_j \neq \pi_j^*(T_jY' + A_jm)$.
- **No Partial Collision.** For any two nodes Y, Y' in the graph, there is no message $m_1, m_2 \in \{0,1\}^n$, such that for any two $i, j \in \{1, \cdots, t\}$, $T_jY + A_jm_1 = T_jY' + A_jm_2$ and $T_iY + A_im_1 = T_iY' + A_im_2$.
- **Free final inputs.** For any two state Y, Y' in the graph and for all $j \in \{1, ..., t\}$, $T_jY \neq T_jY'$.

Overall the simulator samples a y maintaining the permutation property and checks that the following **Bad** events do not occur.

- The properties of tree are not satisfied.
- **Final input collides with previous input:** For any $j \in \{1, \cdots, t\}, T_jY' \in \mathcal{X}(\pi_j^*)$.
- **Partial collision with non-chain inputs:** $\exists m \in \{0,1\}^n$ and $\exists i, j \in \{1, \cdots, t\}$ such that $T_iY' + A_im \in \mathcal{X}(\pi_i^*)$ and $T_jY' + A_jm \in \mathcal{X}(\pi_j^*)$.

For the inverse query of permutation i, the simulator tries to ensure that the output does not create a new node in the graph. For each existing node Y and for all $j \neq i$, consider the set $\Gamma_2(Y, j) = \{x' = T_iY + A_im | m = A_j^{-1}(x_j + T_jY), x_j \in \mathcal{X}(\pi_j^*)\}$. Note that as A is a vector, A_j is an element from the underlying field. On input y simulator samples x from $\{0,1\}^n \setminus \mathcal{X}(\pi_j^*)$, and sets the **Bad** true if $x \in \Gamma_2(Y, j)$ for some node Y and for some $j \neq i$. For a formal description of the simulator we refer the reader to the full version of this paper. Next we state an upper bound for the abort probability of the simulator and we have the following lemma.

Lemma 2. *For any attacker $\mathcal{A}$ interacting with $(S^{\mathcal{R}}, \mathcal{R})$ making at most q queries to the simulator,*

$$\Pr[S^{\mathcal{R}} \rightarrow \perp] \leq \frac{\mathcal{O}(q^4)}{2^n - q}.$$

Interpolation Probability of $\mathcal{OV}^{\mathcal{A}}_{(\mathsf{Luffa}_{\Pi}, \Pi)}$. With an approach similar to Section 4 and Section 5 one can show, for any Consistent Luffa-Irreducible output view $\mathcal{OV}$ we have,

$$\Pr[\mathcal{OV}^{\mathcal{A}}_{(\mathsf{Luffa}_{\Pi}, \Pi)} = \mathcal{OV}] \geq \left(1 - \frac{2t\sigma^2}{2^n}\right)\left(1 - \frac{4q^2}{2^n}\right) \Pr[\mathcal{OV}^{\mathcal{A}}_{(R, S^R)} = \mathcal{OV}].$$

Applying the above inequality together with Lemma 2 to Theorem 1 we prove Theorem 8.

7 Conclusion

In this paper, we proved indifferentiability of domain extension algorithms of Fugue and Luffa. Although, none of them are in the final round, the modes of these hash functions are interesting in their own right. Specifically, domain extension algorithm of Luffa opens up a interesting research direction regarding security efficiency tradeoff. Improving our bound for Luffa remains an interesting open problem too. However, such an analysis seems to need substantial insight.

References

1. Andreeva, E., Mennink, B., Preneel, B.: On the indifferentiability of the grøstl hash function. In: Garay, J.A., De Prisco, R. (eds.) SCN 2010. LNCS, vol. 6280, pp. 88–105. Springer, Heidelberg (2010)
2. Andreeva, E., Mennink, B., Preneel, B.: Security reductions of the second round sha-3 candidates. Cryptology ePrint Archive, Report 2010/381 (2010), `http://eprint.iacr.org/`
3. Aumasson, J.-P., Phan, R.C.W.: Distinguisher for the full final round of fugue-256. In: NIST Second Sha3 Conference (2010)
4. Bertoni, G., Daemen, J., Peeters, M., Van Assche, G.: On the indifferentiability of the sponge construction. In: Smart, N.P. (ed.) EUROCRYPT 2008. LNCS, vol. 4965, pp. 181–197. Springer, Heidelberg (2008)
5. Bhattacharyya, R., Mandal, A., Nandi, M.: Security analysis of the mode of JH hash function. In: Hong, S., Iwata, T. (eds.) FSE 2010. LNCS, vol. 6147, pp. 168–191. Springer, Heidelberg (2010)
6. Canetti, R., Goldreich, O., Halevi, S.: The random oracle methodology. J. ACM 51(4), 557–594 (2004) (revisited)
7. De Canniere, C., Sato, H., Watanabe, D.: Hash Function Luffa: Specification Ver 2.0.1 @Sha3 Zoo (2009)
8. Chang, D., Lee, S.-J., Nandi, M., Yung, M.: Indifferentiable security analysis of popular hash functions with prefix-free padding. In: Lai, X., Chen, K. (eds.) ASIACRYPT 2006. LNCS, vol. 4284, pp. 283–298. Springer, Heidelberg (2006)

9. Chang, D., Nandi, M.: Improved indifferentiability security analysis of chopMD hash function. In: Nyberg, K. (ed.) FSE 2008. LNCS, vol. 5086, pp. 429–443. Springer, Heidelberg (2008)
10. Coron, J.-S., Dodis, Y., Malinaud, C., Puniya, P.: Merkle-damgård revisited: How to construct a hash function. In: Shoup, V. (ed.) CRYPTO 2005. LNCS, vol. 3621, pp. 430–448. Springer, Heidelberg (2005)
11. Halevi, S., Hall, W.E., Jutla, C.S.: The Hash Function "Fugue" @Sha3 Zoo (2009)
12. Halevi, S., Hall, W.E., Jutla, C.S., Roy, A.: Weak ideal functionalities for designing random oracles with application to fugue. Sha3 Zoo (2010)
13. Maurer, U.M., Renner, R.S., Holenstein, C.: Indifferentiability, impossibility results on reductions, and applications to the random oracle methodology. In: Naor, M. (ed.) TCC 2004. LNCS, vol. 2951, pp. 21–39. Springer, Heidelberg (2004)
14. Nandi, M.: A simple and unified method of proving indistinguishability. In: Barua, R., Lange, T. (eds.) INDOCRYPT 2006. LNCS, vol. 4329, pp. 317–334. Springer, Heidelberg (2006)

Analysis of Message Injection in Stream Cipher-Based Hash Functions

Yuto Nakano[1], Carlos Cid[2], Kazuhide Fukushima[1], and Shinsaku Kiyomoto[1]

[1] KDDI R&D Laboratories Inc.
2-1-15 Ohara, Fujimino, Saitama 356-8502, Japan
{yuto,ka-fukushima,kiyomoto}@kddilabs.jp
[2] Information Security Group,
Royal Holloway, University of London
Egham, TW20 0EX – UK
carlos.cid@rhul.ac.uk

Abstract. A common approach for the construction of cryptographic hash functions is to design the algorithm based on an existing symmetric encryption primitive. While there has been extensive research on the design of block cipher-based hash functions, little has been done on the study of design and security of stream cipher-based hash functions (SCH). In this paper we discuss the general construction of stream cipher-based hash functions, devoting special attention to one of the function's crucial components: the message injection function. We define two types of message injection functions, which may be appended to the keystream generator (e.g. a stream cipher) to build an SCH. Based on these constructions, we evaluate the security of simple SCHs whose stream cipher function consists of a LFSR-based filter generator. We see this as an initial step in the more formal study of the security of hash function constructions based on stream ciphers.

Keywords: hash function, stream cipher, collision resistance.

1 Introduction

There has been much recent activity in the area of cryptographic hash function design and analysis: popular hash functions such as MD5 [12] and SHA-1 [10] have been shown to have weaknesses, e.g. lack of collision resistance [16,17], and this has spurred much interest in research in hash functions, culminating in the establishment of the NIST-sponsored SHA-3 competition to select a new hash function standard. As a result, several new designs have been proposed in the past few years.

A common approach for the construction of cryptographic hash functions is to design the algorithm based on an existing symmetric encryption primitive. There are several advantages in this type of approach: it may for instance be possible to derive the security of the hash function from the underlying symmetric algorithm; moreover, such a construction may allow one to use the symmetric

J. Lopez and G. Tsudik (Eds.): ACNS 2011, LNCS 6715, pp. 498–513, 2011.

algorithm for both encryption and as the building block for the hash function. This may be particularly attractive when trying to reduce the total cost of implementation of a cryptographic system in which encryption and hashing are required (for example, implementations in resource-constrained devices).

While there has been extensive research on the design of block cipher-based hash functions, not much has been done on the study of design and security of stream cipher-based functions. The general construction of a stream cipher-based hash function (SCH) was first introduced by Golić [3], as a mode of operation of stream ciphers. An SCH consists of a keystream generator (e.g. a stream cipher) and an additional function, which inputs the message into the internal state of the stream cipher. Thus, we can model a stream cipher-based hash function as a message injection function and a stream cipher function. The stream cipher function is the core component of SCHs and an appropriate algorithm is selected from among existing stream cipher algorithms. The pre-computation/injection function is used to input the message into the internal state of the stream cipher algorithm.

Several *dedicated* hash functions such as Boole [13] and MCSSHA-3 [5] are based on stream ciphers, and although these have been shown to be insecure, they did present good performance on a variety of platforms. It is also not uncommon to encounter in practice *ad hoc* constructions where a stream cipher is used to build a hash-like function. Despite of that, to the authors' best knowledge a well-developed set of design principles and security evaluation criteria for SCHs has yet to be established.

Nakano *et al.* [8] proposed a model for SCHs and showed necessary conditions for the construction of secure stream cipher-based hash functions. They concentrated on the problem of how to construct SCHs that are collision resistant, as this is perhaps the most challenging task for a designer. Their considerations are however not yet sufficient and a proposal for a concrete construction technique remains an open problem. In this paper, we describe the construction of SCHs based on bit-oriented simple stream cipher algorithms and provide the security analysis of possible message injection functions for SCHs. As an initial work in the area, we do not claim to provide an exhaustive discussion of all relevant security and design aspects of SCHs. We only deal here with collision resistance, and an analysis of other security requirements such as pre-image and second pre-image resistance remains to be considered. Moreover, in order to make the discussion simple, we only deal with bit-oriented linear feedback shift registers for the stream cipher. Other stream cipher constructions (such as the ones based on sponge functions) are not discussed in this paper.

The remaining of this paper is organised as follows: in Section 2 we present a brief discussion of related work. Section 3 provides definitions of message injection functions. We analyse the security of the message injection functions in Section 4. In Section 5, as an extension, we consider two LFSR-based SCHs. We present a discussion about the security and efficiency of the message injection functions in Section 6, and a comparison with existing algorithms is made in Section 7. Finally we conclude our paper in Section 8.

2 Related Work

We introduce below the necessary definitions and briefly describe research results of relevance to the construction of stream cipher-based hash functions.

2.1 Security Definitions of Hash Functions

The conventional security requirements for cryptographic hash functions are pre-image resistance, second pre-image resistance, collision resistance [7]; more recently length-extension security [4] has also been proposed as a security requirement for hash functions. Let $\mathcal{H}$ be a hash function, n be the hash output length, M and M' be messages; furthermore, let the symbol $||$ denote the concatenation of data.

Pre-image Resistance: given $h = \mathcal{H}(M)$ for some (unknown, randomly generated) M, finding any M' such that $\mathcal{H}(M') = h$ requires on average the work effort on the order of 2^n hash operations.

Second Pre-image Resistance: given a randomly generated M and $h = \mathcal{H}(M)$, finding any $M' \neq M$ such that $\mathcal{H}(M') = h$ requires on average the work effort on the order of 2^n hash operations.

Collision Resistance: finding M and M' such that $\mathcal{H}(M) = \mathcal{H}(M')$ and $M \neq M'$ requires on average the work effort on the order of $2^{n/2}$ hash operations.

Length-extension Security[1]: given $\mathcal{H}(M)$, the complexity of finding (z, x) such that $x = \mathcal{H}(M||z)$ should be greater than or equal to either the complexity of guessing M itself or 2^n.

2.2 Hash Function Constructions

The great majority of hash functions are of dedicated design. In this section, we briefly discuss the design of hash functions based on symmetric encryption algorithms, as well as some related SHA-3 proposals.

Block Cipher-Based Hash Functions. Preneel *et al.* studied general constructions of block cipher-based hash functions in [11]; they found that 12 out of all 64 possible constructions are secure. Later, Black *et al.* [2] extensively analysed the constructions. Stam [15] extended the work taking pre- and post-processing into consideration. We note that these results on the security of block cipher-based hash functions are based on the assumption that the primitive used is an *ideal block cipher* (rather than a specific algorithm).

Golić's Construction. Golić [3] studied modes of operation for stream ciphers, and showed how to convert a keystream generator into a stream cipher

[1] This requirement has been proposed in the NIST SHA-3 competition.

with memory (SCM), and how to build a hash function from an SCM. When a feedback shift register (FSR) based keystream generator is used, the SCM mode can be easily converted into a hash function by adding the plaintext bit to the feedback bit of the FSR. The SCM mode is clocked m times with an m-bit message and the corresponding m-bit ciphertext is stored in memory; then the SCM mode is clocked another m times with the m-bit ciphertext being input in reverse order. Finally, the SCM mode is clocked αm times where α is a small positive integer (e.g., three), and the last h successive ciphertext bits (or keystream bits) are output as the hash value. As the ciphertext in reverse order is used, the scheme requires an amount of memory that is equivalent to the message size.

SHA-3 Candidates. The NIST-sponsored SHA-3 competition kicked off in late 2008, and is expected to announce the new hash function standard in 2012. In total 64 algorithms were submitted to the competition, although only 56 of them are publicly known. Seven of the original submissions have similar structure to a stream cipher, and are thus of relevance to our work. Below we give a brief description of three of them.

Abacus. The Abacus [14] hash function has four registers (`ra, rb, rc, rd`). The message injection phase processes one byte in each round. First, the values of the four registers and one byte of the message are XORed and the result goes through S-Boxes. In the next step, four counters are combined with registers. Then, a maximum distance separable (MDS) matrix-based function is applied and four bytes are output. Finally, another S-Box operation is applied to the four bytes of registers. Two second pre-image attacks on Abacus were independently proposed by Nikolić *et al.*[9] and Wilson [18]. Wilson also showed a collision attack [18].

Boole. Boole [13] is constructed from a non-linear feedback shift register, input accumulators, and an output filter function. Boole consists of three phases: an input phase, a mixing phase, and an output phase. The state update function of the register, referred to as a cycle, transforms the state S_t into S_{t+1}. A message word is input to three accumulators, which are then updated. The register is then cycled once. After the input phase, the mixing phase is applied: the register is mixed with three accumulators, and the register is cycled 16 times. Finally the hash value is generated in the output phase. A pre-image attack and a collision attack against Boole were proposed by Nikolić [6].

MCSSHA-3. The MCSSHA-3 [5] hash function has a non-linear shift register whose size is the same as the hash value. A message is added with the feedback from the shift register. Before addition of the message, an 8×8 substitution is applied to the feedback. In MCSSHA-3, the message injection is performed every four clocks. Once a message block is input, the shift register is clocked without a message for another three clocks. After the whole message is processed, the h-bit internal state S is obtained, where h denotes the size of a hash value. The register is then clocked $h/2$ times with a $4h$-bit sequence which is a concatenation of the state S. Finally, the internal state is output as a hash value. A collision

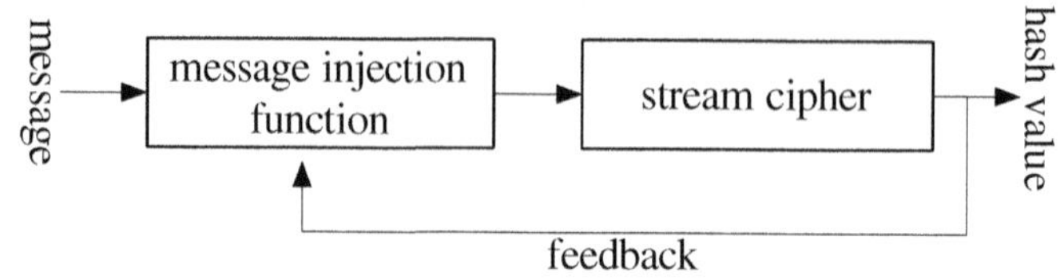

Fig. 1. Model of SCH

attack on MCSSHA-3 was presented by Aumasson and Naya-Plasencia[1]; they also demonstrated a second pre-image attack on MCSSHA-3 and a pre-image attack on the tweaked version: MCSSHA-4.

Model of SCHs. Nakano *et al.* [8] presented a general model of SCH, which is shown in Figure 1. They showed that an SCH can be modelled with two components: a message injection function and a stream cipher function.

A message injection function is appended to a stream cipher to construct an SCH. The message injection function is the component that takes the message and feedback from stream cipher as input, and determines the internal state of the stream cipher. The stream cipher diffuses the message over its internal state. The hash value is then generated as a certain length of the keystream. Generally, keys and IVs (initialization vectors) are set to constant values, usually to zero, and the message is loaded to the internal state of the stream cipher.

SCHs execute three phases: message injection, blank rounds, and hash generation. The message is loaded into the internal state of the stream cipher in the message injection phase. The internal state is updated from the message, previous state, and output feedback in the message injection phase. Blank rounds are the iteration of the state update (e.g. clocking the stream cipher) for diffusing the last input message words over the entire state. In the blank round phase, the internal state is updated from the previous state and output feedback. This phase is similar to the initialization of the stream cipher. Finally the stream cipher outputs a keystream as a hash value in the hash generation phase. During hash generation, only the previous state is used to update the state and the hash value is generated. In [8], the security analysis of SCHs based on the model above was provided, and suggested that secure message injection is a critical stage for achieving collision resistance. In this paper we further consider the problem of how to design a collision resistant message injection for stream cipher-based hash functions.

3 Definition of Message Injection Function

In this section, we present two message injection functions for a bit-oriented stream cipher.

3.1 Stream Cipher Model

Before defining the message injection function, let us define the stream cipher considered in this paper. We deal with a simple stream cipher that consists of

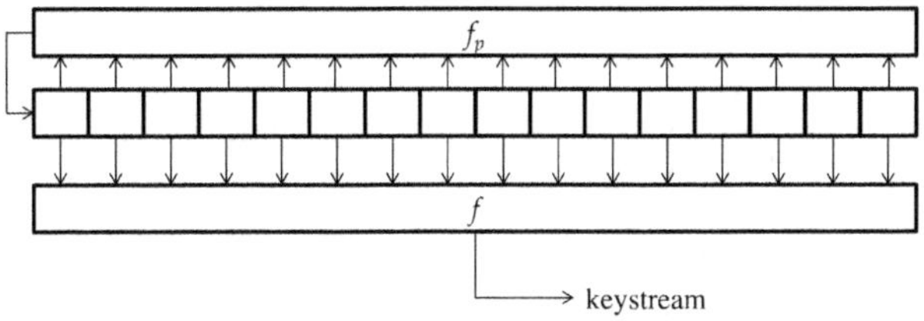

Fig. 2. An LFSR and a filter function

an LFSR and a filter function as in Figure 2, and an extension that consists of two LFSRs and a filter function as in Figure 5. Modern stream ciphers require an initialization operation for loading and mixing the encryption key and initialization vector into the internal state before producing the keystream; in the constructions discussed, the output of the non-linear filter can be fed back to the end of the LFSR as part of the initialization process. We use the initialization operation for message injection of the SCH.

Let l be the length of the LFSR, and f denote the filtering function. We assume that the feedback polynomial of the LFSR is primitive, and that the function f is balanced. Then, the LFSR reaches all internal states except all-zero and the period of state transition is given by $2^l - 1$. By clocking the LFSR $2^l - 1$ times, '1' is fed back 2^{l-1} times and '0' is fed back $2^{l-1} - 1$ times. The probability '1' is fed back and that of '0' are given by

$$\Pr[1 \text{ is fed back}] = \frac{2^{l-1}}{2^l - 1} \approx \frac{1}{2},$$
$$\Pr[0 \text{ is fed back}] = \frac{2^{l-1} - 1}{2^l - 1} \approx \frac{1}{2}.$$

These probabilities can be developed to the probability which feedback has a difference. Let two distinct internal states be S_1 and S_2, and $HW(x)$ be the Hamming weight of a binary string x. We say two internal states have a difference $\Delta(= S_1 \oplus S_2)$ when $HW(\Delta) \geq 1$. We denote the feedback derived from internal states S_1 and S_2 as b_{fdbk1} and b_{fdbk2}, respectively. We also say the feedback of the LFSR has a difference $\Delta_1(= b_{fdbk1} \oplus b_{fdbk2})$ when $HW(\Delta_1) = 1$. Let us denote registers with a difference as '1' and without the difference as '0'.

Remark 1. *The probability that the feedback has a difference is* $1/2$.

Please note that Remark 1 holds if the length l of the LFSR is large enough.

Let g be a message injection function. The message injection function is a map $g : \{0,1\}^l \times \{0,1\}^m \to \{0,1\}^l$, where l and m denote the internal state size and the message block size. Then the message injection function can be given by

$$S'_t = g(S_t, m_t).$$

Let $f_p(x) = c_1x_{t,1} \oplus c_2x_{t,2} \oplus \cdots \oplus c_lx_{t,l}$ be the linear recursion function of the LFSR, then the update of the register can be described as:

$$s_{t+1,i} = \begin{cases} s_{t,i+1} & (1 \leq i \leq l-1), \\ f_p(s_{t,1}, \ldots, s_{t,l}) & (i = l). \end{cases} \quad (1)$$

The keystream z_t is given by

$$z_t = f(d_1s_{t,1}, \ldots, d_ls_{t,l}),$$

where the constants $d_j \in \{0,1\}$ for $1 \leq j \leq l$ choose the registers to be input to the filter function.

We define two message injections as: the message is XORed with the keystream and this result is XORed with feedback of the LFSR; or the message is XORed with the keystream and the result is XORed with the internal state. We describe both injections in more detail in the following sections.

3.2 Inject into Feedback

The SCH with this message injection function is shown in Figure 3. This is the most natural way to inject the message into the internal state, and a subcase of the "inject into the internal state" mode, which we explain later. The construction proposed by Golić applies this method. MD5 and SHA-1 can also be categorized as this type since the step operation of these hash functions update only one chaining variable and the others are just shifted[2]. Once the value of the register is fixed, it is not updated until it is fed back again. Furthermore, the feedback can be controlled by the message. It is possible for an adversary to control the entire internal state. Note that we consider the bit-oriented LFSR only and do not consider the message expansion.

The message injection function can be described as follows.

$$s_{t+1,i} = \begin{cases} s_{t,i+1} & (1 \leq i \leq l-1) \\ f_p(s_{t,1}, \ldots, s_{t,l}) \oplus (f(d_1s_{t,1}, \ldots, d_ls_{t,l}) \oplus m_t) & (i = l). \end{cases} \quad (2)$$

3.3 Inject into the Internal State

We show the scheme of this message injection function in Figure 4. This message injection function updates the internal state by XORing the value of the internal state with the message and the keystream. Since we consider the bit-oriented LFSR only, the same message data is XORed in different positions. This scheme requires selectors $\sigma_i \in \{0,1\}$ that determine the registers to be updated. The message injection function can be described as:

$$s_{t+1,i} = \begin{cases} s_{t,i+1} \oplus \sigma_i(z_t \oplus m) & (1 \leq i \leq l-1) \\ f_p(s_{t,1}, \ldots, s_{t,l}) & (i = l). \end{cases} \quad (3)$$

The security of the message injection depends not only on the stream cipher but also where to inject the message. As we discuss later in Section 4, the number of registers updated by the message is the important factor for the message injection.

[2] Although these hash functions also include a message expansion mechanism.

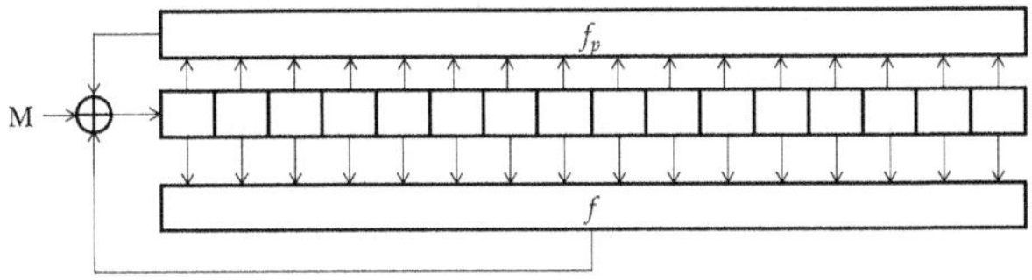

Fig. 3. Inject into Feedback

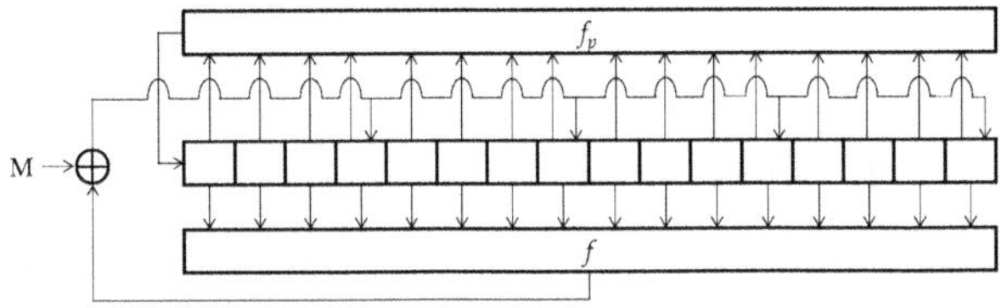

Fig. 4. Inject into the Internal State

4 Security Analysis

In this section, we evaluate the security of SCHs constructed from a simple stream cipher and message injection functions introduced in Section 3 against collision attacks.

4.1 Inject into Feedback

A difference Δ_1 on feedback of the LFSR and/or an output difference Δ_2 of the f function can be easily cancelled out by using a message difference Δm as:

$$\Delta m = \Delta_1 \oplus \Delta_2,$$

and the value in the leftmost register is easily controlled. Assume that the LFSR is clocked n times, where n is large enough when compared to the length of the LFSR. Then the difference in the register is forced out, hence only the difference on the leftmost register should be taken into consideration. Iterating this cancellation of the difference for l times enables an adversary to control the entire internal state of the LFSR. Therefore, the collision is easily generated. This type of message injection cannot provide the collision resistance without the message expansion.

4.2 Inject into the Internal State

Suppose that the same message-dependent data is injected into r positions of the LFSR at regular intervals, then l/r registers can be controlled by the message and a collision of these registers is easily generated. From the fact that the adversary can control l/r bits of the internal state, $l(1-1/r)$ bits of the internal state could have differences. Since feedback has the difference with probability

$1/2$, the filter function must output the difference for $l(1-1/r)/2$ clocks and it must not output the difference for the other $l(1-1/r)/2$ clocks. Suppose that the filter function output the difference with the probability p, then the probability of the collision is given by

$$\Pr[\text{coll}] = [p(1-p)]^{\frac{l(1-1/r)}{2}}.$$

Here we introduce a useful remark for the evaluation of the filter function. Since the filter function is balanced, the difference is output with probability $1/4 + 1/4 = 1/2$.

Remark 2. *If the output of the filter function is balanced, then it propagates the difference with probability* $p = 1/2$.

From Remark 2, we obtain $\Pr[\text{coll}] = 2^{-l(1-1/r)}$. Using the birthday attack, the probability is bounded below by,

$$\Pr[\text{coll}] = 2^{-\frac{l(1-1/r)}{2}}.$$

5 Extension to Two LFSRs

In this section, we consider the extension of the message injection function and its security evaluation to the two-LFSR-based SCH.

5.1 Two-LFSR-Based SCH

The internal state size of LFSR-A and LFSR-B is give by l_a and l_b, respectively. The function f is the same as the stream cipher with one LFSR; the t_f-to-1 balanced filter function. Let f_A and f_B be feedback polynomials of LFSR-A and LFSR-B, and coefficients of each polynomial are given by α_j and β_j. For the sake of the simplicity, we denote $S_t = \bigoplus \alpha_i s_{t,i}$, $U_t = \bigoplus \beta_i u_{t,i}$, and the input to the filter function as S'_t, then the state update of each register can be denoted as

$$s_{t+1,i} = \begin{cases} s_{t,i+1} & (1 \le i \le l_a - 1), \\ f_A(S_t) & (i = l_a), \end{cases}$$

$$u_{t+1,j} = \begin{cases} u_{t,j+1} & (1 \le j \le l_b - 1), \\ f_B(U_t) \oplus f(S'_t) & (j = l_b), \end{cases}$$

We define $\sigma_{A,i}, \sigma_{B,i} \in \{0,1\}$ that determine which registers to be updated by the message. The output of the stream cipher is defined as,

$$z_t = u_{t,1}.$$

Then, we have four choices where to inject the message for this stream cipher as:

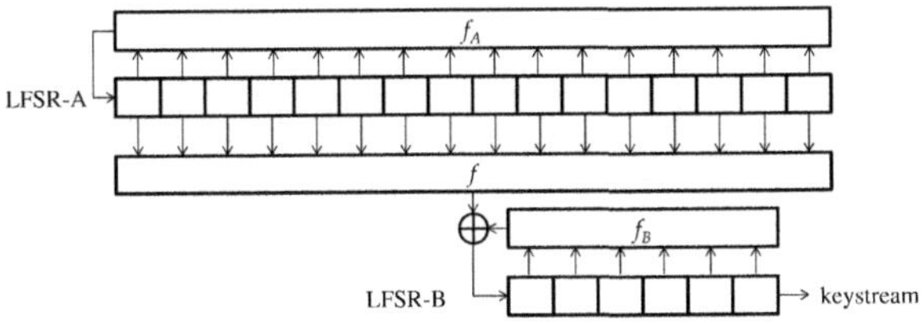

Fig. 5. Two LFSRs and a filter function

$$s_{t+1,i} = \begin{cases} s_{t,i+1} & (1 \le i \le l_a - 1), \\ f_A(S_t) \oplus (z_t \oplus m) & (i = l_a), \end{cases}$$
$$u_{t+1,j} = \begin{cases} u_{t,j+1} & (1 \le j \le l_b - 1), \\ f_B(U_t) \oplus f(S'_t) & (j = l_b), \end{cases} \tag{4}$$

$$s_{t+1,i} = \begin{cases} s_{t,i+1} & (1 \le i \le l_a - 1), \\ f_A(S_t) \oplus (z_t \oplus m) & (i = l_a), \end{cases}$$
$$u_{t+1,j} = \begin{cases} u_{t,j+1} & (1 \le j \le l_b - 1), \\ f_B(U_t) \oplus (z_t \oplus m) \oplus f(S'_t) & (j = l_b), \end{cases} \tag{5}$$

$$s_{t+1,i} = \begin{cases} s_{t,i+1} \oplus \sigma_{A,i}(z_t \oplus m) & (1 \le i \le l_a - 1), \\ f_A(S_t) & (i = l_a), \end{cases}$$
$$u_{t+1,j} = \begin{cases} u_{t,j+1} & (1 \le j \le l_b - 1), \\ f_B(U_t) \oplus f(S'_t) & (j = l_b), \end{cases} \tag{6}$$

$$s_{t+1,i} = \begin{cases} s_{t,i+1} \oplus \sigma_{A,i}(z_t \oplus m) & (1 \le i \le l_a - 1), \\ f_A(S_t) & (i = l_a), \end{cases}$$
$$u_{t+1,j} = \begin{cases} u_{t,j+1} \oplus \sigma_{B,i}(z_t \oplus m) & (1 \le j \le l_b - 1), \\ f_B(U_t) \oplus f(S'_t) & (j = l_b). \end{cases} \tag{7}$$

Two methods described as Eq. (4) and Eq. (5) are essentially the same, the message is XORed with feedback of LFSRs. The internal state is directly updated by the message-dependent data in Eq. (6) and Eq. (7).

5.2 Security Analysis

We first introduce a remark regarding the security evaluation of the LFSR-B. Let Pr[diff. on B cancelled] be the probability that all differences on the LFSR-B are

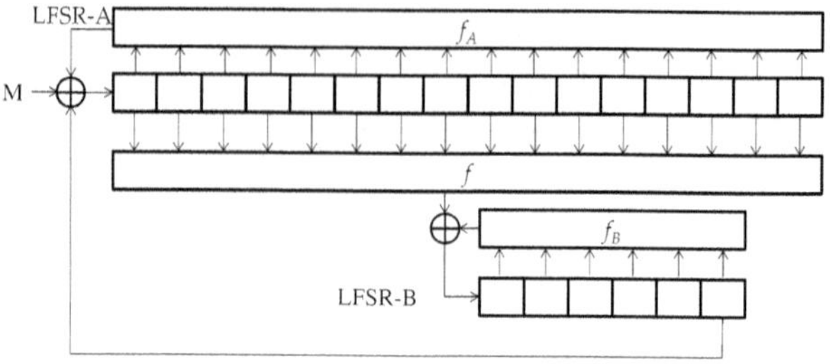

Fig. 6. Inject into Feedback of LFSR-A

cancelled out when no difference is input from the filter function f. We denote the register with the difference as '1' and without as '0'. When the filter function does not output differences, '0's are XORed with feedback of the LFSR-B and the filter function does not influence the LFSR-B, that is, only LFSR-B should be taken into consideration. There exists an integer n for any randomly chosen two internal states U_t and $U_{t'}$ that satisfies $U_t = U_{t'+n}$. The period of the LFSR-B is given by $2^{l_b} - 1$ since its feedback polynomial is primitive. The computational cost for finding such n is given by $2^{l_b/2}$. Hence the probability that difference on the LFSR-B is cancelled is given by $\Pr[\text{diff on B cancelled}] = 2^{-l_b/2}$.

Remark 3. *If the feedback polynomial of the LFSR-B is primitive and the difference is not input into the LFSR-B, then* $\Pr[\textit{diff. on B cancelled}] = 2^{-l_b/2}$.

Inject into Feedback of LFSR-A. This scheme is shown in Figure 6. The LFSR-A is easily controlled by a message, since the message is just XORed with feedback. Once all differences on LFSR-A are cancelled out, no difference is output from the function f. The LFSR-B still has a difference at this point and this difference has to be cancelled out. Hence the probability which two different messages collide is given by greater of the birthday paradox or the probability of the difference on LFSR-B being cancelled out,

$$\begin{aligned}\Pr[\text{coll}] &= \max(2^{-l_b/2}, \Pr[\text{diff. on B cancelled}]) \\ &= 2^{-l_b/2}.\end{aligned}$$

Inject into Feedback of Both LFSRs. Let us consider the case that messages are injected into feedback of both LFSRs. When the message is injected into both LFSRs, the adversary can control either LFSR-A or LFSR-B. Here, we assume that the size of the LFSR-A is larger than that of the LFSR-B, then it is natural for the adversary to control the LFSR-A. After all differences on LFSR-A are cancelled out, no message difference should be injected. However LFSR-B still has differences to be cancelled out. The probability that those differences are cancelled out is the same as the probability of the collision and given as

$$\begin{aligned}\Pr[\text{coll}] &= \max(2^{-l_b/2}, \Pr[\text{diff. on B cancelled}]) \\ &= 2^{-l_b/2}.\end{aligned}$$

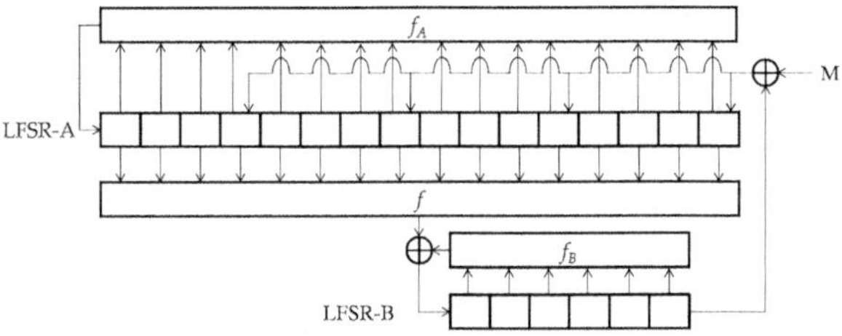

Fig. 7. Inject into the Internal State of LFSR-A

Inject into the Internal State of LFSR-A. Figure 7 shows this scheme. The message is XORed with the output of the LFSR-B and the result is again XORed with the internal state of the LFSR-A. Suppose that r bits of the internal state of LFSR-A are updated by the message-dependent data at regular intervals, then l_a/r bits of the LFSR-A can be controlled. The remaining $l_a(1-1/r)$ bits are unfixed and the probability which the LFSR-A collides is given by

$$\Pr[\text{coll.A}] = 2^{-\frac{l_a(1-1/r)}{2}}.$$

The LFSR-B collides when no difference is input, the difference on LFSR-B is cancelled, or happen to collide. Therefore the LFSR-B collides with the probability,

$$\Pr[\text{coll.B}] = \max((1-p)^{l_b}, \Pr[\text{diff. on B cancelled}], 2^{-l_b/2}) = 2^{-l_b/2}.$$

Hence the probability of the collision is

$$\Pr[\text{coll}] = 2^{-\frac{l_a(1-1/r)+l_b}{2}}$$

Inject into the Internal State of Both LFSRs. We consider that the message-dependent data is injected into both LFSRs: the message is XORed with r bits of the LFSR-A and q bits of the LFSR-B at regular intervals. The adversary can control part of the internal state of either l_a/r bits of LFSR-A or l_b/r bits of LFSR-B. We assume that l_a/r is larger than l_b/r, then it is natural that the adversary tries to control LFSR-A. The remaining $l_a(1-1/r)$ bits of LFSR-A and the entire internal state of LFSR-B are unfixed. Therefore, the adversary can generate the collision with probability,

$$\Pr[\text{coll}] = 2^{-\frac{l_a(1-1/r)+l_b}{2}}.$$

6 Discussion

In this section, we discuss the security and the efficiency of two types of message injection functions. Table 1 summarizes collision probabilities and the number of required operations for each message injection function.

Table 1. Message Injection Functions

Function	Collision Prob.	No. of Operation
Single LFSR		
Inject into Feedback	1	1 *XOR*
Inject into the Internal State	$2^{-[l(1-1/r)]/2}$	r *XORs*
Two LFSRs		
Inject into the Feedback of LFSR-A	$2^{-l_b/2}$	1 *XOR*
Inject into the Feedback of Both LFSRs	$2^{-l_b/2}$	2 *XORs*
Inject into the Internal State of LFSR-A	$2^{-[l_a(1-1/r)+l_b]/2}$	r *XORs*
Inject into the Internal State of Both LFSRs	$2^{-[l_a(1-1/r)+l_b]/2}$	$(r+q)$ *XORs*

6.1 Single LFSR

Inject into feedback only requires an XOR operation for each cycle of the message injection function. Hence this message injection function is lightweight and we believe SCH constructed with this function achieves as a high performance as its original stream ciphers. However the entire internal state can be controlled by the message and the collision is easily generated without the secure message expansion.

Inject into the internal state requires r XOR operations for a cycle of the message injection function. It also requires the selector, which selects registers to be updated with the message. This selector is fixed by the specification of the algorithm, hence does not affect the performance of the SCH. This scheme can be collision resistant by choosing r and l adequately without the message expansion.

6.2 Two LFSRs

Inject into feedback of LFSR-A only requires an XOR operation and inject into feedback of both LFSRs requires two XOR operations. These two injections are lightweight and achieve as a high performance as their original stream ciphers. However, the entire internal state of LFSR-A can be fixed and the computational cost for the collision is reduced from $2^{(l_a+l_b)/2}$ to $2^{l_b/2}$. By making the LFSR-B larger than the size of hash value, the collision resistance can be ensured. The computational cost is not affected if the message is injected into feedback of LFSR-A or both LFSRs.

Inject into the internal state of LFSR-A and both LFSRs requires r and $(r+q)$ operations of XOR, respectively. Similar to the case of single LFSR, the selector has to be introduced to these schemes. These two scheme is the most secure way among six since the adversary can only control the part of the LFSR-A. The adequate three parameters l_a, l_b, and r can ensure the collision resistance. Although inject into both LFSRs require additional q operations of

XOR compared to inject into LFSR-A, probabilities of collisions are the same in two cases.

7 Comparison to Real Algorithms

In this section, we apply our security analysis to real algorithms, which are Abacus, MCSSHA-3, and Boole. Note that we cannot directly compare our evaluation to above algorithms since these algorithms use word-oriented shift registers. However, computational costs shown in attacks on these algorithms meet our estimated ones.

7.1 Abacus

The message injection of Abacus is inject into feedback. Abacus can be considered as the extended version of two-LFSR-based SCH, since Abacus has four registers. The adversary cannot control the entire state, but the largest register can be fixed to zero. This is exploited in the second pre-image attack and collision attack in [9] and [18].

Four registers of Abacus can be classified into registers updated by the message and the others. Suppose the byte-oriented shift register is the extended version of the bit-oriented LFSR, then security analysis of Section 5 is approximately applicable to Abacus. The register where the message is injected corresponds to LFSR-A of Section 5 and the others correspond to LFSR-B. Therefore, the size of the LFSR-B is given by $l_b = 344$ and the lower-bound of the probability for the collision is estimated at 2^{-172} from table 1. This probability is pretty close to the probability of the collision attack given in [18].

7.2 MCSSHA-3

MCSSHA-3 also uses the byte-oriented feedback shift register, and as the same reason of Abacus, our security evaluation is applicable to it. The message injection of MCSSHA-3 is also inject into feedback. Since the message injection of MCSSHA-3 is done every four clocks, the FSR of MCSSHA-3 can be divided into two categories; registers where the message is injected and the others. Hence we can divide the FSR of MCSSHA-3 into two FSRs and our security evaluation for two LFSRs is applicable. From table 1, the lower bound of the collision probability for two LFSRs with inject into feedback is given by $2^{-l_b/2}$. In MCSSHA-3, the size of the registers is given by $l_b = 192$, the probability for collision is estimated at 2^{-96}. This probability is pretty close to the probability of the collision attack given in [1].

7.3 Boole

Boole applies the inject into the internal state. In the security evaluation, we assumed that the message-dependent data is injected into r registers at regular intervals. We cannot directly compare our security analysis and that of Boole since the message injection of Boole does not inject at regular intervals. The FSR of Boole corresponds to the LFSR-A in our evaluation and three accumulators corresponds to the LFSR-B. The message is injected into two registers of the LFSR-A and three registers of the LFSR-B. From our evaluation, the probability of collision is estimated to be $2^{-[512(1-1/2)+96]/2} = 2^{-176}$ since $l_a = 512$, $l_b = 96$, and $r = 2$.

We assume that the same message-dependent data is injected into multiple registers in our evaluation, however, different message-dependent data is injected into each register in Boole. This is the first reason why the probability of the collision derived from our evaluation and shown in [6] has a gap.

In [6], they control the message difference that leads to collision, and the computational cost for collision attack is dramatically reduced to be 2^{33}. This is the second reason of the gap of the probability between ours and that of [6]. Boolean functions used in Boole have a vulnerability and collisions on Boolean functions are easily generated. By exploiting this vulnerability, differences in the accumulators can be cancelled. The difference injected into the register can be also cancelled by the difference which is output from accumulators. As the result, the difference is cancelled efficiently and the computational cost for the collision attack is dramatically reduced to 2^{33}.

8 Conclusion

In this paper, we evaluated the security (mainly collision resistance) and the efficiency of SCHs based on the classic filter generator. We considered a stream cipher consisting of a single LFSR, and then extended it to the case with two LFSRs. We defined two message injection functions for each stream cipher: inject into feedback and inject into the internal state. We constructed six SCHs by appending these message injection functions to the stream ciphers and derived probabilities of collision.

Inject into feedback is lightweight and can achieve high performance, however it cannot be secure without a message expansion. On the other hand, inject into the internal state has the potential to realize the collision resistance without a message expansion. As discussed in Section 4, the security of the resulting SCHs can be increased by making r large. The computational cost of calculating hash values can be optimized to use appropriate r. The construction that injects the message into the internal state is adjustable not only for systems that require a high security level but also for small devices whose resources are highly constrained. Our analysis suggests criteria for parameters, length of FSRs, and the number of message-injecting registers in the design of SCHs. We also compared our security analysis with real algorithms and confirmed that our evaluation appears to be reasonably accurate, especially for inject into feedback function.

In our opinion, there remains many aspects to be researched for achieving a secure design of stream cipher-based stream ciphers; these include analysis of pre-image and second pre-image resistance, extension to other hash function constructions, such as ones based on sponge functions, among others. We see however this paper as an initial step in the more formal study of the security of hash function constructions based on stream ciphers, and hope it will spur further research in this topic.

References

1. Aumasson, J.-P., Naya-Plasencia, M.: Cryptanalysis of the MCSSHA Hash Functions (2009), Presented at WEWoRC 2009, http://131002.net/data/papers/AN09.pdf
2. Black, J., Rogaway, P., Shrimpton, T.: Black-Box Analysis of the Block-Cipher-Based Hash-Function Constructions from PGV. In: Yung, M. (ed.) CRYPTO 2002. LNCS, vol. 2442, pp. 320–335. Springer, Heidelberg (2002)
3. Golić, J.D.: Modes of Operation of Stream Ciphers. In: Stinson, D.R., Tavares, S. (eds.) SAC 2000. LNCS, vol. 2012, pp. 233–247. Springer, Heidelberg (2001)
4. Jonsson, J., Widmayer, C., Kelsey, J.: Public Comments on the Draft Federal Information Processing Standard (FIPS) Draft FIPS 180-2, Secure Hash Standard, SHS (2001), http://www.cs.utsa.edu/~wagner/CS4363/SHS/dfips-180-2-comments1.pdf
5. Maslennikov, M.: Secure hash algorithm MCSSHA-3. Submission to NIST (2008), http://registercsp.nets.co.kr/MCSSHA/MCSSHA-3.pdf
6. Mendel, F., Nad, T., Schläffer, M.: Collision Attack on Boole. In: Abdalla, M., Pointcheval, D., Fouque, P.-A., Vergnaud, D. (eds.) ACNS 2009. LNCS, vol. 5536, pp. 369–381. Springer, Heidelberg (2009)
7. Menezes, A.J., van Oorschot, P.C., Vanstone, S.A.: Handbook of Applied Cryptography. CRC Press, Boca Raton (1996)
8. Nakano, Y., Kurihara, J., Kiyomoto, S., Tanaka, T.: On a Construction of Stream-cipher-based Hash Functions. In: SECRYPT, pp. 334–343 (2010)
9. Nikolić, I., Khovratovich, D.: Second preimage attack on Abacus (2008), http://lj.streamclub.ru/papers/hash/abacus.pdf
10. NIST. Secure hash standard. FIPS180-1 (1995)
11. Preneel, B., Govaerts, R., Vandewalle, J.: Hash Functions Based on Block Ciphers: A Synthetic Approach. In: Stinson, D.R. (ed.) CRYPTO 1993. LNCS, vol. 773, pp. 368–378. Springer, Heidelberg (1994)
12. Rivest, R.: The MD5 message digest algorithm. RFC1321 (1992)
13. Rose, G.G.: Design and primitive specification for Boole. submission to NIST (2008), http://seer-grog.net/BoolePaper.pdf
14. Sholer, N.: Abacus a candidate for SHA-3. submission to NIST (2008), http://ehash.iaik.tugraz.at/uploads/b/be/Abacus.pdf
15. Stam, M.: Blockcipher-Based Hashing Revisited. In: Dunkelman, O. (ed.) FSE 2009. LNCS, vol. 5665, pp. 67–83. Springer, Heidelberg (2009)
16. Wang, X., Yin, Y.L., Yu, H.: Finding collisions in the full SHA-1. In: Shoup, V. (ed.) CRYPTO 2005. LNCS, vol. 3621, pp. 17–36. Springer, Heidelberg (2005)
17. Wang, X., Yu, H.: How to Break MD5 and Other Hash Functions. In: Cramer, R. (ed.) EUROCRYPT 2005. LNCS, vol. 3494, pp. 19–35. Springer, Heidelberg (2005)
18. Wilson, D.: A second-preimage and collision attack on Abacus (2008), http://web.mit.edu/dwilson/www/hash/abacus_attack.pdf

Secure Authenticated Comparisons*

Keith B. Frikken[1], Hao Yuan[2], and Mikhail J. Atallah[3]

[1] Computer Science and Software Engineering, Miami University
frikkekb@muohio.edu
[2] Department of Computer Science, City University of Hong Kong
haoyuan@cityu.edu.hk
[3] Department of Computer Science, Purdue University
mja@cs.purdue.edu

Abstract. In the third-party model for the distribution of data, the data owner provides a third party (referred to as the dealer) with data as well as integrity verification information for that data, in the form of digital signatures that the dealer can use to convince a user of the data's integrity (the dealer is not trusted with the owner's signature keys, which is why it receives pre-signed items). The user's interactions are with the dealer, who is in charge of enforcing access control and confidentiality for the data (i.e., no user should learn more than the outcome of their authorized query). This kind of outsourcing is becoming increasingly important because of its advantageous economics – a dealer who acts as a repository for many owners can achieve economies of scale that are not feasible for the individual owners, and the model allows the data owners to focus on what they do best (the creation and/or acquisition of high-quality data). A problem that arises in the context of outsourced databases (particularly for XML data) is the following: There is a total order Π on n items stored with the dealer, and a user query consists of a pair of items whose relative ordering should be revealed along with a proof that the result is correct. The proof is generated using the dealer's local data (i.e., without bothering the data owner). The main difficulty is achieving efficient storage and query-processing while achieving the desiderata (that the user should learn nothing other than the answer to their query, and that a misbehaving dealer should not be able to convince a user of a wrong ordering). This paper gives a solution that is provably secure under a new assumption and can efficiently generate a very short proof. Furthermore, this scheme is generalized to partial orders that can be decomposed into d total orders. In this case, a user either learns the ordering of the queried items, or learns that they are incomparable.

Keywords: Authenticated Outsourcing, Cryptographic Protocols.

* Portions of this work were supported by National Science Foundation Grants CNS-0915436, CNS-0913875, CNS- 0915843, Science and Technology Center CCF-0939370; by an NPRP grant from the Qatar National Research Fund; by Grant FA9550-09-1-0223 from the Air Force Office of Scientific Research; by sponsors of the Center for Education and Research in Information Assurance and Security; and by a grant from City University of Hong Kong (Project No. 7200218). The statements made herein are solely the responsibility of the authors.

J. Lopez and G. Tsudik (Eds.): ACNS 2011, LNCS 6715, pp. 514–531, 2011.

1 Introduction

Critical data is nowadays accessed through the Internet, often through servers that do not belong to the data owner. The protection of such re-distributed information was explicitly mentioned by Tygar in his early summary of critical open problems in electronic commerce [42]. When important decisions are made based on such data (e.g., business, health-care, scientific, etc), it becomes especially important to provide a user the means of verifying the integrity of the data to which they are entitled; such access rights for data occur because of a user's role in their organization, their government security clearance level, or perhaps because of payment for access. A compelling case was made in the third-party distribution literature, reviewed below in Section 5, for the use of schemes that reduce the trust level required from a third-party dealer. But some thorny issues arise when data distribution is done through a third party dealer who is "moderately trusted" in the sense that, although the data owner relies on the dealer for distribution, the owner does not wish to provide the dealer with the authority to sign on its behalf; this is typically not out of fear for the signature keys (special keys can be created for that purpose), but rather for purposes of data quality control and fear of liability – a compromised dealer should not have the capability of convincing users that corrupted data is pristine. This caution is often not motivated by mistrust of the dealer as an organization, rather, it is a realistic recognition of the fact that networked systems are vulnerable to break-ins, spyware, insider misbehavior, or simply accidents.

The third-party distribution model offers many economic advantages but also a challenge: How does the third-party distributor, who does not have the authority to sign, prove to a user the integrity of the answer to the user's query? There are typically too many possible queries for the owner to pre-sign and store with the dealer a signature for each – the space taken by the data and its signatures on the dealer's system should have size that is close to linear in the size of that data. The problem is conceptually easy when a query asks for a subset of the n data items, as the use of aggregate signatures enables the dealer to convince the user with a single signature (that would be the aggregate, computed by the dealer, of the pre-stored signatures for the individual items in the user's query). But queries about semi-structured data involve comparisons, e.g., they require a proof that node v is to the left of a sibling node w without revealing to the user anything else (e.g., the number of other siblings that are between v and w in the structure); the authors of [28,29] explicitly make the case for the necessity in the XML context of carrying out sibling-order comparisons with the exact properties we require. The problem of comparisons also arises with "flat" data, when there is value-added information computed by the owner and that imposes structure on the individual items. The simplest such structure is a total ordering of the n items stored at the dealer, according to the data owner's proprietary evaluation methodology for the items; there could even be k such total orderings, each of which is based on different evaluation criteria from the others. A user who is comparing the qualifications of two or more of the items stored with the dealer will want to know which is the best for the purpose contemplated. Examples of such criteria include: For publicly traded securities, their current level of attractiveness as an investment, or how they will be affected by some future event; in geopolitical consulting, estimates how one specific hypothetical future would impact various corporate or political entities (or, more boldly, estimates of the likelihood of each of n possible futures to materialize); for individuals or corporations offering services, their levels of proficiency, or future promise, or likelihood of misbehavior; for technologies or products, their effectiveness/reliability/cost for particular tasks. If a user is only authorized to compare two items X and Y, providing the user with the actual score of each would allow the user to compare items other than X and Y (violating disclosure rules, and possibly depriving the data owner of future revenue).

Our contribution: This paper presents an efficient and provably secure scheme for the problem of authenticated secure comparisons that was informally sketched above, and that will be more formally defined in the next section. The scheme requires an amount of storage, at the dealer, that is $O(n \log n)$ where n is the number of items, and the setup and protocols are efficient in that they take a constant amount of communication (hence rounds). We prove that our scheme does not reveal to a user anything other than the outcome of a comparison, that a user can verify the correctness of the answer, and that a rogue dealer is unable to prove a false answer to a query. The security of our scheme is proven in the standard model and is based on a new assumption over bilinear maps, which we call a Computational Linear Splitting Assumption (CLS): That it is hard for an adversary who has g^a, g^b, g^c, g^d to produce g^m and g^n such that $mn = ab + cd$ (note that a decisional version of this assumption is not true). To give evidence of the security of this assumptions, we prove CLS is secure in the generic group model.

2 Preliminaries

2.1 Notation

The security of our approach is in the computational model, and thus we assume the adversaries are probabilistic polynomial time (PPT). A function $\epsilon(n)$ is negligible if for all polynomials p, $\exists N$ such that $\forall n > N$, $\epsilon(n) < \frac{1}{p(n)}$. Given a set S, the notation $s \overset{U}{\leftarrow} S$ corresponds to choosing a value uniformly from S.

2.2 Problem Formulation

In our model we consider three types of entities, a data owner, a data dealer, and users. The data owner has a set of identities which we denote as $\{1, \ldots, n\}$[1] and a permutation $\Pi : \{1, \ldots, n\} \rightarrow \{1, \ldots, n\}$ that defines a total ordering of the identities (based on, e.g., some proprietary information of the data owner). The data owner would like to support queries of the form "Which is greater according to Π, identity i or identity j?". What makes this difficult is that the data owner wants to provide the answers in an outsourced, verified, and zero-knowledge manner. More specifically, the data owner does not want to provide a service that answers queries, and would like instead to outsource the answering of such queries to the data dealer. However, the users of the system do not trust that the data dealer will honestly answer queries, and thus the data dealer must provide a proof that the results are correct ("as good if the user got them directly from

[1] If the identities are not in this domain, then the data owner simply provides a mapping between the real identities and these values.

the owner"). Finally, the queries should be answered in a confidential manner such that the users do not receive additional information other than the answers of their queries as well as what can be inferred from such queries.

Formally, authenticated outsourced secure comparisons consist of three algorithms:

1. $\mathrm{SETUP}(\Pi, 1^\kappa)$ takes as input the ordering Π of the identities and a security parameter 1^κ. The output is (dk, pk) which are respectively the dealer key and the public key.
2. $\mathrm{PROVE}(i, j, pk, dk)$ takes as input identities i and j (such that $\Pi(i) < \Pi(j)$), the public key, pk, and the dealer key, dk. The output is a certificate $cert$.
3. $\mathrm{VRFY}(i, j, pk, cert)$ takes as input identities i and j, the public key, pk, and a certificate $cert$. The output of this algorithm is a binary value b. If $b = 1$ we say the the certificate is accepted, otherwise we say that the certificate is rejected.

2.3 Security Definitions

The security goals of this paper are threefold: i) correctness, ii) protection against dishonest dealer, and iii) zero-knowledge queries.

Correctness. The correctness requirement simply states that if the dealer is honest then the verification algorithm accepts the certificates. More formally, given permutation Π and security parameter κ, then $\forall i, j \in [1, n]$ such that $\Pi(i) < \Pi(j)$:

$$\Pr\big[\mathrm{VRFY}(i, j, pk, cert) = 1 :$$
$$(dk, pk) \leftarrow \mathrm{SETUP}(\Pi, 1^\kappa), cert \leftarrow \mathrm{PROVE}(i, j, dk, pk)\big] = 1.$$

Protection against Dishonest Dealer. This requirement states that a dishonest dealer cannot convince an honest user that $\Pi(i) < \Pi(j)$ when in fact $\Pi(j) < \Pi(i)$. For a specific protocol $\Lambda = (\mathrm{SETUP}, \mathrm{PROVE}, \mathrm{VRFY})$, this notion is captured in the experiment in Figure 1. Protocol Λ is secure against forgery by a dishonest dealer if, for all probabilistic polynomial time adversaries A, $\Pr[\mathbf{Exp}_{\Lambda,A}^{\text{dealer-forge}}(1^\kappa, \Pi)] = 1]$ is negligible in κ.

Zero Knowledge Queries. Let π be an oracle that provides indirect access to the ordering Π; moreover, $\pi(i, j)$ returns 1 if $\Pi(i) < \Pi(j)$ and returns 0 otherwise. Suppose algorithm SIM^π has oracle access to π. More specifically, whenever SIM needs to know the relationship about i and j it queries the oracle for $\pi(i, j)$. After several queries to π, let $\hat{G}$ be the relationship graph defined by these queries. Clearly, if there is no path between i and j (or vica versa) in $\hat{G}$, then either ordering is possible. Furthermore, SIM does not know this relationship. Informally, a scheme has zero-knowledge queries if for every PPT adversary $\mathcal{A}$, there is a PPT simulator SIM^π that can simulate the $\mathcal{A}$'s view in the real protocol.

More formally, in the real protocol $\mathcal{A}$ is given the public key pk, and then can adaptively ask for the comparison and certificate for each of the pairs $(i_1, j_1), \ldots, (i_m, j_m)$. Thus the view of adversary $\mathcal{A}$ is $(pk, res_1, cert_1, \ldots, res_m, cert_m)$ where res_i is the

Experiment $\mathbf{Exp}_{A,A}^{\text{dealer-forge}}(1^\kappa, \Pi)$
$(dk, pk) \leftarrow \text{SETUP}(1^\kappa, \Pi)$
$(i, j, cert) \leftarrow A(1^\kappa, \Pi, dk, pk)$
if $\Pi(i) > \Pi(j)$ and $\text{VRFY}(i, j, pk, cert) = 1$ then return 1
else return 0

Fig. 1. A forgery experiment involving a dishonest dealer

Experiment $\mathbf{Exp}_{A}^{\text{CLS}}(1^\kappa)$
$[q, \mathbb{G}, \mathbb{G}_T, g, e] \leftarrow \text{BGen}(1^\kappa)$
$a, b, c, d \stackrel{U}{\leftarrow} \mathbb{Z}_q$
$(\alpha, \beta) \leftarrow A(q, \mathbb{G}, \mathbb{G}_T, g, e, g^a, g^b, g^c, g^d)$
if $e(\alpha, \beta) = e(g^a, g^b) * e(g^c, g^d)$ then return 1
else return 0

Fig. 2. $\mathbf{Exp}_{A}^{\text{CLS}}(1^\kappa)$

result of the ith comparison, and we denote this view as $\text{VIEW}_{\mathcal{A}}^{\Lambda}(1^\kappa, \Pi)$. In the simulated protocol $\mathcal{A}$ interacts with a simulator SIM^π, which must produce a public key and then adaptively provide comparisons and certificates for pairs $(i_1, j_1), \ldots, (i_m, j_m)$. Let $\text{SIM}_{\mathcal{A}}^{\pi}(1^\kappa)$ be the values output by a simulator when asked queries by adversary $\mathcal{A}$.

We say that Λ provides perfect zero knowledge queries if for all PPT adversaries $\mathcal{A}$, there exists a PPT simulator SIM such that $\text{VIEW}_{\mathcal{A}}^{\Lambda}(1^\kappa, \Pi)$ and $\text{SIM}_{\mathcal{A}}^{\pi}(1^\kappa)$ are identically distributed.

2.4 Bilinear Maps and Assumptions

We utilize bilinear maps in our protocol. Specifically, we utilize a method $\text{BGen}(1^\kappa)$ that produces $\mathbf{G} = [q, \mathbb{G}, \mathbb{G}_T, g, e]$ where $\mathbb{G}$ and $\mathbb{G}_T$ are groups of order q, q is a prime with κ bits, g is a generator of the group $\mathbb{G}$, and $e : \mathbb{G} \times \mathbb{G} \to \mathbb{G}_T$ is a non-degenerate mapping where the bilinear property holds. That is, $e(g^r, g^s) = e(g, g)^{rs}$.

In this paper we propose a new assumption (which is proven secure in the generic group model in the Appendix), called the Computational Linear Splitting (CLS) assumption. At a high level, CLS states that if an adversary is given $\mathbf{G}, g, g^a, g^b, g^c, g^d$, then it is hard for the adversary to produce g^m and g^n such that $mn = ab + cd$. Clearly, a decisional version of this assumption is not true. That is, given g^m and g^n it is easy to verify whether $mn = ab + cd$. This is done by simply checking whether $e(g^m, g^n) = e(g^a, g^b) * e(g^c, g^d)$. Our protocols take advantage of this gap between the computational and decisional versions of these problems, by requiring the prover to create such a pair of values, which can then be verified by the user. More formally, the CLS experiments is defined in Figure 2. Our assumption is that for all probabilistic polynomial time adversaries A, $\Pr[\mathbf{Exp}_{A}^{\text{CLS}}(1^\kappa) = 1]$ is negligible in κ. It is useful to compare this assumption to the well-known Computational Diffie-Hellman (CDH) assumption, which states that it is difficult to computer g^{ab} when given g^a and g^b. Obviously, if the CDH problem is easy, then so is the CLS problem. However, the converse does not appear to be true since the adversary has the flexibility of choosing any m and n values.

3 Scheme

Initially we introduce two straightforward solutions to the problem, based on signatures. We then proceed to the main construction of the paper.

3.1 Two Straightforward Schemes

Suppose that (KeyGen, Sign, Vrfy) is a standard signature scheme. A simple solution to this problem is to set the public key to the verification key, and $\forall i, j \in [1, n]$ the owner signs the message $i||j$ (resp $j||i$) if $\Pi(i) < \Pi(j)$ (resp $\Pi(j) < \Pi(i)$). The dealer is then simply given all of these signatures. When the user asks for a proof about a pair of values the dealer simply provides the signature from the owner. Now, forging a proof is as hard as forging a signature, and confidentiality follows because all that is revealed is the result. The downside with this approach is that the dealer must store $O(n^2)$ values which limits the scalability of this approach.

Another approach takes the following form: The owner creates commitments for the rank of each items and signs messages of the form $(i||\text{Commit}(\Pi(i)))$ for all $i \in [1, n]$ and sends these signatures to the dealer. When the user asks for the comparison of i and j, the dealer reveals the corresponding commitments and their signatures along with a zero knowledge proof that the value held by one commitment is smaller than the other value. We delay an exact comparison of this approach to our approach until section 5.1.

3.2 A More Interesting Scheme

The two key ideas in our scheme are (i) the use of dummy values each with a number of pre-signed comparison statements about how each dummy value compares relative to some of the identities; (ii) the randomization of the signed statements involving the dummies (so that the user cannot recognize if the same dummy is involved in two comparisons). The dummy values are set up in a sorted binary tree. The root of the tree will be used to prove that any identity in the left half of the tree is smaller than any identity in the right half of the tree. The rest of the internal nodes of the tree are used to prove relationships between nodes in the same subtree. A crucial property that we exploit is that if $\Pi(i) < \Pi(j)$, then there will be exactly one node in the binary tree such that i is in the left subtree of the node and j is in the right subtree; hiding which node is being so used, is necessary to prevent extra information from being revealed.

The algorithms are as follows:

- SETUP Run $\text{BGen}(1^\kappa)$ to obtain $\mathbf{G} = [q, \mathbb{G}, \mathbb{G}_T, g, e]$. Choose $2n+2$ values from $\mathbb{Z}_q$ and denote them by $s, s', s_1^+, \ldots, s_n^+, s_1^-, \ldots, s_n^-$ and $n-1$ values $t_1, \ldots, t_{n-1}$ from $\mathbb{Z}_q^\star$. Let $\hat{T}$ be a complete binary tree built on top of the sorted (according to Π) version of $[1, n]$, that is, the leftmost (resp., rightmost) leaf of $\hat{T}$ is the i such that $\Pi(i) = 1$ (resp., $\Pi(i) = n$). Denote the non-leaf nodes of $\hat{T}$ as $v_1, \ldots, v_{n-1}$. For each non-leaf node v_j of $\hat{T}$ the following is done: For every i in the left subtree[2] of v_j in $\hat{T}$, the owner creates a value $R_{i,j} = g^{\frac{ss_i^+}{t_j}}$. For every i in the right subtree

[2] Where the "left subtree" of a node is the subtree of the node's left child.

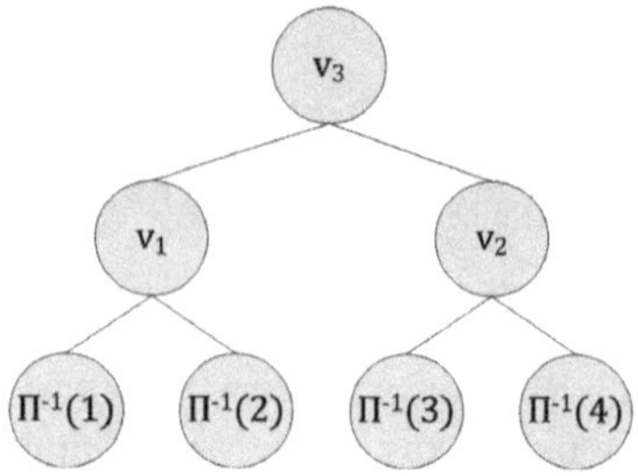

Fig. 3. Example Tree

of v_j in $\hat{T}$, the owner creates a value $S_{i,j} = g^{\frac{s' s_i^-}{t_j}}$. At a high level $R_{i,j}$ (resp. $S_{i,j}$) will be a "certificate" that i is in the left (resp., right) subtree of v_j. Denote all of these R values (resp., S values) for all vertices as the set $\hat{R}$ (resp., $\hat{S}$).
The respective keys produced are as follows:

- **Public Key pk:** $\mathbf{G}, g^s, g^{s'}, g^{s_1^+}, \ldots, g^{s_n^+}, g^{s_1^-}, \ldots, g^{s_n^-}$
- **Dealer Key dk:** $\hat{T}, \Pi, \hat{R}, \hat{S}, g^{t_1}, \ldots, g^{t_{n-1}}$

– PROVE When given i, j, pk, and dk such that $\Pi(i) < \Pi(j)$, the dealer finds the unique vertex $v_k \in \hat{T}$ such that i is in the left subtree of v_k and j is in the right subtree of v_k. Choose a random α from $\mathbb{Z}_q^\star$ and set $cert := ({R_{i,k}}^\alpha * {S_{j,k}}^\alpha, (g^{t_k})^{\alpha^{-1}}) = (g^{\frac{\alpha(s s_i^+ + s' s_j^-)}{t_k}}, g^{\frac{t_k}{\alpha}})$.

– VRFY When given i, j, pk, and $cert = (\beta, \gamma)$, then check if $e(\beta, \gamma) = e(g^s, g^{s_i^+}) * e(g^{s'}, g^{s_j^-})$. Output 1 if this check is true and output 0 otherwise.

Example: Suppose that $n = 4$, the tree created by the above algorithm is shown in Figure 3. For this example, the R and S values that would be part of the dealer key are as follows: $R_{\Pi^{-1}(1),1}$, $R_{\Pi^{-1}(3),2}$, $R_{\Pi^{-1}(1),3}$, $R_{\Pi^{-1}(2),3}$, $S_{\Pi^{-1}(2),1}$, $S_{\Pi^{-1}(4),2}$, $S_{\Pi^{-1}(3),3}$, and $S_{\Pi^{-1}(4),3}$. Let $x = \Pi^{-1}(2)$ and $y = \Pi^{-1}(3)$. To prove that $\Pi(x) < \Pi(y)$, the PROVE protocol would use node v_3 which separates the two nodes in the tree. That is, the certificate would be: $({R_{\Pi^{-1}(2),3}}^\alpha * {S_{\Pi^{-1}(3),3}}^\alpha, (g^{t_3})^{\alpha^{-1}})$.

3.3 Complexity Analysis

The parameters of interest are: i) setup complexity: $O(n \log n)$, ii) size of public key: $O(n)$, iii) size of the dealer key: $O(n \log n)$, iv) complexity of proof: $O(1)$, v) proof size: $O(1)$, and vi) verification cost: $O(1)$. The setup complexity, public key size, dealer key size, proof size, and verification cost all are straightforward from the protocol. It would appear that proof creation requires $O(\log n)$ complexity because it must find the dummy node that separates the leaf nodes corresponding to i and j. However, this node is simply the nearest common ancestor (NCA) of these leaf nodes. It is possible to preprocess the tree in linear time to allow finding the NCA in $O(1)$ time using the data structure of Harel and Tarjan [25]. Note that this pre-processing would be done during the setup.

Reducing the Size of the Public Key to $O(1)$. With a slight increase in the proof size (and in the verification cost), it is possible to reduce the public key size to $O(1)$. The key to this modification is that the querier only needs to know the values $\mathbf{G}, g^s, g^{s'}, g^{s_i^+}$, and $g^{s_j^-}$ when verifying the proof for $\Pi(i) < \Pi(j)$. The linear part of the key is the $+$ and $-$ values. In the modified scheme the public key consists of $\mathbf{G}, g^s, g^{s'}, pk$ where pk is a public key associated with a signature scheme. The owner signs each piece (along with meta-information about the component) of the original public key with its signature key and gives it to the dealer. When the dealer creates the proof, it simply sends the needed pieces (along with their signatures) to the verifier. More formally, the dealer appends the following values to the proof: $(i, +, g^{s_i^+}), Sign_{pk}(i|| + ||g^{s_i^+})$ and $(j, -, g^{s_j^-}), Sign_{pk}(j|| - ||g^{s_j^-})$.

3.4 Proof of Security

The correctness follows from: let $(dk, pk) \leftarrow \text{SETUP}(\Pi, 1^\kappa)$, let i and j be two identities such that $\Pi(i) < \Pi(j)$, and let $(\beta, \gamma) \leftarrow \text{PROVE}(i, j, dk, pk)$. Now, $e(\beta, \gamma) = e(g^{\frac{\alpha(ss_i^+ + s's_j^-)}{t_k}}, g^{\frac{t_k}{\alpha}}) = e(g, g)^{ss_i^+ + s's_j^-} = e(g^s, g^{s_i^+}) * e(g^{s'}, g^{s_j^-})$.

Security against Dishonest Dealer. Suppose that the scheme, Λ, is not secure against a dishonest dealer; that is there exists a polynomial-time adversary $\mathcal{A}$ such that $\Pr[\mathbf{Exp}_{\Lambda,\mathcal{A}}^{\text{dealer-forge}}(1^\kappa, \Pi)] = 1] > \frac{1}{p(\kappa)}$ for some polynomial $p(\kappa)$. We will show how to construct an adversary $\mathcal{B}$ with black box access to $\mathcal{A}$ that solves the CLS with non-negligible probability.

Given an instance of CLS, we are given $\mathbf{G} = [q, \mathbb{G}, \mathbb{G}_T, g, e]$ and g^a, g^b, g^c, and g^d. Algorithm $\mathcal{A}$ will produce a forgery for some pair of values, but as the choice of which pair of values is unknown, algorithm $\mathcal{B}$ will randomly pick two values $\hat{\imath}, \hat{\jmath}$ such that $\Pi(\hat{\jmath}) < \Pi(\hat{\imath})$. We will ensure that the public key and dealer key given to $\mathcal{A}$ will be distributed identically to the real protocol, and since there are only $O(n^2)$ pairs of values, $\mathcal{B}$ will guess which pair that $\mathcal{A}$ will produce a forgery with non-negligible probability.

$\mathcal{B}$ chooses values $r_1^+, \ldots, r_n^+, r_1^-, \ldots, r_n^-, r_1, \ldots, r_{n-1}$ uniformly from $\mathbb{Z}_q^\star$. We implicitly set $s = a$, $s' = c$, $s_{\hat{\imath}}^+ = b$ and $s_{\hat{\jmath}}^- = d$. Note that a forgery proving that $\Pi(\hat{\imath}) < \Pi(\hat{\jmath})$, would require creating two values g^e and g^f such that $e(g^e, g^f) = e(g, g)^{ab+cd}$, which solves the CLS problem.

Moreover, we implicitly set

$$s_k^+ = \begin{cases} cr_k^+ & : \Pi(k) < \Pi(\hat{\imath}) \\ b & : \Pi(k) = \Pi(\hat{\imath}) \\ r_k^+ & : \Pi(k) > \Pi(\hat{\imath}) \end{cases} \quad s_k^- = \begin{cases} r_k^- & : \Pi(k) < \Pi(\hat{\jmath}) \\ d & : \Pi(k) = \Pi(\hat{\jmath}) \\ ar_k^- & : \Pi(k) > \Pi(\hat{\jmath}) \end{cases}$$

Given a tree vertex v_k, denote as ℓ_k the rightmost leaf node in the left subtree of v_k. If $\Pi(\hat{\jmath}) \leq \Pi(\ell_k)$ we set $t_k = ar_k$ and otherwise we set $t_k = cr_k$. It is important to notice that we cannot generate many of the above values, but we can generate all

$\mathcal{B}(1^\kappa, \mathbf{G}, g^a, g^b, g^c, g^d)$
  $\hat{\imath}, \hat{\jmath} \overset{U}{\leftarrow} [1, n]$
  $(pk, dk) \leftarrow \mathrm{PGen}(1^\kappa, \Pi, \hat{\imath}, \hat{\jmath})$
  $(i, j, cert = (\alpha, \beta)) \leftarrow \mathcal{A}(1^\kappa, \Pi, dk, pk)$
  If $i = \hat{\imath}$ and $j = \hat{\jmath}$ output α, β
  else output FAIL

Fig. 4. Algorithm $\mathcal{B}$

values $g^{s_k^+}$, $g^{s_k^-}$, and g^{t_k}. Furthermore, each s_k^+, s_k^-, and t_k are distributed identically to the values chosen in the protocol. All that is left to show is how to compute $R_{k,m}$ and $S_{k,m}$. Let k be an identity in the left subtree of v_m, then we need to be able to compute $R_{k,m} = g^{\frac{ss_k^+}{t_m}} = g^{\frac{as_k^+}{t_m}}$, there are two cases to consider: i) $\Pi(\hat{\jmath}) \leq \Pi(\ell_m)$: In this case $t_m = ar_m$, and thus $R_{k,m} = g^{\frac{as_k^+}{ar_m}} = g^{\frac{s_k^+}{r_m}}$, and ii) $\Pi(\hat{\jmath}) > \Pi(\ell_m)$: Since $\hat{\jmath}$ is after ℓ_m, then $t_m = cr_m$ and $\Pi(k) \leq \Pi(\ell_m) < \Pi(\hat{\jmath}) < \Pi(\hat{\imath})$, thus $s_k^+ = cr_k^+$, and hence, $R_{k,m} = g^{\frac{acr_k^+}{cr_m}} = g^{\frac{ar_k^+}{r_m}}$.

Let k be an identity in the right subtree of v_m, then we need to be able to compute $S_{k,m} = g^{\frac{s's_k^-}{t_m}} = g^{\frac{cs_k^-}{t_m}}$, there are two cases to consider: i) $\Pi(\hat{\jmath}) \leq \Pi(\ell_m)$: In this case $t_m = ar_m$. Also, since k is in the right subtree of v_m, we know that $\Pi(\hat{\jmath}) < \Pi(k)$, thus $s_k^- = ar_k^-$, and hence, $S_{k,m} = g^{\frac{car_k^-}{ar_m}} = g^{\frac{cr_k^-}{r_m}}$, and ii) $\Pi(\hat{\jmath}) > \Pi(\ell_m)$: In this case $t_m = cr_m$, and thus $S_{k,m} = g^{\frac{cs_k^-}{cr_m}} = g^{\frac{s_k^-}{r_m}}$.

Let $\mathrm{PGen}(1^\kappa, \Pi, \hat{\imath}, \hat{\jmath})$ be the procedure described above for computing pk and dk. Algorithm $\mathcal{B}$ works as described in Figure 4.

If $\mathcal{B}$ does not output FAIL and $\mathcal{A}$ succeeds in producing a forgery, then $\mathcal{B}$ will solve the CLS. Now,

$$\Pr[\mathbf{Exp}_{\mathcal{B}}^{\mathrm{CLS}}(1^\kappa) = 1] = \Pr[\mathbf{Exp}_{\Lambda,\mathcal{A}}^{\text{dealer-forge}}(1^\kappa, \Pi) = 1 \wedge \mathcal{B}(1^\kappa, \mathbf{G}, g^a, g^b, g^c, g^d) \neq FAIL]$$

Regardless of $\mathcal{B}$'s choice for $\hat{\imath}$ and $\hat{\jmath}$ the public key and dealer key produced by $\mathcal{B}$ is distributed identically. Thus these two events are independent, and the former's probability is $\geq \frac{1}{p(\kappa)}$ and the latter's probability is $\geq \frac{1}{n^2}$. Thus $\Pr[\mathbf{Exp}^{\mathrm{CLS}_{\mathcal{B}}}(1^\kappa) = 1] \geq \frac{1}{n^2 p(\kappa)}$ which is not negligible. $\square$

Zero Knowledge Queries. The proof of zero knowledge queries rests on the certificates produced by $\mathrm{PROVE}(i, j, pk, dk)$ being distributed identically to tuples of the form $\{g^{r^{-1}(ss_i^+ + s's_j^-)}, g^r) : r \leftarrow \mathbb{Z}_q^\star\}$. This follows because the certificate is of the form $(R_{i,k}{}^\alpha * S_{j,k}{}^\alpha, (g^{t_k})^{\alpha^{-1}}) = (g^{\frac{\alpha(ss_i^+ + s's_j^-)}{t_k}}, g^{\frac{t_k}{\alpha}})$ where $\alpha \overset{U}{\leftarrow} \mathbb{Z}_q^\star$. Let $\rho = \frac{t_k}{\alpha}$, in this case the certificate is simply $(g^{\rho^{-1}(ss_i^+ + s's_j^-)}, g^\rho)$. Now ρ and r are distributed identically, because α is a randomly chosen value uniformly from $\mathbb{Z}_q^\star$. Thus the claim follows.

Given the above observation, the simulator will operate as follows: it will do the same random setup to produce its own pk, dk. The public key will be chosen identically to that in the real protocol (since it is generated in the same manner). A difference is that the simulator knows the values of s, s', s_1^+, s_1^-, etc. And so, the simulator can prove any relationship between any pairs of values. When the client asks for a specific proof, the simulator gets the correct response using its oracle to π and then produces a proof that is distributed identically to the real proof. The specific simulator, $\text{SIM}_{\mathcal{A}}^{\pi}(1^\kappa)$, is described formally below:

Simulator Setup: The simulator does the following steps:

1. $\mathbf{G} \leftarrow \text{BGen}(1^\kappa)$
2. $s, s', s_1^+, \ldots, s_n^+, s_1^-, \ldots, s_n^- \stackrel{U}{\leftarrow} \mathbb{Z}_q$
3. $pk := \mathbf{G}, g^s, g^{s'}, g^{s_1^+}, \ldots, g^{s_n^+}, g^{s_1^-}, \ldots, g^{s_n^-}$
4. The simulator gives $\mathcal{A}$ the public key pk.

Query responses: When $\mathcal{A}$ queries for the relationship between i and j, the simulator uses its oracle access to π to determine the relationship between the two values. Without loss of generality suppose that $\Pi(i) < \Pi(j)$. The simulator then does the following: Choose $r \stackrel{U}{\leftarrow} \mathbb{Z}_q^\star$, and return $(g^{r^{-1}(ss_i^+ + s's_j^-)}, g^r)$.

Now, $\text{VIEW}_{\mathcal{A}}^{A}(1^\kappa, B_1, \ldots, B_n, \Pi)$ consists of pk and several certificates. Now pk is generated by the same process in both the simulator and the real protocol, and thus these are distributed identically. Furthermore, the certificates also have the same distribution. Thus these views are distributed identically. □

4 Extensions

4.1 Min/Max Queries

Consider the following authenticated range minimum query (RMQ) problem: a user wants to query for the position of the minimum element among $\Pi(i), \Pi(i+1), \Pi(i+2), \cdots, \Pi(j)$, where $i < j$ are two query parameters. We use $\text{RMQ}(i, j)$ to denote the position k such that $\Pi(k) = \min_{i \le p \le j} \Pi(p)$.

To return authenticated $k = \text{RMQ}(i, j)$ without leaking Π, a naive method is: For each $p \in [i, k-1] \cup [k+1, j]$, the dealer sends to the user a proof that $\Pi(k) < \Pi(p)$. This naive approach has a complexity of $O(n)$. More efficient solution can be obtained using the technique of Cartesian trees [43]. In the Cartesian tree for Π, each tree node v_i has two keys: i and $\Pi(i)$. The tree is organized in a unique way that: it is a binary search tree under the first keys, and a minimum heap under the second keys. In Gabow, Bentley and Tarjan [18], it was shown that the Cartesian tree for Π can be constructed in $O(n)$ time, and $\text{RMQ}(i, j) = \text{LCA}(i, j)$, where $\text{LCA}(i, j)$ is the least (or lowest) common ancestor of nodes v_i and v_j in the Cartesian tree. Therefore, it is sufficient to prove that v_k (where $k = \text{RMQ}(i, j)$) is the lowest common ancestor of v_i and v_j, with the requirement that the structure of the Cartesian tree is not leaked.

To prove that node v_k is the lowest common ancestor of v_i and v_j, it is equivalent to prove that:

1. v_k is an ancestor of v_i;
2. v_k is an ancestor of v_j;
3. v_k's left child (if it exists) is not an ancestor of v_j;
4. v_k's right child (if it exists) is not an ancestor of v_i.

Property (1) and (2) can be easily tested in the following way [39,26]: A node u is a proper ancestor of a node v if and only if the preorder number [27] of u is smaller than the preorder number [27] of v, and the postorder number of u is bigger than the postorder number of v. Therefore, the owner can define two new mappings $\Pi_1(i) =$ preorder number of v_i, and $\Pi_2(i) =$ postorder number of v_i, and then let the dealer to run the authenticated secure comparison protocol to prove (1) and (2) based on Π_1 and Π_2. Property (3) can be tested by showing that the postorder number of the left child of v_k is smaller than the postorder number of the left child of v_j. Note that we can add one or two dummy children as new leaves to each tree node to make sure each tree node has two children. Similarly, property (4) can be tested by comparing the preorder numbers of the right children of v_k and v_i. To generate efficient proofs for property (3) and (4), let T' be the new Cartesian tree with added dummy leaves, then the owner can define two new mappings Π_3 and Π_4, where $\Pi_3(i)$ is the preorder number of the left child of v_i in T', and $\Pi_4(i)$ is the postorder number of the right child of v_i in T'. The dealer then runs the authenticated secure comparison protocol to prove property (3) and (4) based on Π_3 and Π_4. The complexity for proving the LCA and RMQ is then reduced to $O(1)$.

4.2 Extension to Partial Orders

Given a partial order, it is possible to decompose the partial order into the intersection of a group of t total orders, and the dimension of a partial order is the minimum such t. In this section we describe a technique for extending the system to d-dimensional partial orders where the decomposition into d total orders is known. Unfortunately, computing the dimension of a partial order is NP-complete [44]. However, there are some cases where computing the dimension (and the decomposition) is efficient. For example, trees have dimension at most 2, and the decomposition is straightforward. Also, partial orders whose transitive reduction is a planar graph have dimension at most 3, and the decomposition can be computed in linear time [40]. Furthermore, if the transitive reduction is 4-colorable then the dimension is at most 4 [40]. Thus the following section extends the previous scheme to any partial order where the decomposition into total orders is known, which is significantly more general than the protocol in section 3.2. It is important to note that the following scheme does not need a minimal decomposition, however a non-minimal decomposition leads to a less efficient scheme.

Specifically, we represent the partial order over $[1, n]$ as d total order $\Pi_1, \ldots, \Pi_d$. Moreover, if $\bigwedge_{i=1}^{d} \Pi_i(k) < \Pi_i(j)$, then we say that k is less than j. However, if $\exists i_1, i_2 \in [1, d] : \Pi_{i_1}(k) < \Pi_{i_1}(j)$ and $\Pi_{i_2}(j) < \Pi_{i_2}(k)$, then we say that j and k are incomparable. Proving statements of the form "k is less than j" is straightforward, we utilize d versions of the scheme outlined in section 3.2, one for each of the d dimensions. If "k is less than j', then the dealer will be able to prove this statement for all dimensions. The interesting part of this protocol is creating a proof that two elements are incomparable. If j and k are incomparable, then there will be a pair, Π_{d_1} and Π_{d_2} of partial orders where $\Pi_{d_1}(k) < \Pi_{d_1}(j)$ and $\Pi_{d_2}(j) < \Pi_{d_2}(k)$, and thus it would appear that all that needs to be done is to use the schemes for dimensions d_1 and d_2 to prove both of these statements. Unfortunately, this reveals additional information, namely it would reveal the specific dimensions to the querier. To hide this information we will apply the scheme to all d-dimensions a second time, and furthermore use the same values $s, s', s_1^+, \ldots, s_n^+, s_1^-, \ldots, s_n^-$ for each dimension. With this approach the dealer will be able to prove that "k is less than j" and that "j is less than k" for incomparable values. However, for comparable values the dealer would not be able to generate such proofs.

The scheme requires $O(d)$ versions of the original protocol. Thus the size of the public key is $O(dn)$ and the size of the dealer key is $O(dn \log n)$. The size of a proof that $\Pi(i) < \Pi(j)$ is $O(d)$ and the size of a proof of incompatibility is $O(1)$. Note that a zero-knowledge incompatibility proof based on range proof protocols for commitments would result in a proof size of $O(d)$. Therefore, our scheme has a significant advantage for proving incompatibility efficiently.

An authenticated outsourced private secure comparisons for partial orders consists of five algorithms ($POSETUP, POPROVELT, POPROVEIN, POVRFYLT, POVRFYIN$), which correspond to setup, proving less than, proving incomparability, verifying less than, and verifying incomparability.

Before describing the scheme formally, we define two variations of the Setup protocol in section 3.2. These variations allow us to specify some of the parameters used by the setup algorithm.

1. SETUP($\mathbf{G}, \Pi, 1^\kappa$): This does all of the steps in setup, but uses the group $\mathbf{G}$ instead of creating its own group.
2. SETUP($\mathbf{G}, \Pi, 1^\kappa, s, s', s_1^+, \ldots, s_n^+, s_1^-, \ldots, s_n^-$): This does all of the steps in SETUP, but uses the group $\mathbf{G}$ instead of creating its own group, and uses the values $s, s', s_1^+, \ldots, s_n^+, s_1^-, \ldots, s_n^-$ instead of generating its own parameters.

The actual protocols are as follows:

- POSETUP Run BGen(1^κ) to obtain $\mathbf{G} = [q, \mathbb{G}, \mathbb{G}_T, g, e]$. For each $j = 1$ to d run SETUP($\mathbf{G}, \Pi_j, 1^\kappa$) and denote the keys as (pk_j, dk_j).
  Furthermore, choose $2n+2$ values from $\mathbb{Z}_q$ and denote them by $s, s', s_1^+, \ldots, s_n^+, s_1^-, \ldots, s_n^-$ For each $j = 1$ to d run SETUP($\mathbf{G}, \Pi_j, 1^\kappa, s, s', s_1^+, \ldots, s_n^+, s_1^-, \ldots, s_n^-$) and denote the keys as $(\hat{pk}_j, \hat{dk}_j)$. Note that $\hat{pk}_j$ is the same for all dimensions $j \in [1, d]$, and thus it is denoted simply as $\hat{pk}$. The keys defined by setup are as follows:
  - **Public Key pk:** $pk_1, \ldots, pk_d, \hat{pk}$
  - **Dealer Key dk:** $dk_1, \ldots, dk_d, \hat{dk}_1, \ldots, \hat{dk}_d$
- POPROVELT When given i, j, pk, and dk such that i is less than j. For all $\ell \in [1, d]$, $\Pi_\ell(i) < \Pi_\ell(j)$. Thus for each dimensions $\ell \in [1, d]$, the dealer creates $cert_\ell = \text{PROVE}(i, j, pk_\ell, dk_\ell)$. The certificate is $cert_1, \ldots, cert_d$.
- POVRFYLT When given i, j, pk, and $cert = cert_1, \ldots, cert_d$, the verification algorithm checks whether VRFY($i, j, pk, cert_\ell$) accepts for all dimensions $\ell \in [1, d]$. If so, then this outputs 1 and accepts the certificate, and otherwise this rejects the certificate and outputs 0.

- POPROVEIN When given i, j, pk, and dk such that i is incomparable with j. There exists total orders Π_{d_1} and Π_{d_2} such that $\Pi_{d_1}(i) < \Pi_{d_1}(j)$ and $\Pi_{d_2}(j) < \Pi_{d_2}(i)$. The dealer creates $cert_{i,j} = \text{PROVE}(i, j, \hat{pk}, dk_{d_1})$ and $cert_{j,i} = \text{PROVE}(j, i, \hat{pk}, dk_{d_2})$. The certificate is $(cert_{i,j}, cert_{j,i})$.
- POVRFYIN When given i, j, pk, and $cert = (cert_{i,j}, cert_{j,i})$, the verification algorithm checks whether $\text{VRFY}(i, j, \hat{pk}, cert_{i,j})$ and $\text{VRFY}(j, i, \hat{pk}, cert_{j,i})$. If both accept, then this outputs 1 and otherwise it outputs 0.

5 Related Work

Hacigümüş [24] proposed the database outsourcing framework. Devanbu et al. [15] considered the query result integrity verification problem for basic operators (e.g., Selection, Projection, Union, etc) in outsourced databases, based on the Merkle hash tree technique [33]. Mykletun et al. [34] used signature aggregation approaches (Condensed-RSA and BGLS [3]) to authenticate the SELECT query result. Narasimha and Tsudik [35] used signature aggregation and chaining to verify the query result completeness for basic operators in outsourced databases. Cheng, Pang and Tan [10] also used a similar signature chain approach to verify the completeness of multi-dimensional query results. Their signature chain approaches [35,10] to verify query result completeness leaks some data tuples outside the query result. Pang and Tan [36] proposed a repeated hashing approach to solve the data leakage problem in verifying query result completeness. Their approach can also handle range aggregate operators (like COUNT, SUM and MAX/MIN). Using a similar method, Cheng and Tan [11] considered the problem of authenticating k nearest neighbor queries. Chen et al. [9] addressed the completeness verification problem in a different access control model. Although the scheme of Pang and Tan [36] can be used to authenticate range aggregate query result, their approach leaks information other than the query results. Haber et al. [23] also provided schemes to authenticate query results, but their schemes still leak information other than the query results. Other related work on database outsourcing include authenticate dynamic outsourced databases [30,20], authenticated data structures [19,21,37], and query result freshness [30].

For the secure authenticated comparison problem, the closest work is range proof [32,16,4,31,22,5,45,8]: Given a commitment of a secret, a range proof is a zero-knowledge protocol to prove that the committed secret belongs to a specific range. Using range proof, one can solve the secure authenticated comparison problem in the following way: Let $\text{Commit}(x, r)$ represent a commitment of a secret x, where r is a random number. The commitment scheme can be either Pedersen commitment [38] or Fujisaki-Okamoto commitment [17,14], depending on the range proof protocol used.

Note that those commitment schemes are homomorphic, i.e.,

$\text{Commit}(x_1, r_1)/\text{Commit}(x_2, r_2) = \text{Commit}(x_1 - x_2, r_1 - r_2)$. The owner signs the commitments $\text{Commit}(\Pi(i), r_i)$ of $\Pi(i)$ for every i, makes them public, and then sends r_i's to the dealer. Whenever a user asks for comparing $\Pi(i)$ and $\Pi(j)$, the dealer gives a zero-knowledge proof to convince the user that the secret hidden by $\text{Commit}(\Pi(i), r_i)$ is smaller or bigger than the secret hidden by $\text{Commit}(\Pi(j), r_j)$ using a range proof. For example, to prove that $\Pi(i) < \Pi(j)$, one can prove that the secret in $\text{Commit}(\Pi(j), r_j)/\text{Commit}(\Pi(i), r_i)$ is in the range $[1, n-1]$.

The complexity for generating the range proof (in terms of modular exponentiations and final proof size) is $O(\log n)$ if the range proof protocol is based on the classic bit-by-bit approach [32,16]. Note that it is possible to make the proof size constant using the scheme of Boudot [4], Lipmaa [31], Groth [22], or Yuen et al. [45]. Note that the proof size generated by the protocol of Camenisch, Chaabouni and shelat [5] (later improved by Chaabouni, Lipmaa and shelat [8]) can also be constant for the secure authenticated comparison problem (see the following subsection). In the following section, we argue that the range proof based approach is less efficient than our scheme for the secure authenticated comparison problem.

5.1 A More Exact Comparison with Previous Work

Our protocol is non-interactive *without assuming the random oracle model* [1,7]. Most of the previous protocols [4,31,22,5] are 3-round Σ-protocols [12] if the random oracle model is not allowed. Moreover, they are *honest-verifier* zero-knowledge unless assuming random oracle model or applying some general transformation (e.g., the 4-round general zero-knowledge transformation for certain Σ-protocols by Cramer et al. [13]). The only previous range proof that is non-interactive without random oracle is due to Yuen et al. [45].

Now consider the exact proof sizes of our protocol and the previous ones. We will use $|G|$ to denote the number of bits to represent a group element in G. We follow the parameters in Camenisch, Chaabouni and shelat [5], e.g., $|G| = 256$, $|G_T| = 3072$, and $|Z_p| = 256$ for the bilinear groups G, G_T of prime order p.

In our protocol, only two group elements in G are sent to the user, so the proof size is $2|G| = 512$ bits.

Using the protocol of Camenisch, Chaabouni and shelat [5], the most efficient proof size is achieved when $u = n$ and $l = 1$ in their paper (recall that their basic protocol proves that a committed secret lies in $[0, u^l)$). The solution is sketched below.

- SETUP: The owner sends $n-1$ Boneh-Boyen signatures [2] (secure against chosen message attack) to the dealer, the signatures are for messages $1, 2, 3, \cdots, n-1$. For each $1 \leq i \leq n$, the owner commits the rank $\Pi(i)$ to C_i, signs C_i using any signature scheme, and publishes C_i and its signature.
- PROVE: To show that $\Pi(i) < \Pi(j)$, the dealer can prove in zero knowledge that it possesses a signature for the secret in C_j/C_i. This is sound because C_j/C_i is a commitment of $\Pi(j) - \Pi(i)$, which belongs to the interval $[1, n-1]$. If the dealer wants to fake a proof, then it needs to forge a signature for some integer in $[-n+1, -1]$.

The proof size by this approach is 4608 bits. Note that Camenisch and Lysyanskaya's signature scheme [6] also has an efficient signature possession proof that can be used in a way similar to the above scheme, but it is less efficient than the one by Camenisch, Chaabouni and shelat [5].

Other approaches by Boudot [4], Lipmaa [31], Groth [22] are much less efficient in the query stage, because of the needs to use 3072-bit RSA keys to match the security of our protocol and Camenisch et al.'s [5]. For example, under the random oracle model, Lipmaa's sum of four squares protocol [31] requires about $30720 + \frac{5}{2}\log_2(n-1)$ bits to do the non-negativeness proof. Groth's sum of three squares improvement [22] requires about $23936 + 2\log_2(4n - 3)$ bits. The protocol of Yuen et al. [45] requires 27648 bits, whose size is more than 50 times larger than our 512-bit proof size.

Therefore, our protocol can be viewed as a non-interactive protocol that uses much smaller query-stage proof size, while increasing server space from $O(n)$ to $O(n \log n)$.

6 Conclusion/Future Work

In this paper we give a protocol for outsourced comparisons, where the dealer has to prove that the answers are correct and all that is revealed to the querier is the outcome of the comparison queries. We give a protocol for comparisons over total orders for n items, where the proof size is $O(1)$ and the dealer is required to store $O(n \log n)$ values. The security of our approach is based on a new assumption, the Computational Linear Splitting Assumption, which may be of independent interest.

Acknowledgments

The authors thank the anonymous reviewers for their comments and useful suggestions.

References

1. Bellare, M., Rogaway, P.: Random oracles are practical: A paradigm for designing efficient protocols. In: CCS 1993: Proceedings of the 1st ACM Conference on Computer and Communications Security, pp. 62–73. ACM, New York (1993)
2. Boneh, D., Boyen, X.: Short signatures without random oracles. In: Cachin, C., Camenisch, J.L. (eds.) EUROCRYPT 2004. LNCS, vol. 3027, pp. 56–73. Springer, Heidelberg (2004)
3. Boneh, D., Gentry, C., Lynn, B., Shacham, H.: Aggregate and verifiably encrypted signatures from bilinear maps. In: Biham, E. (ed.) EUROCRYPT 2003. LNCS, vol. 2656, pp. 416–432. Springer, Heidelberg (2003)
4. Boudot, F.: Efficient proofs that a committed number lies in an interval. In: Preneel, B. (ed.) EUROCRYPT 2000. LNCS, vol. 1807, pp. 431–444. Springer, Heidelberg (2000)
5. Camenisch, J.L., Chaabouni, R., Shelat, A.: Efficient protocols for set membership and range proofs. In: Pieprzyk, J. (ed.) ASIACRYPT 2008. LNCS, vol. 5350, pp. 234–252. Springer, Heidelberg (2008)
6. Camenisch, J., Lysyanskaya, A.: A signature scheme with efficient protocols. In: Cimato, S., Galdi, C., Persiano, G. (eds.) SCN 2002. LNCS, vol. 2576, pp. 268–289. Springer, Heidelberg (2003)
7. Canetti, R., Goldreich, O., Halevi, S.: The random oracle methodology. J. ACM 51(4), 557–594 (2004) (revisited)
8. Chaabouni, R., Lipmaa, H., Shelat, A.: Additive combinatorics and discrete logarithm based range protocols. In: Steinfeld, R., Hawkes, P. (eds.) ACISP 2010. LNCS, vol. 6168, pp. 336–351. Springer, Heidelberg (2010), `http://eprint.iacr.org/`
9. Chen, H., Ma, X., Hsu, W., Li, N., Wang, Q.: Access control friendly query verification for outsourced data publishing. In: Jajodia, S., Lopez, J. (eds.) ESORICS 2008. LNCS, vol. 5283, pp. 177–191. Springer, Heidelberg (2008)

10. Cheng, W., Pang, H., Tan, K.: Authenticating multi-dimensional query results in data publishing. In: Proceedings of Data and Applications Security XX, 20th Annual IFIP WG 11.3 Working Conference on Data and Applications Security, pp. 60–73 (2006)
11. Cheng, W., Tan, K.: Authenticating knn query results in data publishing. In: Jonker, W., Petković, M. (eds.) SDM 2007. LNCS, vol. 4721, pp. 47–63. Springer, Heidelberg (2007)
12. Cramer, R.: Modular design of secure yet practical cryptographic protocols. Ph.D. thesis, CWI and University of Amsterdam (1996)
13. Cramer, R., Damgård, I.B., MacKenzie, P.D.: Efficient zero-knowledge proofs of knowledge without intractability assumptions. In: Imai, H., Zheng, Y. (eds.) PKC 2000. LNCS, vol. 1751, pp. 354–373. Springer, Heidelberg (2000)
14. Damgård, I., Fujisaki, E.: A statistically-hiding integer commitment scheme based on groups with hidden order. In: Zheng, Y. (ed.) ASIACRYPT 2002. LNCS, vol. 2501, pp. 125–142. Springer, Heidelberg (2002)
15. Devanbu, P., Gertz, M., Martel, C., Stubblebine, S.G.: Authentic data publication over the Internet. Journal of Computer Security 11(3), 291–314 (2003)
16. Durfee, G., Franklin, M.: Distribution chain security. In: CCS 2000: Proceedings of the 7th ACM conference on Computer and Communications Security, pp. 63–70. ACM, New York (2000)
17. Fujisaki, E., Okamoto, T.: Statistical zero knowledge protocols to prove modular polynomial relations. In: Kaliski Jr., B.S. (ed.) CRYPTO 1997. LNCS, vol. 1294, pp. 16–30. Springer, Heidelberg (1997)
18. Gabow, H.N., Bentley, J.L., Tarjan, R.E.: Scaling and related techniques for geometry problems. In: STOC 1984: Proceedings of the Sixteenth Annual ACM Symposium on Theory of Computing, pp. 135–143. ACM, New York (1984)
19. Goodrich, M., Tamassia, R.: Efficient authenticated dictionaries with skip lists and commutative hashing. Technical Report, Johns Hopkins Information Security Institute (2000)
20. Goodrich, M.T., Tamassia, R., Triandopoulos, N.: Super-efficient verification of dynamic outsourced databases. In: Malkin, T. (ed.) CT-RSA 2008. LNCS, vol. 4964, pp. 407–424. Springer, Heidelberg (2008)
21. Goodrich, M.T., Tamassia, R., Triandopoulos, N., Cohen, R.: Authenticated data structures for graph and geometric searching. In: Joye, M. (ed.) CT-RSA 2003. LNCS, vol. 2612, pp. 295–313. Springer, Heidelberg (2003)
22. Groth, J.: Non-interactive zero-knowledge arguments for voting. In: Ioannidis, J., Keromytis, A.D., Yung, M. (eds.) ACNS 2005. LNCS, vol. 3531, pp. 467–482. Springer, Heidelberg (2005)
23. Haber, S., Horne, W., Sander, T., Yao, D.: Privacy-preserving verification of aggregate queries on outsourced databases. Technical Report of HP Labs, HPL-2006-128 (2006)
24. Hacigümüş, H., Iyer, B., Mehrotra, S.: Providing database as a service. In: Proceedings of the 18th International Conference on Data Engineering, San Jose, CA, February 26 - March 1 (2002)
25. Harel, D., Tarjan, R.E.: Fast algorithms for finding nearest common ancestors. SIAM J. Comput. 13(2), 338–355 (1984)
26. Kannan, S., Naor, M., Rudich, S.: Implicit representation of graphs. SIAM J. Discret. Math. 5(4), 596–603 (1992)
27. Knuth, D.E.: The art of computer programming. fundamental algorithms, vol. I. Addison-Wesley, Reading (1973)
28. Kundu, A., Bertino, E.: A new model for secure dissemination of xml content. IEEE Transactions on Systems Man and Cybernetics-Part C-Applications Reviews 38(3), 292–301 (2008)
29. Kundu, A., Bertino, E.: Structural signatures for tree data structures. Proc. VLDB Endow. 1(1), 138–150 (2008)

30. Li, F.I., Hadjieleftheriou, M., Kollios, G., Reyzin, L.: Dynamic authenticated index structures for outsourced databases. In: Proceedings of the ACM SIGMOD International Conference on Management of Data, Chicago, Illinois, USA, June 27-29, pp. 121–132 (2006)
31. Lipmaa, H.: On diophantine complexity and statistical zero-knowledge arguments. In: Laih, C.-S. (ed.) ASIACRYPT 2003. LNCS, vol. 2894, pp. 398–415. Springer, Heidelberg (2003)
32. Mao, W.: Guaranteed correct sharing of integer factorization with off-line shareholders. In: Imai, H., Zheng, Y. (eds.) PKC 1998. LNCS, vol. 1431, pp. 60–71. Springer, Heidelberg (1998)
33. Merkle, R.C.: Protocols for public key cryptosystems. In: IEEE Symposium on Security and Privacy, pp. 122–134 (1980)
34. Mykletun, E., Narasimha, M., Tsudik, G.: Authentication and integrity in outsourced databases. In: Proceedings of ISOC Symposium on Network and Distributed Systems Security, NDSS 2004 (2004)
35. Narasimha, M., Tsudik, G.: Authentication of Outsourced Databases Using Signature Aggregation and Chaining. In: Li Lee, M., Tan, K.-L., Wuwongse, V. (eds.) DASFAA 2006. LNCS, vol. 3882, pp. 420–436. Springer, Heidelberg (2006)
36. Pang, H., Tan, K.: Verifying completeness of relational query answers from online servers. ACM Trans. Inf. Syst. Secur. 11(2), 1–50 (2008)
37. Papamanthou, C., Tamassia, R., Triandopoulos, N.: Authenticated hash tables. In: CCS 2008: Proceedings of the 15th ACM Conference on Computer and Communications Security, pp. 437–448. ACM, New York (2008)
38. Pedersen, T.P.: A threshold cryptosystem without a trusted party (extended abstract). In: Davies, D.W. (ed.) EUROCRYPT 1991. LNCS, vol. 547, pp. 522–526. Springer, Heidelberg (1991)
39. Santoro, N., Khatib, R.: Labelling and implicit routing in networks. The Computer Journal 28(1), 5 (1985)
40. Schnyder, W.: Planar graphs and poset dimension. Order, 323–343 (1989)
41. Shoup, V.: Lower bounds for discrete logarithms and related problems. In: Fumy, W. (ed.) EUROCRYPT 1997. LNCS, vol. 1233, pp. 256–266. Springer, Heidelberg (1997)
42. Tygar, J.D.: Open problems in electronic commerce. In: Proceedings of 18th ACM PODS (1999)
43. Vuillemin, J.: A unifying look at data structures. Commun. ACM 23(4), 229–239 (1980)
44. Yannakakis, M.: The complexity of the partial order dimension problem. SIAM Journal on Algebraic and Discrete Methods 3(3), 351–358 (1982)
45. Yuen, T., Huang, Q., Mu, Y., Susilo, W., Wong, D.S., Yang, G.: Efficient non-interactive range proof. In: Ngo, H.Q. (ed.) COCOON 2009. LNCS, vol. 5609, pp. 138–147. Springer, Heidelberg (2009)

A Proof of CLS Assumption

Lemma 1. *In the generic group model, the probability that an adversary that performs k operations solves the CLS problem is at most $\frac{5(k+2)^2}{q}$.*

Proof: We keep track of two lists, $L = \{(F_i, \lambda_i)\}$ and $L_T = \{(F_{i,T}, \lambda_{i,T})\}$, of internal/external representation pairs. Initially, we choose five random external representations corresponding to the internal representations: 1, A, B, C, D. The lists are updated according to the procedures listed below:

- *Group Action in* $\mathbb{G}$: Given internal representation F_1 and F_2, compute $F' = F_1 + F_2$. If $(F', \lambda) \in L$ respond with λ. Otherwise choose a new value λ, add (F', λ) to L, and respond with λ.
- *Inversion in* $\mathbb{G}$: Given internal representation F, compute $F' = -F$. If $(F', \lambda) \in L$ respond with λ. Otherwise choose a new value λ, add (F', λ) to L, and respond with λ.
- *Group Action/Inversion in* $\mathbb{G}_T$: Do the same process as for $\mathbb{G}$, except use L_T.
- *Bilinear Map:* Given internal representation F_1 and F_2, compute $F' = F_1 * F_2$. If $(F', \lambda) \in L_T$ respond with λ. Otherwise choose a new value λ, add (F', λ) to L_T, and respond with λ.

Eventually, the adversary outputs a pair of external representations σ_1 and σ_2. If σ_i is not an external representation contained in L, then create a new internal representation $F = E_i$ and add F, σ_i to L. Clearly, all internal representations in L (resp L_T) have degree ≤ 1 (resp 2). We proceed by showing that the above lists are consistent, and then bound the probability that B's answer is a solution to CLS. We assign random values to A, B, C, and D. We also assign random values to E_1 and E_2 if necessary. It was shown in [41] that a d-degree nonzero polynomial in a group of order q will evaluate to 0 with probability at most $\frac{d}{q}$ when the variables of the polynomial. A corollary to this observation is that two non-equal polynomials of degree d will evaluate to the same value for a random assignment of variables with probability at most $\frac{1}{q}$.

The first type of inconsistency would be two internal representations F and G in L such that $F \neq G$, but $F(a,b,c,d,e_1,e_2) = G(a,b,c,d,e_1,e_2)$. By the results in [41], this only happens with probability at most $\frac{1}{q}$. And since there are at most $k+2$ pairs in L, the probability of this happening for any pair is at most $\frac{(k+2)^2}{q}$.

The second type of inconsistency would occur if there is two internal representations F_T and G_T in L_T such that $F_T \neq G_T$ by $F_T(a,b,c,d) = G_T(a,b,c,d)$. Now each polynomial in L_T has degree at most 2, and thus the probability that two specific polynomials cause an inconsistency is at most $\frac{2}{q}$. And since there are at most k pairs in L_T, the probability of this happening for any pair is at most $\frac{2k^2}{q}$.

Finally, we show that it is unlikely that there are two polynomials, F and G, in L such that $F(a,b,c,d,e_1,e_2) * G(a,b,c,d,e_1,e_2) = ab + cd$. This would imply that the answer returned by the adversary does not solve the CLS. All polynomials in L have degree one. Also, there are no two one-degree polynomials over (A,B,C,D,E_1,E_2) such that their product is the polynomial $AB + CD$. Consider the polynomial $F(A,B,C,D,E_1,E_2) * G(A,B,C,D,E_1,E_2) - AB - CD$. By the above, this is not the zero polynomial, and it has at most degree 2. Thus given a random assignment to the variables will produce a value of 0 with at most probability $\frac{2}{q}$. By the union bound, the probability of this happening for any pair is at most $\frac{2(k+2)^2}{q}$.

Thus the probability of any of the three events above happening is at most $\frac{5(k+2)^2}{q}$. □

Public-Key Encryption with Delegated Search

Luan Ibraimi, Svetla Nikova, Pieter Hartel, and Willem Jonker

Faculty of EEMCS, University of Twente, The Netherlands

Abstract. In a public key setting, Alice encrypts an email with the public key of Bob, so that only Bob will be able to learn the contents of the email. Consider a scenario where the computer of Alice is infected and unbeknown to Alice it also embeds a malware into the message. Bob's company, Carol, cannot scan his email for malicious content as it is encrypted so the burden is on Bob to do the scan. This is not efficient. We construct a mechanism that enables Bob to provide trapdoors to Carol such that Carol, given an encrypted data and a malware signature, is able to check whether the encrypted data contains the malware signature, without decrypting it. We refer to this mechanism as *public-key encryption with delegated search* ($\mathcal{PKEDS}$).

We formalize $\mathcal{PKEDS}$ and give a construction based on ElGamal public-key encryption ($\mathcal{PKE}$). The proposed scheme has ciphertexts which are both searchable and decryptable. This property of the scheme is crucial since an entity can search the entire content of the message, in contrast to existing searchable public-key encryption schemes where the search is done only in the metadata part. We prove in the standard model that the scheme is *ciphertext indistinguishable* and *trapdoor indistinguishable* under the Symmetric External Diffie-Hellman (SXDH) assumption. We prove also the *ciphertext one-wayness* of the scheme under the modified Computational Diffie-Hellman (mCDH) assumption. We show that our $\mathcal{PKEDS}$ scheme can be used in different applications such as detecting encrypted malware and forwarding encrypted email.

1 Introduction

Consider an organization, Carol, whose employees use public-key encryption to communicate with other users from outside the organization. All organizational incoming encrypted emails are stored in a server which is managed by Carol. Bob (an employee) can download his encrypted email from the server and decrypt it locally using his private key. Encryption prevents an attacker from learning confidential information, but it opens another problem: the server cannot search the ciphertext for malware. While encryption helps Bob to protect his sensitive data, the hardness of processing the ciphertext without decrypting it, helps the attacker to hide malicious content from the server. Suppose Alice, who resides outside the organization, encrypts a message with the public key of Bob, so that only Bob will be able to learn the contents of the message. Unbeknown to Alice, her computer is infected and embeds malware into the message. Since the malware is encrypted, the server is unable to scan the ciphertext for malicious

J. Lopez and G. Tsudik (Eds.): ACNS 2011, LNCS 6715, pp. 532–549, 2011.

code. A naive solution to detect encrypted malware is for Bob to send his private key to the server. Once the server gets the key, it decrypts the ciphertext and then scans the plain data for a malicious content. However this solution is too risky since the server accesses the plain data and a compromise of the server compromises all Bob's data. Another solution would be to force Bob to scan his email for malicious content. However this approach is not efficient.

In this paper we focus on finding mechanisms which allow the server to search the ciphertext, without decrypting it. Searching encrypted data [BDCOP04] is an attractive technique that might address the aforementioned problem. It allows the server to search encrypted data without learning information about the plain data or the search query. Boneh et al. [BDCOP04] were the first to propose *public-key encryption with keyword search* (a.k.a $\mathcal{PEKS}$ or searchable encryption). It works as follows. Alice creates a ciphertext c_w which encrypts the keyword w and Bob creates a trapdoor t_w for a keyword w. The trapdoor t_w is sent to the server, which on receipt of the searchable ciphertext c_w and the trapdoor t_w, runs the Test function which returns *true* only if both the searchable ciphertext c_w and the trapdoor t_w are associated with the same keyword, otherwise it outputs *false*. $\mathcal{PEKS}$, is only used to encrypt keywords (meta-data) describing the document, while to encrypt the entire document Alice must use a traditional public-key encryption $\mathcal{PKE}$ scheme, where the ciphertext is decryptable but not searchable. This approach is not suitable for some applications, such as detecting encrypted malware, for the following reasons: a) the server can search only inside the $\mathcal{PEKS}$ ciphertext, the other part of the ciphertext created by the $\mathcal{PKE}$ scheme is not searchable, and b) Bob has to stay online - the malware signature database maintained by the server might get updated frequently, therefore Bob has to create trapdoors and send them to the server.

1.1 Our Contribution

In this paper we construct a public-key encryption scheme with delegated search ($\mathcal{PKEDS}$) which has the following properties:

1. Each part of the encrypted data is both searchable and decryptable, unlike in $\mathcal{PEKS}$ where only the metadata part of the ciphertext is searchable. Hence, our scheme can be used alone, without employing an additional $\mathcal{PKE}$ scheme, to provide end-to-end security.
2. Once delegated by Bob, the server is allowed to create any trapdoor without contacting Bob, thus, once the delegation is done, Bob can go offline. We construct a mechanism that enables Bob to provide the server with a master trapdoor t_* such that, given the encrypted data and a word w picked by the server, the server can test whether the word w is in the encrypted data, without decrypting it.
3. The server can answer queries made by Bob. In the proposed scheme, Bob can provide the server with a trapdoor t_w associated with a specific word w such that the server can test whether the word w occurs in the encrypted data, without allowing the server to learn the word w.

We provide a security proof in the standard model and show that the scheme is *ciphertext indistinguishable* and *trapdoor indistinguishable* under the Symmetric External Diffie-Hellman (SXDH) assumption. Note that in our scheme it is *inherently* impossible to achieve these properties against the server (i.e. we can achieve these properties against any adversary excluding the server). The first limitation comes as a result of allowing the server to hold the master trapdoor t_*, from which the server can create any trapdoor associated to any word and break ciphertext indistinguishability security. The second limitation comes from the nature of public-key encryption where an entity (i.e. the server) which holds a trapdoor t_w associated with a specific word w and knows the public key of the receiver can create a valid ciphertext and break trapdoor indistinguishability. This is also observed by Shen, Shi and Waters [SSW09] and to date the property of trapdoor indistinguishability is only achieved in the symmetric key setting where only the secret key holder can create a valid ciphertext. The best that we can achieve against the server is *ciphertext one-wayness* and we show that our scheme is secure in this respect under the modified Computation Diffie-Hellman (mCDH) assumption in the standard model. The construction of the scheme is based on ElGamal private-key encryption ($\mathcal{PKE}$) [ElG85]. Indeed, our scheme can be viewed as an extension of ElGamal, with additional features: it allows the receiver to create trapdoors and the server to search the encrypted data. The proposed construction uses Type-3 pairings [GPS08] which are employed by the server to search the ciphertext and by the receiver to run the TrapGen function in order to generate the trapdoor. The use of Type-3 pairing is crucial, both for running testing functions and for preventing an adversary (other than the server) to break the security of the scheme.

1.2 Related Work

In the public key setting, Boneh et al. [BDCOP04] introduced the first *private-key encryption with keyword search* ($\mathcal{PEKS}$) in which everyone can create a searchable ciphertext but only the owner of a private key can create a trapdoor. The proposed $\mathcal{PEKS}$ scheme [BDCOP04] is based on anonymous identity-based encryption (IBE) as introduced in [BF01]. Abdalla et al. [ABC$^+$05] fix a consistency flaw from [BDCOP04] and provide a transform of an anonymous IBE scheme from Boyen and Waters [BW06] to construct a $\mathcal{PEKS}$ scheme in the standard model. In addition, they show how to extend $\mathcal{PEKS}$ scheme to design a private-key encryption with temporary keyword search.

There are number of improvements to the initial concept of $\mathcal{PEKS}$ in which the search is only done by comparing the keyword of the ciphertext with the keyword of the trapdoor. Boneh and Waters [BW07] propose a scheme which supports conjunctive, subset, and range queries over the keywords. Hwang and Lee [HL07] propose a $\mathcal{PEKS}$ scheme which works in the multiuser setting, where the keyword is encrypted under many public keys for many receivers. Fuhr and Paillier [FP07] propose a decryptable $\mathcal{PEKS}$ scheme with a security proof in the heuristic random oracle model, and Hofeinz and Weinreb [HW08] propose a decryptable $\mathcal{PEKS}$ scheme with a security proof in the standard model.

A similar concept to decryptable $\mathcal{PEKS}$ is the *hybrid model* [BSNS06, ZI07] which integrates $\mathcal{PKE}$ and $\mathcal{PEKS}$ into a single scheme by allowing both schemes to share the same key pair (pk, sk). The difference between the hybrid model and the original $\mathcal{PEKS}$ scheme, is that the first integrates both $\mathcal{PEKS}$ and $\mathcal{PKE}$ schemes into a single scheme, while the later assumes that in addition to $\mathcal{PEKS}$ scheme there is an another separate $\mathcal{PKE}$ scheme. While the hybrid model ties $\mathcal{PEKS}$ and $\mathcal{PKE}$, it does not guarantee any relation between messages encrypted under $\mathcal{PEKS}$ and messages encrypted under $\mathcal{PKE}$ scheme. Particularly, an attacker can always encrypt one message using $\mathcal{PEKS}$ scheme and encrypt a different message using $\mathcal{PKE}$ scheme, in this way causing the server to send emails in which the receivers are not interested. The proposed $\mathcal{PKEDS}$ scheme guarantees this relation since it allows the receiver to decrypt the searchable ciphertext and check whether the keywords indeed describe the original message.

Organization of the Paper. In Section 2 we give a brief description of bilinear groups and complexity assumptions. In Section 3 we define the algorithms of the $\mathcal{PKEDS}$ scheme and formalize the security requirements. In Section 4 we present our $\mathcal{PKEDS}$ construction. In Section 5 we prove its security and in Section 6 we discuss its applications. The last section concludes the paper.

2 Bilinear Groups and Complexity Assumptions

Our construction uses prime order bilinear groups. Let $\mathbb{G}$, Γ and $\mathbb{G}_T$ be groups of prime order p, and let g and γ be generator of $\mathbb{G}$ and Γ, respectively. A pairing (or bilinear map) $\hat{e} : \mathbb{G} \times \Gamma \rightarrow \mathbb{G}_T$ has the following properties [BF01]:

1. Bilinearity: for all $u \in \mathbb{G}$, $v \in \Gamma$ and $a, b \in \mathbb{Z}_p^*$, we have $\hat{e}(u^a, v^b) = \hat{e}(u, v)^{ab}$.
2. Non-degeneracy: $\hat{e}(g, \gamma) \neq 1$.
3. The function $\hat{e}$ can be efficiently computed.

Pairings can be categorized into three types [GPS08]: a) Type-1: is known as symmetric pairing and $\mathbb{G} = \Gamma$, b) Type-2: is known as asymmetric pairing and $\mathbb{G} \neq \Gamma$, but there is an efficiently computable isomorphism $\psi : \mathbb{G} \rightarrow \Gamma$, and c) Type-3: is same as Type-2 except that there is no known efficiently computable isomorphism $\psi : \mathbb{G} \rightarrow \Gamma$.

The ciphertext and trapdoor indistinguishability of the proposed scheme are based on the Symmetric External Diffie-Hellman (SXDH).

Definition 1. ***Symmetric External Diffie-Hellman*** (SXDH) ***Assumption:*** *In Type-3 pairings the Decision Diffie-Hellman Problem (DDH) is intractable both in* $\mathbb{G}$ *and* Γ, *i.e. given a tuple* $(g, g^a, g^b, g^c) \in \mathbb{G}$ *or* $(\gamma, \gamma^a, \gamma^b, \gamma^c) \in \Gamma$ *with* $a, b \in \mathbb{Z}_p$, *decide whether* $c = ab$ *or* $c \in_R \mathbb{Z}_p$.

To prove the one-wayness of the scheme, we use a slightly stronger variant of the CDH assumption which we call it the *modified* CDH (mCDH).

Definition 2. ***Modified Computational Diffie-Hellman*** (mCDH) ***Assumption:*** *Given tuples* $(g, g^a, g^b) \in \mathbb{G}$ *and* $(\gamma, \gamma^b) \in \Gamma$ *with* $a, b \in \mathbb{Z}_p$, *it is hard to compute* g^{ab}.

Note that the mCDH assumption is implied by the BDH-3 [CM09] assumption. Therefore, an algorithm that breaks the mCDH assumption can be converted to an algorithm that breaks the BDH-3 assumption.

3 Description and Security Model of $\mathcal{PKEDS}$ Scheme

Definition 3. *A private-key encryption scheme with delegated search ($\mathcal{PKEDS}$) consists of the following nine algorithms (Setup, $KeyGen_S$, $KeyGen_R$, Encrypt, Delegate, TrapGen, $Test_1$, $Test_2$, Decrypt):*

- $\mathsf{Setup}(\lambda)$: *The setup algorithm is run by a system administrator and takes as input a security parameter λ and outputs public parameters $\mathcal{PP}$.*
- $\mathsf{KeyGen}_S(\mathcal{PP})$: *This key generation algorithm is run by the server and takes as input public parameters $\mathcal{PP}$ and outputs the server's public/private key pair (pk_S, sk_S).*
- $\mathsf{KeyGen}_R(\mathcal{PP})$: *This key generation algorithm is run by Bob (the receiver) and takes as input public parameters $\mathcal{PP}$ and outputs Bob's public/private key pair (pk_R, sk_R).*
- $\mathsf{Encrypt}(pk_R, w)$: *The encryption algorithm is run by Alice (the message sender) and takes as input Bob's public key pk_R and a word w, and outputs a ciphertext c_w.*
- $\mathsf{Delegate}(sk_R, pk_S)$: *The delegate algorithm is run by Bob and takes as input Bob's private key sk_R, the server's public key pk_S, and outputs the master trapdoor t_*.*
- $\mathsf{TrapGen}(sk_R, pk_S, w)$: *The trapdoor generation algorithm is run by Bob and takes as input Bob's private key sk_R, the server's public key pk_S and a word w, and outputs the trapdoor t_w.*
- $\mathsf{Test}_1(c_w, t_*, t_w, sk_S)$: *The first testing algorithm is run by the server and takes as input a ciphertext c_w, a master trapdoor t_*, a trapdoor t_w associated with the word w, and the server's private key sk_S, and outputs* true *if the ciphertext and the trapdoor are associated with the same word, otherwise outputs $\perp$.*
- $\mathsf{Test}_2(c_w, t_*, w, sk_S)$: *The second testing algorithm is run by the server and takes as input a ciphertext c_w, a master trapdoor t_*, a word w, and the server's private key sk_S, and outputs* true *if the ciphertext contains the word w, otherwise outputs $\perp$.*
- $\mathsf{Decrypt}(c_w, sk_R)$: *The decryption algorithm is run by Bob and takes as input a ciphertext c_w and Bob's private key sk_R, and outputs the word w or $\perp$ if c_w is invalid.*

Correctness. We say that $\mathcal{PKEDS}$ is *correct* if for all security parameters $\lambda \in \mathbb{N}$, for all server public/private key pairs produced by KeyGen_S, for all receiver public/private key pairs produced by KeyGen_R, for all ciphertexts c_w produced by Encrypt, for all master trapdoors t_* produced by Delegate and for all trapdoors t_w produced by TrapGen , we should have:

$$\Pr\left[\begin{array}{c} \mathcal{PP} \leftarrow \mathsf{Setup}(\lambda), (pk_S, sk_S) \leftarrow \mathsf{KeyGen}_S(\mathcal{PP}), (pk_R, sk_R) \leftarrow \mathsf{KeyGen}_R(\mathcal{PP}), \\ c_w \leftarrow \mathsf{Encrypt}(pk_R, w), t_* \leftarrow \mathsf{Delegate}(sk_R, pk_S), t_w \leftarrow \mathsf{TrapGen}(sk_R, pk_S, w) : \\ w \leftarrow \mathsf{Decrypt}(c_w, sk_R) \wedge true \leftarrow \mathsf{Test}_1(c_w, t_*, t_w, sk_S) \\ \wedge\ true \leftarrow \mathsf{Test}_2(c_w, t_*, w, sk_S) \end{array}\right] = 1$$

Ciphertext Indistinguishability. In the following we describe the basic security property for a $\mathcal{PKEDS}$ scheme which is ciphertext indistinguishability. This property guarantees that it is infeasible for an adversary (other than the server) to learn any information about any word from the ciphertext. The following definition formally captures this property.

Definition 4. ***(CI-ATK)*** *Let $\mathcal{PKEDS}$ be a private-key encryption with delegated search scheme and let $\mathcal{A}$ be a polynomial-time (PPT) adversary. Let*

$$\mathsf{ADV}^{\mathcal{CI}-ATK}_{\mathcal{A},\mathcal{PKEDS}}(\lambda) \stackrel{def}{=} \Pr\left[\begin{array}{c} \mathcal{PP} \leftarrow \mathsf{Setup}(\lambda), (pk_S, sk_S) \leftarrow \mathsf{KeyGen}_S(\mathcal{PP}), \\ (w_0, w_1, R^*) \leftarrow \mathcal{A}^{\mathcal{O}_1,\mathcal{O}_2,\mathcal{O}_3}(pk_S, \mathcal{PP}), v \leftarrow \{0,1\}, \\ c_{w_v} \leftarrow \mathsf{Encrypt}(pk_{R^*}, w_v), \\ v' \leftarrow \mathcal{A}^{\mathcal{O}_1,\mathcal{O}_2,\mathcal{O}_3}(c_{w_v}, w_0, w_1, pk_{R^*}, pk_S) : v' = v \end{array}\right] - \frac{1}{2}$$

where $w_0 \neq w_1 \wedge |w_0| = |w_1|$, the key-generation oracle $\mathcal{O}_1(R)$ is defined as $\mathsf{KeyGen}_R(\mathcal{PP})$ and returns (pk_R, sk_R) , the registration oracle $\mathcal{O}_2(R)$ is defined as $\mathsf{Delegate}(sk_R, pk_S)$ and returns t_*, the trapdoor generation oracle $\mathcal{O}_3(R, w)$ is defined as $\mathsf{TrapGen}(sk_R, w)$ and returns t_w. We restrict $\mathcal{A}$ such that if $\mathcal{A}$ queries $\mathcal{O}_1$ for the receiver key pair R^*, then $\mathcal{O}_1$ returns only the public key pk_{R^*} of the receiver R^*. We say that $\mathcal{PKEDS}$ is secure from ciphertext indistinguishable attacks if $\mathsf{ADV}^{\mathcal{CI}-ATK}_{\mathcal{A},\mathcal{PKEDS}}$ is negligible for any $\mathcal{A}$.

The scheme does not achieve $\mathcal{CI}$ against the server. Given the challenge ciphertext c_{w_b}, the master trapdoor t_* and words (w_0, w_1), the server runs $\mathsf{Test}_2(c_{w_b}, t_*, w_0, sk_S)$ to check whether the trapdoor and the ciphertext are associated with the same word. If the output of Test_2 is *true* then the server learns that $b = 0$, otherwise it learns that $b = 1$. As $\mathcal{CI}$ against the server is inherently not possible (remember that our focus is to allow the server to search the ciphertext); the best we can achieve against the server is ciphertext one-wayness.

Note. The security model that we consider in this paper is *weaker* than the original security model considered in $\mathcal{PEKS}$ [BDCOP04]. Our model gives more power to the server since the scheme allows the server to generate any trapdoor. This is risky for low entropy messages since the server by trial and error can find out the message. Nevertheless, for high entropy messages this attack is hard. For instance, if the server scans a ciphertext which contains user fingerprints, then its hard for the server to guess the fingerprint and run the trial and error attack. Indeed, our scheme is suitable for situations when the server is managed by an organization which wants to protect their employees from potential malicious senders while avoiding the need of giving the private key to the server.

Trapdoor Indistinguishability. The *trapdoor indistinguishability* property guarantees that it is infeasible for an adversary (except the server) to learn any information about any word from the trapdoor. Baek et al. [BSNS08] observe that $\mathcal{PEKS}$ scheme presented by Boneh et al. [BDCOP04] assumes a secure channel between the server and the receiver. If there is no secure channel, then everyone can break the *trapdoor indistinguishability* property since everyone can play the role of the server. To remove the secure channel between the receiver and the server, Baek et al. [BSNS08] propose a scheme where the sender encrypts the $\mathcal{PEKS}$ ciphertext with the public key of the server, in such a way that only the server who knows the private key can reveal the $\mathcal{PEKS}$ ciphertext. In this paper we take a different approach to achieve *trapdoor indistinguishability* against outside adversaries. The Baek et al. [BSNS08] solution is not suitable in our setting since we allow the receiver to decrypt the $\mathcal{PKEDS}$ ciphertext, otherwise if we encrypt the $\mathcal{PKEDS}$ ciphertext with the server's public key, then the receiver cannot decrypt the ciphertext without getting help from the server. Instead, the role of the server is to search the encrypted data and not to help the receiver to decrypt the ciphertext. To achieve *trapdoor indistinguishability*, we need the secure channel established between the receiver and the server, as assumed in [BDCOP04]. This implies that, instead of encrypting the communication between the sender and the receiver, we encrypt the communication between the receiver and the server. Namely, before sending the trapdoor to the server, the receiver encrypts the trapdoor under the server's public key. Since only the server has the private key, only the server can reveal the trapdoor and search the encrypted data. We capture the property of *trapdoor indistinguishability* through the following definition.

Definition 5. ***(TI-ATK)*** *Let $\mathcal{PKEDS}$ be a private-key encryption with delegated search scheme and let $\mathcal{A}$ be a polynomial-time (PPT) adversary. Let*

$$\mathsf{ADV}^{\mathcal{TI}-ATK}_{\mathcal{A},\mathcal{PKEDS}}(\lambda) \stackrel{def}{=} \Pr\left[\begin{array}{c} \mathcal{PP} \leftarrow \mathsf{Setup}(\lambda), (pk_S, sk_S) \leftarrow \mathsf{KeyGen}_S(\mathcal{PP}), \\ (w_0, w_1, R^*) \leftarrow \mathcal{A}^{\mathcal{O}_1,\mathcal{O}_2,\mathcal{O}_3}(pk_S, \mathcal{PP}), v \leftarrow \{0,1\}, \\ t_{w_v} \leftarrow \mathsf{TrapGen}(pk_{R^*}, w_v), \\ v' \leftarrow \mathcal{A}^{\mathcal{O}_1,\mathcal{O}_2,\mathcal{O}_3}(t_{w_v}, w_0, w_1, pk_{R^*}, pk_S) : v' = v \end{array}\right] - \frac{1}{2}$$

where $w_0 \neq w_1 \wedge |w_0| = |w_1|$ and oracles $\mathcal{O}_1, \mathcal{O}_2, \mathcal{O}_3$ are defined in the same way as in $\mathcal{CI} - ATK$, but the difference is that the adversary is not limited in his queries. We say that $\mathcal{PKEDS}$ is secure from trapdoor indistinguishable attacks if $\mathsf{ADV}^{\mathcal{TI}-ATK}_{\mathcal{A},\mathcal{PKEDS}}$ is negligible for any $\mathcal{A}$.

Under this definition we achieve $\mathcal{TI}$ only for adversaries other than the server. Informally speaking, we achieve this property by not allowing an adversary to search the encrypted data. In particular, we cannot achieve $\mathcal{TI}$ from the server who runs the Test_1 function since the server can guess the word in the following way: the server sends to the challenger two words (w_0, w_1) and the challenger replies to the server by sending t_{w_v} for a random bit $v \in \{0,1\}$. Next, the server chooses a random bit $v' \in \{0,1\}$ and runs $c_{v'} \leftarrow \mathsf{Encrypt}(pk_R, w_{v'})$. Finally, the server run $\mathsf{Test}_1(c_{v'}, t_*, t_{w_v}, sk_S)$ and outputs $v' = v$ if the output of Test_1 is *true*.

Ciphertext One-Wayness. The property of *ciphertext one-wayness* guarantees that it is hard for an adversary to invert the ciphertext and to learn the word even if the adversary holds the server's private key, the master trapdoor and the trapdoor associated with that word, but the adversary does not hold the receiver's private key. The following definition formally captures this attack.

Definition 6. ***(COW-ATK)*** *Let $\mathcal{PKEDS}$ be a private-key encryption with delegated search scheme and let $\mathcal{A}$ be a polynomial-time (PPT) adversary. Let*

$$\mathsf{ADV}^{COW-ATK}_{\mathcal{A},\mathcal{PKEDS}}(\lambda) \stackrel{def}{=} \Pr\left[\begin{array}{c} \mathcal{PP} \leftarrow \mathsf{Setup}(\lambda), (pk_S, sk_S) \leftarrow \mathsf{KeyGen}_S(\mathcal{PP}), \\ R^* \leftarrow \mathcal{A}^{\mathcal{O}_1,\mathcal{O}_2,\mathcal{O}_3}(pk_S, sk_S, \mathcal{PP}), w^* \leftarrow \mathcal{M}(k), \\ c_{w^*} \leftarrow \mathsf{Encrypt}(pk_{R^*}, w^*), t_{w^*} \leftarrow \mathsf{Encrypt}(sk_{R^*}, pk_S, w^*), \\ w' \leftarrow \mathcal{A}^{\mathcal{O}_1,\mathcal{O}_2,\mathcal{O}_3}(c_{w^*}, t_{w^*}, pk_{R^*}, sk_S, pk_S) : w' = w^* \end{array}\right]$$

where oracles $\mathcal{O}_1, \mathcal{O}_2, \mathcal{O}_3$ are defined in the same way as in $\mathcal{CI} - ATK$ with the restriction that $\mathcal{A}$ is not allowed to learn sk_{R^*} from the oracle $\mathcal{O}_1$. We say that $\mathcal{PKEDS}$ is secure from ciphertext one-way attacks if $\mathsf{ADV}^{COW-ATK}_{\mathcal{A},\mathcal{PKEDS}}$ is negligible for any $\mathcal{A}$.

4 A Construction of $\mathcal{PKEDS}$ Scheme

We are now ready to present our construction. When explaining the scheme we consider the single-user setting. The scheme consists of nine algorithms (Setup, KeyGen_S, KeyGen_R, Encrypt, Delegate, TrapGen, Test_1, Test_2, Decrypt) defined as follows:

Setup: On input of the security parameter λ the algorithm outputs public parameters ($\mathcal{PP}$) which contain the description of groups $< \mathbb{G}, \Gamma >$ of order p, the bilinear map $\hat{e} : \mathbb{G} \times \Gamma \rightarrow \mathbb{G}_T$, generators g and γ of groups $\mathbb{G}$ and Γ, respectively.

$\mathsf{KeyGen}_S(\mathcal{PP})$: On input of public parameters $\mathcal{PP}$ the algorithm picks a random $x \in \mathbb{Z}_p$ and outputs the server's key pair:

$$(sk_S, pk_S) = (x, p_s = \gamma^x)$$

$\mathsf{KeyGen}_R(\mathcal{PP})$: On input of public parameters $\mathcal{PP}$ the algorithm picks a random $\alpha, y \in \mathbb{Z}_p$ and outputs the receiver's key pair:

$$(sk_R, pk_R) = ((y, \gamma^\alpha), p_r = g^y)$$

$\mathsf{Encrypt}(pk_R, w)$: On input of the receiver's public key and a word $w \in \mathbb{G}$ the algorithm picks a random $k \in \mathbb{Z}_p$ and outputs the ElGamal ciphertext:

$$c_w = (c_1, c_2) = \left(w \cdot p_r^k, g^k\right)$$

$\mathsf{Delegate}(pk_S, sk_R)$: The algorithm creates a master trapdoor to allow the server to search the encrypted data for any word of his choice. The algorithm picks at random $r_1, r_2 \in \mathbb{Z}_p$ and outputs the master trapdoor:

$$t_* = (t_1, t_2, t_3, t_4) = (\gamma^\alpha \cdot p_s^{r_1}, \gamma^{r_1}, \gamma^{y\alpha} \cdot p_s^{r_2}, \gamma^{r_2})$$

$\mathsf{TrapGen}(sk_R, pk_S, w)$: The algorithm creates a trapdoor to allow the server to search for a specific message w. The algorithm picks a random $\delta \in \mathbb{Z}_p$ and outputs the trapdoor:

$$t_w = (t_5, t_6) = \left(\hat{e}(w, \gamma^{\alpha}) \cdot \hat{e}(p_r, p_s^{\delta}), \gamma^{\delta}\right)$$

$\mathsf{Test}_1(c_w, t_*, t_w, sk_S)$: The algorithm tests whether the ciphertext contains the same message as the trapdoor. The algorithm parses c_w as (c_1, c_2), t_* as (t_1, t_2, t_3, t_4), t_w as (t_5, t_6) and defines:

$$t_7 = \frac{t_1}{t_2^x} \quad , \quad t_8 = \frac{t_3}{t_4^x} \quad , \quad \tilde{a} = \frac{\hat{e}(p_r, t_6^x) \cdot \hat{e}(c_1, t_7)}{t_5} \quad , \quad \tilde{b} = \hat{e}(c_2, t_8).$$

Finally, the algorithm checks whether $\tilde{a} \stackrel{?}{=} \tilde{b}$. If this equation holds, the algorithm outputs *true* indicating that the ciphertext contains the same message as the trapdoor, otherwise it outputs *false*.

$\mathsf{Test}_2(c_w, t_*, w, sk_S)$: The algorithm tests whether the ciphertext contains the word w. The algorithm parse c_w as (c_1, c_2), t_* as (t_1, t_2, t_3, t_4) and defines:

$$t_7 = \frac{t_1}{t_2^x} \quad , \quad t_8 = \frac{t_3}{t_4^x} \quad , \quad \tilde{c} = \hat{e}(c_1, t_7) \quad , \quad \tilde{d} = \hat{e}(c_2, t_8)$$

Finally, the algorithm checks whether $\frac{\tilde{c}}{\tilde{d}} \stackrel{?}{=} \hat{e}(w, t_7)$. If this equation holds, the algorithm outputs *true* indicating that the ciphertext contains the word w, otherwise it outputs *false*.

$\mathsf{Decrypt}(sk_R, c_w)$: On input of the ciphertext and the private key the algorithm outputs:

$$w = \frac{c_1}{c_2^y}.$$

4.1 Efficiency

In Table 1 we count the number of calculations in KeyGen_S, KeyGen_R, Encrypt, Delegate, TrapGen, Test_1, Test_2 and Decrypt. KeyGen_S requires one exponentiation in Γ and KeyGen_R requires one exponentiation in $\mathbb{G}$ and one exponentiation in Γ. Encrypt and Decrypt are same as in ElGamal. Encrypt requires two exponentiations in $\mathbb{G}$ which are independent of the message and can be computed ahead of time and Decrypt requires only one exponentiation in $\mathbb{G}$. Delegate requires five exponentiations in Γ. TrapGen requires one exponentiation in Γ, one exponentiation in $\mathbb{G}_T$ and two pairing operations. Test_1 requires two exponentiations in Γ, one exponentiation in $\mathbb{G}_T$ and three pairing operations. Test_2 requires two exponentiations in Γ and three pairing operations.

5 Security

We now show that the scheme satisfies the correctness property, and is ciphertext indistinguishable, trapdoor indistinguishable and that the scheme offers ciphertext one-wayness.

Table 1. Efficiency of $\mathcal{PKEDS}$

	Exp.($\mathbb{G}$)	Exp.(Γ)	Exp.($\mathbb{G}_T$)	Pairing
KeyGen$_S$	//	1	//	//
KeyGen$_R$	1	1	//	//
Encrypt	2	//	//	//
Delegate	//	5	//	//
TrapGen	//	1	1	2
Test$_1$	//	2	1	3
Test$_2$	//	2	//	3
Decrypt	1	//	//	//

Correctness. Firstly, we show that when a ciphertext is created as a result of running Encrypt on input of the word w and the receiver's public key pk_R, then the same word w is revealed when running Decrypt on input of the ciphertext and the receiver's private key sk_R. This proof is ElGamal encryption/decyption and proceeds as follows:

$$\frac{c_1}{c_2^y} = \frac{w \cdot p_r^k}{g^{rs}} = \frac{w \cdot g^{rs}}{g^{rs}} = w$$

Next, we show the correctness for Test$_1$ algorithm. We observe that:

$$t_7 = \frac{\gamma^\alpha \cdot p_s^{r_1}}{\gamma^{xr_1}} = \gamma^\alpha \qquad t_8 = \frac{\gamma^{y\alpha} \cdot p_s^{r_2}}{\gamma^{xr_2}} = \gamma^{y\alpha}$$

$$\tilde{a} = \frac{\hat{e}(p_r, \gamma^{x\delta}) \cdot \hat{e}(w \cdot p_r^k, \gamma^\alpha)}{\hat{e}(w, \gamma^\alpha) \cdot \hat{e}(p_r, p_s^\delta)} = \hat{e}(p_r^k, \gamma^\alpha) \qquad \tilde{b} = \hat{e}(g^k, \gamma^{y\alpha}) = \hat{e}(p_r^k, \gamma^\alpha)$$

Thus $\tilde{a} = \tilde{b}$ and the output is *true* indicating that the word associated with the ciphertext and the word associated with the trapdoor are the same. Finally, we show the correctness for the Test$_2$ algorithm. We observe that:

$$t_7 = \frac{\gamma^\alpha \cdot p_s^{r_1}}{\gamma^{xr_1}} = \gamma^\alpha \qquad t_8 = \frac{\gamma^{y\alpha} \cdot p_s^{r_2}}{\gamma^{xr_2}} = \gamma^{y\alpha} \qquad \tilde{c} = \hat{e}(w \cdot p_r^k, \gamma^\alpha)$$

$$\tilde{d} = \hat{e}(g^k, \gamma^{y\alpha}) \qquad \frac{\tilde{c}}{\tilde{d}} = \hat{e}(w, \gamma^\alpha)$$

Thus $\frac{\tilde{c}}{\tilde{d}} = \hat{e}(w, t_7)$ and the output is *true* indicating that the ciphertext is an encryption of the word w.

Ciphertext Indistinguishability. When proving this property we will closely follow the security proof of [BGMM05]. In the following we show that our construction is CI-ATK secure as long as the SXDH assumption holds.

Theorem 1. *Suppose that there exists an adversary $\mathcal{A}$ that can break the CI-ATK of the $\mathcal{PKEDS}$ scheme with advantage ε. Then we can construct a reduction $\mathcal{B}$ that breaks the* SXDH *assumption with advantage $(1 - \frac{q}{n})\frac{1}{n}\frac{\varepsilon}{2}$ where q is the number of queries asked by $\mathcal{A}$, and n is the number of receivers in the system.*

Proof. The challenger selects a bilinear map $\hat{e} : \mathbb{G} \times \Gamma \rightarrow \mathbb{G}_T$, and generators g and γ of groups $\mathbb{G}$ and Γ, respectively. Then, it picks at random $a, b \in \mathbb{Z}_p$, computes $T_0 = g^{ab}$ and picks at random $T_1 \in_R \mathbb{G}$. It flips a fair coin $\mu \in_R \{0,1\}$ and gives the SXDH tuple $(g, g^a, g^b, T_\mu) \in \mathbb{G}$ to the reduction $\mathcal{B}$. The goal of $\mathcal{B}$ is to solve the SXDH assumption and acts as $\mathcal{A}$'s challenger as follows:

1. Setup : $\mathcal{B}$ generates the server's key pair $(sk_S, pk_S) = (x, p_s = \gamma^x)$, where $x \in_R \mathbb{Z}_p$ is chosen in the same way as in the scheme. $\mathcal{B}$ publishes $\mathcal{PP}$ and the server's key pair. The distribution of the $\mathcal{PP}$ and the server's key pair is identical to the $\mathcal{PP}$ and the server's key pair of the scheme since g and γ are random generators, and x is a random exponent, all chosen in the same way as in the scheme.
2. KeyGen$_R$ *to* $\mathcal{O}_1$: $\mathcal{B}$ answers the receiver's key generation queries by computing $(sk_R, pk_R) = ((y, \gamma^\alpha), p_r = g^y)$ where $\alpha, y \in_R \mathbb{Z}_p$ are chosen in the same way as in the scheme (each user has different α and y value). If the query is for R^*, $\mathcal{B}$ sets the public key equal to $p_r = g^a$ (this parameter is taken from the SXDH instance). Note that $\mathcal{B}$ does not know the private key of R^* ($\mathcal{B}$ does not know a). The distribution of the receiver's key pair is identical to the distribution of the receiver's key pair of the scheme since g is a random generator, α, a and y are random exponents, all chosen in the same way as in the scheme.
3. Delegate Query *to* $\mathcal{O}_2$: $\mathcal{A}$ requests a master trapdoor for the receiver R. If R is equal to R^*, $\mathcal{B}$ aborts the simulation and returns a guess μ'. Otherwise, if R is not equal to R^*, $\mathcal{B}$ computes the random elements $r_1, r_2 \in \mathbb{Z}_p$ and outputs the master trapdoor $t_* = (t_1, t_2, t_3, t_4) = (\gamma^\alpha \cdot p_s^{r_1}, \gamma^{r_1}, \gamma^{y\alpha} \cdot p_s^{r_2}, \gamma^{r_2})$. When $\mathcal{B}$ does not abort, the distribution of the master trapdoor is identical to the distribution of the master trapdoor in the scheme since r_1 and r_2 are random elements from $\mathbb{Z}_p$, same as in the real scheme.
4. TrapGen Query *to* $\mathcal{O}_3$: $\mathcal{A}$ requests a trapdoor for the pair (R, w) [1]. If R is equal to R^*, $\mathcal{B}$ aborts the simulation and returns a guess μ'. Otherwise, if R is not equal to R^*, $\mathcal{B}$ picks random $\delta \in \mathbb{Z}_p$ and outputs the trapdoor $t_w = (t_5, t_6) = \left(\hat{e}(w, \gamma^\alpha) \cdot \hat{e}(p_r, p_s^\delta), \gamma^\delta\right)$.
   When $\mathcal{B}$ does not abort, the distribution of the trapdoor is identical to the distribution of trapdoor in the scheme since δ is chosen at random from $\mathbb{Z}_p$, same as in the real scheme.
5. Challenge : $\mathcal{A}$ requests a ciphertext for one of the two words w_0 and w_1 generated under the public key of the receiver R. If R is equal to R^*, $\mathcal{B}$ flips a fair binary coin $v \in \{0,1\}$ and outputs the ciphertext $\hat{c}_{w_v} = (c_1, c_2) = (w_v \cdot T_\mu, g^b)$, where g^b and T_μ are parameters taken from SXDH instance. The distribution of the ciphertext is identical to the distribution of the ciphertext in the scheme only if $T_\mu = g^{ab}$. Otherwise, if $T_\mu \in_R \mathbb{G}$, the ciphertext is a random element from $\mathbb{G}$. If R is not equal to R^*, $\mathcal{B}$ aborts the simulation and returns a guess μ'.
6. Guess : At the end of the game, without loss of generality, we assume that $\mathcal{A}$ has ciphertexts for all keywords generated by each user, and has requested

[1] A trapdoor associated with the word w generated by user (receiver) R.

trapdoor queries to oracles $\mathcal{O}_2$ and $\mathcal{O}_3$ generated from all but one user. Therefore, we assume that at the end of the game, in a non-aborted simulation case, $\mathcal{A}$ should have ciphertexts generated by all users for each keyword, and have at least one challenge ciphertext, denoted as $\hat{c}_v$, which is either a valid or invalid ciphertext generated by R^* for which $\mathcal{A}$ does not have the corresponding trapdoor. Lastly, $\mathcal{A}$ outputs a guess v'. If the guess is correct $v' = v$, then $\mathcal{B}$ sets $\mu' = 0$ indicating that $T_0 = g^{ab}$, otherwise $\mathcal{B}$ sets $\mu' = 1$ indicating that $T_1 \in_R \mathbb{G}$.

Suppose $\mathcal{B}$ does not abort (noted as $\overline{abort}$) during the simulation. If $\mu = 0$ then the ciphertext $\hat{c}_v$ is a valid ciphertext generated by user R^* and $\mathcal{A}$ sees an encryption of w_v. In this case we have: $\Pr\left[v' = v|\overline{abort} \wedge \mu = 0\right] = \frac{1}{2} + \varepsilon$. If $\mu = 1$ then the ciphertext $\hat{c}_v$ is random ciphertext for $\mathcal{A}$ (i.e. $\mathcal{A}$ gains no information about w_v). Hence we have: $\Pr\left[v' \neq v|\overline{abort} \wedge \mu = 1\right] = \frac{1}{2}$. Note that the advantage of $\mathcal{B}$ is same as the advantage of $\mathcal{A}$. For the first case when the guess of $\mathcal{A}$ is correct $v' = v$, $\mathcal{B}$ will output $\mu' = 0$ and we have $\Pr\left[\mu' = \mu|\overline{abort} \wedge \mu = 0\right] = \frac{1}{2} + \varepsilon$. For the second case when the guess is not correct $v' \neq v$, $\mathcal{B}$ will output $\mu' = 1$ and we have $\Pr\left[\mu' = \mu|\overline{abort} \wedge \mu = 1\right] = \frac{1}{2}$.

Now assume that $\mathcal{B}$ aborts (noted as $abort$) the simulation when running either TrapGen Query or Challenge phase. In this case $\mathcal{B}$ outputs its guess μ' which is independent of the guess given by $\mathcal{A}$ in Guess phase. Therefore the advantage of $\mathcal{B}$ in the abort case is: $\Pr\left[\mu' = \mu|abort\right] = \frac{1}{2}$. Putting all together we define the overall advantage of the reduction $\mathcal{B}$:

$$\Pr[abort]\Pr\left[\mu' = \mu|abort\right] + \Pr[\overline{abort}](\Pr[\mu = 0]\Pr[\mu' = \mu|\overline{abort} \wedge \mu = 0] +$$
$$\Pr[\mu = 1]\Pr[\mu' = \mu|\overline{abort} \wedge \mu = 1]) - \frac{1}{2} = \frac{\Pr[\overline{abort}]\varepsilon}{2}.$$

Now we have to define exactly the value of $\Pr[\overline{abort}]$ and give the exact overall advantage of $\mathcal{B}$. Let assume that $\mathcal{A}$ makes at most q queries during TrapGen Query phase and there are n users in the system. Since there is only one user for whom $\mathcal{B}$ cannot answer in TrapGen Query phase, the probability that a query causes $\mathcal{B}$ to abort is at most $\frac{1}{n}$. Since $\mathcal{A}$ can make q queries the overall probability that $\mathcal{A}$ aborts during TrapGen Query phase is $\frac{q}{n}$. Thus the probability that $\mathcal{B}$ does not abort in the TrapGen Query is $1 - \frac{q}{n}$. The probability that $\mathcal{B}$ will not abort in the Challenge phase is at least $\frac{1}{n}$. We now conclude that the reduction $\mathcal{B}$ solves the SXDH assumption with advantage at least $(1 - \frac{q}{n})\frac{1}{n}\frac{\varepsilon}{2}$, as required. □

Trapdoor Indistinguishability. We prove that our scheme is TI-ATK secure as long as the SXDH assumption is intractable. Unlike the ciphertext indistinguishability where the reduction had an SXDH instance from the group $\mathbb{G}$, when proving this property the reduction has an SXDH instance from the group Γ.

Theorem 2. *Suppose that there exists an adversary $\mathcal{A}$ that can break the trapdoor indistinguishability of the $\mathcal{PKEDS}$ scheme with advantage ε. Then we can construct a reduction $\mathcal{B}$ that solves the* SXDH *assumption with advantage $\frac{\varepsilon}{2}$.*

Proof. The challenger selects a bilinear map $\hat{e} : \mathbb{G} \times \Gamma \rightarrow \mathbb{G}_T$, and generators g and γ of groups $\mathbb{G}$ and Γ, respectively. Next, the challenger defines $T_0 = \gamma^{ab}$ for a random $a, b \in_R \mathbb{Z}_p$ and picks at random $T_1 \in_R \Gamma$. After flipping a fair coin $\mu \in_R \{0,1\}$, the challenger gives the SXDH tuple $(\gamma, \gamma^a, \gamma^b, T_\mu) \in \Gamma$ to $\mathcal{B}$. The reduction $\mathcal{B}$ solves the SXDH assumption by running $\mathcal{A}$ as a subroutine:

1. Setup: $\mathcal{B}$ sets the server's public key $pk_S = (p_s = \gamma^a)$, where γ^a is taken from the SXDH instance, and implicitly sets the server's secret-key $sk_S = a$. The reduction publishes the $\mathcal{PP}$ and the server's public keys which distribution is identical to the $\mathcal{PP}$ and the server's public keys of the scheme since g and γ are random generators, a is random exponent, all chosen in the same way as in the scheme.
2. KeyGen$_R$ *to* $\mathcal{O}_1$: $\mathcal{B}$ answers receiver's key generation queries by computing $(sk_R, pk_R) = ((y, \gamma^\alpha), p_r = g^y)$ where $\alpha, y \in_R \mathbb{Z}_p$ are chosen in the same way as in the scheme (each user has different α and y value). The distribution of receiver's key pair is identical to the distribution of receiver's key pair of the scheme since g is a random generator, α and y are random exponents, all chosen in the same way as in the scheme.
3. Delegate Query *to* $\mathcal{O}_2$: $\mathcal{A}$ requests a master trapdoor for the receiver R. $\mathcal{B}$ compute random elements $r_1, r_2 \in \mathbb{Z}_p$ and outputs the master trapdoor $t_* = (t_1, t_2, t_3, t_4) = (\gamma^\alpha \cdot p_s^{r_1}, \gamma^{r_1}, \gamma^{y\alpha} \cdot p_s^{r_2}, \gamma^{r_2})$.
The distribution of the master trapdoor is identical to the distribution of the master trapdoor in the scheme since r_1 and r_2 are random elements from $\mathbb{Z}_p$, same as in the real scheme.
4. TrapGen Query *to* $\mathcal{O}_3$: $\mathcal{A}$ requests a trapdoor for the pair (R, w). $\mathcal{B}$ picks random $\delta \in \mathbb{Z}_p$ and outputs the trapdoor t_w associated with the keyword w, $t_w = (t_5, t_6) = \left(\hat{e}(w, \gamma^\alpha) \cdot \hat{e}(p_r, p_s^\delta), \gamma^\delta\right)$. The distribution of the trapdoor is identical to the distribution of the trapdoor in the scheme since δ is randomly chosen from $\mathbb{Z}_p$, same as in the real scheme.
5. Challenge : $\mathcal{A}$ sends two words w_0 and w_1 to $\mathcal{B}$ and asks for a trapdoor generated by user R^*. $\mathcal{B}$ flips a fair coin $v \in_R \{0,1\}$, picks at random $\alpha \in \mathbb{Z}_p$ and implicitly sets $\delta = b$ (where b is an exponent from the SXDH instance) and returns the trapdoor to $\mathcal{A}$: $t_{w_v} = (t_5, t_6) = \left(\hat{e}(w_v, \gamma^\alpha) \cdot \hat{e}(p_r, T_\mu), \gamma^b\right)$, where $(p_r = g^y,\ y)$ is R^*'s public/private key pair.
6. Guess : $\mathcal{A}$ outputs a guess v'.

If $\mu = 0$ and $T_\mu = \gamma^{ab}$, then the generated challenged trapdoor t_{w_v} is a valid trapdoor generated by user R^* and the view of $\mathcal{A}$ is distributed as if it had received the trapdoor from the real scheme. In this case we have: $\Pr[v' = v|\mu = 0] = \frac{1}{2} + \varepsilon$. If $\mu = 1$ and $T_\mu \in_R \Gamma$, then the generated trapdoor t_{w_v} is an invalid trapdoor. In this case we have: $\Pr[v' \neq v|\mu = 1] = \frac{1}{2}$. Putting all together we define the overall advantage of $\mathcal{B}$:

$$(\Pr[\mu = 0] \Pr[\mu' = \mu|\mu = 0] + \Pr[\mu = 1] \Pr[\mu' = \mu|\mu = 1]) - \frac{1}{2} = \frac{\varepsilon}{2}.$$

Ciphertext One-Wayness. In this section we show that our construction is COW-ATK secure in the standard model.

Theorem 3. *The $\mathcal{PKEDS}$ scheme with the message space in $\mathbb{G}$ is $\mathcal{COW}-\mathcal{ATK}$ secure in the standard model assuming* mCDH *is intractable.*

Proof. The challenger selects a bilinear map $\hat{e} : \mathbb{G} \times \Gamma \rightarrow \mathbb{G}_T$, and generators g and γ of groups $\mathbb{G}$ and Γ, respectively. Then, it picks at random $a, b \in \mathbb{Z}_p$, and gives mCDH tuples $(g, g^a, g^b) \in \mathbb{G}$ and $(\gamma, \gamma^b) \in \Gamma$ to the reduction $\mathcal{B}$. The goal of $\mathcal{B}$ is to solve the mCDH assumption and acts as $\mathcal{A}$'s challenger as follows:

1. Setup : $\mathcal{B}$ generates the server's key pair $(sk_S, pk_S) = (x, p_s = \gamma^x)$, where $x \in_R \mathbb{Z}_p$ is chosen in the same way as in the scheme. $\mathcal{B}$ publishes $\mathcal{PP}$ and the server's key pair. The distribution of the $\mathcal{PP}$ and the server's key pair is identical to the $\mathcal{PP}$ and the server's key pair of the scheme since g and γ are random generators, and x is a random exponent, all chosen in the same way as in the scheme.
2. KeyGen$_R$ *to* $\mathcal{O}_1$: $\mathcal{B}$ answers receiver's key generation queries by computing $(sk_R, pk_R) = ((y, \gamma^\alpha), p_r = g^y)$ where $\alpha, y \in_R \mathbb{Z}_p$ are chosen in the same way as in the scheme (each user has different α and y value). If the query is for R^*, $\mathcal{B}$ sets the public key equal to $p_r = g^a$ (this parameter is taken from the mCDH instance). Note that $\mathcal{B}$ does not know the private key of R^* (the reduction does not know a). The distribution of receiver's key pair is identical to the distribution of receiver's key pair of the scheme since g is a random generator, α, a and y are random exponents, all chosen in the same way as in the scheme.
3. Delegate Query *to* $\mathcal{O}_2$: $\mathcal{A}$ requests a master trapdoor for the receiver R. $\mathcal{B}$ computes random elements $r_1, r_2 \in \mathbb{Z}_p$ and outputs the master trapdoor $t_* = (t_1, t_2, t_3, t_4) = (\gamma^\alpha \cdot p_s^{r_1}, \gamma^{r_1}, \gamma^{y\alpha} \cdot p_s^{r_2}, \gamma^{r_2})$. If $R = R^*$ then $y = a$. The distribution of the master trapdoor is identical to the distribution of the master trapdoor in the scheme since r_1 and r_2 are random elements from $\mathbb{Z}_p$, same as in the real scheme.
4. TrapGen Query *to* $\mathcal{O}_3$: $\mathcal{A}$ requests a trapdoor for the pair (R, w). $\mathcal{B}$ picks random $\delta \in \mathbb{Z}_p$ and outputs the trapdoor $t_w = (t_5, t_6) = (\hat{e}(w, \gamma^\alpha) \cdot \hat{e}(p_r, p_s^\delta), \gamma^\delta)$. The distribution of the trapdoor is identical to the distribution of trapdoor in the scheme since δ is chosen at random from $\mathbb{Z}_p$, same as in the real scheme.
5. Challenge : The reduction picks at random $c' \in \mathbb{Z}_p$, computes $\frac{g^{c'}}{g^a} = g^{\bar{c}}$ (thus, $c' = a + \bar{c}$), implicitly sets $w^* = g^{b\bar{c}}$ and outputs the challenge ciphertext $\hat{c}_{w^*} = (c_1, c_2) = (g^{bc'}, g^b)$ and the challenge trapdoor $t_{w^*} = (t_5, t_6) = (\hat{e}(g^{\bar{c}}, \gamma^{b\alpha}) \cdot \hat{e}(p_r, p_s^\delta), \gamma^\delta)$. The challenge ciphertext $\hat{c}_{w^*} = (c_1, c_2) = (g^{bc'}, g^b) = (g^{b\bar{c}} \cdot g^{ab}, g^b)$ is a valid encryption of the message $w^* = g^{b\bar{c}}$ under the public key of R^* and has the same distribution as the ciphertext in the real scheme. The same holds for the trapdoor t_{w^*}.
6. Output : At the end of the game, $\mathcal{A}$ outputs the message w'.

$\mathcal{B}$ checks whether $\hat{e}(w', \gamma) \stackrel{?}{=} \hat{e}(g^a, \gamma^b)$. If so, then $\mathcal{A}$ has decrypted the challenge ciphertext and $\mathcal{B}$ uses the output of $\mathcal{A}$ to solve the mCDH assumption: $g^{ab} = \frac{c_1}{w'}$, which will reach a contradiction and proof the theorem. □

6 Applications

In this section we illustrate two applications of the $\mathcal{PKEDS}$ scheme: to detect encrypted malwares and to forward encrypted emails.

Detecting Encrypted Malware. A polymorphic virus uses encryption to modify its form as it spreads in such a way that different infected files have different byte-strings (each file is encrypted with a different key) [MC00]. The polymorphic virus instance is divided into three parts: the decryption algorithm, the decryption key and the encrypted virus. The decryption algorithm uses the decryption key in order to decrypt and run the virus. The fact that polymorphic viruses store the decryption key within each virus instance, makes them detectable by a virus scanner. As pointed out in the introduction, in this paper we consider attackers who use encryption to hide malware even in a more powerful way. Namely, in our attack scenario an attacker does not include the decryption key within the encrypted data, indeed an attacker does not know the decryption key which belongs to the receiver. Hence, the virus instance that we consider contains only the encrypted malware and the decryption algorithm.

In the proposed $\mathcal{PKEDS}$, the server can use the master trapdoor, a ciphertext and a malware signature, to check whether the ciphertext contains the malware signature, without decrypting it. This kind of detection is known as signature-based detection and is performed by most existing antivirus software packages, which maintain a database with known malware signatures and check whether the scanned data has the same signature as one of the signatures stored in the database. If the signatures match, then a malware is found and the antivirus takes further steps to quarantine, repair or delete the data. In our context we assume that the server has a database with known malware signatures and for each signature, using the master trapdoor t_*, it creates a trapdoor and checks whether the trapdoor and the scanned ciphertext have the same signature. If so, then a malware is detected and the server takes further steps to quarantine or delete the ciphertext, otherwise the ciphertext is clean and is forwarded to the receiver. The crucial property of the scheme is that ciphertexts are both searchable and decryptable, thus the server can search every part of the ciphertext for a possible malware. Note that allowing the server to search the encrypted data does not mean that the server can perform any kind of intrusion detection. For instance, 0-day attacks cannot be detected through signature-based detection.

Roschke et al. [RICM10] propose a technique which detects malicious content in the encrypted data. The solution presented in [RICM10] is based on the IBE [BF01] scheme and suffers from the key escrow property where the compromise of the master secret key compromises the whole system. The conceptual difference between our approach and the approach presented in [RICM10], is that the technique in [RICM10] uses the master secret key of the IBE to decrypt the ciphertext and then uses the virus scanner to scan the plaintext, while in our approach scanning can be done in the encrypted data, without having to decrypt it.

Forwarding Encrypted Emails. The original motivation for a $\mathcal{PEKS}$ scheme is to allow an email server to categorize user encrypted emails based on keywords contained in the message text. Using this property, Bob can create trapdoors t_{work} and t_{family}, and instruct the server to forward his encrypted emails tagged with the word "work" to his secretary and encrypted emails tagged with the word "family" to one of his family members. Waters et al. [WBDS04] showed that $\mathcal{PEKS}$ schemes can also be used to build an searchable audit log which is encrypted. The $\mathcal{PKEDS}$ scheme adds the decryption property to the $\mathcal{PEKS}$ scheme and as such can be used in every application that $\mathcal{PEKS}$ can be used. The $\mathcal{PKEDS}$ scheme has the following additional advantages compared to $\mathcal{PEKS}$:

- $\mathcal{PKEDS}$ can be used alone, without employing an additional $\mathcal{PKE}$ encryption scheme, to send encrypted messages in an open environment. This property inherently brings an additional advantage - each word of the message becomes searchable by the server. For instance, to encrypt a message m with consists from words $w_1, ..., w_k$, the sender generates the ciphertext:
$$(\mathsf{Encrypt}_{\mathcal{PKEDS}}(w_1)||...||\mathsf{Encrypt}_{\mathcal{PKEDS}}(w_k))$$
Where $\mathsf{Encrypt}_{\mathcal{PKEDS}}$ is the encryption algorithm for the $\mathcal{PKEDS}$ scheme. It is clear that to reveal the message m, the receiver has to decrypt each searchable ciphertext separately. From the computational point of view, using $\mathcal{PKEDS}$ alone might be expensive since it requires a number of $\mathcal{PKEDS}$ ciphertexts linear in the number of words in the document. Note that in $\mathcal{PEKS}$ the server can search only for keywords and this might be a problem for scenarios when the original message might contain some words that appear in the receiver's query but do not appear in the keyword list, and as a result the server will not forward these documents to the receiver.
- $\mathcal{PKEDS}$ can be used in the same way as $\mathcal{PEKS}$ is used, namely, use $\mathcal{PKEDS}$ to encrypt only keywords in addition to a non-searchable $\mathcal{PKE}$ scheme which encrypts the original message. For instance, to encrypt a message m with keywords $w_1, ..., w_k$, the sender generates the ciphertext:
$$(\mathsf{Encrypt}_{\mathcal{PKE}}(m)||\mathsf{Encrypt}_{\mathcal{PKEDS}}(w_1)||...||\mathsf{Encrypt}_{\mathcal{PKEDS}}(w_k))$$
Where $\mathsf{Encrypt}_{\mathcal{PKE}}$ is a regular encryption function for the $\mathcal{PKE}$ scheme and $\mathsf{Encrypt}_{\mathcal{PKEDS}}$ is the encryption function for the $\mathcal{PKEDS}$ scheme. Unlike $\mathcal{PEKS}$ which does not guarantee any relation between keywords (encrypted under $\mathcal{PEKS}$) and the original message (encrypted under $\mathcal{PKE}$), $\mathcal{PKEDS}$ guarantees this *relation* since it allows the receiver to decrypt the searchable ciphertext and check whether the keywords indeed describe the original message. Another benefit is that the receiver can categorize her messages according to keywords, unlike in $\mathcal{PEKS}$ where the receiver cannot categorize her messages since the searchable ciphertext is not decryptable and consequently the receiver, without decrypting the ciphertext, does not know which keywords describe the message. From the computational point of view, using $\mathcal{PKEDS}$ in addition to another $\mathcal{PKE}$ scheme would require a number of $\mathcal{PKEDS}$ ciphertexts linear in the number of *keywords* in the message, same as in $\mathcal{PEKS}$.

7 Conclusion

In this work we have presented a private-key encryption with delegated search ($\mathcal{PKEDS}$) with a security proof in the standard model. In the proposed scheme the private key holder creates a master trapdoor t_* and delegates to another entity (i.e. the server) the ability to search ciphertexts intended for the receiver without decrypting it. The main property of the scheme is that ciphertexts are both searchable and decryptable, thus the scheme can be used to search not only for keywords describing the document, but search also for words inside the document. The proposed scheme also allows the receiver to provide the server with a special key (a.k.a. trapdoor t_w) associated with a specific word w, such that it enables the server to test whether the word w is in the ciphertext. As an application, we show how $\mathcal{PKEDS}$ can be used for detecting encrypted malware and for forwarding encrypted email.

Acknowledgments

We thank Qiang Tang and the anonymous reviewers for their valuable suggestions and comments.

References

[ABC⁺05] Abdalla, M., Bellare, M., Catalano, D., Kiltz, E., Kohno, T., Lange, T., Malone-Lee, J., Neven, G., Paillier, P., Shi, H.: Searchable encryption revisited: Consistency properties, relation to anonymous IBE, and extensions. In: Shoup, V. (ed.) CRYPTO 2005. LNCS, vol. 3621, pp. 205–222. Springer, Heidelberg (2005)

[BDCOP04] Boneh, D., Di Crescenzo, G., Ostrovsky, R., Persiano, G.: Public key encryption with keyword search. In: Cachin, C., Camenisch, J.L. (eds.) EUROCRYPT 2004. LNCS, vol. 3027, pp. 506–522. Springer, Heidelberg (2004)

[BF01] Boneh, D., Franklin, M.: Identity-based encryption from the weil pairing. In: Kilian, J. (ed.) CRYPTO 2001. LNCS, vol. 2139, pp. 213–229. Springer, Heidelberg (2001)

[BGMM05] Ballard, L., Green, M., Medeiros, D.B., Monrose, F.: Correlation-resistant storage via keyword-searchable encryption. Technical report, Cryptology ePrint Archive, Report 2005/417 (2005), `http://eprint.iacr.org/2005/417`

[BSNS06] Baek, J., Safavi-Naini, R., Susilo, W.: On the integration of public key data encryption and public key encryption with keyword search. In: Katsikas, S.K., López, J., Backes, M., Gritzalis, S., Preneel, B. (eds.) ISC 2006. LNCS, vol. 4176, pp. 217–232. Springer, Heidelberg (2006)

[BSNS08] Baek, J., Safavi-Naini, R., Susilo, W.: Public key encryption with keyword search revisited. In: Gervasi, O., Murgante, B., Laganà, A., Taniar, D., Mun, Y., Gavrilova, M.L. (eds.) ICCSA 2008, Part I. LNCS, vol. 5072, pp. 1249–1259. Springer, Heidelberg (2008)

[BW06] Boyen, X., Waters, B.: Anonymous hierarchical identity-based encryption (without random oracles). In: Dwork, C. (ed.) CRYPTO 2006. LNCS, vol. 4117, pp. 290–307. Springer, Heidelberg (2006)

[BW07] Boneh, D., Waters, B.: Conjunctive, subset, and range queries on encrypted data. In: Vadhan, S.P. (ed.) TCC 2007. LNCS, vol. 4392, pp. 535–554. Springer, Heidelberg (2007)

[CM09] Chatterjee, S., Menezes, A.: On Cryptographic Protocols Employing Asymmetric Pairings – The Role of Ψ Revisited. Technical report, Cryptology ePrint Archive, Report 2009/480 (2009), `http://eprint.iacr.org/2009/480`

[ElG85] El Gamal, T.: A public key cryptosystem and a signature scheme based on discrete logarithms. In: Blakely, G.R., Chaum, D. (eds.) CRYPTO 1984. LNCS, vol. 196, pp. 10–18. Springer, Heidelberg (1985)

[FP07] Fuhr, T., Paillier, P.: Decryptable searchable encryption. In: Susilo, W., Liu, J.K., Mu, Y. (eds.) ProvSec 2007. LNCS, vol. 4784, pp. 228–236. Springer, Heidelberg (2007)

[GPS08] Galbraith, S.D., Paterson, K.G., Smart, N.P.: Pairings for cryptographers. Discrete Applied Mathematics 156(16), 3113–3121 (2008)

[HL07] Hwang, Y., Lee, P.: Public key encryption with conjunctive keyword search and its extension to a multi-user system. In: Takagi, T., Okamoto, T., Okamoto, E., Okamoto, T. (eds.) Pairing 2007. LNCS, vol. 4575, pp. 2–22. Springer, Heidelberg (2007)

[HW08] Hofheinz, D., Weinreb, E.: Searchable encryption with decryption in the standard model. Technical report, Cryptology ePrint Archive, Report 2008/423 (2008), `http://eprint.iacr.org/2008/423`

[MC00] Morar, J.F., Chess, D.M.: Can Cryptography Prevent Computer Viruses? In: VIRUS, vol. 127 (2000)

[RICM10] Roschke, S., Ibraimi, L., Cheng, F., Meinel, C.: Secure Communication Using Identity Based Encryption. In: De Decker, B., Schaumüller-Bichl, I. (eds.) CMS 2010. LNCS, vol. 6109, pp. 256–267. Springer, Heidelberg (2010)

[SSW09] Shen, E., Shi, E., Waters, B.: Predicate privacy in encryption systems. In: Reingold, O. (ed.) TCC 2009. LNCS, vol. 5444, pp. 457–473. Springer, Heidelberg (2009)

[WBDS04] Waters, B.R., Balfanz, D., Durfee, G., Smetters, D.K.: Building an encrypted and searchable audit log. In: Proceedings of ISOC Network and Distributed System Security Symposium (NDSS 2004), Citeseer (2004)

[ZI07] Zhang, R., Imai, H.: Generic combination of public key encryption with keyword search and public key encryption. In: Bao, F., Ling, S., Okamoto, T., Wang, H., Xing, C. (eds.) CANS 2007. LNCS, vol. 4856, pp. 159–174. Springer, Heidelberg (2007)

Author Index

GPSR Compliance

The European Union's (EU) General Product Safety Regulation (GPSR) is a set of rules that requires consumer products to be safe and our obligations to ensure this.

If you have any concerns about our products, you can contact us on ProductSafety@springernature.com

In case Publisher is established outside the EU, the EU authorized representative is:

Springer Nature Customer Service Center GmbH
Europaplatz 3
69115 Heidelberg, Germany

Batch number: 10844619

Printed by Printforce, the Netherlands